OPTIMIZATION OF DISTRIBUTED
PARAMETER STRUCTURES : VOLUME II

NATO ADVANCED STUDY INSTITUTES SERIES

Proceedings of the Advanced Study Institute Programme, which aims at the dissemination of advanced knowledge and the formation of contacts among scientists from different countries.

The series is published by an international board of publishers in conjunction with NATO Scientific Affairs Division

A	Life Sciences	Plenum Publishing Corporation
B	Physics	London and New York
C	Mathematical and Physical Sciences	D. Reidel Publishing Company Dordrecht and Boston
D	Behavioural and Social Sciences	Sijthoff & Noordhoff International Publishers B.V.
E	Applied Sciences	Alphen aan den Rijn, The Netherlands and Rockville, Md., U.S.A.

Series E: Applied Sciences – No. 50

OPTIMIZATION OF DISTRIBUTED PARAMETER STRUCTURES - VOLUME II

edited by

Edward J. Haug
The University of Iowa,
College of Engineering
Iowa City, IA, 52242, U.S.A.

and

Jean Cea
University of Nice
Department of Mathematics
Nice, France

Springer-Science+Business Media, B.V. 1981

Proceedings of the NATO Advanced Study Institute on
Optimization of Distributed Parameter Structural Systems
Iowa City, Iowa, U.S.A.
May 20 - June 4, 1980

ISBN 978-94-009-8608-4 ISBN 978-94-009-8606-0 (eBook)
DOI 10.1007/978-94-009-8606-0

Originally published by Sijthoff & Noordhoff International Publishers B.V.,
Alphen aan den Rijn, The Netherlands
Softcover reprint of the hardcover 1st edition 1981

NATO-NSF ADVANCED STUDY INSTITUTE ON

OPTIMIZATION OF DISTRIBUTED PARAMETER STRUCTURES

Iowa City, Iowa, United States, 21 May - 4 June, 1980

Director : E. J. Haug, Materials Division, College of Engineering, University of Iowa, Iowa City, Iowa

Co-director: J. Cea, Department of Mathematics, University of Nice, Nice, FRANCE

Scientific Contents of the Advanced Study Institute

The Advanced Study Institute was organized to bring together engineers and mathematicians working in optimization of distributed parameter structures. The principle attention of the Institute was focused on structures described by boundary-value problems, as opposed to finite element models of structures. The diversity of interest and specialization of participants ranged from applications oriented engineers, through mathematically oriented research engineers, to applied mathematicians working in fields related to structural optimization. A key objective of the institute was to promote interaction by engineers and applied mathematicians who have, in the past, taken rather different approaches to structural optimization . As part of this objective, the emerging field of shape optimal design was given a high degree of emphasis in the Institute, to provide a forum for the study of mathematical techniques of shape optimization that have been applied to diverse fields of science and to consider their applicability for structural optimization.

The scientific program began with a review of distributed parameter structural optimization literature (E. Haug) and a survey of design sensitivity analysis methods applicable to structural optimization (E. Haug). Reviews of optimality criterion methods (J.Taylor), variational methods for developing optimality criterion (E. Rozvany), and an analysis of singular problems in optimal design (E. Masur) were presented as an overview of optimality criteria methods. Optimality criteria methods for design of columns for buckling, beams and shafts for vibration, and plates for vibration were presented (N. Olhoff), as was optimization of grids, shells, and arches (G. Rozvany). Mathematical aspects of recently encountered optimal design problems with repeated eigenvalues were presented (B. Rousselet, and K. Choi). Design of plates for minimum deflection and stress (N. Banichuk) and a survey of structural optimization methods under nonconservative loading (R. Plaut) were presented. An in depth treatment of optimization methods and their application for structures under earthquake loads was presented as a sequence of lectures in the program (K. Pister, E. Polak, M. Bhatti). Numerical methods of optimal remodeling (J. Taylor) and gradient projection techniques for static and dynamic problems (E. Haug, J. Arora) were presented. Approximately 40% of lecture time in the Institute was devoted to presentation of basic mathematical ideas of shape optimazation, with applications to numerious fields of applied physics (J. Cea, B. Rousselet, and J. Zolesio). Applications of shape optimal design techniques to structural systems (N. Banichuk) and optimization of shape of elastic bodies contact (R. Benedict) were presented. In addition to the principle lectures of the institute, participants gave contributed papers and lectures, many of which are contained in these proceedings.

LECTURERS

J. Arora, University of Iowa, Iowa City, Iowa, U.S.A.
N. Banichuk, USSR Academy of Sciences, Moscow, USSR
R. Benedict, University of Iowa, Iowa City, Iowa, U.S.A.
M. Bhatti, University of Iowa, Iowa City, Iowa, U.S.A.
J. Cea, University of Nice, Nice, FRANCE
K. Choi, University of Iowa, Iowa City, Iowa, U.S.A.
E. Haug, University of Iowa, Iowa City, Iowa, U.S.A.
E. Masur, University of Illinois, Chicago, Illinois, U.S.A.
N. Olhoff, Technical University of Denmark, Lyngby, DENMARK
K. Pister, University of California, Berkeley, CA., U.S.A.
R. Plaut, Virginia Polytechnic Institute, Blacksburg, VA., U.S.A.
E. Polak, University of California, Berkeley, California, U.S.A.
B. Rousselet, University of Nice, Nice, FRANCE
G. Rozvany, Monash University, Clayton, AUSTRALIA
J. Taylor, University of Michigan, Ann Arbor, Michigan, U.S.A.
J. Zolesio, University of Nice, Nice, FRANCE

CONTRIBUTORS

C. Akkoc, Middle East Technical University, Gaziantep, TURKEY
H. Alper, Bogazici University, Istanbul, TURKEY
D. Anderson, University of Warwick, Coventry, ENGLAND
F. Cheng, University of Missouri-Rolla, Rolla, Missouri, U.S.A.
K. Cheng, Technical University of Denmark, Lyngby, DENMARK
Y. Chun, Villanova University, Villanova, Pensylvania, U.S.A.
C. Cinquini, Universita Di Pavia, Pavia, ITALY
J. Claudon, University of Tokoyo, Tokoyo, JAPAN
M. Delfour, Universite de Montreal, Montreal, CANADA
C. Fleury, Universite de Liege, Liege, BELGIUM
J. Kalker, Delft University of Technology, Delft, NETHERLANDS
P. Kirmser, Kansas State University, Manhattan, Kansas, U.S.A
R. Kohn, New York University, New York, New York, U.S.A.
V. Komkov, West Virginia University, Morgantown, WV., U.S.A.
J. Kruzelecki, Technical University of Cracow, Crakow, POLAND
B. Kwak, The Korea Advanced Institute of Science, Seoul, KOREA
E. Lightfoot, University of Oxford, Oxford, ENGLAND
C. Martin, University of Nebraska, Lincoln, Nebraska, U.S.A.
A. Morris, Royal Aircraft Establishment, Hampshire, ENGLAND
P. Pedersen, The Technical University of Denmark, Lyngby, DENMARK
B. Pierson, Iowa State University, Ames, Iowa, U.S.A.
C. Polizzotto, Universita di Palermo, Palermo, ITALY
G. Sacchi, Istituto di Scienza E Delle Costruzioni, Milano, ITALY
M. Saglam, Bogazici Universitesi, Istanbul, TURKEY
R. Sandstrom, University of Michigan, Ann Arbor, Michigan, U.S.A.
P. Sinha, University of Warwick, Coventry, ENGLAND
J. Sokolowski, Systems Research Institute, Warszawa, POLAND

CONTRIBUTORS (cont)

W. Spillers, Rensselaer Polytechnic Institute, Troy, N.Y., U.S.A.
M. Szata, Technical University of Wroclaw, Wroclaw, POLAND
B. Topping, The University of Edinburgh, Edinburgh, ENGLAND
A. Torkamani, University of Pittsburgh, Pittsburgh, Penn., U.S.A.
J. Whitesell, University of Michigan, Ann Arbor, Michigan, U.S.A.
K. Willmert, Clarkson College, Potsdam, New York, U.S.A.
H. Zehlein, Gesselschaft fur Kernforschung, Karlsruhe, GERMANY

PARTICIPANTS

A. Aly, University of Waterloo, Waterloo, Ontario, CANADA
E. Atimtay, Middle East Technical University, Ankara, TURKEY
A. Bilgutay, Middle East Technical University, Ankara, TURKEY
N. Carmichael, University of Warwick, Coventry, ENGLAND
R. Caron, University of Waterloo, Waterloo, Ontario, CANADA
A. Ghali, University of Calgary, Calgary, CANADA
J. Gierlinski, University of Southampton, Southampton, ENGLAND
D. Grierson, University of California, Los Angeles, Calif., U.S.A.
E. Johnson, Virginia Polytechnic Institute, Blacksburg, VA.,U.S.A.
A. Kildegaard, Aalborg University Center, Aalborg, DENMARK
O. Lev, Merritt Consulting Engineers, Redlands, California, U.S.A.
B. Lysik, Technical University of Wroclaw, Wroclaw, POLAND
J. McNabb, Bradley University, Peoria, Illinois, U.S.A.
C. Ng, David Taylor Res. & Dev. Center, Bethesda, MD, U.S.A.
T. Panzeca, Istituto di Scienza e Delle Costruzioni, Palermo,ITALY
M. Papadrakakis, Athens, GREECE
G. Payre, Universite de Sherbrooke, Sherbrooke, Quebec, CANADA
S. Tang, Ford Research Laboratory, Dearborn, Michigan, U.S.A.
G. Turvey, University of Lancaster, Lancaster, ENGLAND

TABLE OF CONTENTS

PREFACE

These proceedings contain lectures and contributed papers presented at the NATO-NSF Advanced Study Institute on Optimization of Distributed Parameter Structures (Iowa City, Iowa 21 May - 4 June, 1980). The institute was organized by E. Haug and J. Cea, with the enthusiastic help of leading contributors to the field of distributed parameter structural optimization. The principle contributor to this field during the past two decades, Professor William Prager, participated in planning for the Institute and helped to establish its technical direction. His death just prior to the Institute is a deep loss to the community of engineers and mathematicians in the field, to which he made pioneering contributions.

The proceedings are organized into seven parts, each addressing important problems and special considerations involving classes of structural optimization problems. The review paper presented first in the proceedings surveys contributions to the field, primarily during the decade 1970-1980. Part I of the proceedings addresses optimality criteria methods for analyzing and solving problems of distributed parameter structural optimization. Optimality criteria obtained using variational methods of mechanics, calculus of variation, optimal control theory, and abstract optimization theory are presented for numerous classes of structures; including beams, columns, plates, grids, shells, and arches. Optimality criteria and numerical methods based on these criteria represent a reasonably well developed field that is thoroughly covered in Part I. A special topic receiving considerable attention here is the emerging problem of occurrance of repeated eigenvalues in optimal designs, for both buckling and vibration. New results are presented and directions for future effort in this important class of problems are suggested.

Numerical methods for optimal remodeling and direct numerical optimization of structures are presented in Part II of the proceedings. The theme here is iterative optimization, primarily through linearization or other related approximations of the general nonlinear structural optimization problem. Numerical methods for vibration, buckling, displacement, and dynamic constraints are treated. Direct numerical methods for treating repeated eigenvalues are suggested and examples are presented.

Part III is devoted to optimization of structures under earthquake loads. The essentially nonlinear dynamic behavior of such structures is treated in detail and iterative optimization methods that are suitable for solution of such structural optimization problems are presented. A software system that has been used in earthquake structural optimization is presented.

Contributed papers on finite dimensional structural optimization are presented in Part IV. A survey of nonlinear programming methods and their application to large scale structures, using finite element methods, is presented. Numerous special applications found in papers presented in this part of the proceedings.

Part V is devoted to optimization of structures under nonconservative loading and other special problems of distributed parameter structural optimization. A survey of optimization under nonconservative loading is presented. Related problems in flutter optimization and other complex problems of distributed parameter structural optimization are presented.

Part VI is devoted to the shape optimal design problem. Since only limited work in the field of structural shape optimization has been done, lectures in this part of the proceedings focus on shape optimization examples from diverse fields of applied physics. Mathematical and numerical methods for solving such problems are presented to serve as a guide for future work in shape optimal design of structures. The coherence of the mathematical theory of shape optimization and results obtained for problems other than structures show great potential for development of this field in the future.

The final part of the proceedings, Part VII, presents a thorough treatment of design sensitivity analysis of structural systems. Recently developed, rigorous mathematical theories of design sensitivity analysis are presented and their application to structural optimization problems is illustrated. Results reported in this part of the proceedings are intended to serve as a foundation for future work in both optimality criteria and direct iterative methods for distributed parameter structural optimization.

The extent and variety of the lectures and papers presented in these proceedings illustrate the extensive contribution of numerous individuals in preparation and conduct of the Institute. The Institute directors wish to thank all contributors to these proceedings for their substantial effort. Special thanks go to T. Brannian for her efforts in the administrative planning and support of the Institute and for her typing of preliminary manuscripts. The dedicated and tireless support of R. Huff, K. Walters and A. Craven in typing the final copy of the entire manuscript is greatly appreciated. Thanks are also due to Dr. K.K. Choi and graduate assistants H. Lam, J. Hou, Y. Yoo, A. Shabana, A. Belegundo in proofreading of the final text. Finally, without the financial support of the NATO Office of Scientific Affairs and the U.S. National Science Foundation, the Institute and these proceedings would not have been possible. Their support is gratefully acknowledged by all concerned with the Institute.

December 1980 E. J. Haug

J. Cea

Part 5

NONCONSERVATIVE LOADING AND OTHER PROBLEMS

STRUCTURAL OPTIMIZATION UNDER NONCONSERVATIVE LOADING*

Terrence A. Weisshaar† and Raymond H. Plaut§

Department of Aerospace and Ocean Engineering† and
Department of Civil Engineering§
Virginia Polytechnic Institute and State University
Blacksburg, Virginia

ABSTRACT

Basic problems of optimization of structures under nonconservative loads are discussed and literature pertaining to their solution is reviewed. Problems associated with buckling and flutter instability of columns and with flutter instability of aerolastic structures are treated. Emphasis is on continuously varying design formulation, or a distributed parameter model.

1. INTRODUCTION

This paper reviews work on the optimization of elastic structures subjected to nonconservative loads, in which the stability of the structure is the governing factor. The structures considered have a trivial equilibrium configuration for all values of the load. As the load is increased quasi-statically from zero, this configuration becomes unstable at a critical value of the load.

Two types of instability are possible, divergence or flutter. Divergence, sometimes called buckling or static instability, corresponds to a monotonic increase of displacements from the equiliri-um state. Flutter, sometimes called dynamic instability, corresponds to oscillations with increasing amplitude. Most problems involving nonconservative loads exhibit flutter when instability.

*This work has been supported in part by the U.S. National Science Foundation, under Grant No. ENG77-17487.

occurs. Conservative systems (in which the work done by the loads is independent of the path taken by the system) can only become unstable by divergence. Mathematically, the linearized problem for a conservative system is self-adjoint, while that for a nonconservative system is non-self-adjoint. The loads appear as eigenvalues in the equations of motion. One type of nonconservative load is the so-called follower load. This load has a constant magnitude, but acts in a direction that depends upon the configuration of the structure. Tangential loads are the most common example of this type. Aerodynamic loads form another category that has received extensive consideration. The loading parameter in the latter case is related to the speed of the airflow past the structure.

Optimization of these structures occurs when mass distribution of the structure is allowed to vary and the optimal distribution is sought. In the minimum-weight formulation, the critical load is specified and the total mass is to be minimized. In the maximum-load formulation, the total mass is held fixed, while the critical load is to be maximized. Optimization problems involving follower loads are reviewed in Section 2. The types of problems treated in the literature are described, and the basic methods of solution are discussed. Section 3 then describes work involving aerodynamic loads. In Section 4, observations and suggestions for further research are presented, followed by a list of references.

2. OPTIMIZATION WITH FOLLOWER LOADS

References 1 to 14 deal with optimal design of elastic columns subjected to follower loads. Most of these studies involve cantilevered columns, such as that shown in Fig. 2.1. In nondimensional terms, $w(x,t)$ is the transverse deflection at the coordinate x and time t, P is a tangential end load, and $qg(x)$ is a distributed tangential load. The trivial configuration $w(x,t)=0$ is a static equilibrium state for all values of P and q.

Let $m(x)$ denote the mass per unit length along the column, and let $s(m)$ be the bending stiffness. Problems that involve only concentrated masses are not discussed here. Also, consider Kelvin-Voigt internal damping with coefficient η, and external damping, with coefficient β. The linear differential equation of motion is given by [4,6]

$$(sw'')'' + (P + qf)w'' + m\ddot{w} + \eta(s\dot{w}'')'' + \beta\dot{w} = 0 \qquad (2.1)$$

with boundary conditions

$$w = w' = 0 \;, \qquad \text{at } x = 0 \qquad (2.2)$$

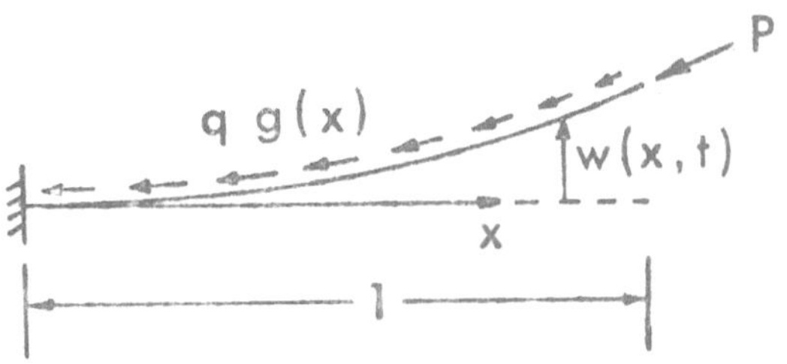

Figure 2.1 Cantilever Subjected to Follower Loads

and

$$sw'' + \eta s\dot{w}'' = (sw'')' + \eta(s\dot{w}'')' = 0 \quad , \qquad \text{at } x = 1 \qquad (2.3)$$

where

$$f(x) = \int_x^1 g(\xi)\, d\xi \qquad (2.4)$$

and primes and dots denote partial differentiation with respect to x and t, respectively.

The case of an end load alone (i.e., $q=0$) is often called Beck's column. Optimization of Beck's column with no damping is carried out in Refs. 2, 3, 5,6,7 and 13, while damping is included in Ref. 6. Hauger's column, for which $P=0$ and $g(x) = 1 - x$, is treated in Ref. 5 with no damping and in Refs. 4, 9, 10, and 11 with damping. The internal damping in Ref. 10 is more general than in Eq. 2.1. Reference 14 considers the case $P = 0$, $g(x) = 1$, and no damping.

In Ref. 1, an anti-tangential end load is considered, whose direction has the slope $\alpha w'(1,t)$, with $\alpha < 0$. Reference 8 treats a column moving in space under a tangential end load P, as a model of a flexible missile. Finally, a simple model of an idealized wing with an engine at its tip is analyzed in Ref. 12. The end load acts in a direction perpendicular to the tangent at the end of the cantilever beam. Bending and torsion are included in the equations. These last three references do not involve damping.

The column subjected to an anti-tangential end load in Ref. 1 becomes unstable by divergence. In that case, as for conservative problems, a static analysis using the equation of equilibrium can be used. For the problems in Refs. 2 to 14, flutter occurs at the critical load and a dynamic analysis is required.

In Ref. 1, the bending stiffness s is assumed to be proportional to the mass per unit length m, as for a column with a rectangular cross section of varying width. A linear relationship s(m) is used in Ref. 2, corresponding to a sandwich cross section with a core of constant height. If the cross sections have similar shape (e.g., circular, or rectangular with constant ratio of height to width), s is proportional to the square of m. This case is considered in Refs. 3 to 7, 9, 11, 13 and 14. Finally, the case of s proportional to the cube of m, corresponding to a rectangular cross section with varying height, is treated in Refs. 4, 9, and 10.

Of Refs. 1 to 14, only Refs. 1 and 3 utilize the minimum-weight formulation (with specified critical load). The others maximize the critical load, subject to a specified total mass.

In order to obtain an optimum design numerically, Refs. 2, 4, 5, 6 and 8 to 14 discretize the mass distribution and the deflection function. A piecewise constant mass distribution is used in Refs. 5, 6, 11, 12, and 14, a piecewise linear shape is used in Ref. 13, a parabolic-type shape is used in Refs. 4, 9, and 10, and a truncated Fourier series is used in Ref. 8. Reference 13 utilizes the finite element method, while the others approximate the deflection function by a truncated series of functions with arbitrary coefficients. Gradient-type iteration techniques are usually employed to obtain the optimum design. They are often based on an adjoint variational principle. Some alternative optimization procedures are presented in Refs. 1, 3, and 7.

As an illustrative example, consider Beck's column with no damping and let

$$w(x,t) = \sum_{n=1}^{\infty} (A_n \cos \omega_n t + B_n \sin \omega_n t)\, u_n(x) \tag{2.5}$$

where $u_n(x)$ represents a vibration mode and ω_n is the corresponding frequency. Then, Eqs. 2.1 to 2.3 yield the following boundary value problem for the modes (with subscripts deleted):

$$(su'')'' + P u'' - \omega^2 m u = 0 \tag{2.6}$$

with boundary conditions

$$u = u' = 0 \,, \quad \text{at } x = 0 \text{ , and } \quad su'' = (su'')' = 0 \,, \quad \text{at } x = 1 \tag{2.7}$$

The adjoint problem, called Reut's problem, with mode v(x), satisfies the same equation as u(x), but has boundary conditions [3,5]

$$v = v' = 0, \quad \text{at } x = 0, \quad \text{and } sv'' + Pv = (sv'')' + Pv' = 0, \quad \text{at } x = 1 \tag{2.8}$$

The functional

$$J = \frac{1}{2} \int_0^1 (su''v'' + Pu''v - \omega^2 muv)\, dx \tag{2.9}$$

is stationary with respect to variations in u and v that satisfy the respective kinematic boundary conditions (i.e., the conditions at x = 0 in Eqs. 2.7 and 2.8).

In Eqs. 2.6, P and ω^2 are eigenvalues. Plots of P versus ω^2 are called characteristic curves. For uniform mass distribution, the first few characteristic curves for Beck's column are depicted in Fig. 2.2. The variable P_c denotes the critical load, which corresponds to the onset of flutter instability. For $0 \leq P < P_c$, all eigenvalues ω^2 are real, distinct, and positive. At P_c, two frequencies coalesce, and then become complex for higher values of the load. There are an infinite number of flutter loads, at which two real frequencies coalesce, and P_c is the lowest. The characteristic curves for the adjoint problem are the same as those in Fig. 2.2.

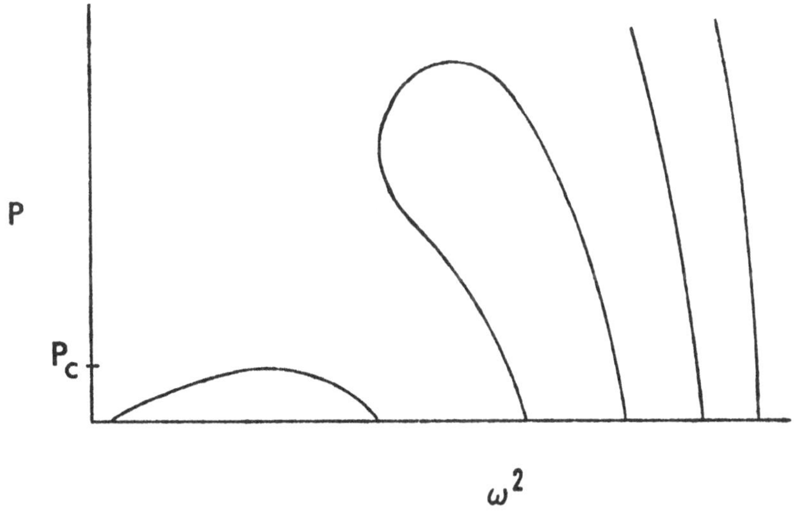

Figure 2.2 Characteristic Curves for Unform Beck's Column [13]

From the adjoint variational principle, a generalized Rayleigh quotient can be formulated for the load, as follows:

$$P = -\frac{\int_0^1 su''v''dx - \omega^2 \int_0^1 muv\,dx}{\int_0^1 u''v\,dx} \tag{2.10}$$

This quotient is stationary with respect to kinematically admissible variations in u and v. At P_c, the slope $dP/d(\omega^2)$ is zero for the appropriate characteristic curve, hence differentiating Eq. 2.10, one obtains the flutter condition [15]

$$\int_0^1 muv\,dx = 0 \tag{2.11}$$

involving the appropriate flutter modes u and v.

Consider the maximum-load formulation, with the total mass fixed, i.e.

$$\int_0^1 m\,dx = 1 \tag{2.12}$$

in nondimensional terms. Define the functional F by

$$F = P + \lambda\left(\int_0^1 m\,dx - 1\right) \tag{2.13}$$

where λ is a Lagrange multiplier and P is defined by Eq. 2.10. If the mass distribution is varied and if a single flutter load is maximized at the optimum design, then the variation of F with respect to m must be zero at that design. This condition yields the pointwise equation [3]

$$\frac{ds}{dm} u''\,v'' - \omega^2\,uv = \text{constant} \tag{2.14}$$

if the denominator in Eq. 2.10 is normalized. Equation 2.14 is called the optimality criterion, or optimality condition.

Let P_{ij} denote the flutter load corresponding to coalescence of the ith and j^{th} frequencies, where the frequencies are numbered in ascending order according to their values at P = 0. For the uniform Beck's column, $P_c = P_{12}$ (Fig. 2.2). Claudon [5] and

Hanaoka and Washizu [13] form a gradient-type expression for P_{12}, based on Eqs. 2.10 and 2.11, and increase P_{12} iteratively, with small changes in the mass distribution. At a particular point in the procedure, however, with the second flutter load P_{34} decreasing while P_{12} is increasing, P_{12} and P_{34} become equal, as shown in Fig. 2.3. Continuation of the procedure to maximize P_{12} would not make sense, because the critical load would switch to P_{34}, i.e. P_{34} would become smaller than P_{12}. Failure to recognize this switching can lead to erroneous results [7]. The optimality criterion given by Eq. 2.14, with u and v corresponding to P_{12}, is based on stationarity of P_{12} and is not valid at the optimum design in this problem.

At this point in the procedure, Hanaoka and Washizu [13] consider maximization of the pair of flutter loads, P_{12} and P_{34}, constraining them to be equal. Such a bimodal optimization was earlier presented by Olhoff and Rasmussen [16] for a conservative buckling problem (see also Refs. 17 to 20). For flutter instability, if two flutter loads are equal at the optimum design, Masur and Mróz [19] derive the optimality criterion.

$$(1-\gamma)\left(\frac{ds}{dm}u''v'' - \omega^2 uv\right) + \gamma\left(\frac{ds}{dm}\hat{u}''\hat{v}'' - \hat{\omega}^2\hat{u}\hat{v}\right) = \text{constant}$$

$$0 \le \gamma \le 1 \quad , \tag{2.15}$$

where u, v, ω correspond to one flutter load and $\hat{u}$, $\hat{v}$, $\hat{\omega}$ to the other. For columns under conservative loads, one can let v = u, $\hat{v} = \hat{u}$, and $\omega^2 = \hat{\omega}^2 = 0$ in Eq. 2.15, to obtain the type of bimodal optimality criterion given in Refs. 16, 18, and 19. A different criterion is presented by Haug [20]. In the conservative case, the mode shapes are not uniquely determined, whereas they are for the flutter case depicted in Fig. 2.3.

As Hanaoka and Washizu [13] numerically increase the equal flutter loads P_{12} and P_{34} simultaneously, it is found that P_c makes another switch, as depicted in Figs. 2.4 and 2.5. In this case, P_c drops suddenly from $P_c = P_{12} = P_{34}$ in Fig. 2.4 to $P_c = P_{23}$ in Fig. 2.5. This type of phenomenon will be called discontinuous switching, whereas the situation in Fig. 2.3 will be called continuous switching. That is, as the mass distribution is varied in a continuous manner, discontinuous switching refers to a change in the set of modes that correspond to the critical load, together with a discontinuous change in the objective function (here P_c), while in continuous switching the objective function does not exhibit a sudden change.

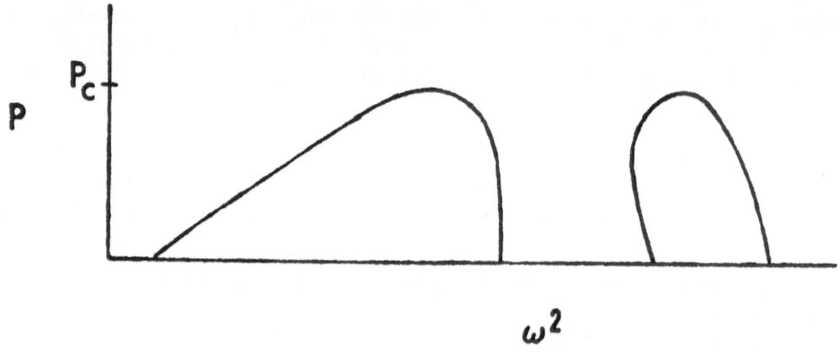

Figure 2.3 Characteristic Curves at Continuous Switching for Beck's Column [13]

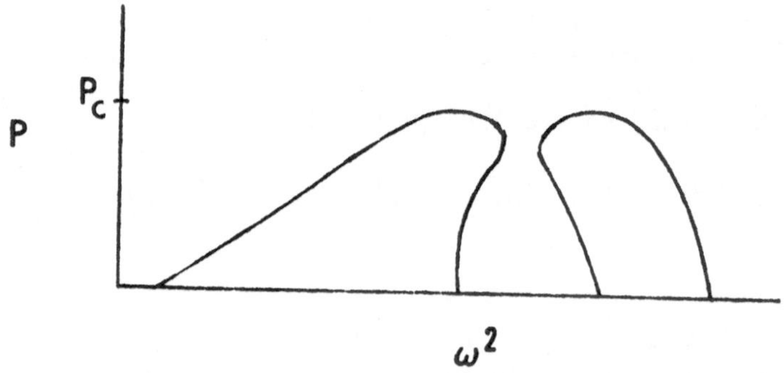

Figure 2.4 Characteristic Curves Before Discontinuous Switching for Beck's Column [13]

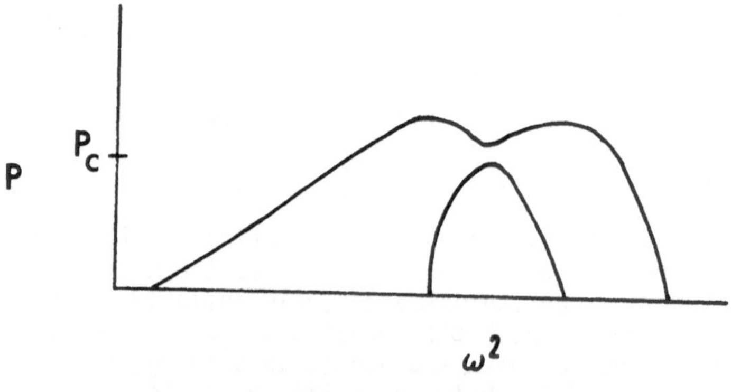

Figure 2.5 Characteristic Curves after Discontinuous Switching for Beck's Column [13]

For the undamped Beck's column, based on these numerical results from Ref. 13, neither a single flutter load nor a pair of equal flutter loads satisfies a stationarity condition at the optimum design, which corresponds to the characteristic curves in Fig. 2.4.

Characteristic curves for undamped problems, with other types of loading than Beck's column, are computed by Claudon in Refs. 5, 12, and 14. For uniform mass distribution, the curves have shapes similar to those in Fig. 2.2. As P_{12} is increased iteratively, switching occurs discontinuously. Typical results are illustrated in Figs. 2.6 and 2.7 for Hauger's column [5]. Here, P_c drops suddenly from P_{12} in Fig. 2.6 to P_{23} in Fig. 2.7. Again, a stationarity condition is not satisfied at the optimum design.

During the transition from Fig. 2.4 to Fig. 2.5 and from Fig. 2.6 to Fig. 2.7, i.e. at the moment of discontinuous switching of this type, the characteristic curves make contact, as sketched in Fig. 2.8. At the contact point A, the mode shapes are uniquely determined [13]. The slopes of the characteristic curves, determined by differentiating Eq. 2.10, become degenerate at A. Therefore, the denominator of Eq. 2.10 and the second integral in the numerator are zero at A (Claudon, private communication, 8/27/76). In addition, since P is finite, the first integral in the numerator must also vanish at A.

Another type of discontinuous switching that can occur during an optimization procedure is described by Sundararajan [8]. It

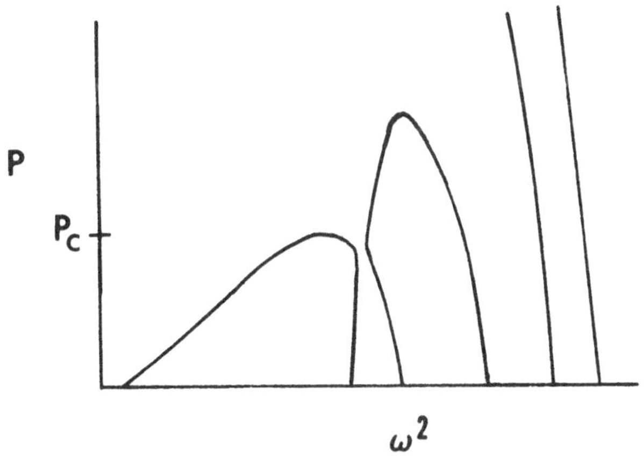

Figure 2.6 Characteristic Curves Before Discontinuous Switching for Hauger's Column [5]

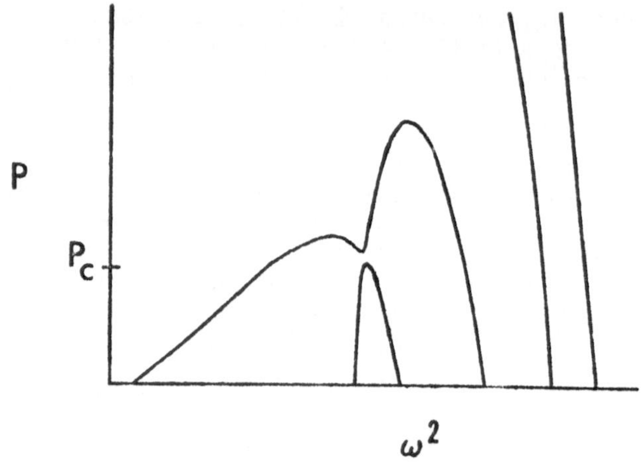

Figure 2.7 Characteristic Curves after Discontinuous Switching for Hauger's Column [5]

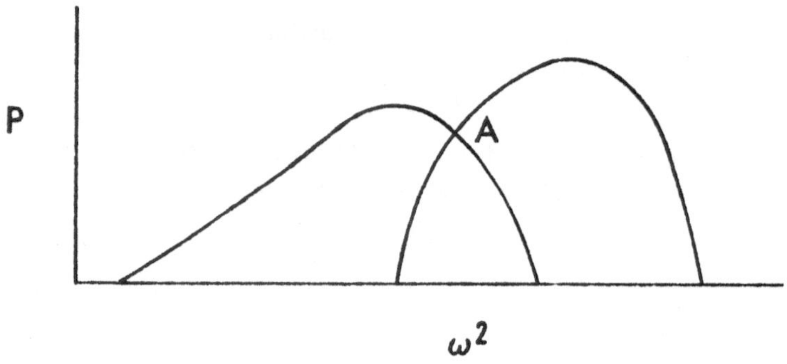

Figure 2.8 Intersection of Characteristic Curves at Discontinuous Switching [5,13]

involves a change in the critical load from a flutter load to a divergence load, or vice versa. For example, one may be increasing the flutter load P_{12}, as in Fig. 2.9, but if the characteristic curve intersects the load axis, as in Fig. 2.10, the critical load shifts abruptly to the value at which the lowest frequency becomes zero, which corresponds to the onset of divergence instability. Alternatively, if a divergence load is being increased in an iterative manner, a transition from Fig. 2.10 to Fig. 2.9 can occur. At the moment of discontinuous switching, the first characteristic curve is tangent to the load axis, say at a point B. For the modes at B, the corresponding frequency of

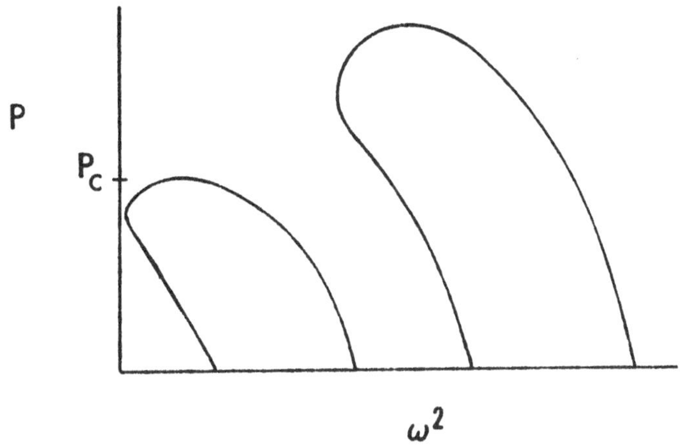

Figure 2.9   Characteristic Curves with Critical Load at Flutter

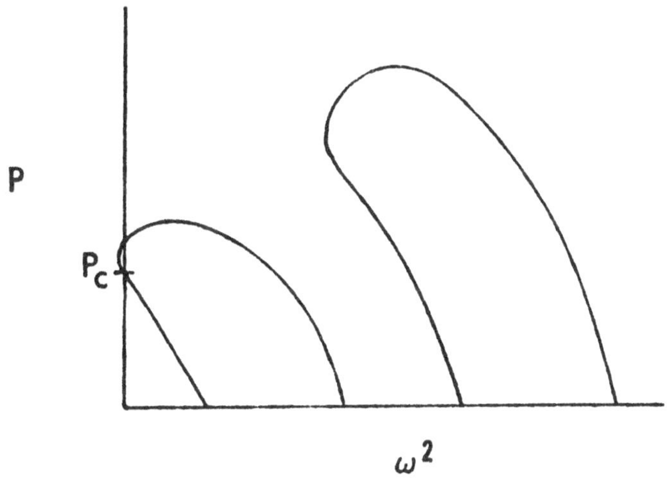

Figure 2.10   Characteristic Curves with Critical Load at Divergence

vibration is zero, the denominator of Eq. 2.10 vanishes (since the slope of the characteristic curve is infinite), and the first integral in the numerator vanishes (since P is finite).

3. AEROELASTIC OPTIMIZATION OF DISTRIBUTED PARAMETER SYSTEMS

In contrast to most work on optimization with follower loads, aeroelastic optimization of structures usually involves a search for minimum-weight structural configurations that satisfy aeroelastic performance requirements, such as constraints on minimum flutter or divergence speeds. Aeroelastic optimization of actual flight structures must, of necessity, deal with discrete parameter models, such as finite element representations. However, the utility of solutions to highly idealized distributed parameter systems should not be underestimated, particularly when it comes to understanding the complicated optimization process itself.

Numerous literature reviews have appeared in the past decade, which place developments in the aeroelastic optimization of distributed parameter systems into historical perspective. Among these reviews, those of Armand [21], Pierson [22], Niordson and Pedersen [23], McIntosh [24], Olhoff [25], and Taylor [26] are particularly useful. In addition, the excellent review by Stroud [27], although focusing attention on discrete parameter systems, is necessary reading for those interested in structural optimization with flutter constraints.

By its nature, the aeroelastic instability of structures is a complicated phenomenon. The aerodynamic forces are nonconservative and are governed by complex phase relationships between the motion of the system and the forces arising from this motion. Perhaps the simplest representation of aerodynamic forces occurs in the classical panel flutter problem [28]. In this problem, quasi-steady piston theory is used to generate the motion dependent forces present on a panel operating in a supersonic airstream. Neglecting damping terms allows one to formulate a problem for which instability occurs through frequency coalescence.

The relative simplicity of the panel flutter equations of motion has led to the popularity of this problem in the literature on aeroelastic optimization. Numerous studies have been published. References 29-33 are typical of the early papers that consider the one-dimensional panel flutter problem, which arises when the extent of the panel in the cross-stream direction is sufficiently large to neglect the effect of boundary conditions on two edges (see Fig. 3.1). In nondimensional terms, similar to Eqs. 2.6 and 2.7, the governing equation for this problem is given by

$$(su'')'' + \lambda u' - \omega^2 mu = 0 \tag{3.1}$$

with boundary conditions

$$u = su'' = 0 \quad , \qquad \text{at } x = 0,1 \tag{3.2}$$

where λ is the airspeed parameter. The flow is in the positive x-direction and the edges $x = 0,1$ are simply supported [34]. The total mass is to be minimized, subject to a specified (nondimensional) flutter speed $\lambda = \lambda_c$, where λ_c is the critical airspeed parameter for a uniform thickness reference panel.

The adjoint problem of Eqs. 3.1 and 3.2, with mode v(x), corresponds to airflow in the negative x-direction and has the differential equation

$$(sv'')'' - \lambda\, v' - \omega^2\, mv = 0 \tag{3.3}$$

and the same boundary conditions as Eq. 3.2. The functional

$$G = \frac{1}{2}\int_0^1 (su''v'' + \lambda\, u'v - \omega^2\, muv)\, dx \tag{3.4}$$

is stationary with respect to variations in u and v satisfying u = v = 0, at x = 0,1. The flutter condition is given by Eq. 2.11 and the optimality criterion by Eq. 2.14 [35].

The most popular approach to the solution of the panel flutter optimization problem stems from methods of optimal control theory. Weisshaar [30,31] relies on a transition matrix, or shooting technique, to solve the nonlinear differential equations (obtained from control theory) that govern optimality of the design. Pierson [32,33] applies gradient projection methods that are also adapted from control theory, with great success.

The optimal design for thickness distribution of the face sheets of a sandwich panel is sketched in Fig. 3.2 [31]. A minimum thickness constraint is included, in order to avoid a solution with points of zero thickness. The design is symmetric about the center, since optimum design for the original problem and for the adjoint problem (with opposite direction of airflow) must be the same [29]. It differs significantly from optimum designs for a conservative, simply supported column with constraints on buckling or fundamental natural frequency, where the thickness is maximum at the center.

A report by Armand [36] develops a framework for the utilization of optimal control theory to solve continuous, two-dimensional structural optimization problems. Librescu and Beiner [37] first formulated the two-dimensional panel flutter optimization problem, but present no solutions. Pierson and Genalo [38] used a gradient projection technique to successfully obtain numerical solutions to some two-dimensional panel flutter problems, with some difficulty.

Notable by its absence is any solution to the optimization of a continuous wing structure with a flutter constraint. Turner [29]

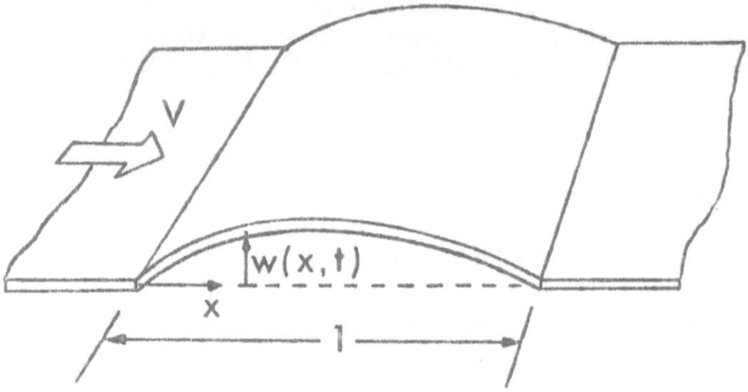

Fiqure 3.1 One-Dimensional Panel Flutter Problem Geometry

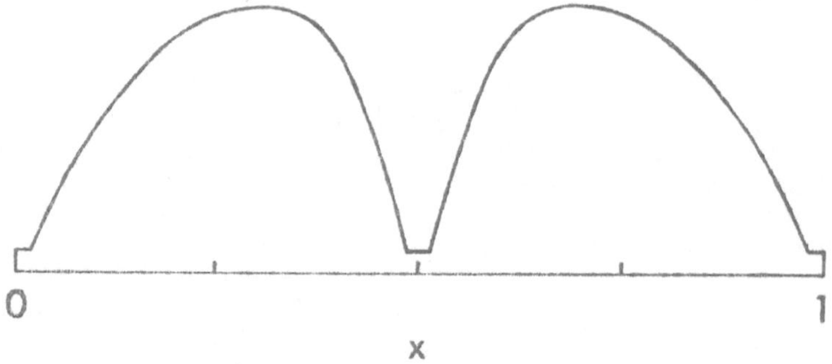

Fiqure 3.2 Optimal Thickness Distribution for One-Dimensional Panel [31]

suggests such a problem, as do McIntosh, Weisshaar, and Ashley [39]. The problem is computationally difficult, but by no means impossible, particularly in light of the development of computational techniques in the decade that has ensued since its proposal.

The more usual case of aerodynamic flutter does not simply involve frequency coalescence, and is much more complicated. The aerodynamic forces are dependent upon such factors as Mach number and the reduced frequency of oscillation $k = \omega b/V$, where ω is the oscillatory frequency, b is the wing semi-chord, and V is the flight speed. Solutions to aeroelastic optimization problems involving continuous structures subjected to such forces (e.g., the continuous wing structure mentioned above) are needed.

Static aeroelastic optimization problems have not been treated extensively in the literature. Divergence of sweptback or unswept aircraft wings is usually not a design problem. However, for recently proposed forward swept wing concepts [40], divergence control becomes extremely important.

Optimization of a straight or unswept wing, with a specified divergence constraint, is discussed by Armand and Vitte [41]. Gwin [42] investigates the swept wing divergence optimization problem with a discrete parameter system. To date, no studies exist that consider the divergence optimization of a distributed parameter forward swept wing with bending-torsion deformation freedom. The interested reader is referred to Bisplinghoff, Ashley, and Halfman [43] for a general discussion of the aeroelastic problem of divergence.

Research in the area of distributed parameter optimization has taken on added importance in the past several years. This stems from efforts to develop more efficient techniques than the mathematical programming methods that are usually applied to solve discrete system optimization problems. One alternative is the formulation of optimality criteria for aeroelastic optimization, coupled with resizing algorithms. A recent review of optimality criteria for several classes of structural optimization problems is presented by Venkayya [44]. Haftka, et al. [45] discuss the use of optimality criteria, together with resizing algorithms, for aeroelastic optimization. McIntosh and Ashley [46] suggest alternative optimality criteria for discrete structures with flutter constraints and obtain favorable computational efficiencies, when compared to mathematical programming results. Clearly, developments in this area are of significance to discrete structure optimization.

Conversely, developments in the optimization of discrete parameter structures have significance to continuous system optimization. In particular, the possibility of mode switching during the optimization procedure is cause for concern. Stroud discusses this problem at length in Reference 27. Rather than examining the character of the solution in the load-frequency domain, as described in Section 2 for undamped follower load problems, the general behavior of the aerodynamic flutter solution is usually investigated in terms of an artificial damping parameter g, as a function of flight speed V. Positive values of g correspond to flight speeds at which flutter will occur.

Figure 3.3 shows a typical V-g curve, in which the lowest flutter speed V_f occurs in Mode 1, while the next flutter speed occurs in Mode 2 and is widely separated from that of Mode 1. In this case, efforts at optimization can be concentrated on Mode 1.

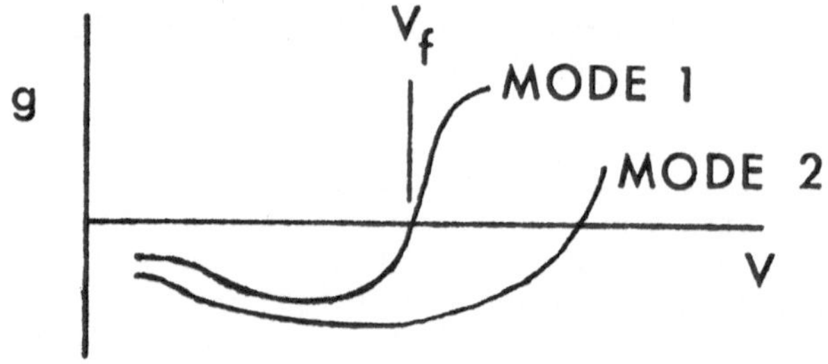

Figure 3.3 Isolated Flutter Mode, No Switching [27]

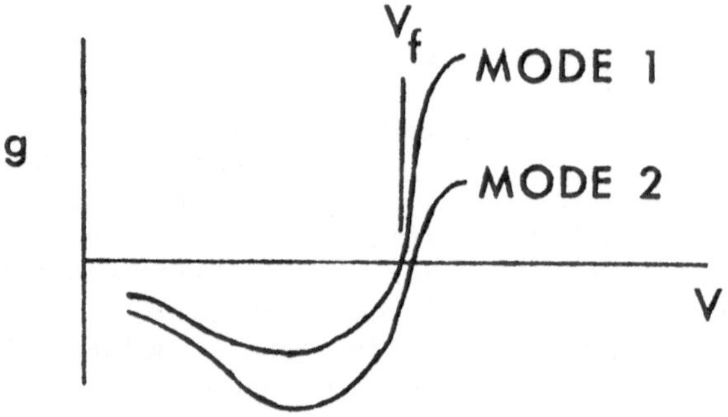

Figure 3.4 Neighboring Modes, Possibility of Continuous Switching [27]

Two closely spaced modes of instability are illustrated in Fig. 3.4. During optimization, Modes 1 and 2 may switch back and forth with one and then the other becoming critical. This would correspond to continuous switching, as defined in Section 2.

Figure 3.5 illustrates a curious, but not uncommon, case of the so-called hump mode. Initially, Mode 2 may be critical. However, as optimization proceeds, the hump in Mode 1 may rise so that it penetrates the region $g > 0$. If this occurs, the flutter speed will suddenly drop. Discontinuous switching occurs in this case, since the flutter speed will be a discontinuous function of changes in the mass distribution.

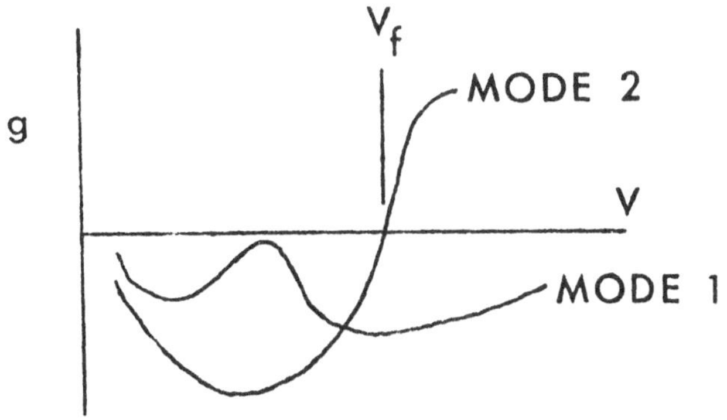

Figure 3.5 Hump Mode, Possibility of Discontinuous Switching [27]

4. OBSERVATIONS AND SUGGESTIONS

For the aeroelastic optimization of actual structures, it is natural to specify that instability should not occur below a certain airspeed and then to minimize the structural weight. Papers involving follower loads, however, usually employ the maximum-load formulation. Mathematically, each of these formulations has advantages and disadvantages. In the first instance, that of weight minimization, the objective function is just the integral of the mass distribution, while in the second it may also involve vibration frequencies and vibration modes (as in Eq. 2.10.). The optimization technique may be more efficient with a simpler objective function. On the other hand, the constraint in the maximum-load formulation, Eq. 2.12, can be satisfied easily by appropriately scaling any design. In contrast, the critical load is obtained from the solution of an eigenvalue problem and cannot be specified explicitly in terms of the mass distribution. Hence, the admissible class of mass distributions for the minimum-weight formulation cannot be identified as simply as in the maximum-load formulation.

The phenomenon of mode switching appears to be an important factor in many optimization problems involving nonconservative loads and may occur continuously or discontinuously. For the buckling of conservative systems, continuous switching has been noted [16], but discontinuous switching would be unlikely. For one thing, a change between divergence and flutter instability (as in Figs. 2.9 and 2.10) is not possible, since only divergence can occur. Also, the characteristic curves for structures such as columns and plates under conservative compressive loads always have negative slopes, as depicted in Fig. 3.6, and then switching can only occur continuously when two characteristic curves intersect at P_c on the load axis. Discontinuous switching would also

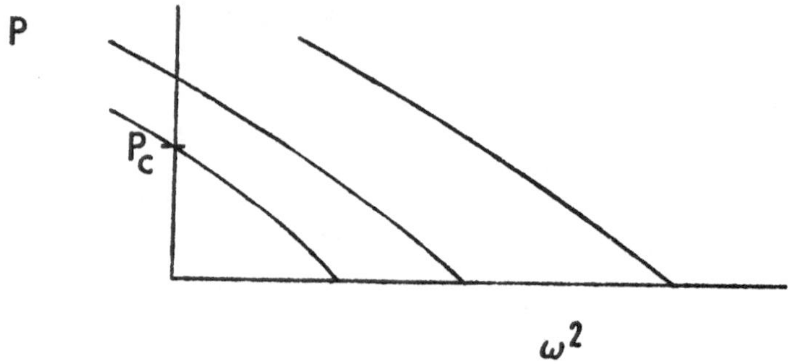

Figure 3.6 Typical Characteristic Curves for Conservative Columns and Plates

be unlikely in the optimization of structures with respect to natural frequencies [47].

As an example of the potential difficulties due to mode switching, consider the undamped Beck's column and mass distributions satisfying Eq. 2.12. The critical load is $P_c = P_{12}$ for one set of these distributions, $P_c = P_{23}$ for another set, $P_c = P_{34}$ for another set, $P_c = P_{12} = P_{34}$ for another set, and so forth. An iterative technique will usually involve the governing instability modes and therefore will be confined to one of these sets at a time. Determination of the transition from one set of mass distributions to another is crucial, and the development of an efficient method for sensing this transition is highly desirable.

In Reference 13, when P_{12} becomes equal to P_{34} (Fig. 2.3), the technique is altered and optimization is carried out among the set of mass distributions for which $P_c = P_{12} = P_{34}$. A similar procedure is used in Ref. 16. However, it may be more profitable to attempt an optimization of P_{34} by itself. Of course, this would only work if the increase in P_{34} were accompanied by a larger increase in P_{12}, so that P_{34} would remain critical. If not, optimization of the double eigenvalue would be appropriate. As noted in Section 2, a double flutter load is generally associated with two unique mode shapes, with different frequencies, and therefore is inherently different from the case of a double divergence load.

During their maximization of the double flutter load, Hanaoka and Washizu [13] find that they can delay the occurrence of discontinuous switching (Figs. 2.4 and 2.5) by adding a constraint with attempts to keep the second and third frequencies separated. An understanding of the behavior of the characteristic curves during the optimization may therefore enable one to add suitable constraints, or to otherwise modify the numerical procedure, in order to improve the final design of the structure.

Due to mode switching, optimality criteria, such as Eqs. 2.14 and 2.15, may not be valid at the optimum design, and hence may be misnomers. They should be stated with the appropriate qualifications. Nevertheless, they may be usefully employed during iterative procedures. Optimality criteria methods have not been used in follower load problems, nor have they been used in many aeroelastic optimization studies. Since they have been successfully applied to a variety of other problems in optimal structural design [44], they might lead to efficient algorithms in nonconservative problems as well.

In follower load investigations, some authors have considered a very restrictive class of mass distributions (e.g., parabolic-type shapes). Other papers have assumed a restrictive class of deflection functions. The proximity of the resulting mass distribution to the actual optimum design of the distributed parameter system is questionable. Also, the condition governing mode switching at the optimum design (if any occurs) may be different in the continuous system. Further study of undamped follower load problems and their characteristic curves is recommended, with the aim of improving the solution techniques and results. Also, an investigation of possible mode switching in damped follower load problems should be undertaken, perhaps utilizing methods already developed in aeroelastic optimization.

The inclusion of damping in the analysis, such as in Eq. 2.1, can significantly alter the critical load, if instability is of the flutter type. It would be interesting to determine if damping also causes a large change in the optimum mass distribution, by comparing optimum designs for undamped and damped analyses of the same structure. Only limited results of this type are available (e.g. see References 6, 11, 33, and 34).

In conclusion, there is a need for continued research on structural optimization under nonconservative loading. Mode switching is not completely understood and recognized. Optimality criteria that include continuous and discontinuous switching are not established. Most numerical techniques are not well developed to handle such switching. Finally, not only do previous solutions to some basic follower load optimization problems need improvement, but some practical aeroelastic problems (such as basic wing flutter optimization) have not yet been solved.

REFERENCES

1. Zyczkowski, M. and Gajewski, A., "Optimal Structural Design in Non-Conservative Problems of Elastic Stability," Instability of Continuous Systems, (H. Leipholz, ed.), Springer-Verlag, Berlin, 1971, pp. 295-301.
2. Plaut, R.H., "On the Optimal Structural Design for a Nonconservative, Elastic Stability Problem," Journal of Optimization Theory and Applications, Vol. 7, 1971, pp. 52-60.
3. Vepa, K., "Generalization of an Energetic Optimality Condition for Non-Conservative Systems," Journal of Structural Mechanics, Vol. 2, 1973, pp. 229-257.
4. Anderson, G.L., "Optimal Design of a Cantilever Subjected to Dissipative and Nonconservative Forces," Journal of Sound and Vibration, Vol. 33, 1974, pp. 155-169.
5. Claudon, J.-L., "Characteristic Curves and Optimum Design of Two Structures Subjected to Circulatory Loads," Journal de Mécanique, Vol. 14, 1975, pp. 531-543.
6. Plaut, R.H., "Optimal Design for Stability Under Dissipative, Gyroscopic, or Circulatory Loads," Optimization in Structural Design, (A. Sawczuk and Z. Mróz, eds.), Springer-Verlag, Berlin, 1975, pp. 168-180.
7. Odeh, F. and Tadjbakhsh, I., "The Shape of the Strongest Column with a Follower Load," Journal of Optimization Theory and Applications, Vol. 15, 1975, pp. 103-118.
8. Sundararajan, C., "Optimization of a Nonconservative Elastic System with Stability Constraint," Journal of Optimization Theory and Applications, Vol. 16, 1975, pp. 355-378.
9. Thomas, C.R., "Mass Optimization of Non-Conservative Cantilever Beams with Internal and External Damping," Journal of Sound and Vibration, Vol. 43, 1975, pp. 483-498.
10. Thomas, C.R., "Stability and Mass Optimization of Non-Conservative Euler Beams with Damping," Journal of Sound and Vibration, Vol. 47, 1976, pp. 395-401.
11. Claudon, J.-L., "Détermination et Maximisation de la Charge Critique d'une Colonne de Hauger en Présence d'Amortissement," Zeitschrift für angewandte Mathematik und Physik, Vol. 29, 1978, pp. 226-236.
12. Claudon, J.-L. and Sunakawa, M., "Optimal Design of a Thin Cantilever Beam with a Transversal Follower Force," (in Japanese), presented at the 21st Lecture Series on Structural Mechanics, Naha, Okinawa, Japan, July, 1979.
13. Hanaoka, M. and Washizu, K., "Optimum Design of Beck's Column," Computers & Structures, to appear.
14. Claudon, J.-L., "Sur l'Optimisation de la Stabilite d'une Colonne Soumise a une Charge Circulatoire," to appear.
15. Plaut, R.H., "Determining the Nature of Instability in Nonconservative Problems," AIAA Journal, Vol. 10, 1972, pp. 967-968.

16. Olhoff, N. and Rasmussen, S.H., "On Single and Bimodal Optimum Buckling Loads of Clamped Columns," International Journal of Solids and Structures, Vol. 13, 1977, pp. 605-614.
17. Prager, S. and Prager, W., "A Note on Optimal Design of Columns," International Journal of Mechanical Sciences, Vol. 21, 1979, pp. 249-251.
18. Masur, E.F. and Mróz, Z., "Singular Solutions in Structural Optimization Problems," Variational Methods in the Mechanics of Solids (S. Nemat-Nasser, ed.), Pergamon Press, to appear.
19. Masur, E.F. and Mróz, Z., "Non-Stationary Optimality Conditions in Structural Design," International Journal of Solids and Structures, Vol. 15, 1979, pp. 503-512.
20. Haug, E.J., "Optimization of Distributed Parameter Structures with Repeated Eigenvalues," New Approaches to Nonlinear Problems in Dynamics (P.J. Holmes, ed.), SIAM Publications, 1980, to appear.
21. Armand, J.-L., "Applications of Optimal Control Theory to Structural Optimization: Analytical and Numerical Approach," Optimization in Structural Design (A. Sawczuk and Z. Mróz, eds.), Springer-Verlag, Berlin, 1975, pp. 15-39.
22. Pierson, B.L., "A Survey of Optimal Structural Design Under Dynamic Constraints," International Journal for Numerical Methods in Engineering, Vol. 4, 1972, pp. 491-499.
23. Niordson, F.I. and Pedersen, P., "A Review of Optimal Structural Design," Proceedings of the 13th International Congress of Theoretical and Applied Mechanics (E. Becker and G. K. Mikhailov, eds.), Springer-Verlag, Moscow, 1973, pp. 264-278.
24. McIntosh, S.C., Jr., "Structural Optimization via Optimal Control Techniques: A Review," Structural Optimization Symposium (L.A. Schmit, ed.), AMD-Vol. 7, American Society of Mechanical Engineers, New York, 1974, pp. 49-64.
25. Olhoff, N., "A Survey of the Optimal Design of Vibrating Structural Elements: Part I: Theory; Part II: Applications," The Shock and Vibration Digest, Vol. 8, No. 8, 1976, pp. 3-10; Vol. 8, No. 9, 1976, pp. 3-10.
26. Taylor, R.F., Development of Stability Methods for Application to Nonlinear Aeroelastic Optimization, Technical Report AFFDL-TR-79-3114, Air Force Flight Dynamics Laboratory, Wright-Patterson Air Force Base, July, 1979.
27. Stroud, W.J., "Automated Structural Design with Aeroelastic Constraints: A Review and Assessment of the State of the Art," Structural Optimization Symposium (L.A. Schmit, ed.), AMD-Vol. 7, American Society of Mechanical Engineers, New York, 1974, pp. 77-118.
28. Bisplinghoff, R.L., and Ashley, H., Principles of Aeroelasticity, Dover, New York, 1975, pp. 416-441.
29. Turner, M.J., "Optimization of Structures to Satisfy Flutter Requirements," AIAA Journal, Vol. 7, 1969, pp. 233-239.

30. Weisshaar, T.A., An Application of Control Theory Methods to the Optimization of Structures Having Dynamic or Aeroelastic Constraints, SUDAAR No. 412, Department of Aeronautics and Astronautics, Stanford University, October, 1970.
31. Weisshaar, T.A., "Aeroelastic Optimization of a Panel in High Mach Number Supersonic Flow," Journal of Aircraft, Vol. 9, 1972, pp. 611-617.
32. Pierson, B.L., "Panel Flutter Optimization by Gradient Projection," International Journal for Numerical Methods in Engineering, Vol. 9, 1975, pp. 271-296.
33. Pierson, B.L., "Aeroelastic Panel Optimization with Aerodynamic Damping," AIAA Journal, Vol. 13, 1975, pp. 515-517.
34. Plaut, R.H., "The Effects of Various Parameters on an Aeroelastic Optimization Problem," Journal of Optimization Theory and Applications, Vol. 10, 1972, pp. 321-330.
35. Plaut, R.H., "Elastic Minimum-Weight Design for Specified Critical Load," SIAM Journal on Applied Mathematics, Vol. 25, 1973, pp. 361-371.
36. Armand, J.-L., Applications of the Theory of Optimal Control of Distributed-Parameter Systems to Structural Optimization, NASA CR-2044, June, 1972.
37. Librescu, L. and Beiner, L., "The Weight Optimization Problem for Supersonic Rectangular Flat Panels with Specified Flutter Speed," Revue Roumaine des Sciences Techniques, Série de Mécanique Appliquée, Vol. 17, 1972, pp. 1087-1102.
38. Pierson, B.L. and Genalo, L.J., "Minimum Weight Design of a Rectangular Panel Subject to a Flutter Speed Constraint," Computer Methods in Applied Mechanics and Engineering, Vol. 10, 1977, pp. 45-62.
39. McIntosh, S.E., Jr., Weisshaar, T.A. and Ashley, H., Progress in Aeroelastic Optimization-Analytical Versus Numerical Approaches, SUDAAR No. 383, Department of Aeronautics and Astronautics, Stanford University, July, 1969.
40. Krone, N.J., "Divergence Elimination with Advanced Composites," AIAA Paper, No. 75-1009, 1975.
41. Armand, J.-L. and Vitte, W.J., Foundations of Aeroelastic Optimization and Some Applications to Continuous Systems, SUDAAR No. 390, Department of Aeronautics and Astronautics, Stanford University, January, 1970.
42. Gwin, L.B., "Optimal Aeroelastic Design of an Oblique Wing Structure," AIAA/ASME/SAE 15th Structures, Structural Dynamics, and Materials Conference, Paper No. 74-349, Las Vegas, Nevada, April, 1974.
43. Bisplinghoff, R.L., Ashley, H. and Halfman, R.L., Aeroelasticity, Addison-Wesley, Reading, Massachusetts, 1955.
44. Venkayya, V.B., "Structural Optimization: A Review and Some Recommendations," International Journal for Numerical Methods in Engineering, Vol. 13, 1978, pp. 203-228.

A METHOD OF DIRECT SOLUTION TO LINEAR INVERSE PROBLEMS

Morteza A.M. Torkamani

Department of Civil Engineering
University of Pittsburgh, Pittsburgh, Pennsylvania

ABSTRACT

A primal-dual decomposition technique is presented for solution of equations of structural mechanics.

1. INTRODUCTION

Many problems in engineering and science are converted to a system of linear simultaneous equations

$$[A]\{x\} = \{y\} \tag{1.1}$$

Such a set of equations may arise by approximation of a continuous relationship, $y(t) = A[t,\tau,x(\tau)]$, by a discrete representation, letting $y_i = y(t_i)$, $x_j = x(\tau_j)$; or by expansion of the continuous functions $y(t)$ and $x(\tau)$ in terms of appropriate sets of orthogonal functions, in which case y_i and x_j represent expansion coefficients.

If the functions $A_i(x_j)$ are linear in x_j, one may write the problem in the form

$$y_i = \sum_{j=1}^{m} A_{ij}\, x_j \quad , \qquad i = 1,\cdots,n \tag{1.2}$$

or in the matrix form as Eq. 1.1. If the functions $A_i(x_j)$ are not strictly linear, but vary smoothly enough, they may be expanded in a Taylor series about some set of initial values, say x_j,

$$y_i = A_i(x_j) + \sum_{j=1}^{m} \left.\frac{\partial A_i}{\partial x_j}\right|_{x_j^0} \Delta x_j + \cdots \tag{1.3}$$

Defining $\Delta y_i \equiv y_i - A_i(x_j^0)$ and ignorning second and higher order terms in Eq. 1.3, one has

$$\Delta Y_i = \sum_{i=1}^{m} \left.\frac{\partial A_i}{\partial x_j}\right|_{x_j^0} \Delta x_j$$

This is the same form as Eq. 1.2, with the substitution of Δy_i for y_i, Δx_j for x_j and

$$\left.\frac{\partial A_i}{\partial x_j}\right|_{x_j^0} = A_{ij}$$

For simplicity, one may proceed using the notation of Eq. 1.1, with the understanding that the above substitution can be made at any stage of the calculations, for a system that results from the perturbation of a quasilinear problem [1].

If Eq. 1.1 is a well-posed problem, the solution to this system of simultaneous equations is obtained by any of a number of well-known, existing techniques [2]. However, in many circumstances Eq. 1.1 cannot be solved directly. This occurs when matrix A is ill conditioned or is a rectangular matrix. There are methods available to handle this situation, such as Orthogonal Decomposition Methods [3] and the Natural Inverse of a Matrix [4], but in any event, the solution obtained for Eq. 1.1 by any direct method may not be acceptable. This is due to the fact that the error introduced to Eq. 1.1 in the conversion from a continuous system to a discrete system was not considered in the solution.

The objective of this paper is to present a method of solution that introduces these errors in the solution. The technique will be compared to the present method of solution, which has been used in many cases for an acceptable solution of Eq. 1.1.

Consider the following optimization problem: Minimize

$$\{x\}^T [c] \{x\}$$

subject to the constraints

$$\left.\begin{aligned} &([A]\{x\} - \{y\})^T([A]\{x\} - \{y\}) \leq e^2 \\ &\{\varepsilon\} = [A]\{x\} - \{y\} \\ &e^2 = \{e\}^T\{e\} \\ &\{\varepsilon\} \leq \{e\} \end{aligned}\right\} \qquad (1.4)$$

where [c] is a banded, positive definite square matrix that varies from problem to problem. In the limit, it might be an identity matrix. The matrix {e} is a vector of upper bounds on acceptable error.

Equation 1.4 may be obtained from a continuous minimization problem after discretization. The Lagrangian function for Eq. 1.4 is

$$L(x,\gamma) = \{x\}^T[c]\{x\} + \gamma^{-1}\{([A]\{x\} - \{y\})^T([A]\{x\} - \{y\}) - e^2\} \qquad (1.5)$$

The gradient of this Lagrangian function set to zero,

$$\nabla L(x,\gamma) = 0 \qquad (1.6)$$

gives

$$\{x\} = ([A]^T[A] + \gamma[c])^{-1}[A]^T\{y\} \qquad (1.7)$$

where γ^{-1} is a Lagrange multiplier. If the Lagrange multiplier γ^{-1} is known, by substituting γ into Eq. 1.7 and solving for {x}, one will obtain the solution of the optimization problem of Eq. 1.4. However, the Lagrange multiplier is not known. The known quantities are the upper bound on acceptable error, {e}, or

$$e^2 = \{e\}^T\{e\}$$

Therefore, there is a need for a method of solution to the primal optimization problem of Eq. 1.4. The algorithm presented in this paper is a technique for finding the optimum solution to Eq. 1.4. This algorithm has the advantage of solving a convex, nonlinear optimization problem by a series of linear programming optimizations.

2. ALGORITHM TO SOLVE CONVEX MINIMIZATION PROBLEMS

It can be shown that the optimization problem of Eq. 1.4 is a convex minimization problem; see the Appendix. Any convex minimization problem has a unique solution. A solution algorithm to solve convex minimization problems is as follows:

Consider the general convex minimization problem: Minimize

$$\left.\begin{array}{l} f(x) \\ \text{Subject to constraints} \\ g(x) \geq 0 \\ x \in R^n \\ x \in X \end{array}\right\} \qquad (2.1)$$

where $f(x)$ is a convex function, and $g(x)$ is an m-vector of concave functions, and X is a set of all feasible solutions.

(1) Pick a feasible solution x^1 such that $g(x^1) \geq 0$, $x^1 \in X$.

(2) Construct the linear programming problem: Minimize

$$\left.\begin{array}{l} \lambda_1 f(x^1) \geq 0 \\ \text{subject to constraints} \\ \lambda_1 = 0 \\ \lambda_1 \geq 0 \end{array}\right\} \qquad (2.2)$$

Solve this linear programming problem for primal solution λ_1 and dual solution π, η. Denote these solutions by $(\lambda^0, \pi^0, \eta^0)$!

(3) Construct the minimization problem: Minimize

$$\left.\begin{array}{l} f(x) - \pi^0 g(x) - \eta^0 \\ \text{Subject to } x \in X \end{array}\right\} \qquad (2.3)$$

Denote by x^2 the optimal solution of Eq. 2.3.

(4) If $f(x^2) - \pi^0 g(x^2) - \eta^0 \geq 0$, $\bar{x} = \lambda_1^0 x^1$ is the optimal solution of Eq. 2.1.

If $f(x^2) - \pi^0 g(x^2) - \eta^0 < 0$, add a new column to the linear programming problem of Eq. 2.2, to obtain the problem: Minimize

$$\left.\begin{aligned} &\lambda_1 f(x^1) + \lambda_2 f(x^2) \\ &\text{Subject to} \\ &\lambda_1 g(x^1) + \lambda_2 g(x^2) \geq 0 \\ &\lambda_1 + \lambda_2 = 1 \\ &\lambda_i \geq 0 \quad , \qquad i = 1,2 \end{aligned}\right\} \qquad (2.4)$$

Denote by $(\lambda_1^1, \lambda_2^1, \pi^1, \eta^1)$ the primal and dual optimal solutions to Eq. 2.4.

(5) Construct the minimization problem: Minimize

$$\left.\begin{aligned} &f(x) - \pi^1 g(x) - \eta^1 \\ &\text{Subject to} \\ &x \in X \end{aligned}\right\}$$

with the new Lagrange multiplier (π^1, η^1) and solve the problem. Denote by x the solution to this minimization problem.

(6) If $f(x^3) - \pi^1 g(x^3) - \eta^1 \geq 0$.

$\bar{x} = \lambda_1^1 x^1 + \lambda_2^1 x^2$ is the optimal solution to Eq. 2.1. If $f(x^3) - \pi^1 g(x^3) - \eta^1 < 0$, add a new column to linear programming of Eq. 2.4 and continue.

For proof of convergence of the algorithm see Refs. 5 and 6. This algorithm will give the optimal solutions to the primal problem of Eq. 2.1 and the dual system of Eq. 2.3 simultaneously.

There are several important technical cases in which the objective is to solve

$$[A]\{x\} = \{Y\} + \{\varepsilon\} \qquad (2.6)$$

A least square error procedure yields

$$\{x\} = ([A]^T[A])^{-1}[A]^T\{y\} \tag{2.7}$$

In order to obtain an acceptable smooth solution, a positive definite banded symmetric matrix is added to the matrix $([A]^T[A])$, such that

$$\{x\} = ([A]^T[A] + \gamma[c])^{-1}[A]^T\{y\} \tag{2.8}$$

where [c] is a banded positive definite symmetric matrix and γ is a small positive number. By applying this technique, the intention is to include some error in the solution, but the result is not satisfactory. In this method it is not possible to control the contribution of the error to the problem in a direct way. A trial and error method should be used to obtain the desired solution. Of course, the solution to Eq. 2.8 depends on the parameter γ and by varying γ one would have a set of solutions, which may or may not be feasible solutions and none of them are the optimal solution to Eq. 1.4.

3. EXAMPLE

The two story frame structure of Fig. 3.1, with element properties $m_1 = m_2 = 1$, $k_1 = k_2 = 4\pi^2$, and $k_3 = k_4 = 6\pi^2$, is subjected to ground excitation $A(t) = 100 \sin(4\pi t)$.

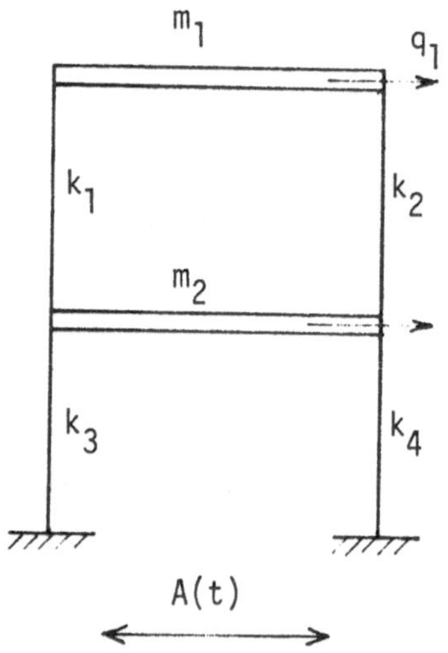

Figure 3.1 Two Story Frame Structure

The absolute acceleration y(t) on the roof level is related to the ground acceleration A(t) by

$$y(t) = \int_0^t A(t - \tau)\, x(\tau)\, d\tau \tag{3.1}$$

where

$$y(t) = 80 \sin(2\pi t) - 100 \sin(4\pi t) + 20 \sin(2\pi\sqrt{6}t)$$

In order to calculate $x(\tau)$, it is assumed that $t_i = i\Delta t$ and $\tau_j = j\Delta\tau$, where $\Delta t = \Delta\tau = 0.05$ and the integral equation of Eq. 3.1 is discretized as

$$y(t_i) = \sum_{j=1}^{t_i} W_j\, A(t_i - \tau_j) x(\tau_j) \Delta\tau \quad , \qquad i = 1,2,... \tag{3.2}$$

where w_j depends on the numerical integration scheme used in the discretization.

In this example, 40 a point time increment was considered. Equation 3.2 resulted in a 40 by 40 simultaneous system of equations. A direct solution to this system of simultaneous equations is shown in Fig. 3.2(b), where the exact solution is shown in Fig. 3.2(a). The optimization technique applied to this problem and further details of each step are given in Ref. 6. The optimal solution is shown in the Fig. 3.2(c).

4. CONCLUSION

A method of solution and an algorithm are presented in this paper for the solution of many problems in engineering and science, by converting to a system of linear simultaneous equations. The method is based on minimizing an objective function, subjected to the upper bound of the error on the constrained equations. The optimization problem is one of convex minimization and a unique solution to the problem exists. The method is applied to a two-story frame structure, subjected to harmonic ground excitation. The optimal solution is in good agreement to the exact solution.

APPENDIX - MATHEMATICAL MINIMIZATION DEFINITIONS AND THEOREMS

If x is a member of the set Ω, one can write $x \in \Omega$. One writes $y\Omega$ if y is not a member of the set Ω. A set Ω in n-space is said to be convex if for every x_1, $x_2 \in \Omega$ and real number α, $0 < \alpha < 1$, the point $\alpha x_1 + (1 - \alpha)x_2 \in \Omega$. Also, a point x in a convex set Ω is said to be an extreme point of Ω if there are not two distinct points x_1 and x_2 in Ω such that $x = \alpha x_1 + (1 - \alpha)x_2$ for some $0 < \alpha < 1$.

A function f(x) defined on a convex set is said to be convex if, for every $x_1, x_2 \in \Omega$ and every α, $0 < \alpha < 1$, $f(\alpha x_1 + (1-\alpha)x_2) \leq \alpha f(x_1) + (1-\alpha)f(x_2)$. If for every $0 < \alpha < 1$ and $x_1 \neq x_2$, $f(\alpha x_1 + (1-\alpha)x_2) < \alpha f(x_1) + (1+\alpha)f(x_2)$, then f(x) is said to be strictly convex. Any function $g_i(x)$ defined on a convex set Ω is said to be concave if the function $f(x) = -g(x)$ is convex. The function $g_i(x)$ is strictly concave if $-g_i(x)$ is strictly convex. A point x in n-space is said to be a convex combination of vectors $y^1, y^2, \cdots, y^k$ in n-space if there exist k positive real number $\lambda_1, \lambda_2, \cdots, \lambda_k$ such that

$$x = \lambda_1 y^1 + \lambda_2 y^2 + \cdots \lambda_k y^k$$

$$\lambda_1 + \lambda_2 + \cdots \lambda_k = 1$$

With these comments one can discuss the primal and dual linear programming problem. For every linear optimization program called a primal:

$$\left.\begin{array}{l} \text{Minimize } cx \\ \text{Subject to } Ax < b \\ \qquad\qquad x \leq 0 \end{array}\right\} \quad \text{(primal)} \qquad \text{(A-1)}$$

there is associated another linear program called a dual

$$\left.\begin{array}{l} \text{Maximize } \lambda b \\ \text{Subject to } \lambda A \leq c \end{array}\right\} \quad \text{(dual)} \qquad \text{(A-2)}$$

where A is a n by m matrix, c is a 1 by m row vector, x is an m by 1 column vector, and λ is a 1 by n row vector. The dual of any linear programming problem can be found by converting the problem to the primal problem shown above. The following are correspondence rules that apply to primal and dual problems:

Primal	Dual
objective form (minimization)	constant terms
constant terms	objective form (maximization)
Relation:	Variable:
(ith) inequality	$\lambda_i \leq 0$
(ith) equation	λ_1 unrestricted in sign
Variables:	Relations:
$x_j \geq 0$	(jth) inequality: $\leq$
x_j unrestricted	(jth) equation: $=$

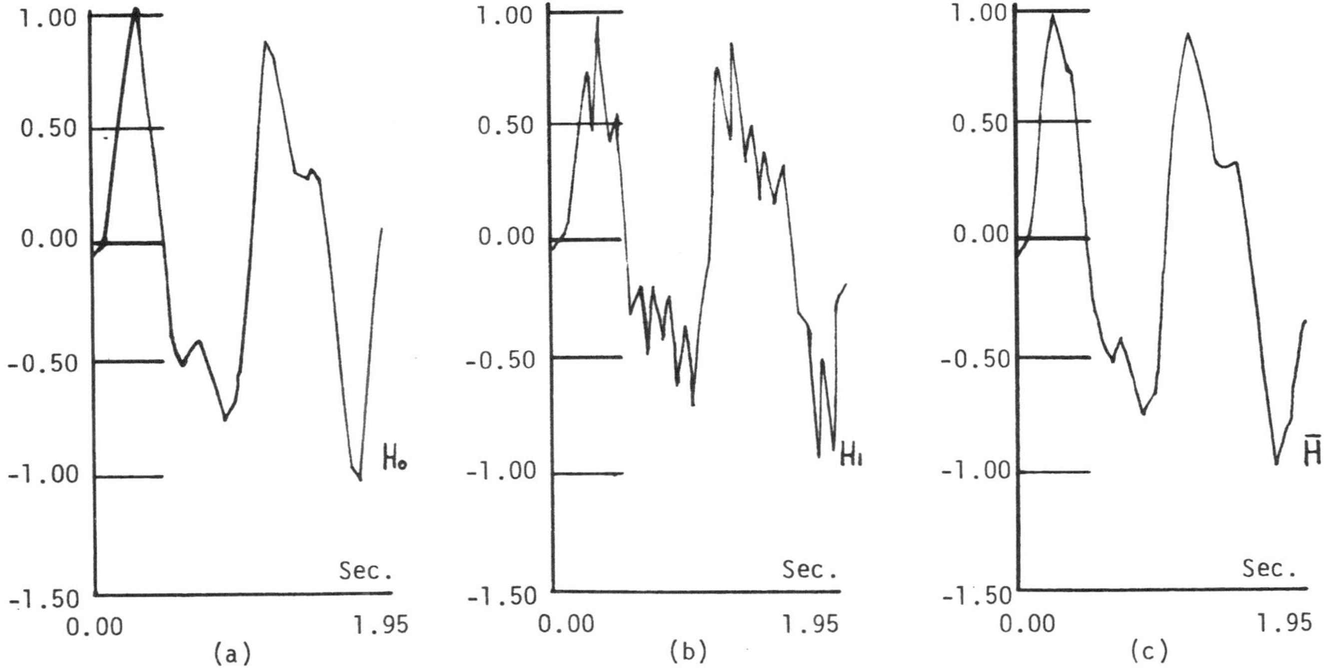

Figure 3.2 (a) Exact Solution,
(b) Direct Solution,
(c) Optimal Solution to Integral Equation

REFERENCES

1. Jackson, D.D., "Interpretation of Inaccurate, Insufficient and Inconsistent Data," Geophys. J.R. Astr. Soc., Vol. 28, No. 2, 1972, pp. 97-110.
2. Bathe, K.J., and Wilson, E. L., Numerical Methods for Finite Element Analysis, Prentice-Hall, Englewood Cliffs, New Jersey, 1976.
3. Jennings, A., Matrix Computation for Engineers and Scientists, John Wiley and Sons, New York, 1977.
4. Lanczos, C., Linear Differential Operators, Van Nostrand Co., London, 1961.
5. Dantzig, G.B., Linear Programming and Extensions, Princeton University Press, Princeton, New Jersey, 1963.
6. Torkamani, M.A.M., and Hart, G.C., Building System Identification Using Earthquake Data, Mechanics and Structures Department, University of California, Los Angeles, UCLA - ENG - 7507, 1975.

MINIMUM-WEIGHT DESIGN OF A ROTATING CANTILEVER BEAM WITH SPECIFIED FLAPPING FREQUENCY

Bion L. Pierson and Michael R. Pouliot

Department of Aerospace Engineering and the Engineering Research Institute, Iowa State University, Ames, Iowa 50011

ABSTRACT

A cantilever beam that rotates at constant angular rate about an axis of rotation normal to the undeflected beam centerline is considered. The problem is to determine the cross-section area distribution along the span that results in the least beam weight, such that a single flapping frequency is the same as for a similarly rotating uniform reference beam. Upper and lower bounds are placed on the area function and a power-law inertia-area relation is assumed. The problem is first formulated as an optimal control problem. Then, a newly-developed Recursive Quadratic Programming algorithm is applied to obtain efficient numerical solutions. Computational results are presented for a variety of conditions. In addition, a stress inequality constraint is imposed and comparative numerical results are given.

1. INTRODUCTION

Optimal control methods have been successfully applied to many one-dimensional, dynamic, continuous, structural optimization problems in recent years (see, for example, Refs. 1 to 6). An interesting survey of this activity through mid-1974 has been given by McIntosh [7]. Both indirect and direct(or gradient function-space) methods have been used in this context.

However, there is now a strong trend in flight path optimization and in other optimal control application areas toward approximating the optimal control problem by a constrained parameter optimization problem (nonlinear programming problem). The motivation for doing so is twofold. The primary advantage is that of reduced computing cost in cases for which high solution accuracy is not a prime concern. Additionally, one is often able to provide considerably more user flexibility in solving modified problems with several different types of constraints. Thus, one of the major objectives of the research presented here has been to explore the application of a promising recent nonlinear programming technique, recursive quadratic programming, to a structural optimal design problem of moderate difficulty, to further evaluate the method.

The basic problem treated here is not new. It has been formulated and solved by Kaza [8] in his 1974 Stanford doctoral dissertation. Kaza's problem is to find the minimum-weight depth distribution of a rotating cantilever beam of rectangular cross-section and specified fundamental flapping frequency. At least at the time of its solution, this problem was regarded as a rather difficult and high-order problem [7]. A few relatively minor extensions of the original problem are provided here and, in addition, a simple stress constraint is imposed on the problem.

2. PROBLEM FORMULATION

2.1 Optimal Control Problem

After assuming harmonic flapping oscillations with frequency ω, the dimensional modal equation of free lateral vibration for a rotating Euler-Bernoulli beam is given by [9]

$$[EIW'']'' - [PW']' - \rho\,\omega^2 AW = 0 \qquad (2.1)$$

Here, E denotes Young's modulus, I denotes the moment of inertia of the cross section, A denotes cross-section area, ρ denotes uniform material density, and W denotes the lateral deflection. Primes denote derivatives with respect to X, distance along the beam. The centrifugal loading at X due to a constant angular velocity Ω about an axis perpendicular to the beam at the clamped end X = 0 is given by

$$P = \rho\,\Omega^2 \int_X^L A(\sigma)\,\sigma\,d\sigma \qquad (2.2)$$

After applying the Leibniz derivative formula, Eq. 2.2 becomes

$$\left.\begin{aligned} P' &= -\rho\,\Omega^2\,A(X)\,X \\ P(L) &= 0 \end{aligned}\right\} \tag{2.3}$$

Next, the inertia-area relation

$$I = CA^k \tag{2.4}$$

is assumed, where C is a dimensional constant. For geometrically similar cross-sections k = 2, and for variable-depth rectangular cross-sections k = 3. The problem can be nondimensionalized by defining the following nondimensional quantities:

$$\left.\begin{aligned} x &= X/L \\ w &= W/L \\ a &= AL/V \end{aligned}\right. \tag{2.5}$$

$$p = P/(\rho\Omega^2 VL) \tag{2.5}$$

Here, L is the fixed beam length and V is a specified characteristic volume, equal to the volume of a uniform cross-section beam of length L that has the same frequency ω as the tapered optimum beam. Substitution of Eqs. 2.4 and 2.5 into Eq. 2.1 yields

$$[a^k w'']'' - \frac{\rho\Omega^2 L^{k+3}}{ECV^{k-1}}[pw']' - \frac{\rho\omega^2 L^{k+3}}{ECV^{k-1}}\,aw = 0 \tag{2.6}$$

Now, the primes denote derivatives with respect to the nondimensional distance x along the beam. In view of Eq. 2.6, the nondimensional frequency parameters

$$\left.\begin{aligned} \gamma &= \frac{\rho\,\omega^2\,L^{k+3}}{ECV^{k-1}} \\ \delta &= (\Omega/\omega)^2 \end{aligned}\right\} \tag{2.7}$$

are introduced. The resulting modal equation may be written as

$$[(a^k w'')' - \delta\gamma p w']' - \gamma a w = 0 \tag{2.8}$$

In a similar fashion, Eq. 2.3 has as its nondimensional counterpart

$$p' = -ax$$
$$p(1) = 0 \tag{2.9}$$

An optimal control problem formulation requires the use of first-order equations of motion. This in turn necessitates a choice of state variables. The selection made here consists of

$$\begin{aligned} y_1 &= w \\ y_2 &= w' \\ y_3 &= a^k w'' \\ y_4 &= (a^k w'')' - \delta\gamma p w' \\ y_5 &= p \end{aligned} \tag{2.10}$$

Thus, the state variables represent deflection, slope, bending moment, shear plus a rotation term, and centrifugal load, respectively. Then, from Eqs. 2.8 to 2.10, the differential constraints become

$$\begin{aligned} y_1' &= y_2 \\ y_2' &= a^{-k} y_3 \\ y_3' &= y_4 + \delta\gamma y_2 y_5 \\ y_4' &= \gamma a y_1 \\ y_5' &= -ax \end{aligned} \tag{2.11}$$

The boundary conditions for this cantilever beam are

$$\begin{aligned} y_1(0) &= y_2(0) = 0 \\ y_3(1) &= y_4(1) = y_5(1) = 0 \end{aligned} \tag{2.12}$$

Since there are more boundary conditions at the free end ($x = 1$) than at the clamped end ($x = 0$), it is advantageous to employ the transformation

$$t = 1 - x \tag{2.13}$$

This reversal in the direction of measuring distance along the beam results in one less terminal state constraint and one less unspecified initial state variable.

The nondimensional area distribution a(t) plays the role of the control function. Upper and lower bounds are imposed on a(t). An additional control parameter is present in the form of an unspecified initial state variable. Since the objective is to achieve the least beam weight, the optimal control problem statement is as follows:

Determine the optimal control function a(t), $0 \leq t \leq 1$, and control parameter σ that minimize

$$J = \int_0^1 a(t)\, dt \tag{2.14}$$

subject to the 5th-order dynamic system

$$\dot{y}_1 = - y_2 \quad , \qquad y_1(0) = 1 \tag{2.15}$$

$$\dot{y}_2 = - a^{-k}(t)\, y_3 \quad , \qquad y_2(0) = \sigma \tag{2.16}$$

$$\dot{y}_3 = - y_4 - \delta\gamma y_2 y_5 \quad , \qquad y_3(0) = 0 \tag{2.17}$$

$$\dot{y}_4 = - \gamma\, a(t)\, y_1 \quad , \qquad y_4(0) = 0 \tag{2.18}$$

$$\dot{y}_5 = (1-t)\, a(t) \quad , \qquad y_5(0) = 0 \tag{2.19}$$

the terminal state constraints

$$\psi_1[y(1)] = y_1(1) = 0 \tag{2.20}$$

$$\psi_2[y(1)] = y_2(1) = 0 \tag{2.21}$$

and the control inequality constraints

$$0 < a_{min} \leq a(t) \leq a_{max}, \qquad 0 \leq t \leq 1 \tag{2.22}$$

where k, δ, γ, a_{min} and a_{max} are specified constants. Here, the dots denote derivatives with respect to transformed distance t.

The value specified for the flapping frequency parameter γ is not arbitrary. Rather, for each value of the rotation frequency parameter δ, γ must be fixed at its corresponding value for the uniform reference beam. The solution to this auxiliary free vibration problem is discussed in Section 4. Other features of interest in this optimal control problem include the nonlinear state variable dependence in Eq. 2.17, the explicit independent-variable dependence in Eq. 2.19, and the presence of the control parameter σ in Eq. 2.16.

Finally, it is shown in the Appendix that the necessary conditions for optimality are invariant under a one-parameter transformation. In particular, the state variables y_1, y_2, y_3, and y_4 may be arbitrarily scaled without affecting the necessary conditions. For convenience, scaling is implemented here by arbitrarily specifying the deflection at the free end of the beam, as shown in Eq. 2.15. Thus, the effect of state variable scaling is to reduce the required number of control parameters from two to one.

In a subsequent version of this problem, the simple axial stress constraint, $P(X)/A(X) \leq S_{max}$, will be imposed. This, of course, represents a state variable inequality constraint. In nondimensional form, this constraint may be taken as

$$s_{max} - y_5/a(t) \geq 0, \qquad 0 \leq t \leq 1 \qquad (2.23)$$

where s_{max} is a specified constant.

2.2 Approximation As A Nonlinear Programming Problem

The key ingredient in the discrete approximation used here for the optimal control problem of Eqs. 2.14 to 2.23 is the representation of the control function a(t) by a function of t and a finite number of parameters. For simplicity and additional reasons noted later, the area distribution is represented by a piecewise-linear time function, with fixed (and not necessarily equally-spaced) t-points. Suppose the beam is partitioned into N segments by the N + 1 points: 0, t_1, t_2, ..., t_{N-1}, 1. Then,

$$a(t) = \begin{cases} \eta_0 + (\eta_1 - \eta_0)t/t_1 \quad , & 0 \le t \le t_1 \\ \eta_1 + (\eta_2 - \eta_1)(t - t_1)/(t_2 - t_1), & t_1 \le t \le t_2 \\ \cdot & \\ \cdot & \\ \cdot & \\ \eta_{N-1} + (\eta_N - \eta_{N-1})(t - t_{N-1})/(1 - t_{N-1}), & t_{N-1} \le t \le 1 \end{cases} \tag{2.24}$$

The N + 1 nodal points η_0, η_1, η_2, ..., η_N serve as problem variables for the nonlinear programming formulation that follows. Because of Eq. 2.24, the performance index of Eq. 2.14 has an exact dependence on the new problem variables,

$$F(\eta_0, \eta_1, \ldots, \eta_N) = \frac{1}{2} \sum_{i=0}^{N-1} (\eta_i + \eta_{i+1})(t_{i+1} - t_i) \tag{2.25}$$

Furthermore, the differential equations of Eqs. 2.15 to 2.19, with Eq. 2.24 are to be numerically integrated from t = 0 to t = 1. Thus, the terminal state constraints of Eqs. 2.20 and 2.21 now take on the role of equality constraints for the nonlinear programming problem,

$$g(\eta_0, \ldots, \eta_N, \sigma) = \begin{bmatrix} y_1(1) \\ y_2(1) \end{bmatrix} = 0 \tag{2.26}$$

The inequality constraints of Eqs. 2.22 and 2.23 are enforced at a finite number of specified t-points. Again, as a result of Eq. 2.24 and the numerical integration process, these inequality constraints may be combined into a single vector constraint that depends on the η_i and σ,

$$h(\eta_0, \ldots, \eta_N, \sigma) \ge 0 \tag{2.27}$$

Finally, after defining a composite (N + 2)-vector of problem variables $z = (\eta_0, \eta_1, \ldots, \eta_N, \sigma)^T$, the approximating nonlinear programming problem takes the general form

$$\left.\begin{array}{l}\text{minimize} \\ \qquad F(z) \\ \text{subject to constraints} \\ \qquad g(z) = 0 \\ \qquad h(z) \geq 0\end{array}\right\} \tag{2.28}$$

The recursive quadratic programming algorithm used here requires first-order gradient information for both the objective function F and the constraint functions g and h. This derivative information is computed using simple first-order forward differences. For example,

$$g_z = \left[\frac{g_i(z_1,\ldots,z_{j-1},z_j+\Delta z_j,z_{j+1},\ldots,z_{N+2})-g_i(z_1,\ldots,z_{N+2})}{\Delta z_j}\right] \tag{2.29}$$

where

$$\Delta z_j = \begin{cases}\varepsilon_{FD}\,|z_j|\,, & |z_j| > 1 \\ \varepsilon_{FD}\,, & |z_j| \leq 1\end{cases} \tag{2.30}$$

Typically, ε_{FD} is chosen as 10^{-5} or 10^{-6}. Provision is also made for using second-order central differences at the user's discretion. Thus, each evaluation of the gradient vector F_z^T and the constraint normal arrays g_z and h_z requires $N+2$ numerical integrations of Eqs. 2.15 to 2.19 when one uses Eq. 2.29 and $2N+4$ integrations when one uses second-order central differences. However, a significant advantage in the choice of the discrete representation of Eq. 2.24 for the control function is that each nodal point perturbation affects only those control values between the nodes adjacent to the perturbed node. Thus, the entire trajectory need not be integrated for every nodal point perturbation. For each perturbation, Eqs. 2.15 to 2.19 are integrated only from the nodal point just prior to the perturbed node. The computational expense for gradient information is therefore reduced by almost half. Also, note that bound-type constraints, such as those corresponding to Eq. 2.22 have constant constraint normals, which do not rely on any numerical integration. A final advantage in the use of Eq. 2.24 is that Eq. 2.22 is satisfied exactly when the upper and lower bounds are enforced on the associated control nodal points.

3. METHOD DESCRIPTION

In this study, a recursive quadratic algorithm is applied to the approximating problem of Eq. 2.28. To simplify an introduction to recursive quadratic programming, consider the nonlinear programming problem of Eq. 2.28, with only equality constraints. First, define the Lagrangian function

$$H(z,\lambda) = F(z) - \lambda^T g(z) \tag{3.1}$$

Then, the first-order necessary conditions for optimality can be written as

$$\left.\begin{aligned} H_z^T &= F_z^T(z^*) - g_z^T(z^*)\ \lambda^* = 0 \\ H_\lambda^T &= -\ g(z^*) = 0 \end{aligned}\right\} \tag{3.2}$$

The unknowns are the elements of the optimal problem vector z^* and the optimal Lagrange multiplier vector λ^*. One solution approach is to apply Newton's method. In that case, the necessary conditions of Eq. 3.2 are linearized about some current estimate of z and λ. The result is

$$\begin{bmatrix} H_{zz}(z,\lambda) & -\ g_z^T(z) \\ -\ g_z(z) & 0 \end{bmatrix} \begin{bmatrix} \Delta z \\ \Delta\lambda \end{bmatrix} = \begin{bmatrix} -F_z^T(z) + g_z^T(z)\lambda \\ g(z) \end{bmatrix} \tag{3.3}$$

where $\Delta z = z^* - z$ and $\Delta\lambda = \lambda^* - \lambda$. The linear system of Eq. 3.3 is solved for Δz and $\Delta\lambda$, which are in turn used to update the estimates z and λ.

Now consider the following particular quadratic programming problem: Minimize

$$F_z(z)\Delta z + \tfrac{1}{2}\Delta z^T B \Delta z$$

with respect to Δz (3.4)

subject to constraints

$$g_z(z)\Delta z + g(z) = 0$$

Quadratic programming problems are of special interest, since well-developed and efficient algorithms are available for their solution. Next, define the Lagrangian function

$$H_q(\Delta z,\lambda_q) = F_z(z)\Delta z + \tfrac{1}{2}\Delta z^T B\Delta z - \lambda_q^T[g_z(z)\Delta z + g(z)] \tag{3.5}$$

so that the necessary conditions for Eq. 3.4 become

$$\left.\begin{aligned}(H_q)^T_{\Delta z} &= F_z^T(z) + B\Delta z - g_z^T(z)\,\lambda_q = 0\\ (H_q)^T_{\lambda q} &= g_z(z)\Delta z + g(z) = 0\end{aligned}\right\} \tag{3.6}$$

A comparison of Eqs. 3.6 and 3.3 reveals that the two sets of equations are identical if $B = H_{zz}(z,\lambda)$ and $\lambda_q = \Delta\lambda + \lambda$. Thus, if $B = H_{zz}(z,\lambda)$, the solution of the quadratic programming problem of Eq. 3.4 yields a vector of Newton corrections to the problem variables. Also, the Lagrange multiplier vector λ_q, which may now be obtained by solving the first of Eqs 3.6 corresponds to the updated Lagrange multiplier of the original Newton iteration. Therefore, one could solve the original nonlinear equality-constrained problem by solving a sequence of quadratic programming problems of Eq. 3.4.

This approach has been developed by Wilson [10] for the case of both equality and inequality constraints by adding to Eq. 3.4 the linearized inequality constraints

$$h_z(z)\Delta z + h(z) \geq 0 \tag{3.7}$$

However, his SOLVER algorithm employs the Hessian H_{zz} for B and therefore suffers all the usual convergence problems associated with a Newton iterative method. An added disadvantage is that second derivatives for both the objective function and the constraint functions are required.

More recently, there has been renewed interest in recursive quadratic programming [11 to 16]. In these second generation recursive quadratic programming algorithms, a variable metric approximation B to the Hessian is used. Thus, a positive definite B matrix can be maintained, and the earlier convergence problems and the burden of providing second derivative information are avoided. The algorithm used here closely follows that of Powell [15,16]. In qualitative terms, the method consists of the following steps:

(1) Select an initial guess for z and set B = I.
(2) Compute function and gradient information: $F(z)$, $F_z(z)$, $g(z)$, $g_z(z)$, $h(z)$, and $h_z(z)$.
(3) Solve the quadratic programming problem, with the Hessian replaced by B, for the direction of search Δz.
(4) Conduct a one-dimensional search along Δz by minimizing an auxiliary performance index. This step-size selection procedure is used to enhance convergence from poor estimates of z.
(5) Update z and test for convergence.
(6) If the termination criteria are satisfied, stop. Otherwise, update B with a modified BFGS (Broyden-Fletcher-Goldfarb-Shanno) recursion formula and return to step 2.

This method has proven to be especially attractive for problems with computationally expensive function and gradient evaluations. This, of course, is precisely what one encounters in discrete approximations to optimal control problems.

4. SOLUTION OF THE UNIFORM REFERENCE BEAM EIGENVALUE PROBLEM

Before one can solve the minimum-weight problem, it is necessary to solve the free vibration problem for the uniform reference beam. That is, one wishes to specify the rotation frequency parameter δ and find the corresponding values of γ, for which the following t-varying, linear, two-point, boundary value problem is satisfied:

$$\left.\begin{aligned}
y_1(0) &= 1, & \dot{y}_1 &= -y_2 &&, & y_1(1) &= 0\\
& & \dot{y}_2 &= -y_3 &&, & y_2(1) &= 0\\
y_3(0) &= 0, & \dot{y}_3 &= -\delta\gamma t(1-\tfrac{1}{2}t)y_2 - y_4 && & &\\
y_4(0) &= 0, & \dot{y}_4 &= -\gamma y_1 && & &
\end{aligned}\right\} \qquad (4.1)$$

Note that the area distribution a(t) in Eqs. 2.15 to 2.19 has been set equal to one and the resulting quadrature in Eq. 2.19 has been performed analytically. The arbitrary boundary condition, $y_1(0) = 1$, serves to scale the mode shapes.

This eigenvalue problem has been posed and solved numerically, by applying the recursive quadratic programming algorithm of Section 3 to the following auxiliary, equality-constrained parameter optimization problem: Minimize

$$F(\gamma,\sigma_R) = \frac{1}{2}y_1^2(1) \tag{4.2}$$

subject to the constraint

$$g(\gamma,\sigma_R) = y_2(1) = 0 \tag{4.3}$$

where $y_1(1)$ and $y_2(1)$ are obtained from a numerical integration of the initial value problem

$$\left.\begin{aligned} y_1(0) &= 1, & \dot{y}_1 &= -y_2 \\ y_2(0) &= \sigma_R, & \dot{y}_2 &= -y_3 \\ y_3(0) &= 0, & \dot{y}_3 &= -\delta\gamma t(1-\tfrac{1}{2}t)y_2 - y_4 \\ y_4(0) &= 0, & \dot{y}_4 &= -\gamma y_1 \end{aligned}\right\} \tag{4.4}$$

and where δ is specified.

Representative solution values for the fundamental mode are listed in Table 4.1. As expected, the flapping frequency increases monotonically as the rotation speed is increased. These are the specified (δ,γ)-pairs for the main problem.

TABLE 4.1 SOLUTION VALUES FOR THE UNIFORM REFERENCE BEAM EIGENVALUE PROBLEM (FUNDAMENTAL MODE)

δ	γ	σ_R
0.00	12.36236	1.37651
0.02	12.66455	1.37554
0.06	13.31497	1.37346
0.10	14.03489	1.37120
0.20	16.22187	1.36447
0.40	23.46892	1.34383
0.50	30.06912	1.32700
0.60	41.44072	1.30171
0.70	64.70696	1.26120
0.80	129.01577	1.19437

5. NUMERICAL RESULTS

Several minimum-weight solutions obtained with the recursive quadratic programming algorithm are now presented. These results have been run on a Digital Equipment VAX-11 time-sharing computer in FORTRAN IV-plus. Double-precision arithmetic is used throughout. The numerical integration is performed by a standard fourth-order Runge-Kutta method, with 100 fixed and uniform integration steps.

5.1 Effect of Rotation Speed

The first solution is for 11 control nodes with $\delta = 0.02$, $k = 2$, $a_{min} = 0.1$, and $a_{max} = 5.0$. The initial control is $a(t) = 1.0$. Upper and lower bounds on area are enforced at all nodal points. Thus, there are 2 + 11 + 11 = 24 constraints and 11 + 1 = 12 problem variables. The solution requires 17 iterations and results in a minimum mass ratio of 0.23707, which represents a 76.3% weight reduction over the uniform reference beam with the same fundamental flapping frequency. The upper-bound area constraint is not active.

The same problem, with 21 control nodes (44 constraints and 22 problem variables), requires 27 iterations and results in a minimum mass ratio of 0.23706. Thus, the N = 11 case produces a very accurate solution estimate for this class of problems. However, to be conservative, all subsequent solutions are for N = 21.

As the rotation speed is increased, the minimum mass ratio is decreased. Optimum area distributions for three different δ values are presented in Fig. 5.1. The reduction of minimum mass ratio with increasing rotation speed is shown in Fig. 5.2. In general, the problems become more difficult to solve as δ, and therefore γ, become larger.

5.2 Effect Of Lower Area Bound

Since the optimal area distributions all reach the lower bound a_{min} at the free end of the beam, the value chosen for a_{min} clearly affects the minimum mass ratio. Solutions for several a_{min} values are given in Fig. 5.3, for the case of $\delta = 0.02$ and $k = 2$. The corresponding minimum mass ratio dependence on a_{min} is shown in Fig. 5.4.

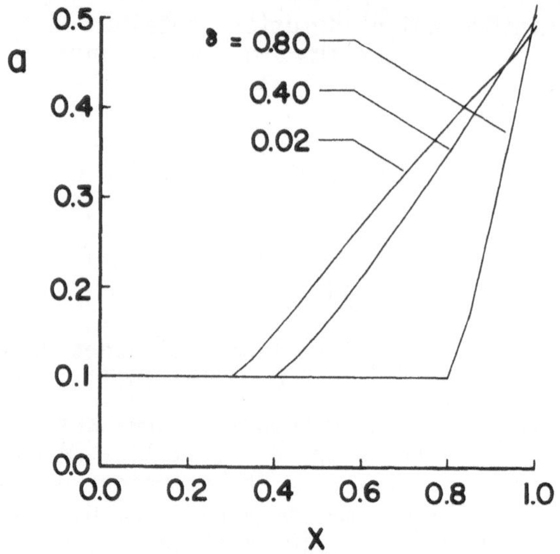

Figure 5.1 Optimal Area Distribution Dependence On Rotational Speed For k = 2 And a_{min} = 0.1

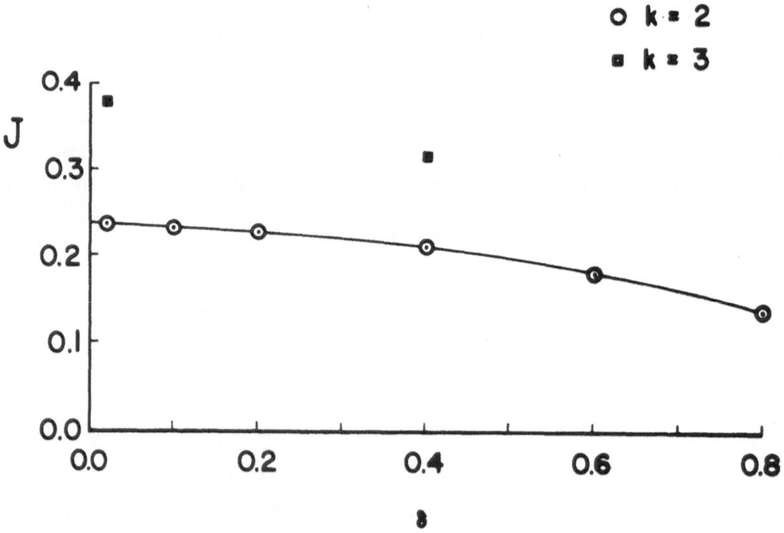

Figure 5.2 Minimum Mass Ratio Variation With Rotational Speed For k = 2 And a_{min} = 0.1

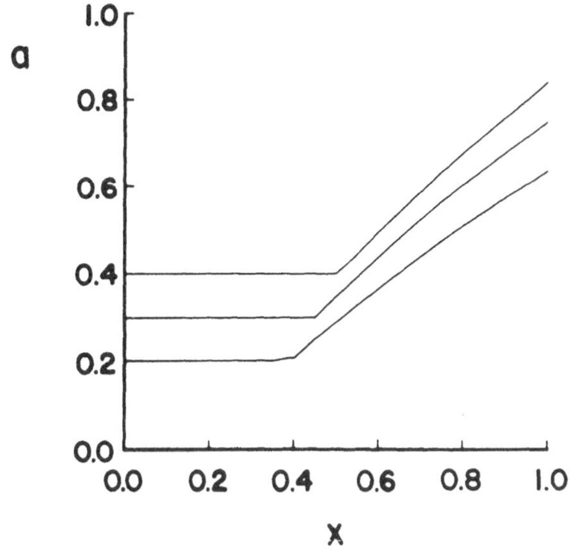

Figure 5.3 Optimal Area Distributions For Several Area Lower Bounds, k = 2, And $\delta = 0.02$

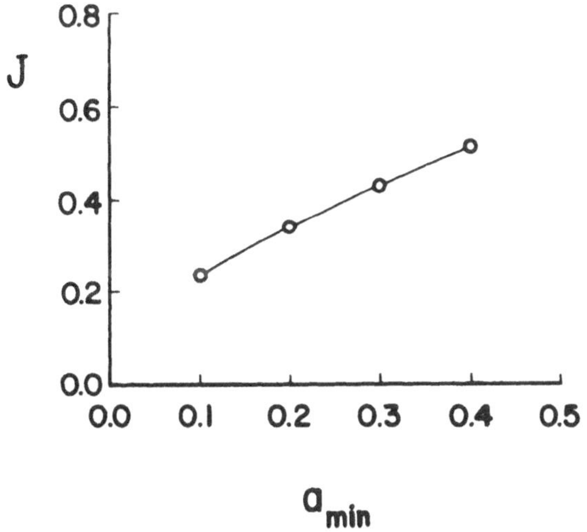

Figure 5.4 Minimum Mass Ratio Variation With a_{min} For k = 2 And $\delta = 0.02$

For the same value of δ, these problems become more sensitive for smaller specified a_{min}. This is because the right-hand side of the differential constraint of Eq. 2.16 becomes unbounded as $a(t)$ approaches zero.

5.3 Active Upper Area Bound

Again let $\delta = 0.02$, $k = 2$, and $a_{min} = 0.1$. Then, set $a_{max} = 0.4$. In the previous solution, with $a_{max} = 5.0$, the largest value of a is $a(1) = 0.4847$. The two optimum area distributions may be compared in Fig. 5.5. With the active upper bound on area, the minimum mass ratio is increased 1.5% from the previous value to 0.24056.

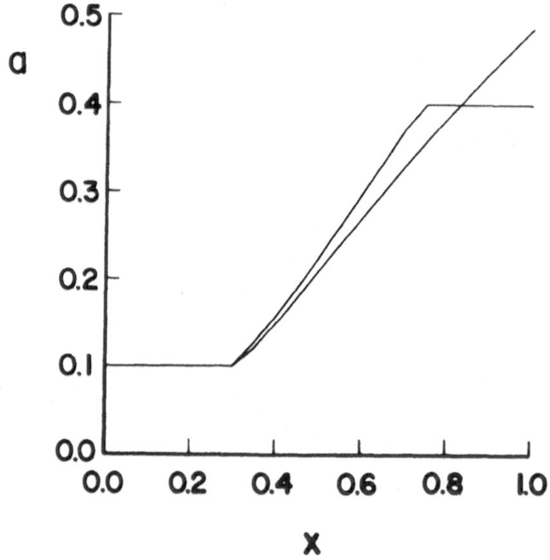

Figure 5.5 Optimal Area Distributions With And Without An Active Upper Bound Constraint For $k = 2$, $\delta = 0.02$, And $a_{min} = 0.1$

5.4 Rectangular Cross-Sections With Variable Depth

The solutions presented thus far have been for geometrically similar cross-sections ($k = 2$). Two solutions are obtained for $k = 3$. As may be noted in Fig. 5.2, a smaller mass reduction is possible in this case. The optimum area distributions are qualitatively similar to their $k = 2$ counterparts.

5.5 Stress Inequality Constraint

The state variable inequality constraint of Eq. 2.23 is now imposed on the problem. For the case $\delta = 0.02$, $k = 2$, and $a_{min} = 0.1$, s_{max} is set at 0.22. The stress constraint is enforced at 8 interior points from $t = 0.20$ through $t = 0.55$, in uniform 0.05 increments. The optimum stress distributions, with and without the stress inequality constraint, are shown in Fig. 5.6. The peak stress reduction of 13.7% is achieved, with a mass penalty of only 0.42% and a rather minor change in the optimum area distribution.

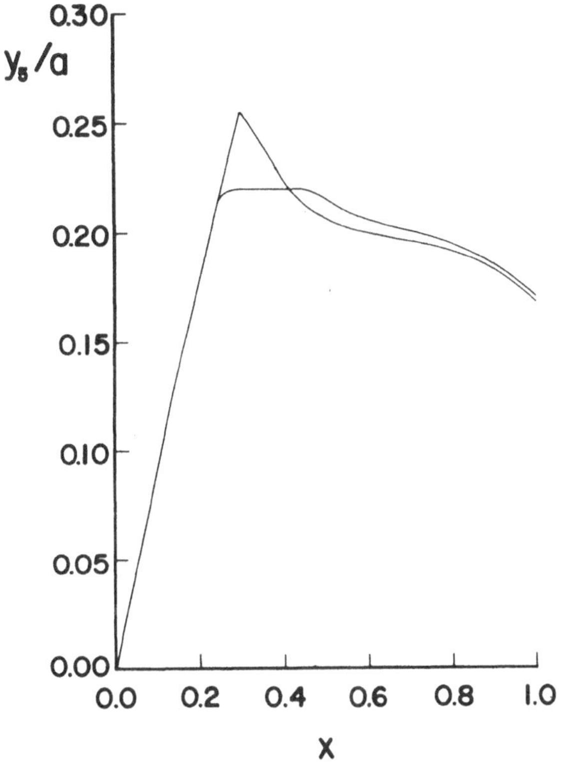

Figure 5.6 Optimal Stress Distributions With And Without The State Variable Inequality Constraint $y_5/a = 0.22$ For $k = 2$, $\delta = 0.02$, And $a_{min} = 0.1$

It is interesting to note that if the stress constraint is enforced at 17 interior points from t = 0.2 to t = 1.0 in uniform 0.05 increments, the number of iterations (quadratic programming problems) required is reduced from 35 to 18, even though the solution remains the same. In each case, an initial unity area distribution is used.

6. DISCUSSION

These initial numerical tests indicate that the combined recursive quadratic programming/discrete variable approach to continuous structural optimization problems is very promising. Relative accuracy appears to be good, even for a small number of control nodes. The most pleasing aspect is the flexibility one has in implementing a wide variety of problem constraints.

It remains to obtain a direct comparison with a continuous optimal control algorithm, such as gradient projection [4], applied to this rotating beam problem. Recent experience on some flight path optimization problems suggests a reduction in total computing time by a factor of 3 to 10, if no state variable inequality constraints are present.

Several extensions to this problem are possible. It would be of interest to examine the case of a pinned, rather than clamped beam. One could also constrain higher-mode flapping frequencies. If more than one frequency is to be fixed, a separate set of differential constraints of Eqs. 2.15 to 2.18 must be included for each mode. It may also be useful to include the effects of a tip mass and/or terms for rotary inertia and shear effects.

Finally, it should be noted that the recursive quadratic programming algorithm itself currently exists in an experimental status. In particular, several modifications are anticipated that will prevent degradation of convergence speed when constraint normals of two or more active constraints become nearly linearly dependent.

APPENDIX: STATE VARIABLE SCALING

The first-order necessary conditions for the optimal control problem of Eqs. 2.14 to 2.21, for unbounded control, consist of the following state and costate equations, associated boundary conditions, and optimality condition:

$$\dot{y}_1 = -y_2 \qquad , \quad y_1(1) = 0$$

$$\dot{y}_2 = -a^{-k}\, y_3 \qquad , \quad y_2(1) = 0$$

$$y_3(0) = 0, \quad \dot{y}_3 = -y_4 - \delta\gamma y_2 y_5$$

$$y_4(0) = 0, \quad \dot{y}_4 = -\gamma a y_1$$

$$y_5(0) = 0, \quad \dot{y}_5 = (1-t)a$$

$$\lambda_1(0) = 0, \quad \dot{\lambda}_1 = \gamma a \lambda_4$$

$$\lambda_2(0) = 0, \quad \dot{\lambda}_2 = \lambda_1 + \delta\gamma y_5 \lambda_3$$

$$\dot{\lambda}_3 = a^{-k}\, \lambda_2 \qquad , \quad \lambda_3(1) = 0$$

$$\dot{\lambda}_4 = \lambda_3 \qquad , \quad \lambda_4(1) = 0$$

$$\dot{\lambda}_5 = \delta\gamma y_2 \lambda_3 - (1-t)a \qquad , \quad \lambda_5(1) = 0$$

$$a^{k+1} = \lambda_2 y_3 / [1 - (1-t)\lambda_5 - \gamma y_1 \lambda_4]$$

Note that if one makes the following substitutions for arbitrary β, these necessary conditions remain invariant:

$$\left.\begin{array}{l} \beta y_i \rightarrow y_i \\ \lambda_i/\beta \rightarrow \lambda_i \end{array}\right\} \quad , \quad i = 1,\ 2,\ 3,\ 4$$

This result follows from the fact that the state and costate equations are linear and homogeneous in the y_i and λ_i, $i = 1,\ldots,4$, and have zero boundary conditions. Further, the integrand of the performance index of Eq. 2.14 is independent of the first four state variables. This state variable scaling is most conveniently accomplished by simply specifying an additional initial state variable, as is done in Eq. 2.15.

REFERENCES

1. Haug, E.J., Pan, K.C., and Streeter, T.D., "A Computational Method For Optimal Structural Design II: Continuous Problems," Int. J. Num. Meth. Engrg., Vol. 9, 1975, pp. 649-667.
2. Hornbuckle, J.C., and Boykin, W.H., Jr., "Equivalence Of A Constrained Minimum Weight And Maximum Column Buckling Load Problem With Solution," J. Appl. Mech., Vol. 45, 1978, pp. 159-164.
3. Johnson, E.H., Rizzi, P., Ashley, H., and Segenreich, S.A., "Optimization Of Continuous One-Dimensional Structures Under Steady Harmonic Excitation," AIAA J., Vol. 14, 1976, pp. 1690-1698.
4. Pierson, B.L., "Panel Flutter Optimization By Gradient Projection," Int. J. Num. Meth. Engrg., Vol. 9, 1975, pp. 271-296.
5. Pierson, B.L., "An Optimal Control Approach To Minimum-Weight Vibrating Beam Design," J. Struct. Mech., Vol. 5, 1977, pp. 147-178.
6. Weisshaar, T.A., "Aeroelastic Optimization Of A Panel In Mach Number Supersonic Flow," J. Aircraft, Vol. 9, 1972, pp. 611-617.
7. McIntosh, S.C., Jr., "Structural Optimization Via Optimal Control Techniques: A Review," Structural Optimization Symposium (L.A. Schmit, Jr., ed.), ASME Applied Mechanics Symposia Series, AMD-Vol. 7, 1974, pp. 49-64.
8. Kaza, K.R.V., Rotation In Vibration, Optimization, And Aeroelastic Stability Problems, Ph.D. dissertation, Department of Aeronautics and Astronautics, Stanford University, Stanford, California, 1974, Chapter 5.
9. Washizu, K., Variational Methods In Elasticity And Plasticity, Second Edition, Pergamon Press, Oxford, 1975, pp. 306-307.
10. Wilson, R.B., A Simplicial Method For Convex Programming, Ph.D. dissertation, Harvard University, Cambridge, Mass., 1963.
11. Biggs, M.C., "Constrained Minimization Using Recursive Equality Quadratic Programming," Numerical Methods for Nonlinear Programming (F.A. Lootsma, ed.), Academic Press, New York, 1972.
12. Biggs, M.C., "A Numerical Comparison Between Two Approaches To The Nonlinear Programming Problem," Towards Global Optimization, Vol. 2 (L.C.W. Dixon and G.P. Szego, eds.), North-Holland, Amsterdam, 1978.
13. Han, S.P., "Superlinearly Convergent Variable Metric Algorithms For General Nonlinear Programming Problems," Math. Prog., Vol. 11, 1976, pp. 263-282.

14. Han, S.P., "A Globally Convergent Method For Nonlinear Programming," J. Optim. Theory Applics, Vol. 22, 1977, pp. 297-309.
15. Powell, M.J.D., "A Fast Algorithm For Nonlinearly Constrained Optimization Calculations," The State of the Art in Numerical Analysis (D. Jacobs, ed.), Academic Press, New York, 1977.
16. Powell, M.J.D., "Algorithms For Nonlinear Constraints That Use Lagrangian Functions," Math. Prog., Vol. 14, 1978, pp. 224-248.

INTERACTION BETWEEN HIGH-SPEED VEHICLES AND FLEXIBLE GUIDEWAYS

P.K. Sinha

Department of Engineering Science, University of Warwick, Coventry, England

ABSTRACT

Dynamic interaction between flexible guideways and a single-degree of freedom suspension system are considered in this paper. Some results relative to the guideway parameters and ride comfort and stability are presented.

1. VEHICLE-GUIDEWAY INTERACTION

The general vehicle-guideway interaction problem is schematically represented in Fig. 1.1. As the vehicle moves along the guideway, it is acted upon by external forces and by suspension system forces that cause linear and rotational accelerations of the vehicle. The suspension reacts with the vehicle body and guideway surface motions, where the latter include geometric irregularities and elastic deformations that produce suspension forces acting on the vehicle and on the guideway. The guideway beams deflect dynamically in response to the moving, unsteady suspension forces, as well as the reaction forces and moments generated at the supports. Finally, support motion is determined by the support and foundation dynamic characteristics and guideway beam forces and moments. Simplifying assumptions are usually made for tractable analysis of this strongly coupled, non-linear, time-varying system. Some analytical and computer simulation results for a single-point moving vehicle are presented in this paper. For additional detail, see Refs. 1 to 5.

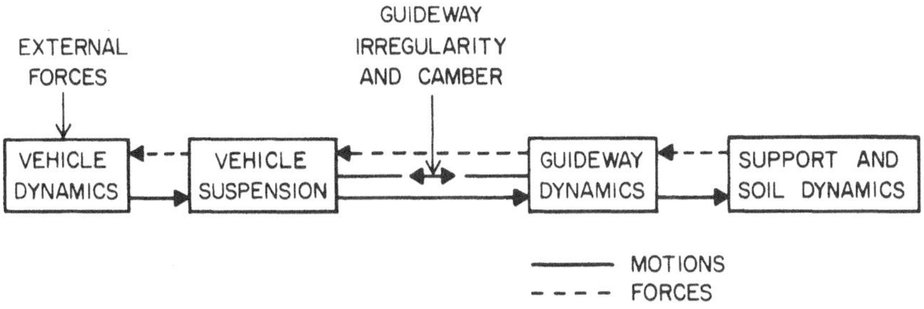

Figure 1.1. Vehicle-Suspension-Guideway Interaction
[—— motions ------ forces]

2. SINGLE-DEGREE OF FREEDOM SYSTEM

A simplified model, with a single-point suspension, is shown in Fig. 2.1. Only motion in the vertical plane is considered, assuming that the external forces on the guideway due to irregularities that would excite lateral and torsional motions are small compared with vehicle weight and vertical plane irregularities, respectively. The vehicle is assumed to be a rigid body of mass m_2 and moment of inertia J about its center of mass. Suspension elements are assumed to include a primary suspension acting directly against the guideway, distributing its force over a pad length* d, and a secondary suspension consisting of a spring and damper or including active elements.

The need for two suspension systems is based on two conflicting requirements; good tracking and acceptable ride comfort. Reliable tracking demands small clearance between the suspension magnet and the guideway, while good ride comfort (defined as low vertical acceleration levels) requires weak coupling of the passenger cabin module to the guideway. When the secondary suspension is passive, the higher modes of vibration of the guideway propagate easily to the passenger cabin. To improve ride quality, the passenger module may be equipped with active suspension (which responds only to absolute velocity).

---

*For magnetically suspended vehicles, $\ell/5 < d < 2\ell/3$

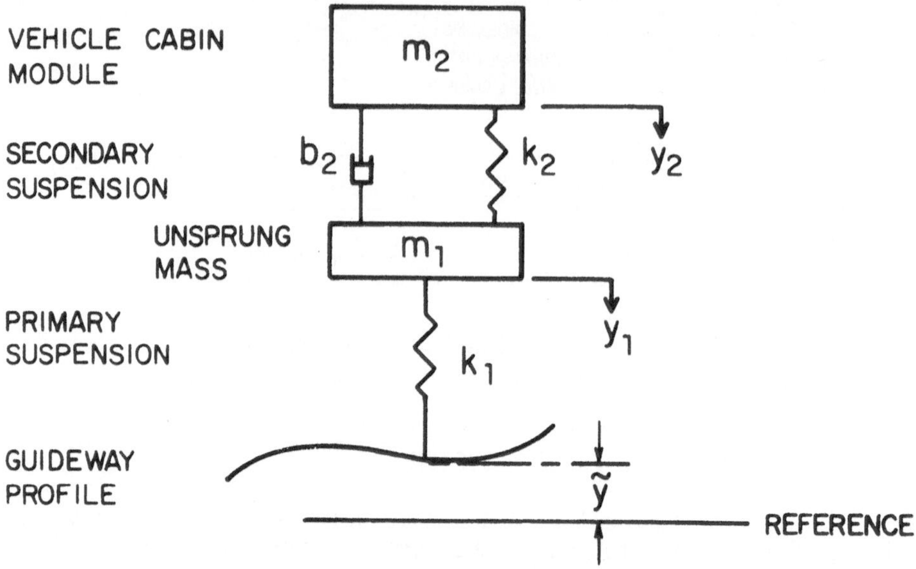

Figure 2.1. Single Point Suspension System

2.1 Suspension Dynamics

The equations describing suspension dynamics in Fig. 2.1 may be expressed as

$$\begin{aligned} m_2\ddot{y}_2(t) &= k_2\,[y_1(t) - y_2(t)] + b_2\,[\dot{y}_1(t) - \dot{y}_2(t)] \\ m_2\ddot{y}_2(t) + m_1\ddot{y}_1(t) &= k_i\,[\tilde{y}(t) - y_2(t)] \end{aligned} \qquad (2.1)$$

where $\tilde{y}(t)$ is the guideway profile, which is given by

$$\tilde{y}_1(t) = y(t,vt) + y_0(vt) \quad , \quad \text{for } \frac{d}{\ell} << 1$$

and

$$\begin{aligned} \tilde{y}_1(t) &= \int_0^{\ell} [y_0(x) + y(t,x)]w(t,x)dx \\ &= \frac{d}{2}\int_{v-d/2}^{v+d/2} [y_0(x) + y(t,x)]dx \quad , \quad \text{otherwise} \end{aligned} \qquad (2.2)$$

and where

$$w(t,\cdot) = \text{pressure distribution function}$$

$$= \begin{cases} 1 \ , & \text{for } (vt - d/2) \leq x \leq (vt + d/2) \\ 0 & \text{elsewhere} \end{cases} \qquad (2.3)$$

The interacting suspension force f(t) is given by

$$f(t) = (m_1 + m_2)\, g - [m_1\ddot{y}_1(t) + m_2\ddot{y}_2(t)] \qquad (2.4)$$

2.2 Beam Dynamics

A simplified model for a guideway is given by the Bernoulli-Euler partial differential equation

$$EI\,\frac{\partial^4 y(t,x)}{\partial x^2} - T\,\frac{\partial^2 y(t,x)}{\partial x^2} + ky(t,x) + b\frac{\partial y(t,x)}{\partial t} + \rho A\frac{\partial^2 y(t,x)}{\partial t^2}$$

$$= p(t,x) = f(t)\; w(t,x) \qquad (2.5)$$

where E is the elastic modulus, I is the moment of inertia, ρ is the mass density, and A is the cross-sectional area of the beam. The term $T\,\frac{\partial^2 y(\cdot)}{\partial x^2}$ is included to account for axial tension T that may be present in the guideway. The term $ky(\cdot)$ represents a distributed elastic restraint (elastic foundation) that is zero for a discretely supported beam. The damping coefficient b represents a dissipative force acting between the beam and ground, and is usually approximated to include all damping effects in the structure. Equation 2.5 also assumes a uniform beam, whose plane sections remain plane with deflection $y(\cdot)$ remaining uniformly normal to the plane of bending. The Bernoulli-Euler representation of the guideway is an acceptable model in the linear dynamic analysis, providing vehicle speeds remain less than ten percent of sonic velocity [$= \sqrt{E/\rho}$] of the guideway stress waves and the ratio of the guideway span width to span length is less than 0.7.

The boundary and initial conditions for Eq. 2.5 are defined by the characteristics of the supports and the coupling between successive beam spans. The simplest case occurs for simply supported beams with single spans and rigid supports. These are given by

$$y(t,0) = y(t,\ell) = EI\,\frac{\partial^2 y(t,0)}{\partial x^2} = EI\,\frac{\partial^2 y(t,\ell)}{\partial x^2}\,,$$

$$\text{for } t \geq 0 \tag{2.6}$$

$$\left.\begin{aligned} y(0,x) &= f_0(x) \\ \frac{\partial y(0,x)}{\partial t} &= g_0(x) \end{aligned}\right\}, \text{ for } x\ (0,\ell) \tag{2.7}$$

where $f_0(\cdot)$ and $g_0(\cdot)$ are specified initial beam displacement and velocity, respectively.

3. INTERACTION ANALYSIS

In a strongly coupled system, Eqs. 2.1 and 2.5 must be solved simultaneously, with boundary and initial conditions given by Eqs. 2.6 and 2.7. In general, there are two types of analytical techniques that are applicable in such problems: (a) direct methods using finite difference numerical algorithms, and (b) modal analysis.

3.1 Direct Numerical Solution

One method of obtaining a direct solution of Eqs. 2.1 and 2.5, based on a Lagrange multiplier technique, is briefly described here. In this method the vehicle and the guideway, represented in terms of generalized coordinates, are treated independently. The coupling is imposed through Lagrange multipliers. The Lagrange's equations are

$$\frac{d}{dt}\left(\frac{\partial T_\nu}{\partial \dot{q}_\nu}\right) - \frac{\partial (T_\nu - V_\nu)}{\partial q_\nu} + \frac{\partial D_\nu}{\partial q_\nu} = f_\nu + \frac{\partial e}{\partial q_\nu}\lambda \tag{3.1}$$

$$\frac{d}{dt}\left(\frac{\partial T_g}{\partial \dot{q}_g}\right) - \frac{\partial (T_g - V_g)}{\partial q_g} + \frac{\partial D_g}{\partial q_g} = f_g + \frac{\partial e}{\partial q_g}\lambda \tag{3.2}$$

for the vehicle and guideway, respectively, where

$e = b_\nu q_\nu + b_g q_g + \alpha$

q = coordinate vector

f = force vector

T = kinetic energy, = $\frac{1}{2}\dot{q}^T m\dot{q}$, m = mass matrix

V = potential energy = $\frac{1}{2}\dot{q}^T k\dot{q}$, k = stiffness matrix

D = viscous dissipation = $\frac{1}{2} q^T C\dot{q}$, C = damping matrix

e = 0 is the system of constraint equations, which enforce contact between vehicle and guideway through suitable choice of b_ν, b_g, and α.

λ = Lagrange's vector multiplier

and a suffix ν denotes vehicle parameters, and a suffix g denotes guideway parameters. The expressions for T, V, and D given above are for linear suspension systems.

Combining the above expressions, Eqs. 3.1 and 3.2 may be expressed as a vector differential equation of the form

$$M\ddot{q} + C\dot{q} + Kq = F + b\lambda \tag{3.3}$$

where $q = [q_\nu \; q_g]^T$ is the system coordinate vector, $M = \text{diag}\,[M_\nu, M_g]$, $C = \text{diag}\,[C_\nu, C_g]$, and $K = \text{diag}\,[K_\nu, K_g]$. The constraint equation is of the form

$$Bq = h \tag{3.4}$$

for some time-dependent vector h. If the transverse velocity and the location of the vehicle on the guideway is known for all t, Eqs. 3.3 and 3.4 may be solved as

$$\ddot{q} = M^{-1}\,[F + B - C\dot{q} - Kq] \tag{3.5}$$

and

$$h = (BM^{-1}B)^{-1}\,[\ddot{h}-\ddot{B}q - 2\dot{B}\dot{q} + BM^{-1}(\dot{q} + BM^{-1}Kq)] \tag{3.6}$$

If q and $\dot{q}$ are known at time $t = t_1$, h may be computed from Eq. 3.6 and $\ddot{q}$ from Eq. 3.5. Numerical integration of $\ddot{q}$ would then yield q and $\dot{q}$ at $t = t_1+\Delta$, for $\Delta>0$. This procedure is then repeated in each successive step of numerical integration.

The advantage of this method is that highly accurate solutions can be generated, even for very complicated multi-vehicle, multi-span systems. Furthermore, beams having non-uniform or non-linear characteristics and vehicles with non-linear suspension systems can be analyzed. The main disadvantage is that the computer time required may be large and may limit the number of cases that may be considered in a design analysis.

3.2 Modal Analysis

The first step in generating a solution of Eqs. 2.1 and 2.5 is to find the solution of the boundary value problem of Eqs. 2.5, 2.6, and 2.7, under the assumption that $p(\cdot) = 0$. For the sake of brevity, b is assumed to be zero (it is of the order of 0.05 for most beam structures). It is then apparent from Eq. 2.5 that the eigenvalues $\{\omega_n\}_{n=1}^{\infty}$ for the homogeneous boundary value problem of Eqs. 2.5 and 2.6 are

$$\omega_n = \left[\frac{EI}{\rho A}\left(\frac{n\pi}{\ell}\right)^4 + \frac{T}{\rho A}\left(\frac{n\pi}{\ell}\right)^2 + \left(\frac{K}{\rho A}\right)\right]^{\frac{1}{2}} \tag{3.7}$$

and the corresponding eigenfunctions $\{y_n\}_{n=1}^{\infty}$ are

$$y_n(x) = \sin\left(\frac{n\pi x}{\ell}\right) \tag{3.8}$$

The second step is to assume that any solution $y(t,x)$ of Eqs. 2.5 and 2.6 can be expanded in terms of the eigenfunctions, i.e.

$$y(t,x) = \sum_{n=1}^{\alpha} a_n(t)y_n(x) \tag{3.9}$$

where $a_n(t)$, n=1,2,... are modal amplitudes that are independent of x. Substituting Eq. 3.9 into Eq. 2.5 and using uniform convergence of the series in Eq. 3.9, one has

$$\sum_{n=1}^{\alpha} [\ddot{a}_n(t) + \omega_n^2 a_n(t)]\, Y_n(x) = \frac{p(t,x)}{\rho A}$$

which by orthogonality of $\{Y_n\}_{n=1}^{\infty}$ in $[0, \ell]$ gives

$$\ddot{a}_n(t) + \omega_n^2 a_n(t) = \frac{2}{\ell g A} \int_0^\ell p(t,x) \sin\left(\frac{n\pi x}{\ell}\right) dx \tag{3.10}$$

which satisfies the initial conditions:

$$\left.\begin{aligned} a_n(o) &= \frac{2}{\ell} \int_0^\ell f_0(x) \sin\left(\frac{n\pi x}{\ell}\right) dx \\ a_n(o) &= \frac{2}{\ell} \int_0^\ell g_0(x) \sin\left(\frac{n\pi x}{\ell}\right) dx \end{aligned}\right\} \tag{3.11}$$

Under weak vehicle-guideway coupling, i.e. when the pressure distribution is dominated by static considerations, Eqs. 3.10 and 3.11 can be easily solved, the solution being

$$a_n(t) = a_n(0) \cos(\omega_n t) + \omega_n^{-1} a_n(0) \sin(\omega_n t) + \omega_n^{-1} \int_0^\ell \sin[\omega_n(t-\tau)]\, p_n(\tau)\, d\tau$$

where

$$p_n(t) = \frac{2}{\rho A} \int_0^\ell p(t,x) \sin\left(\frac{n\pi x}{\ell}\right) dx$$

The case of weak dynamic coupling between the vehicle and the guideway is of significance, primarily due to the acceleration constraints of the passenger module. In this case, the guideway displacement can be computed and used as a known input to the vehicle suspension, to determine vehicle heave accelerations. When the weak coupling assumption does not apply, Eq. 3.10 will have to be solved simultaneously with Eqs. 2.1 and 2.5 by using a point to point numerical integration algorithm.

In the modal analysis method, only a finite number of modes is generally used to approximate the solution. The number of modes required for a given degree of approximation is dependent on the frequency content of the pressure distribution function $p(\cdot)$ and on its time-varying nature. For beams excited by a distribution $p(\cdot)$ of constant shape and moving with a constant velocity v, the number of modes required is related to the cross frequency ratio $|V_c = T_1/(\ell/v)|$, i.e. to the ratio of the period of the fundamental mode $[T_\ell = \frac{2\ell^2}{\pi}(\frac{\rho A}{EI})^{\frac{1}{2}}]$ and the time required to traverse a span. For $V_c \leq 2$ (applicable for $v \leq 350$ mph) the contribution of the m-th mode to the beam deflection is proportional to m^{-4}, which shows rapid convergence of the series expansion of Eq. 3.9.

4. COUPLED INTERACTION

The constant force model mentioned above can be extended to study the coupled vehicle-suspension-guideway interaction, using the one-dimensional vehicle model discussed in the previous sections (Fig. 2.1). The response of vehicle vertical acceleration level (ride comfort) and guideway deflection provides an indication of the extent of coupling. This depends on the following parameters, in addition to the span crossing frequency V_c:

(1) primary-to-secondary suspension stiffness ratio, $k_r = k_1/k_2$

(2) span damping ratio $\xi_1 = b_m/4\pi f_m \rho A$, where b_m is the mth modal damping ratio and $f_m = m^2 f_1$ (term neglected earlier to illustrate the analytical method of modal solution).

(3) vehicle-to-guideway mass ratio $m_r = m_2/\rho A\ell$

(4) guideway-to-vehicle/suspension natural frequency ratio $\Omega = \frac{T}{2\pi}\sqrt{k_2/m_2}$

(5) vehicle /suspension damping ratio $\xi_2 = b_2/2\sqrt{k_2/m_1}$

Figures 4.1 and 4.2 show the variation of maximum vehicle acceleration $\ddot{Y}_2$ and mid-span deflections, respectively, for two values of vehicle damping ratio. These results illustrate that an optimum damping ratio exists, which will maximize ride comfort over any given speed range.

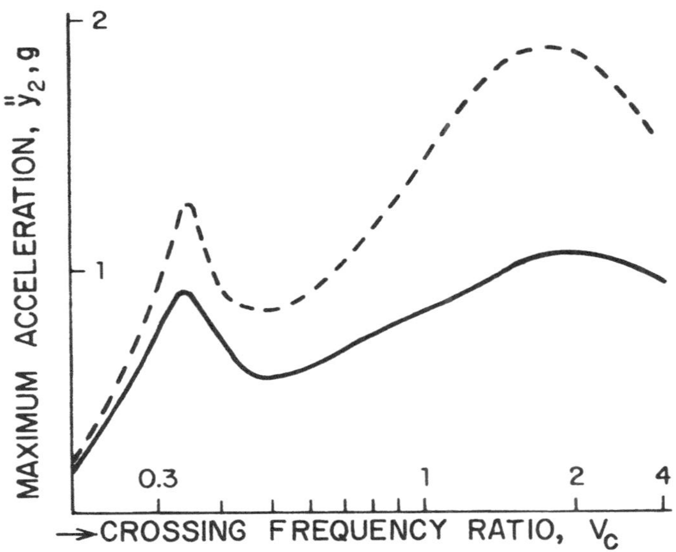

Figure 4.1 Variation of maximum vehicle acceleration

$\left[k_2/k_1 = 10,\ m_2\rho Al = 0.25,\ 2\pi f_1/[k_2/m_2]^{\frac{1}{2}} = 3 \right.$

$\left. b_2/2[k_2/m_2]^{\frac{1}{2}} : \text{——— } 0.25 \text{ ----- } 0.5 \right]$

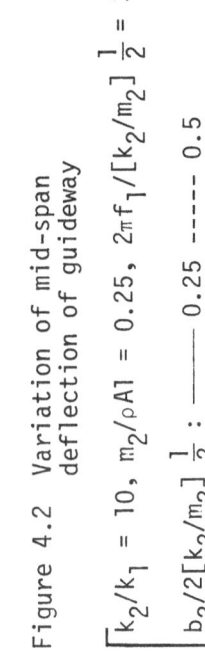

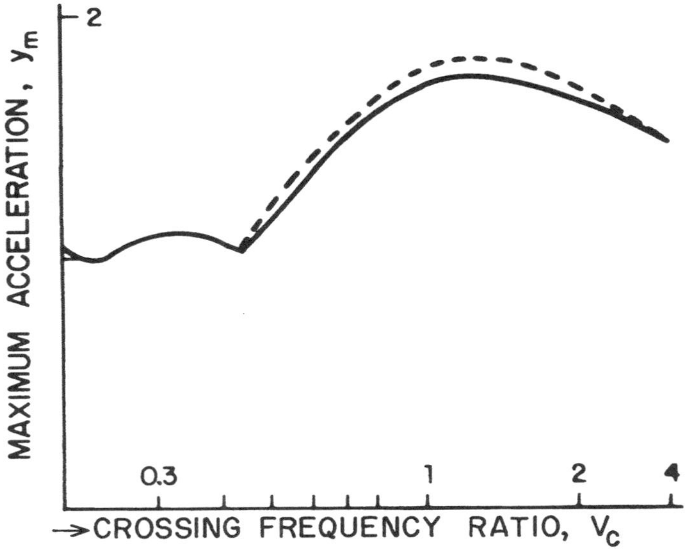

Figure 4.2 Variation of mid-span deflection of guideway

$\left[k_2/k_1 = 10,\ m_2/\rho Al = 0.25,\ 2\pi f_1/[k_2/m_2]^{\frac{1}{2}} = 3 \right.$

$\left. b_2/2[k_2/m_2]^{\frac{1}{2}} : \text{——— } 0.25 \text{ ----- } 0.5 \right]$

5. STABILITY

A direct method of analyzing the stability of the coupled vehicle-guideway system is based on the reduction of ordinary differential equations to a set of linear algebraic equations. Assuming EI to be independent of x and T=b=0, Eqs. 2.1, 2.4, and 2.5 give

$$EI \frac{\partial^4 y(t,x)}{\partial x^4} + \rho A \frac{\partial^2 y(t,x)}{\partial x^2} = \left[(m_1+m_2)g - m_1 \frac{\partial^2 y_1(t)}{\partial t^2} - m_2 \frac{\partial^2 y_2(t)}{\partial t^2}\right] \quad (5.1)$$

where, from sections 3 and 4,

$$\left.\begin{aligned} y(t,x) &= \sum_{n=1}^{\alpha} a_n(t) \sin\left(\frac{n\pi x}{\ell}\right) \\ w(t,x) &= \frac{2}{\pi} \sum_{n=1}^{\alpha} \left[\frac{\sin\left(\frac{n\pi d}{\ell}\right)}{\left(\frac{n\pi d}{\ell}\right)}\right] \sin\left(\frac{n\pi vt}{\ell}\right) \sin\left(\frac{n\pi x}{\ell}\right) \end{aligned}\right\} \quad (5.2)$$

Combining Eqs. 5.1 and 5.2, one has

$$\frac{d^2 a_n(t)}{dt^2} + \left(\frac{\ell w_1}{\pi v}\right)^2 x^4 a_n(t) = \left[(m_1+m_2)g - m_2 \frac{d^2 y_2(t)}{dt^2} - m_1 \frac{d^2 y_1(t)}{dt^2}\right] \times \frac{2}{\pi} \frac{\sin(n\pi d/\ell)}{(n\pi d/\ell)} \sin\left(\frac{n\pi vt}{\ell}\right) \quad (5.3)$$

Similarly,

$$\tilde{y}(t) = \sum_{n=1}^{\alpha} \left[a_n(t) + y_{0n}\right] \frac{\sin(n\pi d/\ell)}{(n\pi d/\ell)} \sin(n\pi vt/\ell) \quad (5.4)$$

Equations 5.3 and 5.4 suggest that a non-zero pad (d>0) has a considerable influence in reducing high-order harmonics, the reduction factor being

$$\frac{\sin(n\pi d/\ell)}{(n\pi d/\ell)} \ll 1 , \qquad \text{for } n \gg 1$$

Assuming a marginally stable solution,

$$y_2(t) = \sum_{n=1}^{\alpha} [b_n \cos (n\pi t/\ell) + \bar{b}_n \sin (n\pi d/\ell)] \tag{5.5}$$

From Eq. 2.1, solving for $y_1(t)$, one has

$$y_1(t) = \sum_{n=1}^{\alpha} \left[(\alpha_1 b_n + \beta_1 \bar{b}_n) \cos(\frac{n\pi vt}{\ell}) + (\alpha_2 b_n + \beta_2 \bar{b}_n) \sin(\frac{n\pi vt}{\ell})\right] \tag{5.6}$$

where α_i and β_i, i=1,2, are constants. Combining Eqs. 2.5, 5.5, and 5.6, and substituting the solution of Eq. 2.5 into Eq. 2.2, one has

$$\tilde{y}(t) = \sum_{j,n} \left[(\gamma_{jn} b_j + \gamma_{jn} b_j) \cos (n\pi vt/\ell) + (\delta_{jn} b_j + \bar{\delta}_{jn} \bar{b}_j) \sin (n\pi vt/\ell)\right] \tag{5.7}$$

Finally, substituting Eqs. 5.5, 5.6, and 5.7 into Eq. 2.1 and equating the coefficients of the cosine and sine terms to zero, one obtains $D'b'=0$ and $D''b''=0$, where D' and D'' are infinite dimensional matrices of constant coefficients and $b'^T = [b_1, \bar{b}_i]$ and $b''^T = [b_2, \bar{b}_i]$, i=1,.... Thus, Eq. 5.5 is a solution of Eqs. 2.1 to 2.5 if and only if

$$\det(D') = \det(D'') = 0 \tag{5.8}$$

This method of determining the boundary of the relative stability region for the system in Fig. 2.1 is suitable for parametric studies. From Eq. 5.8, the relative stability boundaries may easily be obtained. Such a boundary is depicted in Fig. 5.1, which illustrates that a higher guideway first natural frequency has a stabilizing influence, as compared with the composite system in Fig. 2.1.

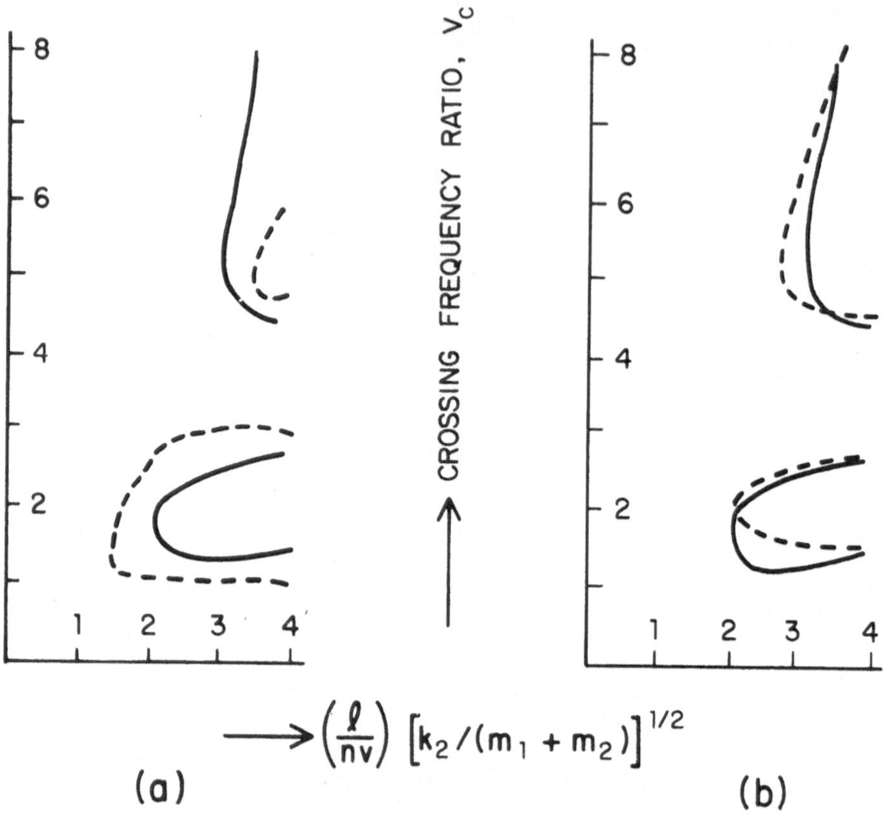

Figure 5.1. Effect Of Vehicle-Guideway Parameters On Stability [a. mass ratio: $(m_1+m_2)/\rho A\ell$; ——— 0.5, ----- 1.0 and b. damping ratio: $b_2/2[k_2(m_1+m_2)]^{1/2}$; ——— 0.3, ----- 0.6]

REFERENCES

1. Manouni, L., High Speed Ground Transportation System; A Review, Internal Report, University of Warwick, March 1980.
2. Richardson, H.H., and Wormley, D.N., "Transportation Vehicle/ Beam-Elevated Guideway Dynamic Interactions: A State-Of-The-Art Review," J. Dyn. Syst. Meas. & Contr., Vol. 96, 1974, pp. 169-179.
3. Blejwas, T.E., Feng, C.C., and Ayre, R.S., "Dynamic Interaction Of Moving Vehicles And Structures," J. Sound & Vibr., Vol. 67, 1979, pp. 513-521.

4. Cheung, Y.I., and Genin, J., "Stability Of A Vehicle On A Multispan Simply Supported Guideway," J. Dyn. Syst. Meas. & Contr., Vol. 100, 1978, pp. 326-332.
5. Sinha, P.K., "Magnetic Suspension For Low-Speed Vehicles," J. Dyn. Syst. Meas. & Contr., Vol. 100, 1978, pp. 333-342.

OPTIMAL PLASTIC DESIGN OF BEAMS FOR WORKHARDENING ADAPTATION

Castrenze Polizzotto and Teotista Panzeca

University of Palermo, Italy

ABSTRACT

This paper deals with optimal design of rigid-plastic beams that are subjected to loadings that are allowed to vary inside a given loading domain, the real loading history being unknown. The optimum design is one that is able to adapt to the loadings, which means that deformation eventually stops being produced and then the structure behaves as a purely rigid body. Bounds on deformation are also provided and suitably included in the optimization problem. The optimality conditions are discussed in order to point out some features of the optimum design. It is shown that the present design criterion provides designs that are more economic than the corresponding limit designs when constraints on deformation are imposed.

1. INTRODUCTION

Workhardening adaptation is a process that a rigid-plastic workhardening structure may experience under the action of variable loadings. When workhardening adaptation occurs, plastic deformation stops being produced and the structure behaves like a purely rigid body, while the amount of plastic deformation produced is limited.

---

*The present paper is part of a research project sponsored by the National (Italian) Research Council, C.N.R., PAdIS Committee.

Workhardening adaptation, a concept first introduced by Prager [1,2] and subsequently studied by others [3,4], is comparable with shakedown of elastic-plastic structures, but while shakedown is mainly caused by self-stresses, workhardening adaptation is due only to the workhardening behavior of the structure. Prager considered quasi-static loads and a kinematic workhardening law, using the term Bauschinger adaptation. Extensions of the theory to dynamic loads and to other workhardening laws were given in Refs. 3 to 6. Second-order geometric effects were also considered [4,7]. Moreover, in analogy with shakedown theory, bounding techniques were developed [6,7] for the assessment of plastic deformation produced during the adaptation process.

Workhardening adaptation as a design criterion was proposed by Prager [8] for truss-like structures with a kinematic workhardening law and by Polizzotto et al. [9] for a wider class of discrete-type structures and workhardening laws. Constraints on deformation were also included in the framework of the optimization problem [10,11], but these constraints are the source of nonlinearity and nonconvexity of the problem.

The present paper is devoted to optimal design of beams, using the workhardening adaptation criterion. For simplicity, consider the bending moment theory of plasticity and suppose that the workhardening is linearly kinematic. The loading is allowed to vary inside a given loading domain, but the real loading history is unknown. The displacements are treated as infinitesimal. Constraints on deformation are also considered in the framework of the optimization problem. The paper is divided into two parts. In the first (Sections 2 and 3) a sufficient criterion for adaptation and also bounds on (plastic) deformation (displacement, strain, and strain intensity) are given. In the second part (Sections 4 to 7) the problem of optimal plastic designs of beams is treated.

2. THE WORKHARDENING ADAPTATION PROBLEM

A rigid-plastic, linearly workhardening beam of length L (Fig. 2.1a), supported at the extreme points and having a mass per unit length $m = m(x)$, is loaded by the load $p = p(x,t)$. Let M, T, w, and k denote the bending moment, the shear force, the deflection, and the curvature, as functions of location x $(0 \leq x \leq L)$ and time t $(t \geq 0)$. Differentiation with respect to x is indicated by primes and with respect to t is indicated by dots.

Equilibrium and compatibility conditions are

$$M'' - m\ddot{w} + p = 0 \ , \quad x \in (0,L) \ , \quad t \geq 0 \tag{2.1}$$

(a)

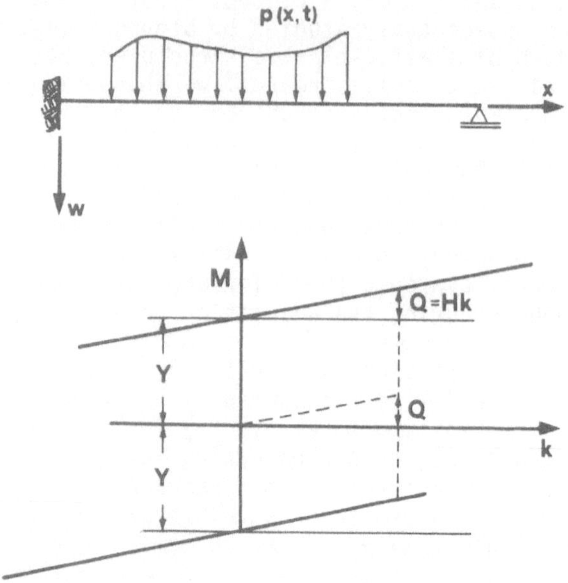

(b)

Figure 2.1 Rigid-Plastic Linearly Workhardening Beam, Dynamically Loaded [(a) Geometrical Sketch, (b) Bending Moment - Curvature Diagram]

$$k = -w'' , \quad x \in (0,L) , \quad t \geq 0 \tag{2.2}$$

to which one must add the boundary conditions

$$\left.\begin{array}{l}\text{either } w = 0 , \quad \text{or } T \text{ is assigned} \\ \text{either } w' = 0, \quad \text{or } M \text{ is assigned}\end{array}\right\} \begin{array}{l}\text{at } x = 0 \text{ and } x = L, \\ \quad t \geq 0\end{array} \tag{2.3}$$

and the initial conditions

$$w = 0 , \quad \dot{w} = \dot{w}_0, \quad x \in (0,L) \text{ and } t = 0 \tag{2.4}$$

where the initial velocity $\dot{w}_0 = \dot{w}_0(x)$ is assigned over all the beam.

The plastic behavior of the beam is specified by the M-k diagram of Fig. 2.1(b), where the virgin yield bending moment $Y = Y(x)$ and the workhardening stiffness $H = H(x)$ are shown. A convenient mathematical description of flow-law plasticity, accounting for rigid recovery, is as follows:

$$\phi_1 = M - Q - Y, \quad \phi_2 = -(M - Q) - Y \tag{2.5}$$

$$\phi_\alpha \le 0, \quad \dot{\lambda}_\alpha \ge 0, \quad \alpha = 1,2 \tag{2.6}$$

$$\phi_\alpha \dot{\lambda}_\alpha = \dot{\phi}_\alpha \dot{\lambda}_\alpha = 0, \quad \text{(the summation rule is applied on } \alpha) \tag{2.7}$$

$$\dot{k} = \dot{\lambda}_1 - \dot{\lambda}_2 , \tag{2.8}$$

where $\phi_\alpha, \alpha = 1,2$, play the role of plastic potentials, $\dot{\lambda}_\alpha$, $\alpha = 1,2$, are the so-called plastic activation coefficients, and Q is a stress parameter called "workhardening moment".

For the sake of greater compactness, set

$$\underset{\sim}{N} = [1, -1], \ \underset{\sim}{Y} = [Y,Y]^T \tag{2.9}$$

$$\underset{\sim}{\phi} = [\phi_1, \phi_2]^T, \ \dot{\underset{\sim}{\lambda}} = [\dot{\lambda}_1, \dot{\lambda}_2]^T \tag{2.10}$$

so that Eqs. 2.5 to 2.8 can be given in the matrix form

$$\underset{\sim}{\phi} = \underset{\sim}{N}^T(M - Q) - \underset{\sim}{Y} \tag{2.11}$$

$$\underset{\sim}{\phi} \le \underset{\sim}{0}, \quad \dot{\underset{\sim}{\lambda}} \ge \underset{\sim}{0}, \quad \underset{\sim}{\phi}^T \dot{\underset{\sim}{\lambda}} = \dot{\underset{\sim}{\phi}}^T \dot{\underset{\sim}{\lambda}} = 0 \tag{2.12}$$

$$\dot{k} = \underset{\sim}{N}\dot{\underset{\sim}{\lambda}} \tag{2.13}$$

The workhardening moment $Q = Q(x,t)$, which meansures the displacement of the yield domain (Fig. 2.1(b)), is by hypothesis linearly linked to the curvature k, i.e.

$$Q = Hk \tag{2.14}$$

The mechanical problem of the beam under given loads and initial velocities is governed by Eqs. 2.1 to 2.4 and Eqs. 2.11 to 2.14. When the beam, after an initial phase in which some limited amount of plastic work has been done, stops producing further plastic deformation and thus behaves as a purely rigid body, one says that there is adaptation, or that the structure is able to adapt. Sufficient criteria for adaptation have been given for solid bodies [2,5], as well as for discrete structures [1,3,4,6,7] and for quasi-static, as well as for dynamic loads. Bounding techniques have been given for discrete structures [6,7].

A unified formulation of adaptation criteria and bounding techniques for beams is presented in this paper. To this purpose, it is helpful to introduce the quantity

$$\eta = \int_0^L \underline{R}^T \underline{\lambda}\, dx \tag{2.15}$$

where $\underline{R}$ is a given two-component, time-independent vector function and $\underline{\lambda}$ is the plastic strain intensity vector, i.e.

$$\underline{\lambda} = \int_0^t \dot{\underline{\lambda}}\, dt \tag{2.16}$$

From the definition of Eq. 2.15, $\eta = \eta(t)$ proves to be a measure of plastic deformation at time t. The kind of plastic deformation depends on the way the vector function $\underline{R}$ is specified in (0,L). A number of different cases are possible, for instance:

(1) If $R_1 = R_2 = 1$, the deformation measure takes the form

$$\eta = \int_0^L (\lambda_1 + \lambda_2)dx = \int_0^t \int_0^L ||\dot{k}||dxdt \tag{2.17}$$

and η proves to be a nonnegative and nondecreasing function of time, which stops increasing only if plastic deformation stops being produced. Thus, if the integral in Eq. 2.17 is limited for any $t > 0$, adaptation occurs.

(2) If $\underline{R} = \underline{N}^T\bar{M}$, where $\bar{M} = \bar{M}(x)$ is a given bending moment distribution, the quantity η, by virtue of Eq. 2.13, becomes

$$\eta = \int_0^L \bar{M}\; k\; dx \tag{2.18}$$

which is a measure of the plastic curvature k at t.

(3) Finally, if $\underline{R} = \underline{N}^T\bar{M}$, as in (2), and in addition $\bar{M}$ is in equilibrium with some given load, $\bar{P}$, then by the virtual work principle

$$\eta = \int_0^L \bar{P}\; w\; dx \tag{2.19}$$

which is a measure of the deflection w at time t.

For structural safety, not only must adaptation occur, but also deformation, expressed in one of the forms (1), (2), or (3) shown above, must be limited. All these requirements imply that the integral in Eq. 2.15 must be limited.

3. BOUNDING AND ADAPTATION THEOREMS

Let $P_i = P_i(x)$, $i \in I_n \equiv (1,\ldots,n)$, be a set of n independent loading conditions for the beam of Section 2 and let Π be the (convex) hull associated with this set of loads. Let $\eta = \eta(t)$ be a deformation measure that is specified as in Eq. 2.15.

Theorem 3.1 (Bounding Theorem): If there exist a time-independent subsequent yield moment distribution over the beam, of the form

$$\tilde{\underline{Y}} = \underline{Y} + \underline{N}^T\tilde{Q} - \omega\underline{R}, \quad \omega > 0 \tag{3.1}$$

and n bending moment distributions $\tilde{M}_i$ that satisfy

$$\left.\begin{array}{l} \tilde{M}_i'' + P_i = 0 \\ \underline{N}^T(\tilde{M}_i - \tilde{Q}) - \underline{Y} + \omega\underline{R} \leq \underline{0} \end{array}\right\} \quad x \in (0,L),\ i \in I_n \tag{3.2}$$

where $\tilde{M}_i$ satisfy the mechanical boundary conditions, then the deformation measure $\eta(t)$ associated with any loading history inside Π is bounded and

$$\eta(t) \leq \frac{1}{\omega}\left[K_0 + \frac{1}{2}\int_0^L H^{-1}\tilde{Q}^2 dx\right], \quad \text{for all } t > 0 \tag{3.3}$$

where

$$K_0 = \frac{1}{2}\int_0^L m\dot{w}_0^{\,2}\, dx \tag{3.4}$$

is the intial kinetic energy.

For the proof of this theorem, see the Appendix. An adaptation criterion is derived from Thm. 3.1 by setting $\underline{R} = \underline{0}$.

Theorem 3.2 (Adaptation Theorem): If there exist a time-dependent subsequent yield moment distribution over the beam, of the form $\tilde{\underline{Y}} = \underline{Y} + N^T\tilde{Q}$, and n bending moment distributions $\tilde{M}_i$ that satisfy

$$\left.\begin{array}{l}\tilde{M}_i'' + P_i = 0 \\ \underline{N}^T(\tilde{M}_i - \tilde{Q}) - \underline{Y} \leq \underline{0}\end{array}\right\} \quad x \in (0,L), \quad i \in I_n \tag{3.5}$$

where $\tilde{M}_i$ satisfy the mechanical boundary conditions, then the structure is able to adapt under loads ranging within the domain Π' associated with the reduced loads P_i/s, $s > 1$, while it may be the limit of adaptation for $s = 1$ (hence $\Pi' \equiv \Pi$).

For the proof, divide Eqs. 3.5 by $s > 1$ and set $\hat{M}_i = \tilde{M}_i/s$ and $\hat{Q} = \tilde{Q}/s$. Thus, one can write

$$\left.\begin{array}{l}\hat{M}_i'' + \frac{1}{s} P_i = 0 \\ \underline{N}^T(\hat{M}_i - \hat{Q}) - \underline{Y} + \frac{s - 1}{s} \underline{Y} \leq \underline{0}\end{array}\right\} \quad x \in (0,L), \quad i \in I_n \tag{3.6}$$

which are equivalent to Eqs. 3.5, but have the form of Eqs. 3.2, provided

$$\underline{R} = \underline{Y}, \quad \omega = \frac{s - 1}{s} \tag{3.7}$$

As a result, the inequality in Eq. 3.3 becomes

$$\int_0^L \underline{Y}^T \underline{\lambda}\, dx \leq \frac{s}{s - 1} \left[\frac{1}{2} \int_0^L m\dot{w}_0^2\, dx + \frac{1}{2} \int_0^L H^{-1} \hat{Q}^2\, dx \right] \tag{3.8}$$

which, until $s > 1$, implies that adaptation occurs. This completes the proof.

For a given structure subjected to given loads P_i, which is not at the adaptation limit, there is some freedom in satisfying Eqs. 3.2. Therefore one can search for the solution $(\tilde{Q}, \tilde{M}_i, \omega)$ that produces the most stringent bound.

4. OPTIMAL DESIGN

A design, i.e. a yield moment distribution $Y = Y(x)$ specified throughout the beam is by definition safe if all the deformation measures η_j of a given set prove to be no greater than corresponding upper limites U_j, i.e.

$$\eta_j = \int_0^L \underline{R}_j^T \underline{\lambda}\, dx \leq U_j \,, \quad \text{for all } j \in I_\ell \equiv (1,\ldots,\ell) \tag{4.1}$$

where the vectors $\underline{R}_j$ are suitably specified in (0,L).

Since each deformation measure may be referred to different limit states of the structure, one may introduce safety factors γ_j, $j = 1,\ldots,\ell$, so that a specific loading domain Π_j may be associated with η_j. These domains are specified by loads P_{ij} that are given by

$$P_{ij} = P_f + \gamma_j Z_i \,, \quad \text{for all } i \in I_n \text{ and } j \in I_\ell \tag{4.2}$$

where P_f is a fixed load and Z_i are alternative loads.

In view of Thm. 3.1, the inequalities in Eq. 4.1 are satisfied if the following conditions are satisfied:

$$\left.\begin{array}{l} M''_{ij} + P_{ij} = 0 \,, \quad \omega_j > 0 \\[2ex] \underline{N}^T(M_{ij} - Q_j) - \underline{Y} + \omega_j \underline{R}_j \leq \underline{0} \\[2ex] \frac{1}{2}\int_0^L m\dot{w}_0^2\, dx + \frac{1}{2}\int_0^L H^{-1} Q_j^2\, dx - \omega_j U_j \leq 0 \end{array}\right\} x \in (0,L),\ i \in I_n,\ j \in I_\ell \tag{4.3}$$

where the bending moments M_{ij} satisfy the mechanical boundary conditions. Equations 4.3 are thus a conservative definition of a safe design. The latter is also able to adapt, if for $j = 1$ one chooses $\underline{R}_1 = \underline{0}$, $\omega_1 = 1$, and $U_1 >> 0$.

Among the set of safe designs, the optimum design is one of minimum cost, the unit cost being proportional to the yield moment Y. Assume that Y is not allowed to be less than a minimal value Y_0, i.e.

$$Y = Y_0 + y, \quad y = y(x) \geq 0, \quad x \in (0,L) \tag{4.4}$$

Moreover, let the fixed load P_f be design dependent, i.e.

$$P_f = a + by \tag{4.5}$$

where b is the self-weight load per unit yield moment.

Therefore, denoting the cost per unit length by G(y), the optimal design problem is as follows

$$\min C = \int_0^L G(y)dx \tag{4.6}$$

subject to the constraints in Eqs. 4.2 to 4.5, where H and m are given functions of y.

If, for the sake of simplicity, from the set of deformation measures of Eq. 4.1, one retains only the first one, then designs are obtained that are only able to adapt. In addition, assuming that G = cy and $\gamma_1 = 1$, the optimal design problem is as follows:

$$\left.\begin{array}{l}\text{Minimize}\\ \quad C = \int_0^L cy\, dx\\ \text{subject to } y \geq 0 \text{ and for } x \in (0,L) \text{ and } i \in I_n\\ \quad M_i'' + Z_i + a + by = 0\\ \quad \underline{N}^T(M_i - Q) - \underline{e}(Y_0 + y) \leq \underline{0}\end{array}\right\} \tag{4.7}$$

where M_i satisfy the mechanical boundary conditions and $e = [1,1]^T$. Observe that the formulation of Eq. 4.7 is a direct consequence of Thm. 3.2.

An admissible design, i.e. a design $y = y(x) \geq 0$ that satisfies the constraints in Eqs. 4.7, is safe with regards to the polyhedral loading domain Π specified by the given relevant loading conditions $P_i = P_f + Z_i$. In order to characterize the optimum design, one proceeds as usual by applying the Lagrange multiplier method. Introducing the multipliers $\dot{u}_i$ (velocities) and $\dot{\mu}_i \geq 0$ (plastic strain rate coefficients), one obtains the following Euler-Lagrange equations:

for equilibrium,

$$M_i'' + P_f + Z_i = 0, \quad x \in (0,L),\ i \in I_n$$

for compatibility,

$$\dot{\kappa}_i = -\dot{u}_i'', \quad x \in (0,L),\ i \in I_n$$

$$\left.\begin{array}{l}\text{and at } x = 0 \text{ and } x = L\text{, with } i \in I_n, \\ \quad \text{either } \dot{u}_i = 0\text{, or } T_i \text{ is specified} \\ \quad \text{either } \dot{u}_i' = 0\text{, or } M_i \text{ is specified}\end{array}\right\} \tag{4.8}$$

$$\left.\begin{array}{l}\text{with yield conditions for } x \in (0,L) \text{ and } i \in I_n \\ \quad \underline{\phi}_i = \underline{N}^T(M_i - Q) - \underline{e}(Y_0 + y) \\ \quad \underline{\phi}_i \leq \underline{0}, \quad \dot{\underline{\mu}}_i \geq \underline{0}, \quad \underline{\phi}_i^T \dot{\underline{\mu}}_i = 0 \\ \text{and curvature rates} \\ \quad \dot{\kappa}_i = \underline{N}\dot{\underline{\mu}}_i \;, \quad x \in (0,L), \quad i \in I_n\end{array}\right\} \tag{4.9}$$

the resultant mechanism

$$\left.\begin{array}{l}\dot{\underline{\mu}}_r = \sum\limits_{i=1}^{n} \dot{\underline{\mu}}_i \;, \\ \dot{\kappa}_r = \sum\limits_{i=1}^{n} \dot{\kappa}_i = 0, \quad \dot{u}_r = \sum\limits_{i=1}^{n} \dot{u}_i = 0\end{array}\right\}, \; x \in (0,L) \tag{4.10}$$

and finally the uniformity conditions (β = arbitrary constant)

$$\left.\begin{array}{l}\psi = \underline{e}^T\dot{\underline{\mu}}_r - \beta c, \\ \psi \leq 0, \; y \geq 0, \; \psi y = 0\end{array}\right\} \; x \in (0,L) \tag{4.11}$$

Under the action of every load $P_i = P_f + Z_i$, the optimal beam deforms according to a mechanism $\dot{u}_i$, to which the curvature rate $\dot{\kappa}_i$ and the plastic strain rate intensity vector $\dot{\underline{\mu}}_i$ are associated. The sum of these mechanisms $\dot{u}_r$, which is called the resultant mechanism, proves to be zero (as a result of the workhardening law being kinematic [3,5]), but $\dot{\underline{\mu}}_r \neq 0$. Therefore, a plastic work density D_r is associated with the resultant mechanism, i.e.

$$D_r = \sum_{i=1}^{n} M_i\dot{\kappa}_i = (Y_0 + y)\underline{e}^T\dot{\underline{\mu}}_r \tag{4.12}$$

which is easy to show using Eq. 4.9. The optimality conditions in Eq. 4.11 show that the gradient of D_r, per unit yield moment, i.e the quantity

$$\frac{\partial D_r}{\partial y} = \underline{e}^T \dot{\underline{\mu}}_r = \sum_{i=1}^{n} ||\kappa_i|| \tag{4.13}$$

proves to be proportional to the cost gradient c, in those segments of the beam where a design growth is required, i.e. where $y > 0$. This result is the known Principle of Uniform Energy Dissipation, by Drucker and Shield [12-14], and for this reason the above optimality conditions are called uniformity conditions [15,16].

Figure 4.1 shows a region of the positive quadrant of the $(\dot{\mu}_{r1},\dot{\mu}_{r2})$-plane, bounded by the straight line of equation $\mu_{r1} + \mu_{r2} = \beta c$, $(\beta = 1)$, whose external normal is the vector $\underline{e} = \partial \underline{Y}/\partial y$. According to the relationships in Eqs. 4.11, the vector $\dot{\underline{\mu}}_r$ is allowed to vary only within this region. When the vector touches the boundary, in the relevant section of the beam there is design growth, $y > 0$ and the corresponding plastic resistance increment vector, given by $\Delta \underline{Y} = \underline{e}y$, lies along the external normal $\underline{e}$ (the latter can always be thought of as a unit vector). If, on the other hand, $\dot{\underline{\mu}}_r$ does not touch the boundary, in the relevant section there is no design growth and $y = 0$.

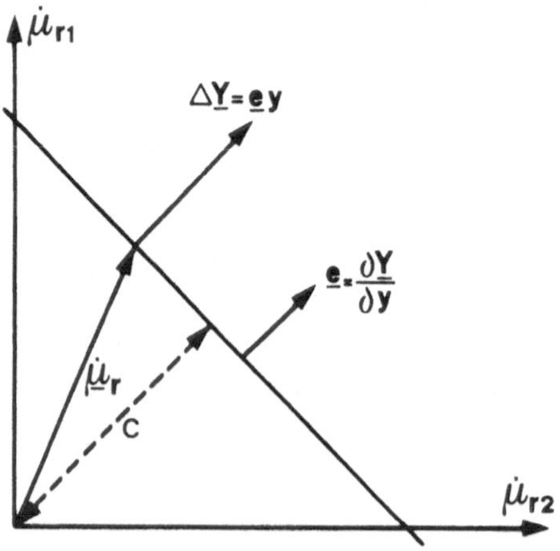

Figure 4.1 $\dot{\underline{\mu}}_r$-Space, Growing Domain, and Growth-Law

As a result of what has been said about Fig. 4.1, one sees that this figure is a geometric picture of an analogy that can be envisaged between the plastic flow-law and a design growth-law [15,16].

On the basis of the Euler-Lagrange equations, one can show that the optimum design is not affected by the fixed load $P_f = a + by$. To show this, let ΔP_f be an arbitrary variation of P_f, specified throughout the beam, and let ΔM be a bending moment field that is in equilibrium with ΔP_f and that satisfies the mechanical boundary conditions. Substituting $M + \Delta M$ and $Q - \Delta M$ for M and Q, respectively, together with all the other quantities left unchanged, all the Euler-Lagrange equations are still satisfied, which proves the statement. Thus, one can say that, when the workhardening is kinematic, any fixed load can be sustained by the beam without addition of any other material to that strictly required by the variable loads.

5. THE WORKHARDENING ADAPTATION DESIGN VS THE STANDARD LIMIT DESIGN

It is evident from the formulation in Eqs. 4.7 that if $H = 0$, and hence $Q = 0$, throughout the beam, the optimal design problem for workhardening adaptation transforms into the traditional problem for plastic collapse. Calling the two optimum costs C and C_p, one has

$$C \leq C_p \tag{5.1}$$

and the workhardening thus produces more economic optimum designs.

The workhardening adaptation design (WHAD) shows properties that are similar to those of the standard limit design (SLD). However, in the latter, the ability of the structure to sustain the loads is provided by the material employed for structural modelling. In the former, a greater or smaller percentage of this material can be saved, because of the strengthening produced by plastic deformation, which is required for adaptation to occur. The amount of material saved depends upon the workhardening law and also upon the range and shape of the loading domain. For instance, if the workhardening law is isotropic, this amount would theoretically be 100% [9].

A comparison of workhardening adaptation designs with the corresponding standard limit designs shows considerable material saving, which however must be greatly reduced when constraints on deformation are taken into account. However, it seems to be reasonable to conjecture that the reduced saving is still worth enough to make the WHAD compatitive with respect to the SLD [9].

For the formulation of the WHAD to be more rational, constraints on deformation should be included in the framework of the optimization problem, in the form shown in Eqs. 4.3, but the resulting problem is more difficult to solve [10,11]. For a simpler procedure it seems to be sufficient to use the formulation in Eqs. 4.7 and make a subsequent check on deformation of the optimum structure, by using the bounding technique provided in Section 3.

The simple rigid-plastic idealization, usually thought of as being applicable when elastic strains are expected to be small in comparison with plastic strains, has received much consideration in the literature. Solutions to static as well as to dynamic problems have been provided and used with some confidence in engineering practice. However, in the authors' opinion, reliability of the optimum designs provided by the present theory has not yet been well assessed and there are still some theoretical and practical aspects of this theory to be better clarified. Among them, a main point is the formulation of suitable fatigue and serviceability safety criteria for an adequate limitation of the deformation required by adaptation.

6. EXAMPLES

As a first example, consider a clamped beam that must sustain a point load $P\tau$, $(-\frac{1}{2} \leq \tau \leq 1)$, in the middle section (Fig. 6.1(a)). In Fig. 6.1(b), the optimum WHAD and SLD profiles are shown. The design variable for the WHAD proves to be equal to 3/4 of the design variable for the SLD, at every section. Figures 6.1(b_1) and 6.1(b_2) show the collapse mechanisms for the two optimum designs, precisely a single collapse mechanism for the SLD and two collapse mechanisms for the WHAD. These latter being opposite to one another, so their sum (resultant mechanism) is zero, as is required by the optimality conditions.

If one introduces the constraint of uniform cross section (y = constant), the WHAD and SLD profiles are as shown in Fig. 6.1(c) and the collapse mechanisms are as shown in Fig. 6.1(c_1).

A second example is considered, as shown in Fig. 6.2(a). A clamped, uniform beam of length 3a must sustain two alternative loads, $P\tau_1$ and $P\tau_2$, with $0 \leq \tau_1 \leq 1$ and $0 \leq \tau_2 \leq 1$, applied at the one third points of the beam. The relevant loading conditions are the points (1,0), (0,1), and (0,0) of Fig. 6.2(b). Obviously, the optimum beam is also able to adapt when the loading is allowed to vary within the convex hull Π shown in Fig. 6.2(b). The SLD yield moment is Pa/3, while the WHAD yield moment is Pa/6. The

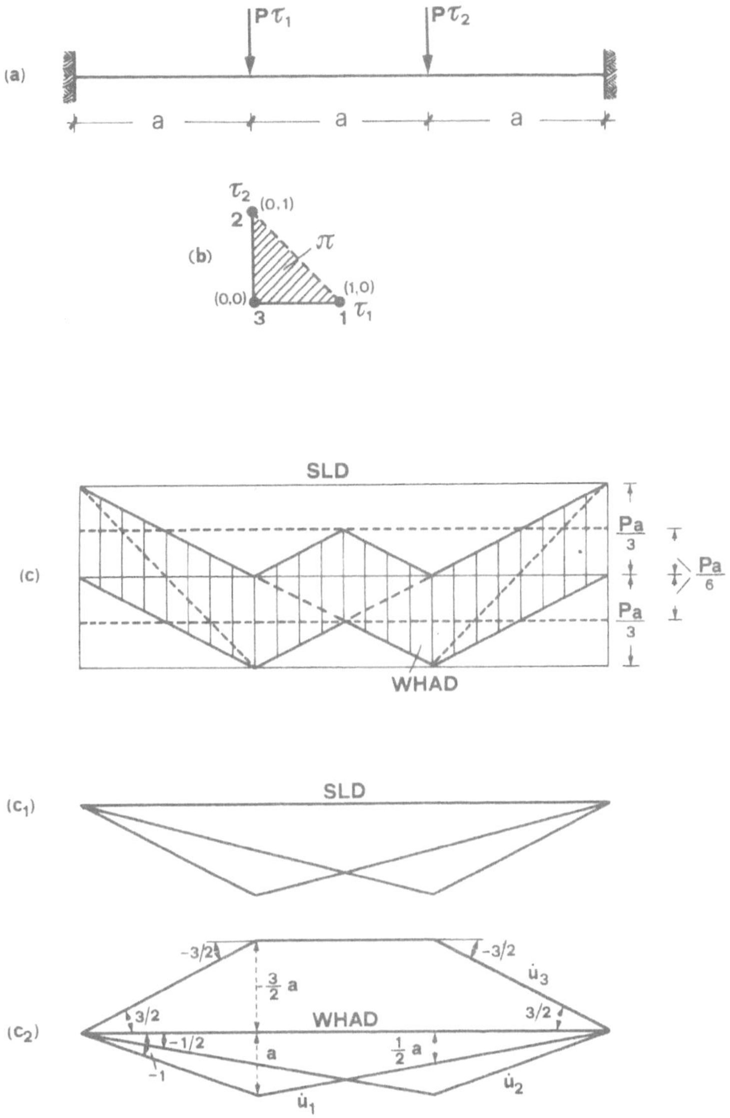

Figure 6.1 Standard Limit Design (SLD) and Workhardening Adaptation Design (WHAD) of a Clamped Beam. [(a) Geometry and Loading Scheme, (b) Non-uniform SLD and WHAD Profiles and (b_1, b_2) Relevant Collapse Mechanisms, (c) Uniform SLD and WHAD Profiles, and (c_1) Relevant Collapse Mechanisms]

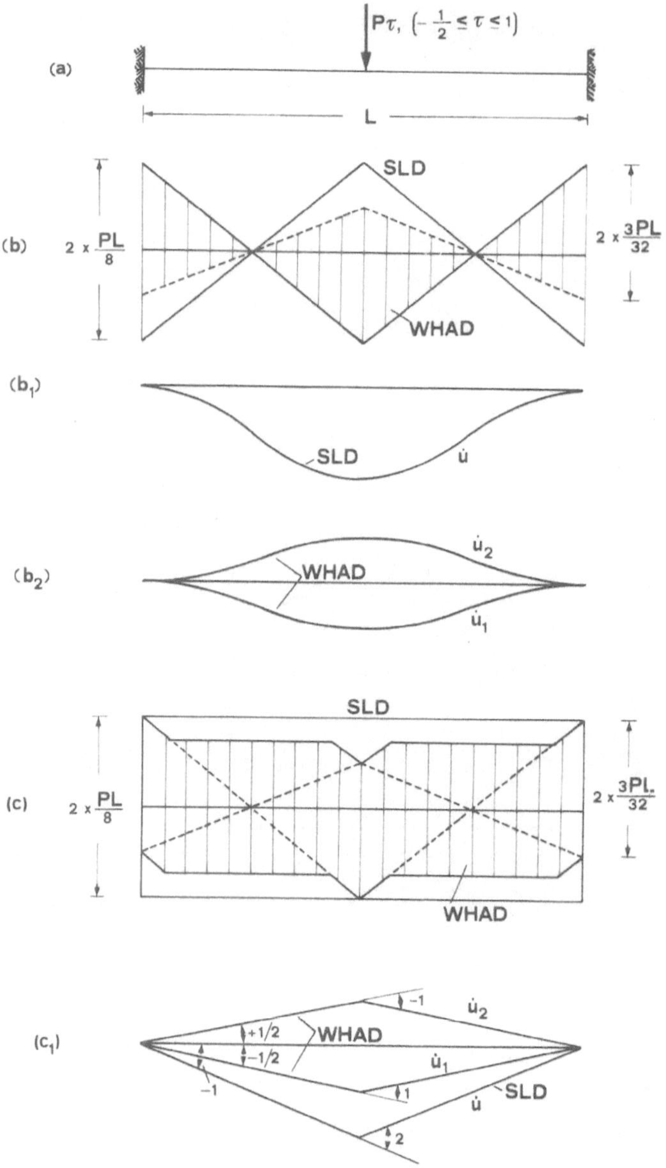

Figure 6.2 Standard Limit Design (SLD) and Workhardening Adaptation Design (WHAD) of a Clamped Beam. [(a) Geometry and Loading Scheme, (b) Loading Domain Π, (c) Uniform SLD and WHAD Profiles, and (c_1,c_2) Relevant Collapse Mechanisms]

SLD shows only two collapse mechanisms, (Fig. 6.2(c_1)), while the WHAD shows three such mechanisms, whose sum (resultant mechanism) proves to be zero (Fig. 6.2(c_2)).

7. CONCLUSION

In the present paper, rigid-plastic kinematically workhardening beams subjected to variable loads have been considered. Criteria for adaptation, as well as bounding principles have been discussed. Finally the optimal design problem has been treated, using the workhardening adaptation concept as a design criterion.

It has been shown that constraints on deformation can be included in the design formulation, even though the case in which such constraints are not included was studied in more detail. In this case, the optimum design is only able to adapt, and the deformation is to be checked a posteriori by using the bounding technique provided in this paper.

The Workhardening Adaptation Design (WHAD) shows some features that are quite similar to those of the Standard Limit Design (SLD). Comparing the WHAD with the corresponding SLD, the former is seen to be more economic than the latter and the material saved can be considerable, also when limitations on deformation are imposed.

The paper points out the need for a greater future understanding of a few aspects of the present theory, in particular the reliability of the WHAD in connection with serviceability and fatigue safety criteria of engineering practice.

APPENDIX

In this Appendix a proof of Thm. 3.1 is given.

Observe that any loading history inside Π can be given in the form

$$P = \sum_{i=1}^{n} \tau_i P_i, \quad \text{for all} \quad t \geq 0 \tag{A.1}$$

where

$$\tau_i = \tau_i(t) \geq 0 , \quad \text{for all} \quad i \in I_n \equiv (1,\ldots,n) \tag{A.2}$$

and

$$\sum_{i=1}^{n} \tau_i = 1 \quad , \qquad \text{for all } t \geq 0 \tag{A.3}$$

Therefore, multiplying Eqs. 3.2 by τ_i and summing with respect to i gives

$$\left.\begin{array}{l} \tilde{M}'' + P = 0 \\ \underline{N}^T(\tilde{M} - \tilde{Q}) - \underline{Y} + \omega\underline{R} \leq \underline{0} \end{array}\right\} x \in (0,L) \tag{A.4}$$

plus the mechanical boundary conditions on the bending moment, $\tilde{M}$, which is given by

$$\tilde{M} = \sum_{i=1}^{n} \tau_i \tilde{M}_i \tag{A.5}$$

Multiplying the second of Eqs. A.4 by $\dot{\lambda}$, which is the plastic activation coefficient vector associated with the relevant loading history, one has recalling Eq. 2.13 that

$$(\tilde{M} - \tilde{Q})\dot{k} - \underline{Y}^T\dot{\underline{\lambda}} + \omega\underline{R}^T\dot{\underline{\lambda}} \leq \underline{0} \tag{A.6}$$

From this, substracting the third of Eqs. 2.12, which in explicit form is

$$(M - Q)\dot{k} - \underline{Y}^T\dot{\underline{\lambda}} = 0 \tag{A.7}$$

yields

$$(\tilde{M} - M)\dot{k} - (\tilde{Q} - Q)\dot{k} + \omega\underline{R}^T\dot{\underline{\lambda}} \leq 0 \tag{A.8}$$

which, through an integration over the interval (0,L), becomes

$$\int_0^L (\tilde{M} - M)\dot{k}\, dx - \int_0^L (\tilde{Q} - Q)\dot{k}\, dx + \omega \int_0^L \underline{R}^T\dot{\underline{\lambda}}\, dx \leq 0 \tag{A.9}$$

Now observe that, since $\tilde{M}$ is in statical equilibrium with P and M is in dynamic equilibrium with the same load, while $\dot{k}$ is a compatible curvature rate distribution, the first integral in Eq. A.9 is

$$\int_0^L (\tilde{M} - M)\dot{k}\, dx = \int_0^L m\, \ddot{w}\, \dot{w}\, dx \tag{A.10}$$

Moreover, by Eq. 2.14 differentiated with respect to t, the second integral in Eq. A.9 is

$$\int_0^L (\tilde{Q} - Q)\dot{k}\, dx = \int_0^L H^{-1}(\tilde{Q} - Q)\dot{Q}dx \tag{A.11}$$

Therefore, Eq. A.9 can be given the form

$$\int_0^L \underline{R}^T \dot{\underline{\lambda}}\, dx = \dot{\eta} \underline{<} - \frac{1}{\omega}\dot{B} \tag{A.12}$$

where B = B(t) is the positive definite functional

$$B = \frac{1}{2}\int_0^L m\, \dot{w}^2\, dx + \frac{1}{2}\int_0^L H^{-1}(Q - \tilde{Q})^2\, dx \tag{A.13}$$

An integration over the time interval $(0,t_1)$, observing that $\underline{\lambda}(0) = \underline{0}$ and hence $\eta(0) = 0$, now gives

$$\eta(t_1) \underline{<} \frac{1}{\omega} B(0) - \frac{1}{\omega} B(t_1) \underline{<} \frac{1}{\omega} B(0) \tag{A.14}$$

which, since t_1 is any subsequent time, coincides with Eq. 3.3. Thus, the theorem is proved.

REFERENCES

1. Prager, W., "Bauschinger Adaptation of Rigid, Workhardening Trusses," Mech. Res. Commun., Vol. 1, 1974, pp. 253-256.
2. Prager, W., "Adaptation Bauschinger d'un Solide Plastique à Écrouissage Cinématique," C. R. Acad. Sci., Ser. B 280, 1975, pp. 585-587.
3. Polizzotto, C., "Workhardening Adaptation of Rigid-Plastic Structures," Meccanica, Vol. 10, No. 4, 1975, pp. 280-288.
4. König, J.A. and Maier, G., "Adaptation of Rigid-Workhardening Discrete Structures Subjected to Load and Temperature Cycles and Second-Order Geometric Effects," Computer Meth. Appl. Mech. Eng., Vol. 8, 1976, pp. 37-50.
5. Polizzotto, C., "Adaptation of Rigid-Plastic Continua Under Dynamic Loadings," J. Struct. Mech., Vol. 6, No. 3, 1978, pp. 319-329.
6. Polizzotto, C., "On Workhardening Adaptation of Discrete Structures Under Dynamic Loadings," Archives of Mechanics, Vol. 32, No. 1, 1980, pp. 81-99.

7. Mazzarella, C. and Panzeca, T., "On Workhardening Adaptation of Discrete Structures Subjected to Dynamic Forces in the Presence of Second-Order Geometric Effects," J. Struct. Mech. (accepted for publication).
8. Prager, W., "Optimal Plastic Design of Trusses for Bauschinger Adaptation," Omaggio a Carlo Ferrari, Libreria Editrice Universitaria Levrotto & Bella, Torino, 1974.
9. Polizzotto, C., Mazzarella, C., and Panzeca, T., "Optimum Design for Workhardening Adaptation," Computer Meth. Appl. Mech. Eng., Vol. 12, No. 2, 1977, pp. 129-144.
10. Polizzotto, C. and Mazzarella, C., "Structural Optimization Based on the Workhardening Adaptation Concept," Structural Control, (ed. H.H.E. Leipholz), North-Holland, 1980.
11. Mazzarella, C. and Polizzotto, C., "Optimum Design of Rigid-Workhardening Structures with Constraints on Deformation," Engineering Structures, Vol. 2, No. 3, 1980, pp. 138-146.
12. Drucker, D.C. and Shield, R.T., "Design for Minimum Weight," Proc. Int. Congress. Appl. Mech. 5, 1956, pp. 212-222.
13. Save, M.A., "A Unified Formulation of the Theory of Optimal Plastic Design with Convex Cost Functions," J. Struct. Mech. Vol. 1, No. 2, 1972, pp. 247-276.
14. Rozvany, G.I.N., Optimal design of flexural systems, Pergamon Press, 1976.
15. Polizzoto, C., "Optimum Plastic Design Under Combined Stresses," Int. J. Solids Structures, Vol. 11, 1975, pp. 539-553.
16. Polizzotto, C., "Optimum Plastic Design for Multiple Sets of Loads," Meccanica, Vol. 9, 1974, pp. 206-213.

OPTIMIZATION OF SHELLS UNDER COMBINED LOADINGS VIA THE CONCEPT OF UNIFORM STABILITY

Jacek Krużelecki

Institute of Mechanics and Machine Design
Technical University of Cracow, Poland

ABSTRACT

A problem of optimization of a cylindrical shell, subjected to bending and torsional loads, is treated in this paper. The shape and thickness of the shell are used as design variables and a minimum weight design objective is adopted. Constraints on both strength and stability are imposed. Problems with several geometric shapes and two materials are solved.

1. INTRODUCTION

It is generally known that, when subjecting a bar to torsion or bending, only the layers distant from the bar axis may be fully utilized. Retaining only these layers consequently leads to thin walled bars, or cylindrical shells. The cross section of such a shell may be fully determined by two functions of one variable. The first determines the position of the center line of the cross section of the shell $y=y(x)$, whereas the other is the change of wall thickness $h=h(x)$. The functions y and h thus serve as two design variables. Optimal design of thin walled cross section of a shell is investigated in Ref. 1 (large twisting, small bending), 2, and 3 (pure bending).

Minimum weight of the shell is taken here as the criterion of optimality. With the assumption of constant values of both twisting and bending moments, M_t and M_b, along the axis of the shell, the cross section will be uniform. Then, assuming that the material is homogeneous, the criterion of minimum weight of the shell reduces to the criteria of minimum cross sectional area.

Such an optimization problem should be stated under constraints connected with strength, stability, and elastic stiffness, but here only two are considered, which ensure safety of the construction. The first is the strength condition, its form based on the Huber-Hencky strength hypothesis,

$$\sigma_0 \geq \sqrt{\sigma^2 + 3\tau^2} = \sigma_{red} \tag{1.1}$$

where σ_0 denotes the allowable stress and σ_{red} denotes the reduced stress. The second constraint is the stability condition, which may be written in the general form

$$M_0 j \leq J[h(x), \rho(x), E_i, \ell, M_b/M_t] \tag{1.2}$$

where J denotes a functional depending on the functions h(x) and $\rho(x)$ (radius of curvature), the elastic constants denoted by a common symbol E_i, the length of the shell ℓ, and the assumed ratio of moments M_b/M_t. The symbol M_0 stands for a certain intensity of moments (e.g. reduced moment) and j stands for an assumed degree of safety with respect to the stability of the wall. In approximations that are often used in the theory of stability of shells, the stability condition may be replaced by a local condition, refering to particular points of the shell. The local buckling constraint used here is

$$\frac{i\sigma}{\sigma_c} + {\frac{j\tau}{\tau_c}}^{2} \leq 1 \tag{1.3}$$

which is an approximation of the exact buckling condition for a shell of considerable length, where

$$\sigma_c = \beta E \frac{h}{\rho}, \quad \tau_c = \alpha E \left(\frac{h}{\rho}\right)^{3/2} \tag{1.4}$$

are critical stresses for pure bending and twisting, respectively, and i and j are degrees of safety.

Assuming the condition of local stability to be satisfied at any point of the shell (active constraint) and the strength condition at a dangerous point only (passive constraint elsewhere) one obtains a shell that is called the shell of uniform stability. If the active condition is only the strength one constraint, then one has the shell of uniform strength.

Four combinations of these constraints, as applied to four possible combinations of design variables, are shown in Tables 1.1(a) to 1.1(d) where a and p denote active and passive conditions respectively.

TABLE 1.1 COMBINATIONS OF ACTIVE CONSTRAINTS

(a) Optimization of y(x) and h(x)

		STABILITY CONDITION a	STABILITY CONDITION p
STRENGTH CONDITION	a	differential equation 1	single variational optimization 2
STRENGTH CONDITION	p	single variational optimization 3	double variational optimization 4

(b) Optimization of y(x) for prescribed h(x)

		STABILITY CONDITION a	STABILITY CONDITION p
STRENGTH CONDITION	a	X	algebraic equation 2
STRENGTH CONDITION	p	differential equation 3	single variational optimization 4

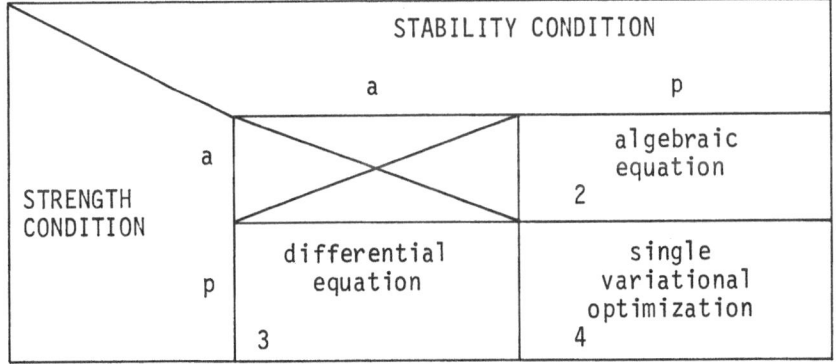

(c) Optimization of h(x) for prescribed y(x)

		STABILITY CONDITION	
		a	p
STRENGTH CONDITION	a		algebraic equation 2
	p	algebraic equation 3	single variational optimization 4

(d) Both functions y(x) and h(x) prescribed

		STABILITY CONDITION	
		a	p
STRENGTH CONDITION	a		
	p		parametrical optimization 4

In Table 1.1, the type of mathematical problem for each case under consideration is shown. All combinations were considered. However, it turns out that some combinations do not lead to any sensible solution, e.g. both constraints passive.

2. FORMULATION OF THE PROBLEM

Consider a cylindrical shell loaded by bending moment M_b and twisting moment M_t, as shown in Fig. 2.1. One seeks the optimum shape of the cross section of the shell, represented by two functions (two design variables), the middle line of the profile $y = y(x)$ and the wall thickness $h = h(x)$. As the criterion of optimality, one takes the minimization of the cross sectional area for given values of external loading (M_b, M_t).

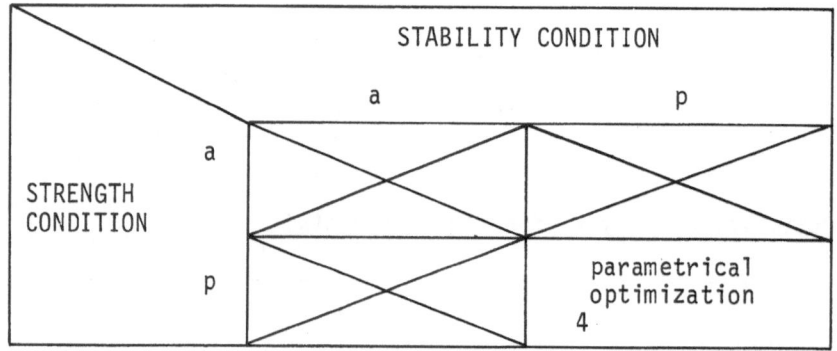

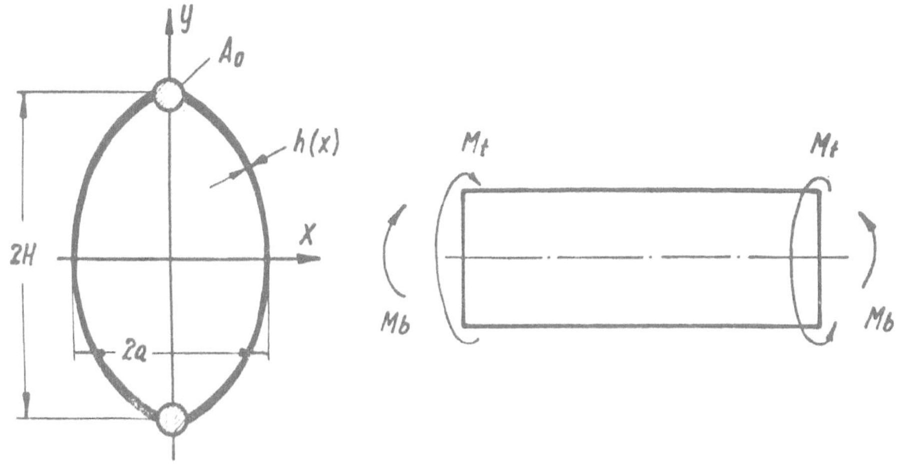

Figure 2.1 Cylindrical Shell

Assume, as shown in Fig. 2.1, that some part of material is concentrated in an outer fibers (rib). This situation is typical for pure bending. For sufficiently large values of M_b, in a combined state of loading, such a rib should appear and improve the cross section.

The functionals that are of interest in this formulation are expressed as follows:

$$A = \oint h ds + 2A_o = \text{minimum} \tag{2.1}$$

$$M_b = \oint \sigma y h ds + 2A_o H \sigma \Big|_{y=H} = \text{constant} \tag{2.2}$$

$$M_t = \oint \rho h (y - xy') \frac{ds}{(1 + y'^2)^{1/2}} = \text{constant} \tag{2.3}$$

where $ds = \sqrt{1 + y'^2}\, dx = \sqrt{1 + x'^2}\, dy$ and $\sigma\big|_{y=H}$ is the maximal normal stress.

Assume that the Bernoulli hypothesis of plane cross sections is valid. Then the normal stress will be distributed as follows:

$$\sigma = \frac{y}{H} \sigma\Big|_{y=H} \tag{2.4}$$

and the shear stress, in conformity with the hydrodynamic analogy, will be

$$\tau h = c \tag{2.5}$$

where c depends on M_t.

The problem of optimization is now stated, under constraints of strength of Eq. 1.1 and local stability of Eq. 1.3. It is convenient to introduce the following dimensionless quantities, variables, and parameters: $\eta = y/H$, $\xi = x/H$, $k = \sigma\big|_{y=H}/\sigma_o$, $g = h/H$, $\rho^* = \rho/H$, $a = A/(\frac{M}{\sigma_o})^{2/3}$, $a_o = A_o/(\frac{M}{\sigma_o})^{2/3}$, $\chi = H/(\frac{M}{\sigma_o})^{1/3}$, $M = \sqrt{M_b^2 + M_t^2}$. Then, Eqs. 2.1 to 2.5 can be rewritten in the form

$$a = 4\chi^2 \int_0^1 g\sqrt{1 + \xi'^2}\, d\eta + 2a_o = \text{minimum} \tag{2.6}$$

$$\frac{M_b}{M} = 4k\, \chi^3 \int_0^1 g\eta^2\sqrt{1 + \xi'^2}\, d\eta + 2a_o k\chi = \text{constant} \tag{2.7}$$

$$\frac{M_t}{M} = -\frac{8}{\sqrt{3}}\, \psi^{5/3} \left(\frac{j\sigma_o}{\sqrt{3}\alpha E}\right)^{2/3} \phi^{5/2}\chi^3 \int_0^1 \xi'\eta\, d\eta = \text{constant} \tag{2.8}$$

where $\psi = \left(\frac{i\sigma_o}{\beta E}\right)^{3/2} / \left(\frac{j\sigma_o}{\sqrt{3}\alpha E}\right)$, $\phi = \left(\frac{jc}{\alpha E\chi}\sqrt[3]{\frac{\sigma_o}{M}}\right)^{2/5} / \left(\frac{i\sigma_o}{\beta E}\right)$, and the constraint on strength has the form

$$g \geq \frac{i\sigma_o}{\beta E}\, \frac{\psi\, \phi^{5/2}}{(1 - k^2\eta^2)^{1/2}} \tag{2.9}$$

and the constraint on buckling is as follows:

$$g^5 - \frac{i\sigma_o}{\beta E}\, k\eta\rho^* g^4 - \left(\frac{i\sigma_o}{\beta E}\right)^5 \phi^5 \rho^{*3} \geq 0 \tag{2.10}$$

or

$$\rho^{*3} + \frac{\eta k g^4}{\left(\frac{i\sigma_o}{\beta E}\right)} \rho^* - \frac{g^5}{\left(\frac{i\sigma_o}{\beta E}\right)^5 \phi^5} \leq 0 \tag{2.11}$$

If one applies Eq. 2.10 or Eq. 2.11, in the form of an equality, at any point of the shell and the strength condition of Eq. 2.9 at a dangerous point only (passive constraint elsewhere), one obtains a shell of uniform stability. On the other hand, in a shell of uniform strength, the strength condition of Eq. 2.9 is active at any point.

At first, from Eq. 2.8 one determines dimensionless height of the cross section as

$$\chi = -\frac{3^{1/6}}{2} \frac{\psi^{-5/9}\left(\frac{j\sigma_o}{\sqrt{3}\,\alpha\,E}\right)^{-2/9} \phi^{-5/6}}{\left[1 + \left(\frac{M_b}{M_t}\right)^2\right]^{1/6} \left[\int_0^1 \xi' \eta d\eta\right]^{1/3}} \tag{2.12}$$

The concentrated area a_o, evaluated from Eq. 2.7, and the global area of the cross section are as follows:

$$2a_o = \frac{2}{3^{1/6}} \frac{\psi^{5/9}\left(\frac{j\sigma_o}{\sqrt{3}\,\alpha\,E}\right)^{2/9} \phi^{5/6}}{\left[1 + \left(\frac{M_b}{M_t}\right)^2\right]^{1/3}} \left[\int_0^1 \xi'\eta\, d\eta\right]^{1/3}$$

$$\times \left[-\frac{M_b}{M_t} - \frac{\sqrt{3}\,k}{2\psi\,\phi^{5/2}} \frac{\int_0^1 \bar{g}\eta^2 \sqrt{1 + \xi'^2}\, d\eta}{\int_0^1 \xi'\eta\, d\eta}\right] \tag{2.13}$$

$$a = \frac{2}{3^{1/6}} \left(\frac{i\sigma_o}{\sqrt{3}\,\alpha E}\right)^{2/9} \frac{\psi^{5/9}\,\phi^{5/6}}{\left[1 + \left(\frac{M_b}{M_t}\right)^2\right]^{1/3} k} \left[\int_0^1 \xi'\eta\, d\eta\right]^{1/3}$$

(equation continued)

$$\times \left[-\frac{M_b}{M_t} - \frac{\sqrt{3}\ k}{2\psi\ \phi^{5/2}} \times \frac{\int_0^1 \bar{g}\eta^2 \sqrt{1+\xi'^2}\, d\eta - \int_0^1 \bar{g}\sqrt{1+\xi'^2}\, d\eta}{\int_0^1 \xi'\eta\, d\eta} \right], \tag{2.14}$$

where $\bar{g} = g/\left(\frac{i\sigma_o}{\beta E}\right)$. Equations 2.13 and 2.14 hold for $a_o > 0$. When $a_o = 0$, from the Eq. 2.13, one can evaluate the ratio of moments

$$\frac{M_b}{M_t} = -\frac{\sqrt{3}}{2}\frac{k}{\psi\ \phi^{5/2}} \frac{\int_0^1 \bar{g}\eta^2\sqrt{1+\xi'^2}\, d\eta}{\int_0^1 \xi'\eta\, d\eta} \tag{2.15}$$

and Eq. 2.14 can be reduced to

$$a = \frac{3^{1/3}\left(\frac{j\sigma_o}{\sqrt{3}\ \alpha E}\right)^{2/9}}{\left[1+\left(\frac{M_b}{M_t}\right)^2\right]^{1/3}\psi^{4/9}\phi^{5/3}} \frac{\int_0^1 \bar{g}\eta\sqrt{1+\xi'^2}\, d\eta}{\left[\int_0^1 \xi'\eta\, d\eta\right]^{2/3}} \tag{2.16}$$

Equations 2.12 to 2.16 are valid for each case under consideration.

Further considerations depend on the problem being investigated. In this paper, the following problems of optimization are investigated:

(1) circular and elliptic profile of constant thickness (D4)

(2) circular profile of uniform strength (C2)

(3) circular and elliptic profile of uniform stability (C3)

(4) constant thickness profile of uniform stability (B3)

(5) profile of uniform strength and stability (A1)

(6) profile of uniform strength (variational problem) (A2)

(7) profile of uniform stability (variational problem) (A3)

where symbols in brackets denote the type of mathematical problem given in tables.

More detailed analysis of two problems of optimal design of cross sections are now presented, namely; constant thickness and circular profile of uniform stability.

3. UNIFORM STABILITY CROSS SECTION OF CONSTANT THICKNESS

In the case of a shell of uniform stability, the buckling constraint of Eq. 1.8 must be satisfied in the form of an equality at each point of the shell. The solution of Eq. 2.11 leads to the differential equation

$$\rho^* = \frac{g^{5/3}}{2^{1/3}\,\phi^{5/3}}\left(\frac{i\sigma_o}{\beta E}\right)^{-5/3}\left[\sqrt[3]{1+\sqrt{1+\frac{4\,g^2\eta^3k^3}{2+\left(\frac{i\sigma_o}{\beta E}\right)^2\phi^5}}}\right.$$

$$\left.+\sqrt[3]{1-\sqrt{1+\frac{4\,g^2\eta^3k^3}{2+\left(\frac{i\sigma_o}{\beta E}\right)^2\phi^5}}}\right] \tag{3.1}$$

where

$$\rho^* = \frac{(1+\xi'^2)^{3/2}}{|\xi''|} = \frac{(1+\xi'^2)^{3/2}}{|\eta''|}$$

which describes the cross section under consideration. The unknown dimensionless thickness g can be evaluated from the strength condition of Eq. 2.9, which is to be satisfied at dangerous points only. In view of constant shear stress τ (h = constant) and maximum normal stress σ for $\eta = 1$, the dangerous point is $\eta_o = 1$. Thus one has

$$g = \frac{i\sigma_o}{\beta E}\;\frac{\psi\,\phi^{5/2}}{\sqrt{1-k^2}} \tag{3.2}$$

Substitution of Eq. 3.2 into Eq. 3.1 leads to

$$\rho^* = \frac{\psi^{5/3}\ \phi^{5/2}}{2^{1/3}\ (1 - k^2)^{1/2}} \left[\sqrt[3]{1 + \sqrt{1 + \frac{4\psi^2\ k^3\ \eta^3}{2 + (1 - k^2)}}} \right.$$

$$\left. + \sqrt[3]{1 - \sqrt{1 + \frac{4\psi^2\ k^3\ \eta^3}{2 + (1 - k)^2}}} \right] \tag{3.3}$$

with the following boundary conditions

$$\left. \begin{aligned} \eta = 1, \quad \eta' = 0 \\ \eta = 0, \quad \xi' = 0 \end{aligned} \right\} \tag{3.4}$$

The differential equation of Eq. 3.3 was integrated numerically, starting from $\eta = 1$, using the Runge-Kutta method. Equation 3.3 depends on two parameters k and ϕ. For various given values of k, parameters ϕ were sought which satisfy the second boundary condition of Eq. 3.4.

In the case of $a_0 > 0$, for a given ratio M_b/M_t, one seeks a pair of parameters k and ϕ (evaluated previously from the boundary condition) that gives the minimal value of the area of Eq. 2.14. This procedure holds only for $a_0 > 0$. If a_0, evaluated from Eq. 2.13, is less than zero ($a_o < 0$) it means that the optimum cross section does not have a rib. Then, one can apply an inverse procedure. For the previously evaluated pair of k and ϕ, one evaluates M_b/M_t and the remaining quantities.

The optimum quantities are shown in Figs. 3.1 to 3.8, denoted as curves.

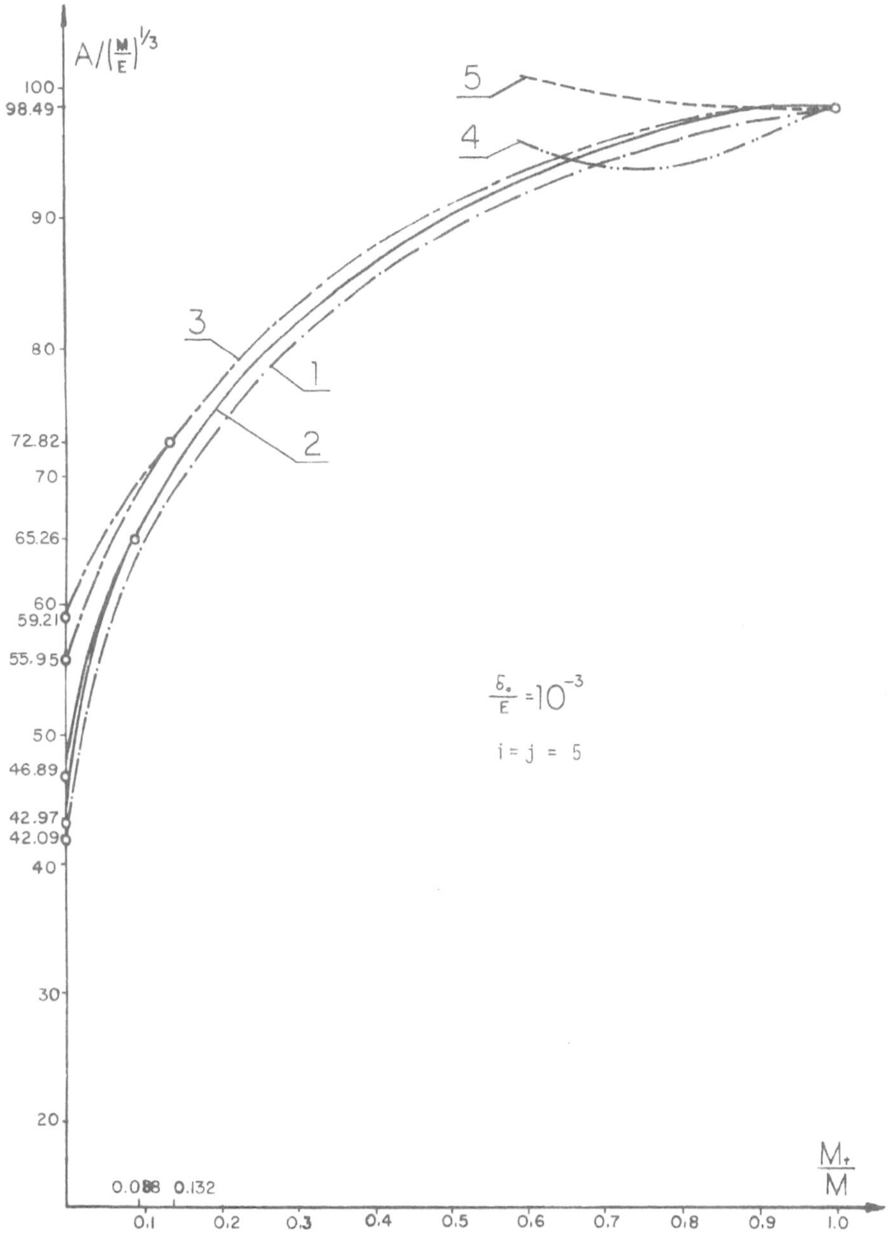

Figure 3.1 Optimum Value of $a/(\frac{M}{E})^{1/3}$ vs. $M_t/M[\sigma_0/E = 10^{-3}]$

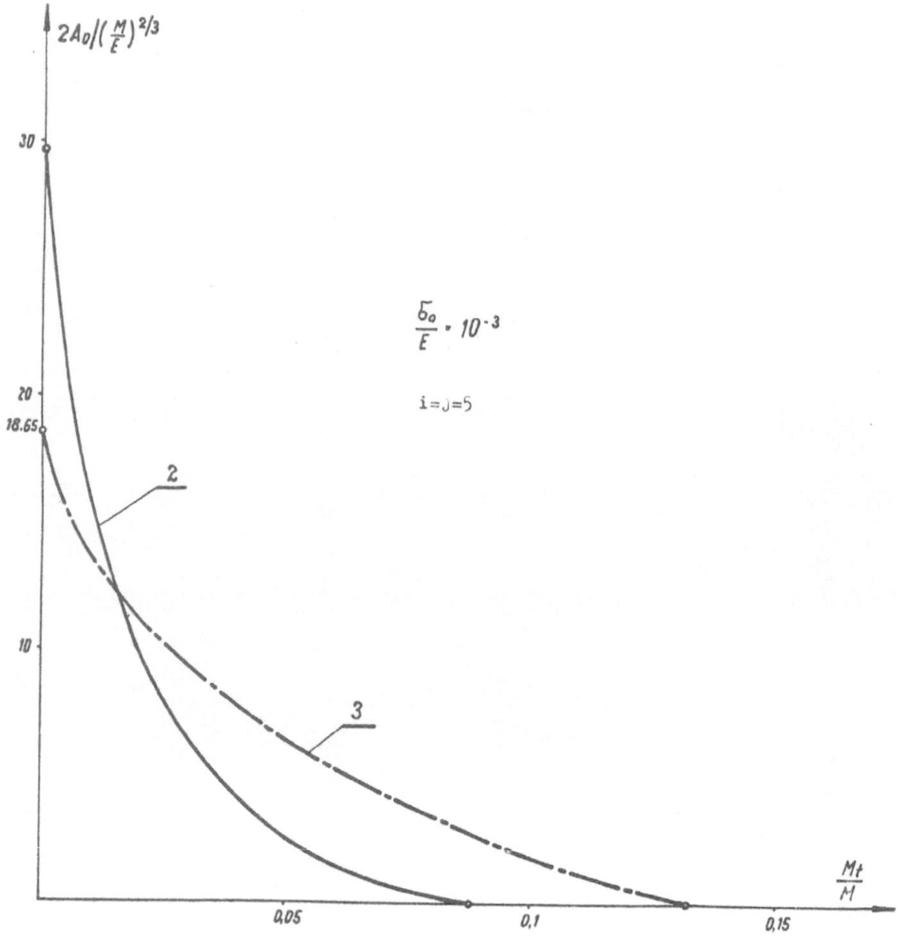

Figure 3.2 Optimum Value of $2A_0/(\frac{M}{E})^{1/3}$ vs. M_t/M $[\sigma_0/E = 10^{-3}]$

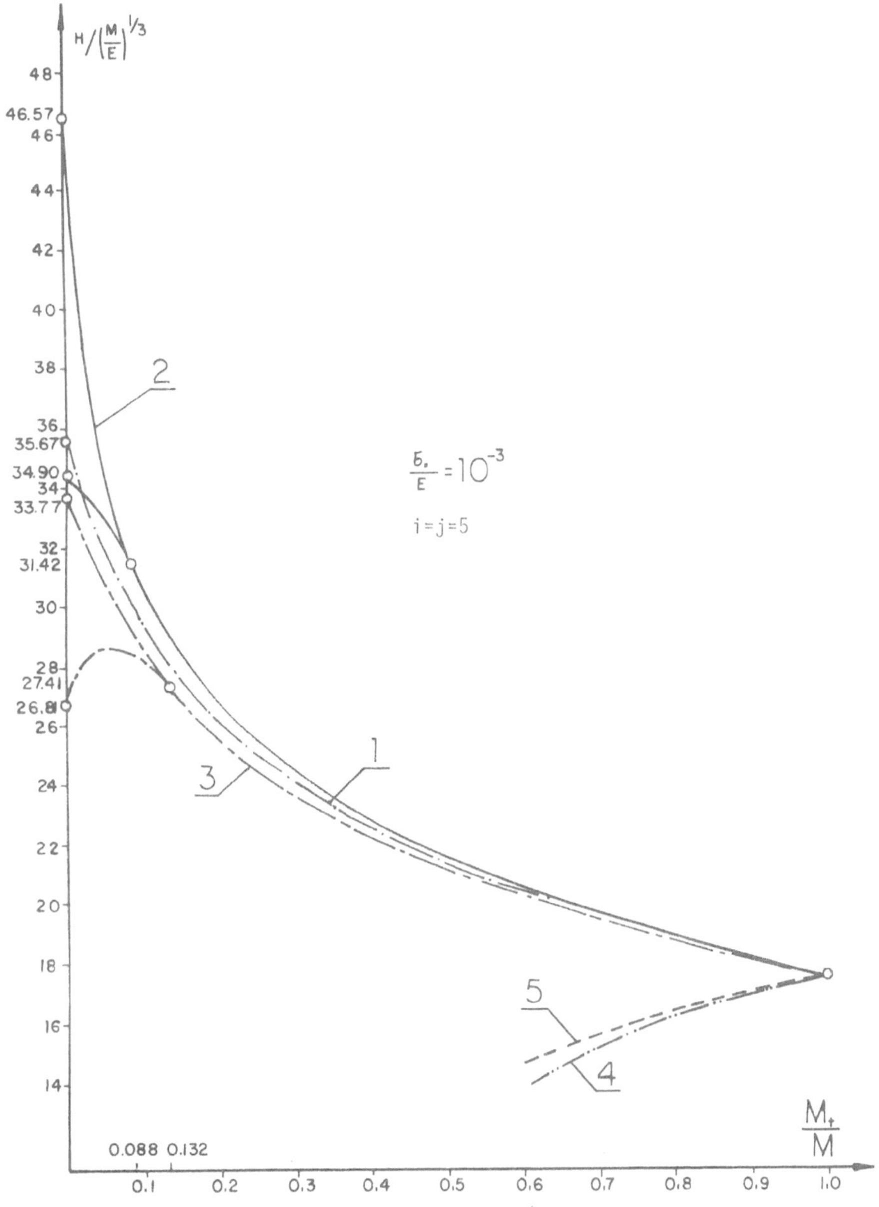

Figure 3.3 Optimum Value of $H/(\frac{M}{E})^{1/3}$ vs. M_t/M [$\sigma_0/E = 10^{-3}$]

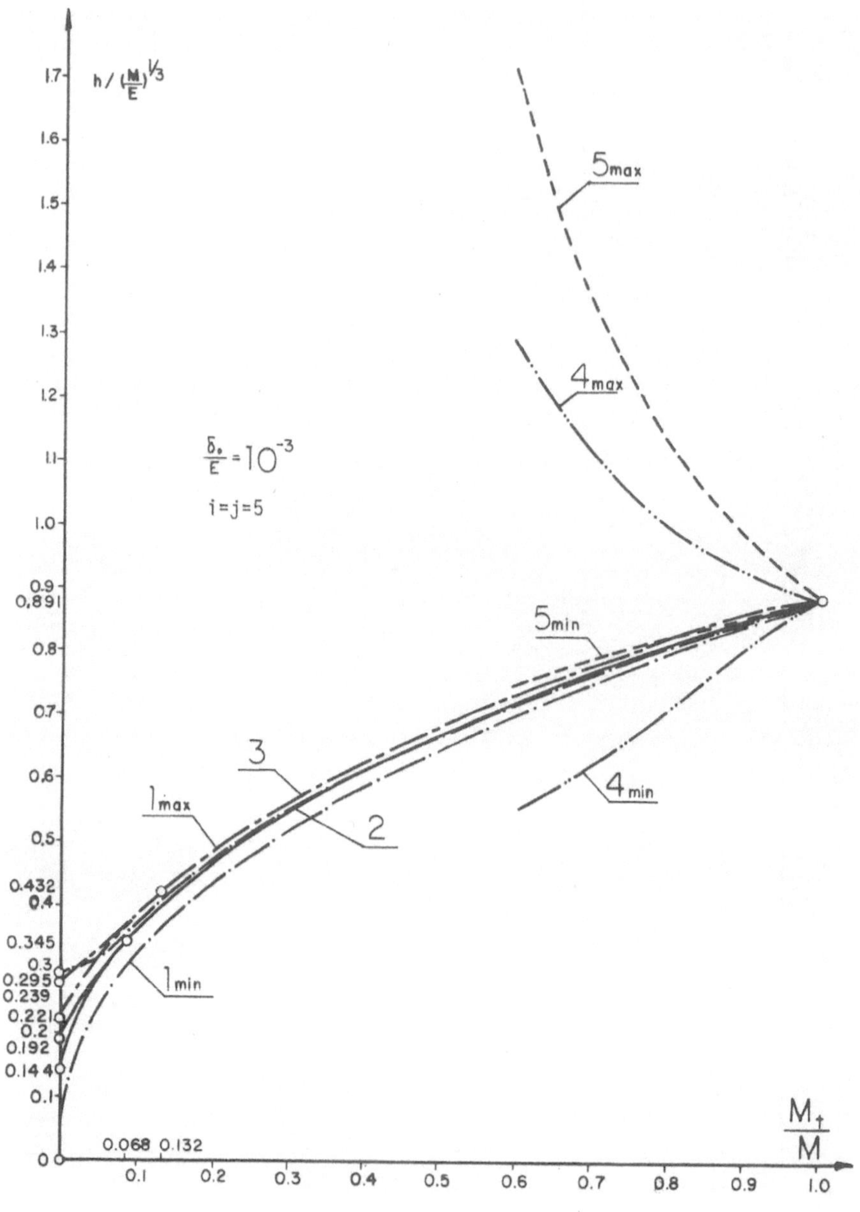

Figure 3.4 Optimum Value of $h/(\frac{M}{E})^{1/3}$ vs. M_t/M $[\sigma_0/E = 10^{-3}]$

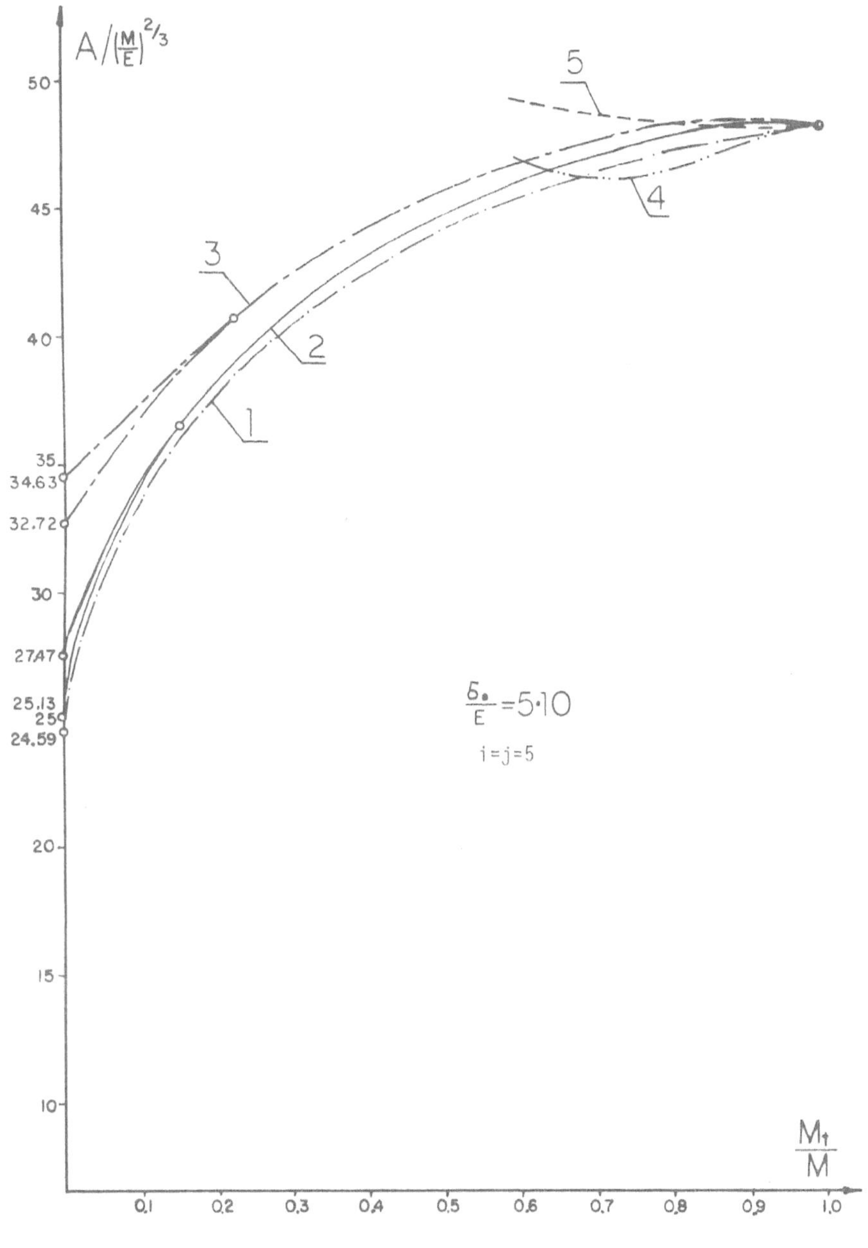

Figure 3.5 Optimum Value of $a/(\frac{M}{E})^{1/3}$ vs. M_t/M $[\sigma_0/E = 5 \times 10^{-3}]$

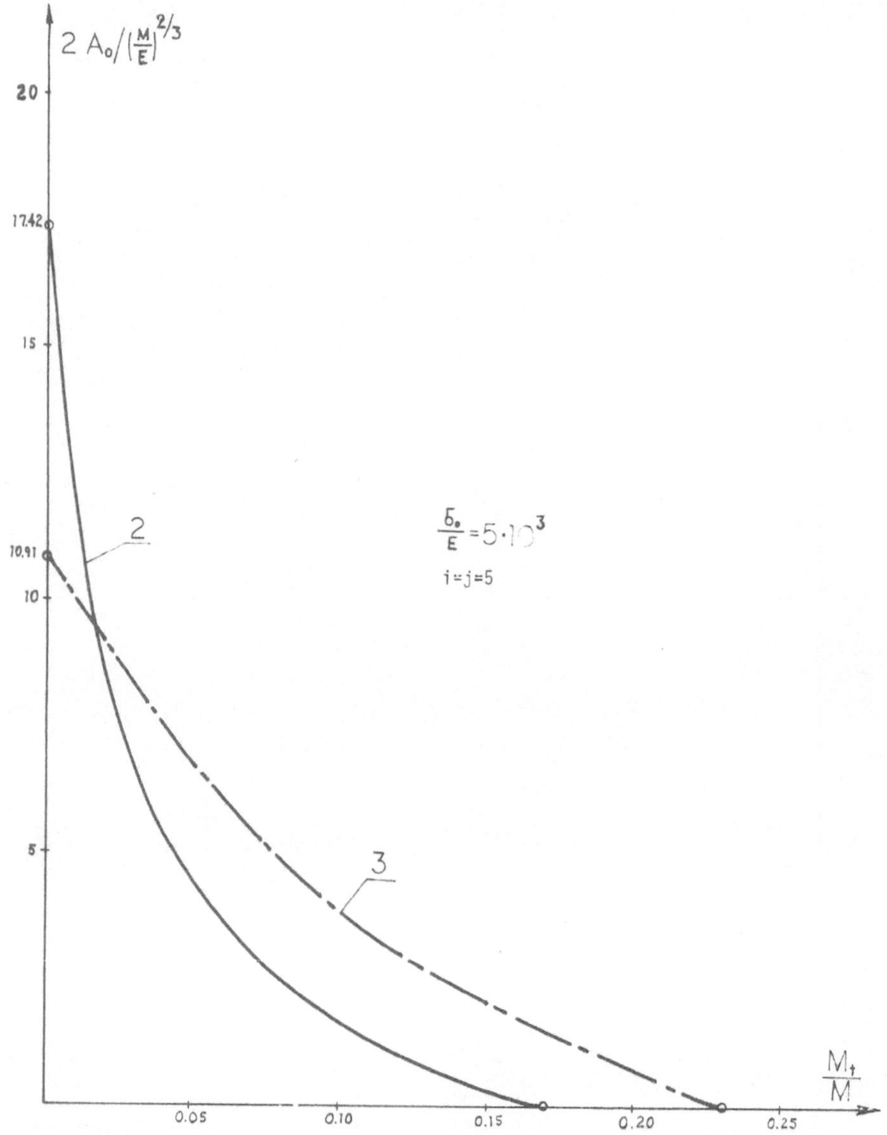

Figure 3.6 Optimum Value of $2A_0/(\frac{M}{E})^{2/3}$ vs. M_t/M [$\sigma_0/E = 5 \times 10^{-3}$]

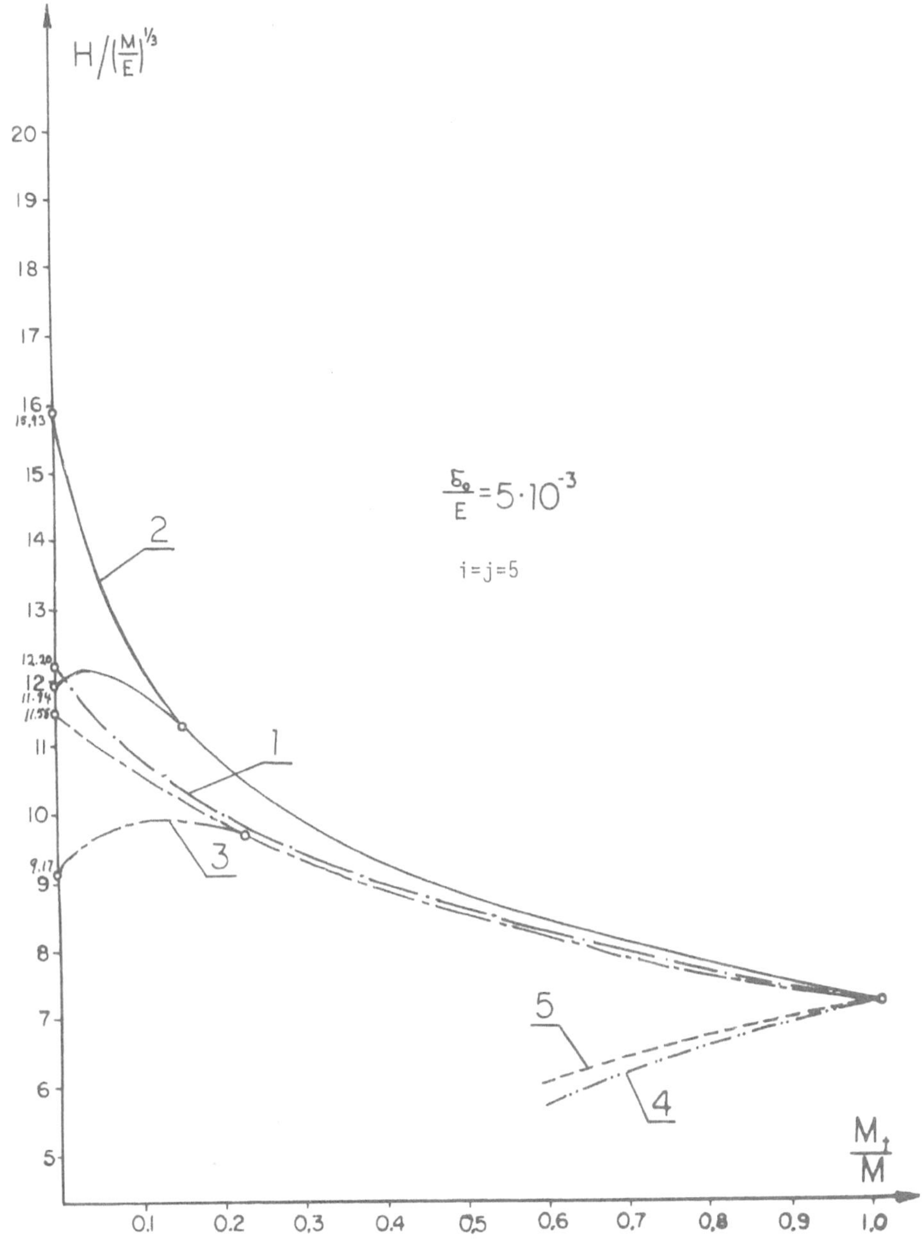

Figure 3.7 Optimum Value of $H/(\frac{M}{E})^{1/3}$ vs. M_t/M [$\sigma_0/E = 5 \times 10^{-3}$]

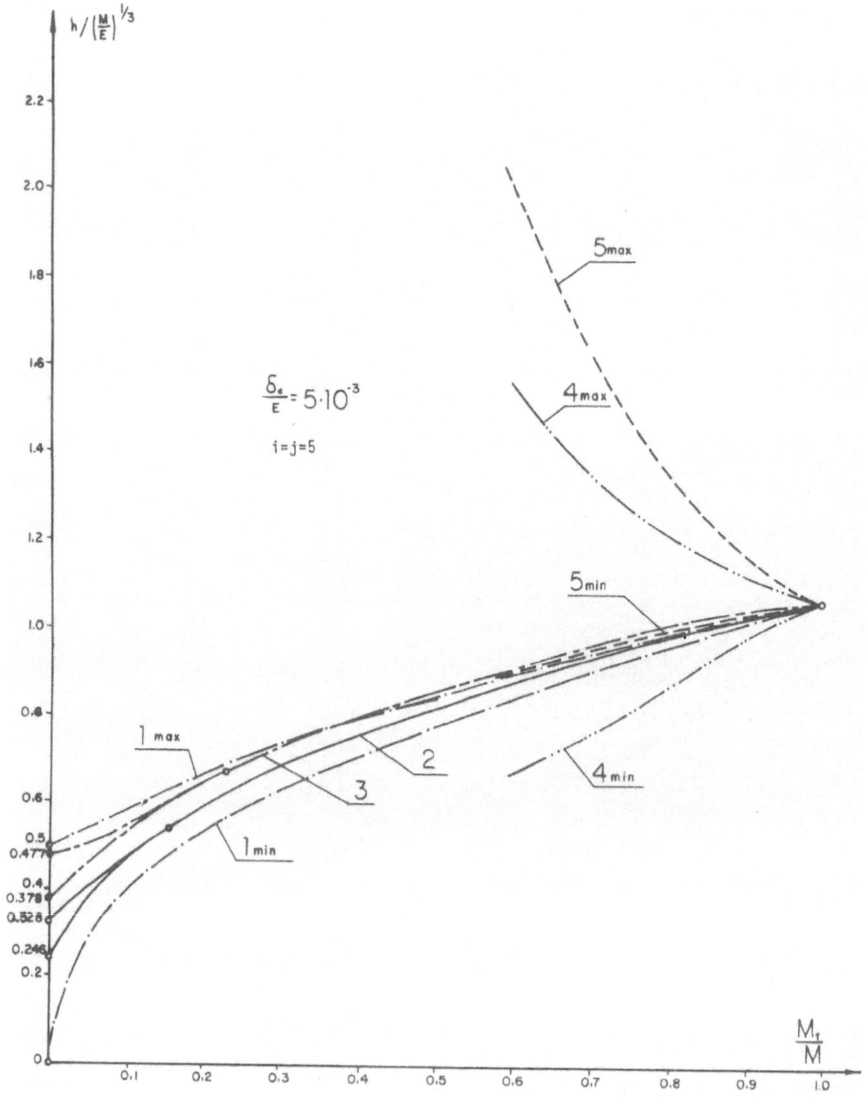

Figure 3.8 Optimum Value of $h/(\frac{M}{E})^{1/3}$ vs. M_t/M [$\sigma_0/E = 5 \times 10^{-3}$]

4. THE CIRCULAR CROSS SECTION OF UNIFORM STABILITY

For a prescribed shape of a cross section, one can obtain the uniform stability by appropriate design of a wall thickness. In this case, one may rewrite the stability condition of Eq. 2.10, determining g, in the shorter form

$$g^5 - C\eta g^4 - D = 0 \tag{4.1}$$

where $C = \frac{i\sigma_0}{\beta E} k\rho^*$ and $D = \rho^{*3}(\frac{i\sigma_0}{\beta E})\ \phi^5$. In case of general loadings ($k, \phi \neq 0$), the analytical solution of Eq. 4.1 is impossible, so it was decided to approximate the exact solution of Eq. 4.1 by the formula

$$g = C\eta + D^{1/5} + \psi_1 \tag{4.2}$$

where the first term in Eq. 4.2 describes the wall thickness for pure bending and the second term for pure twisting.

Substituting Eq. 4.2 into Eq. 4.1, after manipulation one has

$$\psi_1 = - \frac{C\eta\ D^{1/5}}{C\eta + \frac{5}{4} D^{1/5}} \tag{4.3}$$

The accuracy of Eq. 4.2 increases in the whole domain of loadings when one introduces into Eq. 4.3 two parameters γ and β, which should be evaluated numerically from a condition of minimum error (in comparison with Eq. 4.1), so with $r = \beta = 3/2$

$$\psi_1 = - \frac{3}{2} \frac{C\eta\ D^{1/5}}{C\eta + \frac{3}{2} D^{1/5}} \tag{4.4}$$

and errors are smaller than 7%. Finally, the approximation formula is

$$g = \frac{i\sigma_0}{E} \frac{k^2\eta^2 z^7 + k\eta\phi z^5 + 1.5\phi^2 z^3}{k\eta z^2 + 1.5\phi} \tag{4.5}$$

where $z^5 = \rho^*$. Equation 4.5 describes the wall thickness of the cross section of uniform stability, for each case under consideration.

For a circular profile,

$$\eta^2 + \xi^2 = 1 \tag{4.6}$$

and the corresponding radius of curvature is $\rho^* = 1$. It turns out that, in this case, the concentrated area is $a_0 = 0$.

The substitution of Eq. 4.5 into Eqs. 2.15, 2.16, and 2.12 leads to the following formulae:

$$\frac{M_b}{M_t} = \frac{2\sqrt{3}}{\pi}\ \frac{k}{\psi\phi^{5/2}} \left[\frac{2}{3}k - \frac{\pi}{8}\phi + \frac{9\phi^2}{4k} - \frac{27\pi\phi^3}{16k^2} + \frac{81\phi^4}{16k^2} J_1 \vee J_2 \vee J_3\right] \tag{4.7}$$

$$a = 4\left(\frac{\sqrt{3}}{2\pi}\right)^{2/3}\left(\frac{j\sigma_0}{\sqrt{3}\alpha E}\right)^{2/9} \frac{k - \frac{\pi}{4}\phi + \frac{9}{4}\phi^2 J_1 \vee J_2 \vee J_3}{\left[1 + \left(\frac{M_b}{M_t}\right)^2\right]^{1/3} \psi^{4/9}\phi^{5/3}} \tag{4.8}$$

$$\chi = \left(\frac{\sqrt{3}}{2\pi}\right)^{1/3} \frac{\left(\frac{j\sigma_0}{\sqrt{3}\alpha E}\right)^{-2/9} \psi^{-5/9}\phi^{-5/6}}{\left[1 + \left(\frac{M_b}{M_t}\right)^2\right]^{1/6}} \tag{4.9}$$

where

$$J_1 = \frac{2}{\sqrt{\frac{9}{4}\phi^2 - k^2}} \arctan \frac{\sqrt{\frac{9}{4}\phi^2 - k^2}}{\frac{3}{2}\phi + k}, \text{ if } \frac{3}{2}\phi > k \tag{4.10}$$

$$J_2 = \frac{1}{\sqrt{k^2 - \frac{9}{4}\phi^2}} \ln \frac{\left(\frac{3}{2}\phi + k - \sqrt{k^2 - \frac{9}{4}\phi^2}\right)\left(k + \sqrt{k^2 - \frac{9}{4}\phi^2}\right)}{\left(\frac{3}{2}\phi + k + \sqrt{k^2 - \frac{9}{4}\phi^2}\right)\left(k - \sqrt{k^2 - \frac{9}{4}\phi^2}\right)}, \text{ if } k > \frac{3}{2}\phi \tag{4.11}$$

$$J_3 = \frac{3\phi}{k\left(k + \frac{3}{2}\phi\right)}, \text{ if } k = \frac{3}{2} \tag{4.12}$$

The unknown parameter ϕ is to be evaluated from the strength condition. Introducing $\rho^* = 1$ and $\eta = 1$ (dangerous point) into Eq. 3.3, one has

$$\phi = \frac{2^{2/15}(1 - k^2)^{1/3}}{\psi^{2/3}} \left[\sqrt[3]{1 + \sqrt{1 + \frac{4\psi^2 k^3}{2 + (1 - k^2)}}} + \sqrt[3]{1 - \sqrt{1 + \frac{4\psi^2 k^3}{2 + (1 - k^2)}}}\right]^{-2/5} \tag{4.13}$$

The optimum quantities for the circular cross section are shown in Figs. 3.1 to 3.8, denoted as curves.

Besides the two cases treated in the foregoing, the appropriate quantities are shown in Figs. 3.1 to 3.8 for the following optimum cross sections:

(1) constant thickness circular profile (curve 3),

(2) uniform strength circular profile (curve 5),

(3) uniform strength and stability profile (curve 4).

For the last two profiles, the region of permissible loadings is limited by the condition of thin walled cross section.

Results are presented for two materials, characterized by the ratio of material constants σ_0/E.

REFERENCES

1. Mazurkiewicz S., and Zyczkowski M., "Optymalne Ksztaltowanie Przekroju Preta Cienkościennego Jednocześnie Skrecanego i Zginanego", Rozpr. Inz., Vol. 2, No. 14, 1966, pp. 199-213.

2. Zyczkowski M., "Optimale Formen des Dummwandingen Geschlossenen Querschnittes Eines Balkenz bei Berucksichtigung von Stabilitatsbedingungen," Z. angew. Math. Mechanik, Vol. 48, 1968, pp. 455-462.
3. Zyczkowski M. and Kruzelecki J., "Optimal Design of Shells with Respect to Their Stability," Optimization in Structural Design, Springer-Verlag, Berlin 1975, pp. 229-247.

PROCESS DESCRIPTION PARAMETER CHANGE IN DIMENSIONAL BASE OPTIMIZATION*

Mieczyslaw Szata

Institute of Materials Science and Applied Mechanics
Technical University of Wroclaw, Wroclaw, Poland

ABSTRACT

Problems connected with construction of optimal mathematical models for a process, are considered here, based on techniques of dimensional analysis. The Pi-theorem is used in a parametric form. A method of transformation of the exponents when, dimensional base changes are given, is introduced. This transformation makes it easier to construct an algorithm for a dimensional base change.

1. INTRODUCTION

Consider a process in which the quantities $Z_1, Z_2, \ldots, Z_s, Z$ are considered as elements of a dimensional space π, spanned by a fixed system of units $E_1, E_2, \ldots, E_n \in \pi$. One can express the quantities $Z_1, Z_2, \ldots, Z_s, Z$ in terms of the basis E_i in the following way:

$$\left.\begin{aligned} Z_\ell &= \zeta_\ell \prod_{k=1}^{n} E_k^{\,z_{\ell k}}, \quad \ell = 1,\ldots,s; \\ Z &= \zeta \prod_{k=1}^{n} E_k^{\,z_k} \end{aligned}\right\} \tag{1.1}$$

*This work was sponsored by the Marie Curie-Sklodowska Foundation Grant No. J-F7F 010-P.

where $z_{\ell k}, z_k \in R$ and $\zeta_i, \zeta \in R^+$ with R the set of real numbers and R^+ the set of real positive numbers.

A formalism of dimensional analysis presented by Drobot [1] and developed by Kasprzak and Lysik [2] is applied here. The fundamental theorem of dimensional analysis, known as the Pi-theorem, deals with the following function, whose arguments and values form the elements of the space π:

$$Z = \Phi(Z_1, Z_2, \ldots, Z_s), \quad Z, Z_1, Z_2, \ldots, Z_s \in \pi \tag{1.2}$$

According to the approach presented in Refs. 1 and 2, one partitions arguments of the function of Eq. 1.2 into dimensionally independent arguments, $A_1, \ldots, A_m$ (dimensional base) and dimensionally dependent arguments $B_1, \ldots, B_r$, $m+r=s$. According to the Pi-theorem [1], one can write Eq. 1.2 in the form

$$\left.\begin{aligned} Z &= f(\phi_1, \phi_2, \ldots, \phi_r) \prod_{i=1}^{m} A_i^{a_i} \\ B_j &= \phi_j \prod_{i=1}^{m} A_i^{a_{ji}}, \quad j=1,\ldots,r \\ A_i &= \alpha_i \prod_{k=1}^{n} E_k^{z_{ik}}, \quad i=1,\ldots,m \end{aligned}\right\} \tag{1.3}$$

where a_{ji}, $a_i \in R$ and $\phi_j \in R^+$.

Using this formalism, it is possible to construct mathematical models for physical processes. The change of dimensional base (as is shown for example in Ref. 2) has significant influence on the accuracy of the process description. A parametric description of the process being considered is very useful in the construction of a base change algorithm. Such a description is presented in Section 2. The important problem in formulation of an algorithm of dimensional base change is that of determining the transformation between the parameters connected with two different choices of dimensional bases. In the present paper the objective is to determine this transformation. Section 3 is devoted entirely to this problem.

2. A PARAMETRIC DESCRIPTION OF A PROCESS

Assuming that the first m arguments of the function of Eq. 1.2, i.e.

$$(Z_1, Z_2, \ldots, Z_p, \ldots, Z_m) \tag{2.1}$$

form a dimensionally independent set (a basis), one can present the description of the process considered in the following manner:

$$\left.\begin{aligned} \zeta_p &= \alpha_p, \quad p=1,\ldots,m \\ \zeta_{m+q} &= \phi_q \prod_{i=1}^{m} \alpha_i^{a_{qi}}, \quad q=1,\ldots,r \\ \zeta &= f(\phi_1, \phi_2, \ldots, \phi_r) \prod_{i=1}^{m} \alpha_i^{a_i} \end{aligned}\right\} \tag{2.2}$$

where exponents $a_{ji}, a_i \in R$, $i=1,\ldots,m$; $j=1,\ldots,r$ satisfy the following sets of equations

$$\left.\begin{aligned} \sum_{i=1}^{m} a_{ji} z_{ik} &= z_{m+j,k}, \\ \sum_{i=1}^{m} a_i z_{ik} &= z_k, \quad j=1,\ldots,r \\ & \qquad\quad k=1,\ldots,n \end{aligned}\right\} \tag{2.3}$$

while $\alpha_1, \alpha_2, \ldots, \alpha_m, \phi_1, \phi_2, \ldots, \phi_r$ are variables in this parametric description of the function of Eq. 1.2.

When another base

$$(Z_1, Z_2, \ldots, Z_{p-1}, Z_{m+q}, Z_{p+1}, \ldots, Z_m) \tag{2.4}$$

is chosen from among arguments of the function of Eq. 1.2, differing from the base of Eq. 2.1 in that the argument Z_p has been changed to Z_{m+q} (previously it was a dimensionally dependent argument), one obtains another parametric expression for the same process, namely,

$$\left.\begin{aligned}
&\zeta_1 = \overset{*}{\alpha}_1,\dots,\zeta_{p-1} = \overset{*}{\alpha}_{p-1},\ \zeta_p = \overset{*}{\phi}_q \prod_{i=1}^{m} \overset{*}{\alpha}_i^{\overset{*}{a}_{qi}},\\
&\zeta_{p+1} = \overset{*}{\alpha}_{p+1},\dots,\\
&\zeta_m = \overset{*}{\alpha}_m,\ \zeta_{m+1} = \overset{*}{\phi}_1 \prod_{i=1}^{m} \overset{*}{\alpha}_i^{\overset{*}{a}_{1i}},\dots,\zeta_{m+q}\\
&\quad = \overset{*}{\alpha}_p,\dots,\zeta_{m+r} = \overset{*}{\phi}_r \prod_{i=1}^{m} \overset{*}{\alpha}_i^{\overset{*}{a}_{ri}}\\
&\zeta = \overset{*}{f}(\overset{*}{\phi}_1,\overset{*}{\phi}_2,\dots,\overset{*}{\phi}_r) \prod_{i=1}^{m} \overset{*}{\alpha}_i^{\overset{*}{a}_i}
\end{aligned}\right\} \tag{2.5}$$

where exponents $\overset{*}{a}_{ji},\overset{*}{a}_i \in R$, satisfy a set of equations

$$\left.\begin{aligned}
&\sum_{i=1}^{m} \overset{*}{a}_{ji} z_{ik} = z_{m+j,k}, \qquad j=1,\dots,q-1,q+1,\dots,r\\
&\sum_{i=1}^{m} \overset{*}{a}_{qi} z_{ik} = z_{pk}, \qquad k=1,\dots,n\\
&\sum_{i=1}^{m} \overset{*}{a}_i z_{ik} = z_k
\end{aligned}\right\} \tag{2.6}$$

and $\overset{*}{\alpha}_1,\overset{*}{\alpha}_2,\dots,\overset{*}{\alpha}_m,\overset{*}{\phi}_1,\overset{*}{\phi}_2,\dots,\overset{*}{\phi}_r$ are variables of a parametric description of the process described in the base of Eq. 2.4. Bases of Eqs. 2.1 and 2.4 differ only in one element. This results in an easier notation, without any loss of generality.

Note that the choice of the dimensional base from among the arguments $Z_1,Z_2,\dots,Z_s$ of the function of Eq. 1.2 affects any numerical description of the process considered which uses an adequate form of right members of a parametric notations of Eqs. 2.2 and 2.5, and of the quantities of Eqs. 2.3 and 2.4. Therefore, the acceptance of a new base can be described using relationships between a parametrization $(\alpha_1,\alpha_2,\dots,\alpha_m,\phi_1,\phi_2,\dots,\phi_r)$ and $(\overset{*}{\alpha}_1,\overset{*}{\alpha}_2,\dots,\overset{*}{\alpha}_m,\overset{*}{\phi}_1,\overset{*}{\phi}_2,\dots,\overset{*}{\phi}_r)$. From the relations established between the quantities of Eqs. 2.3 and 2.6, the following relationships between the solutions of the corresponding system of algebraic equations can be obtained:

$$\left.\begin{array}{ll}\left.\begin{array}{l}\overset{*}{a}_{ji} + a_{ji} + \overset{*}{a}_{qi}a_{jp} = 0\\ a_{qi} + \overset{*}{a}_{qi}\, a_{qp} = 0\\ \overset{*}{a}_i - a_i - \overset{*}{a}_{qi}a_p = 0\end{array}\right] & , i=1,\ldots,m,\ i\neq p\\ \left.\begin{array}{l}\overset{*}{a}_{jp} - \overset{*}{a}_{qp}a_{jp} = 0\\ 1 - \overset{*}{a}_{qp}\, a_{qp} = 0\\ \overset{*}{a}_p - \overset{*}{a}_{qp}a_p = 0\\ a_{qp} \neq 0 \neq \overset{*}{a}_{qp}\end{array}\right] & , j=1,\ldots,r,\ j\neq q\end{array}\right\} \quad (2.7)$$

The system of equations in Eq. 2.7 does not immediately follow from Eqs. 2.3 and 2.6. The technique for obtaining these relationships is the purpose of this work. It is presented in Section 3. But, first it is helpful to make the following observation: Consider $\zeta_1,\zeta_2,\ldots,\zeta_s$ given by Eqs. 2.2 and 2.5 and use Eqs. 2.7. Upon transition from parametric description given by Eqs. 2.2 and 2.3, involving the base of Eq. 2.1, to the parametric description given by Eqs. 2.5 and 2.6, involving the base of Eq. 2.4, the following transformation of parameters is obtained:

$$\left.\begin{array}{l}\alpha_1 = \overset{*}{\alpha}_1,\ldots,\alpha_{p-1} = \overset{*}{\alpha}_{p-1},\ \alpha_p = \overset{*}{\phi}_p \prod\limits_{\substack{i=1\\ i\neq p}}^{m} \overset{*}{\alpha}_i^{-\frac{a_{qi}}{a_{qp}}}\ \overset{*}{\alpha}_p^{\frac{1}{a_{qp}}}\\ \alpha_{p+1} = \overset{*}{\alpha}_{p+1},\ldots,\ \alpha_m = \overset{*}{\alpha}_m,\ \phi_1 = \overset{*}{\phi}_1\ \overset{*}{\phi}_q^{-a_{1p}},\ldots,\\ \phi_q = \overset{*}{\phi}_q^{-a_{qp}},\ldots,\ \phi_r = \overset{*}{\phi}_r\overset{*}{\phi}_q^{-a_{rp}},\ \phi = \overset{*}{\phi}\ \overset{*}{\phi}^{-a_p}\end{array}\right\} \quad (2.8)$$

which affects the value of the function of Eq. 1.3.

3. EXPONENT TRANSFORMATION INVOLVED IN THE BASE CHANGE

The method of determining the exponents $\overset{*}{a}_{ji}$, using exponents a_{ji}, is presented here. In the classical approach of dimensional analysis, applied for example in Refs. 3 and 4, the algorithm for calculation of exponents leads to sets of homogeneous linear equations. The formalism of these calculations, based on the theory presented in Ref. 1, applies also to sets of linear equations, but these equations are no longer homogeneous. It follows from the method presented here that in some cases it is necessary to combine elements of both the approaches.

One can write the exponents $z_{\ell k}$ and z_k introduced by Eq. 1.1 ($\ell=1,\ldots,s$; $k=1,\ldots,n$) in the form of a $\underline{D}$ - matrix, called the dimensional matrix. The transpose matrix of $\underline{D}$, $\underline{D}^T$ has the following form:

$$\underline{D}^T = \begin{array}{c} \backslash \\ Z_1 \\ Z_2 \\ \vdots \\ Z_s \\ Z \end{array} \begin{array}{c} \begin{array}{cccc} E_1 & E_2 & \cdots & E_n \end{array} \\ \begin{bmatrix} z_{11} & z_{12} & \cdots & z_{1n} \\ z_{21} & z_{22} & \cdots & z_{2n} \\ \cdots & \cdots & \cdots & \cdots \\ z_{s1} & z_{s2} & \cdots & z_{sn} \\ z_1 & z_2 & \cdots & z_n \end{bmatrix} \end{array} \tag{3.1}$$

Exponents $z_{\ell 1},\ldots,z_{\ell n}$ of the variable z_ℓ form ℓ-th row of the $\underline{D}^T$ matrix, for $\ell=1,\ldots,s$. In the last row one has exponents of Z function values (with respect to a fixed system of units E_k, $k=1,\ldots,n$).

In the k-th column of the $\underline{D}^T$ matrix, there are exponents $z_{1k},\ldots,z_{sk},z_k$ of arguments and function values in a fixed measure unit E_k, $k=1,\ldots,n$. The form of equations defining exponents a_{ji} and a_i evidently depends on the choice of quantities forming the dimensional base. Thus, for the base formed by the first m arguments $Z_1,Z_2,\ldots,Z_m$, exponents a_{ji} and a_i ($j=1,\ldots,r$; $i=1,\ldots,m$) are defined by Eqs. 2.3. For the base described by Eq. 2.4, one can calculate exponents $\overset{*}{a}_{ji}$ and $\overset{*}{a}_i$ from Eq. 2.6. It is easy to check that Eqs. 2.3 can be written in the matrix form

$$\underline{D}\ \underline{X} = \underline{0} \qquad (3.2)$$

where $\underline{0}$ is the matrix of zero elements and the $\underline{X}$ matrix is

$$\underline{X} = \begin{array}{c|ccccc|} Z \backslash j & 1 & 2 & \dots & r & r+1 \\ \hline Z_1 & -a_{11} & -a_{21} & \dots & -a_{r1} & -a_1 \\ Z_2 & -a_{12} & -a_{22} & \dots & -a_{r2} & -a_2 \\ \vdots & \vdots & \vdots & & \vdots & \vdots \\ Z_m & -a_{1m} & -a_{2m} & \dots & -a_{rm} & -a_m \\ Z_{m+1} & 1 & 0 & \dots & 0 & 0 \\ Z_{m+2} & 0 & 1 & \dots & 0 & 0 \\ \vdots & \vdots & \vdots & & \vdots & \vdots \\ Z_{m+r} & 0 & 0 & \dots & 1 & 0 \\ Z & 0 & 0 & \dots & 0 & 1 \end{array} \qquad (3.3)$$

The $\underline{X}$ matrix is constructed according to the following basic principle: Every column corresponds to one dimensionally dependent quantity. The first column corresponds to the first dimensionally dependent quantity, the second column - to the second one, etc. Since, in the first base, Z_{m+1} is the first dimensionally dependent quantity, unity is contained in row (m+1) of the first column of the $\underline{X}$ matrix.

As is known from the Pi-theorem, one can express any dimensionally dependent quantity by an exponent product of base quantities, with exponents a_{ji} (i=1,2,...,m; j=1,2,...,r). Therefore, exponents a_{1i} (i=1,2,...,m) are in the first column of the $\underline{X}$ matrix, whose rows correspond to quantities forming the dimensional base. The remaining rows of the first column contain zeroes.

In the same way, one forms the j-th column (j=1,...,r). This column corresponds to the j-th dimensionally dependent quantity. In the j-th column, one inserts unity in row (m+j). In the first m rows of this column, one places exponents $a_{j1}, a_{j2}, \dots, a_{jm}$. The remaining positions are filled with zeros.

Equation 3.2, for a fixed j value, can be written in the form

$$\underline{D}\,\underline{X}_j = \underline{0}, \quad j=1,2,\ldots,r \tag{3.4}$$

where $\underline{X}_j$ is a column vector (the j-th column of the $\underline{X}$ matrix) and $\underline{0}$ is a vector containing only zero elements. It is possible to show (see e.g. Ref. 3) that vectors X_j can be obtained by solving the following matrix equation

$$\underline{D}\,\underline{X} = \underline{0}\,, \quad j=1,\ldots,r \tag{3.5}$$

where $\underline{X}$ is an unknown column vector with dimension identical to $\underline{X}_j$ (i.e. of a length r+m+1=s+1). As is known (see e.g. Ref. 1), the dimensional base of a process described by Eq. 1.2 has m quantities, if and only if the rank of the dimensional $\underline{D}$-matrix is equal to m. For considerations here, it is assumed that the dimensional base has m dimensional quantities, thus the $\underline{D}$-matrix has a rank m. Then, Eq. 3.5 has r+m+1-m=r+1 linearly independent solutions. The $\underline{X}$ matrix has the same number of columns, in which all the linearly independent solutions are included. As one can see, the vectors forming the fundamental set of solutions of the homogeneous linear system of equations of Eq. 3.5 constitutes the columns of the X-matrix. Following the terminology of Chen [4], the $\underline{X}$-matrix is called the complete $\underline{X}$-matrix of the $\underline{D}$-matrix.

Consider now the matrix equation of Eq. 3.5 in particular the k-th row of the $\underline{D}$-matrix (k=1,2,...,n) and the j-th column of the $\underline{X}$-matrix (j=1,2,...,r)

$$\sum_{i=1}^{m} -a_{ji}\, z_{ik} + z_{m+j,k} = 0, \quad k=1,\ldots,n,\ j=1,\ldots,r \tag{3.6}$$

Taking the product of the k-th row of the $\underline{D}$-matrix (k=1,2,...,n) and the (r+1)-th column of the $\underline{X}$-matrix, one obtains

$$\sum_{i=1}^{m} -a_i z_{ik} + z_k = 0, \quad k=1,\ldots,n \tag{3.7}$$

Comparing Eqs. 3.6 and 3.7 with Eq. 2.3, one sees that they are identical. Thus, the systems of equations of Eqs. 2.3 and 3.2 are equivalent. Assume that the values of exponents a_{ji} and a_i for the first base are known. One can compute them by solving, for instance, Eqs. 2.3. Values of exponents are substituted into the $\underline{X}$-matrix of Eq. 3.3.

It is now to be shown that one can obtain values of exponents $\overset{*}{a}_{ji}$ and $\overset{*}{a}_{i}$ for the second base choice, without actually solving Eqs. 3.6 but by using only the $\underline{X}$-matrix structures. Consider the case in which the natural numbers p and q can take any values within the following bounds

$$1 \le p \le m$$

$$1 \le q \le n-m=r$$

First, one constructs the $\underline{X}$ and $\overset{*}{\underline{X}}$ matrices. In the $\underline{X}$-matrix, defined by the Eq. 3.3, one notes the p-th and (m+q)-th rows. Their elements are closely connected with the change of the p-th element in the base with the q-th dimensionally dependent element (Eq. 2.4). The complete $\underline{X}$ matrix is

$$\underline{X} = \begin{array}{l} 1 \\ 2 \\ \vdots \\ p \\ \vdots \\ m \\ m+1 \\ m+2 \\ \vdots \\ m+q \\ \vdots \\ m+r \\ m+r+1 \end{array} \begin{bmatrix} -a_{11} & -a_{21} & \cdots & -a_{q1} & \cdots & -a_{r1} & -a_{1} \\ -a_{12} & -a_{22} & \cdots & -a_{q2} & \cdots & -a_{r2} & -a_{2} \\ \vdots & \vdots & & \vdots & & \vdots & \vdots \\ -a_{1p} & -a_{2p} & \cdots & -a_{qp} & \cdots & -a_{rp} & -a_{p} \\ \vdots & \vdots & & \vdots & & \vdots & \vdots \\ -a_{1m} & -a_{2m} & \cdots & -a_{qm} & \cdots & -a_{rm} & -a_{m} \\ 1 & 0 & \cdots & 0 & \cdots & 0 & 0 \\ 0 & 1 & \cdots & 0 & \cdots & 0 & 0 \\ \vdots & \vdots & & \vdots & & \vdots & \vdots \\ 0 & 0 & & 1 & & 0 & 0 \\ \vdots & \vdots & & \vdots & & \vdots & \vdots \\ 0 & 0 & \cdots & 0 & \cdots & 1 & 0 \\ 0 & 0 & \cdots & 0 & \cdots & 0 & 1 \end{bmatrix} \qquad (3.8)$$

The complete $\overset{*}{\underline{X}}$-matrix, for the second base, has the form

$$
\overset{*}{\underline{X}} = \begin{array}{c|ccccccc|}
i\backslash j & 1 & 2 & \cdots & q & \cdots & r & r+1 \\
\hline
1 & -\overset{*}{a}_{11} & -\overset{*}{a}_{21} & \cdots & -\overset{*}{a}_{q1} & \cdots & -\overset{*}{a}_{r1} & -\overset{*}{a}_{1} \\
2 & -\overset{*}{a}_{12} & -\overset{*}{a}_{22} & \cdots & -\overset{*}{a}_{q2} & \cdots & -\overset{*}{a}_{r2} & -\overset{*}{a}_{2} \\
\vdots & \vdots & \vdots & & \vdots & & \vdots & \vdots \\
p & 0 & 0 & & 1 & \cdots & 0 & 0 \\
\vdots & \vdots & \vdots & & \vdots & & \vdots & \vdots \\
m & -\overset{*}{a}_{1m} & -\overset{*}{a}_{2m} & \cdots & -\overset{*}{a}_{qm} & \cdots & -\overset{*}{a}_{rm} & -\overset{*}{a}_{m} \\
m+1 & 1 & 0 & \cdots & 0 & \cdots & 0 & 0 \\
m+2 & 0 & 1 & \cdots & 0 & \cdots & 0 & 0 \\
\vdots & \vdots & \vdots & & \vdots & & \vdots & \vdots \\
m+q & -\overset{*}{a}_{1p} & -\overset{*}{a}_{2p} & \cdots & -\overset{*}{a}_{qp} & \cdots & -\overset{*}{a}_{rp} & -\overset{*}{a}_{p} \\
\vdots & \vdots & \vdots & & \vdots & & \vdots & \vdots \\
m+r & 0 & 0 & \cdots & 0 & \cdots & 1 & 0 \\
m+r+1 & 0 & 0 & \cdots & 0 & \cdots & 0 & 1
\end{array} \tag{3.9}
$$

The $\underline{X}$ and $\overset{*}{\underline{X}}$ matrices contain two different sets of fundamental solutions of the same homogeneous linear system of equations of Eq. 3.5. Thus, the relationship between the $\underline{X}$ and $\overset{*}{\underline{X}}$ matrices can be written in the form

$$
\underline{X} = \overset{*}{\underline{X}}\ \underline{T} \tag{3.10}
$$

where T is a nonsingular matrix.

Consider now the transformation matrix $\underline{T}$. One may construct an auxiliary matrix $\underline{T1}$, composed of rows p, m+1,m+2,...,m+q-1, m+q+1,...,m+r,m+r+1 of the $\overset{*}{\underline{X}}$ matrix. Its construction begins by defining the matrix

$$\overset{*}{\underline{T1}} = \begin{bmatrix} 0 & 0 & \dots & 0 & 1 & 0 & \dots & 0 & 0 \\ 1 & 0 & \dots & 0 & 0 & 0 & \dots & 0 & 0 \\ 0 & 1 & \dots & 0 & 0 & 0 & \dots & 0 & 0 \\ \vdots & \vdots & & \vdots & \vdots & \vdots & & \vdots & \vdots \\ 0 & 0 & \dots & 1 & 0 & 0 & \dots & 0 & 0 \\ \vdots & \vdots & & \vdots & \vdots & \vdots & & \vdots & \vdots \\ 0 & 0 & \dots & 0 & 0 & 0 & \dots & 1 & 0 \\ 0 & 0 & \dots & 0 & 0 & 0 & \dots & 0 & 1 \end{bmatrix} \tag{3.11}$$

The $\overset{*}{\underline{T1}}$ matrix can be obtained from the identity matrix $\underline{I}$ by interchanging the first and p-th rows. The $\overset{*}{\underline{T1}}$-matrix obtained in this way has the same rows as the $\underline{I}$-matrix, but arranged in a different order.

Then, one constructs the $\underline{T1}$-matrix. It is formed from rows $p, m+1, m+2, \dots, m+q-1, m+q+1, \dots, m+r, m+r+1$ of the $\underline{X}$-matrix and has the form

$$\underline{T1} = \begin{array}{c} 1 \\ 2 \\ 3 \\ \vdots \\ q \\ q+1 \\ \vdots \\ r \\ r+1 \end{array} \begin{bmatrix} -a_{1p} & -a_{2p} & \dots & -a_{q-1,p} & -a_{qp} & -a_{q+1,p} & \dots & -a_{rp} & -a_{p} \\ 1 & 0 & \dots & 0 & 0 & 0 & \dots & 0 & 0 \\ 0 & 1 & \dots & 0 & 0 & 0 & \dots & 0 & 0 \\ \vdots & \vdots & & \vdots & \vdots & \vdots & & \vdots & \vdots \\ 0 & 0 & \dots & 1 & 0 & 0 & \dots & 0 & 0 \\ 0 & 0 & \dots & 0 & 0 & 1 & \dots & 0 & 0 \\ \vdots & \vdots & & \vdots & \vdots & \vdots & & \vdots & \vdots \\ 0 & 0 & \dots & 0 & 0 & 0 & \dots & 1 & 0 \\ 0 & 0 & \dots & 0 & 0 & 0 & \dots & 0 & 1 \end{bmatrix} \tag{3.12}$$

Equation 3.10 displaying the relationship between the $\underline{X}$ and $\underline{\overset{*}{X}}$ matrices, is also true for submatrices of the $\underline{X}$ and $\underline{\overset{*}{X}}$ matrices, formed from the same numbered rows. Hence, one can write

$$\underline{T1} = \underline{\overset{*}{T1}}\,\underline{T} \tag{3.13}$$

Thus, one calculates $\underline{T}$ as

$$\underline{T} = \underline{\overset{*}{T1}}^{-1}\,\underline{T1} \tag{3.14}$$

It is easy to notice that the $\underline{\overset{*}{T1}}$-matrix has properties similar to the identity matrix

$$\underline{\overset{*}{T1}}^{-1} = \underline{\overset{*}{T1}}^{T}$$

where $\underline{\overset{*}{T1}}^{T}$ is the transpose of $\underline{\overset{*}{T1}}$.†

Hence, one notes that

$$\underline{T} = \underline{\overset{*}{T1}}^{T}\,\underline{T1} \tag{3.16}$$

The $\underline{T}$-matrix has the same rows as the $\underline{T1}$-matrix but in a different order. This order change is identical with the change of order of the rows of the T-matrix, which would result in formulation of the identity matrix $\underline{I}$.

The form of the $\underline{T}$-matrix obtained from Eq. 3.16 is

$$\underline{T} = \begin{array}{c} \\ 1 \\ 2 \\ \vdots \\ q-1 \\ q \\ q+1 \\ \vdots \\ r+1 \end{array}
\begin{array}{c} \begin{array}{cccccccccc} 1 & 2 & \dots & q-1 & q & q+1 & \dots & r & r+1 \end{array} \\
\begin{bmatrix} 1 & 0 & \dots & 0 & 0 & 0 & \dots & 0 & 0 \\ 0 & 1 & \dots & 0 & 0 & 0 & \dots & 0 & 0 \\ \vdots & \vdots & & \vdots & \vdots & \vdots & & \vdots & \vdots \\ 0 & 0 & \dots & 1 & 0 & 0 & \dots & 0 & 0 \\ -a_{1p} & -a_{2p} & \dots & -a_{q-1,p} & -a_{q,p} & -a_{q+1,p} & \dots & -a_{rp} & -a_{p} \\ 0 & 0 & \dots & 0 & 0 & 1 & \dots & 0 & 0 \\ \vdots & \vdots & & \vdots & \vdots & \vdots & & \vdots & \vdots \\ 0 & 0 & \dots & 0 & 0 & 0 & \dots & 0 & 1 \end{bmatrix} \end{array} \tag{3.17}$$

†As it is known, a matrix with these properties makes scalar products invariant, i.e. for all $x,y \in R^3$, $(\underline{\overset{*}{T1}}x, \underline{\overset{*}{T1}}y) = (x,y)$.

4. CONCLUSIONS

The procedure determining the exponents of a transformation connected with a base change is presented in Section 3. This procedure makes it possible to construct an algorithm for a dimensional base change, using only a parametric description of the process considered in Eq. 1.3. This algorithm is then used for the construction of a mathematical model of this process. It enables one to conduct an investigation of various relationships between properties of the process description, considered as a function the changed dimensional base, even before any measurements are carried out. A similar analysis of parameters in the dimensional process description has been carried out in Refs. 3, 4, and 5. For instance, Chen [4] has derived an optimization of dimensionless products, using the so called B-matrix as a code-matrix. In practice, this optimization leads to the optimization of the B-matrix structure. The B-matrix-equivalent given in this work is the X-matrix, defined by Eq. 3.3. Happ [5], in a manner similar to Chen [4], assumes an identical optimization criterion consisting of three conditions, namely:

(1) All elements are integers. This means that the exponents a_i and a_{ji} are integers.
(2) The maximum number of B-matrix elements should take the zero value. This means that the X-matrix defined by Eq. 3.3, consisting of exponents a_i and a_{ji} that are zeroes and ones, should have the maximum number of zero entries, which means the maximum number of zero values for the a_i and a_{ji} exponents.
(3) The sum of absolute values of B-matrix elements ought to be minimal. In this paper, the same remark concerns the X-matrix.

As one easily sees, the postulate (1) is not in agreement with the formalism of Drobot's dimensional analysis [1]. In Drobot's scheme, and also in Carlson's scheme [3], all exponents are real numbers. Chen [4] limits their values to the set of integers. However, it is necessary to remember that the formulation of the Pi-theorem, as applied by Chen, yields no clue concerning the method for the selection of exponents, except one - product of dimensional quantities raised to the powers of unknown exponents have to be dimensionless. Therefore, the limitation of exponent values only to the set of integers does not reduce the domain of application, in agreement with Drobot's scheme.

In this case, the form of dimensionless products may be different. This implies a different numerical function, but the associated dimensional function describing the process considered may be the same. A very important problem is the so called optimum form of dimensionless products (according to Chen), and in the case of Drobot's scheme - the change of a dimensional base. Both procedures have a similar meaning. However, the criterion previously formulated in Chen's scheme does not have sufficient motivation. It simply requires that the B-matrix satisfies some postulates, the meaning of which is not even intuitively clear.

In Drobot's scheme the base choice may be made in many ways. The quality of an experiment description with a fixed form of the f function depends on the choice of the bases. This fact has been noticed by Kokar [6], who has proposed a procedure for securing the optimum base choice. This procedure is sensible, with regard to the criterion, which concerns the accuracy of a process description. Thus, it satisfies all physically intuitive requirements and it is clear. According to this procedure the optimization of the dimensional base choice can not be carried out a priori, but must be based on experimental results. Hence, the solution of this problem requires a process description using all dimensional bases.

REFERENCES

1. Drobot, S., "On the Foundations of Dimensional Analysis," Studia Mathematica, vol. 14, 1953, pp. 84-89.
2. Kasprzak, W., Lysik, B., and Szata, M., "Problems of the Selection of a Dimensional Base in Experiment Design and Planning," Bull. Acad. Polon. Sci., Ser. Sci. Techn., vol. 25, No. 8, 1977, pp. 247-255.
3. Carlson, D.E., "On Some New Results in Dimensional Analysis," Arch. Rational Mech. Anal., 1978, pp. 191-210.
4. Chen, W.K., "Algebraic Theory of Dimensional Analysis," Journal of the Franklin Institute, vol. 292, No. 6, 1971, pp. 403-422.
5. Happ, W.W., "Computer-Oriented Procedures for Dimensional Analysis," Journal of Appl. Phys., vol. 38, No. 10, 1967, pp. 3918-3926.
6. Kokar, M., "A System Approach to a Search of Laws of Empirical Theories," Current Topics in Cybernetics and Systems, Springer-Verlag, Berlin, 1978.

QUANTITATIVE STABILITY ANALYSIS AND LOAD DOMAINS

Pauli Pedersen

Department of Solid Mechanics, Technical University of Denmark, DK2800 Lyngby, Denmark

In the lecture by Plaut [1], the notion of discontinuous switching of the load of instability was used. A few comments, based on Ref. 2, may be helpful in understanding the nature of these non-conservative problems. As will be seen, the discontinuities are due to the fact that one focuses on critical loads, rather than on stable/unstable load domains. Furthermore, a stability analysis from a quantitative point of view will show that it is possible to pass through a domain of moderate

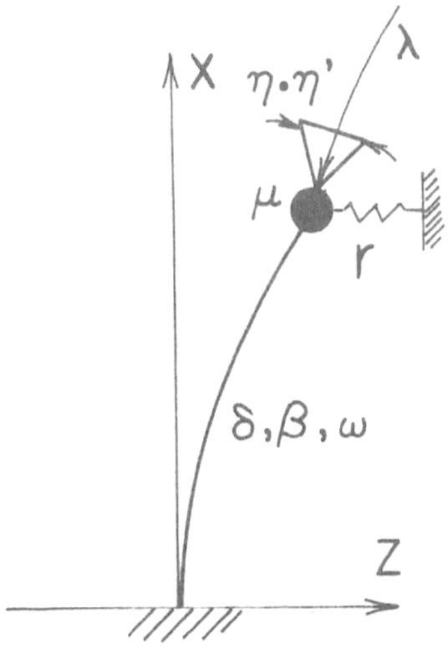

Figure 1 Extended Follower Force Problem

instability, just as rotors pass through critical domains of instability. Details are given in Ref. 2 and here only the simple example of Fig. 1 is discussed to illustrate the idea.

The time dependent displacement function $\tilde{y}(x,t)$ is represented in separated form by

$$\tilde{y}(x,t) = y(x)e^{\phi t} = z(x)e^{(\alpha+i\omega)t} = y(x)e^{\alpha t}e^{i\omega t},$$

and one has three possible results:

(1) ϕ^2 is real and $<0 \Rightarrow \alpha = 0 \Rightarrow$ stability

(2) ϕ^2 is real and $>0 \Rightarrow \omega = 0 \Rightarrow$ divergence

(3) ϕ^2 is complex $\Rightarrow \alpha > 0,\ \omega \neq 0 \Rightarrow$ flutter

The characteristic curves for a nonconservative and almost conservative problem are shown in Fig. 2. Thus, the critical load λ_1 separates the stable domain from the domain of instability by divergence. With increasing load, the column becomes more and more unstable (α increasing).

The characteristic curves for a moderately non-conservative problem, as shown in Fig. 3, have changed in nature. The

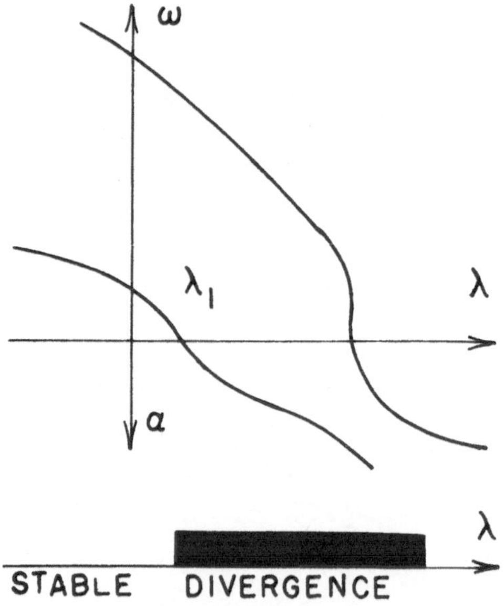

Figure 2 Result for $\eta < .354$

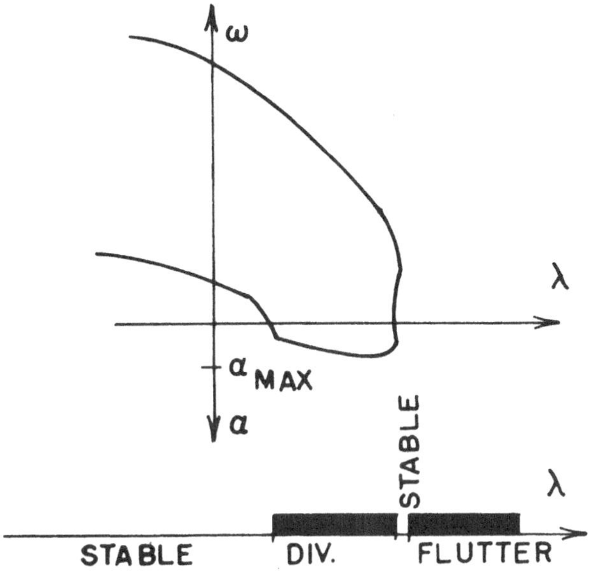

Figure 3 Result for $.354 < \eta < .5$

instability is now limited (α_{max}) and a domain of stability separates the domain of instability by divergence and flutter. When the domain of divergence is small, $\eta \to 0.5$ and α_{max} is moderate, it should be possible to pass through to the second domain of stability. The limiting case of $\eta = 0.5$ is shown in Fig. 4. When the non-conservative component of the force takes over, one gets the results of Fig. 5. The domain of instability by divergence has then disappeared.

The usual way of showing the critical load, as a function of the non-conservative component of the force, is given in Fig. 6. This gives an impression of a discontinuity in stability of the structure. However, from Figs. 3, 4, and 5, one sees that what appeared to be a discontinuity may be better understood by considering the behavior plotted against the load domain.

REFERENCES

1. Weisshaar, T.A. and Plaut, R.A., "Structural Optimization Under Non-conservative Loading", Optimization of Distributed Parameter Structures (Ed. E.J. Haug and J. Cea), Sijthoff & Nordhoff, Alphen aan den Rijn, Netherlands, 1980.
2. Pedersen, P., "Influence of Boundary Conditions on the Stability of a Column Under Non-conservative Load", Int. J. Solids Structures, Vol. 13, 1977, pp. 445-455.

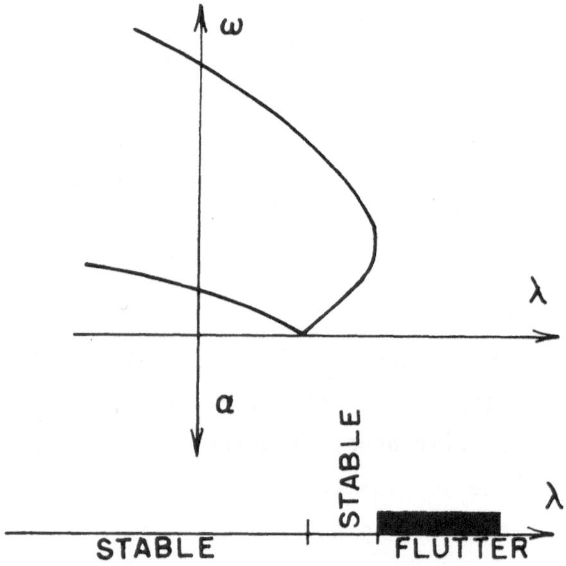

Figure 4 Result for η = .5

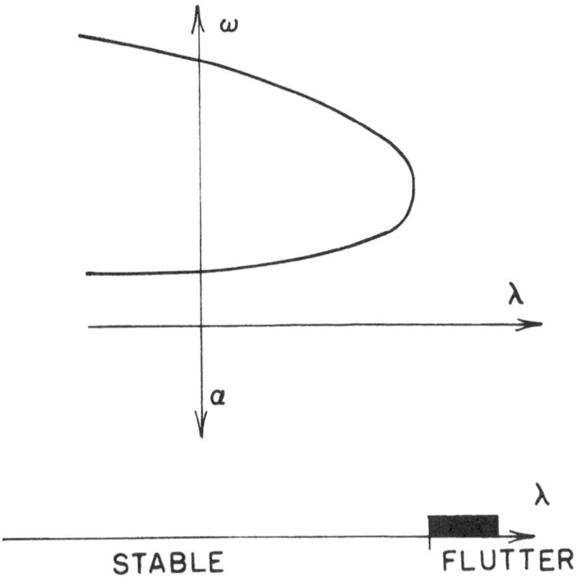

Figure 5 Result for $\eta > .5$

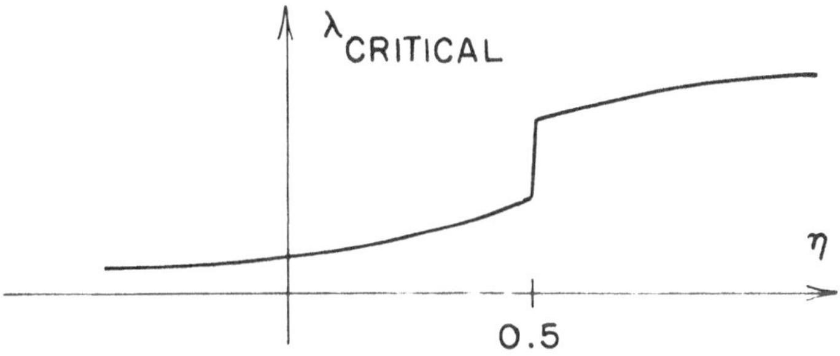

Figure 6 Discontinuous Presentation

Part 6

SHAPE OPTIMAL DESIGN

OPTIMALITY CONDITIONS AND ANALYTICAL METHODS OF SHAPE OPTIMIZATION

Nick V. Banichuk

Institute of Problems of Mechanics, USSR Academy of Sciences, Moscow, USSR

ABSTRACT

Necessary extremum conditions for optimal design problems with unknown boundaries and integral functionals are derived. More complicated optimization problems with local functionals (such as stress intensity) are also discussed. Analytical methods for seeking the shape of elastic bodies that are based on the perturbation theory, conformal mapping, and integral and functional equations are presented.

1. INTRODUCTION

Shape optimization is the basic problem in the theory of optimal structural design. In the majority of works concerned with optimal structural design, thickness distribution of beams, rods, trusses, plates is considered as the unknown control function. Problems of finding thickness distribution for thin-shelled structures have the property that coefficients of equations depend on control functions that describe the thickness distribution. This is due to the approximate nature of the basic equations, averaged with respect to one spatial variables (thickness). For determining optimal shape of elastic bodies in the general case, it is necessary to solve problems with unknown domain of definition of the equilibrium equations. The shape of the region of integration of these equations is not specified beforehand, but is to be determined from the condition that a certain integral functional attains an extremal value. These are the so-called problems with unknown boundaries. As well as having a direct practical application, such problems are of great interest from a purely mathematical angle, i.e. from the point of view of developing effective analytical methods for optimizing mechanical systems that are described by elliptical partial differential equations.

Apparently Saint-Venant was the first to discuss such problems in connection with finding the cross-sectional shape of an elastic bar of maximum torsional rigidity [1]. The assumption that among all bars with simply-connected and convex cross-sections, the circular bar has the maximum torsional rigidity was made in Ref. 1. It was confirmed by some calculations and comparisons. Some works (see Refs. 2 and 3) were devoted to proof of this hypothesis. Employing the symmetrization theorem, Polya [2] proved optimality of the bar with circular cross-section. The results of this investigation are presented, for example, in Ref. 3.

Shape optimization problems with unknown free boundary and integral performance criteria were considered by Wasiutinski [4] and Prager [5]. Free boundaries were considered to be varied and optimality conditions were obtained, requiring that the density of elastic strain energy is constant along the optimum boundary.

Another class of self-adjoint optimization problems with unknown boundaries was studied in Ref. 6. In Refs. 6 to 9, optimality conditions for unknown shapes of isotropic and anisotropic bars of multiply-connected cross-section were obtained. A perturbation technique was applied in Refs. 6 and 8 for solving the problem of maximizing rigidity of a bar. Complex variable methods were used for study of the same problem in Refs. 7 and 10. Rigorous analytic solution for anisotropic bars of maximal torsional rigidity and for multi-purpose isotropic bars were obtained in Refs. 7, 11, and 12.

The problem with unknown boundary arising from optimization of vibration frequencies of membranes was considered in Ref. 13. Necessary extremum conditions were obtained in this paper and the influence of membrane contour variations on the natural frequencies of transverse vibrations was evaluated. Note that it was suggested by Rayleigh that, among all membranes of a given area, the circular membrane has the minimum fundamental frequency. An analogous statement was also made concerning the vibration of the clamped plates. These statements were confirmed by Rayleigh's calculation of fundamental frequency variation for almost circular membranes and by Hadamard's study of clamped plates (see Ref. 3).

Problems with unknown boundaries and local functionals have proved to be more complicated. Minimization of the maximum stress intensity and minimization of the maximum elastic displacements are typical examples of optimization problems with unknown boundaries. Methods of solution of these problems are intensively developing now (see Ref. 15). The optimality of holes with equal-stress contours was proved for plane problems of the theory of elasticity [16-19] (plane stress and plane deformation). Similar results for hole shapes in plates working in bending was established in Ref. 17. To find the shape of equal stress holes, the method of conformal mapping was used in Ref. 20 and the integral equation method was used in Refs. 19 and 21. Results concerning stress intensity minimization of variable diameter, elastic bars under torsion are presented in Ref. 16. More general numerical methods of optimization for problems with unknown boundaries, based on nonlinear programming, were applied in Ref. 22.

In discussing problems with unknown boundaries, it is important to make the following observation. Optimization problems with given domains and optimal design problems with unknown boundaries are considered in different ways and different methods have been used for their solution. However, this separation is to some extent conditional. In some cases, it is possible to use a mapping of the domain with unknown boundaries onto a given canonical domain. Then it is not directly the shape of the domain, but the mapping function defined on the canonical domain, that must be found. The coefficients of equations and integrands of functionals of the reduced problem then depend on these mapping functions and their derivatives. Thus, problems with unknown boundaries are reduced to optimization problems with given domains. However, in some cases it is very difficult to find the mapping and this reduction is noneffective, whereas the method of boundary variation often appears to be convenient and permits one to establish some general properties of optimal solutions. Hence, it is worth developing both approaches for solving optimization problems with unknown boundaries.

2. OPTIMIZATION PROBLEMS WITH INTEGRAL PERFORMANCE CRITERIA

Basic formulas for the first variation of the performance functional, which is stipulated by the variation of the boundary of domain, are first presented and extremum conditions are obtained. Then, these conditions are applied to optimization problems with unknown boundaries and integral functionals.

2.1 Variation of Performance Functionals and Extremum Conditions

Let the vector-function u be defined on the domain Ω with boundary Γ in s-dimensional space. Let J denote the performance criteria to be optimized and J_j $(j = 1,\dots,r)$ be integral functionals depending on u. Equality type constraints are imposed on the values of J_j. The boundary Γ is composed of three parts Γ_0, Γ_1, and Γ_2. The function u is subjected to boundary conditions on Γ_2 and is not given on Γ_1. The part Γ_0 is considered to be given and the other part $\Gamma_1 + \Gamma_2$ is to be determined from the extremum condition for J. Basic relationships of the problem have the form

$$J = F(J_1, J_2, \ldots, J_r)$$

$$J_j = \int_\Omega f_j(x, u, u_x)\,d\Omega \tag{2.1}$$

$$F_i(J_1, J_2, \ldots, J_r) = C_i$$

where f_j and F_i $(i=1,\ldots,k;\ j=1,\ldots,r)$ are given functions of their arguments, C_i are given constants, $x = \{x_1, x_2, \ldots, x_s\}$, $u = \{u_1(x), u_2(x), \ldots, u_m(x)\}$, and $u_x = \{u_{1x_1}, \ldots, u_{mx_s}\}$.

Consider the problem of finding the extremum of functional J, with respect to u and Γ, under the given constraints. For this purpose, one introduces into consideration the function $\Phi = F - \sum_{i=1}^{k} \lambda_i F_i$ and writes the expression for the first variation δJ in terms of variations of the boundary $\Gamma_1 + \Gamma_2$ and the function u, in the following way [15]:

$$\delta J = \int_\Omega \sum_{\ell=1, i=1}^{r,m} \frac{\partial \Phi}{\partial J_\ell} \left(\frac{\partial f_\ell}{\partial u_i} - \sum_{j=1}^{s} \frac{\partial}{\partial x_j} \frac{\partial f_\ell}{\partial u_{ix_j}} \right) \delta u_i\, d\Omega$$

$$+ \int_{\Gamma_1+\Gamma_2} \sum_{\ell=1}^{r} \frac{\partial \Phi}{\partial J_\ell} \left[(n, \delta r) f_\ell + \sum_{i=1}^{m} (Z_{\ell i}, n)(\delta u_i - \nabla u_i, \delta r)) \right] d\Gamma$$

where $n = \{n_1, \ldots, n_s\}$, $n_j = \partial x_j / \partial n$, $||n||=1$, $\delta r = \{\delta x_1, \ldots, \delta x_s\}$ $\nabla u_i = \{u_{ix_1}, \ldots, u_{ix_s}\}$, $Z_{\ell i} = \{\partial f_\ell / \partial u_{ix_1}, \ldots, \partial f_\ell / \partial u_{ix_s}\}$, and the expressions $(Z_{\ell i}, n)$, $(n, \delta r)$, and $(\nabla u_i, \delta r)$ denote the scalar product of vectors. Assume that the function u satisfies Euler equations in the domain Ω. Then the integral over Ω in the expression for δJ vanishes.

One has $\delta u = 0$ on the boundary Γ_2, where the function u is given. For the part of the boundary Γ where δu is arbitrary, one obtains the system of transversality conditions $\sum_{\ell=1}^{r} \Phi_{J\ell}(Z_{\ell i},n) = 0 (i=1,\ldots,m)$. Taking into account that $(\delta u)_{\Gamma_2} = 0$ and supposing that the transversality conditions are satisfied, the expression for the first variation of the optimized functional reduces to

$$\delta J = \int_{\Gamma_1} \sum_{\ell=1}^{r} \frac{\partial \Phi}{\partial J_\ell} f_\ell(n,\delta r)d\Gamma + \int_{\Gamma_2} \sum_{\ell=1}^{r} \frac{\partial \Phi}{\partial J_\ell} \left[(n,\delta r)f_\ell - \sum_{i=1}^{m} (Z_{\ell i},n)(\nabla u_i,\delta r)\right] d\Gamma$$

Consider a boundary variation of the form $\delta r = tn$, where t is a scalar parameter. Using the condition $\delta J = 0$ and the arbitrariness of t, one obtains the optimality conditions for the unknown boundaries as

$$\sum_{\ell=1}^{r} \left(\frac{\partial \Phi}{\partial J_\ell} f_\ell\right)_{\Gamma_1} = 0 \qquad (2.2)$$

$$\left(\sum_{\ell=1}^{r} \frac{\partial \Phi}{\partial J_\ell} \left\{f_\ell - \sum_{i=1}^{m} (Z_{\ell i},n)(\nabla u_i,n)\right\}\right)_{\Gamma_2} = 0 \qquad (2.3)$$

Quantities Φ_{J_ℓ} are determined for the values of the functionals $J_1,\ldots,J_r$, corresponding to the extremal of the variational problem of Eq. 2.1. Therefore, both the Euler equations defining the function u in Ω and the relationships of Eqs. 2.2 and 2.3 are integro-differential equations.

In the particular case when $(u)_{\Gamma_2}$ = const, the vectors ∇u_i defined at points $x \in \Gamma_2$ are parallel to the normal vector n and therefore $(Z_{\ell i},n)(\nabla u_i,n) = (Z_{\ell i},\nabla u_i)$. The optimality condition for boundary Γ_2, where the function u is constant, takes the form

$$\left(\sum_{\ell=1}^{r} \frac{\partial\Phi}{\partial J_\ell}\left(f_\ell - \sum_{i=1,j=1}^{m,s} \frac{\partial f_\ell}{\partial u_{ix_j}} u_{ix_j}\right)\right)_{\Gamma_2} = 0 \qquad (2.4)$$

If f_ℓ are homogeneous functions of order β, with respect to variables u_{ix_j}, then according to the Euler theorem for homogeneous functions, the subtracted term in Eq. 2.4 is equal to βf_ℓ and consequently one again obtains the optimality condition of Eq. 2.2.

Note that in more general cases, when the first m_1 components of the vector-function u are given on boundary Γ_2 and the other components are not fixed and are determined as part of the solution of the extremal problem, the optimality condition for the unknown boundary takes the form of Eq. 2.3, with the only exception that summation over i is taken from 1 to m_1, rather than to m.

2.2 A Minimum Compliance Problem

Assume that an elastic body occupies the domain Ω with boundary $\Gamma = \Gamma_u + \Gamma_q + \Gamma_f$. Displacements u are given on Γ_u and loads q are given on Γ_q. The part of the boundary Γ_f is free of load. The problem under consideration consists in minimization of the compliance,

$$J = \frac{1}{2}\int_{\Gamma_q} qu\,d\Gamma_q \longrightarrow \min_{\Gamma_f} \qquad (2.5)$$

with respect to Γ_f. The isoperimetric condition that the volume of the domain Ω is given, is imposed on the variations of Γ_f. One may use the principle of minimum potential energy to reformulate the problem. According to the variational principle, the minimum of the functional

$$\Pi = \frac{1}{2}\int_{\Omega} \sigma_{ij}\varepsilon_{ij}d\Omega - \int_{\Gamma_q} uqd\Gamma$$

is realized for the actual distribution of displacement vector u, under the condition that the function u is prescribed on Γ_u. It is assumed that the stresses σ_{ij} and the deformations ε_{ij} are expressed in terms of displacements u, by means of the kinematic conditions and Hook's law. For the actual distribution of dis- tribution of displacements, one has $J = -\Pi$. Taking into account this equality and the variational principle, it is possible to express the performance criterion in terms of the functional Π, by means of formula $J = -\min_u \Pi$. Hence, it is possible to reduce the problem of Eq. 2.5 to successive determination of the minimum and maximum of the functional

$$J_* = \min_{\Gamma_f} J = \min_{\Gamma_f}(-\min_u \Pi) = -\max_{\Gamma_f} \min_u \Pi$$

The external maximum with respect to Γ_f is sought, subject to the isoperimetric condition that the volume of domain Ω is given, while the internal minimum with respect to u is sought, subject to the boundary condition on Γ_u. The functional includes the equilibrium equation, as its Euler equation, and thus they needn't be considered explicitly.

Taking into account that the displacement vector is not given on Γ_f, one may use Eq. 2.2 to obtain the necessary optimality condition. According to the problem under consideration, $F = \Pi$, $F_1 = V = 1$, $\Phi = \Pi - \lambda V$, and $V \equiv \int d\Omega$. One thus has

$$(\varepsilon_{ij}\sigma_{ij})_{\Gamma_f} = \lambda \tag{2.6}$$

By λ here and in the following one denotes the unknown constant that is to be determined in solution of the optimization problem.

The optimality condition of Eq. 2.6, meaning that the density of elastic deformation energy is constant on Γ_f, was obtained in Refs. 4 and 5. For elastic bodies under plane stress or plane deformation, the optimality condition is reduced to that of constant stress $\sigma_s^2 = \lambda$ on Γ_f. Here, n and s denote the normal and tangent directions to the contour and σ_s is the normal component of the stress tensor acting on the plane orthogonal to s. Thus the equal-stress condition is satisfied for optimal boundaries that are free of load.

2.3 A Bar of Maximum Torsional Rigidity

Consider the problem of an anisotropic bar under torsion [26]. Let the bar lie along the axis z in a rectangular cartesian coordinate system xyz and be subjected to torsional moments at its ends. Let Ω denote the cross-section of the bar perpendicular to the z axis and Γ the boundary of the region Ω. One may introduce a stress function $\phi(x,y)$ and express through it the non-vanishing components τ_{xz} and τ_{yz} of the stress tensor ($\tau_{xz} = \theta\phi_y$ and $\tau y_z = -\theta\phi_x$, where θ is the angle of twist per unit length of the bar). The torsional rigidity K is calculated from

$$K = 2 \int_\Omega \phi d\Omega$$

It may be noted, that the twisting moment M, the torsional rigidity K, and the angle of twist θ are related through $M = K\theta$.

The optimization problem considered here consists of seeking the contour Γ bounding the region Ω that satisfies the isoperimetric condition that the cross-sectional area is given and is such that the torsional rigidity of the bar achieves a maximum value, i.e. $K \longrightarrow \max_\Gamma$.

The stress function ϕ, as is well known, is determined by solving the Dirichlet problem for the torsion equation. This equation plays the differential constraint role in the optimization problem. The differential constraint can be excluded, if one takes into account that the actual stress function ϕ reduces the functional

$$\Pi = \int_{\Omega} (a\phi_x^2 + b\phi_y^2 - 2c\phi_x\phi_y - 4\phi)d\Omega$$

to a minimum. One can then use the equality $K = -\Pi$ [6]. The minimum is achieved with respect to the functions ϕ, satisfying the boundary condition $(\phi)_\Gamma = 0$. Here a, b, and c denote the elastic constants of the anisotropic material of the bar. Thus, one obtains the following optimization problem:

$$J_* = \max_\Gamma K = -\min_\Gamma \min_\phi \Pi \qquad (2.7)$$

One may now derive an optimality condition for determining the best shape of the contour Γ. In the case under consideration, one puts $F = \Pi$, $F_1 = S = 1$, $\Phi = F - \lambda S$, $S \equiv \int_\Omega d\Omega$. Taking into account the constancy of the function ϕ on the varied contour Γ and using Eq. 2.4, one has

$$(a\phi_x^2 + b\phi_y^2 - 2c\phi_x\phi_y)_\Gamma = \lambda \qquad (2.8)$$

This condition was obtained in Ref. 6. For isotropic bars the necessary optimality condition takes the simple form

$$\left(\frac{\partial\phi}{\partial n}\right)_\Gamma^2 = \lambda$$

2.4 A Membrane with Minimum Natural Frequency

Consider now (in nondimensional variables) the problem with unknown boundary arising from optimization of fundamental vibration frequency of a membrane clamped on the plane contour Γ. One part of the contour is given, but the other part is to be determined from the extremum condition for the fundamental frequency ($J = \omega$). It is assumed that the area of domain Ω, bounded by the contour Γ, is equal to unity. For this problem

$$J = F(J_1, J_2) \equiv J_1/J_2$$

$$J_1 = \int_\Omega (u_x^2 + u_y^2)d\Omega$$

$$J_2 = \int_\Omega u^2 d\Omega$$

$$F_1 = S = 1$$

$$S = \int_\Omega d\Omega$$

The amplitude deflection function u satisfies the boundary condition $(u)_\Gamma = 0$. Using Eq. 2.4, one derives the optimality condition for the unknown part of the boundary as

$$u_x^2 + u_y^2 - \omega^2 u^2 = \lambda$$

which has been obtained in Ref. 13.

3. OPTIMIZATION PROBLEMS WITH LOCAL PERFORMANCE CRITERIA

3.1 Elastic Body Weakened By a Hole

Consider the loading of elastic body weakened by a hole. Let Ω denote a domain that is occupied by the material, while Γ is the boundary of the body. For each point $x \in \Omega + \Gamma$, one may characterize the state of stress by a function f of stress tensor invariants I_1, I_2, and I_3, i.e. $f = f(I_1, I_2, I_3)$. Define F to be a function whose attainment at the point $x \in \Omega + \Gamma$ of a given value k^2 (the constant k^2 is a material characteristic) means that the material is in the limit state at x. Deformations of the material are elastic if the inequality $f < k^2$ is satisfied. Violation of this inequality is treated in different mechanics theories as the appearance of a plasticity zone, domains of inelastic strains, brittle fracture, and other effects. Henceforth, the equality $f = k^2$ will be interpreted as a plasticity condition. It is hence assumed that the expression for f, which is represented in terms of the stress tensor components, is a homogeneous function with a homogeneity index of 2.

For a given Γ, let the boundary value problem of elastic equilibrium be solved for certain sufficiently small values of the external forces and let the stresses $\sigma_{ij}^{0}(x)$ be found. Then, a set of points $\Omega_0(\Omega_0 \subset \Omega + \Gamma)$ can be determined where the maximum of the function f is realized: $J = (f)_{\Omega_0} \max_x f$. If the values of the external forces increase in proportion to a certain parameter p, then the plastic state will first be achieved at the points $x \in \Omega_0$, for the value

$$p_0 = \frac{k}{\sqrt{J}}$$

Evidently, the smaller the value of J, the larger the loads (larger values of p_0) at which the onset of plastic strains occurs in the body. Hence, extension of the range of loads for which the strains are elastic and plasticity zones do not occur in the body is achieved by minimizing the quantity J. One thus arrives at the following optimization problem: Determine the shape of the boundary Γ for which the minimum of the maximum is achieved for the quantity J in the domain $\Omega + \Gamma$, i.e.,

$$J_* = \min_\Gamma J = \min_\Gamma \max_{x\in\Omega} f \tag{3.1}$$

The problem of Eq. 1.9 is among the optimization problems with local performance criteria, because of the local nature of the functional to be optimized. In seeking the minimum with respect to Γ, it is assumed that the desired contour cannot shrink to a point, i.e. the absence of a hole is not allowed. The shape of the hole contour plays the role of the control function and the equilibrium equations enter the optimization problem as differential constraints. The problem formulated will be studied for the cases of tension and bending of infinite plates that are weakened by a hole.

Before investigating the optimality conditions in the problem formulated, note a property of harmonic functions which is used later. In the x-y plane, let there be n holes bounded by the closed contours Γ_i, i=1,...,n. Consider a family of harmonic functions χ that are continuous in $\Omega + \Gamma$ and tend to a given positive constant A at infinity, in a domain Ω bounded by the contour $\Gamma = \Sigma\Gamma_i$. Let g denote the boundary values of χ on Γ, i.e. $(\chi)_\Gamma = g$. For any function χ from this family, according to the maximum principle (see Ref. 23, for instance), the following inequality holds $|\chi(x,y)| \le \max_{\xi\eta} |g(\xi,\eta)|$, where $(x,y) \in \Omega$ and $(\xi,\eta) \in \Gamma$. In particular, it follows that $A \le \max_{\xi\eta}|g|$. If the strict equality is realized in this relation, then the function χ is identically equal to the constant A. Hence, the minimum of the functional $\max_{(\xi,\eta)\in\Gamma} |g(\xi,\eta)|$ with respect to g is achieved on the unique function $g(\xi,\eta) \equiv A$ ($\chi(x,y) \equiv A$) of the family under consideration, and its value equals A, i.e.

$$\min_g \max_{\xi\eta} |g| = A \tag{3.2}$$

This property of harmonic functions will now be used to estimate stresses on hole boundaries.

3.2 Infinite Plate (Plane Elasticity) with Hole

Consider the plane problem of elasticity theory on the state of stress of an infinite plate weakened by a hole. Let Ω denote a domain in the x-y plane that is occupied by the plate material, while Γ is the hole boundary. Assume that the plate is under tension, while the hole contour Γ is free of applied load. The boundary conditions on Γ are $\sigma_n = 0$ and $\tau_n = 0$ and the conditions at infinity are $(\sigma_x)_\infty = \sigma_1$, $(\sigma_y)_\infty = \sigma_2$, and $(\tau_{xy})_\infty = 0$, where σ_1 and σ_2 are given positive constants and n and s denote the normal and tangent directions to the contour of Γ. For this problem, the third stress tensor invariant is equal to zero and the function f is represented in the form $f = f(I_1,I_2)$, where $I_1 = \sigma_x + \sigma_y$ and $I_2 = \tau_{xy}^2 - \sigma_x\sigma_y$. The expressions $f = I_1^2 + 3I_2$ and $f^2 = I_1^2 + 4I_2$ corresponds to the Mises and Tresca plasticity criteria, respectively.

First consider the expression $I_1^2 + 3I_2$ as f, which equals (to the accuracy of a factor) the square of the shear stress intensity. One may introduce the auxiliary function $\chi = \sigma_x + \sigma_y$, which is known to be harmonic [24]. Taking into account the boundary conditions and the equality $\sigma_x + \sigma_y = \sigma_n + \sigma_s$, one has

$$\left.\begin{array}{ll} \Delta\chi = 0 \quad , & (x,y) \in \Omega \\ (\chi)_\Gamma = \sigma_s \ , & (\chi)_\infty = \sigma_1 + \sigma_2 \end{array}\right\} \qquad (3.3)$$

where Δ is the Laplace operator. Furthermore, using the invariance of the expression for f, relative to transfer from the x-y axes to the coordinates n-s and the boundary conditions, one arrives at the following formula for the boundary values of f : $(f)_\Gamma = \sigma_s^2$. Applying the property of harmonic functions noted in Section 3.1 to the function χ defined by Eq. 3.3 and comparing the expressions for the boundary values of f and Γ, one concludes that the minimum of the maximum values of $|\chi|$ and f on the contour Γ is achieved if and only if the stress σ_s is constant along the contour, with

$$(\sigma_s)_\Gamma = \sigma_1 + \sigma_2 \qquad (3.4)$$

The existence of an equal stress contour Γ satisfying Eq. 3.4 is assumed here. Some methods of determining hole shapes with equal stress boundaries are discussed in Sections 5 and 6. Note [20] that for the plane with one hole, the contours Γ satisfying Eq. 3.4 form a one-parameter family of ellipses $x^2\sigma_1^{-2} + y^2\sigma_2^{-2} = \lambda^2$, where λ^2 is a parameter.

It is now to be shown that Eq. 3.4 is a necessary and sufficient condition for optimality of the contour Γ of the hole in the problem of Eq. 3.1. To show this, it will be sufficient to prove that for contours Γ satisfying Eq. 3.4, the maximum of the function f in the domain $\Omega + \Gamma$ is reached on the contour Γ.

One may use the complex representation of the stress tensor components, in terms of the potentials $\Phi(z)$ and $\Psi(z)$ of Kolosov-Muskhelishvili [24]

$$\sigma_x + \sigma_y = 2\,[\Phi(z) + \overline{\Phi(z)}]$$

$$\sigma_y + \sigma_z + 2i\tau_{xy} = 2\,[\overline{z}\Phi'(z) + \Psi(z)]$$

where $z = x + iy$ and $\overline{z} = x - iy$. In compliance with Eq. 3.4, the function $\chi = \sigma_x + \sigma_y = \sigma_1 + \sigma_2$ and, therefore, $\Phi'(z) = 0$. The second equation of Kolosov-Muskhelishvili is converted into $\sigma_y - \sigma_x + 2i\tau_{xy} = 2\Psi(z)$ and multiplied by its complex-conjugate $\sigma_y - \sigma_x - 2i\tau_{xy} = 2\overline{\Psi(z)}$. Taking into account the equality

$\sigma_x + \sigma_y = \sigma_1 + \sigma_2$, one has $Q = (\sigma_1 + \sigma_2)^2 + 4(\tau_{xy}^2 - \sigma_x\sigma_y) = 4\bar{\Psi}\Psi$. Note that the function Q differs from the expression for f only by the factor in the second term. One can thus express f, in terms of Ψ and $\bar{\Psi}$, as

$$f = 3\Psi\bar{\Psi} + \frac{1}{4}(\sigma_1 + \sigma_2)^2$$

Furthermore, one can represent the potentials Ψ and $\bar{\Psi}$ in the form $\Psi = \phi + i\psi$ and $\bar{\Psi} = \phi - i\psi$, where the functions ϕ and ψ satisfy the Cauchy-Riemann conditions. The function f now has the form $f = 3(\phi^2 + \psi^2) + \frac{1}{4}(\sigma_1 + \sigma_2)^2$. One may now apply the Laplace operator to the expression obtained. Performing elementary manipulations and taking account of the Cauchy-Riemann conditions, as well as the resulting equality $\Delta\phi = \Delta\psi = 0$, it is easy to show that $\Delta f = 12(\nabla\phi)^2 \geq 0$.

Therefore, f does not achieve its maximum value at internal points of the domain $\Omega + \Gamma$. Comparing the values $(f)_\Gamma = (\sigma_1 + \sigma_2)^2$ and $(f)_\infty = \sigma_1^2 + \sigma_2^2 - \sigma_1\sigma_2$ results in the deduction that the maximum of f is reached on Γ. The optimality of equal stress contours is thus proved.

Note that the condition of Eq. 3.4 is both necessary and sufficient for a global optimum. Furthermore, note that the proof of the optimality was independent of the connectedness of the domain $\Omega + \Gamma$. Hence a system of n equally stressed holes will be optimal.

If f is given by the expression $f = I_1^2 + 4I_2$, which corresponds to the Tresca plasticity criterion, then it can also be shown that the optimality condition of Eq. 3.4 remains true. The more general dependence of f on the stress tensor invariants I_1 and I_2 are examined in Ref. 18 and the optimality of equally stressed holes is proved.

Note that the optimality condition takes the form $(\sigma_s)_\Gamma = \sigma_1 + \sigma_2 + \sigma_0$ when a constant pressure $\sigma_0 \geq 0$ is applied to the hole boundary, i.e. $(\sigma_n)_\Gamma = -\sigma_0$.

3.3 Infinite Plate (Bending) with Hole

Similar problems are encountered when one seeks the shapes of holes in infinite plates that are subjected to bending. Assume that a plate is bent by moments $(M_x)_\infty = M_1 > 0$, $(M_y)_\infty = M_2 > 0$, and $(M_{xy})_\infty = 0$ at infinity and the hole contour is free of loads, i.e. $M_n = 0$, $M_{ns} = 0$, and $Q_n = 0$. The optimization problem examined here involves finding the shape of the contour Γ of a hole for which the minimum of the maximal value of the second invariant of the stress tensor deviator over the region occupied by the plate material, is realized; i.e.

$$\left.\begin{aligned} J_* &= \min_\Gamma J = \min_\Gamma \max_{xy} \max_\zeta f \\ f &= f_1 + f_2 \\ f_1 &= 144\,\zeta^2 h^{-6}\,[(M_x + M_y)^2 + 3(M_{xy}^2 - M_x M_y)] \\ f_2 &= \frac{27}{4}\,h^{-6}(h^2 - 4\zeta^2)^2\,[Q_x^2 + Q_y^2] \end{aligned}\right\} \qquad (3.5)$$

where $(x,y) \in \Omega + \Gamma$ and the variable ζ is measured along the normal to the middle surface of the plate and varies in the limits $-h/2 \le \zeta \le h/2$. The function f depends explicitly on ζ and implicitly on (x,y) through the quantities M_x, M_y, M_{xy}, Q_x, and Q_y.

One first examines the auxiliary problem of minimizing the maximum value of f_1 on the contour Γ. It can be shown [17] that $(f_1)_\Gamma = aM_s^2$, where $a = 144\zeta^2 h^{-6}$. Introduce now the auxiliary function $\chi = M_x + M_y$, which is harmonic and satisfies the following conditions: $\Delta\chi = 0$ in Ω, $(\chi)_\Gamma = M_s$, and $(\chi)_\infty = M_1 + M_2$, analogous to Eq. 3.3. If one considers these relations, the formula $(f_1)_\Gamma = aM_s^2$, and the property of harmonic functions in Eq. 3.2, one comes to the conclusion that the minimum of the maximal of f_1 is reached on the contour Γ, if and only if

$$(M_s)_\Gamma = M_1 + M_2 \qquad (3.6)$$

Thus, the equality of Eq. 3.6 is a necessary and sufficient optimality condition for the auxiliary problem.

Consider further the problem of minimizing the maximum value of f_1 in the region $\Omega + \Gamma$. It is to be proved that the optimality condition for this problem is also expressed by Eq. 3.6. To this purpose, one may study the behavior of the function

f_1 in the region $\Omega + \Gamma$ and show that for contours satisfying the condition of Eq. 3.6, the function f_1 reaches its maximum value on the boundary of the hole. Complex representations [25] of the bending moments M_x, M_y, and M_{xy} and the shear forces Q_x and Q_y in terms of the two analytic functions $\Phi(z)$ and $\Psi(z)$

$$M_x + M_y = -2D(1 + \nu)[\Phi(z) + \overline{\Phi(z)}]$$

$$M_y - M_x + 2iM_{xy} = 2D(1 - \nu)[\bar{z}\Phi'(z) + \Phi(z)]$$

$$Q_x - i\, Q_y = -4D\Phi'(z)$$

Using these representations and performing calculations analogous to those made for the plane problem of elasticity theory, one obtains $\Delta f_1 \geq 0$. Therefore, f_1 cannot reach its maximum value in the interior points of the region $\Omega + \Gamma$. It follows from a direct comparison of the values $a(M_1 + M_2)^2$ and $a((M_1 + M_2)^2 - 3M_1M_2)$ that are taken by the function f_1, respectively, on Γ and at infinitely distant point of the region, that the maximum of f_1 is reached on the boundary Γ. Thus, under the condition of Eq. 3.6, the maximum of f_1 is reached on Γ. But the maximal value of f_1 on Γ reaches its minimum if and only if the condition of Eq. 3.6 is satisfied. Therefore, the necessary and sufficient optimality condition is also given by Eq. 3.6 in the problem of minimizing the maximal value of f_1 on the region $\Omega + \Gamma$.

Returning to the basic problem of minimizing the maximum value of f, note that $Q_x - iQ_y = 4D\Phi'(z) = 0$ and, consequently, $Q_x = Q_y = 0$, for holes satisfying Eq. 3.6. In this case, $f_2 = 0$ for all $(x,y) \in \Omega + \Gamma$ and the following equality holds: $f = f_1$. For holes of arbitrary shape, $f = f_1 + f_2 \geq f_1$. Taking into account that the minimum of the maximum value of f_1 on the region $\Omega + \Gamma$ is reached if and only if Eq. 3.6 is satisfied, one conclues that Eq. 3.6 is also the necessary and sufficient optimality condition for the problem of Eq. 3.5.

4. SOLUTION OF THE OPTIMIZATION PROBLEM BY A PERTURBATION TECHNIQUE

An effective approach to solving two-dimensional optimization problems with unknown boundaries is a perturbation method. Application of this method involves power series expansion of both state variables (displacements, strains, stresses, etc.) and control functions describing the shape of a body. A shape optimization problem is thus reduced to a set of more simple problems, the solutions of which permits one to determine the best shape with any desired accuracy. It should be noted that in applications of this method to shape optimization, several degenerations are possible. They lead to certain purely mathematical problems that have not yet been solved. No general comments or singular cases are considered here. This discussion is presented only to illustrate the perturbation method for the example of a well-posed problem of cross-sectional shape optimization of a twisted bar.

Consider a homogeneous, isotropic bar with a doubly-connected cross-section that is subjected to a twisting torque. Let Γ_0 and Γ denote, respectively, the inner and the outer boundaries of the region Ω. The nomenclature for the main variables is the same as in Section 2. In the following, it is convenient to employ a curvilinear coordinate system s - t related to the contour Γ_0. The coordinate t of a point $P \in \Omega$ is the distance along the normal 0_2P from the point P to the contour Γ_0, while s is the distance along the contour measured from a certain fixed point, 0_1 to the point 0_2. Let $h = h(s)$ be the equation of the contour Γ, $R = R(s)$ be the radius of curvature of the contour Γ_0, and ℓ_0 be its length. It is assumed that the bar is thin-walled and that the minimum radius of curvature is of the order of the length ℓ_0 of the contour Γ_0, i.e.

$$\max_s h(s) = H << \ell_0$$

$$\min_s R(s) \sim \ell_0$$

Thus, $H/\ell_0 = \varepsilon$ is a small parameter ($\varepsilon << 1$). In other words, consideration is limited to only weakly curved contours.

The inner contour Γ_0 is considered to be given, but the outer boundary is to be determined. The optimization problem solved here consists of seeking a function $h(x)$ that satisfies the isoperimetric condition of constant cross-sectional area of the bar and maximizes the torsional rigidity K.

Define the new variables,

$$s' = \frac{s}{\ell_0}, \quad t' = \frac{t}{H}, \quad h' = \frac{h}{H}, \quad \phi' = \frac{\phi}{H\ell_0}, \quad S_0' = \frac{S_0}{\ell_0^2}$$

$$S' = \frac{S}{H\ell_0}, \quad R' = \frac{R}{\ell_0}, \quad K' = \frac{K}{H\ell_0^3}, \quad C' = \frac{C}{H\ell_0}, \quad \lambda' = \frac{\lambda}{\ell_0}$$

The equations for the stress function ϕ, optimality condition for the unknown boundary (see Section 2), boundary condition for ϕ on Γ_0 and Γ, Bredt condition, isoperimetric condition of constant cross-sectional area of the bar, and formula for torsional rigidity take the form (primes are suppressed)

$$\left.\begin{aligned}
&(T\phi_t)_t + \varepsilon^2(T^{-1}\phi_s)_s = -2\varepsilon T \\
&T = 1 + \frac{\varepsilon t}{R} \\
&[\nabla\phi(s,h)]^2 = \lambda^2 \\
&\phi(s,0) = C \\
&\phi(s,h) = 0 \\
&\int_0^1 \phi_t(s,0)ds = -2S_0 \\
&\int_0^1 \left(h + \frac{\varepsilon h^2}{2R}\right)ds = S \\
&K = 2\left(CS_0 + \varepsilon \int_0^1 \int_0^h T\phi dtds\right)
\end{aligned}\right\} \qquad (4.1)$$

The solution of this problem is sought in the form of series expansions with respect to the small parameter ε,

$$\phi = \phi^0 + \varepsilon\phi^1 + \varepsilon^2\phi^2 + \ldots$$

$$h = h^0 + \varepsilon h^1 + \varepsilon^2 h^2 \ldots \tag{4.2}$$

$$K = K^0 + \varepsilon K^1 + \varepsilon^2 K^2 + \ldots$$

For finding the zeroth, first and second order approximate solution, it is sufficient to substitute the expansion of Eq. 4.2 into the relations of Eq. 4.1 and to equate the coefficients of like powers of ε. The resulting boundary-value problems serve to determine the unknown functions. Thus, for determining the unknown functions of zeroth order, one has

$$\left.\begin{aligned} &\phi^0_{tt} = 0 \\ &\phi^0(s,0) = C^0 \\ &\phi^0(s,h^0) = 0 \\ &\phi^0_t(s,h^0) = -\lambda^0 \\ &\int_0^1 \phi^0_t(s,0)ds = -2S_0 \\ &\int_0^1 h^0 ds = S \end{aligned}\right\} \tag{4.3}$$

Similarly, the first order approximation is the solution of the boundary-value problem

$$\left.\begin{aligned} &\phi^1_{tt} = -2 - \frac{1}{R}\phi^0_t \\ &\phi^1(s,0) = C^1 \\ &\phi^1(s,h^0) = \lambda^0 h^1 \\ &\phi^1_t(s,h^0) = -\lambda^1 \\ &\int_0^1 \phi^1_t(s,0)ds = 0 \end{aligned}\right\} \tag{4.4}$$

$$\int_0^1 h^1 ds = -\frac{1}{2}\int_0^1 \frac{(h^0)^2}{R} ds$$

and the second order approximation is the solution of the problem

$$\left.\begin{aligned}
&\phi^2_{tt} = -\frac{1}{R}[(t\phi^1_t)_t + 2t] - \phi^0_{ss}\\
&\phi^2(s,0) = C^2\\
&\phi^2(s,h^0) = \lambda^1 h^1 + \lambda^0 h^2\\
&\phi^2_t(s,h) = (2 - \frac{\lambda^0}{R})h^1 - \lambda^2\\
&\int_0^1 \phi^2_t(s,0)ds = 0\\
&\int_0^1 h^2 ds = -\int_0^1 \frac{h^0 h^1}{R} ds
\end{aligned}\right\} \tag{4.5}$$

From the zeroth, first, and the second order approximate solutions, one can define the torsional rigidity of the bar (the functional of the variational problem) as

$$K = K^0 + \varepsilon K^1 + \varepsilon^2 K^2 = 2C^0 S_0 + 2\varepsilon\left(\int_0^1\int_0^{h^0} \phi^0 dtds + C^1 S_0\right)$$

$$+ 2\varepsilon^2\left(\int_0^1\int_0^{h^0} \phi^1 dtds + C^2 S_0\right) + 0(\varepsilon^3)$$

Proceed now with the solution of the above boundary-value problems, starting with the zeroth order terms. From the equation and boundary conditions of Eq. 4.3 (the first line), one finds the function ϕ^0. Using the expression obtained for ϕ^0 and the optimality condition of Eq. 4.3 (the second line), one determines h^0. Thus, one has $\phi^0 = C^0(1 - t/h^0)$ and $h^0 = C^0/\lambda^0$. Substituting the functions obtained into the isoperimetric equality of Eq. 4.3 (the third line) and performing elementary transformations, one finds the constants λ^0 and C^0. Finally, the zeroth order solution takes the form

$$h^0 = S$$

$$\phi^0 = 2SS_0 \; (1 - \frac{t}{S})$$

$$K^0 = 4SS_0^2$$

$$\lambda^0 = 2S_0$$

$$C^0 = 2SS_0$$

Thus, in the zeroth order approximation for hollow bars with fairly shallow (large radius of curvature) inner contours, the optimum bar has constant wall thickness.

Similarly, the solution of the boundary-value problem of Eq. 4.4 furnishes the quantities h^1, ϕ^1, C^1, λ^1, and K^1. For the sake of brevity, only the final result is presented, which is

$$\phi^1 = t^2 \; (\frac{S_0}{R} - 1) + 2SS_0\left(\int_0^1 \frac{ds}{R} - \frac{1}{R}\right) + S^2 - 2S_0S^2 \int_0^1 \frac{ds}{R}$$

$$h^1 = - \frac{S^2}{2R}$$

$$K = 4S_0S^2 \left(1 - S_0 \int_0^1 \frac{ds}{R}\right)$$

$$C^1 = S^2 \left(1 - 2S_0 \int_0^1 \frac{ds}{R}\right)$$

$$\lambda^1 = 2S \left(1 - S_0 \int_0^1 \frac{ds}{R}\right)$$

Thus, to within terms of the order ε^2, one has

$$h = h^0 + \varepsilon h^1 = S(1 - \varepsilon \frac{S}{2R})$$

or in terms of the original dimensional quantities

$$h = S\ell_0^{-1}(1 - S/2R\ell_0)$$

From the formula for h, it is evident that the wall thickness of the optimum bar decreases as one moves along the inner contour in the direction of increasing curvature.

In order to determine the terms of order ε^2, it is necessary to integrate Eq. 4.5 and to find the arbitrary constants of integration, from the boundary and the isoperimetric conditions. Finally, one gets

$$h^2 = -\frac{S^3}{2R^2}\left(3 + \frac{R}{S_0} - 4R^2\int_0^1 \frac{ds}{R^2} - \frac{R^2}{S_0}\int_0^1 \frac{ds}{R}\right)$$

Then, the value K^2 is determined. Using the expressions for K^0, K^1, and K^2 and returning to the original dimensioned quantities, one gets the following expression for the torsional rigidity of the optimum bar:

$$K = \frac{4SS_0^2}{\ell_0^2} + \frac{4S^2S_0}{\ell_0}\left(1 - \frac{S_0}{\ell_0^2}\int_0^{\ell_0} \frac{ds}{R}\right)$$

$$+ \frac{4S^3}{3\ell_0^6}\left[3S_0^2\left(\int_0^{\ell_0} \frac{ds}{R}\right)^2 + S_0^2\ell_0 \int_0^{\ell_0} \frac{ds}{R}\right.$$

$$\left. - 4S_0\ell_0^2 \int_0^{\ell_0} \frac{ds}{R} + \ell_0^4\right]$$

Compare now the torsional rigidity K of the optimally designed bar with that of a bar of constant thickness (h(s) = const), having the same inner contour and cross-sectional area K_C. Performing similar calculations as for the optimization problem, one obtains, to the same degree of accuracy, an expression for K_C. Using this expression and the expression for the torsional rigidity of the optimum bar, one obtains

$$\Delta K = K - K_C = \frac{S^3 S_0^2}{\ell_0^6}\left[\ell_0 \int_0^{\ell_0} \frac{ds}{R^2} - \left(\int_0^{\ell_0} \frac{ds}{R}\right)^2\right]$$

Employing the Schwarz inequality, it is easy to show that the expression within the square brackets is always positive. Therefore, for any inner contour Γ_0 with $\min_s R(s) \sim \ell_0$, the following inequality is true $\Delta K \geq 0$. The equality sign holds for the case in which Γ_0 is a circle (R = const is the radius of the circle). In this case, the thin-walled optimum bar is of constant thickness.

The possibilities of a perturbation technique are not exhausted by the optimization problem considered in this Section. Likewise, the perturbation technique may be used to solve problems of optimizing thickness distributions of a thin-walled structure. Various other constraints can be accounted for within the framework of the technique used.

5. METHODS OF COMPLEX VARIABLE FUNCTION THEORY

The application of methods of complex variable function theory turns out to be effective in seeking optimum shapes. Some methods for two-dimensional problems with unknown boundaries, based on conformal mapping, are presented in Refs. 7, 10, and 20.

5.1 Shape of a Bar of Greatest Torsional Rigidity

Consider the problem of determining the shape of a bar of greatest torsional rigidity. The case of a simply connected cross-sectional domain Ω (the hole is absent) with contour Γ is first examined. To determine the contour Γ, one may introduce a function f(z) that is regular in the domain Ω, so that

$$\text{Re } f(z) = \phi(x,y) + \frac{1}{2}(x^2 + y^2)$$

where z = x + iy. Taking the optimality condition $(\phi_t)_\Gamma = -2$ into account, one obtains the boundary condition for f as

$$f'(t) = \bar{t} - i\lambda \frac{dt}{|dt|}$$

If $z = \omega(\zeta)$ is the conformal mapping of the circle $|\zeta| < 1$ onto the unknown domain Ω, then the expression $\bar{\omega}(\tau) - \lambda\tau\bar{\omega}'(\tau)$ ($\tau = \exp i\theta$, $0 < \theta < 2\pi$) is to be the boundary value of the regular function on the unit circle. The function $w(\zeta) = -C\zeta$ satisfies this condition. Hence, the circular bar has the maximal torsional rigidity. This result is in accordance with the theorem formulated by Sent-Venant in 1856 and proved by Polya and Szego in 1946.

As an example of more complicated problems, consider torsion of a bar having a cylindrical hole of given shape. Let the cross-section of twisted bar occupy a doubly-connected domain Ω that is bounded by the contour $\Gamma_0 + \Gamma$, where Γ_0 is the given inner and Γ the unknown outer contour. It is required to find the external contour Γ, satisfying the isoperimetric condition of constant cross-section area of the bar and maximizing the torsional rigidity K.

By introduction of the complex torsional function, the optimization problem is reduced to the problem of seeking the domain Ω , where a univalent analytic function f(z) is defined, which satisfies the condition

$$f(t) + \bar{f}(t) = 2C + t\bar{t}$$

on the contour Γ_0 and the condition

$$f(t) = \frac{1}{2} t\bar{t} + \frac{1}{2i} \int_{t_0}^{t} (\bar{t}dt - td\bar{t}) - i\lambda \int_{t_0}^{t} |dt|$$

on the unknown boundary Γ.

The inverse boundary value problem of determining the doubly-connected domain Ω with given inner contour Γ_0 and unknown outer contour Γ can be solved by means of mappings. The mappings of the exterior of the unit circle $|\xi| > 1$ and $|\zeta| > 1$ in ξ and ζ -planes, respectively, onto the exterior of the contours Γ_0 and Γ in the z-plane are used for solving the problem in Refs. 7 and 10. These mappings are realized by means of the functions

$$z = \omega_1(\zeta) = B\zeta \sum_{k=0}^{n} b_k \zeta^{-k}$$

$$z = \omega_2(\xi) = A\xi \sum_{k=0}^{n} a_k \xi^{-k}$$

where $a_0 = b_0 = 1$ and A and B are real constants that define the scale. The coefficients a_i are known, but the quantities b_i are to be determined. The parameter $H = A/B$ defines the relative size of the section. A system of n nonlinear equations was obtained in Ref. 10, in order to determine the unknown coefficients b_j and to seek the unknown contour Γ. This system was solved by the Newton method for known values of a_j and H. The calculations were performed and presented in Ref. 10 for $0 < H < 1$, for inner contours having forms of an ellipse, square, and rectangle. Note that the technique used, based on two mappings, was propósed and applied to some problems of the theory of elasticity by Sherman [30].

5.2 Shape of Equal Stress Holes in a Plate (Plane Stress)

The method of determination of equal stress hole contours in a perforated plate, based on conformal mapping and reducing to a standard Dirichlet problem for the exterior of the same number of parallel slits on a parametric plane, presented in Ref. 20 is described.

Determination of the stress state of a perforated plane, weakened by equal stress holes, is reduced to determination of a complex potential $\Psi(z)$. The function $\Psi(z)$ has the assymptotics $\beta + O(z^{-2})$ when $z \to \infty$ $(\beta = \frac{1}{2}(\sigma_2 - \sigma_1))$. Taking into account the optimality condition $(\sigma_s)_\Gamma = \sigma_0 + \sigma_1 + \sigma_2$, boundary condition $(\sigma_n)_\Gamma = -\sigma_0$, and the known relation [24]

$$\sigma_s - \sigma_n + 2i\tau_{sn} = e^{2i\theta}(\sigma_y - \sigma_x + 2i\tau_{xy})$$

one may write the boundary condition for the function Ψ as

$$e^{2i\theta}\Psi(z) = \mu \qquad z \in \Gamma$$

where $\mu = \frac{1}{2}(\sigma_1 + \sigma_2 + 2\sigma_0)$ and θ is the angle between the external normal to the outline and the x-axis. One may now use the conformal mapping $z = \omega(\zeta)$ of the exteriors of n parallel slits in the ζ-plane onto the exteriors of n equal stress contours in plane z. Note that every n-connected domain, including the infinitely remote point, can be mapped conformally onto the exterior of some n slits that are parallel to the real axis, with correspondence of points at infinity [27]. For $n \geq 3$, this mapping is unique, if the following behavior of the mapping function $\omega(\zeta)$ at infinity: $\omega(\zeta) = \zeta + O(1)$, as $\zeta \to \infty$ is specified.

Substituting the expression for $\exp(2i\theta) = \omega'(\zeta)/|\omega'(\zeta)|$, see Ref. 20, into the boundary condition, one obtains

$$-\psi(\zeta)\,\omega'(\zeta) = \mu\overline{\omega'(\zeta)} \tag{5.1}$$

where $\psi(\zeta) = \Psi(\omega(\zeta))$ and Γ_ζ is the boundary of the slits in the ζ-plane. The functions $\psi(\zeta)$ and $\omega(\zeta)$ are to be determined from the boundary value problem of Eq. 5.1. Taking the real and imaginary parts in Eq. 5.1, one obtains

$$\left.\begin{array}{lll} \mathrm{Re}\, T'(\zeta) = 0, & \zeta \in \Gamma_\zeta, & T'(\zeta) \equiv (\psi(\zeta) + \mu)\omega'(\zeta) \\ \mathrm{Im}\, N'(\zeta) = 0, & \zeta \in \Gamma_\zeta, & N'(\zeta) \equiv (\psi(\zeta) - \mu)\omega'(\zeta) \end{array}\right\} \tag{5.2}$$

The functions $T'(\zeta)$ and $N'(\zeta)$ are analytic everywhere in the exterior of the slits Γ_ζ. They are bounded in a neighborhood of any infintely remote point, since the functions $\psi(\zeta)$ and $\omega'(\zeta)$ are bounded as $\zeta \to \infty$. One may require that all the hole contours Γ be smooth. Under this additional condition, the function $\psi(\zeta)$ is bounded everywhere, with the exception of the ends of the slits Γ_ζ, in whose neighborhoods $\omega'(\zeta)$ evidently has a power singularity of order 1/2. According to this condition, the analytic functions $T'(\zeta)$ and $N'(\zeta)$ are bounded in the entire ζ-plane, with the exception of the ends of the slits Γ_ζ, at which they have a power-law singularity of order 1/2. The boundary value problems of Eq. 5.2 are classical Dirichlet problems for the exterior of slits, where the solution of the problems is sought in the class of functions bounded at infinity and having a power singularity at the ends of the slits. After the functions $T(\zeta)$ and $N(\zeta)$ have been found, the desired functions $\psi(\zeta)$ and $\omega(\zeta)$ are determined by using

$$\left.\begin{array}{l} \omega(\zeta) = \dfrac{1}{2\mu}\,[T(\zeta) - N(\zeta)] + C_0 \\[2ex] \psi(\zeta) = \dfrac{T'(\zeta) - N'(\zeta)}{2\omega'(\zeta)} \end{array}\right\} \tag{5.3}$$

where C_0 is an arbitrary constant.

In the case of one hole in the z-plane, there will be one slit on the ζ-plane which can be considered the slit (-1,+1) along the real axis, without loss of generality. For $\zeta \to \infty$ one has $\omega(\zeta) = C_1\zeta + O(\zeta^{-1})$. The quantity C_1 can be considered real. According to the Rieman theorem, this condition, together with assignment of the slit length, exhausts the possible arbitrariness in the description of the conformal transformation of the two given domains. The solution of the boundary value problems of Eq. 5.2 for the exterior of the slit mentioned is the following:

$$T'(\zeta) = \frac{C_1\zeta(\beta + \mu) + d_1}{\sqrt{\zeta^2 - 1}}$$

$$N'(\zeta) = C_1(\beta - \mu) + \frac{id_2}{\sqrt{\zeta^2 - 1}}$$

The real constants d_1 and d_2 are arbitrary. Integrating the expressions for T' and N' and using the first formula of Eq. 5.3, with $C_0 = 0$, one obtains

$$\omega(\zeta) = \frac{C_1}{2}\left(m_1\zeta + m_2\sqrt{\zeta^2-1}\right) + \frac{d_1 - id_2}{2\mu}\,\ln\left(\zeta + \sqrt{\zeta^2-1}\right)$$

where (5.4)

$$m_1 = 1 - \beta/\mu$$

$$m_2 = 1 + \beta/\mu$$

The function $\omega(\zeta)$ should be unique in the exterior of the slit (-1,+1). According to the solution given in Eq. 5.4, this condition is satisfied only if $d_1 = d_2 = 0$, which is henceforth assumed. The equation of the hole contour in parametric form is obtained from Eq. 5.4 as

$$x = \frac{1}{2}\,C_1m_1t\ , \quad y = \pm\, C_1m_2\,\sqrt{1 - t^2}$$

where t is a real parameter ($-1 < t < 1$).

Consider now the case of two holes. One maps the exterior of the holes conformally onto the exterior of two slits (λ_1,λ_2) and (λ_3,λ_4) along the real axis ξ of the ζ plane. The coefficient C_1 can be considered real and positive. The solution of the Dirichlet problems of Eq. 5.2 is

$$T'(\zeta) = \frac{C_1\zeta^2(\mu + \beta) + d_1\zeta + d_2}{\sqrt{(\zeta - \lambda_1)(\zeta - \lambda_2)(\zeta - \lambda_3)(\zeta - \lambda_4)}}$$

$$N'(\zeta) = C_1(\beta - \mu) + \frac{i(d_3\zeta + d_4)}{\sqrt{(\zeta - \lambda_1)(\zeta - \lambda_2)(\zeta - \lambda_3)(\zeta - \lambda_4)}}$$

where d_1, d_2, d_3, and d_4 are real arbitrary constants. In this case, the contours of both holes are symmetric, relative to the abscissa and ordinate axes.

For definiteness, consider one hole to be located entirely in the left half-plane $(x < 0)$, and the other to be symmetrically in the right half-plane $(x > 0)$. In the symmetric case under consideration, one can, without limiting the generality, set $\lambda_1 = -2$, $\lambda_2 = -1$, $\lambda_3 = 1$, $\lambda_4 = 2$, and $d_1 = d_3 = d_4 = 0$. The constant C_0 is selected from the condition that $\omega(0) = 0$. One now determines the function

$$\omega(\zeta) = \frac{C_1}{2\mu}\left\{(\mu - \beta)\zeta + 2(\mu + \beta)\, E\left(\arcsin\zeta, \frac{1}{2}\right) - \left[(\mu + \beta) - 2 + \frac{1}{2} d_2\right] F\left(\arcsin\zeta, \frac{1}{2}\right)\right\}$$

where F and E are elliptic integrals of the first and second kinds, respectively.

Note that an effective exact solution is found by this method, not only for the case of one and two holes, but also for the case of periodic and doubly-periodic series of holes.

6. THE METHOD OF INTEGRAL EQUATIONS

The method of determining equally stressed hole contours in a perforated plate, based on the reduction to integral equations, is developed in Ref. 19.

As is known [28], a canonical domain obtained from the ζ-plane by removing n circles, can be mapped onto any n-connected domain of the complex z-plane, with a point at infinity. When $n > 2$, the mapping $\omega(\zeta)$, which has the form $\omega(\zeta) = C\zeta + \omega_0(\zeta)$, where $\omega_0(\zeta)$ is bounded at infinity, depends on 3n real parameters, six of which (e.g. one circumference, one fixed point on this circumference, and a center of another circumference) can be specified arbitrarily, and C is a scale multiplier. Consequently, a system of contours of equal stress, if it exists, forms a (3n - 6)-parameter family. The limits of variation of the parameters can be found from geometrical considerations. The presence of symmetry may lead to a reduction in the number of parameters.

One has the following relations [24] for determining the stress components at the boundary Γ:

$$\left.\begin{aligned} &\sigma_r + \sigma_\theta = 4\mathrm{Re}\Phi(\zeta) \\ &\sigma_\theta - \sigma_r + 2i\tau_{r\theta} = \frac{2(\zeta - \zeta_k)^2}{\tau_k^2\,\overline{\omega_0'(\zeta)}}\left(\omega(\zeta)\Phi'(\zeta) + \omega'(\zeta)\Psi(\zeta)\right) \end{aligned}\right\} \tag{6.1}$$

Here σ_r, σ_θ, and $\tau_{r\theta}$ denote the normal and shear stress in a polar coordinate system, with a pole at the center ζ_k of a circle of radius r_k and a boundary Γ_k, $k = 1,\ldots,n$. If a homogeneous state of stress with the stress components $\sigma_x = \sigma_1$, $\sigma_y = \sigma_2$, and $\tau_{xy} = 0$ is given at infinity, then $\Phi(\zeta)$ and $\Psi(\zeta)$ have the form $\Phi(\zeta) = \frac{1}{4}(\sigma_1 + \sigma_2) + \Phi_0(\zeta)$ and $\Psi(\zeta) = \frac{1}{2}(\sigma_2 - \sigma_1) - \Psi_0(\zeta)$, where $\Phi_0(\zeta)$ and $\Psi_0(\zeta)$ are holomorphic and have asymptotics of order $O(z^{-2})$ when $z \to \infty$. Assuming that $\sigma_r = -\sigma_0$ at all contours, one sets $(\sigma_\theta)_\Gamma = \sigma_1 + \sigma_2 + \sigma_0$. In this case, $\Phi(\zeta) = \frac{1}{4}(\sigma_1 + \sigma_2)$ and the second formula of Eq. 6.1 reduces to the relation

$$\frac{\mu C r_k^2}{(\zeta-\zeta_k)^2} + \frac{\mu r_k^2 \omega_0'(\zeta)}{(\zeta-\zeta_k)^2} = C\beta + \omega_0'(\zeta)\Psi_0(\zeta) \tag{6.2}$$

Consider now the second term in the left-hand side of Eq. 6.2. Taking into account the fact that $r_k^2/(\zeta - \zeta_k)^2 = -d\bar{\zeta}/d\zeta$, when $\zeta \in \Gamma_\zeta$, one can write this term in the form

$$\frac{r_k^2\omega_0'(\zeta)}{(\zeta - \zeta_k)^2} = -\lim_{\Delta\zeta\to 0} \frac{\overline{\Delta\zeta}}{\Delta\zeta} \; \frac{\overline{\omega_0(\zeta + \Delta\zeta) - {}_0(\zeta)}}{\overline{\Delta\zeta}} = -\frac{d\omega_0}{d\zeta}$$

Substituting this relation into Eq. 6.2 and integrating, one obtains

$$\Lambda(\zeta) + \overline{\omega_0(\zeta)} = -C\left(\frac{\beta}{\mu}\zeta + \frac{r_k^2}{\zeta - \zeta_k}\right) + d_k, \quad \zeta \in \Gamma_k, \qquad k = 1,\ldots,n \tag{6.3}$$

The function $\Lambda(\zeta)$ is holomorphic and $\Lambda'(\zeta) = \omega_0'(\zeta)\Psi_0(\zeta)/\mu$ and d_k are arbitrary constants. Varying d_k, if necessary, one can reach the state for which the bounded functions $\Lambda(\zeta)$ and $\omega_0(\zeta)$ decrease at infinity. To solve the boundary-value problem, one writes, following Sherman [29], $\Lambda(\zeta)$ and $\omega_0(\zeta)$ in terms of the Cauchy-type integrals

$$\Lambda(\zeta) = \frac{1}{2\pi i}\int_\Gamma \frac{u(t)}{t-\zeta}\,dt, \quad \omega_0(\zeta) = \frac{1}{2\pi i}\int_\Gamma \frac{\bar{u}(t)}{t-\zeta}\,dt \tag{6.4}$$

The expressions in Eq. 6.3 decrease at infinity. Substituting Eq. 6.4 into Eq. 6.3, one obtains

$$u(\zeta) + \frac{1}{2\pi i}\int_\Gamma u(t)\,d\ln\frac{t-\zeta}{\bar{t}-\zeta} - d_k = -C\left(\frac{\beta}{\mu}\zeta + \frac{r_k^2}{\zeta-\zeta_k}\right) \tag{6.5}$$

The constants d_k can now be determined from the relation

$$d_k = -\frac{1}{2\pi r_k}\int_{\Gamma_k} u(t)\,ds, \quad ds = |dt| \tag{6.6}$$

Equation 6.5 represents a Fredholm-type equation with a real symmetric kernel. Separating the real and imaginary parts, one obtains a pair of integral equations, relative to a double layer potential of a modified Dirichlet problem, in the class of bounded continuous functions. Since the above equations have unique solutions, Eqs. 6.5 and 6.6 also have a solution for any arbitrary value of the right-hand side. When $n = 1$, Eq. 6.5 has the obvious solution $\Lambda(\zeta) = \zeta$ and $\omega_0(\zeta) = \beta/\mu\zeta$. The function $\omega = C(\zeta + \beta/\mu\zeta)$ coincides with the function in Eq. 5.4, if in addition one maps the outside of the unit circle onto the outside of the segment [-1,1], using the Joukowski function $\frac{1}{2}(\zeta + \zeta^{-1})$.

When $n > 1$, one can solve Eq. 6.5 using the numerical method of least squares. The calculation of optimum forms of cyclically and symmetrically distributed holes ($n = 2,3,4$, and 6) were performed in Ref. 19.

REFERENCES

1. Saint-Venant, B. de, "Memoire Sur la Torsion des Prismes," Memoires presentes par divers savants a l'Academie des Sciences, t. 14, 1856, 233-560.
2. Polya, G., "Torsional Rigidity, Principal Frequency, Electrostatic Capacity and Symmetrization," Quart. Appl. Math., Vol. 6, No. 3, 1948.
3. Polya, G. and Szego, G., Isoperimetric Inequalities in Mathematical Physics, Princeton University Press, New Jersey, 1951.
4. Wasiutinski, Z., "On the Congruency of the Forming According to the Minimum Potential Energy with that According to the Equal Strength," Bull. Acad. Polonaise Sci., Ser. Sci. Techn., Vol. 8, No. 6, 1960.
5. Prager, W., "Optimality Criteria Derived from Classical Extremum Principles," Solid Mech. Study No. 1, Solid Mech. Division, Univ. of Waterloo, April, 1969.
6. Banichuk, N.V., "On a Variational Problem with Unknown Boundaries and Determination of the Optimal Shapes of Elastic Bodies," PMM, Vol. 39, No. 6, 1975.
7. Kurshin, L. M., "On the Problem of Determining the Shape of a Rod Section of Maximum Torsional Stiffness," Dokl. Acad. Nauk, SSSR, Vol. 223, No. 3, 1975.
8. Banichuk, N.V., "Optimization of Elastic Bars in Torsion," Int. J. Solids Structures, Vol. 12, No. 4, 1976.
9. Gurvitch, E.L., "On Isoperimetric Problems for Domains with Partly Known Boundaries," Journ. of Optimizat. Theory and Appl., Vol. 20, No. 1, 1976.
10. Kurshin, L.M. and Onoprienko, P.N., "Determination of Shape of Doubly Connected Cross-Sections of Bars of Maximum Torsional Rigidity," PMM, Vol. 40, No. 6, 1976.
11. Banichuk, N.V., "Extremum Problem for a System with Distributed Parameters and Determination of the Optimum Properties of an Elastic Medium," Dokl. Akad. Nauk SSSR, Vol. 242, No. 5, 1978.

12. Banichuk, N.V. and Karihaloo, B.L., "Minimum-Weight Design of Multi-Purpose Cylindrical Bars," Int. J. Solids Structures, Vol. 12, No. 4, 1976.
13. Hutchinson, J.W. and Niordson, F.I., "Designing Vibrating Membranes," Mechanics of a Continuous Medium and Kindred Problems of Analysis. "Nauka", Moscow, 1972.
14. Lord Rayleigh, Theory of Sound. Dover Publications, New York, 1945.
15. Banichuk, N.V., Optimization of Shape of Elastic Bodies, "Nauka", Moscow, 1980.
16. Wheeler, L., "On the Role of Constant-Stress Surfaces in the Problem of Minimizing Elastic Stress Concentration," Int. J. Solid Structures, Vol. 12, No. 11, 1976.
17. Banichuk, N.V., "Optimizating Hole Shape in Plates Working in Bending," Izv. AN SSSR, MTT Mechanic of Solids, No. 1, 1977.
18. Banichuk, N.V., "Optimality Conditions in the Problem of Seeking the Hole Shapes in Elastic Bodies," PMM, Vol. 41, No. 5, 1977.
19. Vigdergauz, S.B., "Integral Equation for the Inverse Problem of the Plane Theory of Elasticity," PMM, Vol. 40, No. 3, 1976.
20. Cherepanov, G.P., "Inverse Problems of a Plane Theory of Elasticity," PMM, Vol. 38, No. 6, 1974.
21. Vigdergauz, S.B., "On a Case of the Inverse Problem of Two-Dimensional Theory of Plasticity," PMM, Vol. 41, No. 5, 1977.
22. Tvergaard, V., "On the Optimum Shape of a Fillet in a Flat Bar with Restrictions," Proc. IUTAM Symp. Optimization in Structural Design, Springer-Verlag, Berlin, 1975.
23. Bladimirov, V.S., Equations of Mathematical Physics. "Nauka," Moscow, 1976.
24. Muskhelishvili, N.I., Some Fundamental Problems of the Mathematical Theory of Elasticity, "Nauka", Moscow, 1966.
25. Savin, G.N., Stress Concentration Around Holes (in Russian), Gostekhizdat, Moscow - Leningrad, 1951.
26. Lekhnitskii, S.G., Theory of Elasticity of an Anisotropic Solid (in Russian), Nauka Press, Moscow, 1977.
27. Keldysh, M.V., "Conformal Mapping of Multiconnected Domains on a Canonical Domain," Uspekhi Matem. Nauk, No. 6, 1939.
28. Courant, R., Dirichlet's Principle, Conformal Mapping and Minimal Surfaces, N.Y., Interscience, 1950.
29. Sherman, D.I., "On a Method of Considering Boundary-Value Problems of the Theory of Functions and Two-Dimensional Problems of Elasticity Theory," Mechanics of Continuous Media and Kinder Problems of Analysis, "Nauka", Moscow, 1972.
30. Sherman, D.I., "On a Method of Solving Some Torsion, Bending and Plane Theory of Elasticity Problems for Multiple Connected Domains," Prikl. Mekhan., Vol. 3, No. 4, 1957.

PROBLEMS OF SHAPE OPTIMAL DESIGN

Jean Cea

Département de Mathématiques, Université de Nice,
06034, Nice Cedex, France

ABSTRACT

This paper presents numerous problems of shape optimal design that arise in structural mechanics, acoustics, electric fields, fluid flow, and other areas of engineering and applied science. Several methods of characterizing domains are identified. Examples are used to clearly illustrate the essential elements of shape optimal design problems. A variational formulation of the state equations of the systems is presented throughout, in preparation for development of numerical methods of solution.

1. INTRODUCTION

1.1 Optimal Design Problems

The focus of this paper is on a family of control problems or identification problems in which the variable to be controlled or identified is a geometrical domain, i.e. a subset of $\mathbb{R}^n$. Such problems can be stated as follows: Let Ω be an open set of $\mathbb{R}^n$, with boundary Γ; from Ω (and given data) one constructs a solution y_Ω, the state function, of a state equation $E(\Omega, y_\Omega)$; from Ω and y_Ω one constructs a cost function $J(\Omega)$ (in certain cases J can depend on certain eigenvalues of a given operator in Ω); finally, given a family $\mathcal{a}$ of domains Ω, one wishes to minimize $J(\Omega)$ where $\Omega \in \mathcal{a}$.

General Scheme. Consider first the following mappings, which were described verbally above:

$$\Omega \xrightarrow[\text{state equation}]{} y_\Omega \xrightarrow[\text{cost function}]{} J(\Omega)$$

One may now consider two basic problems, as follows:

Problem 1. Find a domain $\Omega^* \in \mathcal{a}$ that minimizes $J(\Omega)$ over the family $\mathcal{a}$, i.e. for all $\Omega \in \mathcal{a}$,

$$J(\Omega^*) \leq J(\Omega)$$

or if one wishes to maximize $J(\Omega)$, one seeks a domain $\Omega^* \in \mathcal{a}$ such that for all $\Omega \in \mathcal{a}$

$$J(\Omega^*) \geq J(\Omega)$$

Problem 2. Minimize $J(\Omega)$, $\Omega \in \mathcal{a}$, i.e. to find the greatest lower bound of $J(\Omega)$ for $\Omega \in \mathcal{a}$,

$$\operatorname*{Inf}_{\Omega \in \mathcal{a}} J(\Omega)$$

even though there may not be a domain $\Omega^* \in \mathcal{a}$ that yields the infimum.

Particular Scheme. In some problems, Ω is defined by a parameter u that belongs to a given set U. In these cases one has the mappings

$$u \longrightarrow \Omega_u \longrightarrow y_u \longrightarrow J(u)$$

and problems 1 and 2 may be stated in the following form:

Problem 1'. Find $u^* \in U$ such that for all $u \in U$,

$$J(u^*) \leq J(u)$$

Problem 2'. Minimize $J(u)$, i.e. find

$$\operatorname*{Inf}_{u \in U} J(u)$$

Usually the problem 1 is very hard and very few theoretical results are known (see papers of Chenais [1] and Koenig-Zolesio [2]). For the second problem, one would like to find a sequence Ω_n of domains such that

$$\lim_{n\to\infty} J(\Omega_n) = \operatorname{Inf}_{\Omega\in\mathcal{A}} J(\Omega)$$

In many cases, one can construct a sequence Ω_n for which, one can prove only that

$$J(\Omega_{n+1}) < J(\Omega_n)$$

The reasons for this difficulty are explained in the following.

1.2 Limitation of Numerical Methods in Shape Optimal Design

An example is first given in which the functional J is not convex (in fact this is the usual case). For this reason, classical optimization methods are not assured of finding a general extremum, but only to a local extremum. The following example is due to J. Cea and R. Glowinski:

Geometry. The domain Ω is an interval on the real line,

$$\Omega = \Omega_a = \{x \mid x \in \mathbb{R},\ 0 < x < a\}$$

where a is the parameter that defines Ω.

State Equation. The governing state equations are

$$\begin{cases} -\,y_a''(x) = 2,\ x \in (0,a) \\ y_a'(0) = 0 \\ y_a(a) = 0 \end{cases}$$

Cost Function. The cost function to be minimized is

$$J(a) = \int_{\Omega_a} (y_a(x) - 1)^2\, dx = \int_0^a (y_a(x) - 1)^2\, dx$$

One can verify that the solution of the state equations is

$$y_a(x) = a^2 - x^2$$

and that

$$J(a) = \frac{8}{15} a^5 - \frac{4}{3} a^3 + a$$

The graph of J is shown in Fig. 1.1, where

$$a_1 = \sqrt{\frac{3}{4}\left(1 - \frac{1}{\sqrt{3}}\right)}$$

$$a_2 = \sqrt{\frac{3}{4}\left(1 + \frac{1}{\sqrt{3}}\right)}$$

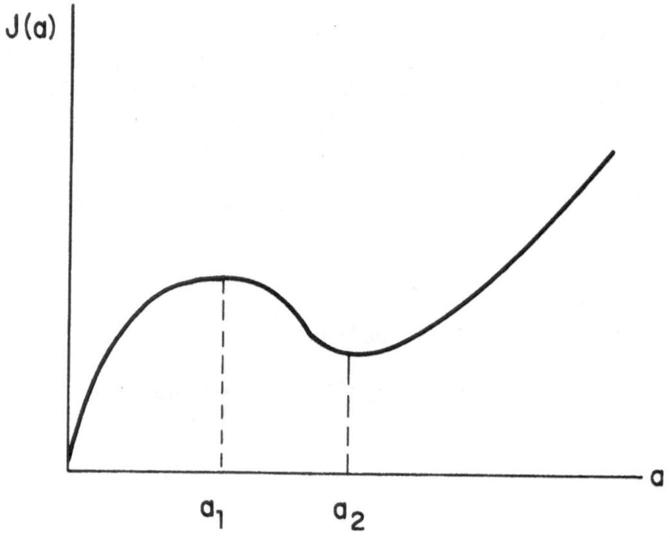

Figure 1.1 Graph of J vs a

If one introduces the constraint

$$\alpha_1 \leq a \leq \alpha_2$$

where $0 \leq \alpha_1 < \alpha_2 < +\infty$, it is obvious that the usual optimization methods are not sure to converge to the optimal solution.

1.3 First Model Problem

Geometry. The domain is $\Omega \in \mathbb{R}^n$, with Γ = boundary of Ω.

State Equation. Let f be a given function in $L^2(\mathbb{R}^n)$. The state equation, in variational form is to find $y_\Omega \in H^1(\Omega)$ so that

$$\int_\Omega (\nabla y_\Omega \cdot \nabla \phi + y_\Omega \phi)\,dx = \int_\Omega f\phi\,dx$$

for all $\phi \in H^1(\Omega)$.

Cost Function. Let y_g be a given function in $L^2(\mathbb{R}^n)$. The cost function is defined as a measure of the difference between the given function y_g and the solution y_Ω of the state equation,

$$J(\Omega) = \frac{1}{2}\int_\Omega |y_\Omega - y_g|^2\,dx$$

Model Problem 1. The first model problem is to minimize $J(\Omega)$, i.e.

$$\operatorname*{Inf}_{\Omega} J(\Omega)$$

Note that the state equations, written in differential equation form, are formally

$$\begin{cases} -\Delta y_\Omega + y_\Omega = f, & \text{in } \Omega \\ \dfrac{\partial y_\Omega}{\partial n} = 0, & \text{on } \Gamma \end{cases}$$

where n is the unit exterior normal to Γ. This is a Neumann problem in a variable domain Ω.

1.4 Second Model Problem

Geometry. The domain Ω is

$$\Omega = \Omega_0 \cup \gamma \cup \Omega_1 \subset \mathbb{R}^n$$

where, as shown in Fig. 1.2, Γ is the boundary of Ω

Γ is fixed

γ is variable

Ω_0 and Ω_1 are variable domains.

It will be convenient to associate the decomposition Ω_0, γ, Ω_1 of Ω with Ω.

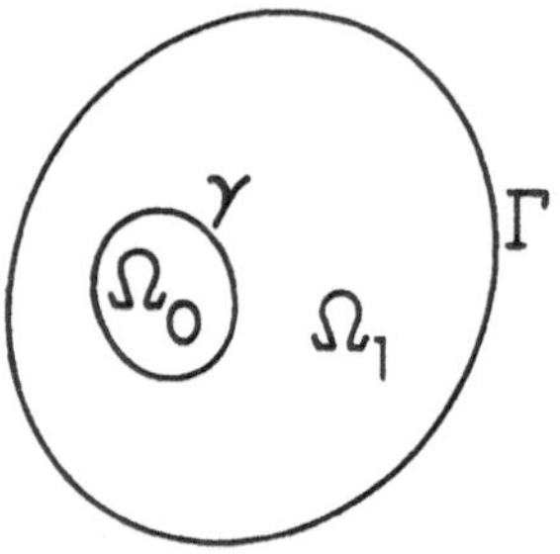

Figure 1.2 Domain for Model Problem 2

State Equation. The following notation is helpful: If $\psi \in H^1(\Omega)$, then $\psi_i = \psi|_{\Omega_i}$ = the restriction of ψ to Ω_i, i=0, 1. The variational form of the state equations is to find $y \in H^1(\Omega)$ (in fact $y = y_\Omega$) so that

$$\int_{\Omega_0} \nabla y_0 \cdot \nabla \phi_0 \, dx + k \int_{\Omega_1} \nabla y_1 \cdot \nabla \phi_1 \, dx + \int_\Omega y\phi dx = \int_\Omega f\phi dx$$

for all $\phi \in H^1(\Omega)$, where f is given in $L^2(\Omega)$ and k is given $k > 1$, for example.

Cost Funtion.. The cost function is again a measure of the difference between a given function $y_g \in L^2(\Omega)$ and the solution y_Ω of the state equation,

$$J(\Omega) = \frac{1}{2} \int_\Omega |y_\Omega - y_g|^2 \, dx$$

Model Problem 2. The second model problem is to minimize $J(\Omega)$, i.e.

$$\inf_{\Omega} J(\Omega)$$

Note that the solution $y = y_\Omega$ of the state equations satisfies formally the following equations:

$$\begin{cases} -\Delta y_0 + y_0 = f_0, & \text{in } \Omega_0 \\ -k\Delta y_1 + y_1 = f_1, & \text{in } \Omega_1 \\ y_0 = y_1, & \text{on } \gamma \\ \dfrac{\partial y_0}{\partial n} = k\,\dfrac{\partial y_1}{\partial n}, & \text{on } \gamma \\ \dfrac{\partial y_1}{\partial n} = 0, & \text{on } \Gamma \end{cases}$$

where n is a convenient unit exterior vector to Ω_0 and to Ω. This is a transmission problem with variable domains Ω_0, Ω_1 but $\Omega_0 \cup \Omega_1$ is fixed.

1.5 The Present Study

In Section 2, many examples of optimal design problems are studied, most related to partial differential equations. Alternate methods of defining a subset of $\mathbb{R}^n$ are then reviewed in Section 3. Finally, in Section 4, methods of taking the derivative of $J(\Omega)$ with respect Ω are studied.

Numerical methods based on these techniques are applied to the two model problems in a companion paper [3]. In order to be clear, these model problems have been chosen to be very simple.

2. EXAMPLES OF PROBLEMS OF SHAPE OPTIMIZATION

2.1 Different Kinds of Optimal Design Problems

A variety of applications, from diverse fields of science and engineering, lead to shape optimization problems. Some basic kinds of problems that arise are the following:

(1) One may search for a best shape among an admissible set of shapes.
(2) An object has a given shape and one wishes to find a best location of this object.
(3) An object has a part Γ_0 of its boundary that can be reached for physical measurement, but the other part Γ_1 is not accessible. One wishes to identify the part Γ_1 using experimental data and a set of state equations. This is called shape identification.

(4) Let $\Gamma \cup \Sigma$ be the boundary of an open set Ω of $\mathbb{R}^n$ with Γ known and Σ to be found, and let A, B, and C be operators. One seeks y and Σ such that:

$$Ay = 0 \ , \text{ in } \Omega$$

$$By = 0 \ , \text{ on } \Gamma$$

$$\left.\begin{aligned} By &= 0 \\ Cy &= 0 \end{aligned}\right\} , \text{ on } \Sigma$$

If $\Gamma \cup \Sigma$ is given, the problem $Ay = 0$ in Ω, $By = 0$ on $\Gamma \cup \Sigma$ is well posed. The condition $Cy = 0$ on Σ is satisfied only for a good choice of Σ. The optimal design problem is then to find Σ such that $Cy = 0$ on Σ, where y is a solution of the state equations $Ay = 0$ on Ω, $By = 0$ on $\Sigma \cup \Gamma$. This is called a free boundary-value problem.

(5) A set Ω is now a subset of $\mathbb{R}^{n\times 1}$, defined as the set of points (x,z), where $x \in \Omega_0$ is a given set in $\mathbb{R}^n$ and $z \in]0, u(x)[$, where u(x) is the thickness of Ω at the point x of Ω_0. One seeks a best thickness, based on an optimality criteria.

2.2 Concrete Examples of Shape Optimal Design

Example 1: Electrical Field and Insulator. In this example shown in Fig. 2.1, Δ is an electrical wire that produces an electrical field, in particular in a fixed domain Ω_2. One wants to find a best shape Ω_0 for the insulating material such that the field in Ω_2 is minimum.

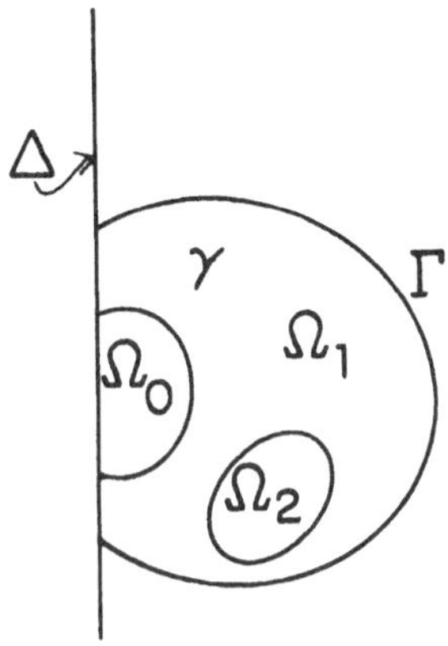

Figure 2.1 Electrical Field and Insulator

Geometry. The boundary Γ can be very far from Ω_0 and Ω_2 and $\Omega = \Omega_0 \cup \gamma \cup \Omega_1$. The set Ω is fixed, but the decomposition of Ω is variable, so Ω_0 and Ω_1 are variables.

State Equation. The boundary-value problem governing the electric field is

$$- \nabla(\varepsilon_i \nabla u_i) = f_i \;, \quad \text{in } \Omega_i,\; i = 0,1$$

$$\left.\begin{aligned} u_0 &= u_1 \\ \varepsilon_0 \frac{\partial u_0}{\partial n} &= \varepsilon_1 \frac{\partial u_1}{\partial n} \end{aligned}\right\}, \text{ on } \gamma$$

and some conditions are given on Δ and Γ. The functions and parameters f_0, f_1, ε_0, and ε_1 are given.

Constraint. The volume of insulating material of Ω_0 is given as

$$\int_{\Omega_0} dx = V_0$$

Cost Functions. Alternate cost functions for this problem are

$$J_1(\Omega) = J_1(\Omega_0,\Omega_1) = \underset{x \in \Omega_2}{\text{Max}} \; |\nabla u_1(x)|$$

and

$$J_2(\Omega) = \int_{\Omega_2} |\nabla u_1(x)|^2 \, dx.$$

Problem. The shape optimal design problem is to find γ to minimize J_1 or J_2.

References. For additional detail, see Ref. 4.

Example 2: Heating and Air-conditioning. In this example, shown in Fig. 2.2, T is the temperature in the room of fixed domain Ω_1. The heat, for example, comes from a domain Ω_0. One wants to find the best shape of Ω_0 (or if the shape is given, the best location of Ω_0) such that the flow of heat received by Ω_1 through γ is maximum or minimum.

Geometry. The geometry of the problem is shown in Fig. 2.2, where $\Omega = \Omega_0 \cup \gamma \cup \Omega_1$, Ω is given as a fixed domain, Ω_0 is variable (and thus Ω_1 also). Thus Ω is decomposed to Ω_0, Ω_1, and γ.

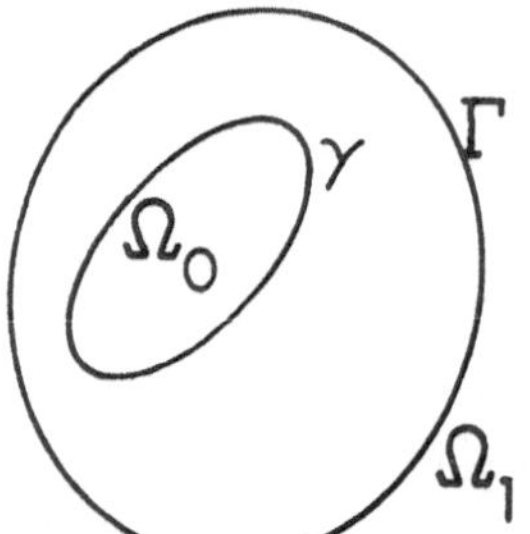

Figure 2.2 Heating and Air-Conditioning

State Equation. The temperature T satisfies the boundary-value problem

$$\begin{cases} \Delta T = 0, & \text{in } \Omega_1 \\ T = \alpha, & \text{on } \gamma \\ T = \beta, & \text{on } \Gamma \end{cases}$$

where α and β are given.

Cost Function. The cost function for this problem is

$$J(\Omega) = \int_\gamma \frac{\partial T}{\partial n}\, d\sigma$$

Problem. The shape optimal design problem is to find γ to maximize or minimize J.

References. For additional detail, see Refs. 4 and 5.

Example 3: Electrical Capacitor. In this problem shown in Fig. 2.3, y is a potential. One wants to find the best shape γ of the boundary of a void, such that the capacity J of Ω is minimum.

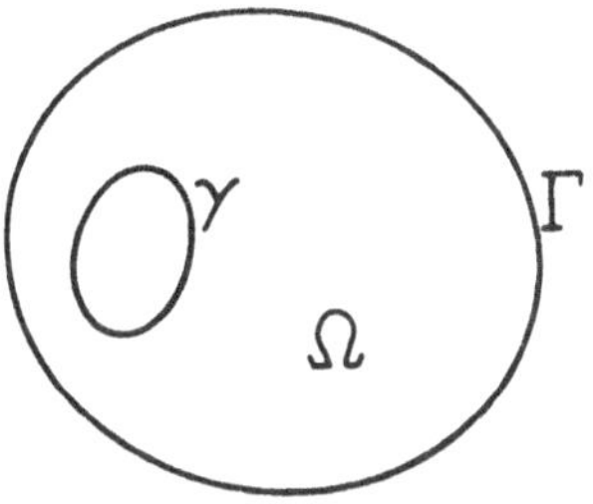

Figure 2.3 Electrical Capacitor

Geometry. The boundary Γ is given and γ is variable.

State Equation. The potential y satisfies the boundary-value problem

$$\begin{cases} \Delta y = 0, & \text{in } \Omega \\ y = \alpha, & \text{on } \gamma \\ y = \beta, & \text{on } \Gamma \end{cases}$$

where α and β are given numbers.

Constraints. The volume of Ω is given

$$\int_{\Omega} dx = V$$

Cost Function. The cost function is the capacity,

$$J(\Omega) = \frac{1}{4\pi} \int_{\Omega} |\nabla y|^2 \, dx$$

Problem. The shape optimal design problem is to find γ to minimize $J(\Omega)$.

References. For more detail, see Refs. 6-9.

Example 4: Elastic Bars in Torsion (I). In this problem, shown in Fig. 2.4, ϕ is a stress function for torsion of a bar that is made of nonhomogeneous material. The objective is to find the shape of a domain Ω with given area, to maximize the torsional stiffness of the bar.

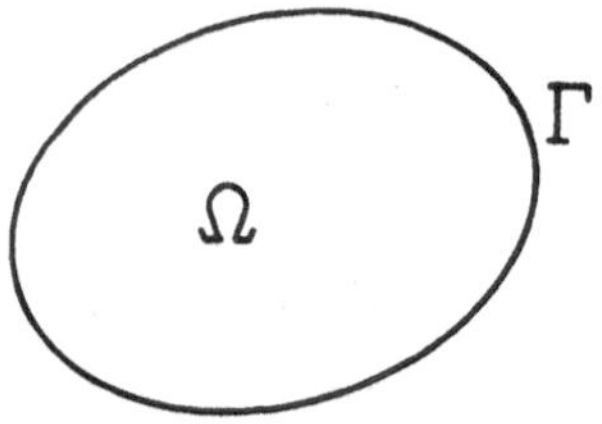

Figure 2.4 Simply Connected Bar Cross Section

Geometry. The cross section Ω of Fig. 2.4 is the design variable.

State Equations. The stress function ϕ satisfies the boundary-value problem

$$\begin{cases} \left(a\phi_x - c\phi_y\right)_x + \left(b\phi_x - c\phi_y\right)_y + m = 0, \text{ in } \Omega \\ \phi = 0, \text{ on } \Gamma \end{cases}$$

where $(x,y) \in \mathbb{R}^2$, a,b, and c, $\in C^1(\mathbb{R}^2)$, $a > 0$, $ab - c^2 > 0$, and $m \in \mathbb{R}$. The material property functions a, b, and c are given. In the case of a homogeneous bar,

$$\begin{cases} -\Delta\phi = 2, \text{ in } \Omega \\ \phi = 0, \text{ on } \Gamma \end{cases}$$

Constraints. The area of the cross-section is given as

$$\int_\Omega dxdy = S$$

where S is a given constant.

Cost Function. The torsional stiffness of the bar is

$$J(\Omega) = \int_\Omega \phi \; dxdy$$

Problem. The shape optimal design problem is to find Ω to maximize J.

References. For more detail, the reader is referred to Refs. 10 and 11.

Example 5: Elastic Bars in Torsion (II). In this problem, shown in Fig. 2.5, ϕ is a stress function for torsion of a multiply connected bar that is made of a homogeneous material. The objective is to find the shape of a hole Ω_0 of given area that lies within a fixed boundary Γ and maximizes torsional stiffness of the bar.

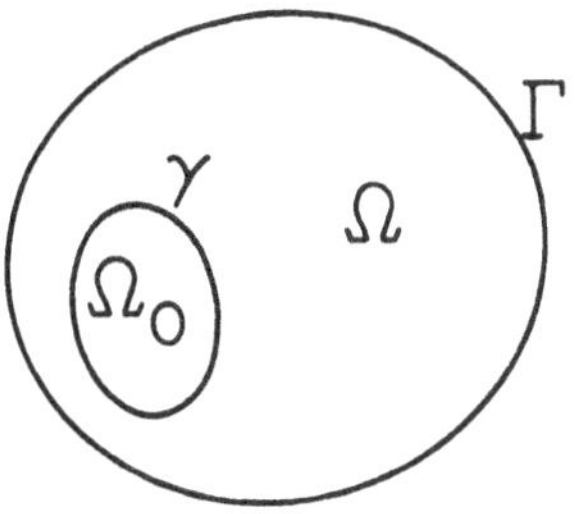

Figure 2.5 Multiply Connected Bar Cross Section

Geomerty. The hole Ω_0 shown in Fig. 2.5 is of variable shape. Its boundary γ must lie within the fixed outer boundary Γ.

State Equations. The stress function ϕ satisfies the boundary-value problem

$$\begin{cases} -\Delta\phi = 2, & \text{in } \Omega \\ \phi = 0, & \text{on } \Gamma \\ \phi = c, & \text{on } \gamma \\ \int_\gamma \frac{\partial\phi}{\partial n}\, dx = -2 \int_{\Omega_0} dxdy \end{cases}$$

where ϕ and c are the unknowns. We have the constraint:

Constraint. The area of the hole Ω_0 is given as

$$\int_{\Omega_0} dxdy = S$$

where S is a given constraint.

Cost Function. The torsional stiffness of this bar is

$$J(\Omega) = \int_\Omega \phi\, dxdy + c \int_{\Omega_0} dxdy$$

Problem. The shape optimization problem is to find γ to maximize J, in order to minimize torsional displacement.

References. For more detail, see Refs. 10 and 11.

Example 6: Elastic Bars in Torsion (III). In this problem, shown in Fig. 2.6, ϕ is a stress function for torsion of a bar made of two materials. The subdomain Ω_0 of one material is fixed. The subdomain Ω_1 of the other material, with given area, is to be found to maximize torsional stiffness of the bar.

Geometry. The two given materials occupy subdomains Ω_0 and Ω_1, with $\Omega = \Omega_0 \cup \gamma \cup \Omega_1$. The outer boundary Γ of Ω_1 is variable.

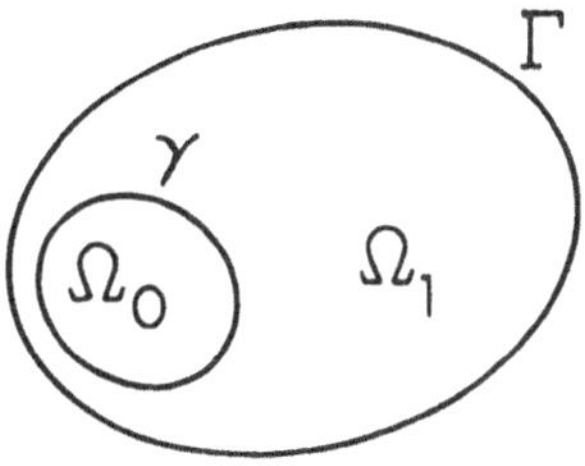

Figure 2.6 Cross Section of Bar Made of Two Materials

State Equations. The stress function ϕ satisfies the boundary-value problem

$$\begin{cases} -\Delta\phi_i = 2\,G_i, \quad i = 0,1, & \text{in } \Omega_i \\ \phi_1 = 0, & \text{on } \Gamma \\ \left.\begin{aligned} \phi_1 &= \phi_0 \\ G_1 \frac{\partial\phi_1}{\partial n} &= G_0 \frac{\partial\phi_0}{\partial n} \end{aligned}\right\}, & \text{on } \gamma \end{cases}$$

where G_0 and G_1 are given moduli of rigidity of the two materials.

Constraints. The area of the subdomain Ω_1 is given as

$$\int_{\Omega_1} dxdy = S_1$$

where S_1 is a given constant.

Cost Function. The torsional stiffness of the bar is

$$J(\Omega) = \sum_{i=0}^{1} \int_{\Omega_i} \phi_i \; dxdy$$

Problem. The shape optimization problem is to find Γ to maximize J.

References. For additional detail, see Refs. 10 and 11.

<u>Example 7: Hole Shapes in Elastic Bodies.</u> In this problem, shown in Fig. 2.7, ϕ is a stress function for in plane deformation of an infinite elastic slab, with a hole Ω_0. The shape of the hole Ω_0 is to be selected to minimize the maximum value of a stress measure, throughout Ω.

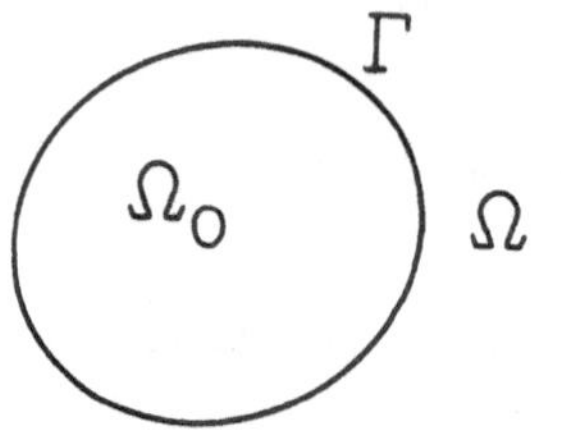

Figure 2.7 Infinite Planar Solid with A Hole

Geometry. The domain Ω is $\mathbb{R}^2$ with the hole Ω_0 removed, i.e. $\Omega = \mathbb{R}^2/\bar{\Omega}_0$. The boundary Γ of Ω_0 is variable.

State Equation. The stress function ϕ satisfies

$$\Delta^2\phi = 0, \quad \text{in } \Omega$$

Furthermore, some boundary conditions are given on Γ and at the infinity.

Cost Function. A failure criteria, for example the von Mises criteria, is defined as the cost function. Let I_1 and I_2 be stress invariants, defined by

$$I_1 = \phi_{y,y} + \phi_{x,y}$$

$$I_2 = (-\phi_{xy})^2 - \phi_{yy}\phi_{xx}$$

and let F be a given function. Failure occurs at a point if $F(I_1,I_2) \geq k$, where k is known. The cost function is thus defined as the maximum value of F over Ω,

$$J(\Omega) = \max_{(x,y)\in\Omega} F(I_1(x,y),\ I_2(x,y))$$

Problem. The shape optimization problem is to find Γ to minimize J, in order to avoid failure, i.e.,

$$\min_{\Gamma} \max_{(x,y)\in\Omega} F(I_1, I_2)$$

Reference. For more detail, see Ref. 12.

Example 8: Optimal Total Drag of A Conical Body in A Hypersonic Flow. In this problem, shown in Fig. 2.8, a hypersonic flow of gas occurs in the x direction. The shape of the conical body is to be selected to minimize total drag.

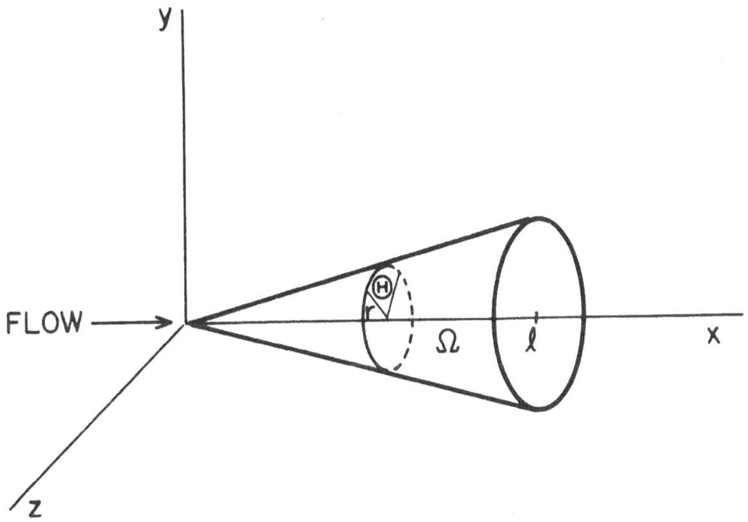

Figure 2.8 Conical Body

Geometry. The body Ω is conical with given length and base area. The unknown (or design variable) is the transversal contour. The equation of the contour in the coordinates x, r, and θ is

$$r = r(x, \theta)$$

or more precisely

$$r = \frac{x}{\ell} r(\theta)$$

where ℓ is the given length. One takes (x,y) as a plane of symmetry. The unknown or design variable is the function $r(\theta)$.

State Equation. The body is in an hypersonic flow, with undisturbed flow direction along the x axis. The state equations are equations of steady state hypersonic flow, which may be found in Ref. 13.

Cost Function. The cost function here is the total drag (the sum of the pressure drag and the skin-friction drag), which can be expressed as

$$J(r) = \int_0^{\ell} \int_0^{\pi} F(r(\theta), r'(\theta), x)\, d\theta dx$$

where F is known [13].

Problem. The shape optimization problem is to choose $r(\theta)$ to minimize $J(r)$.

Reference. For details of this problem, see Ref. 13.

Example 9: An Eigenvalue Problem(1). In this problem, shown in Fig. 2.9, domain modifications are considered to control the eigenvalues of a general second order, divergence form operator.

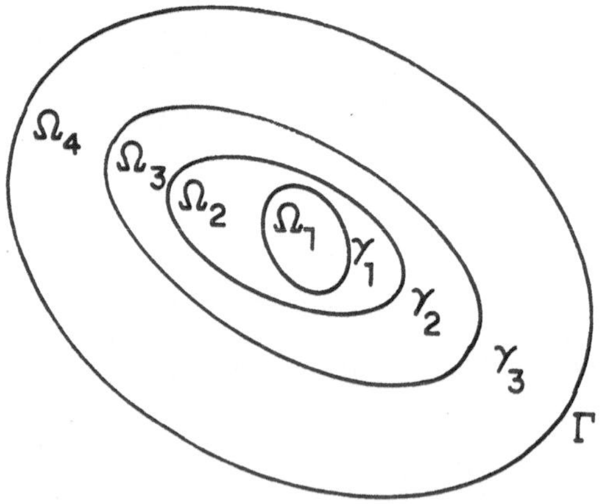

Figure 2.9 Domain for Eigenvalue Problem (1)

Geometry. Two alternatives exist for shape design variables:

(1) The subdomains Ω_1, Ω_2, and Ω_3 may be fixed and Ω_4 is variable, i.e. Γ is unknown.
(2) The subdomains Ω_1 and Ω_2 and boundary Γ may be fixed and γ_3 is variable.

In either case one denotes by Ω the union of the sets Ω_1, γ_1, Ω_2, γ_2, Ω_3, γ_3, and Ω_4.

State Equation. The state y is defined in $\bar{\Omega}$, $y_i = y\big|_{\Omega_i}$, $i = 1,\ldots,4$, as the solution of

$$-\nabla(a_i \nabla y_i) + Ly = \lambda Ky, \quad i = 1, \ldots, 4$$

where transmission conditions are given on γ_1, γ_2, and γ_3 and a boundary condition is given on Γ. The constants are a_i are given and the global operators L and K are defined by matrices, with coefficients that are constants in every Ω_i. Here, λ is an eigenvalue and y is the associated eigenfunction.

Cost Function. Suppose that there exists a unique eigenvalue $\lambda(\Omega)$, which is real and has the minimum module of all eigenvalues. The cost function is

$$J(\Omega) = \frac{1}{2} \, |\lambda(\Omega) - 1|^2$$

Problem. The shape optimal design problem is to choose Ω_4 or γ_3 (geometries (1) and (2), respectively) to minimize $J(\Omega)$.

Reference. This problem is similar, in a simply case, to optimal design of a nuclear reactor, treated in Ref. 14.

Example 10: An Eigenvalue Problem (2). In this problem, shown in Fig. 2.10, the shape of a simply connected domain is sought to control the first N eigenvalues of the Laplace operator, with Dirichlet boundary conditions.

Geometry. The variable here is Ω or Γ.

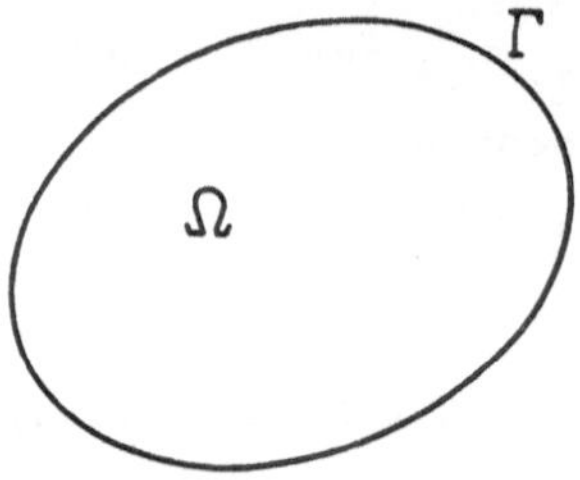

Figure 2.10 Domain for Eigenvalue Problem (2)

State Equation. The boundary-value problem is

$$\begin{cases} -\Delta y = \lambda y, & \text{in } \Omega \\ y = 0, & \text{on } \Gamma \end{cases}$$

where y is an eigenfunction corresponding to the eigenvalue λ.

Cost Function. Let $\mu_1, \ldots, \mu_N$ be N given real numbers

$$0 < \mu_1 < \mu_2 < \ldots < \mu_N$$

Suppose that the N first eigenvalues satisfy

$$0 < \lambda_1 < \lambda_2 < \ldots < \lambda_N$$

The cost function, defined as a measure of the deviation of the λ_i from the corresponding μ_i, is

$$J(\Omega) = \frac{1}{2} \sum_{i=1}^{N} |\lambda_i - \mu_i|^2$$

Problem. The shape optimization problem is to minimize J. This problem leads to a numerical solution of a problem posed by Kac [15].

References. For more detail, see Refs. 15, 16, and 17.

Example 11: A Free Boundary-Value Problem (1). In this problem, shown in Fig. 2.11, a free boundary-value problem of determining the interior boundary γ, where excess data are given, is reduced to a free boundary shape optimal design problem.

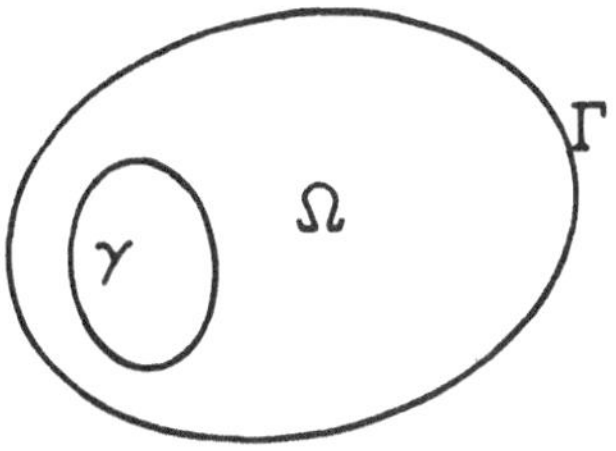

Figure 2.11 Domain of Free Boundary Problem (1)

Geometry. The part Γ of the boundary is given and the part γ is variable. Hence Ω is variable in $\mathbb{R}^n$.

State Equation. The free boundary-value problem that is to be solved for the function y and boundary γ is

$$\begin{cases} \Delta y = 0, & \text{in } \Omega \\ y = 0, & \text{on } \Gamma \\ y = 1, & \text{on } \gamma \\ \left|\frac{\partial y}{\partial n}\right| = Q, & \text{on } \gamma \end{cases}$$

where the given function Q is defined on all of ^{n}. When the geometry is given, the problem with the three first equations is well posed. One can transform this problem into an optimal design problem. Let the state equations for the shape optimization problem be

$$\begin{cases} \Delta y = 0, & \text{in } \Omega \\ y = 0, & \text{on } \Gamma \\ y = 1, & \text{on } \gamma \end{cases}$$

Cost Function. The cost function of J of the optimization problem is

$$J(\Omega) = \frac{1}{2}\int_\gamma (|\nabla y - Q)^2 d\sigma$$

where y is the solution of the state equation of the shape optimization problem.

Problem. The minimization of J leads to the solution of the free boundary-value problem.

Reference. For more detail, see Ref. 18.

Example 12: A Free Boundary-Value Problem (2). In this problem, shown in Fig. 2.12, a free boundary-value problem of determining an exterior boundary, where excess data are given, is reduced to a free boundary shape optimal design problem.

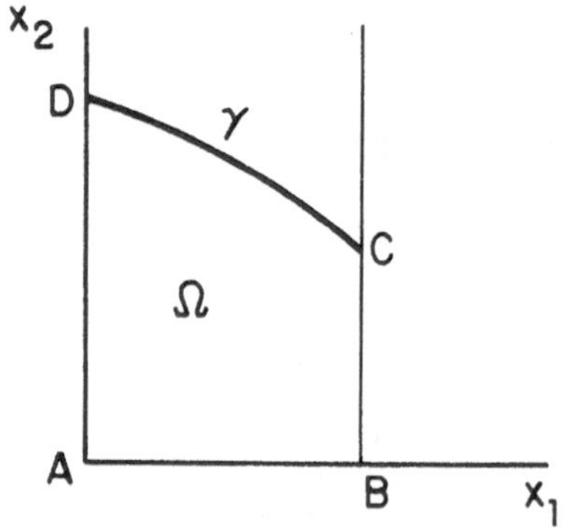

Figure 2.12 Domain for Free Boundary-Value Problem (2)

Geometry. The domain Ω is bounded by 3 lines ($x_1 = 0$, $x_1 = 1$, $x_2 = 0$) and by a curve γ whose equation $x_2 = z(x_1)$, where z is an unknown function. Thus, $\Omega = \Omega_z$

State Equations. The free boundary-value problem to be solved is

$$\begin{cases} \Delta y = 0 & \text{in } \Omega \\ y = 0 & \text{on } BC, \gamma, DA \\ y = f & \text{on } AB \\ \dfrac{\partial y}{\partial n} = g & \text{on } \gamma \end{cases}$$

If the geometry is fixed, the problem with the first 3 equations is well posed, as is the problem with all but the second equation enforced on γ. In the free boundary-value problem, one wishes to determine both γ and y. This problem may be converted to a free boundary optimal design problem. For example, the state equation can be defined as

$$\begin{cases} \Delta y = 0 & \text{in } \Omega \\ y = 0 & \text{on AD, BC} \\ y = f & \text{on AB} \\ \dfrac{\partial y}{\partial n} = g & \text{on } \gamma \end{cases}$$

Cost Function. The cost function for the shape optimization problem is taken as

$$J(\Omega) = \frac{1}{2} \int_{\gamma} |y|^2 \, d\sigma$$

Problem. If $J(\Omega) = 0$ then $y|_{\gamma} = 0$ and all boundary conditions of the free boundary-value problem are satisfied. The shape optimization problem is then to minimize J.

Reference. For more detail, see the paper of Baiocchi on dam spillways [19].

<u>Example 13: Acoustic Horn</u>. In this problem, shown in Fig. 2.13, the objective is to choose the shape of an acoustic horn to maximize the quality of the emitted waves.

Geometry. Two kinds of domains shown in Fig. 2.13 are considered. A part Γ of the boundary is fixed and the part γ is variable.

State Equation. The state variable y satisfies the Helmholtz equation

$$\Delta y + k^2 y = 0, \quad \text{in } \Omega$$

some boundary condition on $\Gamma \cup \gamma$, and some radiation condition at the infinity.

Cost Function. With $y = y(x,z)$ one defines the cost function as

$$J(\Omega) = \frac{1}{2} \int_{-b}^{+b} \left| \frac{\partial y}{\partial x} (x,a) \right|^2 dx$$

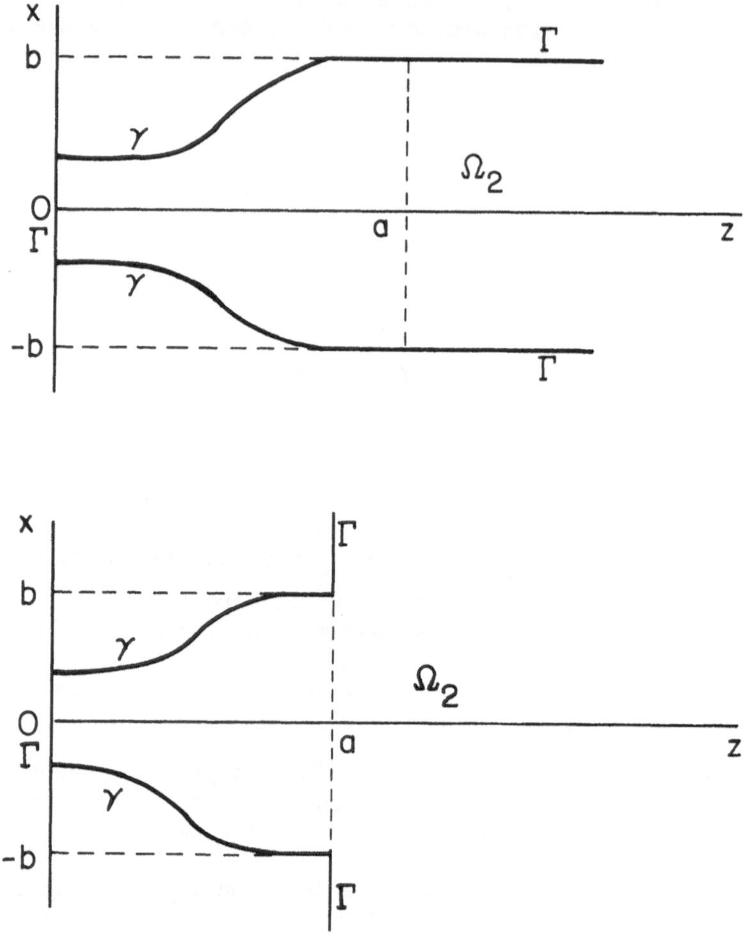

Figure 2.13 Acoustic Horn

Problem. The shape optimization problem is to find γ to minimize J. In fact, what one wants is $y(x,a) \approx$ constant for all $x \in [-b,+b]$, because in this case the quality of the radiation is better.

Reference. This problem is due to a personal communication of M. Bolomey.

<u>Example 14: Vesicle Shape</u>. In this problem, shown in Fig. 2.14, one is interested in the shapes of closed fluid membranes, such as those formed by lecithin in water. Attention

is limited here to the case of rotationally symmetric shapes. The enclosed volume and the area of the surface Γ are given.

Geometry. The boundary Γ of a simply connected domain Ω, shown in Fig. 2.14, is to be found.

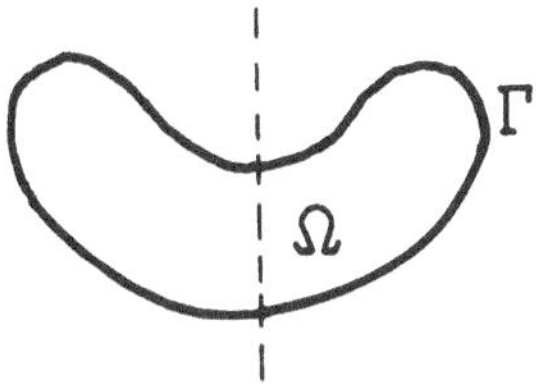

Figure 2.14 Vesicle Shape

Constraints. The volume of Ω and surface area of Γ are given as

$$\int_{\Omega} dx = V$$

$$\int_{\Gamma} ds = S$$

Cost Function. The total elastic curvature is the cost function, which is given by

$$J(\Omega) = \int_{\Gamma} g_c \, d\sigma$$

with $g_c = \alpha(c_1 + c_2 - c_0)^2 + \beta\, c_1\, c_2$

where α and β are elastic constants, c_1 and c_2 are the two principle curvatures, and c_0 is the spontaneous curvature, which can be supposed constant on Γ.

Problem. The shape optimization problem is to find the equilibrium shape of a vesicle, which reduces to minimization of J.

Reference. For details, see Ref. 20.

Example 15: Optimum Thickness. In this problem, shown in Fig. 2.15, the thickness of an object that has a fixed shape is to be found.

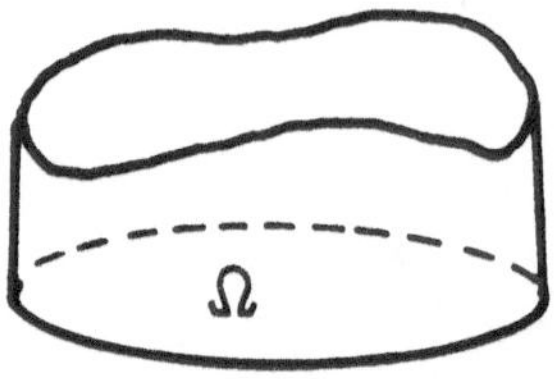

Figure 2.15 Optimum Thickness

Geometry. The domain $\Omega \in \mathbb{R}^n$ is fixed, but the function $u : \Omega \rightarrow \mathbb{R}$, to be determined defines a subdomain of $\Omega \times \mathbb{R}$, $\{(x,t) : x \in \Omega,\ t \in \mathbb{R},\ 0 < t < u(x)\}$, as shown in Fig. 2.15.

State Equation. Let $V = H_0^1(\Omega)$ and f be a given function in $L^2(\Omega)$. The variational formulation of the state equations is to find $y \in V$ so that

$$\int_\Omega u\,(\nabla y \cdot \nabla\phi)\,dx = \int_\Omega f \cdot \phi\,dx$$

for all $\phi \in V$. Formally, this is equivalent to the boundary-value problem

$$\begin{cases} -\displaystyle\sum_{i=1}^{n} \frac{\partial}{\partial x_i}\left(u\,\frac{\partial y}{\partial x_i}\right) = f, & \text{in } \Omega \\ y = 0, & \text{on } \Gamma \end{cases}$$

Constraints. It is required that

$$\alpha \le u(x) \le \beta$$

for all $x \in \Omega$ and

$$\int_\Omega u(x)\,dx = \gamma$$

where α, β, and γ are positive given numbers.

Cost Function.

$$J(u) = \int_\Omega u \, |\nabla y|^2 \, dx = \int_\Omega f \cdot y \, dx$$

Problem. The shape optimization problem is to find the function u(x) on the fixed domain Ω to minimize J.

Reference. For details, see Ref. 21.

Example 16: A Nozzle. In this problem, shown in Fig. 2.16, the shape of a nozzle is to be found to optimize a steady state measure of performance.

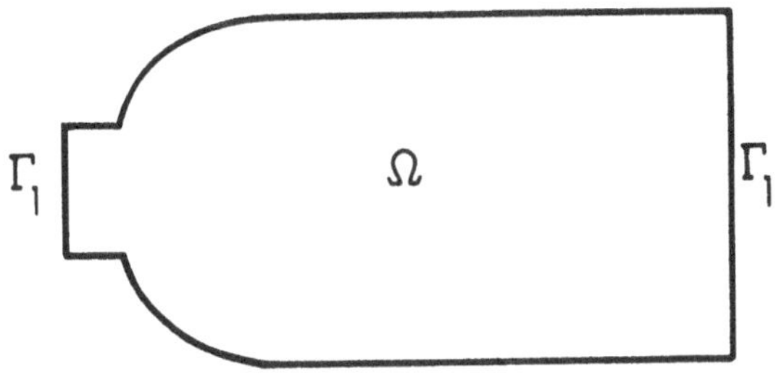

Figure 2.16 Nozzle

Geometry. The boundaries Γ of the nozzle shown in Fig. 2.16 are to be determined, to define the shape of the nozzle.

State Equations. The boundary value problem determining the state variable is

$$\begin{cases} \Delta y = 0, & \text{in } \Omega \\ y = y_0, & \text{on } \Gamma_1 \\ \dfrac{\partial y}{\partial n} = 0, & \text{on } \Gamma \end{cases}$$

Cost Function. The cost function is

$$J(\Omega) = \int_\Gamma (|\nabla y|^2 - f)^m \, d\sigma$$

where y_0 and f are two given functions and m is an even number.

Problem. The shape optimization problem is to minimize $J(\Omega)$, when Ω belongs to an admissible set of domains.

Reference. For details, see Ref. 22.

3. DEFINITION OF DOMAINS

3.1 Definition by a Graph

Suppose the unknown boundary of a domain Ω is defined by a graph. To be specific, consider the following two simple examples in $\mathbb{R}^2$:

Example 3.1. The part γ of the boundary shown in Fig. 3.1 is the unknown (or variable). If the equation of γ is $x_2 = z(x_1)$, the true unknown is the function z and one has $\Omega = \Omega_z$ and $J(\Omega) = J(z)$.

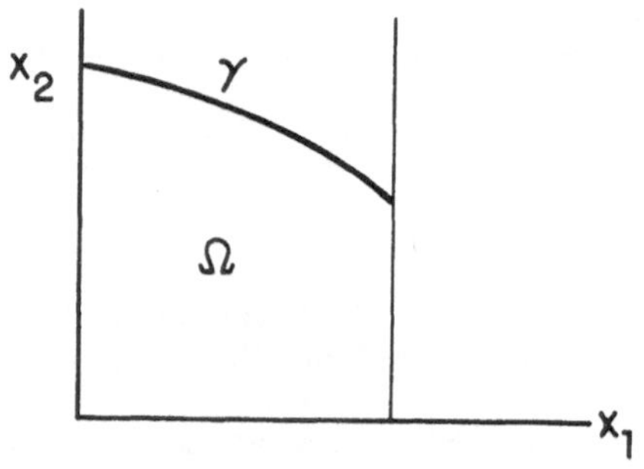

Figure 3.1 Variable Boundary in Euclidean Coordinates

Example 3.2. In polar coordinates, as shown in Fig. 3.2, the boundary Γ is defined by the function g, with $r = g(\theta)$, $0 \le \theta \le 2\pi$. Then, $\Omega = \Omega_g$ and $J(\Omega) = J(g)$.

3.2 Definition as the Image of a Fixed Domain

Let Ω be a fixed domain in $\mathbb{R}^n$ and F a transformation from $\mathbb{R}^n$ to $\mathbb{R}^n$, which depends on some parameter or function $u \in U$, denoted by

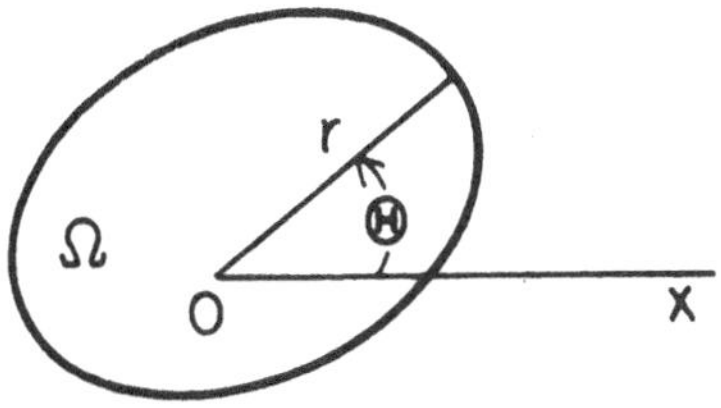

Figure 3.2   Variable Boundary in Polar Coordinates

$$F : \mathbb{R}^n \longrightarrow \mathbb{R}^n$$

$$X \rightarrow F(u,X)$$

$$\Omega \rightarrow \Omega_u = F(u,\Omega)$$

The set U can be a subspace of a finite or infinite dimensional space. The variable defining the set Ω is now u. Consider now the following concrete examples:

Example 3.3. Choose F as the following mapping from the (X_1,X_2) plane to the (x_1,x_2) plane:

$$x_1 = F_1(a_1, a_2, c_1, c_2; X_1, X_2) \equiv c_1 + a_1X_1$$

$$x_2 = F_2(a_1, a_2, c_1, c_2; X_1, X_2) \equiv c_2 + a_2X_2$$

Here, $u = (a_1,a_2,c_1,c_2) \in \mathbb{R}^4$ and Ω is the disk $(X_1)^2 + (X_2)^2 < 1$ of $\mathbb{R}^2$. Its image in the (x_1,x_2) plane is the ellipse shown in Fig. 3.3. In this case, domains Ω_u are restricted to be in a family of ellipses.

Example 3.4. The Example 3.1 belongs also to this family by defining

$$x_1 = F(z; X_1, X_2) \equiv X_1$$

$$x_2 = F(z; X_1, X_2) \equiv X_2\ z(X_1)$$

where z defines the graph γ in the (x_1,x_2) plane, as shown in Fig. 3.4. In this case the parameter z belongs to an infinite dimensional function space.

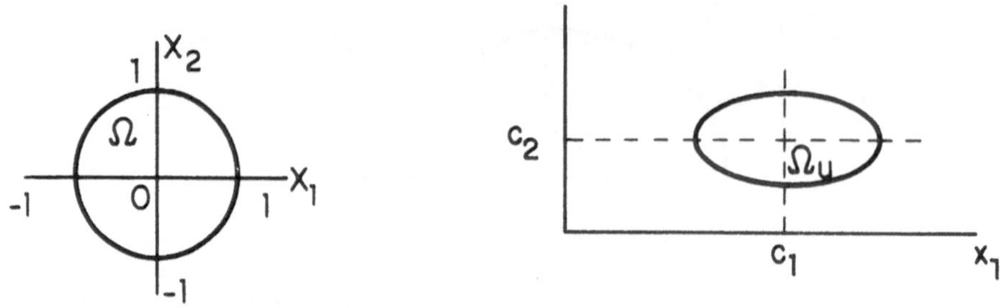

Figure 3.3 Mapping of Example 3.3

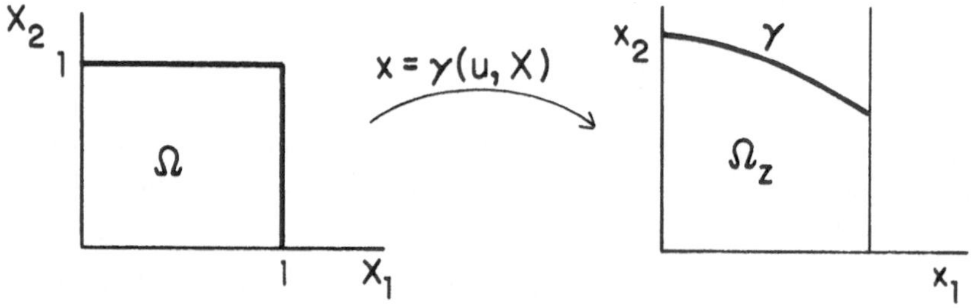

Figure 3.4 Alternate Definition of the Domain of Example 3.1

3.3 Domain Defined by a Velocity Field and an Initial Position

Suppose only one parameter t defines the transformation F, as shown in Fig. 3.5. This parameter can be thought of as the time, so the mapping is given by

$$F : \mathbb{R}^n \longrightarrow \mathbb{R}^n$$

$$X \longrightarrow x = F(X,t)$$

$$\Omega \longrightarrow \Omega_t = F(\Omega,t)$$

It is important to study carefully this case because when the classical optimization techniques are used, one needs some directional derivative: these derivatives can be thought as derivatives of a function which depends only on one parameter.

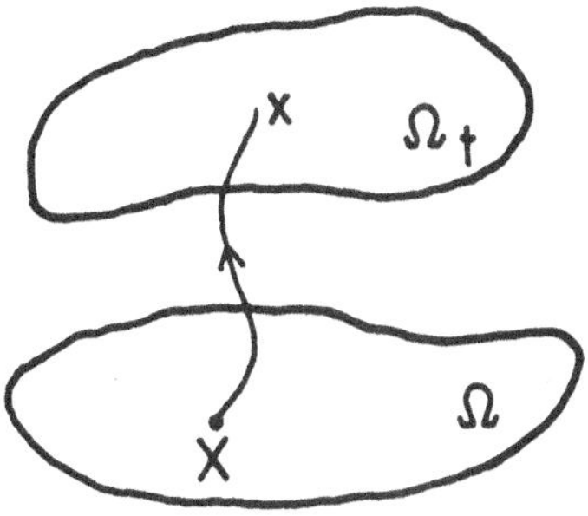

Figure 3.5 One Parameter Family of Mappings

It is convenient to think about Ω_t as a continuous medium, or a continuum of particles. At the initial time t = 0, it was at Ω. One can follow the trajectories of points X that at the initial time t = 0 were in Ω. The initial position X is now at $x = F(X,t)$. Using a different notation,

$$x = x(t) = x(X,t) = F(X,t),\ X \in \Omega,\ x \in \Omega_t \tag{3.1}$$

Thinking of t as time, one can define particle velocity as

$$\frac{dx}{dt} = \frac{dx(t)}{dt} = \frac{dF(X,t)}{dt} = \frac{\partial F(X,t)}{\partial t}$$

since X does not depend on t. One can also express this velocity in terms of the position of the particle at time t, if one supposes that F^{-1} exists. Thus,

$$X = F^{-1}(x,t) = X(x,t)$$

is equivalent to the mapping of Eq. 3.1 and the velocity of the particle at $x = F(x,t)$ is

$$v(x,t) = \frac{dx}{dt} = \frac{\partial F}{\partial t}\left(F^{-1}(x,t),t\right)$$

The trajectory of the particle that was at X at the initial time is now defined by

$$\left.\begin{aligned} \dot{x}(t) &= v(x,t) \\ x(0) &= X \end{aligned}\right\} \tag{3.2}$$

where $\dot{x} = \frac{dx}{dt}$. Thus, if F is given, one can construct the velocity v. Conversely, if the velocity field is given, one can define F by

$$F(X,t) = x(X,t)$$

where x is the solution of Eq. 3.2. The domain Ω_t can now be defined as

$$\Omega_t = F(\Omega,t) = \{x \in \mathbb{R}^n : \text{there exists } X \in \Omega \text{ such that } x = x(t),$$

$$\text{with } \dot{x}(s) = v(x(s),s),\ 0 < s \leq t,\ x(0) = X\} \tag{3.3}$$

In a neighborhood of t = 0, under certain regularity hypothesis, one has

$$F(X,t) = F(X,0) + t\,\frac{\partial F}{\partial t}(X,0) + \ldots$$

$$= X + t\,v(X,0) + \ldots$$

Let $\theta(X) = v(X,0)$, so that

$$F(X,t) = [I + t\theta](X) + \ldots$$

Many authors [6,7,8,16,17,23,24,25] have studied the case in which

$$F(X,t) = X + t\,\theta(X) = [I + t\theta](X)$$

The transformation F is then a perturbation of the identity in the direction θ.

4. DERIVATIVE OF THE COST FUNCTION

4.1 The Meaning of the Derivative

Suppose that $J : V \to \mathbb{R}$, where V is a real Hilbert space, for example $\mathbb{R}^n$. The Gateaux (directional) derivative of J at the point u in the direction v is defined by

$$DJ(u,v) = \lim_{t\to 0} \frac{J(u + tv) - J(u)}{t}$$

One can also write this in a different way. Suppose $u(t)$ is defined by

$$\dot{u}(t) = v$$

$$u(0) = u$$

that is, $u(t) = u + tv$. If one puts

$$j(t) = J(u(t))$$

one has

$$\left.\begin{aligned} \frac{d}{dt} j(t) = \dot{j}(t) &= \lim_{t\to 0} \frac{J(u(t)) - J(u(0))}{t} \\ &= \lim_{t\to 0} \frac{J(u + tv) - J(u)}{t} = DJ(u,v) \end{aligned}\right\} \tag{4.1}$$

Thus, the Gateaux derivative involves the position u and the velocity v at $t = 0$.

The same thing will now be done when u is not an element of a Hilbert space, but is a domain. To give a sense to the derivative of J at Ω_t with the velocity v, suppose v is given and Ω_t is defined by

$$\Omega_t = \{x \in \mathbb{R}^n : \text{there exists } X \in \Omega_0 \text{ such that } x = x(t), \text{ with}$$

$$\dot{x}(s) = v(x(s),s),\ 0 < s \leq t,\ x(0) = X\}$$

The cost function in the shape optimization problem is defined by the mapping

$$\Omega_t \rightarrow y_t \rightarrow j(t) = J(\Omega_t)$$

where y_t is the solution of state equations on Ω_t and one wants to find the derivative $\dot{j}(t)$ of $j(t)$.

In the examples of Section 2, many state equations, cost functions, and constraints were of the following kinds:

$$\int_{\Omega_t} A(y_t(x),\ \phi(x))dx + \int_{\Gamma_t} B(y_t(x),\ \phi(x))d\sigma = 0$$

$$\int_{\Omega_t} C(y_t(x))dx = m$$

$$\int_{\Gamma_t} D(y_t(x))d\sigma = m_1$$

$$j(t) = \int_{\Omega_t} E(y_t(x))dx + \int_{\Gamma_t} F(y_t(x))d\sigma$$

All these expressions are of one of the following types:

$$\int_{\Omega_t} P(x,t)dx$$

$$\int_{\Gamma_t} Q(x,t)d\sigma$$

through some functions y, ϕ, ... The functions P and Q are defined on B_t or its boundary, where

$$B_t = \{(x,s) \in \mathbb{R}^{n+1} : s \in [0,t] \text{ and } x \in \Omega_s\}$$

In continuum mechanics, it is usual to take the derivative of these expressions with respect to t. For example

$$\frac{d}{dt}\int_{\Omega_t} P(x,t)dx = \int_{\Omega_t} \frac{\partial}{\partial t} P(x,t)dx + \int_{\Omega_t} \operatorname{div}(P(x,t)v(x,t))dx \tag{4.2}$$

or

$$\frac{d}{dt}\int_{\Omega_t} P(x,t)dx = \int_{\Omega_t} \frac{\partial}{\partial t} P(x,t)dx + \int_{\Gamma_t} P(x,t) < v(x,t),\ n(x,t) > d\sigma \tag{4.3}$$

where div is the divergence operator, n is the unit exterior normal to Γ_t, v is the speed, and $<,>$ is the scalar product in $\mathbb{R}^n$.

In some cases, it is difficult to give a sense to $\frac{\partial}{\partial t} P(x,t)$. Many expressions of derivative of this kind do exist, so attention is restricted here to these cases, because it will be sufficient for the model problems studied here. In the work of J.P. Zolesio there are many examples of such derivatives.

Equations 4.2 or 4.3 will now be applied to the model problems of Section 1. Note that Joseph [26] has used this technique in order to get the derivatives of eigenvalues.

4.2 Analysis of Model Problem 1

The state equation in variational form is to find $y \in H^1(\Omega_t)$ so that

$$\int_{\Omega_t} (\nabla y_t \cdot \nabla \phi + y_t \, \phi) dx = \int_{\Omega_t} f \phi \, dx \tag{4.4}$$

for all $\phi \in H^1(\Omega_t)$, where Ω_t is as shown in Fig. 4.1.

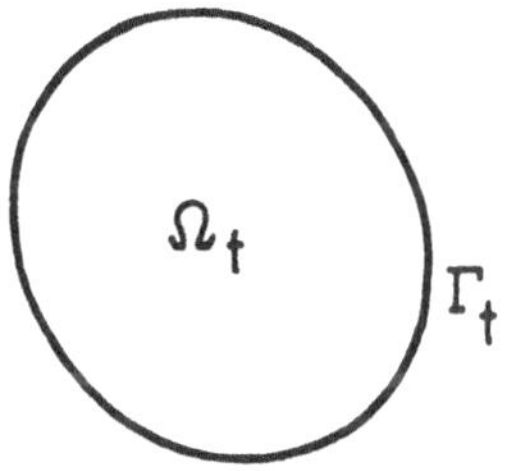

Figure 4.1 Domain Ω_t for Model Problem 1

In order to eliminate certain problems related to the definition of ϕ in Ω_t, which is variable with t, suppose that ϕ is the restriction to Ω_t of a function $\phi \in H^1(\mathbb{R}^n)$.

The cost function, as a function of t, is

$$j(t) = J(\Omega_t) = \frac{1}{2}\int_{\Omega_t} |y_t - y_{g,t}|^2 \, dx$$

where $y_g \in H^1(\mathbb{R}^n)$ and $y_{g,t} = y_g\big|_{\Omega_t}$.

Denote by y' the partial derivative of y with respect to t, i.e. $y'(x,t) = \frac{\partial}{\partial t} y(x,t)$. This can have a sense for $x \in \Omega_t$, because y is defined in B_T, $T \geq t$. One can in fact prove that $y'(\cdot,t)$ exists and is in $H^1(\Omega_t)$. Taking the derivative of the two sides of Eq. 4.4, using Eq. 4.3, for example, one has

$$\int_{\Omega_t} (\nabla y_t' \cdot \nabla\phi + y_t'\phi) \, dx + \int_{\Gamma_t} (\nabla y_t \cdot \nabla\phi + y_t\phi) < v, n > d\sigma$$

$$= \int_{\Gamma_t} f\phi < v,n > d\sigma \tag{4.5}$$

Also,

$$\dot{j}(t) = \int_{\Omega_t} (y_t - y_{g,t})y_t' \, dx + \int_{\Gamma_t} \frac{1}{2}(y_t - y_{g,t})^2 < v,n > d\sigma \tag{4.6}$$

Introduce now the adjoint state $p_t \in H^1(\Omega_t)$, defined as the solution of the variational equation

$$\int_{\Omega_t} (\nabla\psi \cdot \nabla p_t + \psi p_t) dx = \int_{\Omega_t} (y_t - y_{g,t})\psi \, dx \tag{4.7}$$

for all $\psi \in H^1(\Omega_t)$. Choosing $\psi = y_t'$ in Eq. 4.7 and $\phi = p_t$ in Eq. 4.5, one gets

$$\int_{\Omega_t} (y_t - y_{g,t})y_t' \, dx = \int_{\Omega_t} (\nabla y_t' \cdot \nabla p_t + y_t' \, p_t) \, dx$$

$$= \int_{\Gamma_t} [-(\nabla y_t \cdot \nabla p_t + y_t p_t) + f p_t] < v,n > d\sigma$$

and finally, from Eq. 4.6, one has

$$\dot{j}(t) = \int_{\Gamma_t} \left\{- (\nabla y_t \cdot \nabla p_t + y_t p_t) + f p_t + \frac{1}{2} |y_t - y_{g,t}|^2 \right\} < v,n > d\sigma \tag{4.8}$$

This is the derivative of J at Ω_t with the velocity v, which is linear, and continuum in some sense, with respect to v.

4.3 Analysis of Model Problem 2

In this problem $\Omega = \Omega_t = \Omega_{0,t} \cup \gamma_t \cup \Omega_{1,t}$, as shown in Fig. 4.2, where Ω_t is decomposed into three parts; $\Omega_{0,t}$, γ_t, and $\Omega_{1,t}$. Here, γ_t is variable, hence $\Omega_{0,t}$ and $\Omega_{1,t}$ are also variables. The total set Ω, however does not vary with time.

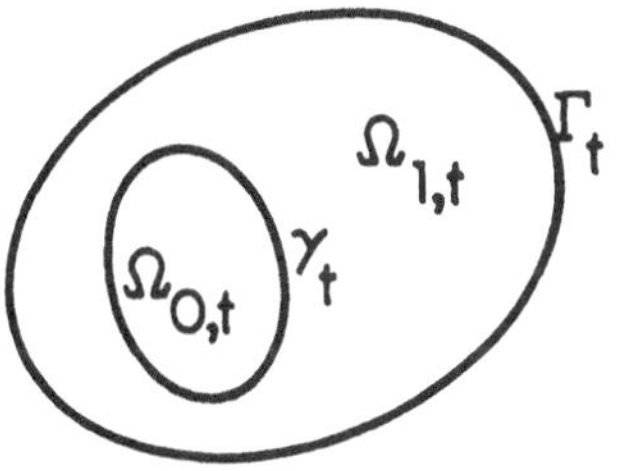

Figure 4.2 Domain Ω for Model Problem 2

The state equation, in variational form, is to find $y_t \in H^1(\Omega)$, i.e. $y_{i,t} = y_t|_{\Omega_{i,t}}$, i = 0,1, so that

$$\int_{\Omega_{0,t}} \nabla y_{0,t} \cdot \nabla \phi_0 \, dx + k \int_{\Omega_{1,t}} \nabla y_{1,t} \cdot \nabla \phi_1 \, dx + \int_\Omega y_t \, \phi \, dx = \int_\Omega f \, \phi \, dx \tag{4.9}$$

for all $\phi \in H^1(\Omega)$.

The cost function is

$$j(t) = J(\Omega_t) = \frac{1}{2} \int_\Omega |y_t - y_g|^2 \, dx \tag{4.10}$$

Denote by y_t' the derivative $\frac{\partial}{\partial t} y_t$. One can prove that this derivative exists and is in $H^1(\Omega)$. Taking the derivatives of the two sides of Eqs. 4.9 and 4.10 one has

$$\int_{\Omega_{0,t}} \nabla y_{0,t}' \cdot \nabla \phi_0 \, dx + k \int_{\Omega_{1,t}} \nabla y_{1,t}' \cdot \nabla \phi_1 \, dx + \int_\Omega y_t' \, \phi \, dx$$

$$+ \int_{\gamma_t} \nabla y_{0,t} \cdot \nabla \phi_0 < v,n > d\sigma + \int_{\gamma_t} k \, \nabla y_{1,t} \cdot \nabla \phi_1 < v,-n > d\sigma = 0 \quad (4.11)$$

where n is the unit normal to γ_t, exterior to $\Omega_{0,t}$.

Since $\Omega = \Omega_t$ does not vary with t,

$$\dot{j}(t) = \int_\Omega (y_t - y_g) \, y_t' \, dx \quad (4.12)$$

One now defines the adjoint state $p_t \in H^1(\Omega)$ as the solution of the variational equation

$$\int_{\Omega_{0,t}} \nabla \psi_0 \cdot \nabla p_{0,t} \, dx + k \int_{\Omega_{1,t}} \nabla \psi_1 \cdot \nabla p_{1,t} \, dx + \int_\Omega \psi \, p_t \, dx$$

$$= \int_\Omega (y_t - y_g) \psi \, dx \quad (4.13)$$

for all $\psi \in H^1(\Omega)$. Using Eq. 4.13 with $\psi = y_t'$ and Eq. 4.11 with $\phi = p_t$, one has in Eq. 4.12

$$\dot{j}(t) = \int_\Omega (y_t - y_g) y_t' \, dx = \int_{\Omega_{0,t}} \nabla y_{0,t}' \cdot \nabla p_{0,t} \, dx$$

$$+ k \int_{\Omega_{1,t}} \nabla y_{1,t}' \cdot \nabla p_{1,t} \, dx + \int_\Omega y_t' \, p_t \, dx$$

$$= - \int_{\gamma_t} \nabla y_{0,t} \cdot \nabla p_{0,t} < v,n > d\sigma + \int_{\gamma_t} k \nabla y_{1,t} \cdot \nabla p_{1,t} < v,n > d\sigma$$

or

$$\dot{j}(t) = \int_{\gamma_t} \{-\nabla y_{0,t} \cdot \nabla p_{0,t} + k\ \nabla y_{1,t} \cdot \nabla p_{1,t}\} < v,n > d\sigma \qquad (4.14)$$

Under some regularity hypothesis, y_t and p_t satisfy the boundary-value problems

$$\left.\begin{aligned}
&-\Delta y_{0,t} + y_{0,t} = f_{0,t}, && \text{in } \Omega_{0,t} \\
&-k\Delta y_{1,t} + y_{1,t} = f_{1,t}, && \text{in } \Omega_{1,t} \\
&\left.\begin{aligned} & y_{0,t} = y_{1,t} \\ & \frac{\partial y_{0,t}}{\partial n} = k\,\frac{\partial y_{1,t}}{\partial n} \end{aligned}\right\}, && \text{on } \gamma_t \\
&\frac{\partial y_{1,t}}{\partial n} = 0, && \text{on } \Gamma
\end{aligned}\right\} \qquad (4.15)$$

and

$$\left.\begin{aligned}
&-\Delta p_{0,t} + p_{0,t} = y_{0,t} - y_{g,0,t}, && \text{in } \Omega_{0,t} \\
&-k\Delta p_{1,t} + p_{1,t} = y_{1,t} - y_{g,1,t}, && \text{in } \Omega_{1,t} \\
&\left.\begin{aligned} & p_{0,t} = p_{1,t} \\ & \frac{\partial p_{0,t}}{\partial n} = k\,\frac{\partial p_{1,t}}{\partial n} \end{aligned}\right\}, && \text{on } \gamma_t \\
&\frac{\partial p_{1,t}}{\partial n} = 0, && \text{on } \Gamma
\end{aligned}\right\} \qquad (4.16)$$

From these conditions, one sees that

$$\nabla y_{0,t} \cdot \nabla p_{0,t} = k^2\ \nabla y_{1,t} \cdot \nabla p_{1,t}, \qquad \text{on } \gamma_t$$

Thus

$$\dot{j}(t) = -\left(1 - \frac{1}{k}\right)\int_{\gamma_t} \nabla y_{0,t}\cdot\nabla p_{0,t} < v,n > d\sigma$$

$$= -(k^2 - k)\int_{\gamma_t} \nabla y_{1,t}\cdot\nabla p_{1,t} < v,n > d\sigma \qquad (4.17)$$

If $k = 1$, y does not depend on the decomposition of Ω and $J(\Omega_t)$ = constant, since $\dot{j}(t) = 0$, as one sees in Eq. 4.14.

One may note that the derivative of a functional that depends on a variable domain Ω is an old problem. See for example the work of Rayleigh [66], Hadamard [28], Garabedian and Schiffer [29], Joseph [26], and more recently many papers of the French school with Bendali and Djaadane [30], Cea [31-34], Chenais [1], Dervieux [23-25], Gioan [33,34], Glowinski [36-38], Koenig [2], Michel [33,34], Morel [35], Morice [39], Murat [6-8], Palmerio [23,25], Pironneau [40-43], Rousselet [16,17], Simon [6-8], and Zolesio [45-48]. For selected references dealing with problems of shape optimal design, see Refs. 49-70.

REFERENCES

1. Chenais, D., "On the Existence of a Solution in a Domain Identification Problem," J. Math. Anal. Appl., Vol. 52, No. 2, 1975.
2. Koenig, M. and Zolesio, J.P., "Localisation d'un Objet de Forme Convexe Donnee," C.R. Acad. Sci., Paris, 1972, pp. 276, 880-852.
3. Cea, J., "Numerical Methods of Shape Optimal Design," Optimization of Distributed Parameter Structures (Eds. E.J. Haug and J. Cea), Sijthoff & Noordhoff, Alphen aan den Rijn, Netherlands, 1980.
4. Cea, J., Colloque Annuel d'Analyse Numerique, France, 1972.
5. Gonzalez de Paz, R.B., Personal Communication, 1980.
6. Murat, F. and Simon, J., Quelques Résultats sur le Contrôle par un Domaine Géométrique, Publication du L.A. 189, Univ. Paris VI, 1974.
7. Murat, F. and Simon, J., Sur le Contrôle par un Domaine Géométrique, Publication du L.A. 189, Univ. Paris VI, 1976.
8. Murat, F. and Simon, J., "Etude de Problèmes d'Optimal Design," Proceedings of the 7th IFIP Conference, Springer Verlag, Lecture Notes in Computer Sciences, No. 41, 1976, pp. 54-62.

9. Gurvitch, E.L., "On Isoparametric Problems for Domains with Partly Known Boundaries," J.O.T.A., Vol. 20, No. 1, 1976, pp. 65-79.
10. Banichuk, N.V., "Optimization of Elastic Bars in Torsion," Int. J. Solids and Structures, Vol. 12, 1976, pp. 275-286.
11. Banichuk, N.V., "On a Two-Dimensional Optimization Problem in Elastic Bar Torsion Theory," MTT, Vol. 11, No. 5, 1976, pp. 45-52.
12. Banichuk, N.V., "Optimality Condition in the Problem of Seeking the Hole Shapes in Elastic Bodies," PMM, Vol. 41, No. 5, 1977, pp. 920-925.
13. Miele, A. and Hull, D.G., "Sufficiency Proofs for the Problem of the Optimum Transversal Contour," SIAM J. Appl. Math., Vol. 15, No. 2, 1967, pp. 466-477.
14. Dervieux, A., Palmerio, B., and Rousselet, B., "Dessin Optimal d'une Reacteur Nucleaire par une Méthode de Gradient," to appear.
15. Kac, M., "Can One Hear the Shape of a Drum?" Amm. Math. Mthly., Vol. 73, 1966, pp. 1-23.
16. Rousselet, B., Probléme Inverse de Valeurs Propres; Optimization Techniques, Springer Verlag, Lecture Notes in Computer Sciences, Vol. 2, No. 41, 1976, pp. 77-85.
17. Rousselet, B., "Sur les Problemes de Valeurs Propres," C.R. Acad. Sci. Paris, Serie A, Vol. 283, 1976, p. 507.
18. Daniljuk, I.I., "Sur une Classe de Fonctionnelles Intégrales à Domaine Variable d'Intégration," Actes, Congrès Internat. Math., 1970, Vol. 2, pp. 703-715.
19. Baiocchi, C., "Su un Problema a Frontiera Connesso a Questioni di Idraulica," Ann. Mat. Pura Appl., Vol. 92, No. 4, 1972, pp. 107-127.
20. Deuling, H.J. and Helfrich, W., "The Curvature Elasticity of Fluid Membranes - A Catalogue of Vesicle Shapes," J. de Phys., Vol. 37, Nov. 1976, pp. 1335-1345.
21. Cea, J. and Malanowski, K., "An Example of a Max-Min Problem in Partial Differential Equations," SIAM J. Control, Vol. 8, No. 3, 1970, pp. 305-316.
22. Angrand, F., Glowinski, R., Periauf, J., Perrier, P., Poirier, G., and Pironneau, O., "Optimum Design For Potential Flows," Communication to Calgary Conference, June, 1980.
23. Dervieux, A. and Palmerio, B., "Une Formule d'Identification dans des Problèmes d'Identification de Domaines," C.R. Acad. Sci. Paris, Série A, Vol. 280, 1975, pp. 1697-1700 and 1761-1764.
24. Dervieux, A. and Palmerio, B., Hadamard's Variational Formula for a Mixed Problem and an Application to a Problem Related to a Signorini Like Variational Inequality, Rapport Laboria, IRIA, Rocquencourt, 1979.
25. Dervieux, A. and Palmerio, B., Une Formule de Hadamard dans les Problèmes d'Optimal Design, Springer Verlag, Lecture Notes in Computer Science, Vol. 40, 1976.

26. Joseph, D.D., "Parameter and Domain Dependence of Eigenvalues of Elliptic Partial Differential Equations," Arch. Rat. Mech. Anal., Vol. 24, 1967.
27. Ramakrishnan, V. and Francavilla, A., "Structural Shape Optimization Using Penalty Functions," J. Struct. Mech., Vol. 3, No. 4, 1975, pp. 403-432.
28. Hadamard, J., "Mémoire sur le Probleme d'Analyse Relatif à l'Equilibre des Plaques Élastiques Encastrées (1908)," oeuvres de J. Hadamard, C.N.R.S., Paris, 1968.
29. Garabedian, P.R. and Schiffer, M., "Convexity of Domain Functionals, J. d'Analyse Math., Vol. 3, 1953, 246-344.
30. Bendali, A. and Djaadane, A., Thèse-Université Alger, 1975.
31. Cea, J., Une Méthode Numerique Pour La Recherche d'un Domaine Optimal, Publication IMAN-P2, Nice, 1975.
32. Cea, J., Identification de Domaines, Springer Verlag, Lecture Notes in Computer Science, Vol. 3, 1973.
33. Cea, J., Gioan, A., and Michel, J., "Quelques Résultats sur l'Identification de Domaines," Calcolo, III-IV, 1973.
34. Cea, J., Gioan, A., and Michel, J., Adaption de la Méthode du Gradient à un Problème d'Identification de Domaine, Springer Verlag, Lecture Notes in Computer Science, Vol. 11, 1974.
35. Morel, P., "Utilisation en Analyse Numerique de la Formule de Derivation d'Hadamard," RAIRO, R-2, 1973, pp. 115-119.
36. Glowinski, R. and Marrocco, A., "Analyse Numérique du Champ Magnétique d'un Alternateur par Éléments Finis et Surrelaxation Ponctuelle non Linéare, Comp. Meth. in Appl. Mech. and Eng., Vol. 3, No. 1, Jan. 1974.
37. Glowinski, R. and Marrocco, A., "Finite Element Approximation and Iterative Methods of Solution for 2-D Nonlinear Magnetostatic Problems," Proceedings of COMPUMAG Conf., Oxford, 1976, pp. 112-125.
38. Glowinski, R. and Marrocco, A., "Numerical Solution of Two-Dimensional Magnetostatic Problems by Augmented Lagrangian Methods," Comp. Meth. in Appl. Mech. and Eng., to appear.
39. Morice, P., "Une Méthode d'Optimisation de Forme de Domaine," Proc. Congress, IFIP-IRIA, Springer-Verlag, 1974.
40. Pironneau, O., "Variational Method for the Numerical Solution of Free Boundary Problems and Optimum Design Problems," in Control Theory of Systems Governed by P.D.E., (Ed., A. Asis), Academic Press, New York, 1977.
41. Pironneau, O., Sur les Problèmes d'Optimisation de Structure en Mécanique des Fluides, Thèse de Doctorat, Pariv VI, 1976.
42. Pironneau, O., "On Optimum Design in Fluid Mechanics," J. Fluid Mech., Vol. 64, 1974, pp. 97-111.
43. Pironneau, O., "Optimum Design with Lagrangian Finite Elements: Design of an Electromagnet," Comp. Math. in Appl. Mech. and Eng., Vol. 15, 1978, pp. 207-308.
44. Pironneau, O. and Saguez, C., Asymptotic Behaviour of Solutions of P.D.E. with Respect to the Domain, IRIA Laboria Report No. 218, 1977.

45. Zolesio, J.P., Identification de Domaine par Deformation, These, Universite de Nice, 1979.
46. Zolesio, J.P., "Un Resultat d'Existence de Vitesse Convergente," C.R. Acad. Sci. Paris, Serie A, Vol. 283, 1976, p. 855.
47. Zolesio, J.P., "Localisation du Support d'un Controle Optimal," C.R. Acad. Sci. Paris, Serie A, Vol. 284, 1977, p. 791.
48. Zolesio, J.P., An Optimal Design Procedure for Optimal Control Support, Lecture Notes in Economical and Mathematical Systems, No. 14, pp. 207-233, 1977.
49. Bamberger, A., Chavent, G., and Lailly, P., Etude Mathématique et Numérique d'un Problème Inverse Pour l'Equation des Ondes à une Dimension, Rapport Interne, No. 14, Ecole Polytechnique, January 1977.
50. Begis, D. and Glowinski, R., "Application de la Méthode des Éléments Finis à l'Approximation d'un Problème de Domaine Optimal," Applied Math. and Optimization, Vol. 2, No. 2, 1975, pp. 130-169.
51. Benson, D.C., "An Elementary Solution of a Variational Problem of Aerodynamics," J. of Opt. Theory and Appl., Vol. 1, No. 2, 1967.
52. Bhaivikatti, S.S. and Ramakrishnan, C.V., "Optimum Design of Fillets in Flat and Round Tension Bars," ASME paper, No. 77-DET-45, 1977.
53. Cea, J., Bardos, C., Grisvard, P., and Lhomne, B., "Some Numerical Applications of the Green Formula," to appear.
54. Chavent, G., Analyse Fonctionnelle et Identification de Coefficients Répartis dans les Équations aux Dérivées Partielles, Thèse de Doctorat, Paris.
55. Francavilla, A., Ramakrishnan, C.V., and Zienkievicz, O.C., "Optimization of Shape to Minimize Stress Concentration," J. Strain Anal., Vol. 10, 1975, pp. 63-70.
56. Friedman, A., "Free Boundary Problems for Parabolic Equations, (II) Evaporation or Condensation of a Liquid Drop," J. Math. Mech., Vol. 9, 1960, pp. 19-66.
57. Friedman, A., "Free Boundary Problems for Parabolic Equations (III) Dissolution of a Gas Bubble in Liquid," J. Math. Mech., Vol. 9, 1960, pp. 327-345.
58. Friedman, A., "The Stefan Problem in Several Space Variables," Trans. Amer. Math. Soc., Vol. 133, 1968, pp. 51-87.
59. Garabedian, P.R. and Spencer, D.C., "Extremal Methods in Cavitational Flows," J. Rat. Mech. Anal., Vol. 1, 1952, pp. 359-409.
60. Kagiwada, H.H. and Kalaba, R.E., "A Practical Method for Determining Green's Functions Using Hadamard's Variational Formula," J. of Opt. Theory and Appl., Vol. 1, No. 1, 1967.

61. Kristensen, E.S. and Madsen, N.M., "On the Optimal Shape of Fillets in Plates Subjected to Multiple in Plane Loading Cases," Int. J. for Num. Meth. in Engr., Vol. 10, 1976, pp. 1007-1011.
62. Lions, J.L., "On the Optimal Control of Distributed Parameter Systems," Techniques of Optimization, (Ed., A. Balakrishnan), Academic Press, New York, 1972.
63. Lions, J.L., Sur Quelques Questions d'Analyse, de Mécanique et de Contrôle Optimal, Le Presses de l'Université de Montréal, 1976.
64. Marrocco, A. and Pironneau, O., "Optimum Design with Lagrangian Finite Elements - Design of an Electro-Magnet," Comp. Meth. in Appl. Mech. and Eng., Vol. 15, 1978, pp. 217-308.
65. Mignot, F., Murat, F., and Puel, J.P., "Variation d'un Point de Retournement par Rapport au Domaine," Comm. on P.D.E., Vol. 4, No. 11, 1979.
66. Rayleigh, J.W., The Theory of Sound, 2nd Ed., Cambridge University Press, 1894-1896.
67. Rubinstein, L.I., "The Stefan Problem," A.M.S., Translations of Mathematical Monographs, Vol. 27, Providence, R.I., 1971.
68. Troesch, B.A. and Troesh, H.R., "Eigenfrequencies of an Elliptic Membrane," Math of Comp., Vol. 27, No. 126, 1973, pp. 755-772.
69. Tvergaard, V., "On the Optimum Shape of a Fillet in a Flat Bar with Restrictions," Proc. IUTAM Symposium on Structural Design, Warsaw, Springer-Verlag, New York, 1973.
70. Zienkiewicz, O.C. and Campbell, J.S., "Shape Optimization and Sequential Linear Programming," Optimum Structural Design, (Ed. R.H. Gallagher and O.C. Zienkiewicz), Wiley, New York, 1973.

NUMERICAL METHODS OF SHAPE OPTIMAL DESIGN

Jean Cea

Département de Mathématiques, Université de Nice
06034 Nice Cedex, France

ABSTRACT

This paper presents numerical methods for shape optimal design and applications to problems of engineering and applied science. Methods presented include gradient methods, fixed point methods, Green function methods, and duality methods. It is shown that derivative formulas derived in companion papers can be used directly to obtain expressions needed to implement these optimization methods for solution of concrete problems of shape optimal design.

1. THE (CONTINUOUS) GRADIENT METHOD

1.1 The Hilbertian Case

Let V be a real Hilbert space and $J : V \to \mathbb{R}$ the cost function. One seeks $u^* \in V$ such that

$$J(u^*) \leq J(v), \text{ for all } v \in V$$

Suppose first that there are no constraints. The idea is to approach u^* by a one parameter family $u(t)$, $t \geq 0$, where $u(t)$ is the solution of the ordinary differential equation

$$\left.\begin{aligned} \dot{u}(t) &= v(u(t),t) \\ u(0) &= u_0 \text{ given} \end{aligned}\right\} \qquad (1.1)$$

The objective now is to choose the velocity v in such a way that

$$\lim_{t\to\infty} u(t) = u^*$$

In order to find the good choice for v, put

$$j(t) = J(u(t))$$

and take the derivative we get

$$\dot{j}(t) = \left\langle \nabla_u J(u(t)), \frac{du(t)}{dt} \right\rangle = < G_t, v_t >_V \qquad (1.2)$$

where $< , >_V$ is the scalar product in V, v_t is the velocity at the time t ($v_t(u) = v(y,t)$), and G_t is the gradient of J. Let B(t) be a uniformly bounded and coercive operator defined on V. One can introduce it with a family of bilinear forms $b(t,\phi,\psi)$ that satisfy

$$b(t,\phi,\phi) \geq \alpha||\phi||^2, \text{ for all } \phi \in V, \ \alpha > 0$$

$$|b(t,\phi,\psi)| \leq M \ ||\phi|| \cdot ||\psi||, \text{ for all } \phi \text{ and } \psi \in V, \ M > 0$$

The form b is related to the operator B by the condition

$$\langle B(t)\phi,\psi\rangle_V = b(t,\phi,\psi), \text{ for all } \phi \text{ and } \psi \in V$$

If one chooses $v_t \in V$ as the solution of

$$b(t,v_t,\psi) = - \langle G_t,\psi\rangle_V, \text{ for all } \psi \in V \qquad (1.3)$$

(or v_t is the solution of $B(t)v_t = -G_t$), one gets from Eq. 1.3, with $\psi = v_t$ and Eq. 1.2,

$$\dot{j}(t) = \langle G_t,v_t\rangle_V = - b(t,v_t,v_t) \leq - \alpha \ ||v_t||^2$$

Hence with this choice of v_t, j decreases. Furthermore,

$$j(t) = j(0) + \int_0^t \dot{j}(s)ds \leq j(0) - \alpha \int_0^t ||v_s||^2 \ ds$$

or

$$j(t) + \alpha \int_0^t ||v_s||^2 \, ds \leq j(0), \text{ for all } t \geq 0 \tag{1.4}$$

Usually, from Eq. (1.4) and some hypothesis on J, one gets

$$\int_0^\infty ||v_s||^2 \, ds < + \infty$$

$$\lim_{s\to\infty} ||v_s|| = 0$$

$$\lim_{s\to\infty} ||G_s|| = 0$$

and finally

$$\lim_{t\to+\infty} u(t) = u^*$$

$$G_t(u^*) = 0$$

The two important steps. The preceding analysis may be viewed as the following two step computation:

(1) Computation of the gradient, or of the linear and continuous form

$$v \to \langle G_t, v \rangle_V$$

which has been used in Eq. (1.3).

(2) Computation of the velocity v_t using an elliptic operator $B(t)$ and the previous linear form. If $B(t) \equiv I$ one gets the classical continuous gradient method,

$$\begin{cases} \dot{u}(t) = -G(u(t)) = -\nabla_u J(u(t)) \\ u(0) = u_0 \text{ given} \end{cases}$$

$$\dot{j}(t) = -||G(u(t))||^2$$

1.2 Shape Optimal Design Case

For shape optimal design, the variable is not u but Ω, which can now be considered as a nondenumerable continuum of particles. One can define Ω_t, by defining a velocity for every particle,

$$\left.\begin{aligned} \dot{x}(t) &= v(x(t),t) \\ x(0) &= X \in \Omega_0 \end{aligned}\right\} \tag{1.5}$$

Now the space of velocities is to be selected. One defines V_t on Ω_t, for example as

$$V_t = (H^m(\Omega_t))^n \tag{1.6}$$

where m depends on the regularity one wants for the velocity.

In the model problems of Ref. 1, it has been seen that

$$\dot{j}(t) = \int_{\Omega_t} \operatorname{div}(C_t\ v_t)dx = \int_{\Gamma_t} C_t < v_t, n > d\sigma \tag{1.7}$$

with a convenient C_t. One can also write this as

$$\dot{j}(t) = \langle G_t, v_t \rangle_{V_t} \tag{1.7'}$$

where G_t is in some sense the gradient of J at time t. Now, one chooses a family of bilinear forms that are uniformly bounded and coercive, for example

$$b(t,\phi,\psi) = <\phi,\psi>_{V_t} = <\phi,\psi>_{(H^m(\Omega_t))^n} \tag{1.8}$$

and choose $v_t \in V_t$ by

$$b(t,v_t,\phi) = - \langle G_t, \phi \rangle_{V_t}, \text{ for all } \phi \in V_t \tag{1.9}$$

This is, for example, find $v_t \in (H^m(\Omega_t))^n$

$$\langle v_t,\phi\rangle_{(H^m(\Omega_t))^n} = \int_{\Gamma_t} C_t < \phi, n > d\sigma, \tag{1.10}$$

for all $\phi \in (H^m(\Omega_t))^n$.

With this choice, one has

$$\dot{j}(t) = \langle G_t, v_t\rangle_{V_t} - b(t,v_t,v_t) \leq - \alpha \; ||v_t||^2 \tag{1.11}$$

and j is decreasing, as desired.

It is important to remember the two steps carried out:

(1) Compute $\dot{j}(t)$, which gives G_t

(2) Compute v_t

1.3 Application to Model Problem No. 1 [1]

To illustrate the continuous gradient method, it is applied to Model Problem No. 1 of Ref. 1. Here, the state of the system is $y_t \in H^1(\Omega_t)$, satisfying the variational state equation

$$\int_{\Omega_t} (\nabla y_t \cdot \nabla \phi + y_t \phi) dx = \int_{\Omega_t} f \; \phi \; dx, \text{ for all } \phi \in H^1(\Omega_t)$$

and the cost function is $j(t) = \frac{1}{2} \int_{\Omega_t} |y_t - y_{g,t}|^2 \; dx$

The variable domain Ω_t is as shown in Fig. 1.1. There are no constraints in this problem.

If the adjoint state is defined as $p_t \in H^1(\Omega_t)$ satisfying the variational equation

$$\int_{\Omega_t} (\nabla \psi \cdot \nabla p_t + \psi \; p_t) \; dx = \int_{\Omega_t} (y_t - y_{g,t}) \; \psi \; dx, \text{ for all}$$

$$\psi \in H^1(\Omega_t)$$

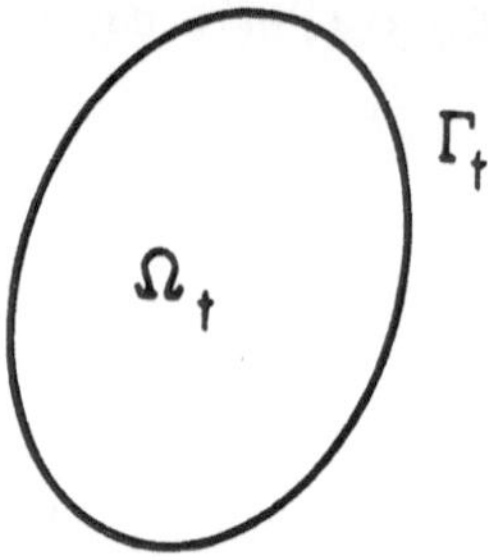

Figure 1.1 Variable Domain for Model Problem No. 1

one has

$$\dot{j}(t) = \int_{\Gamma_t} C_t \langle v_t, n_t \rangle \, d\sigma$$

where

$$C_t = -(\nabla y_t \nabla p_t + y_t p_t - f p_t) + \frac{1}{2} |y_t - y_{g,t}|^2$$

Now, if $V_t = (H^m(\Omega_t))^n$ and $b(t,\phi,\psi) = \langle \phi,\psi \rangle_{(H^m(\Omega))^n}$ one can choose $v_t \in V_t$ as the solution of the variational equation

$$b(t,v_t,\phi) = \int_{\Gamma_t} C_t \langle \phi, n_t \rangle \, d\sigma, \text{ for all } \phi \in V_t$$

One recalls that

$$\Omega_t = \{x \in \mathbb{R}^n : \text{there exists } X \in \Omega_0 \text{ such that } x = x(t),$$

$$\text{where } \dot{x}(s) = v(x(s),s),\ s \geq 0,\ x(0) = X\}$$

Conclusion. Note that if one suppresses the test functions ϕ and ψ arising in the variational equations, one gets relations of the following kind:

$$\begin{cases} E(\Omega_t, y_t) = 0 \\ F(\Omega_t, y_t, p_t) = 0 \\ H(\Omega_t, y_t, p_t, v_t) = 0 \\ \Omega_t \text{ defined using } v_s,\ 0 \le s \le t \end{cases}$$

which is coupled system as seen by the mappings

$$\Omega_t \rightarrow y_t \rightarrow p_t \rightarrow v_t \rightarrow \Omega_t$$

Discretization of the time t. Let τ be a step in the time t, $\tau > 0$, and denote $\Omega_n = \Omega_{n\tau}$, $y_n = y_{n\tau}$, ... One can construct Ω_{n+1} from Ω_n in the following way: For a given Ω_n

$$E(\Omega_n, y_n) = 0$$

determines y_n

$$F(\Omega_n, y_n, p_n) = 0$$

determines p_n

$$H(\Omega_n, y_n, p_n, v_n) = 0$$

determines v_n, and Ω_{n+1} is the set of points x_{n+1} where

$$x_{n+1} = x_n + \tau\, v(x_n, n\tau),\ x_n \in \Omega_n$$

In this way, one has in fact discretized the equations $\dot{x} = v$ by the Euler method.

Discretization of the Space Variables. In specific examples, the equations $E(\Omega, y) = 0$, ..., are usually partial differential equations. If one solves these equations by the Finite Element Method, it will be necessary to associate a triangulation $\mathcal{T}_n$ to every Ω_n. In fact, one can transport this triangulation from Ω_n to Ω_{n+1}. Suppose a,b,c is a triangle of $\mathcal{T}_n$. One can construct a triangle A,B,C of $\mathcal{T}_{n+1}$, as shown in Fig. 1.2, as follows

$$x_A = x_a + \tau\, v(x_a, n\tau)$$

$$x_B = x_b + \tau\, v(x_b, n\tau)$$

$$x_C = x_c + \tau\, v(x_c, n\tau)$$

Obviously, trouble can arise with this transportation method. If so, one must define a new triangulation by another way.

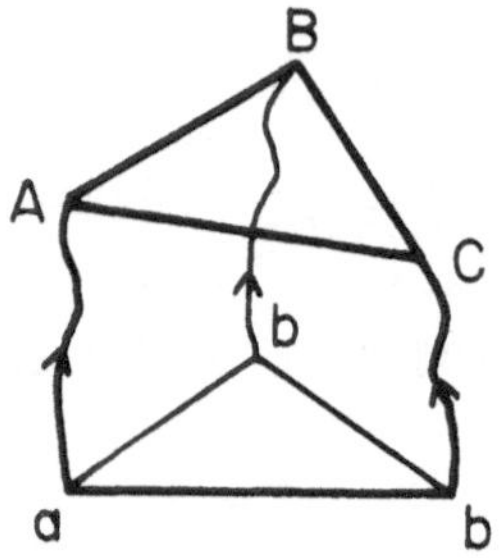

Figure 1.2 Transportation of Triangulation

In many cases, the transportation method has worked very well, particularly when Ω_n is not too far from the solution.

2. THE CONTINUOUS GRADIENT METHOD IN PARTICULAR CASES

2.1 The Domain Defined By A Finite Number of Parameters

Consider now Model Problem No. 2 of Ref. 1 and suppose Ω_0 is a disk of a given radius r and center $a = (a_1, a_2)$ that is the unknown or the variable of the problem, as shown in Fig. 2.1. The state equation for $y \in H'(\Omega)$ is

$$\int_{\Omega_0} \nabla y_0 \cdot \nabla \psi_0 \, dx + \int_{\Omega_1} k \, \nabla y_1 \cdot \nabla \phi_1 + \int_{\Omega} y \, \phi \, dx = \int_{\Omega} f \, \phi \, dx,$$

for all $\phi \in H^1(\Omega)$

The cost function is

$$J(a) = J(a_1, a_2) = J(\Omega) = \frac{1}{2}\int_{\Omega} |y - y_g|^2 \, dx$$

In this problem, the cost function J depends on two real variables a_1 and a_2, through the solution of a partial differential equation. The first task is to compute $\frac{\partial J}{\partial a_1}$ and $\frac{\partial J}{\partial a_2}$, that is to compute the gradient of J.

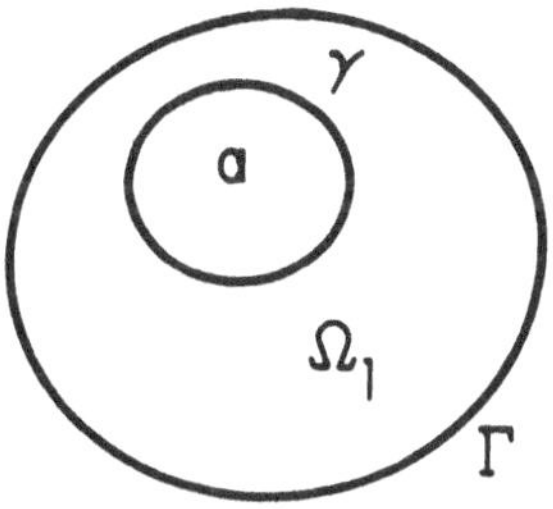

Figure 2.1 Domain for Model Problem No. 2

One introduces the time t and

$$a(t) = a + tb$$

so

$$j(t) = J(a(t))$$

It is well known that

$$\dot{j}(t) = \nabla_a J(a(t)) \cdot \frac{da(t)}{dt} = \nabla_a J \cdot b$$

or

$$\dot{j}(t) = \frac{\partial J}{\partial a_1} b_1 + \frac{\partial J}{\partial a_2} b_2$$

and, in particular,

$$\dot{j}(0) = \frac{\partial J}{\partial a_1}(a)\, b_1 + \frac{\partial J}{\partial a_2}(a)\, b_2 \tag{2.1}$$

Note that $\dot{j}$ is a linear form with respect to b_1 and b_2, so the coefficients of b_1 and b_2 are the derivatives of J. Another way of computating j(t) will be found, which will lead to the derivative of J.

One can find a velocity v_t defined on $\Omega_t = \Omega_{0,t} \cup \gamma_t \cup \Omega_{1,t}$ that is equal to zero on Γ and equal to b on $\overline{\Omega_{0,t}}$. For example, $v(x,t) = b$ for all $x \in \Omega_{0,t} \cup \gamma_t$ and for $i = 1,2$,

$$\Delta v_i(x,t) = 0, \text{ for all } x \in \Omega_{1,t}$$

$$v_i(x,t) = b_i, \text{ for all } x \in \gamma_t$$

$$v_i(x,t) = 0, \text{ for all } x \in \Gamma$$

There is no need to compute v, even though it exists.

In Ref. 1, it was shown that

$$\left.\begin{aligned} \dot{j}(t) &= \int_{\gamma_t} C_t < v,n > d\sigma \\ C_t &= -\left(1 - \frac{1}{k}\right) \nabla y_{0,t} \cdot \nabla p_{0,t} \end{aligned}\right\} \quad (2.2)$$

in polar coordinates, this is

$$\dot{j}(t) = \int_0^{2\pi} C_t[a_1(t) + r\cos\theta,\ a_2(t) + r\sin\theta]$$

$$\times [b_1 \cos\theta + b_2 \sin\theta]\, r\, d\theta$$

or, in a simplified notation,

$$\dot{j}(t) = b_1\, \alpha_1(t) + b_2\, \alpha_2(t) \quad (2.3)$$

where

$$\alpha_1(t) = r \int_0^{2\pi} C_t[a_1(t) + r\cos\theta,\ a_2(t) + r\sin\theta] \cos\theta\, d\theta$$

$$\alpha_2(t) = r \int_0^{2\pi} C_t[a_1(t) + r\cos\theta,\ a_2(t) + r\sin\theta] \sin\theta\, d\theta$$

If one compares Eqs. 2.1 and 2.3, one sees that

$$\frac{\partial J}{\partial a_1} = \alpha_1(0)$$

$$\frac{\partial J}{\partial a_2} = \alpha_2(0)$$

2.2 Domain Defined By A Level Curve

Suppose one is again working with Model Problem No. 2, but with the domain Ω_0 now given by a level curve, as shown in Fig. 2.2(a). The level curve γ_t is defined by a C^1 simple regular function θ, i.e. a C^1 function such that $|\nabla\theta| > 0$, except at one point. The geometry is as shown in Fig. 2.2(b), where

$$\Omega_{0,t} = \{x \in \mathbb{R}^2 : \theta(x) < t\},\ 0 < t < T$$

One supposes that

$$\overline{\Omega_{0,T}} \cap \Gamma = \phi \text{ (null set)}$$

Thus $\Omega_t = \Omega_{0,t} \cup \gamma_t \cup \Omega_{1,t}$ which gives the decomposition in 3 parts.

In this geometry, Γ is fixed and γ_t is variable. The equation of γ_t in the (x_1,x_2) plane is $\theta(x_1,x_2) = t$. In this problem, when θ is given, the problem depends only on one parameter t. The derivative with respect to this parameter is given by Eq. 2.2, but in this case, one has (see Ref. 2)

$$\vec{n} = \frac{\nabla\theta}{|\nabla\theta|},\quad V = \frac{\nabla\theta}{|\nabla\theta|^2}$$

and then

$$\langle V,n\rangle_{\mathbb{R}^2} = \frac{1}{|\nabla\theta|}$$

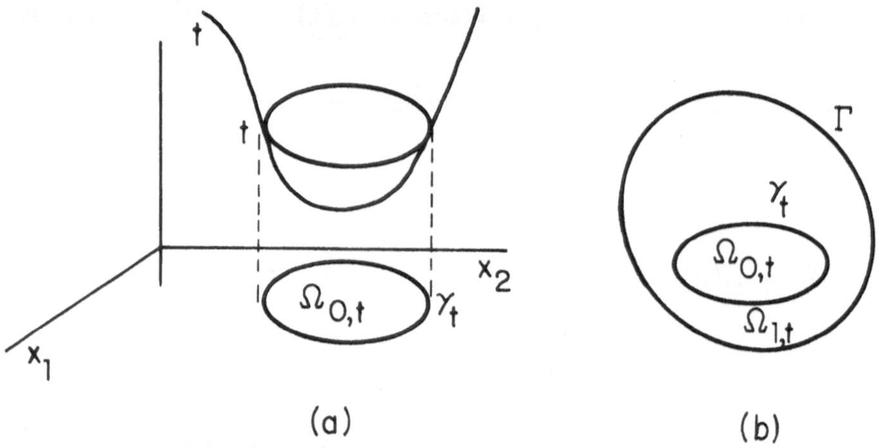

Figure 2.2 Domain Defined by Level Curve

and finally

$$\dot{j}(t) = \int_{\gamma_t} C_t \frac{1}{|\nabla\theta|} d\sigma$$

2.3 Domain Defined By An Infinite Number of Parameters

Consider again Model Problem No. 1, but with a set Ω defined using a function z and its graph γ, as shown in Fig. 2.3. Suppose C and D are fixed (for simplicity) the state function $y \in H'(\Omega)$ is given as the solution of

$$\int_{\Omega} (\nabla y \ \nabla\phi + y\phi) \ d\sigma = \int_{\Omega} f \ \phi \ dx, \text{ for all } \phi \in H^1(\Omega)$$

The cost function is

$$J(z) = J(\Omega) = \frac{1}{2}\int_{\Omega} |y - y_g|^2 \ dx$$

In this case z belongs to an infinite dimensional space.

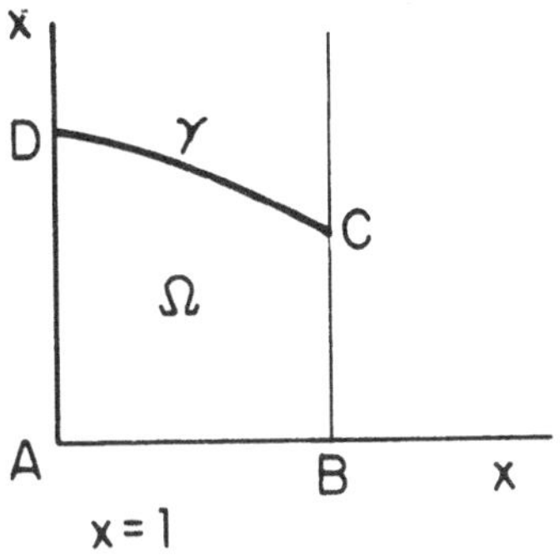

Figure 2.3 Domain of Model Problem 1, Defined by A Function

One may now compute the directional derivative of J at z in the direction μ. If $z_t = z + t\mu$, (with $\mu(0) = \mu(1) = 0$, one has

$$DJ(z,\mu) = \lim_{t\to 0} \frac{J(z+t\mu) - J(z)}{t}$$

$$= \lim_{t\to 0} \frac{J(z_t) - J(z_0)}{t}$$

or

$$DJ(z,\mu) = \dot{j}(0) \tag{2.4}$$

with

$$j(t) = J(z_t)$$

Here also one can find a second way to compute $\dot{j}$. There exists a velocity v_t such that

$$v_t = 0, \text{ on DA, AB, and BC}$$

$$v_t(x) = [0,\mu(x)], \text{ for all } x \in [0,1] (\text{i.e. on } \gamma_t)$$

In Ref. 1 it is shown that

$$\dot{j}(t) = \int_{\gamma_t} C_t \langle v_t, n_t \rangle \, d\sigma$$

where

$$C_t = -(\nabla y_t \cdot \nabla p_t + y_t p_t - f\, p_t) + \frac{1}{2}\, |y_t - y_{g,t}|^2$$

For $t = 0$, on γ one has

$$\vec{n} = \frac{1}{\sqrt{1 + z'^2}} [-z', 1]$$

$$\vec{v} = [0, \mu]$$

Hence,

$$\dot{j}(0) = \int_0^1 C_0 \frac{\mu}{\sqrt{1 + z'^2}} \sqrt{1 + z'^2}\, dx = \int_0^1 C_0\, \mu\, dx \qquad (2.5)$$

Then with Eqs. 2.4 and 2.5, one sees that

$$DJ(z,\mu) = \int_0^1 C_0\, \mu\, dx$$

In each of the cases analyzed so far, the directional derivative of the functional J has been computed. From these, it is possible to use many optimization techniques.

3. A FIXED POINT METHOD

3.1 Derivative With Respect To A Characteristic Function

Consider again Model Problem No. 2 on the domain of Fig. 3.1, with $y \in H'(\Omega)$ defined as the solution of

$$\int_{\Omega_0} \nabla y_0 \cdot \nabla \phi_0\, dx + \int_{\Omega_1} k \nabla y_1 \cdot \nabla \phi_1\, dx + \int_\Omega y\, \phi\, dx = \int_\Omega f\, \phi\, dx,$$

for all $\phi \in H'(\Omega)$

Here, $\Omega = \Omega_0 \cup \gamma \cup \Omega_1$ and the cost function is

$$J(\Omega) = \frac{1}{2} \int_\Omega |y - y_g|^2\, dx$$

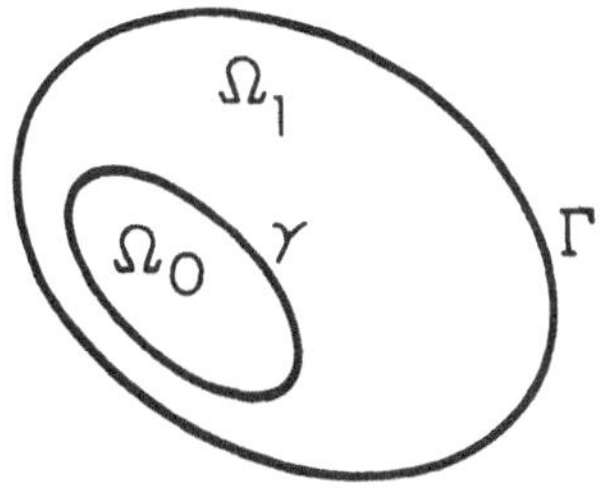

Figure 3.1 Domain for Model Problem No. 2

Denote by u the characteristic function of Ω_0. One can then write the state equation as

$$\int_\Omega u\nabla y\cdot\nabla\phi\, dx + \int_\Omega k(1-u)\nabla y\cdot\nabla\phi\, dx + \int_\Omega y\phi\, dx = \int_\Omega f\phi\, dx$$

or

$$(1-k)\int_\Omega u\nabla y\ \nabla\phi\, dx + k\int_\Omega \nabla y\ \nabla\phi\, dx + \int_\Omega y\ \phi\, dx = \int_\Omega f\ \phi\, dx$$

One may introduce the 3 multilinear forms

$$\left.\begin{aligned} a(u,y,\phi) &= (1-k)\int_\Omega u\ \nabla y\cdot\nabla\phi\, dx \\ b(y,\phi) &= k\int_\Omega \nabla y\cdot\nabla\phi\, dx + \int_\Omega y\ \phi\, dx \\ L(\phi) &= \int_\Omega f\ \phi\, dx \end{aligned}\right\} \qquad (3.1)$$

and the state equation becomes, find $y\in H^1(\Omega)$ to satisfy

$$a(u,y,\phi) + b(y,\phi) = L(\phi),\ \text{for all}\ \phi\in H^1(\Omega) \qquad (3.2)$$

Changing u to $u + \delta u$ gives a change $y \longrightarrow y + \delta y$, where $u + \delta u$ is the characteristic function of open set denoted by $\Omega_0 + \delta\Omega^+ - \delta\Omega^-$ ($\delta\Omega^+\subset\Omega_1$, $\delta\Omega^-\subset\Omega_0$), as shown in Fig. 3.2. Now, $y + \delta y\in H^1(\Omega)$ is the solution of

$$a(u + \delta u, y + \delta y, \phi) + b(y + \delta y, \phi) = L(\phi), \text{ for all } \phi \in H'(\Omega) \tag{3.3}$$

after simplification, this is

$$a(u, \delta y, \phi) + b(\delta y, \phi) = - a(\delta u, y, \phi) - a(\delta u, \delta y, \phi) \tag{3.4}$$

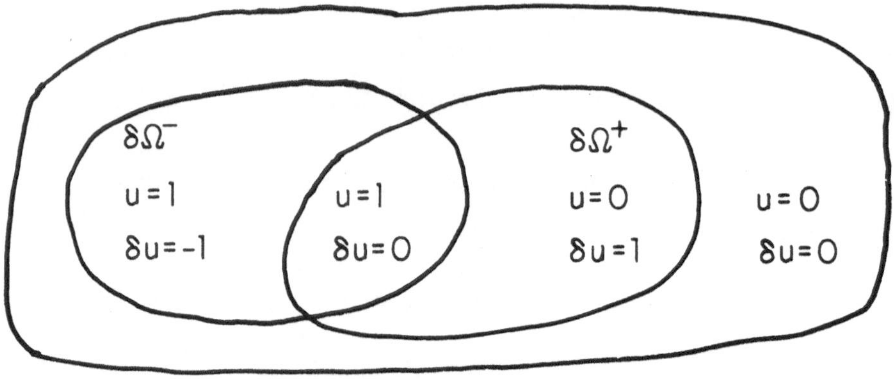

Figure 3.2 Perturbation of Domain

One has also

$$J(u + \delta u) = \frac{1}{2} \int_\Omega |y + \delta y - y_g|^2 \, dx$$

$$= J(u) + \int_\Omega (y - y_g) \delta y \, dx + \frac{1}{2} \int_\Omega (\delta y)^2 \, dx \tag{3.5}$$

Defining an adjoint state $p \in H^1(\Omega)$ as the solution of

$$a(u, \psi, p) + b(\psi, p) = \int_\Omega (y - y_g) \psi \, dx, \text{ for all } \psi \in H^1(\Omega) \tag{3.6}$$

and using Eqs. 3.6 and 3.4 one gets:

$$\int_\Omega (y - y_g) \delta y \, dx = a(u, \delta y, p) + b(\delta y, p)$$

$$= - a(\delta u, y, p) - a(\delta u, \delta y, p)$$

and

$$J(u + \delta u) = J(u) - a(\delta u, y, p) - a(\delta u, \delta y, p) + \frac{1}{2} \int_{\Omega} (\delta y)^2 \, dx \tag{3.7}$$

If $\int_{\Omega} |\delta u| \, dx$ is small, then

$$T_1(u, \delta u) = - a(\delta u, y, p)$$

appears as the term of order 1 and

$$T_2(u, \delta u) = - a(\delta u, \delta y, p) + \frac{1}{2} \int_{\Omega} (\delta y)^2 \, dx$$

is the term of order 2. More explicitly,

$$T_1(u, \delta u) = -(1 - k) \int_{\Omega} \delta u \, \nabla y \cdot \nabla p \, dx = -(1 - k) \int_{\delta \Omega^+} \nabla y \cdot \nabla p \, dx$$

$$+ (1 - k) \int_{\delta \Omega^-} \nabla y \cdot \nabla p \, dx \tag{3.8}$$

Suppose $k > 1$ and u is the optimum ($J(u) = \operatorname{Inf}_V J(v)$), then $T_1(u, \delta u)$ must be nonnegative, for any small δu or any small $\delta\Omega^+$ and $\delta\Omega^-$. This implies, from Eq. 3.8, that

$$\left.\begin{aligned} \nabla y \cdot \nabla p \geq 0, \quad \text{in } \Omega_1 \\ \nabla y \cdot \nabla p \leq 0, \quad \text{in } \Omega_0 \end{aligned}\right\} \tag{3.9}$$

and under certain regularity hypothesis,

$$\nabla y \cdot \nabla p = 0, \quad \text{on } \gamma \tag{3.10}$$

Note that the problem of Eqs. 3.2, 3.6, and 3.10 appears as a generalized free boundary value problem.

Note also that one can very easily get the condition of Eq. 3.10 using the derivative of $j(t) = J(\Omega_t)$, as has been seen in Ref. 1. Recall that

$$\dot{j}(t) = - \left(1 - \frac{1}{k}\right) \int_{\gamma_t} \nabla y_{0,t} \cdot \nabla p_{0,t} \, \langle v, n \rangle \, d\sigma$$

(Equation continued on next page)

$$= -(k^2 - k) \int_{\gamma_t} \nabla y_{1,t} \cdot \nabla p_{1,t} \langle v,n \rangle \, d\sigma$$

If j(t) is the optimum, then for any velocity field v, $\dot{j}(t) = 0$ and then

$$\nabla y_{0,t} \cdot \nabla p_{0,t} = \nabla y_{1,t} \cdot \nabla p_{1,t} = 0$$

which is Eq. 3.10.

Intuitively, one can also derive Eq. 3.9. First of all, the sign of $\nabla y \cdot \nabla p$ is constant in Ω_0 and Ω_1. Suppose one is near the optimum and suppose that on AB (see Fig. 3.3) one has $\nabla y_{0,t} \cdot \nabla p_{0,t} > 0$. If on AB one chooses v in the direction of Ω_0, then

$$-(1 - \frac{1}{k}) \int_{AB} \nabla y_{0,t} \cdot \nabla p_{0,t} \langle v,n \rangle \, d\sigma$$

is negative. One can thus decrease j by eliminating the part of γ where $\nabla y_{0,t} \cdot \nabla p_{0,t} > 0$. That means at the optimum, one has $\nabla y_{0,t} \cdot \nabla p_{0,t} \leq 0$ in Ω_0.

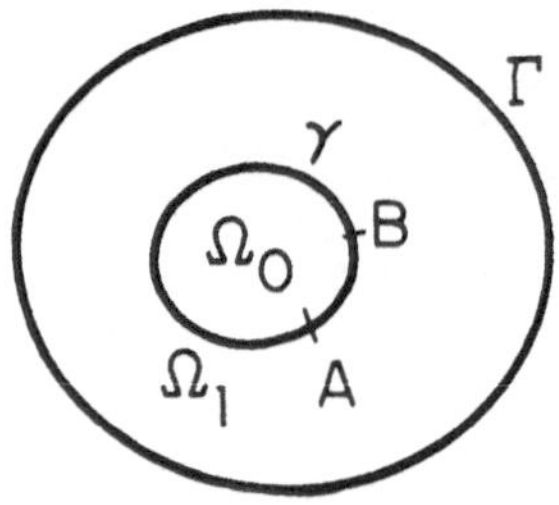

Figure 3.3 Partition of γ

3.2 A Fixed Point Method

Let

$$G_u = -(1 - k) \nabla y \cdot \nabla p$$

(y and p are functions of x and depend on u). From Eq. 3.9, $u = u_{opt}$ if and only if

$G_u(x) \leq 0$, for all $x \in \Omega_{0,opt}$

$G_u(x) \geq 0$, for all $x \in \Omega_{1,opt}$

Define the transformation T by:

$u \to Tu$

$Tu(x) = 1$, if and only if $G_u(x) \leq 0$

$Tu(x) = 0$, if and only if $G_u(x) > 0$

One knows that at the optimum

$$u_{opt} = Tu_{opt}$$

That is, u is a fixed point of T. In order to approximate this fixed point, it is classical to consider the following iterative method:

u_0 given

$u_{n+1} = Tu_n$, $n = 0, 1, \ldots$

3.3 A Numerical Point of View

If $\Omega_{0,n}$ (and $\Omega_{1,n}$) are known, one has to compute y_n, p_n, $G_n = -(1 - k)\, \nabla y_n \cdot \nabla p_n$, and define $\Omega_{0,n+1}$ as the set of x such that $G_u(x) \leq 0$.

To compute y and p, one needs a finite element grid that is variable with n. In fact, let $\mathcal{T}$ be a fixed triangulation of Ω and denote the triangles by Δ_i, $i = 1, \ldots, N$. The iterative method will be limited to $\Omega_{0,n}$ and $\Omega_{1,n}$ that are the union of these triangles. Suppose I_n and J_n satisfy

$$\begin{cases} I_n \cup J_n = \{1,\ldots,N\} \\ I_n \cap J_n = \phi \end{cases}$$

Define $\Omega_{0,n}$ and $\Omega_{1,n}$ as

$$\Omega_{0,n} = \bigcup_{i \in I_n} \Delta_i$$

$$\Omega_{1,n} = \bigcup_{i \in J_n} \Delta_i$$

One Step of the Iterative Method. The method of computing I_{n+1} (and J_{n+1}) from I_n (and J_n) is now explained. Compute y_n, p_n, G_n, and for all i, $\int_{\Delta_i} G_n(x)dx$. If $\int_{\Delta_i} G_n(x)dx \le 0$, then $i \in I_{n+1}$, otherwise $i \in J_{n+1}$. If $I_{n+1} = I_n$ the process is finished and one can propose $\Omega_{0,opt} \simeq \Omega_{0,n+1}$. In general, one has better results if he proposes

$$\Omega_{0,opt} \simeq \{x : x \in \Omega,\ G_{n+1}(x) < 0\}$$

In many examples, good results have been obtained with few iterations ($n \le 4$). One can still improve the process. If $I_n = I_{n+1}$, define a new triangulation, with smaller triangles, particularly around γ_{n+1}.

3.4 The Fixed Point Method In the Case of Model Problem No. 1

Recall that in model problem No. 1, one has

$$\dot{j}(t) = \int_{\Gamma_t} C_t \langle v_t, n_t \rangle \, d\sigma$$

$$C_t = -(\nabla y_t \cdot \nabla p_t + y_t p_t - f p_t) + \frac{1}{2} |y_t - y_{g,t}|^2$$

and obviously if $C_t = 0$ on Γ_t one has a critical domain, which could be the optimum domain. But here, the situation is as follows: Σ_t is the set of points where $C_t = 0$. One wants to use the method previously employed to prolongate C_t outside of Ω_t, as shown in Fig. 3.4.

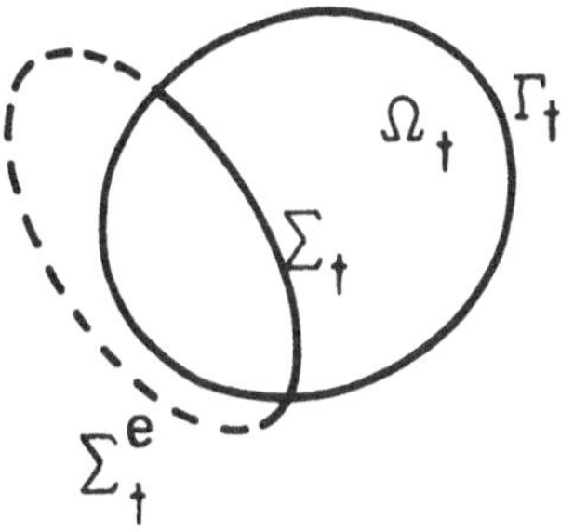

Figure 3.4 Extension of Ω_t

This can be done in the following way. Suppose one knows that Ω_t is included in a fixed domain D of boundary ∂B, as shown in Fig. 3.5. Define $\Omega_{1,t}$ such that

$$D = \Omega_t \cup \Gamma_t \cup \Omega_{1,t}$$

One now defines C_t everywhere in D, since C_t is yet only defined in Ω_t. In $\Omega_{1,t}$ one puts

$$-\Delta C_t = 0, \quad \text{in } \Omega_{1,t}$$

$$C_t = C_t \text{ (which is defined in } \bar{\Omega}_t\text{), on } \Gamma_t$$

$$C_t = 0, \quad \text{on } \partial B$$

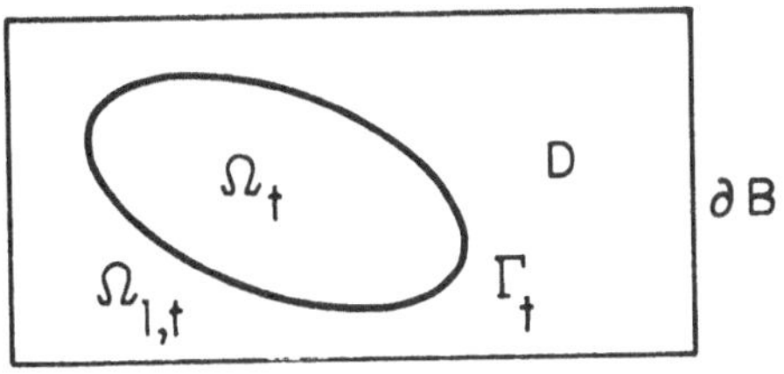

Figure 3.5 Fixed Domain Containing Ω_t

Now one has a function C_t defined on a fixed domain and he can use the fixed point method. If one does not know D, take at each time t a closed domain $D = D_t$ such that $\overline{\Omega}_t \subset D_t$. For a detailed development of this method, see Refs. 3 and 4.

4. TRANSPORTATION INTO A FIXED DOMAIN

The method of transportation into a fixed domain is illustrated with Model Problem No. 1. Suppose the domains Ω_u are the images of a fixed domain Ω by transformation $F(\cdot,u)$, as shown in Fig. 4.1, i.e.

$$x = F(X,u) = x(X,u)$$

$$X = F^{-1}(x,u) = X(x,u)$$

$$\Omega_u = F(\Omega,u)$$

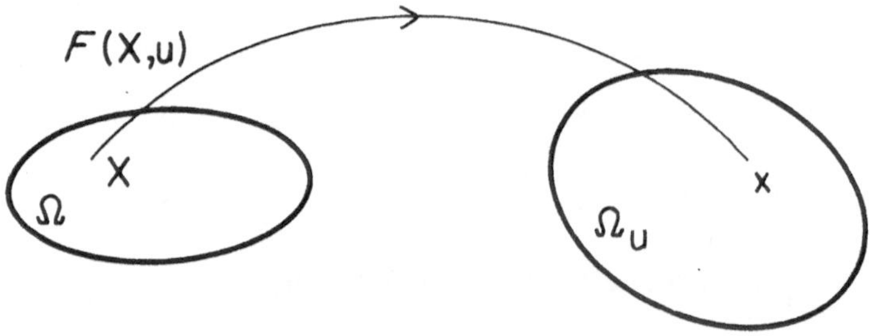

Figure 4.1 Transportation of Domain

The state function $y \in H^1(\Omega_u)$ satisfies

$$\int_{\Omega_u} (\nabla y \cdot \nabla \phi + y\phi)\, dx = \int_{\Omega_u} f\, \phi\, d\sigma, \text{ for all } \phi \in H^1(\Omega_u) \tag{4.1}$$

and the cost function is

$$J(u) = J(\Omega_u) = \int_{\Omega_n} |y - y_g|^2\, d\sigma \tag{4.2}$$

In Eqs. 4.1 and 4.2 the integrals will be transformed to the fixed domain Ω. To do this, one needs to define functions in the fixed space as images of functions defined in Ω_u. Let

$$Z(X,u) = z(x(X,u),u), \text{ for all } X \in \Omega$$

$$z(x,u) = Z(X(x,u),u), \text{ for all } x \in \Omega_u$$

For simplicity, one forgets for the moment the parameter u, and gets

$$\frac{\partial z}{\partial x_i}(x) = \sum_j \frac{\partial Z}{\partial X_j}(X(x)) \frac{\partial X_j}{\partial x_i}(x) = \nabla_X Z(X(x)) \cdot \frac{\partial X}{\partial x_i}(x)$$

and Eq. 4.1 becomes

$$\sum_{i=1}^{n} \int_{\Omega_u} \left\{ \sum_j \frac{\partial Y}{\partial X_j}(X(x)) \frac{\partial X_j}{\partial x_i}(x) \right\} \left\{ \sum_k \frac{\partial \Phi}{\partial X_k}(X(x)) \frac{\partial X_k}{\partial x_i}(x) \right\} dx$$

$$+ \int_{\Omega_u} Y(X(x))\Phi(X(x))dx = \int_{\Omega} F(X(x))\cdot\Phi(X(x))dx$$

or changing the variable x to X,

$$\sum_{i=1}^{n} \int_{\Omega} \left\{ \sum_j \frac{\partial Y(X)}{\partial X_j} \frac{\partial X_j}{\partial x_i}(x(X)) \right\} \left\{ \sum_k \frac{\partial \Phi(X)}{\partial X_k} \frac{\partial X_k}{\partial x_i}(x(X)) \right\} K(X)\,dX$$

$$+ \int_{\Omega} Y(X)\Phi(X)K(X)\,dX = \int_{\Omega} F(X)\Phi(X)K(X)\,dX \qquad (4.3)$$

here K is the Jacobian of the transformation, i.e.

$$K(X) = \det \frac{\partial F}{\partial X} = \det \frac{\partial x}{\partial X} = \frac{D(x_1,\ldots,x_n)}{D(X_1,\ldots,X_n)}$$

Now Eq. 4.3 can be written as

$$\int_{\Omega} \left\{ \sum_{j,k} A_{j,k}(X) \frac{\partial Y}{\partial X_j}(X) \cdot \frac{\partial \Phi(X)}{\partial X_k} + A_0(X)Y(X)\Phi(X) \right\} dX$$

$$= \int_{\Omega} F(X)K(X)\Phi(X)\,dX \qquad (4.4)$$

with

$$A_{j,k}(X) = \nabla_x X_j\ (x(X)) \cdot \nabla_x X_k\ (x(X))\ K(X)$$

$$A_0(X) = K(X)$$

Note that Φ is defined from $\phi \in H^1(\Omega_u)$:

$$\Phi(X) = \phi(x(X))$$

If one chooses ϕ as the image of $\psi \in H^1(\Omega)$,

$$\phi(x) = \psi(X(x))$$

one gets

$$\Phi(X) = \phi\ (x(X)) = \psi(X(x(X))) = \psi(X)$$

Then, $\Phi = \psi \in H^1(\Omega)$

This proves that in Eq. 4.4, one can take Φ arbitrarily in $H^1(\Omega)$ (under certain regularity hypothesis satisfied by F). The same argument proves that $Y \in H^1(\Omega)$. In the same way, one obtains

$$J(u) = \frac{1}{2}\int_{\Omega_u} |y - y_g|^2\ dx = \frac{1}{2}\int_{\Omega} |Y(X) - Y_g(X)|^2\ K(X)dX \tag{4.5}$$

One is now in a fixed domain Ω. The problem is now to minimize $J(u)$, where Y is defined by the state equation of Eq. 4.4. This is a classical control problem. The control u appears in F, then in $A_{j,k}$, A_0, K, F, and Y_g. That is, the control u appears in the coefficients of the equations and in the given functions F and Y_g. Begis and Glowinski [5] have used this method in the case shown in Fig. 4.2. Here, γ is the graph of $x_2 = u(x_1)$, i.e.

$$x_1 = X_1, \qquad \text{so } X_1 = x_1$$

$$x_2 = X_2\ u(X_1), \text{ so } X_2 = \frac{x_2}{u(x_1)}$$

Example 16 of Ref. 1 has been solved by this method.

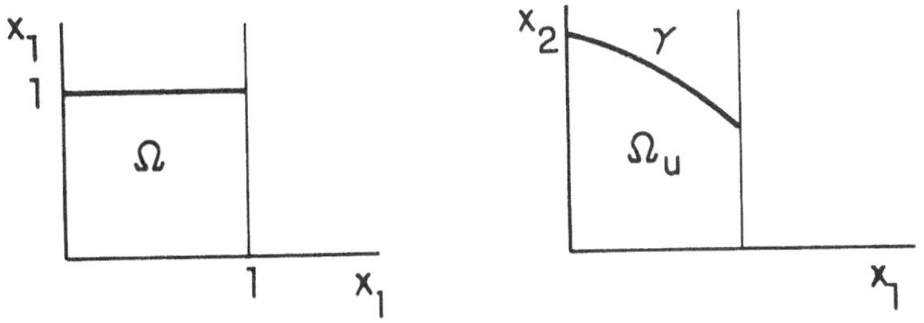

Figure 4.2 Transportation to Fixed Domain

5. FREE BOUNDARY-VALUE PROBLEMS

One can formulate free boundary-value problems as optimal design problems and then, use the previously studied methods. However, one can also use a specific method to solve this kind of problem. The method presented in this section was originally proposed by Cea, Bardos, Grisvard, and Lhomme in Ref. 6.

5.1 Problems Related to the Laplace Equation

Here Ω denotes a subset of a strip B (in $\mathbb{R}^2$),

$$B = \{(x,y) : 0 < x < 1,\ y > 0\}$$

The upper boundary of Ω is the curve γ shown in Fig. 5.1, whose equation in $y = z(x)$. One looks for two functions u and z such that

$$\Delta u = 0, \quad \text{in } \Omega \tag{5.1}$$

$$u = 0, \quad \text{on } \Gamma_1 \cup \Gamma_3 \cup \gamma \tag{5.2}$$

$$u = f, \quad \text{on } \Gamma_2 \tag{5.3}$$

$$\frac{\partial u}{\partial n} = g, \quad \text{on } \gamma \tag{5.4}$$

The functions f and g are given. This is suggested by water dam problems [7] and by problems concerning flow through a jet engine.

Let G be a Green function for the Laplace operator, satisfying homogeneous Dirichlet conditions on the boundary of the strip B (the asymptotic conditions on G as $y \to \infty$ are irrelevant here. Thus, G is such that

$$\left.\begin{aligned} \Delta_{(x,y)}\, G(x,y;x',y') &= \delta(x-x',y-y'), \text{ in } B \\ G(x,y;x',y') &= 0 \qquad\qquad\qquad , \text{ on } \delta B \end{aligned}\right\} \tag{5.5}$$

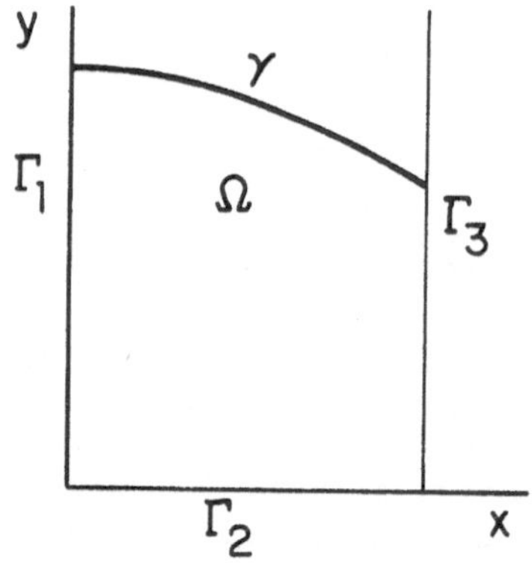

Figure 5.1 Domain for Free Boundary-Value Problem

For instance, G can be defined by the following series

$$G(x,y;x',y') = \sum_{n\geq 1} \frac{1}{n\pi} [e^{-n\pi|y-y'|} - e^{-n\pi|y+y'|}] \sin n\pi x \sin n\pi x' \tag{5.6}$$

When $(x,y) \in \gamma$ (i.e. $y = z(x)$, $x \in (0,1)$), Greens formula leads to

$$\frac{1}{2} u(x,y) = \int_{\Omega} G(x,y;x',y')\, \Delta u(x',y')\,dx'dy'$$
$$+ \int_{\partial\Omega} \left\{ \frac{\partial G}{\partial n_{x',y'}} (x,y;x',y') \cdot u(x',y') - \frac{\partial u}{\partial n_{x',y'}} (x',y') \cdot G(x,y;x',y') \right\} d\sigma_{x',y'}$$

Then, the boundary conditions on u and G imply that for all $x \in (0,1)$,

$$0 = \int_{\Gamma_2} u(x',0)\ [-\frac{\partial G}{\partial y'}\ (x,z(x);\ x',0)]dx$$

$$- \int_{\gamma} \frac{\partial u}{\partial n_{x',y'}}\ (x',y')\cdot G(x,y;x',y')\ d\sigma_{x',y'}$$

where $y = z(x)$ and $y' = z(x')$ on γ. Finally, one has, for all $x \in (0,1)$,

$$\int_0^1 g(x')\ G(x,z(x);x',z(x'))\ \sqrt{1 + (z'(x'))^2}\ dx'$$

$$- \int_0^1 f(x')\ \frac{\partial G}{\partial y'}\ (x,z(x);x',0)dx' = 0 \qquad (5.7)$$

Thus, one is led to an integro-differential equation in which the unknown is the function z.

Numerical Results. For numerical tests, one proceeds as follows:

(1) The curve is γ given by means of the function $x \to z(x)$. Then one applies the finite element method (on triangles, using first order polynomials) to solve the problem of Eqs. 5.1, 5.2, and 5.3, with f given (namely $f(x) = x(1-x)$). Use of Eq. 5.4 has been made to numerically calculate g (always with a mesh-size equal to 1/18).
(2) Next, starting from f and g, one tries to re-obtain z by means of Eq. 5.7. For this purpose, the interval (0.1) is divided into subintervals of length h. Then, the identity of Eq. 5.7 is considered only at the nodes j h ($j = 1,2,\ldots$) and the integrals have been approximated by finite sums. The nonlinear system thus obtained is then solved by the Newton-Raphson method. In numerical tests, the initial value for the unknown function z is the linear function that coincides with the exact solution z at 0 and 1. For measuring errors, one introduces

$$EAM = \max_i |z_c(ih) - z(ih)|$$

$$EAM = \frac{\sum_i |z_c(ih)-z(ih)|}{\sum_i 1}$$

$$ERM = Max \left| \frac{z_c(ih)-z(ih)}{z(ih)} \right|$$

where z_c is the function calculated numerically. Here, NINR is the number of steps (sweeps) in the Newton-Raphson method.

Numerical results using this algorithm are given in Table 5.1.

TABLE 5.1 NUMERICAL RESULTS FOR LAPLACE FREE BOUNDARY-VALUE PROBLEM

z	$z(x) = 2 - x^2$		$z(x) = \frac{1}{2} \cos \pi x + \frac{3}{2}$	
h	1/18	1/6	1/18	1/6
NINR	4	3		
EAM	0.080	0.155	0.071	0.190
EAm	0.030	0.065	0.028	0.086
ERM	0.072	0.118	0.034	0.178

One can apply the method outlined here, even when the goemetry is less convenient than in the above example. Consider as an illustration the case shown in Fig. 5.2. Here, one chooses G such that with P = (x,y), Q = (x',y'),

$$\Delta_P \, G(P;Q) = \delta_{(P-Q)}$$

$$G(P;Q) = 0, \text{ for all } P \in AB \cup BC$$

Then, when one writes down the Green formula for a point $Q \in \gamma$, he has an additional unknown, the normal derivative of u on AD, $\left.\frac{\partial u}{\partial n}\right|_{AD}$. An extra equation is obtained by writing down the Green formula for a point $Q \in AD$. Finally one obtains a system of nonlinear equations in the functions z (which defines γ) and $\frac{\partial u}{\partial n}$ on AD.

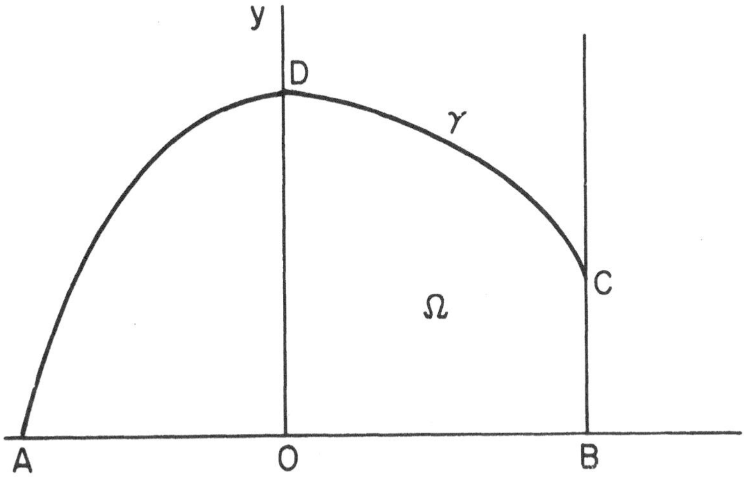

Figure 5.2 Free Boundary-Value Problem with More General Boundary

5.2 Problems Related to the Heat Equation

Let γ be the plane curve shown in Fig. 5.3, whose equation is $x = s(t)$, $t \geq 0$, with $s(0) = a > 0$, where a is a given number. The problem is to find u and s so that

$$\frac{\partial u}{\partial t}(x,t) = \frac{\partial^2 u}{\partial x^2}(x,t), \text{ for } t > 0,\ 0 < x < s(t) \tag{5.8}$$

$$u(0,t) = 0, \text{ for all } t > 0 \tag{5.9}$$

$$u(x,0) = f(x), \text{ for all } x \in (0,a) \tag{5.10}$$

$$u(s(t),t) = 0, \text{ for all } t > 0 \tag{5.11}$$

$$\frac{\partial u}{\partial x}(s(t),t) = \phi(s(t),s'(t)), \text{ for all } t > 0 \tag{5.12}$$

Here, ϕ is a given function. For instance the classical Stefan problem corresponds to

$$\phi(s(t);s'(t)) = -s'(t)$$

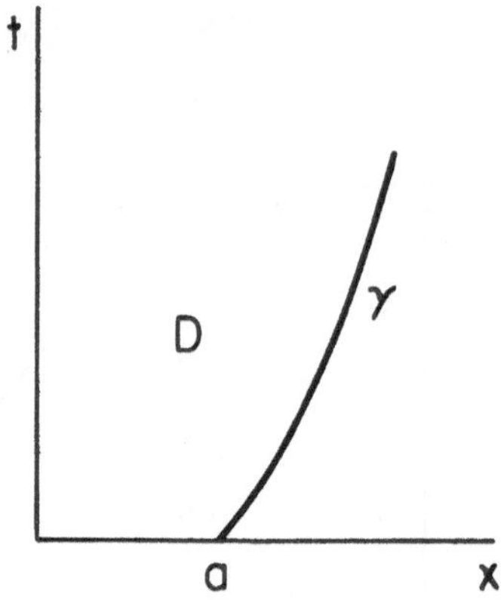

Figure 5.3 Geometry for Heat Equation

See for instance Refs. 8 to 11 for detailed treatments of this problem. When

$$\phi(s(t);s'(t)) = \frac{\lambda}{2\pi\, s(t)}$$

with a given λ, one is dealing with the problem solved by O. Sero-Guillaume [12].

For an appropriate choice of the function of ϕ, the initial data f is to be consistent with Eqs. 5.9 and 5.11 (i.e. $f(0) = 0$ and $f(a) = 0$). Let G be the Green function satisfying

$$\left.\begin{aligned} &-G'_{\tau}(y,\tau;x,t) - G''_{y^2}(y,\tau;x,t) = \delta(y-x,\tau-t) \\ &G(y,\tau;x,t) = 0, \text{ for all } \tau > t \\ &G(y,\tau;x,t) = 0, \text{ on } y = 0 \end{aligned}\right\} \tag{5.13}$$

One easily checks that

$$G(y,\tau;x,t) = F(y,\tau;x,t) - F(y,\tau;-x,t) \tag{5.14}$$

where

$$F(y,\tau;x,t) = Y(t-\tau)\left[\frac{1}{\sqrt{4\pi(t-\tau)}}\right]e^{-\frac{(x-y)^2}{4(t-\tau)}} \tag{5.15}$$

and Y is the heavyside function.

Recall the Green formula

$$\iint_D [G(u'_\tau - u''_{y^2}) - u(-G'_\tau - G''_{y^2})]d\tau\, dy$$

$$= \int_{\partial\Omega} [(uG'_y - u'_y G) \cos(\vec{x}, \vec{y}) + uG \cdot \cos(\vec{x}, \vec{\tau})]\, ds \qquad (5.16)$$

where the domain D is bounded by the curve γ, the x-axis, and t-axis; $\vec{n}$ denotes the unit normal vector to ∂D, pointing outward; and $\vec{y}$ and $\vec{\tau}$ are unit vectors parallel to the x-axis and the t-axis, respectively. Denote by D_t the open set $\{(y,\tau) : 0 < \tau < t, 0 < y < s(t)\}$ and by γ_t the curve $x = s(t)$, $t \in (0,\tau)$. Writing down the identity of Eq. 5.16 at the point $(x = s(t),t)$, one gets

$$\frac{1}{2} u(s(t),t) = \iint_{D_t} G(y,\tau;s(t),t)[u'_\tau(y,\tau) - u''_{y^2}(y,\tau)]\, dy\, d\tau$$

$$+ \int_0^t u(0,\tau)\; G'_y(0,\tau;s(t),t)\, d\tau$$

$$+ \int_0^a u(y,0)\; G(y,0;s(t),t)\, dy \qquad (5.17)$$

$$- \int_{\gamma_t} \Big\{[u(s(\tau),\tau)G'_y(s(\tau),\tau;s(t),t)$$

$$- u'_y(s(\tau),\tau)G(s(\tau),\tau;s(t),t)] \cos(\vec{x},\vec{y})$$

$$+ u(s(\tau),\tau)G(s(\tau),\tau;s(t),t) \cos(\vec{x},\vec{\tau})\Big\}\, d\gamma$$

Due to the conditions on u and G, the Green formula of Eq. 5.16 leads to the condition that for all $t > 0$,

$$\int_0^t \phi(s(t),s'(t))\; G(s(\tau),\tau;s(t),t)\, d\tau$$

(Equation continued on next page)

$$= \int_0^a f(y)\ G(y,0;s(t),t)\ dy \tag{5.18}$$

Here again, an integro-differential equation is derived, where the only unknown is the function s.

A method for solving the problem of Eqs. 5.8 to 5.12 is as follows:

(1) Search for a solution s of Eq. 5.18.
(2) Solve the problem of Eqs. 5.8 to 5.11 with s given.

6. PARTIAL ELIMINATION OF THE DOMAIN

Usually, the state equation is a partial differential equation $E(\Omega,y_\Omega) = 0$, where Ω is variable. In the methods presented thus far, one solves the optimal design problem by an iterative method. Hence, one has to solve many problems of the form $E(\Omega,y_\Omega) = 0$. This could be very expensive. In some cases, one can cut down this cost by using the following techniques: Cut Ω in two parts, as shown in Fig. 6.1 and suppose that only the part Ω_2 is variable. Instead of solving a problem in Ω, solve a problem in Ω_2. To do this, however, one needs a boundary condition on Σ. For problems of order 2, this condition will in general be a relation between $y|_\Sigma$ and $\frac{\partial y}{\partial u}\Big|_\Sigma$

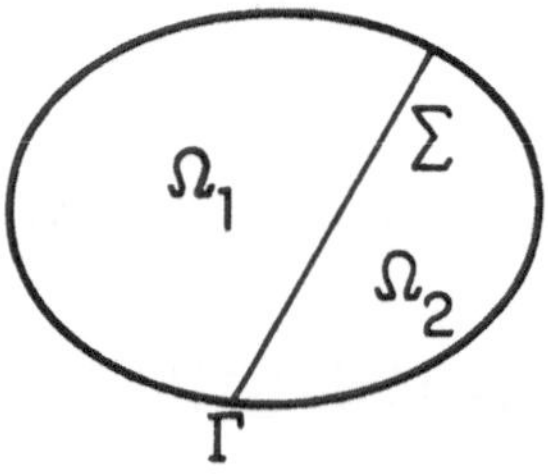

Figure 6.1 Subdivision of Ω

Consider now an example, in which y satisfies the boundary-value problem

$$\left.\begin{aligned} -\Delta y + y &= f, \text{ in } \Omega_1 \cup \Sigma \cup \Omega_2 = \Omega \\ \frac{\partial y}{\partial u} &= 0 \qquad \text{in } \Gamma = \partial\Omega \end{aligned}\right\} \tag{6.1}$$

where f is given. The domain is shown in Fig. 6.2.

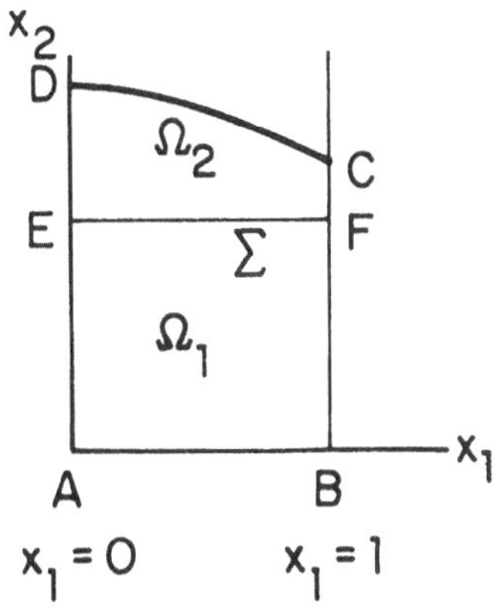

6.2 Partition of Domain for Partial Elimination

Let

$$y_{|EF} = \phi$$

one then has

$$-\Delta y + y = f, \qquad \text{in } \Omega_1$$

$$\frac{\partial y}{\partial u} = 0, \qquad \text{on EA, AB, and BF}$$

$$y = \phi, \qquad \text{on } \Sigma$$

Then, y is well defined and there exist 2 operators A and B such that

$$y = A\phi + Bf$$

From this, one has

$$\left.\frac{\partial y}{\partial n}\right|_{EF} = A\phi + Bf$$

But with $\phi = y|_{EF}$,

$$\left.\frac{\partial y}{\partial n}\right|_{EF} = Ay\Big|_{EF} + Bf\Big|_{\Omega_1} \qquad (6.2)$$

For some geometries, it is possible to explicitly write the operators A and B. This can be done for the example where Ω_1 is a rectangle and the equation is $-\Delta y + y = f$.

Once Eq. 6.2 has been obtained, one has to solve the following problem in Ω_2:

$$\left.\begin{array}{ll} -\Delta y + y = f, & \text{in } \Omega_2 \\ \frac{\partial y}{\partial u} = 0, & \text{on } FC \cup CD \cup DE \\ -\left.\frac{\partial y}{\partial u}\right|_{EF} = \Delta y\Big|_{EF} + Bf\Big|_{\Omega_1} & \end{array}\right\} \qquad (6.3)$$

In the third equation of Eq. 6.3 one has a - sign, because $\vec{n}$ has been changed to $-\vec{n}$. This relation is independent of the shape of Ω_2. It depends only on Ω_1 (and the operators and f). The number of variables in Eq. 6.3 could be much smaller than in Eq. 6.1. Note that for the adjoint state one has the same kind of relation, possibly exactly the same! For more detail, see Ref. 6.

7. SHAPE OPTIMAL DESIGN WITH CONSTRAINTS

In this final section it is shown how to adapt some classical methods of nonlinear programming to the problems studied here. Suppose one has the most general case when Ω_t is defined by a velocity field, the unknown being the velocity. There is no general proof for the convergence of the methods that are given now. It will be possible only to adapt to our problems some classical results of the optimization theory in Hilbert spaces.

7.1 Gradient Projection Method

Suppose one has computed the derivative of $j(t) = J(\Omega_t)$ and has found

$$\dot{j}(t) = \int_{\Gamma_t} C_t < v_t, n_t > d\sigma$$

In order to study only the case of active constraints, suppose for example that there are two equality constraints,

$$\left.\begin{aligned} K_1(\Omega) &= 0 \\ K_2(\Omega) &= 1 \end{aligned}\right\} \tag{7.1}$$

To be specific, take

$$K_1(\Omega) = \int_\Omega dx,$$

$$K_2(\Omega) = \int_\Omega y\ dx$$

and introduce, for i = 1 and 2, the functions $k_i(t) = K_i(\Omega_t)$. Using the same technique for j, one computes their derivatives, as

$$\dot{k}_i(t) = \int_{\Gamma_t} D_{i,t} < v_t, n_t > d\sigma$$

Because of Eq. 7.1, the constraints are

$$\dot{k}_i(t) = \int_{\Gamma_t} D_{i,t} < v_t, n_t > d\sigma = 0 \tag{7.2}$$

Suppose now one has discretized the time t and has found $\Omega_n (= \Omega_{n\tau})$. One needs to compute v_n in order to construct Ω_{n+1}. With no constraints, one has only to solve

$$b_n(v_n, \psi) = -\int_{\Gamma_n} C_n < \psi, n > d\sigma, \text{ for all } \psi \in V_n$$

However, Eq. 7.2 must be satisfied. One way to do this is to project the solution v_n of Eq. 7.3 to the domain of the constraints of Eq. 7.2. This will give the new direction of descent. Another way is to solve the variational inequality problem (or quadratic programming)

$$\text{Min.}\ \frac{1}{2} b_n(\psi,\psi) + \int_{\Gamma_n} C_n < \psi, n > d\sigma$$

under the constraints

$$\int_{\Gamma_t} D_{i,t} < \psi, n > d\sigma = 0, \ i = 1,2$$

7.2 Penalty Method

If, for example, one has the constraint

$$\int_\Omega du = m$$

One can replace the cost function J by

$$J(\Omega) + \frac{1}{\varepsilon}\left[\int_\Omega dx - m\right]^2$$

with a very small $\varepsilon > 0$.

7.3 Duality

Suppose one has a state constraint

$y(x) \le c$, for all $x \in \Omega_2$, Ω_2 fixed, c given

and a domain constraint

$$\int_\Omega dx = m, \quad m \text{ given.}$$

One can introduce Lagrange multipliers so that the Lagrangian will be

$$L(\Omega;\mu,\beta) = J(\Omega) + \int_{\Omega_2} \mu(x)\,(y(x) - c)dx + \beta\left[\int_{\Omega} dx - m\right]$$

with $\mu(x) \geq 0$ a.e. in Ω_2, $\mu \in L^2(\Omega_2)$, and $\beta \in \mathbb{R}$

Necessary conditions of optimality now give us the following:

(1) The derivative of the mapping $\Omega \to L(\Omega,\mu,\beta)$ will be zero, or Ω minimizes the cost function $L(\Omega,\mu,\beta)$.
(2) $y(x) - c \leq 0$, for all $x \in \Omega_2$

$$\mu(x) \geq 0$$

$$(y(x) - c)\ \mu(x) = 0$$

(3) $\int_{\Omega} dx = m$

(in fact one should write $\mu(y-c)$ with μ a measure in Ω_2, instead of

$$\int_{\Omega_2} \mu(y-c)dx$$

To solve this kind of problem, in nonlinear programming one very effective method is the Uzawa Method.

<u>Uzawa Method.</u> This is only the gradient projection method for the dual problem. One defines $J^*(\mu,\beta) = \min_{\Omega} L(\Omega,\mu,\beta)$ and wants to maximize $J^*(\mu,\beta)$, with $\mu \geq 0$. As usual, if the problem $\min_{\Omega} L(\Omega,\mu,\beta)$ has a unique solution, one can expect that [Ref. 2 for example]

$$\frac{\partial J^*}{\partial \beta}(\mu,\beta) = + \int_{\Omega} dx - m$$

$$\frac{\partial J^*}{\partial \mu}(\mu,\beta) = [y - c]$$

The algorithm to go from the step n to step n+1 is

$$\left.\begin{aligned}&\beta_{n+1} = \beta_n + \rho\left[\int_\Omega dx - m\right]\\&\mu_{n+1} = \mu_n + \rho[y_n - c]\\&\Omega_{n+1} \text{ minimizes } L(\Omega, \mu_{n+1}, \beta_{n+1}),\\&\text{without constraints.}\end{aligned}\right\} \quad (7.4)$$

In fact, instead of computing Ω_{n+1} from Ω_n and μ_{n+1} and β_{n+1} one can do only one or a few steps in the iterative method, which lead from Ω_n to Ω_{n+1}.

For the UZAWA method related with partial differential equations, see for example Ref. 13.

REFERENCES

1. Cea, J., "Problems of Shape Optimal Design", Optimization of Distributed Parameter Structures (Eds. E.J. Haug and J. Cea), Sijthoff & Noordhoff, Alphen aan den Rijn, Netherlands, 1980.
2. Zolesio, J.P., "The Material Derivative (Or Speed) Method For Shape Optimization", Optimization of Distributed Parameter Structures (Eds. E.J. Haug and J. Cea), Sijthoff & Noordhoff, Alphen aan den Rijn, Netherlands, 1980.
3. Hadamard, J., "Mémoire sur le Probleme d'Analyse Relatif à l'Équilibre des Plaques Elastiques Encastrées (1908)", oeuvres de J. Hadamard, C.N.R.S., Paris, 1968.
4. Joseph, D.D., "Parameter and Domain Dependence of Eigenvalues of Elliptic Partial Differential Equations", Arch. Rat. Mech. Anal., Vol. 24, 1967.
5. Pironneau, O., Sur les Problèmes d'Optimisation de Structure en Mécanique des Fluides, Thèse de Doctorat, Pariv VI, 1976.
6. Pironneau, O. and Saguez, C., Asymptotic Behaviour of Solutions of P.D.E. with Respect to the Domain, IRIA Laboria Report No. 218, 1977.
7. Deuling, H.J. and Helfrich, W., "The Curvature Elasticity of Fluid Membranes - A Catalogue of Vesicle Shapes", J. de Phys., Vol. 37, Nov. 1976, pp. 1335-1345.
8. Rousselet, B., Probléme Inverse de Valeurs Propres; Optimization Techniques, Springer Verlag, Lecture Notes in Computer Sciences, Vol. 2, No. 41, 1976, pp. 77-85.
9. Rubinstein, L.I., "The Stefan Problem", A.M.S., Translations of Mathematical Monographs, Vol. 27, Providence, R.I., 1971.

10. Troesch, B.A. and Troesh, H.R., "Eigenfrequencies of an Elliptic Membrane", Math of Comp., Vol. 27, No. 126, 1973, pp. 755-772.
11. Troesch, B.A. and Troesh, H.R., "Eigenfrequencies of an Elliptic Membrane," Math of Comp., Vol. 27, No. 126, 1973, pp. 755-772.
12. Sero-Guillaume, O., "Probleme a Frontiere Libre de Type Stefan," These, INP. Lorraine, Nance, France, 1978.
13. Cea, J., Lectures on Optimization - Theory and Algorithms, Springer-Verlag, 1978.

THE MATERIAL DERIVATIVE (OR SPEED) METHOD FOR SHAPE OPTIMIZATION

Jean-Paul Zolesio

Départment de Mathématiques, Université de Nice,
06034, Nice-Cedex, France

ABSTRACT

In this paper, basic results in domain optimization are presented, with emphasis on calculation of the Eulerian semi-derivative of a cost function $J(\Omega)$ that is to be minimized by selecting the domain Ω. Results presented here are treated in greater detail in the author's thesis, "Optimisation de Domaine par Deformation" (Univ. of Nice, 1979), where references are made to the work of J. Cea, B. Rousselet, D. Chenais, B. Palmerio, A. Dervieux, F. Murat, J. Simon, O. Pironeau, A. Bendali, A.M. Micheletti, Glowinsky, J. Giroire, and others who have worked on shape optimal design. The problem of large deformations of the domain is also treated, i.e. the problem of the choice of the speed (or domain deformation field) V that leads to a domain differential equation, for which existence results are given and some nondifferentiable problems are considered.

1. DEFORMATIONS OF DOMAINS

Let V be a regular, n-dimensional vector field defined on $[0,1] \times \mathcal{U}$, where $\mathcal{U}$ is an open neighborhood of a bounded set Ω. Suppose V belongs to $C^0(I, C^k(\mathcal{U}, \mathbb{R}^n))$, i.e. for any t, all the space derivatives of V are continuous (up to the order k) and the mapping $t \longrightarrow V(t)$ is continuous for the topology given by uniform convergence on any compact subset of $\mathcal{U}$ of these derivatives (k is an integer greater or equal to one).

The differential equation and initial conditions

$$\left.\begin{aligned} \frac{d}{dt}\, x(t,X) &= V(t,x(t,X)) \\ x(0,X) &= X \end{aligned}\right\} \qquad (1.1)$$

uniquely define the mapping

$$T_t(V) : X \longrightarrow x(t,X)$$

This mapping has regularity properties stated in the following theorem 1 :

Theorem 1.1 One can find an open set $\mathcal{U}_1$ such that $\bar{\Omega} \subset \mathcal{U}_1 \subset \mathcal{U}$ and an interval I_1, $0 \in I \subset I_1$ such that for any $t \in I_1$, the mapping $T_t(V)$ is a homeomorphism from $\mathcal{U}_1$ onto an open subset 0_t of $\mathcal{U}$. Further, $T_t(V)$ and its inverse mapping $T_t(V)^{-1}$ have C^k regularity.

The geometry of the transformation $T_t(V)$ and the neighborhoods of Theorem 1.1 are shown in Fig. 1.1. Deformations of the set Ω by the field V are denoted $\Omega_t(V) = T_t(V)(\Omega)$. For simplicity, one simply writes Ω_t and T_t.

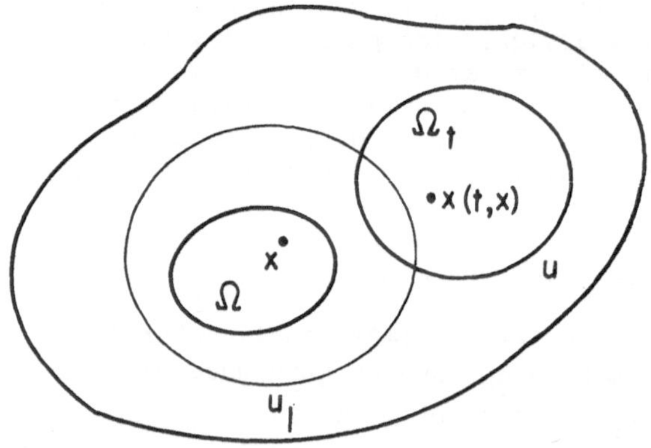

Figure 1.1 Transformation $T_t(V)$ from Ω to Ω_t

2. DERIVATIVES OF FUNCTIONS DEFINED ON Ω_t

2.1 Pointwise Derivatives

Let $y(\Omega_t)$ be a regular function defined on Ω_t. Then $y(\Omega_t)\circ T_t$ is defined on Ω and, for any X in Ω, the derivative

$$\dot{y}(\Omega)(X) = \frac{d}{dt}\, y(\Omega_t)\circ T_t \Big|_{t=0} = \frac{d}{dt}\, y(\Omega_t)(x(t,X)) \Big|_{t=0} \tag{2.1}$$

is (if it exists) the Eulerian or material derivative at $X \in \Omega$.

Of course if $y(\Omega_t)$ is the restriction to Ω_t of a function $Y(t,x)$ defined on $I_1 \times \mathcal{U}_1$, i.e., $y(\Omega_t)(X) = Y(t,x), x \in \Omega_t$, then one has

$$\dot{y}(\Omega)(X) = \frac{\partial Y}{\partial t}(0,X) + \nabla Y(0,X)\cdot V(0,X) \tag{2.2}$$

2.2 Derivatives in Normed Vector Spaces

Let Ω be a regular open set with C^k regularity; i.e., its boundary Γ is a compact manifold of class C^k in $\mathbb{R}^n$ (n=2 or 3). The open set Ω_t has the same regularity. The Sobolev space H^m, for $m \leq k$, are preserved by the diffeomorphism T_t, i.e.

$$H^m(\Omega) = \{z_t\circ T_t: \text{ with } z_t \in H^m(\Omega_t)\}$$

Let $y(\Omega_t)$ be an element of the vector space $H^m(\Omega_t)$ and consider the Eulerian derivative $y(\Omega)$ as the limit (if it exists) in $H^m(\Omega)$ of the differential quotient $[y(\Omega_t)\circ T_t - y(\Omega)]/t$, as t goes to zero, i.e.

$$\lim_{t\to 0} \| [y(\Omega_t)\circ T_t - y(\Omega)]/t - \dot{y}(\Omega) \|_{H^m(\Omega)} = 0$$

Note that this limit has a sense in a Sobolev space setting, whereas for $y(\Omega_t) \in H^m(\Omega_t)$, the pointwise derivative of Section 2.1 would be meaningless.

In the first examples to be given, $\dot{y}(\Omega)$ depends linearly and continuously on the stationary field $V(0)$. In fact, it is proved in Ref. 1 that if $\dot{y}(\Omega)$ depends continuously on V, then it just depends on $V(0)$.

If $m > n/2$, then by the Sobolev imbedding theorem [2], the vector space $H^m(\Omega_t)$ is a topological subspace of $C^0(\Omega_t)$ and then $\dot{y}(\Omega)$ is also the limit of the differential quotient of Section 2.1, in the $C^0(\bar{\Omega})$ topology (uniform convergence on the compact set $\bar{\Omega}$), which is stronger than the material derivative. But if $m \leq n/2$, the function $y(\Omega)$ is just defined almost everywhere on Ω and the pointwise derivative has no sense.

3. DERIVATIVE OF DOMAIN FUNCTIONALS

Let $J(\Omega)$ be a real number associated with any regular domain Ω (for example the measure of Ω). The Eulerian upper semi-derivative of J at Ω, in the direction of the field V is the real number (finite or infinite) given by (in general see Ref. 3)

$$\bar{d}J(\Omega;V) = \limsup_{t\to 0} \, [J(\Omega_t) - J(\Omega)]/t \tag{3.1}$$

In general, this derivative is shown [1] to depend only on the vector field $V(0)$, as stated in the following theorem.

<u>Theorem 3.1.</u> Let the mapping $V \to \bar{d}J(\Omega;V)$ be finite and continuous, then $\bar{d}J(\Omega;V) = \bar{d}J(\Omega;V(0))$

If the limit exists and is finite in the definition of $\bar{d}J$, one gets the Eulerian semi-derivative of J at Ω, in the direction V, and writes $dJ(\Omega;V)$.

Consider now the case in which the mapping $V \to dJ(\Omega;V)$ is linear and continuous. By restriction to the subspace of test functions (with compact support and continuous derivatives up to the order k) $\mathcal{D}^k(\mathcal{U},\mathbb{R}^n)$, this mapping defines a vector distribution $G(\Omega)$ that is called the gradient of J at Ω; i.e. for all $V \in \mathcal{D}^k(\mathcal{U},\mathbb{R}^n)$, $dJ(\Omega;V) = \langle G(\Omega),V\rangle_{\mathcal{D}^k(\mathcal{U},\mathbb{R}^n)' \times \mathcal{D}^k(\mathcal{U},\mathbb{R}^n)}$ where $\mathcal{D}^k(\mathcal{U},\mathbb{R}^n)'$ is the dual of $\mathcal{D}^k(\mathcal{U},\mathbb{R}^n)$ and $\langle\ ,\ \rangle$ denotes the duality pairing.

Now, if V has its support in the open set Ω or in the complement of Ω, then for t small enough one has $\Omega_t = \Omega$ and

$dJ(\Omega;V) = 0$, i.e. $\langle G(\Omega),V\rangle = 0$. Thus the restriction of the distribution $G(\Omega)$ to Ω (and to the complement of $\bar{\Omega}$) is equal to zero and the gradient is a finite order distribution, with support on the boundary Γ of Ω. When the open set Ω is of regularity C^{k+1}, the following two essential properties for the gradient have been demonstrated [1]:

(1) $G(\Omega)$ is supported by the manifold Γ, which is a curve for n=2 and a surface for n=3 and the transverse order of $G(\Omega)$ is equal to zero, which means that the computation of the number $\langle G(\Omega),V\rangle$ just involves the tangential derivatives of V on Γ but not any transverse derivative. This property shows that $G(\Omega)$ is an extension to the open set $\mathcal{U}$ of a vector distribution $g(\Omega)$ on the manifold S, i.e.

$$G(\Omega) = {}^t\gamma_\Gamma g(\Omega)$$

where γ is the restriction mapping (trace) to Γ and ${}^t\gamma_\Gamma$ it's transpose.

(2) The normal field n on Γ is a field of regularity C^k on the manifold Γ (n is taken going out of Ω). The field n can be extended to a field N_0 in a neighborhood $\mathcal{U}$ of Γ, N_0 having the regularity C^k in $\mathcal{U}$.

For any test function V belonging to $\mathcal{D}^k(\mathcal{U},\mathbb{R}^n)$ one may decompose V as

$$V = V_n + V_T$$

where

$$V_n = (V\cdot N_0)N_0$$

$$V_T = V - V_n$$

The field V_T is tangent to the boundary Γ. It is efficient now to prove that $\Omega_t(V_T) = \Omega$ for any t, so that $\langle G(\Omega),V_T\rangle = 0$ and

$$\langle G(\Omega),V\rangle = \langle G(\Omega),(V\cdot N_0)N_0\rangle$$

for any V. By definition of a k-order vector distribution G multiplied by a C^k regular vector field, $G(\Omega) = \big(G(\Omega)\cdot N_0\big)N_0$.

These results may be summarized in the form of a theorem, which is proved in Ref. 1.

Theorem 3.2. Let J possess a gradient $G(\Omega)$ at any domain Ω of C^k regularity and let Ω be a C^{k+1} regular domain. Then there exists a scalar distribution $g_n(\Omega)$ on the manifold Γ, of order $\leq k$, such that

$$G(\Omega) = {}^t\gamma_\Gamma(g_n(\Omega)n) \tag{3.2}$$

In other words, one has the Hadamard formula

$$\bar{d}J(\Omega;V) = \int_\Gamma g_n(\Omega)V(0)\cdot n \, ds \tag{3.3}$$

for any field V belonging to $C^0(I,C^k(\mathcal{U},\mathbb{R}^n))$, where the integral on Γ is, by extension, the notation for the bilinear duality form on Γ between the finite order distributions on Γ and the regular functions on Γ.

In several cases to be treated, $g_n(\Omega)$ is in fact an integrable function on Γ, so the integral is really an integral on Γ, ds being the classical measure for hypersurface in Euclidian $\mathbb{R}^n$ space. Examples are given, however, in which the distribution $g_n(\Omega)$ actually involves tangential derivatives of $v = V(0)\cdot n$ on S.

The neighborhoods $\mathcal{U}$ of Ω and Γ considered were arbitrary. This means that the derivative $\bar{d}J(\Omega;V)$ actually only depends on the germ of V(0) on Γ, i.e. the image of V(0) in the inductive limit $Z_k(\Gamma) = \varinjlim_{\mathcal{U}\supset\Gamma} C^k(\mathcal{U},\mathbb{R}^n)$.

The gradient may be considered as an element of the topological dual vector space $Z_k(\Gamma)'$ and $Z_k(\Gamma)$ as the tangent space to the topological space of regular domains at Ω.

Several different presentations of the gradient have been developed in Refs. 4, 5, 6, and 7, but, in fact, as seen here the gradient is actually unique for a differentiable domain functional J.

In many cases, the gradient may be written as

$$G = {}^t\gamma_\Gamma(g_n(\Omega)n)$$

where $g_n(\Omega)$ is in fact a regular function on Γ. Now, $g_n(\Omega)$ is the restriction to Γ of a function g defined in Ω, or in a neighborhood of Γ. Then,

$$G = {}^t\gamma_\Gamma(g_n(\Omega)n) = {}^t\gamma_\Gamma(\gamma_\Gamma g)n \\ = g\nabla X_\Omega \tag{3.4}$$

where X_Ω is the characteristic function of Ω,

$$X_\Omega(x) = \begin{cases} +1 , & \text{if } x \in \Omega \\ 0 , & \text{if } x \in \Omega^c \end{cases}$$

It may be shown that if Ω is regular,

$$X_\Omega \in H^s(\mathbb{R}^n) \text{, for all } s \text{ with } s < \tfrac{1}{2}$$

The function $g(\Omega)$, when it exists, is not unique. In fact, g is any extension of $g_n(\Omega)$, but the gradient G is unique. To fix the terminology, one says that:

(1) G is the gradient, a vectorial distribution on $\mathbb{R}^n$.

(2) g_n is the scalar gradient, a distribution on the surface Γ.

(3) g is a distributed formulation of the gradient.

Now, many numerical methods may use such a formulation of the gradient.

4. SOME BASIC RESULTS FOR GRADIENT COMPUTATION

Let V be a regular field belonging to $C^0(I,C^k(U,\mathbb{R}^n))$ (from Refs. 1 and 6).

4.1 Derivatives at t=0

Let DT_t be the jacobian matrix (differentiation with respect to X) of the mapping T_t, then $DT_t\big|_{t=0} = I$, the identity matrix,

and

$$\frac{d}{dt} DT_t \Big|_{t=0} = DV(0) \tag{4.1}$$

This result can be seen from the equation of variation [8]

$$\frac{d}{dt} DT_t = \left(DV(t) \circ T_t\right) DT_t$$

where $DV(t)$ is the jacobian of $V(t)$ with respect to x. For the transpose and inverse matrices, one has

$$\left.\begin{aligned} \frac{d}{dt} {}^tDT_t \Big|_{t=0} &= {}^tDV(0) \\ \frac{d}{dt} {}^tDT_t^{-1} \Big|_{t=0} &= - {}^tDV(0) \end{aligned}\right. \tag{4.2}$$

Note that $({}^tDT_t)^{-1} = {}^t\left((DT_t)^{-1}\right) = {}^tDT_t^{-1}$.

For determinants,

$$\left.\begin{aligned} \frac{d}{dt} \det (DT_t) \Big|_{t=0} &= \operatorname{div} V(0) \\ \frac{d}{dt} \det (DT_t^{-1}) \Big|_{t=0} &= - \operatorname{div} V(0) \end{aligned}\right\} \tag{4.3}$$

Let Γ and Γ_t be the boundaries of Ω and Ω_t and n be the normal vector field on Γ (exterior to Ω). Then n_t on Γ_t is given by

$$n_t(x(t,X)) = ||{}^tDT_t^{-1}(X)\ n(X)||^{-1}\ {}^tDT_t^{-1}(X)\ n(X) \tag{4.4}$$

The Eulerian derivative of n is obtained using

$$\frac{d}{dt} ||\ {}^tDT_t^{-1} n|| \Big|_{t=0} = - \langle DV(0)n, n \rangle$$

as

$$\dot{n} = \frac{d}{dt} n_t \circ T_t \Big|_{t=0} = \langle DV(0)n, n \rangle n - {}^tDV(0)n \tag{4.5}$$

4.2 Transport of Derivatives

For any function y that is regular enough,

$$(\nabla y) \circ T_t = {}^tDT_t^{-1} \; \nabla(y \circ T_t) \tag{4.6}$$

$$(\Delta y) \circ T_t = J_t^{-1} \; \mathrm{div} \; (J_t DT_t^{-1} \; {}^tDT_t^{-1} \; \nabla(y \circ T_t)) \tag{4.7}$$

where $J_t = \det (DT_t)$ and the ∇ and Δ operators are with respect to x or X, as dictated by the context. Also,

$$\frac{d}{dt} \left(J_t DT_t^{-1} \; {}^tDT_t^{-1} \right) \Big|_{t=0} = \mathrm{div} \; V(0)I - (DV(0) + {}^tDV(0)) \tag{4.8}$$

4.3 Eulerian Derivative for a Gradient

Let $y(\Omega_t)$ be a regular function defined on Ω_t or on Γ_t, then, from Eqs. 4.2 and 4.6,

$$\dot{\widehat{\nabla y}}(\Omega) = \frac{d}{dt} \left((\nabla y(\Omega_t) \circ T_t) \right) \Big|_{t=0}$$

$$= \frac{d}{dt} \left({}^tDT_t^{-1} \; \nabla(y(\Omega_t) \circ T_t) \right) \Big|_{t=0} \tag{4.9}$$

$$\dot{\widehat{\nabla y}}(\Omega) = - {}^tDV(0) \; \nabla y(\Omega) + \nabla(\dot{y}(\Omega))$$

Now, if $y(\Omega_t)$ is the restriction to $t \times \Omega_t$ (or $t \times \Gamma_t$) of a regular function $Y(t,x)$ defined on $I \times \mathcal{U}$, then one gets another expression,

$$\dot{\widehat{\nabla y}}(\Omega) = \frac{d}{dt} (\nabla Y(t) \circ T_t) \Big|_{t=0}$$

$$= \nabla \left(\frac{\partial Y}{\partial t} (0) \right) + D^2 Y(0) V(0) \tag{4.10}$$

where $D^2Y(0)$ is the second derivative matrix:

$$D^2Y(0)_{ij} = \frac{\partial^2 Y}{\partial x_i \partial x_j}(0)$$

4.4 Unitary Extension N_0 of the Normal Field n to a Neighborhood of Γ

Locally, Γ is defined as $\Gamma = \{x \in \mathcal{U}: c_n(x) = 0\}$, where $c:\mathcal{U} \longrightarrow B = \{z \in \mathbb{R}^n: |z| < 1\}$,

$$c(\mathcal{U} \cap \Gamma) = B_0 = \{z \in B: z_n = 0\}$$

Let $e_n = (0,0,\ldots,0,1)$ and $h = c^{-1}_{\cdot}$ Then, for $x \in \Gamma$,

$$n(x) = ||{}^tDh^{-1}(c(x))e_n||^{-1}\ {}^tDh^{-1}(c(x)\ e_n \qquad (4.4')$$

If Γ is covered by p open sets of the form $\mathcal{U}$ and if r_i is a regular function with compact support in $\mathcal{U}_i$, with $1 = r_1 + r_2 + \ldots + r_p$ on Γ, then

$$M_0(x) = \sum_{i=1}^{p} r_i(x)||{}^tDh_i^{-1}\left(c_i(x)\right)e_n||^{-1}\ {}^tDh_i^{-1}(c(x))\,e_n$$

is a nonunitary, regular extension of n to $\mathcal{U}_1 \cup \mathcal{U}_2 \cdots \cup \mathcal{U}_p = \mathcal{U}$. By restriction of $\mathcal{U}$ (eventually) one may consider M_0 not equal to zero on $\mathcal{U}$ and

$$N_0(x) = ||M_0(x)||^{-1}\ M_0(x)$$

is a unitary extension of n to a neighborhood of Γ. Of course, such an extension is not unique, but it will be shown that on some derivatives of N_0 do not depend on the choice of N_0.

If Ω and Γ have C^k regularity, that is the mappings c_i are C^k, then N_0 and n have C^{k-1} regularity.

4.5 Special Expressions for Vector Fields V Proportional to N_0 (See Ref. 1)

Let Ω have C^{k+1} regularity and v be a C^k regular function, so that $V = vN_0$ has C^k regularity. Then

$$DV = N_0\ {}^t\nabla v + v\ DN_0$$

but N_0 is a unitary field, so

$$ {}^tDN_0\ N_0 = 0$$

Thus,

$$ {}^tD(vN_0)N_0 = \nabla v\left({}^tN_0N_0\right) = \nabla v$$

and for $V = vN_0$, one has the Eulerian derivative on Γ of the normal field, given by

$$\dot{n} = \langle n,\ {}^tDVn\rangle n - {}^tDVn = \langle \nabla v, n\rangle n - \nabla v \tag{4.11}$$

that is $\dot{n}$ is the tangential component of $-\nabla v$ on Γ.

Some expressions that are often used in the following are now obtained. A direct calculation shows that for $V = vN_0$, one has on Γ

$$\left\langle DVn, n\right\rangle = \frac{\partial v}{\partial n}$$

is independent on the choice of N_0,

$$\text{div } V = \text{div } (vN_0) = \frac{\partial v}{\partial n} + v \text{ div } N_0$$

It will also be shown that div (N_0) is also independent, on Γ, of the choice of N_0. Finally, on Γ,

$$\text{div } V - \left\langle DVn, n\right\rangle = v \text{ div } (N_0) \tag{4.12}$$

4.6 Boundary Integration

The superficial measure on Γ may now be written using cofactor matrix notation. For a mapping h, one writes

$$M(h) = \det (Dh)\ {}^tDh^{-1}$$

Then, for any function f with support in $\mathcal{U}$, one has

$$\int_\Gamma f\, ds = \int_{B_0} f\circ h\, ||M(h)e_n||\, dz'$$

The cofactor matrix has the property

$$M(T_t\circ h) = M(T_t)\circ h M(h)$$

Now, the normal field n on Γ may be written (compare with Eqs. 4.4 and 4.4')

$$n = ||M(h)\, e_n||^{-1}\, M(h)e_n \tag{4.13}$$

so that

$$||M(T_t\circ h)e_n|| = ||M(T_t)\circ h\; M(h)e_n||$$

$$= ||M(h)e_n||\;\; ||M(T_t)n||\circ h$$

and one gets the change of variables in a boundary integral as

$$\int_{\Gamma_t} f\, ds_t = \int_{B_0} f\circ T_t\circ h||M(T_t\circ h)e_n||dz'$$

$$= \int_{B_0} \{f\circ T_t||M(T_t)n||\}\circ h||M(h)e_n||dz'$$

that is,

$$\int_{\Gamma_t} f\, ds_t = \int_\Gamma f\circ T_t||M(T_t)n||\, ds \tag{4.14}$$

Now, it is easy to see that the derivative, at t = 0, of the weight function in the last integral is given by (using Eqs. 4.1 and 4.3)

$$\frac{d}{dt}\, ||M(T_t)n||\Big|_{t=0} = \operatorname{div} V(0) - \langle DV(0)n,n\rangle \tag{4.15}$$

and one gets the derivative of the integral as

$$\frac{d}{dt}\int_{\Gamma_t} f\, ds_t = \int_\Gamma \{\dot{f} + f[\operatorname{div} V(0) - \langle DV(0)n,n\rangle]\}ds$$

A second example is given in Section 5, with many other expressions for this derivative.

5. THE DERIVATIVES OF SOME SIMPLE DOMAIN FUNCTIONALS AND IMMEDIATE CONSEQUENCES

5.1 Derivative of an Integral over Ω_t

Let $y(\Omega_t)$ be a regular function given on Ω_t and

$$J_1(\Omega_t) = \int_{\Omega_t} y(\Omega_t)\, dx = \int_{\Omega} y(\Omega_t)\circ T_t J_t dx \tag{5.1}$$

Then, with 2.1 and 4.3

$$dJ_1(\Omega;V) = \int_{\Omega} (\dot{y}(\Omega) + y(\Omega)\, \mathrm{div}\, V(0))\, dx \tag{5.2}$$

Now, if $y(\Omega_t)$ is the restriction to $t \times \Omega_t$ of a regular function Y defined on $I \times \mathcal{U}$, then with Eq. 2.2,

$$dJ_1(\Omega;V) = \int_{\Omega} \frac{\partial Y}{\partial t}(0,x)\, dx + \int_{\Omega} \mathrm{div}\,(y(\Omega)V(0))\, dx$$

and if Ω is regular enough, by Stoke's formula

$$dJ_1(\Omega;V) = \int_{\Omega} \frac{\partial Y}{\partial t}(0,x)\, dx + \int_{\Gamma} y(\Omega)(V(0)\cdot n) ds \tag{5.3}$$

An immediate and well known consequence is:

$$\frac{d}{dt}\int_{\Omega_t} dx = \int_{\Gamma_t} V(t)\cdot n_t\, ds_t = \int_{\Omega_t} \mathrm{div}\, V(t)\, dx$$

If the field V verifies $\mathrm{div}\, V(t) = 0$ for any t, then for any domain Ω the measure of Ω_t is constant and $\det(DT_t(V)) = +1$.

As a first application, let u be a regular function defined on $\mathbb{R}^n$, then

$$\frac{d}{dt}\int_{\Omega_t} u\, y(\Omega_t)\, dx = \int_{\Omega} u \frac{\partial Y}{\partial t}(0)\, dx + \int_{\Omega} \mathrm{div}\,(yV(0))\, dx$$

Let Y_1 and Y_2 be two regular extensions of $y(\Omega)$. Then,

$$\int_\Omega u \frac{\partial Y_1}{\partial t}\, dx = \int_\Omega u \frac{\partial Y_2}{\partial t} (0)\, dx$$

for any function u. This proves that

$$Y' = \frac{\partial Y}{\partial t} (0)\Big|_\Omega \quad \text{or} \quad \frac{\partial Y}{\partial t} (0)\Big|_\Gamma \tag{5.4}$$

is independent of the choice of the regular extension $Y(t,x)$ of $y(\Omega_t)$, hence they are well defined.

Such an extension $Y(t,x)$ always exists when $y(\Omega)$ and Γ are regular enough (see the Hypotheses in Section 5.1). In a general way, even when $y(\Omega_t)$ is just defined on the boundary Γ_t, one considers

$Y'(x) = \frac{\partial Y}{\partial t} (0,x)$, $x \in \Omega$ or $x \in \Gamma$ where Y is any regular extension of $y(\Omega)$.

5.2 Derivative of an Integral Over Γ_t

Let $y(D_t)$ be given on Γ_t and let

$$J_2(\Omega_t) = \int_{\Gamma_t} y(\Omega_t)\, ds_t = \int_S y(\Omega_t)\circ T_t ||M(T_t)n||ds \tag{5.5}$$

where the second equality is obtained with Eq. 4.14 and with Eqs. 2.1 and 4.14, one has

$$dJ_2(\Omega:V) = \int_\Gamma \left\{\dot{y}(\Omega) + y(\Omega)[\operatorname{div} V(0) - \langle DV(0)n,n\rangle]\right\} ds$$

The general method for gradient computation (that is given at Section 10.1) can now be used. Suppose that $V \longrightarrow \dot{y}(\Omega)$ is linear and continuous, then the last expression, for $dJ_2(\Omega;V)$, proves that J_2 possesses a gradient at Ω. Then by Theorem 3.2, if Ω has C^{k+1} regularity, it is sufficient to take the field as $V = vN_0$, which is a stationary field, proportional to an extension N_0 of the normal field n on Γ. In this particular case, with the expressions given previously, one gets (see Eq. 4.12)

$$\operatorname{div} V - \langle DVn, n \rangle = v \operatorname{div}(N_0), \text{ on } \Gamma$$

Then,

$$dJ_2(\Omega;V) = \int_\Gamma [\dot{y}(\Omega) + y(\Omega) v \operatorname{div}(N_0)]\, ds$$

Now suppose that N_0^1 and N_0^2 are two unitary extensions of the field n. Then, by subtraction of the two equalities obtained by writing the foregoing equation with N_0^1 and N_0^2, one has

$$0 = \int_\Gamma y(\Omega) v [\operatorname{div}(N_0^1) - \operatorname{div}(N_0^2)]\, ds$$

which is true for any $y(\Omega)$ and v on Γ. This shows that div (N_0) is on Γ, independent on the choice of N_0

In fact, it is now shown that

$$\operatorname{div}(N_0) = H \equiv \text{the mean curvature of } \Gamma$$

The proof of this result is given first for n = 2, i.e. when Γ is locally the graph of a regular function, say f (suppose $f'>0$ and Ω lies below the graph of f). The normal is given by

$$N(x,f(x)) = [1 + f'(x)^2]^{-\frac{1}{2}}[-f'(x) + 1]$$

One may take

$$N_0(x,y) = n(x,f(x))$$

Then a direct calculation gives

$$\operatorname{div}(N_0)(x,f(x)) = -(1 + f'(x)^2)^{-3/2} f''(x)$$
$$= H(x) \tag{5.6}$$

which completes the proof for n = 2.

When n = 3, Γ is a regular surface. For $x \in \Gamma$, consider the $\mathbb{R}^3$ orthonormal basis $\{e_1, e_2, n\}$, where e_i is a tangent vector to Γ at x, specifically an eigenvector for the Gauss mapping (see Ref. 9)

$$d_x n(e_i) = -k_i e_i$$

where k_1 and k_2 are the extreme normal curvatures of Γ at x, e_1 and e_2 being the main direction curvatures. Then, in a neighborhood of x, Γ may be written in the coordinate $x_1e_1 + x_2e_2 + zn$ as the graph of a mapping $z = f(x_1,x_2)$.

Since the divergence operator is invariant under translation and rotation, one can write div N_0 in the coordinate (x,z). As an extension N_0 of n, take

$$N_0(x_1, x_2, z) = n(x_1,x_2,f(x_1,x_2))$$

By definition of the Gauss mapping,

$$\frac{\partial}{\partial x_i} N_0(x_1,x_2,z) = \langle d_x n(e_i),e_i\rangle = -k_i$$

and

$$\text{div } N_0 = -(k_1 + k_2)$$

Let

$$H = -(k_1 + k_2) \tag{5.7}$$

then

$$\text{div } N_0 = H$$

This completes the proof for n = 3.

Note that with Eqs. 4.12, 5.6, and 5.7, one sees that if the field V is, on Γ proportional to the normal field n on Γ, then one has

$$\text{div } V - \langle DVn,n\rangle = vH \tag{5.8}$$

where $v = V \cdot n$, i.e. $V = vn$ on Γ. Now, the choice of n going outward of the domain Ω is the choice of the orientation of the boundary Γ. If one changes the orientation of Γ, i.e. changes n to -n, then the first member of Eq. 5.8 does not change, while v is changed to -v. This proves that the definitions of Eqs. 5.6

and 5.7 for the mean curvature H depend on the choice made for the orientation of Γ. If one changes this orientation, H has to be changed to -H.

Finally, if $V \longrightarrow \dot{y}(\Omega)$ is linear and continuous,

$$dJ_2(\Omega;V) = \int_\Gamma [\dot{y}(\Omega) + Hy(\Omega)V(0)\cdot n]\, ds \tag{5.9}$$

where H is the mean curvature of Γ given by Eq. 5.6 for n = 2 and Eq. 5.7 for n = 3. Now, if $y(\Omega_t)$ is the restriction to $t \times \Gamma_t$ of $Y(t,x)$ defined on $I \times \mathcal{U}$, then

$$dJ_2(\Omega;V) = \int_\Gamma \frac{\partial Y}{\partial t}(0,x)dx + \int_\Gamma \{\operatorname{div}[y(\Omega)V(0)] - y(\Omega) \langle DV(0)n,n\rangle\} ds$$

and if $V \longrightarrow \left.\dfrac{\partial Y}{\partial t}\right|_{t=0}$ is linear and continuous,

$$dJ_2(\Omega;V) = \int_\Gamma \left\{\frac{\partial Y}{\partial t}(0,x) + \left[\frac{\partial Y}{\partial n}(\Omega) + Hy(\Omega)\right] V(0).n\right\} ds$$

5.3 Other Applications

Integration by parts on Γ. Writing the equality between the two last derivatives, one gets, for any regular and stationary field V and function Y on $\mathbb{R}^n$,

$$\int_\Gamma V\cdot\nabla Y\, ds = -\int_\Gamma Y(\operatorname{div} V - \langle DVn,n\rangle)\, ds + \int_\Gamma \left(\frac{\partial Y}{\partial n} + HY\right) V\cdot n\, ds \tag{5.10}$$

As previously for div N_0, this equality shows that, for any regular field V defined on a neighborhood of Γ, the expression div V - $\langle DVn,n\rangle$ only depends on the value of V on Γ, i.e. it is independent of the choice of an extension of $V|_\Gamma$.

Now, if V is a tangent field to Γ, then $V\cdot n = 0$ and

$$\int_\Gamma V\cdot\nabla Y\, ds = -\int_\Gamma Y(\operatorname{div} V - \langle DVn,n\rangle)\, ds \tag{5.11}$$

This equality may be regarded as the integration by parts formula on Γ, since $V\cdot\nabla Y = \frac{\partial}{\partial V} Y$ is a derivative of V on Γ, i.e. a derivative in a tangential direction on Γ.

As a first application, one sees that the measure of the surface Γ_t is preserved if and only if $\int_{\Gamma_t} H_t V(t)\cdot n_t \, ds = 0$.

The Surface Divergence (See also Ref. 10). If V is proportional to the normal field n on Γ, $V = vn$, then one obtains in Eq. 5.8, with Eqs. 4.12, 5.6, and 5.7,

$$\operatorname{div} V - \langle DVn, n\rangle = vH$$

Now, in general it has been seen that on Γ, $\operatorname{div} V - \langle DVn, n\rangle$ only depends on the restriction of V to Γ, i.e. that

$$\operatorname{div}_\Gamma V = \operatorname{div} V - \langle DVn, n\rangle \tag{5.12}$$

is in fact a differential operator on the surface Γ, involving only tangent derivatives of V.

Let ϕ be given on Γ, $\phi \in H^{1/2}(\Gamma)$. One can find ψ defined on Ω such that

$$\psi = \phi \quad , \text{ on } \Gamma$$

$$\frac{\partial\psi}{\partial n} = 0 \quad , \text{ on } \Gamma$$

Then, by the integration by parts formula of Eq. 5.10,

$$\int_\Gamma (\operatorname{div} V - \langle DVn, n\rangle)\, \phi \, ds = -\int_\Gamma V\cdot\nabla\psi \, ds + \int_\Gamma \left(\frac{\partial\psi}{\partial n} + H\phi\right) V\cdot n \, ds$$

But, $\frac{\partial\psi}{\partial n} = 0$ so $\nabla\psi$ is a tangent vector to Γ, which is independent of the choice of ψ and

$$\nabla_\Gamma \phi = \nabla\psi \tag{5.13}$$

Then,

$$\int_\Gamma (\mathrm{div}_\Gamma\ V)\phi\ ds = -\int_\Gamma V\cdot\nabla_\Gamma\phi\ ds + \int_\Gamma H\phi V\cdot n\ ds \tag{5.14}$$

Note that if ψ is defined in a neighborhood of Γ, then on Γ, one has

$$\nabla_\Gamma\psi = \nabla\psi - \frac{\partial\psi}{\partial n} n \tag{5.13'}$$

i.e. $\nabla_\Gamma\psi$ is the tangential component of $\nabla\psi$. Now, for any function ψ, one has

$$\int_\Gamma (\mathrm{div}_\Gamma\ V)\psi ds = -\int_\Gamma \left(\nabla\psi - \frac{\partial\psi}{\partial n} n\right) V\, ds + \int_\Gamma H\psi V\cdot n\ ds \tag{5.14'}$$

One may call $\mathrm{div}_\Gamma V$, defined by Eq. 5.12, the superficial divergence on Γ and $\nabla_\Gamma\psi$, defined by Eq. 5.13 or Eq. 5.13' the superficial gradient on Γ.

<u>Beltromi-Laplace Operator on Γ.</u> Let ϕ be defined on Γ and put

$$\Delta_\Gamma\phi = \mathrm{div}_\Gamma(\nabla_\Gamma\phi) \tag{5.15}$$

Then, for any function g defined on Γ

$$\int_\Gamma (\Delta_\Gamma\phi) g\ ds = -\int_\Gamma \nabla_\Gamma\psi\cdot\nabla_\Gamma g\, ds + \int_\Gamma Hg\ \langle\nabla_\Gamma\psi, n\rangle\ ds$$

But $\nabla_\Gamma\phi, n = 0$, so

$$\int_\Gamma (\Delta_\Gamma\phi)\ g\ ds = -\int_\Gamma \nabla_\Gamma\phi\cdot\nabla_\Gamma g\ ds \tag{5.16}$$

<u>The Operator $\mathrm{div}_\Gamma(\mu\nabla_\Gamma\phi)$.</u> Let μ, ϕ, and g be defined on Γ. Then

$$\int_\Gamma \mathrm{div}_\Gamma\ (\mu\nabla_\Gamma\phi)\ g\ ds = -\int_\Gamma \mu\nabla_\Gamma\phi\cdot\nabla_\Gamma g ds + \int_\Gamma g\, H\langle\mu\nabla_\Gamma\phi, n\rangle\ ds$$

where $\langle\mu\nabla_\Gamma\phi, n\rangle = 0$, so

$$\int_\Gamma \mathrm{div}_\Gamma(\mu\nabla_\Gamma\phi)\ g\ ds = -\int_\Gamma \mu\nabla_\Gamma\phi\cdot\nabla_\Gamma\ g\ ds \tag{5.17}$$

Now, if ψ and h are regular functions defined in a neighborhood of Γ,

$$\nabla_\Gamma\psi\cdot\nabla_\Gamma h = \left(\nabla\psi - \frac{\partial\psi}{\partial n}\, n\right)\cdot\left(\nabla h - \frac{\partial h}{\partial n}\, n\right)$$

$$= \nabla\psi\cdot\left(\nabla h - \frac{\partial h}{\partial n}\, n\right)$$

$$= \left(\nabla\psi - \frac{\partial\psi}{\partial n}\, n\right)\cdot\nabla h$$

$$= \nabla\psi\cdot\nabla h - \frac{\partial\psi}{\partial n}\frac{\partial h}{\partial n}$$

and one gets

$$\int_\Gamma \mathrm{div}_\Gamma\,(\mu\nabla_\Gamma\psi)h\; ds = -\int_\Gamma \mu\left(\nabla\psi\cdot\nabla h - \frac{\partial\psi}{\partial n}\frac{\partial h}{\partial n}\right) ds \qquad (5.17')$$

and also, for any h,

$$\int_\Gamma \mathrm{div}_\Gamma\,(\mu n)h\; ds = -\int_\Gamma \mu\,\langle n, \nabla_\Gamma h\rangle\; ds + \int_\Gamma hH\,\mu ds$$

but, $\langle n, \nabla_\Gamma h\rangle = 0$, so

$$\mathrm{div}_\Gamma\,(\mu n) = H\mu \qquad (5.18)$$

Note that if ψ and h are defined in a neighborhood of Γ, then

$$\nabla_\Gamma\psi\cdot\nabla h = \left(\nabla\psi - \frac{\partial\psi}{\partial n}\, n\right)\cdot\nabla h$$

$$= \left(\nabla\psi - \frac{\partial\psi}{\partial n}\, n\right)\cdot\left(\nabla h - \frac{\partial h}{\partial n}\, n\right)$$

$$= \nabla_\Gamma\psi\cdot\nabla_\Gamma h$$

If ψ is constant on Γ, then $\nabla_\Gamma\psi = 0$.

Now, if V is a regular vector field on Γ, one may put $V_\Gamma = V - (V.n)n$, i.e. V_Γ is the tangential component of V. Then with $V = V_\Gamma + \mu n$, $\mu = \langle V, n\rangle$, one has

$$\mathrm{div}_\Gamma\,(V) = (V\cdot n)H + \mathrm{div}_\Gamma\,(V_\Gamma) \qquad (5.19)$$

and, in particular, when V is a gradient vector of a function h, that is defined in a neighborhood of Γ, one gets

$$\operatorname{div}_\Gamma (\nabla h) = \frac{\partial h}{\partial n} H + \Delta_\Gamma h$$

Using $\operatorname{div}_\Gamma V = \operatorname{div} V - \langle DVn, n\rangle$, one gets

$$\operatorname{div}_\Gamma (\nabla h) = \Delta h - \frac{\partial^2 h}{\partial n^2}$$

where

$$\frac{\partial^2 h}{\partial n^2} = \langle D^2 hn, n\rangle, \text{ on } \Gamma$$

Finally, one gets

$$\Delta_\Gamma h = \Delta h - H \frac{\partial h}{\partial n} - \frac{\partial^2 h}{\partial n^2}, \text{ on } \Gamma \tag{5.20}$$

Now, with $V = \mu\nabla h$,

$$\operatorname{div}_\Gamma (\mu\nabla h) = \mu H \frac{\partial h}{\partial n} + \operatorname{div}_\Gamma (\mu\nabla_\Gamma h)$$

$$= \operatorname{div} (\mu\nabla h) - \langle D(\mu\nabla h)n, n\rangle$$

But,

$$D(\mu\nabla h) = \nabla\mu \, {}^t\nabla h + \mu D^2 h$$

and one gets

$$\langle D(\mu\nabla h)n, n\rangle = \frac{\partial\mu}{\partial n}\frac{\partial h}{\partial n} + \mu \frac{\partial^2 h}{\partial n^2}$$

and then

$$\operatorname{div}_\Gamma (\mu\nabla_\Gamma h) = \operatorname{div} (\mu\nabla h) - \mu H \frac{\partial h}{\partial n} - \frac{\partial\mu}{\partial n}\frac{\partial h}{\partial n} - \mu \frac{\partial^2 h}{\partial n^2} \tag{5.21}$$

<u>Expressions for the Material Derivative</u> $\dot{\widehat{\frac{\partial y}{\partial n}}}$. For a field proportional to the normal field n (exterior to Ω on Γ), $V = vn$, it was shown in the foregoing that (with Eqs. 4.11, 5.13 and 5.13')

$$\dot{n} = -\nabla_\Gamma v$$

Now,

$$\frac{\dot{\widehat{\partial y}}}{\partial n} = \dot{n}\nabla y + n\,\dot{\widehat{\nabla y}}$$

For the material derivative $\dot{\widehat{\nabla y}}$, the following two expressions are obtained:

(1) Using the material derivative $\dot{y}$,

$$\dot{\widehat{\nabla y}} = -{}^{t}DV\nabla y + \nabla(\dot{y})$$

(2) Using $Y' = \left.\frac{\partial Y}{\partial t}\right|_{t=0,\,x\in\Gamma}$ where Y is any regular extension of y

$$\dot{\widehat{\nabla y}} = \nabla(Y') + D^2YV$$

Using the second expression (2) and a field $V = vn$ on Γ, one gets (with 5.4 and 5.13')

$$\frac{\dot{\widehat{\partial y}}}{\partial n} = -\nabla v\cdot\nabla_\Gamma y + \frac{\partial}{\partial n}Y' + v\frac{\partial^2 y}{\partial n^2} \tag{5.22}$$

where

$$\frac{\partial^2 y}{\partial n^2} = \langle D^2Yn, n\rangle$$

Now, if y is constant on Γ,

$$\frac{\dot{\widehat{\partial y}}}{\partial n} = \frac{\partial}{\partial n}Y' + v\frac{\partial^2 y}{\partial n^2}$$

Using the first expression (1), one gets

$$\frac{\dot{\widehat{\partial y}}}{\partial n} = -\nabla v\cdot\nabla_\Gamma y - \langle D\grave{V}n, \nabla y\rangle + \frac{\partial}{\partial n}(\dot{y})$$

But, for a field $V = vn$ on Γ,

$$ {}^t DVn = \nabla v, \text{ on } \Gamma $$

and, with

$$ \nabla y = \nabla_\Gamma y + \frac{\partial y}{\partial n} n $$

one gets

$$ \langle DVn, y \rangle = \langle DVn, \nabla_\Gamma y \rangle + \frac{\partial v}{\partial n}\frac{\partial y}{\partial n} $$

and

$$ \dot{\widehat{\frac{\partial y}{\partial n}}} = -\nabla_\Gamma v \cdot \nabla_\Gamma y - \frac{\partial v}{\partial n}\frac{\partial y}{\partial n} - \langle DVn, \nabla_\Gamma y \rangle + \frac{\partial}{\partial n}(\dot{y}) $$

Finally, with Eq. 5.13', this may be written in the form

$$ \dot{\widehat{\frac{\partial y}{\partial n}}} = -\nabla v \nabla y - \langle DVn, \nabla_\Gamma y \rangle + \frac{\partial}{\partial n}(\dot{y}) \tag{5.23} $$

6. EXAMPLES OF QUADRATIC COST FUNCTIONS

As an example of an integral cost function over Ω, let

$$ J_3(\Omega) = \int_\Omega (y(\Omega) - Z_g)^2 \, dx $$

where Z_g is a regular function given on $\mathbb{R}^n$. Then,

$$ dJ_3(\Omega;V) = \int_\Omega \{2[y(\Omega) - Z_g][\dot{y}(\Omega) - \nabla Z_g \cdot V(0)] $$
$$ + [y(\Omega) - Z_g]^2 \operatorname{div} V(0)\} \, dx $$

This shows that J_3 possesses a gradient, if $V \longrightarrow \dot{y}(\Omega)$ is linear and continuous. If $y(\Omega_t)$ is the restriction to $t \times \Omega_t$ of $Y(t,x)$, then

$$dJ_3(\Omega{:}V) = 2\int_\Omega [y(\Omega) - Z_g] \frac{\partial Y}{\partial t}(0,x)dx$$

$$+\int_\Gamma [y(\Omega) - Zg]^2 V(0).n\ ds$$

As an example of an integral cost function over Ω, but involving the gradient of the state variable, let

$$J_4(\Omega) = \int_\Omega ||\nabla(y(\Omega) - Z_g)||^2\ dx$$

To simplify the calculus, suppose that $y(\Omega_t)$ is the restriction to $t \times \Omega_t$ of $Y(t,x)$ and that $V \longrightarrow \frac{\partial Y}{\partial t}(0,x)$ is linear and continuous. Then, using the example of Section 5.1,one gets (see Eq. 5.3)

$$dJ_4(\Omega;V) = 2\int_\Omega \nabla[y(\Omega) - Z_g]\cdot\nabla\left(\frac{\partial}{\partial t} Y(0,x)\right)dx$$

$$+\int_\Gamma [y(\Omega) - Z_g]^2\ V(0)\cdot n\ ds$$

As an example of an integral cost function over Γ, let

$$J_5(\Omega) = \int_\Gamma [y(\Omega) - Z_g]^2\ ds$$

Then, with Eq. 5.9,

$$dJ_5(\Omega;V) = 2\int_\Gamma [y-Z_g][\dot{y}(\Omega) - \nabla Z_g \cdot V(0)]\ ds$$

$$+\int_\Gamma [y - Z_g]^2 HV(0)\cdot n\ ds$$

Now, if $V \longrightarrow \dot{y}(\Omega)$ is linear and continuous, and if $y(\Omega_t)$ is the restriction to $t \times \Omega_t$ of $Y(t,x)$, then

$$dJ_5(\Omega;V) = 2\int_\Gamma [y(\Omega) - Z_g] \frac{\partial Y}{\partial t}(0)\ ds$$

$$+\int_\Gamma \left\{2[y(\Omega) - Z_g] \frac{\partial}{\partial n}[y(\Omega) - Z_g] + H[y(\Omega) - Z_g]^2\right\}$$

$$\times\ V(0)\cdot n\ ds$$

Finally, as a more complicated example of an integral cost function over Γ, with the integrand depending on the gradient of the state variable, let

$$J_6(\Omega_t;V) = \int_{\Gamma_t} \left[\frac{\partial}{\partial n_t} y(\Omega_t) - Z_g\right]^2 ds \tag{6.1}$$

following the second example given at 5.9 we have

$$dJ_6(\Omega;V) = 2\int_\Gamma \left[\frac{\partial}{\partial n} y(\Omega) - Z_g\right]\left[\dot{n}\nabla y + n\dot{\widehat{\nabla y}} - Z_g\right] ds$$
$$+ \int_\Gamma \left[\frac{\partial}{\partial n} y(\Omega) - Z_g\right]^2 \left[\operatorname{div} v(0) - \langle DV(0)n,n\rangle\right] ds \qquad . \tag{6.2}$$

But,

$$\dot{n} = \langle Dvn,n\rangle\, n - {}^tDVn$$

$$\dot{\widehat{\nabla y}} = {}^tDV\nabla y(\Omega) + \nabla(y(\Omega))$$

$$Z_g = \nabla Z_g \cdot V$$

so that

$$dJ_6(\Omega;V) = 2\int_\Gamma \left[\frac{\partial}{\partial n} y(\Omega) - Z_g\right]\Big[\langle DVn,n\rangle\, n\cdot\nabla y(\Omega) - \langle DV\nabla y(\Omega),n\rangle$$
$$- \langle DVn,\nabla y(\Omega)\rangle + n\nabla(y(\Omega)) - \nabla Z_g\cdot V(0)\Big] ds$$
$$+ \int_\Gamma \left[\frac{\partial}{\partial n} y(\Omega) - Z_g\right]^2 \left[\operatorname{div} V(0) - \langle DV(0)n,n\rangle\right] ds$$

Suppose that $V \longrightarrow y(\Omega)$ is linear and continuous. Then the last expression shows that J_6 possesses a gradient at Ω. Consider $V = vN_0$ and, to simplify the calculations, let $y(\Omega_t)$ be the restriction to $t \times \Omega_t$ of $Y(t,x)$. Then,

$$\dot{n} = (\nabla v\cdot n) - \nabla v$$

$$\dot{\widehat{\nabla y}}(\Omega) = \left(\frac{\partial Y}{\partial t}\right) + D^2YV$$

and

$$dJ_6(\Omega;V) = 2\int_\Gamma \left[\frac{\partial}{\partial n}y(\Omega) - Z_g\right]\left[\frac{\partial v}{\partial n}\frac{\partial y}{\partial n} - \nabla v \cdot \nabla y + \frac{\partial}{\partial n}\left(\frac{\partial Y}{\partial t}\right)\right.$$

$$\left. + \langle D^2 Yn,n\rangle v - \frac{\partial}{\partial n} Z_g v\right] ds$$

$$+ \int_\Gamma \left[\frac{\partial}{\partial n} Y(\Omega) - Z_g\right]^2 \left[\mathrm{div}\ V(0) - \langle DV(0)n,n\rangle\right] ds$$

where the last term in the second integral is just vH.

With $\mu = \frac{\partial y}{\partial n} - Z_g$ and $\nabla_\Gamma v \nabla_\Gamma y = \nabla v \nabla y - \frac{\partial v}{\partial n}\frac{\partial y}{\partial n}$ one has

$$\int_\Gamma \mu \nabla_\Gamma v \nabla_\Gamma y\ ds = \int_\Gamma \mathrm{div}_\Gamma\ (\mu \nabla_\Gamma y) v\ ds$$

With $v = \langle V,n\rangle$, one gets

$$dJ_6(\Omega;V) = \int_\Gamma \left\{2\ \mathrm{div}_\Gamma \left[\left(\frac{\partial y}{\partial n} - Z_g\right)\nabla_\Gamma\ y\right] + \left[\frac{\partial^2 y}{\partial n^2} - \frac{\partial}{\partial n} Z_g\right]\left[\frac{\partial y}{\partial n} - Z_g\right]\right\} v\ ds$$

$$+ \int_\Gamma \left[\frac{\partial y}{\partial n} - Z_g\right]^2 Hv\ ds + 2\int_\Gamma \left[\frac{\partial y}{\partial n} - Z_g\right]\frac{\partial}{\partial n} Y'\ ds \qquad (6.3)$$

Note that Eq. 5.26 could have been derived directly from Eqs. 5.25, 5.22, and 5.17.

Now, if y is constant on the boundary Γ, then $y = \frac{\partial y}{\partial n} n$ and

$$\mathrm{div}_\Gamma \left[\left(\frac{\partial y}{\partial n} - Z_g\right)\nabla y\right] = \mathrm{div}_\Gamma\ (\mu n)$$

with

$$\mu = \left(\frac{\partial y}{\partial n} - Z_g\right)\frac{\partial y}{\partial n}$$

and one gets

$$\mathrm{div}_\Gamma\ (\mu n) = H\mu$$

Thus,

$$dJ_6(\Omega;V) = \int_\Gamma \left\{2\left[\frac{\partial y}{\partial n} - Z_g\right]\frac{\partial y}{\partial n} + \left[\frac{\partial y}{\partial n} - Z_g\right]^2\right\} Hv\, ds$$

$$+ \int_\Gamma \left[\frac{\partial^2 y}{\partial n^2} - \frac{\partial}{\partial n} Z_g\right]\left[\frac{\partial y}{\partial n} - Z_g\right] v\, ds$$

$$+ 2\int_\Gamma \left[\frac{\partial y}{\partial n} - Z_g\right]\frac{\partial}{\partial n} Y'\, ds \qquad (6.4)$$

7. A COST FUNCTION ARISING FROM AN INTEGRAL EQUATION FORMULATION FOR EXTERIOR PROBLEMS

Let f_t be a regular function defined on $\Gamma_t \times \Gamma_t$. One may write

$$\dot{f}_t(x,y) = \frac{d}{dt} f(T_t(x),T_t(y))\Big|_{t=0}$$

i.e. the Eulerian or material derivative simultaneously in the two variables x and y. If

$$h_t(y) = \int_{\Gamma_t} f_t(x,y)\, ds(x)$$

then following example 2 one has (see 5.9)

$$\dot{h}_t(y) = \int_{\Gamma_t} \dot{f}_t(x,y)\, ds(x) + \int_{\Gamma_t} f_t(x,y)\, H_t(x)\, v(x)\, ds(x)$$

where H_t is the mean curvature of Γ_t and $v = V(0)\, n_t$ on Γ_t.

Let

$$J_7(\Gamma;\Gamma) = \iint_{\Gamma_t \times \Gamma_t} f_t(x,y)\, ds(x)ds(y) \qquad (7.1)$$

then

$$dJ_7(\Gamma;V) = \int_\Gamma \dot{h}\, ds + \int_\Gamma Hh_0 v\, ds$$

where h_0 is defined on $\Gamma = \Gamma_0$, i.e. at $t = 0$. Thus,

$$dJ_7(\Gamma;V) = \iint_{\Gamma\times\Gamma} \dot{f}(x,y)\, ds(x)ds(y) + \iint_{\Gamma\times\Gamma} [f(x,y) + f(y,x)]\, H(x)v(x)\, ds(x)ds(y)$$

8. A FIRST CLASS OF DOMAIN FUNCTIONALS WITHOUT GRADIENT

In studying domain functionals that do not have a gradient, the following well known optimization result [11] is useful:

Theorem 8.1. Let W be a compact set, $F:\mathbb{R} \times W \longrightarrow \mathbb{R}$ such that:

(i) $(t,w) \longrightarrow F(t,w)$ is lower semicontinuous on $\mathbb{R} \times W$

(ii) for all $w \in W$, $t \longrightarrow F(t,w)$ is differentiable

(iii) $(t,w) \longrightarrow \frac{\partial}{\partial t} F(t,w)$ is lower semicontinuous on $\mathbb{R} \times W$

For any t, let

$$W(t)^* = \{u \in W: j(t) = \inf_{w \in W} F(t,w) = F(t,u)\}$$

Then the function j(t) is semi differentiable on the right and

$$\lim_{s \to 0^+} (j(t+s) - j(t))/s = \inf_{w \in W(t)^*} \frac{\partial}{\partial t} F(t,w)$$

This result, simply stated, is that the derivative of a minimum is the minimum of the derivative. An analogous result holds for a maximum.

Now, if $y(\Omega_t)$ is a function defined on Ω_t, say $y(\Omega_t)$ belongs to $C^0(\Omega_t)$, consider the functional

$$J_8(\Omega_t) = \inf_{x \in \bar{\Omega}_t} [y(\Omega_t)(x)]$$

This may be written as

$$J_8(\Omega_t) = \inf_{x \in \bar{\Omega}} y(\Omega_t) \circ T_t(X)$$

Taking $W = \Omega$ and $F(t,x) = y(\Omega_t) \circ T_t(X)$, one has

$$\frac{\partial}{\partial t} F(t,X)\Big|_{t=0} = \dot{y}(\Omega)(X)$$

Let

$$\Omega^* = \{X \in \bar{\Omega} : J_8(\Omega) = y(\Omega)(X)\}$$

be the set of minimizing points for J_8. Then by the proposition,

$$dJ_8(\Omega;V) = \inf_{X \in \Omega^*} \dot{y}(\Omega)(X)$$

Note that even if $V \longrightarrow y(\Omega)$ is linear, $V \longrightarrow dJ_8(\Omega;V)$ is not linear (it is linear of course if Ω^* reduces to a single point, as occurs in some particular cases).

Now, replacing $y(\Omega)$ by a function of $y(\Omega)$, for example $[y(\Omega) - Z_g]^2$, $|\nabla y(\Omega)|^2$, ... one easily gets the derivatives for the infimum (or supremum) on the compact set Ω_t.

A different problem arises when the extremum is taken over a fixed compact set A. Let A be a compact set in the open set Ω. Then for t small enough, $A \subset \Omega_t$. Let

$$J_9(\Omega_t) = \inf_{x \in A} y(\Omega_t)(x)$$

If one supposes that $y(\Omega_t)$ is the restriction to $t \times \Omega_t$ of a regular function $Y(t,x)$, then

$$J_9(\Omega_t) = \inf_{x \in A} Y(t,x)$$

and (taking $W = A$ and $F = Y$ in the proposition), with

$$A^* = \{x \in A : J_9(\Omega) = y(\Omega)(x)\}$$

the proposition yields

$$dJ_9(\Omega;V) = \inf_{x \in A^*} \frac{\partial Y}{\partial t}(0,x)$$

Note that dJ_8 and dJ_9 are different, for in the former case in which $y(\Omega)$ is the restriction of Y, $\dot{y}(\Omega) = \frac{\partial Y}{\partial t}(0) + \nabla Y(0)\cdot V(0)$ and one should have for J_8:

$$dJ_8(\Omega;V) = \inf_{X\in\Omega^*} \frac{\partial Y}{\partial t}(0,X) + \nabla Y(0,X)\cdot v(0,X)$$

Turn now to the functional obtained as the infimum (or supremum) on the boundary Γ_t of Ω_t. Let

$$J_{10}(\Gamma_t) = \inf_{X\in\Gamma_t} \frac{\partial y}{\partial n_t}(\Omega_t)$$

with

$$\Gamma^*_t = \{x \in \Gamma_t : n_t(x)\cdot\nabla y(\Omega_t)(x) = J_{10}(\Omega_t)\}$$

Then,

$$dJ_{10}(\Gamma;V) = \inf_{X\in\Gamma^*} \dot{\widehat{\frac{\partial y}{\partial n}}}(\Omega)(X)$$

that is,

$$dJ_{10}(\Gamma;V) = \inf_{X\in\Gamma^*} \langle DV(0)n,n\rangle \frac{\partial y}{\partial n}(\Omega) - \langle(DV(0) + {}^tDV(0))n,\nabla y\rangle + \nabla(\dot{y}(\Omega))$$

For a field V that is proportional to n on Γ, V may be written as $V = \acute{v}N_0$ in a neighborhood of Γ. With the expression of $\dot{\widehat{\frac{\partial y}{\partial n}}}(\Omega)$ for this situation (see Eq. 5.23), one gets

$$dJ_{10}(\Gamma;V) = \inf_{X\in\Gamma^*} -\nabla_\Gamma y \nabla_\Gamma v + \frac{\partial}{\partial n} Y' + \frac{\partial^2 y}{\partial n^2} v$$

9. NON DIFFERENTIABLE DOMAIN FUNCTIONALS OBTAINED WITH NON DIFFERENTIABLE INTEGRANDS

Consider now domain functionals in which extremality technique of Section 8 is combined with methods for regular functionals. For example, let

$$J_{11}(\Omega_t) = \int_{\Omega_t} | y(\Omega_t) | dx$$

which is the same as

$$J_{11}(\Omega_t) = \sup_{u \in W} \int_{\Omega_t} u \circ T_t^{-1} \, y(\Omega_t) \, dx$$

where

$$W = \{u: -1 \leq u(x) \leq 1 \text{ a.e. } x \text{ in } \Omega\}$$

is a compact set in the weak- topology. Taking

$$F(t,u) = \int_{\Omega_t} u \circ T_t^{-1} \, y(\Omega_t) \, dx$$

and noting that

$$\dot{\widehat{u \circ T_t^{-1}}} = \frac{d}{dt} u \circ T_t^{-1} \circ T_t \Big|_{t=0} = 0$$

Then, as for J_1 in Section 5.1,

$$\frac{\partial}{\partial t} F(0,u) = \int_{\Omega} (u\dot{y}(\Omega) + uy(\Omega) \operatorname{div} V(0)) \, dx$$

and with

$$W^* = \left\{ u \in W : J_{11}(\Omega) = \int_{\Omega} uy(\Omega) \, dx \right\}$$

one has, from Theorem 8.1

$$dJ_{11}(\Omega; V) = \sup_{u \in W^*} \int_{\Omega} (u\dot{y}(\Omega) + uy(\Omega) \operatorname{div} V(0)) \, dx$$

Now u belongs to W^* if and only if u = -1 where $y(\Omega) < 0$ and u = +1 where $y(\Omega) > 0$. Then, one may write

$$y(\Omega)^{-1}(0) = \{x \in \Omega : y(\Omega)(x) = 0\}$$

and consider the function sgn defined by

$$\mathrm{sgn}(t) = \begin{cases} +1 & \text{if } t>0 \\ 0 & \text{if } t=0 \\ -1 & \text{if } t<0 \end{cases}$$

In order to write $dJ_{11}(\Omega;V)$ one now notes that the $u \in W^*$ that yields the foregoing supremum must equal $\mathrm{sgn}(\dot{y}(\Omega) + y(\Omega)\,\mathrm{div}\,V(0))$ on $y^{-1}(\Omega)(0)$. Thus,

$$dJ_{11}(\Omega;V) = \int_\Omega \mathrm{sgn}\,(y(\Omega))\,[\dot{y}(\Omega) + y(\Omega)\,\mathrm{div}\,V(0)]\,dx$$

$$+\int_{y(\Omega)^{-1}(0)} |\,\dot{y}(\Omega) + y(\Omega)\,\mathrm{div}\,V(0)\,|\,dx$$

Now, it is clear that J_{11} possesses a gradient at Ω if and only if the measure of $y(\Omega)^{-1}(0)$ is equal to zero and $V \longrightarrow y(\Omega)$ is linear and continuous.

As another example, consider the constraint

$$\frac{\partial}{\partial n_t} y(\Omega_t) \leq Z_g, \text{ on } \Gamma_t$$

where Z_g is given. This may be written as

$$J_{12}(\Omega_t) = \int_{\Gamma_t} \left(\frac{\partial}{\partial n_t} y(\Omega_t) - Z_g\right)^+ ds = 0$$

where $(\;)^+$ is the positive part of the associated argument. One may, as before, write

$$J_{12}(\Omega_t) = \sup_{u \in W} \int_{\Gamma_t} u \circ T_t^{-1}\left(\frac{\partial}{\partial n_t} y(\Omega_t) - Z_g\right) ds$$

where W is given by

$$W = \{u: 0 \le u(x) \le 1, \text{ a.e. } x \text{ in } \Gamma\}$$

Then, if

$$W^* = \{u \in W: u \text{ yield } J_{12}(\Omega_t)\}$$

Then, as for J_2 of Section 5.2, one applies Theorem 8.1 to write

$$dJ_{12}(\Omega;V) = \sup_{u \in W^*} \int_\Gamma \left\{ u\left[\dot{\widehat{\frac{\partial}{\partial n} y(\Omega)}} - \nabla Z_g \cdot V(0)\right] \right.$$

$$\left. + Hu\left[\frac{\partial}{\partial n} y(\Omega) - Z_g\right] V(0)\cdot n \right\} ds$$

That is, with the same considerations on W^* as in the treatment of J_{11}, one has

$$dJ_{12}(\Omega;V) = \int_{\left\{x\in\Gamma: \frac{\partial}{\partial n} y(\Omega)(x) > Z_g(x)\right\}} \left\{ \dot{\widehat{\frac{\partial}{\partial n} y(\Omega)}} - \nabla Z_g \cdot V(0) + H\left[\frac{\partial}{\partial n} y(\Omega) - Z\right] V(0)\cdot n \right\} ds$$

$$+ \int_{\left\{x\in\Gamma: \frac{\partial}{\partial n} y(\Omega)(x) = Z_g(x)\right\}} \left\{ \dot{\widehat{\frac{\partial}{\partial n} y(\Omega)}} - \nabla Z_g \cdot V(0) + H\left[\frac{\partial}{\partial n} y(\Omega) - Z_g\right] V(0)\cdot n \right\}^+ ds$$

For other examples of non differentiable domain functionals, see Refs. 1, 12, 13 and 14, where the domain functional $J(\Omega)$ may itself be regular, but the function $g(\Omega)$, being the solution of a well posed variational inequality in the domain Ω, possesses a material derivative $\dot{y}(\Omega)$ and a derivative Y', but they are not linear in the vector field V.

10. CONTROL OF THE SOLUTION OF A PARTIAL DIFFERENTIABLE EQUATION

Let $y(\Omega_t)$ be the solution of a boundary-value problem in the geometry Ω_t and consider the gradient of a cost function

$$J(\Omega_t) = J(\Omega_t, y(\Omega_t))$$

For simplicity, attention is limited here to linear elliptic problems.

In all examples of cost functions with a gradient, i.e. functionals J_1, J_2, ..., and J_7, the Eulerian derivative $dJ(\Omega;V)$ has two additional terms:

(1) one is linear on $v = V(0) \cdot n$, the normal component of the speed

(2) the second contains $\frac{\partial}{\partial t} Y$

One has now to consider this second term and to give its linear expression in v, the first term being unchanged.

10.1 A General Method for the Computation of the Gradient of $J(\Omega)$

Two hypotheses are now needed, involving $y(\Omega_t)$ as an element of the sobolev space $H^m(\Omega_t)$:

(1) Hypothesis H1. $\dot{y}(\Omega)$ exists and the mapping $V \longrightarrow \dot{y}(\Omega)$ is linear and continuous

(2) Hypothesis H2. The domain Ω is regular enough so that there exists an extension (linear and continuous) operator P, such that

$$P: H^m(\Omega) \longrightarrow H^m(\mathbb{R}^n)$$

In each example treated, H1 is verified using the implicit function theorem (the complete proof is given for only one example). The hypothesis H2 is true for regular open sets for k large enough, but this problem is not treated in detail here.

Consider the function

$$Y(t,x) = [P(y(\Omega_t) \circ T_t)] \circ T_t^{-1}(x)$$

Then, $y(\Omega_t)$ appears as the restriction to t x Ω_t of the function Y, which is defined on $\mathbb{R} \times \mathbb{R}^n$ and is regular by H1.

Define

$$Y' = \frac{\partial Y}{\partial t}(0)$$

10.2 The Dirichlet Problem

Let f belong to $L^2(\mathbb{R}^n)$ and g to $H^{3/2}(\mathbb{R}^n)$ and $y(\Omega_t)$ be the solution of the problem

$$\left.\begin{aligned} -\Delta y(\Omega_t) &= f, \text{ in } \Omega_t \\ y(\Omega_t) &= g, \text{ on } \Gamma_t \end{aligned}\right\}$$

Then, $y(\Omega_t)$ belongs to $H^2(\Omega_t)$ and one has

$$\int_{\Omega_t} -\Delta Y(t)u\,dx = \int_{\Omega_t} fu\,dx$$

for any regular function u. Taking the derivatives with respect to t, one gets at t = 0

$$\int_{\Omega} -\Delta Y'u\,dx + \int_{\Gamma} -\Delta Y u V(0)\cdot n\,ds = \int_{\Gamma} fuV(0)\cdot n\,ds$$

That is,

$$-\Delta(Y') = 0, \text{ in } \Omega$$

On the other hand, one has, throughout $\mathbb{R}^n$,

$$Y' = P\,\dot{y}(\Omega) - \nabla(py(\Omega))\cdot V(0)$$

and on Ω simply

$$Y' = \dot{y}(\Omega) - \nabla y(\Omega)\cdot V(0)$$

But, on Γ,

$$\dot{y}(\Omega) = \nabla g\cdot V(0)$$

Then, on Γ,

$$Y' = \nabla[g - y(\Omega)]\cdot V(0)$$

But, $g - y(\Omega) = 0$ on Γ. Thus the gradient is in the normal direction, i.e.

$$\nabla[g - y(\Omega)] = \frac{\partial}{\partial n}[g - y(\Omega)]\,n$$

That is, on Γ,

$$Y' = \frac{\partial}{\partial n}[g - y(\Omega)]V(0)\cdot n$$

and the restriction of Y' to Ω appears as the solution of a Dirichlet problem (as does $y(\Omega_t)$).

10.3 The Neumann Problem

Let f belong to $L^2(\mathbb{R}^n)$ and g to $H^{1/2}(\mathbb{R}^n)$ and let $y(\Omega_t)$ be the solution of the problem

$$\left.\begin{aligned} &-\Delta y(\Omega_t) + y(\Omega_t) = f, \text{ in } \Omega_t \\ &\frac{\partial}{\partial n_t} y(\Omega_t) = g, \text{ on } \Gamma_t \end{aligned}\right\}$$

The solution $y(\Omega_t)$ belongs to $H^2(\Omega_t)$ and one has

$$\int_{\Omega_t} Yu\, dx + \int_{\Omega_t} \nabla Y\cdot\nabla u\, dx = \int_{\Omega_t} fu\, dx + \int_{\Gamma_t} gu\, ds$$

for any $u \in H^1(\mathbb{R}^n)$.

Differentiating the last expression with respect to t, one gets

$$\begin{aligned} &\int_\Omega Y'u\, dx + \int_\Omega \nabla Y'\cdot\nabla u\, dx + \int_\Gamma (\nabla Y\cdot\nabla u)V(0)\cdot n\, ds \\ &\quad + \int_\Gamma (Yu)V(0)\cdot n\, ds \\ &= \int_\Gamma (fu)V(0)\cdot n\, ds + \int_\Gamma \left[\frac{\partial}{\partial n}(gu) + Hgu\right] V(0)\cdot n\, ds \end{aligned}$$

Taking u with support in Ω, one gets

$$Y' - \Delta Y' = 0, \text{ in } \Omega$$

and using the Green formula

$$\int_\Gamma \frac{\partial}{\partial n} Y'u\, ds = \int_\Gamma \left[-\nabla y(\Omega)\cdot\nabla u + fu + \frac{\partial}{\partial n}(gu) + Hgu\right] V(0)\cdot n\, ds$$

This relation holds for any u that is regular on Γ. Then, taking u as the restriction to Γ of U, with $\frac{\partial}{\partial n} U = 0$ on Γ, one gets

$$\int_\Gamma \frac{\partial}{\partial n} Y'U \, ds = \int_\Gamma -\nabla_\Gamma U \nabla_\Gamma y v \, ds + \int_\Gamma (f + \frac{\partial}{\partial n} g + Hg) Uv \, ds$$

where $v = V(0)\cdot n$. Note that the first integral on the right can be written as

$$\int_\Gamma U \, \mathrm{div}_\Gamma \, (v \nabla_\Gamma y) \, ds.$$

Since relation holds for any function U on Γ, one gets on Γ,

$$\frac{\partial}{\partial n} Y' = \mathrm{div}_\Gamma \, (v \nabla_\Gamma y) + (Hg + \frac{\partial}{\partial n} g + f) v$$

and Y' is the solution of a Neumann problem, just as y was.

10.4 Gradient Computation

One has now to compute the second term of $dJ(\Omega;B)$, which involves $Y' = \frac{\partial Y}{\partial t}(0,x)$. Seven domain functionals with gradients and two basic elliptic problems have thus far been studied. This gives 14 examples of gradient calculus. For simplicity, just two examples are treated here. One must compute the term involving Y' as a linear term in $v = V(0)\cdot n$. For this, the classical method of the adjoint equation is used, whose solution is noted p.

<u>Dirichlet Problem with J_6 As Cost Function</u>. The state $y(\Omega_t) \in H^2(\Omega_t)$ is the solution of

$$\left.\begin{aligned} -\Delta y(\Omega_t) &= f \;, & &\text{in } \Omega_t \\ y(\Omega_t) &= g \;, & &\text{on } \Gamma_t \end{aligned}\right\}$$

Looking to $dJ_6(\Omega;V)$ given in Section 6, one considers the term

$$j = \int_\Gamma \left[\frac{\partial y}{\partial n} - Z_g\right] \frac{\partial}{\partial n} Y' \, ds$$

Let $p(\Omega)$ be the solution of the problem

$$\left.\begin{aligned} -\Delta p(\Omega) &= 0 \ , && \text{in } \Omega \\ p &= \frac{\partial y}{\partial n}(\Omega) - Z_g \ , && \text{on } \Gamma \end{aligned}\right\}$$

By the Green formula, one gets, taking $V = vN_0$,

$$j = \int_\Gamma p \frac{\partial}{\partial n} Y' \, ds = \int_\Omega \nabla p \cdot \nabla Y' \, dx = \int_\Gamma \nabla p \cdot nY' \, ds$$

$$= \int_\Gamma \frac{\partial}{\partial n}(g - y) \frac{\partial p}{\partial n} v \, ds$$

and the gradient of J_6 is given by

$$g_n(\Omega) = 2 \operatorname{div}_\Gamma \left[\left(\frac{\partial y}{\partial n} - Z_g\right) \nabla_\Gamma y\right] + \left(\frac{\partial^2 y}{\partial n^2} - \frac{\partial}{\partial n} Z_g\right)\left(\frac{\partial y}{\partial n} - Z_g\right)$$

$$+ \left(\frac{\partial y}{\partial n} - Z_g\right)^2 H + 2 \frac{\partial}{\partial n}(g - y) \frac{\partial p}{\partial n}$$

Neumann Problem with J_3 as Cost Function. The state $y(\Omega_t) \in H^2(\Omega_t)$ is the solution of

$$\left.\begin{aligned} -\Delta y(\Omega_t) + y(\Omega_t) &= f \ , && \text{in } \Omega_t \\ \frac{\partial y}{\partial n_t}(\Omega_t) &= g \ , && \text{on } \Gamma_t \end{aligned}\right\}$$

and

$$J_3(\Omega_t) = \int_{\Omega_t} [y(\Omega_t) - Z]^2 \, dx$$

is a quadratic cost function, studied in Section 6. One has now to calculate the term

$$j = 2 \int_\Omega [y(\Omega) - Z] Y' \, dx$$

as a linear term on $v = V(0) \cdot n$.

Consider p as the solution of the adjoint problem

$$\left.\begin{aligned} p(\Omega) - \Delta p(\Omega) &= 2(y - Z) \quad , && \text{in } \Omega \\ \frac{\partial p}{\partial n}(\Omega) &= 0 \quad , && \text{on } \Gamma \end{aligned}\right\}$$

Then, using $p(\Omega)$ one gets, using the Green formula,

$$j = \int_\Omega (-\Delta p + p)Y' \, dx = \int_\Omega (p(\Omega)Y' + \nabla p(\Omega)\cdot\nabla Y') \, dx$$

$$= \int_\Gamma \frac{\partial}{\partial n} Y' p(\Omega) \, ds$$

Using the weak formulation for $\frac{\partial}{\partial n} Y'$, one has

$$j = \int_\Gamma \left[-\nabla_\Gamma y(\Omega)\cdot\nabla_\Gamma p(\Omega) + fp(\Omega) + \frac{\partial g}{\partial n} p(\Omega) + Hgp(\Omega)\right] v \, ds$$

Using the remaining expressions for the scalar gradient, one has

$$g_n(\Omega) = (y(\Omega) - z)^2 - \nabla_\Gamma p(\Omega)\nabla_\Gamma y(\Omega) + p(\Omega)\left(Hg + \frac{\partial g}{\partial n} + f\right)$$

Note that if one considers the Neumann problem

$$\left.\begin{aligned} -\Delta y(\Omega) &= f \quad , && \text{in } \Omega \\ \frac{\partial y}{\partial n}(\Omega) &= g \quad , && \text{on } \Gamma \end{aligned}\right\}$$

then $y(\Omega)$ is just defined up to an additive constant. Then, one cannot consider J_3 as a cost function, but may consider J_4 or J_6. For example, with J_4 one has

$$J(\Omega) = J_4 = \frac{1}{2}\int_\Omega ||\nabla(y(\Omega) - z)||^2 \, dx$$

and

$$dJ(\Omega;V) = j + \frac{1}{2}\int_\Gamma ||\nabla(y(\Omega) - z)||^2 \, V(0)\cdot n \, ds$$

where

$$j = \int_\Gamma \nabla(y(\Omega) - z)\cdot\nabla Y' \, ds = \int_\Gamma \frac{\partial y'}{\partial n}(y(\Omega) - z)\, ds$$

Thus, no adjoint equation is needed and one gets directly, with $v = V(0)\cdot n$,

$$j = \int_\Gamma (y - z)\left[\mathrm{div}_\Gamma (v\nabla_\Gamma y) + \left(f + Hg + \frac{\partial g}{\partial n}\right)v\right] ds$$

and the scalar gradient is given by

$$g_n = \nabla_\Gamma y(\Omega)\nabla_\Gamma(z - y(\Omega)) + \left(f + \frac{\partial g}{\partial n} + Hg\right)(y(\Omega) - z)$$

$$+ (y(\Omega) - z)^2$$

11. CONTROL BY THE GEOMETRY OF THE SOLUTION OF A TRANSMISSION PROBLEM

Let $\mathcal{O}$ be a regular open set in $\mathbb{R}^n$ and Σ its boundary, as shown in Fig. 11.1. Consider the diffeomorphisms T_t of $\mathcal{O}$ onto itself, which is a geometric constraint. If the field $V(t)$ is equal to zero or tangent on Σ, then $T_t = T_t(V)$ is a diffeomorphism with C^k regularity of $\mathcal{O}$ onto itself. Then, if Ω is a domain in $\mathcal{O}$, $\bar{\Omega}_t$ is also in $\mathcal{O}$.

Let $\Omega_t^c = \mathcal{O}\setminus(\Omega_t \cup \Gamma_t)$ which is a regular open set in $\mathcal{O}$ with boundary $\Sigma \cup \Gamma_t$, Let $y(\Omega) = (y_i, y_e)$ be the solution of the problem

$$\left.\begin{aligned} -\Delta y_i &= f, && \text{in } \Omega \\ -\Delta y_e &= g, && \text{in } \Omega^c \\ \frac{\partial}{\partial n} y_e &= 0, && \text{on } \Sigma \end{aligned}\right\}$$

with the transmission conditions

$$\left.\begin{aligned} \frac{\partial y_i}{\partial n} - k\frac{\partial y_e}{\partial n} &= h, && \text{on } \Gamma \\ y_i - y_e &= \mu, && \text{on } \Gamma \end{aligned}\right\}$$

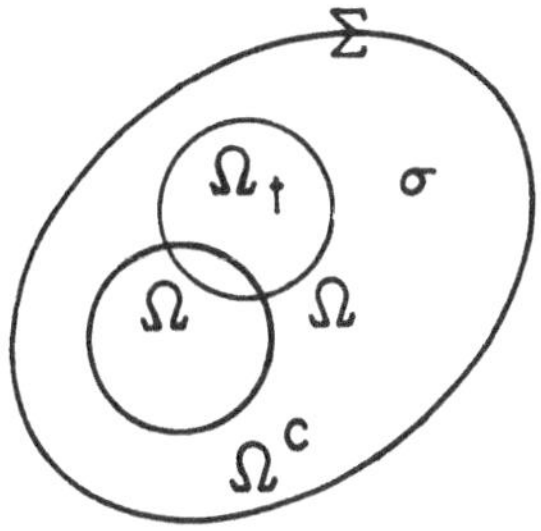

Figure 11.1 Domain $\mathcal{O}$

where n is an outward normal to Ω on Γ, k is a real number, and f, g, h and μ are given functions on $\mathcal{O}$.

Let Y_i and Y_e be regular extensions of y_i and y_e. Now, $y_i - y_e = \mu$, on Γ, so the material derivative gives

$$\dot{\overbrace{y_i - y_e}} = \dot{\mu} = \nabla\mu \cdot V = Y'_i - Y'_e + \nabla(y_i - y_e) \cdot V$$

Thus, on Γ,

$$Y'_i - Y'_e = -\nabla(y_i - y_e) \cdot V + \nabla\mu \cdot V$$

$$= -\nabla_\Gamma(y_i - y_e) \cdot V - \frac{\partial}{\partial n}(y_i - y_e) V \cdot n + \nabla\mu \cdot V$$

Since

$y_i - y_e = \mu$, on Γ, $\nabla_\Gamma (y_i - y_e) = \nabla_\Gamma \mu$, while by the transmission conditions on Γ,

$$\frac{\partial}{\partial n}(y_i - y_e) = h + (k - 1) \frac{\partial}{\partial n} y_e$$

Thus, on Γ, one has

$$Y'_i - Y'_e = (-\nabla_\Gamma \mu + \nabla\mu) \cdot V + \left[(1 - k) \frac{\partial y_e}{\partial n} - h\right] V \cdot n$$

$$= \left[(1 - k) \frac{\partial y_e}{\partial n} + \frac{\partial \mu}{\partial n} - h\right] V \cdot n$$

For any regular function u on $\mathcal{O}$, one has

$$\int_{\Omega_t} \nabla y_t^i \nabla u \, dx + k \int_{\Omega_t^c} \nabla y_t^e \nabla u \, dx$$

$$= \int_{\Omega_t} fu \, dx + \int_{\Omega_t^c} gu \, dx + \int_{\Gamma_t} hx \, ds$$

Taking the derivative with respect to t, at t = 0, one gets

$$\int_{\Omega} \nabla Y_i' \cdot \nabla u \, dx + \int_{\Gamma} \nabla y_i \cdot \nabla u \, v \, ds$$

$$+ k\left(\int_{\Omega^c} \nabla Y_e' \cdot \nabla u \, dx - \int_{\Gamma} \nabla y_e \cdot \nabla u \, v \, ds\right)$$

$$= \int_{\Gamma} fu \, v \, ds - \int_{\Gamma} gu \, v \, ds + \int_{\Gamma} \left(\frac{\partial}{\partial n}(hu) + Hhu\right) v \, ds$$

Choosing u first with its support in Ω and then in Ω^c one gets

$$\left.\begin{aligned} \Delta Y_i' &= 0 , && \text{in } \Omega \\ \Delta Y_e' &= 0 , && \text{in } \Omega^c \end{aligned}\right\}$$

Using the Green formula

$$\int_{\Gamma} \left(\frac{\partial}{\partial n} Y_i' - k \frac{\partial}{\partial n} Y_e'\right) u \, ds = -\int_{\Gamma} (\nabla y_i - k\nabla y_e)\nabla u \, v \, ds$$

$$+ \int_{\Gamma} \left[(f - g)u + \frac{\mu}{\partial n}(hu) + Hhu\right] v \, ds$$

But, on Γ, $y_i = y_e + \mu$, so

$$\nabla_\Gamma y_i = \nabla_\Gamma y_e + \nabla_\Gamma \mu$$

and

$$\nabla y_i = \nabla_\Gamma y_i + \nabla_\Gamma \mu + (\nabla y_i \cdot n)n$$

$$\nabla y_e = \nabla_\Gamma y_e + (\nabla y_e \cdot n)n$$

Thus, on Γ,

$$\nabla y_i - k\nabla y_e = (1 - k)\nabla_\Gamma y_e + hn + \nabla_\Gamma \mu$$

since

$$\frac{\partial y_i}{\partial n} - k\,\frac{\partial y_e}{\partial n} = h$$

and one gets

$$\int_\Gamma \left(\frac{\partial}{\partial n} Y'_i - k\,\frac{\partial}{\partial n} Y'_e\right) u\, ds$$

$$= (k - 1)\int_\Gamma \nabla_\Gamma\, y_e\cdot\nabla_\Gamma u\; v\; ds - \int_\Gamma \nabla_\Gamma\mu\cdot\nabla_\Gamma u\; v\; ds$$

$$+ \int_\Gamma \left(f - g + \frac{\partial h}{\partial n} + Hh\right) u\; v\; ds$$

and, finally, one sees that $Y' = (Y'_i, Y'_e)$ is the solution of the following transmission problem:

$$\left.\begin{array}{lll} \Delta Y'_i = 0 , & & \text{in } \Omega \\ \Delta Y'_e = 0 , & & \text{in } \Omega^c \\ \dfrac{\partial}{\partial n} Y'_e = 0 , & & \text{on } \Sigma \end{array}\right\}$$

with the transmission conditions on Γ

$$Y'_i - Y'_e = \left[(1 - k)\,\frac{\partial y_e}{\partial n} + \frac{\partial \mu}{\partial n} - h\right] V\cdot n$$

$$\frac{\partial}{\partial n} Y'_i - k\,\frac{\partial}{\partial n} Y'_e = (1 - k)\ \mathrm{div}_\Gamma\ (v\nabla_\Gamma y_e) + \mathrm{div}_\Gamma\ (v\nabla_\Gamma\mu)$$

$$+ \left(f - g + \frac{\partial h}{\partial n} + Hh\right) v$$

With these calculations, one is ready to compute the gradient of the cost function. Consider for simplicity the cost function $J = J_3$ of Section 6, i.e.

$$J(\Omega) = \int_\Omega (y_i(\Omega) - z)^2\, dx$$

One now has only to compute, as a linear expression in v, the term

$$j = 2\int_\Omega (y_i(\Omega) - z)\, Y'_i\, dx$$

Consider the adjoint equation for $p = p(\Omega)$, which is the solution of the transmission problem

$$\left.\begin{aligned} -\Delta p_i &= 2(y_i(\Omega) - z)\ , && \text{in } \Omega \\ -\Delta p_e &= 0\ , && \text{in } \Omega^c \\ \frac{\partial p_e}{\partial n} &= 0\ , && \text{on } \Sigma \end{aligned}\right\}$$

with the transmission conditions

$$\left.\begin{aligned} \frac{\partial p_i}{\partial n} - k\frac{\partial p_e}{\partial n} &= 0\ , && \text{on } \Gamma \\ p_i &= p_e\ , && \text{on } \Gamma \end{aligned}\right\}$$

Then,

$$j = \int_\Omega -\Delta p_i Y'_i\, dx + k\int_{\Omega^c} - \Delta p_e Y'_e\, dx$$

$$= \int_\Gamma \left(\frac{\partial}{\partial n} Y'_i p_i - \frac{\partial p_i}{\partial n} Y'_i\right) ds$$

$$-k\int_\Gamma \left(\frac{\partial}{\partial n} Y'_e p_e - \frac{\partial p_e}{\partial n} Y'_e\right) ds$$

and since $p_i = p_e$, on Γ,

$$j = \int_\Gamma \left(\frac{\partial}{\partial n} Y'_i - k\frac{\partial}{\partial n} Y'_e\right) p\, ds - \int_\Gamma \left(\frac{\partial p_i}{\partial n} Y'_i - k\frac{\partial p_e}{\partial n} Y'_e\right) ds$$

Using the transmission conditions for Y' on Γ, the first term may be directly computed. For the second term one writes

$$\frac{\partial p_e}{\partial n} Y'_e = \frac{\partial p_e}{\partial n} Y'_i + \frac{\partial p_e}{\partial n}\left[(k - 1)\frac{\partial y_e}{\partial n} - \frac{\partial \mu}{\partial n} + h\right] v$$

Then,

$$j = \int_\Gamma \left[(1 - k)\ \mathrm{div}_\Gamma\ (v \nabla y_e) + (f - g + \frac{\partial h}{\partial n} + Hh)v + \mathrm{div}_\Gamma (v \nabla_\Gamma \mu)\right] p\, ds$$

$$+ \int_\Gamma \frac{\partial p_e}{\partial n} \left[(k - 1)\ \frac{\partial y_e}{\partial n} - \frac{\partial \mu}{\partial n} + h\right] ds$$

and one gets the scalar gradient

$$g_n = (k - 1) \left[\nabla\ y_e \cdot \nabla p + \frac{\partial y_e}{\partial n} \frac{\partial}{\partial n} p_e\right] - \nabla_\Gamma \mu \cdot \nabla_\Gamma p$$

$$+ \frac{\partial p_e}{\partial n} \left(h - \frac{\partial \mu}{\partial n}\right) + p\left(f - g + \frac{\partial h}{\partial n} + Hh\right) + (y_i - z)^2$$

That is,

$$g_n = (k - 1) \nabla y_e \cdot \nabla p_e - \nabla_\Gamma \mu \nabla_\Gamma p + \frac{\partial p_e}{\partial n} \left(h - \frac{\partial \mu}{\partial n}\right)$$

$$+ p(f - g + \frac{\partial h}{\partial n} + Hh) + (y_i - z)^2$$

Using the transmission condition on Γ in the reverse way (eliminating p_e and y_e instead of p_i and y_i) one would get analogous formulas involving $\nabla y_i \cdot \nabla p_i$ and $\frac{\partial p_i}{\partial n}$, instead of $\nabla y_e \cdot \nabla p_e$ and $\frac{\partial p_e}{\partial n}$. These formulas would, of course, be equivalent to the foregoing results.

12. VERIFICATION OF HYPOTHESIS H1 AND SUMMARY FOR SECOND ORDER PROBLEMS

The linearity and continuity of $V \longrightarrow \dot{y}(\Omega)$ (Hypothesis H1) classically using the implicit function theorem

Let R be a continuously differentiable mapping from $I \times A$ into F, where $I = [0,t]$, A is an open set of a Banach space E, and F is another Banach space. Let (t_0, y_0) be a point in $I \times A$ such that $R(t_0, y_0) = 0$ and $D_2R(t_0, y_0)$ is invertable, then locally there exists a continuously differentiable mapping u from I to E such that $u(t_0) = y_0$ and $R(t, u(t)) = 0$. Furthermore, $u'(t) = -D_2R(t, u(t))^{-1} D_1R(t, u(t))$.

TABLE 12.1 SUMMARY FOR 2nd ORDER PROBLEMS (LaPlace Equation)

	Dirichlet	Neumann
EQUATION	$-\Delta y = f\,,\quad \text{in } \Omega$ $y = g\,,\quad \text{on } \Gamma$	$-\Delta y + y = f,\quad \text{in } \Omega$ $\frac{\partial y}{\partial n} = g,\quad \text{on } \Gamma$
DERIVATIVE EQUATION	$-y' = 0\,,\quad \text{in } \Omega$ $y' = \frac{\partial}{\partial n}(g - y)v\,,\quad \text{on } \Gamma$	$-\Delta y' + y' = 0,\quad \text{in } \Omega$ $\frac{\partial}{\partial n} y' = \text{div}_\Gamma(v\nabla_\Gamma y)$ $+ \left(Hg + \frac{\partial}{\partial n} g + f\right)v$
COST	$J = J_6 = \int_\Gamma \left(\frac{\partial y}{\partial n} - z\right)^2 ds$	$J = J_3 = \int_\Omega (y - z)^2\, dx$
ADJOINT EQ.	$-\Delta p = 0\,,\quad \text{in } \Omega$ $p = 2\left(\frac{\partial y}{\partial n} - z\right),\quad \text{on } \Gamma$	$-\Delta p + p = 2(y - z),\quad \text{in } \Omega$ $\frac{\partial p}{\partial n} = 0\,,\quad \text{on } \Gamma$
GRADIENT $G = {}^t\gamma_s(g_n n)$	$g_n = 2\,\text{div}_\Gamma(\mu\nabla_\Gamma y)$ $+ \mu \frac{\partial \mu}{\partial n} + \mu^2 H$ $+ 2\frac{\partial}{\partial n}(g - y)\frac{\partial p}{\partial n}$ where $\left(\mu = \frac{\partial y}{\partial n} - z\right)$	$g_n = (y - z)^2$ $- \nabla_\Gamma p \cdot \nabla_\Gamma y$ $+ p\left(Hg + \frac{\partial g}{\partial n} + f\right)$

Neumann	Transmission
$-\Delta y = f, \text{ in } \Omega$ $y = g, \text{ on } \Gamma$	$-\Delta y_i = f, \text{ in } \Omega, \quad -\Delta y_e = g, \text{ in } \Omega^c,$ $\frac{\partial}{\partial n} y_e = 0 \text{ on } \Sigma,$ and $\frac{\partial}{\partial n} y_i - k \frac{\partial}{\partial n} y_e = h$ and $y_i - y_e = \mu, \text{ on } \Gamma$
$-\Delta y' = 0, \text{ in } \Omega$ $\frac{\partial}{\partial n} y' = \text{div}_\Gamma (v \nabla_\Gamma y)$ $+ \left(Hg + \frac{\partial g}{\partial n} + f\right) v$	$\Delta y_i' = 0, \text{ in } \Omega, \ \Delta y_e' = 0, \text{ in } \Omega^c,$ $\frac{\partial}{\partial n} y_e' = 0 \text{ on } \Sigma,$ and $y_i' - y_e' \left[(1-k) \frac{\partial}{\partial n} y_e + \frac{\partial \mu}{\partial n} - h\right] v$ and $\frac{\partial}{\partial n} y_i' - k \frac{\partial}{\partial n} y_e' = (1-k) \text{ div}_\Gamma (v \nabla_\Gamma y_e) + \text{div}_\Gamma (v \nabla_\Gamma \mu)$ $+ \left(f - g + \frac{\partial}{\partial n} h + Hh\right) v, \text{ on } \Gamma$
$J = J_4 = \int_\Omega \lVert \nabla (y - z) \rVert^2 \, dx$	$J = J_3 = \int_\Omega (y_i - z)^2 \, dx$
NONE	$-\Delta p_i = 2(y_i - z), \text{ in } \Omega \quad -\Delta p_e = 0, \text{ in } \Omega^c,$ $\frac{\partial p_e}{\partial n} = 0 \text{ on } \Sigma,$ $p_i = p_e, \text{ on } \Gamma$ $\frac{\partial}{\partial n} p_i - k \frac{\partial}{\partial n} p_e = 0, \text{ on } \Gamma$
$g_n = (y - z)^2$ $+ \nabla_\Gamma y \cdot \nabla_\Gamma (z - y)$ $+ (y - z) \left(Hg + \frac{\partial g}{\partial n} + f\right)$	$g_n = (k-1) \nabla y_e \cdot \nabla p - \nabla_\Gamma \mu \cdot \nabla_\Gamma p$ $+ \frac{\partial p_e}{\partial n} \left(h - \frac{\partial h}{\partial n}\right) + (y_i - z)^2$ $+ p \left(f - g + \frac{\partial h}{\partial n} + Hh\right)$

For the Dirichlet problem, for example, one has

$$u(t) = [y(\Omega_t) - g]\circ T_t(V)$$

belongs to $H^2(\Omega)$ for any t and u(t) is defined, on the fixed open set Ω, as the solution of the following problem: find $u(t) \in E = H^2(\Omega) \cap H_0^1(\Omega)$ such that

$$R(t,u(t)) = 0 \quad , \qquad \text{in } F = L^2(\Omega)$$

where the mapping R is given by

$$R(t,u) = A(t)u - f\circ T_t(V) - A(t)(g\circ T_t(V))$$

with A(t) the transported operator (as was seen at 4.7).

$$A(t)u = J_t^{-1} \operatorname{div} (J_t DT_t^{-1}\, {}^tDT_t^{-1} \nabla u)$$

and $J_t = \det(DT_t)$. In this example R is linear in u, so it is equal to D_2R, which is invertible, since the Dirichlet problem is well posed in $H^2(\Omega) \cap H_0^1(\Omega)$ for f in $L^2(\Omega)$.

Now one must verify that $D_1R(t,u(t))$ is in $L^2(\Omega)$, and it is just an exercise to verify that $D_1R(t,u(t))$ is (in $L^2(\Omega)$) linear and continuous in V(t). Finally, with $\dot{y}(\Omega) = u'(0) + \nabla g\cdot V(0)$, one sees that $\dot{y}(\Omega)$ is linear in V(0), that is H1.

Note that the expression of $A' = \frac{\partial}{\partial t} A(t)\Big|_{t=0}$, which is very complicated (it can only be used for constant coefficients), has not been used. Also, the expression of u' given by the implicit function theorem is just used for proving the existence of u'.

In order to show the unity of the foregoing method, results obtained here for four basic second order problems are summarized in Table 12.1.

13. EXTERIOR PROBLEMS USING AN INTEGRAL EQUATION FORMULATION ON THE SURFACE Γ

13.1 Exterior Dirichlet Problem for the Laplace Operator

For convenience, suppose n = 3. Then the problem

$$\left.\begin{array}{ll} \Delta y(\Omega) = 0 \ , & \text{in } \Omega^c \\ y(\Omega) = g \ , & \text{on } \Gamma \end{array}\right\}$$

has a unique solution $y(\Omega)$ in

$$W^1(\Omega^c) = \{y:(1 + |x|)^{-1} y \in L^2(\Omega^c), \frac{\partial y}{\partial n_i} \in L^2(\Omega^c)\}$$

where $\Omega^c = \mathbb{R}^n \setminus (\Omega \cup \Gamma)$ and one supposes $0 \in \Omega$, Ω is a regular bounded deomain. The boundary integral method consists of calculating an auxiliary unknown q (a potential), such that $y(\Omega)$ is given by [see Ref. 18]

$$y(\Omega)(x) = \frac{1}{4\pi} \int_\Gamma q(x) ||x - u||^{-1} \, ds(u)$$

for regular data g and Γ, where $q = q(\Omega)$ is the solution of the boundary coercitive problem for $q \in H^{-1/2}(\Gamma)$ and all $r \in H^{-k}(\Gamma)$,

$$\iint_{\Gamma\times\Gamma} q(x)r(u)||x - u||^{-1} \, ds(u) \, ds(x) = \int_\Gamma g(u)r(u) \, ds(u)$$

Denote by b(q,r) the bilinear form in the first member. To control the solution $y(\Omega_t)$ of the exterior problem, one writes the cost function in terms of $q(\Omega_t)$. For example, if A is a compact set in Ω^c and if the cost function is

$$J(\Omega) = \int_A (y(\Omega) - Z_g)^2 \, dx$$

it may be rewritten as

$$J(\Omega) = \iint_{\Gamma\times\Gamma} q(x)q(y)I_2(x,y)ds(x)ds(y) + \int_\Gamma q(x)I_1(x)ds(x) + \int_A Z_g^2 \, dx$$

where I_1 and I_2 are two regular functions that are independent on Ω_t and Γ_t. In the example, one has

$$I_2(x,y) = \frac{1}{16\pi^2} \int_A ||x - z||^{-1} ||y - z||^{-1} \, ds(z)$$

$$I_1(x) = \frac{1}{4\pi} \int_A ||x - y||^{-1} \, ds(y)$$

Now for regular data, $q(\Omega_t)$ belongs to $H^{1/2}(\Gamma_t)$ and the following theorem has been proved in Ref. 1.

Theorem 13.1. Let g belong to $H^3(\mathbb{R}^n)_{loc}$ and $k \geq 4$, then $q = \frac{d}{dt} q(\Omega_t) \circ T_t(V) \Big|_{t=0}$ is linear and continuous in $V(0) \in H^{1/2}(\Gamma)$.

Using the general method for gradient computation , with $V = vN_0$, one differentiates the last expression for $J(\Omega)$ with respect to t and at t = 0 gets

$$dJ(\Omega;V) = \int_\Gamma \Big\{ [\dot{q}(x) + (qHv)(x)] \Big(\int_\Gamma q(y) [I_2(x,y) + I_2(y,x)] \, ds(y) \Big) \Big\} \, ds(x) + \int_\Gamma q[\dot{I}_1 + HI_1 v] \, ds + \iint_{\Gamma\times\Gamma} q(x) q(y) \dot{I}_2(x,y) \, ds(x) \, ds(y) + \int_\Gamma \dot{q} I_1 \, ds$$

Denote the kernel

$$K(x,y) = ||x - y||^{-1}$$

Then with the notation introduced for J_7, one has

$$\dot{K}(x,y) = -||x - y||^{-3} \langle V(x) - V(y), x - y \rangle$$

By differentiation with respect to t of the problem of which $q(\Omega_t)$ is solution, one gets

$$\frac{d}{dt} \iint_{\Gamma_t \times \Gamma_t} q(\Omega_t) r_t(y) K(x,y) \, ds \, ds = \frac{d}{dt} \int_{\Gamma_t} g r_t \, ds$$

where $r_t = u \circ T_t^{-1}$. With u regular on Γ, $\dot{r} = 0$, and at $t = 0$ one gets

$$b(\dot{q},u) = \int_\Gamma \left(\frac{\partial g}{\partial n} + Hg\right) uv \, dx$$

$$- \int_\Gamma \left\{ H(x) \left(\int_\Gamma K(x,y)[q(x)u(y) + u(x)q(y)] dx(y) \right) v(x) \right\} ds(x)$$

$$+ \iint_{\Gamma\times\Gamma} v(x) \, ||x-y||^{-3} \, \langle n(x), x-y \rangle$$

$$\times \, [q(x)u(y) + u(x)q(y)] \, ds(x) \, ds(y)$$

Introduce now the adjoint equation for $p = p(\Omega)$,

$$b(p,r) = \int_\Gamma r(x) \left(\int_\Gamma p(y)[I_2(x,y) + I_2(y,x)] ds(y) + I_1(x) \right) ds(x)$$

for all $r \in H^{-1/2}(\Gamma)$. Thus $dJ(\Omega;V)$ is equal to $b(q,p)$ + linear terms in v, i.e.

$$dJ(\Omega;V) = \int_\Gamma \left(\frac{\partial g}{\partial n} + Hg\right) pv ds + \iint_{\Gamma\times\Gamma} v(x) \, ||x-y||^{-3} \langle n(x), x-y \rangle$$

$$\times \, (pq + qp) ds \, ds$$

$$- \int_\Gamma H(x) \left(\int ||x-y||^{-1} \, (pq + qp) \, ds(y) \right) v(x) \, ds(x)$$

$$+ \int_\Gamma (qHv)(x) \left(\int_\Gamma q(x) \, [I_2(x,y) + I_x(y,x)] ds(y) \right) ds(x)$$

$$+ \iint_{\Gamma\times\Gamma} q(x) \, q(y) \, \dot{I}_2(x,y) \, ds \, ds$$

$$+ \int_\Gamma q(\dot{I}_1 + HI_1 v) \, ds$$

Now, $\dot{I}_2$ and $\dot{I}_1$ are linear terms in v. Neglecting their contributions, which are trivial calculations, one gets the gradient given for $x \in \Gamma$ by

$$g_n(x) = \left[\frac{\partial}{\partial n} g(x) + H(x)g(x)\right]p(x)$$

$$+\int_\Gamma ||x - y||^{-3}\langle n(x), x-y\rangle[p(x)q(y) + q(x)p(y)]\, ds(y)$$

$$-H(x)\int_\Gamma ||x - y||^{-1}[p(x)q(y) + q(x)p(y)]\, ds(y)$$

$$+ q(x)H(y)\int_\Gamma q(y)[I_2(x,y) + I_2(y,x)]\, ds(y)$$

$$+ q(x)H(x)I_1(x)$$

13.2 Exterior Dirichlet Problem for the Helmoltz Equation

It is well known that when going from the wave equation to the Helmoltz equation, one loses the uniqueness of the solution. To recover this uniqueness, it is necessary to introduce a Sommerfeld radiation condition, which express the fact that the distrubance propagates to infinity, without ever coming back. The Dirichlet problem has a unique solution, Ref. 16, setting

$$\left.\begin{aligned} & y(\Omega_t) \in \left\{ y \in H^1(\Omega^c)_{loc} : \frac{\partial y}{\partial r} - iky \in L^2(\Omega^c)\right\} \\ & y(\Omega_t) + \bar{K}^2 y(\Omega_t) = 0 \quad , \qquad \text{in } \Omega_t^c \\ & y(\Omega_t) = g \quad , \qquad \text{on } \Gamma_t \end{aligned}\right\}$$

14. GLOBAL (OR LARGE) DEFORMATION OF DOMAINS

14.1 The Large Deformation Problem

Thus only gradient and derivative of gradient of functionals $J(\Omega)$ have been calculated. For this, one needed only a local deformation of the open set Ω, using a field V defined in a neighborhood $\mathcal{U}$ of Ω and for t small enough, having just defined the deformed domain $\Omega_t = T_t(V)(\Omega)$ for t small.

Now consider V such that Ω_t can be defined for any t, $t \in \mathbb{R}^+$. Let

$$C_\infty^{0,k} = \{V: V \in C^0(\mathbb{R}^+, C^k(\mathbb{R}^n,\mathbb{R}^n) \cap L^\infty(\mathbb{R}^+, C^k(\mathbb{R}^n,\mathbb{R}^n))\}$$

From Ref. 1, one has the following theorem.

Theorem 14.1. Let $V \in C_\infty^{0,k}$. Then, for any $t \geq 0$, the mappings $T_t(V)$ and $T_t(V)^{-1}$ are C^k diffeomorphisms from $\mathbb{R}^n$ to $\mathbb{R}^n$ and for any subset Ω of $\mathbb{R}^n$, $\Omega_t = T_t(V)(\Omega)$ is defined.

14.2 Derivative Following a Global Deformation

For any $t \geq 0$, let V_t and V^{-t} be the fields defined by

$$V_t(r) = V(t + r)$$

$$V^{-t}(r) = -V(t - r)$$

For the fields just defined, one has

$$T_{t+r}(V) = T_r(V_t)\circ T_t(V)$$

and

$$T_t(V)^{-1} = T_t(V^{-t})$$

Note that if V is an autonomous field, then

$$T_t(V)^{-1} = T_t(-V)$$

since $V^{-t} = -V$.

Now if $j(t) = J(\Omega_t(V))$, for any domain functional J, then

$$dj(t) = \lim_{r\to 0^+} [\, j(t + r) - j(t)]/r$$

$$= \lim_{r\to 0^+} [\, J(T_r(V_t)(\Omega_t)) - J(\Omega_t)]/r$$

$$= dJ(\Omega_t;V_t)$$

Now, if $W \rightarrow dJ(\Omega_t;W)$ is linear and continuous on $C^k(u_t,\mathbb{R}^n)$ (u_t is an arbitrary neighborhood of Γ_t), i.e. if J possesses a gradient at Ω_t, then

$$dj(t) = \langle G(\Omega_t),V_t(0)\rangle = \langle G(\Omega_t),V(t)\rangle$$

since $V_t(0) = V(t)$. Note that dj(t) is the right semi-derivative.

The left semi-derivative is

$$\lim_{r\to 0^+} [j(t - r) - j(t)]/(-r) = [J(\Omega_{t-r}) - J(\Omega_t)]/(-r)$$

Recall that

$$\Omega_{t-r} = T_{t-r}(v)(\Omega)$$

One may now manipulate to obtain the relations

$$T_t(V) = T_{(t-r)+r}(V) = T_r(V_{t-r})\circ T_{t-r}(V)$$

and

$$T_{t-r}(V) = T_r(V_{t-r})^{-1}\circ T_t(V)$$

But,

$$T_r(V_{t-r})^{-1} = T_r(W)$$

where $W(s) = -V_{t-r}(r - s) = -V(t - s)$

Then,

$$T_{t-r}(V) = T_r(V^t)\circ T_t(V)$$

and

$$\Omega_{t-r} = T_r(V^t)(\Omega_t)$$

The defining equation for the left semi-derivative may now be written as

$$[J(\Omega_{t-r}) - J(\Omega_t)]/(-r) = -\{J\ T_r(V^t)(\Omega_t)\ ] - J(\Omega_t)\}/r$$

$$\xrightarrow[r \to 0^+]{} -dJ(\Omega_t;V^t)$$

and if J possesses a gradient at Ω_t, one gets the left semi-derivative as

$$-dj(t;-1) = -dJ(\Omega_t;V^t)$$

$$= -<G(\Omega_t),V^t(0)>$$

$$= <-G(\Omega_t),-V(t)>$$

$$= <G(\Omega_t),V(t)>$$

$$= dj(t;+1)$$

One has now the following result: if J possesses a gradient at any Ω_t, then j(t) is differentiable and $j'(t) = <G(\Omega_t),V(t)>$.

14.3 Construction of a Global Deformation

As a first example without geometrical constraint on the open sets Ω, suppose that either:

(H) J possesses a gradient $G(\Omega)$ at any regular domain, Ω of regularity C^k and $\Omega \longrightarrow G(\Omega)$ is continuous.

or, the upper gradient being the closed convex set defined by $IJ(\Omega) = \{G : dJ(\Omega;V) \leq <G,V(0)>, \text{ for all } V\}$.

(H') The upper gradient $IJ(\Omega)$ is neither empty and one has built a continuous selection $\Omega \longrightarrow G(\Omega)$ of the multi-valued mapping $\Omega \longmapsto IJ(\Omega)$.

To minimize J, the idea is to take the field V(t), at any t, in the opposite direction to $G(\Omega_t)$. Now, $G(\Omega_t)$ is a distribution with support on the boundary Γ_t, while V(t) must be a regular function, smooth enough in a neighborhood of the closure $\bar{\Omega}_t$ of the domain Ω_t.

Now, one has to regularize $G(\Omega_t)$. To do this, consider the operator A_m on $\mathbb{R}^n$,

$$A_m u = (I - \Delta)^m u$$

where Δ is the Laplace operator and m is an integer. One may use the well known property that if G is a distribution with compact support, then one can find $r > 0$ such that $G \in H^{-r}(\mathbb{R}^n, \mathbb{R}^n)$ (r may be determined with the order of G). Now, A_m^{-1} is an isomorphism from H^{-r} on to H^{-r+2m} (supposing $m > r$), so by the Sobolev imbedding Theorem [2], for m large enough one has

$$H^{-r+2m}(\mathbb{R}^n, \mathbb{R}^n) \subset C^k(\mathbb{R}^n, \mathbb{R}^n)$$

for $m > k + \frac{n}{2}$.

Suppose now that $V \in C_\infty^{0,k}$ is a solution, for any $t > 0$, of

$$V(t) = -A_m^{-1} G(\Omega_t(V))$$

in a neighborhood of $\Gamma_t(V)$. Now, A_m possesses an elementary solution E_m, given by

$$E_m(r) = \frac{(2)^m}{(m-1)!} r^{m-\frac{n}{2}} J_{\frac{n}{2}-m}(2\pi r)$$

where n is the dimension n = 2 or 3 and J_i is the Bessel function of index i. The equation for V(t) may now be written in the form

$$V(t,x) = -(E * G(\Omega_t))(x)$$

with x in a neighborhood of Γ_t. Using the general expression for the gradient (see Theorem 1.1 of Section 1),

$$G(\Omega_t) = {}^t\gamma_{\Gamma_t}(g_{n_t} n_t)$$

one gets

$$V(t,x) = -\int_{\Gamma_t} g_{n_t}(y) n_t(y) E_m(x - y)\, ds_t(y)$$

for all $t > 0$ and for x in a neighborhood of Γ_t.

Now, under the Hypothesis H or H', one has

$$J(\Omega_t(V)) = J(\Omega) + \int_0^t dJ(\Omega_s; V(s))ds \le J(\Omega) + \int_0^t < G(\Omega_s), V(s) > ds$$

Equality occurs when a gradient G exists. For an upper gradient (the case H'), one has just the majorization, i.e.

$$J(\Omega_t) \le J(\Omega) - \int_0^t < A_m^{-1} G(\Omega_s), G(\Omega_s) > ds$$

Let $m = 2p$, p an integer. Then,

$$< A_m^{-1} G, G > = ||G||^2_{H^{-m}(\mathbb{R}^n, \mathbb{R}^n)}$$

If J is bounded from below, for example $J(\Omega) \ge 0$, then one gets

$$\int_0^\infty ||G(D_t)||^2_{H^{-m}(\mathbb{R}^n, \mathbb{R}^n)}\, dt < \infty$$

which implies

$$||G(D_t)||^2_{H^{-m}(\mathbb{R}^n, \mathbb{R}^n)} \to 0 \qquad \text{as } t \to \infty$$

in measure, which is taken as a weak necessary optimality condition. This condition may be harder in some examples, essentially when the gradient is uniformly continuous.

Now the choice, $V(t) = -E * G(\Omega_t)$, in a neighborhood of Γ_t, is a domain differential equation.

14.4 Domain Differential Equation

Following the last comments of Section 1, the data of the field $V(t)$ in any neighborhood of Γ_t, is the derivative at t of the mapping $t \longmapsto \Omega_t$. Denote $V(t) = \frac{\partial}{\partial t}\Omega_t$, then the differential equation of Section 14.3 for V may be written

$$\frac{\partial}{\partial t}\Omega_t = -E{*}G(\Omega_t)\ \ t > 0,\ \Omega_0 = \Omega \qquad \text{(DDE)}$$

There is at this time no uniqueness result for the solutions of the domain differential equation DDE, but several existence results may be given for the speed V in the vector space $C_\infty^{0,k}$. These results need much more functional analysis [1], but the fundamental point is underlined here. It is efficient to prove the existence on [0,1], where one may use the Leray-Schauder fixed point theorem for the following mapping:

$$V \to [t \to \Omega_t(V)] \to [t \to G(\Omega_t(V))] \to [t \to A^{-1}G(\Omega_t(V))]$$

A fixed point for this mapping is obtained in a convex, closed subset of $C^0([0,1],C^k(\mathbb{R}^n,\mathbb{R}^n))$

A fundamental observation is that the field $V = -E_m{*}G(\Omega_t)$ is not, on Γ_t, proportional to the normal field n_t on Γ_t. To get a field V that is proportional to the normal field on Γ_t, one could try to use the scalar distribution g_{n_t} on Γ_t, instead of the gradient $G(\Omega_t) = {}^t\gamma_{\Gamma_t}(g_{n_t} n_t)$ and look for a field

$$V(t) = v(t)n_t \ , \qquad \text{on } \Gamma_t$$

with $v(t)$ a scalar function in a neighborhood of Γ_t, that is a solution of the following equation:

$$\left.\begin{aligned} & v(t) + E{*}g_{n_t}(\Omega_t(V)) \\ & V(t) = v(t)n_t \end{aligned}\right\}$$

Now, one meets a difficulty for the existence of solutions for this equation. If V has C^k regularity, then the boundaries $\Gamma_t(V)$ are C^k manifold (if $\Gamma_0 = \Gamma$ was C^k), but the normal field n_t and any local extension N_t is only C^{k-1} and one cannot get a fixed point in $C^0(I,C^k)$ of the mapping

$$V \to [t \to \Gamma_t(V),\Omega_t(V)] \to [t \to n_t, v(t) = -E{*}g_{n_t}]$$

$$\to [t \to V(t) = v(t)n_t]$$

14.5 Optimal Global Deformation Using A Field Z Proportional to the Normal Field

It was seen that the domain equation DDE furnishes fields V that are not proportional to the normal field, but now from Ref. 1, one has a theorem that enables construction of a normal field.

Theorem 14.2. Let Ω be a C^{k+1} regular domain and V a vector field of C^{k+1} regularity, i.e. $V \in C^0(I,C^{k+1}(\mathbb{R}^n,\mathbb{R}^n))$. Let $\Gamma_t(V) = T_t(V)$ and N(t) be any extension of the normal field to $\mathbb{R}^n$ (or to a neighborhood u of Γ_t), $N(t,x) \in C^0(\mathbb{R}^+,C^k(\mathbb{R}^n,\mathbb{R}^n))$. Then, the field

$$Z = (V\cdot N)N$$

is of regularity C^k and it constructs the same domains (of regularity C^{k+1}) as the vector field V, i.e.

$$\left.\begin{array}{l} \Omega_t(V) = \Omega_t(Z) \\ \Gamma_t(V) = \Gamma_t(Z) \\ t \geq 0 \end{array}\right\}$$

Now, let V be a solution in $C_\infty^{0,k+1}$ of the domain differential equation DDE. Taking m large enough, the previous existence result assures a solution V in $C_\infty^{0,k+1}$. The field Z = (V N)N is given on the boundary Γ_t by

$$Z(t,x) = - [E_m * G(\Omega_t)(x) \cdot n_t(x)] n_t(x)$$

for all $t \geq 0$ and $x \in \Gamma_t(Z)$.

That is,

$$Z(t,x) = - \left[\int_{\Gamma_t(Z)} g_{n_t}(y) E(x-y) n_t(x) \cdot n_t(y) ds(y) \right] n_t(x)$$

Note that in the integral on Γ_t, the following scalar product appears:

$$n_t(x) \cdot n_t(y)$$

which implies that if $Z(t,x) = z(t,x) n_t(x)$, for $x \in \Gamma_t(Z)$, the scalar function z is not a solution for the scalar equation discussed before, but is defined by

$$Z(t,x) = z(t,x) n_t(x), x \in \Gamma_t(Z)$$

with

$$z(t,x) = - \int_{\Gamma_t(Z)} g_{n_t}(y) E_m(x-y) n_t(x)\ n_t(y) ds(y)$$

15. GLOBAL DEFORMATION FOR A DOMAIN FUNCTIONAL J THAT IS INVARIANT WITH TRANSLATION OF THE DOMAIN

The example developed in the previous section furnished a general method to get weak optimal conditions when minimizing a domain functional J. The elementary solution E_m is, however, complicated for the computation, even though many expansions for E_m are given in the literature. In fact, in the previous section, if one is concerned with the problem without constraints on the geometrical domains Ω_t, the operator A_m and its elementary solution E_m can be replaced by any operator having the same properties. For example, consider the problem for $V(t) = (V_1(t), \ldots, V_N(t))$ for any $t \_ 0$, $1 \leq i \leq N$,

$$\int_{\mathbb{R}} \Delta V_i(t) \Delta h dx = - \int_{\Gamma_t(V)g} g_{n_t}(x) n_{t,i}(x) h(x) ds(x)$$

for all functions $h \in H^2(\mathbb{R}^n)$. Giroire has shown [15] that this problem is well posed and possesses a unique solution in a Sobolev-like space W if and only if $G_i = g_{n_t} n_{t,i}$ has the special property

$$\int_{\Gamma_t} G_i ds = 0$$

Taking the speed $V = (0,\ldots 0,1,0,\ldots,0)$, i^{th} one sees that this condition may be written $\langle G(\Omega_t),V\rangle = 0$ for any constant field. But, a constant field gives for $T_t(V)$ a translation, so the condition for the problem to be well posed is that J is invariant with domain translations, i.e.

$$J(\Omega+a) = J(\Omega), \text{ for any } a \in \mathbb{R}^n$$

Now, the interpretation of the problem of Section 14.3, whose solution is V, is

$$\left.\begin{array}{lll} \Delta^2 V_i(t) = 0 \ , & \text{in } \Omega_t \cup \Omega_t^c & \\ \left[\frac{\partial}{\partial n_t} V_i(t)\right] = g_{n_t}(t) n_{t,i} \ , & & \text{on } \Gamma_t(V) \\ \left[V_i(t)\right] = 0 \ , & \text{on } \Gamma_t(V) & \end{array}\right\}$$

where [q] is the jump of q when going through $\Omega_t(V)$, i.e.

$$[q] = q\Big|_{\Gamma_t = \partial\Omega_t} - q\Big|_{\Gamma_t = \partial\Omega_t^c}$$

that is $\Delta^2 V_i(t) = - {}^t\gamma_{\Gamma_t}(g_{n_t} n_{t,i})$ in $\mathcal{D}'(\mathbb{R}^n,\mathbb{R}^n)$

where g_{n_t} is the scalar gradient distribution at Ω_t (see Eq. 3.4) and $n_{t,i}$ is the ith component of the normal field n_t.

The elementary solution for this problem is simple,

$$L_n(r) = \begin{cases} r^2 \log r \,, & \text{when } n=2 \\ r^2 \,, & \text{when } n=3 \end{cases}$$

Now for all $t \geq 0$, V is solution of

$$V(t,x) = -\int_{\Gamma_t(V)} L_n(x-y) g_{n_t}(y) n_t(y) ds(y)$$

and the field Z associated is

$$Z(t,x) = -\left[\int_{\Gamma_t(Z)} L_n(x-y) g_{n_t}(y) n_t(y)\; n_t(x) ds(y)\right] n_t(x)$$

for all $x \in \Gamma_t(Z)$.

Then,

$$\Gamma_t = \Gamma_t(V) = \Gamma_t(Z)$$

$$\Omega_t = \Omega_t(V) = \Omega_t(Z)$$

$$J(\Omega_t) \leq J(\Omega) - \int_0^t \left[\int_{\mathbb{R}^n} \sum_i (V_i(t,x))^2 dx\right] dt$$

where K is a real (positive) number and $y(\Omega_t)$ is now a complex function. Now, the result of J. Giroire is the following (see Ref. 16).

If K is not an eigenvalue for the Laplace operator in $H_0^1(\Omega_t)$, then $y(\Omega_t)$ is related to the potential $q(\Omega_t)$ as in the previous case. Simply, one has to change the kernel $||x-y||^{-1}$ for the elementary solution of the Hemoltz equation (n=3) to

$$K(x,y) = e^{iK\,||x-y||} / ||x-y||$$

For the gradient computation one just changes the expression of the bilinear form b(p,r), using this kernel instead of $||x-y||^{-1}$ and, in the expression of the gradient g_n, one changes the contribution of $\dot{K}$, that is an exercise. For more details see Ref. 11 and Ref. 1.

REFERENCES

1. Zolesio, J.P., Identification de domaines par deformations, Thesis, Nice University, 1979.
2. Adams, Sobolev Spaces, Academic Press, New York, 1975.
3. Penot, J.P., "Calcul sons differentiel," J. Funct. Anal., 27, No. 2, 1977, pp. 248-276.
4. Zolesio, J.P., Doctor Speciaty Thesis, Nice University, 1973.
5. Derveux, A. and Palmeria, B., "Une formule de Hadamard," C.R. Acad. Sci., Vol. 280, Paris, 1975, p. 1667 and p. 1761.
6. Murat, F. and Simon, J., Sur le controle par un domaine geometrique, Publication du L.A., 189, University of Paris VI, 1976.
7. Haug, E.J. and Rousselet, B., "Design Sensitivity Analysis of Shape Variations," Optimization of Distributed Parameter Structures, (Eds. E.J. Haug and J. Cea), Sijthoff and Noordhoff, Alphen aan den Rijn, Netherlands, 1980.
8. Diendonne, J., Treatise on Analysis, Vol. 1, Academic Press, New York, 1972.
9. Carmo, D.O., Differential Geometry of Curves and Surfaces, Prentice Hall, Englewood Cliffs, N.J., 1976.
10. Simon, J., Publication du L.A. 189, Univ. Paris VI, 1980.
11. Dusselrer and Zolesio, J.P., to appear.
12. Zolesio, J.P., An Optimal Design Procedure for Optimal Control Support, Lecture notes in Economical and Mathematical Systems, No. 14, 1977, pp. 207-233.
13. Delfour, M., Payre, G. and Zolesio, J.P., "Design of a Mass-Optimized Thermal Diffuser," Optimization of Distributed Parameter Structures (Eds. E.J. Haug and J. Cea), Sijthoff and Noordhoff, Alphen aan den Rijn, Netherlands, 1980.
14. Sokolowsky, J. and Zolesio, J.P., to appear.
15. Giroire, J., Formulation Variationelle par Equations Integraes de Problemes aux Limites Exterieurs, Ecole Polytechnique, Paris, Publication No. 6.
16. Giroire, J., Helmoltz Equation, Ecole Polytechnique, Paris, Publication No. 35.
17. Zolesio, J.P., "Domain Variational Formulation for Free Boundary Problems," Optimization of Distributed Parameter Structures (Eds. E.J. Haug and J. Cea), Sijthoff and Noordhoff, Alphen aan den Rihn, Netherlands, 1980.
18. Mikhlin, S.G., Mathematical Physics; An Advanced Course, North-Holland Publishing Company, 1976.

DOMAIN VARIATIONAL FORMULATION FOR FREE BOUNDARY PROBLEMS

Jean-Paul Zolésio

Departement de Mathematiques, Université de Nice,
06034 Nice Cedex, France

ABSTRACT

This paper concerns the problem of finding a domain Ω, with boundary Γ, in which the solutions (or a solution) u of a variational problem $A_\Omega\, u = f$, in Ω, and $u = 0$, on Γ, satisfy a supplementary condition on Γ, say a condition involving the normal derivative $\frac{\partial u}{\partial n}$ on Γ. The curve Γ is then the free boundary and Ω (or Γ) is the unknown of the problem. A domain functional $J(\Omega)$ is introduced in the following way: The solutions u in Ω are variational ones, i.e. they are solutions of the optimization problem

$$\inf \{E_\Omega(u) : u \in H(\Omega)\}$$

Where E_Ω is an energy term, i.e. a generalization of a convex elliptic quadratic function (in some examples E_Ω is nonconvex, or nondifferentiable) and $H(\Omega)$ is a vector space of functions, or a part of it. In general the domain functional J considered is of the form

$$J(\Omega) = \mathrm{Inf}\, \{E_\Omega(u) : u \in H(\Omega)\} + J_0(\Omega)$$

and all the stationary domains of J furnish solutions for the free boundary problem.

Motivated by free boundary problem in the Grad equations, arising from plasma physics, some considerations on the levels curves $u^{-1}(t)$ of a regular function f are developed. These are regular curves generated by the autonomous field $\pm|\nabla u|^{-2}\ \nabla u = V$, and if u is assumed to be zero on the boundary Γ of a domain Ω, the search for a solution $A_\Omega\ u = f$, in Ω may be considered as the search for the field V generating the level curves of u. The level curves of u can be parametrized by $s = \beta(u)$ and $u = u^*(s)$,s being considered as an independent variable and u* is the monotone rearrangement of u. The search for u is then seen to involve two steps: (1) the search for the geometry of the family of the level curves of u, i.e. the function $\beta(u)$, then (2) the search for u*. The search for the geometry is presented here as a generalization of domain optimization problems. The deformation of the open set Ω is realized by a diffeomorphism T taken in the form $T_t(V)$, but the gradient of the cost functional is now a distribution with distributed support. The construction of u is also presented as the construction of the surface $\Gamma_t = u^{-1}(t)$, then the determination of the normal component of the field that generates the domain is seen as a domain differential equation in $t = u^*(s)$.

1. FREE BOUNDARY PROBLEM : A FIRST EXAMPLE OF DOMAIN VARIATIONAL PROBLEMS

1.1 A Basic Problem

Let f be a regular function given on $\mathbb{R}^n$ and consider the problem of finding a domain Ω, with boundary Γ and u the solution of

$$\left.\begin{aligned} -\Delta u &= f, \text{ in } \Omega \\ u &= 0, \text{ on } \Gamma \\ \frac{\partial u}{\partial n} &= 0, \text{ on } \Gamma \end{aligned}\right\} \tag{1.1}$$

A first approach is to consider the cost functions:

$$J_a(\Omega) = \int_\Gamma \left(\frac{\partial y}{\partial n}\right)^2 ds$$

with $-\Delta y = f$, in Ω, $y = 0$, on Γ, or

$$J_b(\Omega) = \int_\Gamma w^2 \, ds$$

with $-\Delta w = f$, in Ω, $\frac{\partial w}{\partial n} = 0$, on Γ. Equations 1.1 are satisfied if one finds Ω to minimize J_a (or J_b) such that J_a (or J_b) is equal to zero.

This technique possesses two difficulties:

(1) one does not need to minimize J_a (or J_b), but to get a zero value, which is the absolute minimum.
(2) The gradients of J_a and J_b are complicated. They involve the mean curvature H of Γ (see for example Ref. 1).

One now introduces a domain functional $J(\Omega)$ whose stationary domains (the domains Ω such that $dJ(\Omega;V) = 0$) are weak solutions of the free boundary problem of Eq. 1.1. If Γ is smooth enough to define the normal field n, then the solution u of the homogeneous Dirichlet problem is the solution for the homogeneous Neumann problem. Note that no general result on the existence of stationary domains is given.

1.2 The Free Boundary Problem As A Domain Variational Problem

The method given here is valid for any elliptic boundary problem with Dirichlet conditions. The solution of the Dirichlet problem is obtained by minimizing an energy functional, i.e. $y(\Omega)$ is the unique solution, in $H_0^1(\Omega)$ of the optimization problem

$$\inf \{E(y) : y \in H_0^1(\Omega)\}$$

where

$$\left.\begin{aligned} E(y) = E_\Omega(y) &= \int_\Omega [(1/2) \, ||\nabla y||^2 - fy] \, dx \\ \text{or in a more compact notation,} & \\ E_\Omega(y) &= a_\Omega(y,y) - L_\Omega(y) \end{aligned}\right\} \quad (1.2)$$

where a_Ω is an elliptic bilinear form and L_Ω is a linear form, both depending on the domain Ω.

Now consider the domain functional

$$J(\Omega) = \inf_{y \in H_0^1(\Omega)} E_\Omega(y) \tag{1.3}$$

and one has the following theorem.

Theorem 1.1. The domain functional J possesses a gradient at any open set Ω. For a regular domain Ω it is given by

$$dJ(\Omega;V) = -(1/2) \int_\Gamma \left(\frac{\partial}{\partial n} y(\Omega)\right)^2 (V(0)\cdot n)\ ds$$

Further, if Ω is a stationary point for J (e.g. a local extremum) and Γ be regular enough, then Γ is a solution of the free boundary problem of Eq. 1.1.

Before proving the theorem, note that in contrast with J_a and J_b,

(1) the gradient of J is simple
(2) one does not have to minimize J to zero, but simply to find a stationary point (a domain) of J.

Proof. One can write

$$J(\Omega) = \inf \int_\Omega [(1/2)\ ||\nabla y||^2 - fy]\ dx = -(1/2) \int_\Omega fy(\Omega)\ dx$$

Writing $y(\Omega_t)$ as the restriction to $t\times\Omega_t$ of a function $Y(t,x)$, as in the general method [1], one gets

$$dJ(\Omega;V) = -(1/2) \int_\Omega fY'\ dx -(1/2) \int_\Gamma fy(\Omega)\ (V(0)\cdot n)\ ds$$

one could then introduce the adjoint equation solution $p(\Omega)$ (here $p(\Omega)$ is just $-\frac{1}{2}y(\Omega)$) to get an expression for the gradient of J.

Instead, the proof here essentially uses Theorem 2.1 of Ref. 2, which states that the derivative of a minimum is the minimum of the derivative. Let V be a regular field and $\Omega_t = T_t(V)(\Omega)$. Define, for $w \in H_0^1(\Omega)$

$$F(t,w) = E_{\Omega_t}\ (w \circ T_t^{-1})$$

Then,

$$J(\Omega_t) = \inf_{y\in H_0^1(\Omega_t)} E_{\Omega_t}(y) = \inf_{w\in H_0^1(\Omega)} E_{\Omega_t}(w\circ T_t^{-1})$$

since the mapping $w \to w\circ T_t^{-1}$ is one-to-one, and on to $H_0^1(\Omega_t)$, for any t. Then,

$$J(\Omega_t) = \inf_{w\in H_0^1(\Omega)} F(t,w)$$

Now, as was done for eigenvalue (see Ref. 2), it is easy to show that the solution $y(\Omega_t)\circ T_t$ of this infimum is in $H^2(\Omega)$ and is continuously dependent on t. Thus, for $0 \le t \le 1$, it is enough to consider the infimum, not on all the vector space $H_0^1(\Omega)$, but on a ball K of $H_0^1(\Omega)\cap H^2(\Omega)$ centered at $y(\Omega)$. Now, K is compact for the strong topology of $H_0^1(\Omega)$, so Theorem 2.1 of Ref. 2 gives

$$dJ(\Omega;V) = \frac{\partial}{\partial t} F(t, y(\Omega))\Big|_{t=0}$$

$$= \frac{\partial}{\partial t}\left\{(1/2)\int_{\Omega_t} ||\nabla(y(\Omega)\circ T_t^{-1})||^2 dx - \int_{\Omega_t} f y(\Omega)\circ T_t^{-1} dx\right\}\Big|_{t=0}$$

The function $y(\Omega)$ may be extended to a function in $H^2(\mathbb{R}^n)$ (still noted $y(\mathbb{R})$) (see Ref. 1, Sections 5 and 10) then, let $Y(t,x) = ||\nabla y(\Omega)\circ T_t^{-1}(x)||^2$ which is regular and defined on $\mathbb{R}\times\mathbb{R}^n$. By differentiation at t=0, (as for J_1 in Ref. 2, Section 5), with $y = y(\Omega)$ and $V = V(0)$ one gets

$$dJ(\Omega;V) = -\int_\Omega < \nabla y, \nabla(\nabla y\cdot V) > dx + (1/2)\int_\Gamma ||\nabla y||^2 (V\cdot n)\, ds$$

$$+\int_\Omega f\,(\nabla y\cdot V)\, dx - \int_\Gamma f\, y\, (V\cdot n)\, ds$$

But, $y = 0$ on Γ and

(Equation continued on next page)

$$-\int_{\Omega} < \nabla y, \nabla(\nabla y \cdot V) > dx = \int_{\Omega} \Delta y \ (\nabla y \cdot V) \ dx - \int_{\Gamma} \frac{\partial y}{\partial n} (\nabla y \cdot V) \ ds$$

$$= - \int_{\Omega} f \ (\nabla y \cdot V) \ dx - \int_{\Gamma} (\frac{\partial y}{\partial n})^2 \ (V \cdot n) \ ds$$

since, on Γ, $y = 0$, so $\nabla y = \frac{\partial y}{\partial n} n$. Thus,

$$dJ(\Omega;V) = - \frac{1}{2} \int_{\Gamma} (\frac{\partial y}{\partial n})^2 \ (V \cdot n) \ ds$$

This completes the proof.

2. A VARIATIONAL DOMAIN PROBLEM WITH AN EXISTENCE RESULT

2.1 The Problem (a two fluid problem)

Let D be a (fixed) regular domain in $\mathbb{R}^n$ with boundary S. One seeks a domain Ω, $\Omega \subset D$, as shown in Fig. 2.1, with positive measure, $0 < |\Omega| < |D|$, whose boundary is such that

$$S \cap \Gamma = \phi$$

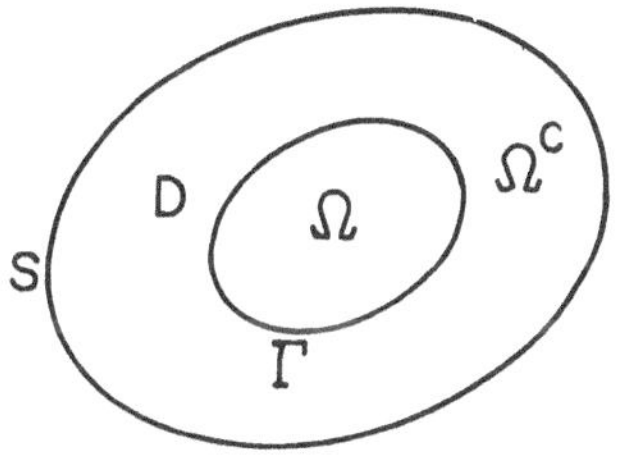

Figure 2.1 Domain For Two Fluid Problem

Let $y(\Omega)$ and $z(\Omega)$ be solutions of the following problems:

(1) the inner problem in Ω (f is a positive bounded function, $f \in L^{\infty}(\Omega)$):

$$-\Delta y = f, \qquad \text{in } \Omega$$
$$y = 0 \ , \qquad \text{on } \Gamma$$

(2) the exterior problem (posed in $\Omega^c = D \setminus (\Omega \cup \Gamma)$), (2.1)

$$-\Delta z = 0, \text{ in } \Omega^c$$

$$z = -1, \text{ on } S$$

(3) with the coupling condition;

$$\frac{\partial z}{\partial n} = \frac{\partial y}{\partial n}, \text{ on } \Gamma$$

$$z = y, \text{ on } \Gamma$$

By the maximum principle, y is positive in Ω and z is negative in Ω^c. In other words, if one considers u defined on D by

$$u(x) = \begin{cases} y(x), & x \in \Omega \\ z(x), & x \in \Omega^c \end{cases}$$

u is the solution of the problem

$$-\Delta u = \begin{cases} f, & \text{in } \{x \in D : u(x) = 0\} \text{ a.e.} \\ 0, & \text{in } \{x \in D : u(x) = 0\} \text{ a.e.} \end{cases} \tag{2.2}$$

Then Ω appears as the domain of positivity for u and the free boundary Γ as the level curve

$$\Gamma = u^{-1}(0) \tag{2.3}$$

2.2 The Domain Variational Formulation For The Free Boundary Γ

Let

$$R(\Omega^c) = \{z \in H^1(\Omega^c) : z = 0 \text{ on } \Gamma,\ z = -1 \text{ on } S\} \tag{2.4}$$

which is a closed convex set in $H^1(\Omega^c)$. Then the domain functional is

$$J(\Omega) = \inf_{z \in R(\Omega^c)} \frac{1}{2} \int_{\Omega^c} ||\nabla z||^2 dx + \inf_{z \in H_0^1(\Omega)} \int_\Omega \left[\frac{1}{2} ||\nabla y||^2 - fy\right] dx \tag{2.5}$$

Exactly as for the previous example, one gets the following theorem.

Theorem 2.1. The domain functional J possesses a gradient at any regular domain Ω, given by

$$dJ(\Omega;V) = \frac{1}{2}\int_{\Gamma}\left\{\left[\frac{\partial}{\partial n}z(\Omega^c)\right]^2 - \left[\frac{\partial}{\partial n}y(\Omega)\right]^2\right\}(V(0)\cdot n)\ ds$$

If Ω is a stationary point of J and if Γ is smooth enough, then Γ is a solution for the free boundary problem.

Proof. The proof is the same as for Thm 1.1. Note simply that for z in Ω^c, the normal vector exterior to Ω^c, on Γ is -n, this minus sign leads to the difference in the gradients and not to the sum. Now, $y(\Omega)$ is positive on Ω, so $\frac{\partial}{\partial n}y(\Omega) \leq 0$ on Γ. Similarly, $z(\Omega^c)$ is negative on Ω^c, so $\frac{\partial}{\partial(-n)}z(\Omega^c) \geq 0$ on Γ, i.e. $\frac{\partial}{\partial n}z(\Omega) \leq 0$ on Γ. Thus, Γ is the solution of the free boundary problem and the proof is complete.

2.3 On The Existence And Regularity Of The Free Boundary

For u belonging to $H^1(D)$ and u = -1 on S, consider

$$E(u) = \int_D \left(\frac{1}{2}\,||\nabla y||^2 - f\,u^+\right)dx$$

Using once more the Theorem 8.1 of Ref. 1 (as in the examples of Section 9) one gets the right semi-derivative at u in the direction $g, g \in H_0^1(D)$

$$dE(u;g) = \int_D \nabla U\cdot\nabla g\ dx - \int_{\{x/u>0\}} fg\,dx - \int_{\{x/u=0\}} fg^+\ dx$$

Now it is classical that E reaches its minimum at a u (not unique since E is not convex). The necessary condition for optimality leads to the inequality $E(u;g) \geq 0$, for any g in $H_0^1(D)$. Taking g positive and negative, one gets

$$f\,1_{u\geq 0} \leq -\Delta u \leq f\,1_{u>0} \qquad \text{a.e.}\ x \text{ in } D$$

Then, f being a bounded function, Δu belongs to $\overset{\infty}{L}(D)$ and it is classical that u belongs to $C^1(D)$. Now, if f is strictly positive one has

$$1_{u \geq 0} \leq 1_{u > 0}$$

(inequality between two characteristic functions) that is, if $\Gamma = \{x \in D : u(x) = 0\}$, then the measure of $\Gamma = 0$. Let $\Omega = \{x \in D : u(x) > 0\}$. Then Γ is its boundary and is a solution (non unique) of the free boundary problem, which is a stationary point for J.

Under the hypothesis

$$||\nabla u(x)|| > 0, \text{ for all } x \in \Gamma$$

Γ is a C^1 regular manifold, with an orientation given by the normal vector field exterior to Ω on Γ,

$$n(x) = -(||\nabla u||^{-1} \nabla u)(x), \; x \in S$$

This technique of differentiation of $J(\Omega)$ as an infimum (using the Theorem 5.1 of Ref. 2) does not require that $y(\Omega_t)$ is differentiable in t, but simply that it is continuous in t. This property will be useful in the next example, where $y(\Omega_t)$ is not differentiable, but $J(\Omega_t)$ is differentiable, because it is an infimum.

3. A NON DIFFERENTIABLE VARIATIONAL DOMAIN PROBLEM (WITH AN EXISTENCE RESULT)

3.1 The Free Boundary Problem [3,4,5]

Retain the notation Ω, Γ, D, S, y, z, u of Section 2. For any measurable function u on D, consider the operator

$$\overline{\beta}_D(u)(x) = \text{measure } \{y \in D : u(y) \leq u(x)\}$$

This operator $\overline{\beta}_D(u)$ is nonlinear, nonlocal, and nonmonotone. It arises in some plasma physics problems and in some questions about the poloidal coordinates to be encountered in the next sections.

Consider the "step of u at x", defined by

$$P(u,x) = \{y \in D : u(x) = u(y)\} \tag{3.1}$$

If $|P(u,x)| = 0$ one says that u has no step at x and one has

$$\Omega = \{x \in D : u(x) > 0\},$$

$$\Gamma = \{x \in D : u(x) = 0\} = u^{-1}(0) = P(u,y), \text{ for all } y \in \Gamma \tag{3.2}$$

The free boundary problem is; find $u \in C^1(\overline{D}) \cap H^2(D)$, such that

$$\left.\begin{aligned} -\Delta u(x) &= \begin{cases} \overline{\beta}_D(u)(x), & \text{a.e. } x \in \Omega \\ 0, & \text{a.e. } x \in \Omega^c \end{cases} \\ u &= -1, \text{ on } S, \\ |\Gamma| &= 0,\ |\Omega| > 0, \text{ and for any } x \in \Omega,\ |P(u,x)| = 0 \end{aligned}\right\} \tag{3.3}$$

Here, Γ is the free boundary. As before, put $y = y(\Omega) = u_{|\Omega}$ and $z = z(\Omega^c) = u_{|\Omega^c}$, so y is positive in Ω and z is negative in Ω^c. Now, the coupling condition is less explicit than in the previous example.

The global regularity of u implies $\frac{\partial y}{\partial n} = \frac{\partial z}{\partial n}$, on Γ, but for $x \in \Omega$,

$$\overline{\beta}_D(u)(x) = \overline{\beta}_\Omega(y)(x) + |\Omega^c|$$

Then, $y(\Omega)$ is the solution of the inner problem

$$\left.\begin{aligned} -\Delta y &= \overline{\beta}_\Omega(y) + |\Omega^c|, \quad \text{in } \Omega \\ y &= 0, \quad \text{on } \Gamma \end{aligned}\right\} \tag{3.4}$$

and $z(\Omega)$ is the solution of the exterior problem

$$\left.\begin{aligned} -\Delta z &= 0, \quad \text{in } \Omega^c \\ z &= 0, \quad \text{on } \Gamma,\ z = -1, \quad \text{on } S \end{aligned}\right\} \tag{3.5}$$

These two problems are coupled by $|\Omega^c|$.

2.2 The Existence Results

For the inner problem of Eq. 3.4 in Ω (for a fixed domain Ω and a given number $|\Omega^c|$), let

$$F(y) = \int_\Omega ||\nabla y||^2\, dx - 2\,|\Omega^c| \int_\Omega y dx - |\Omega| \int_\Omega y dx$$

$$- \iint_{\Omega \times \Omega} (y(x) - y(u))^+ \, dxdu \qquad (3.6)$$

Theorem 3.1. F reaches its minimum in $H_0^1(\Omega)$ and any local minimum of F in $H_0^1(D)$ is a solution of the inner problem of Eq. 3.4.

For the free boundary problem, let

$$H(w) = \int_D ||\nabla w||^2 dx - |D| \int_D (w-1)^+ dx - \iint_{D\times D} [(w(x)-1)^+$$

$$- (w(u)-1)^+]^+ dxdu \qquad (3.7)$$

Theorem 3.2. H reaches its minimum in $H_0^1(D)$ and for any local minimum w of H in $H_0^1(D)$, $u = w - 1$, $\Gamma = \{x \in D : w = 1\}$, $y = u_{|\Omega}$, and $z = u_{|\Omega^c}$ furnish a solution of the free boundary problem of Eqs. 3.1, 3.2, and 3.3, if Γ is smooth enough.

3.3 The Variational Formulation Of The Free Boundary Problem

One may introduce the domain functional whose stationary points furnish the free boundary. With

$$R(\Omega^c) = \{z \in H^1(\Omega^c) : z = 0 \text{ on } \Gamma,\ z = -1 \text{ on } S\}$$

$$G(\Omega) = \{y \in H_0^1(\Omega) : y \geq 0 \text{ a.e. in } \Omega\}$$

consider $y = (w-1)_{|\Omega}$ and $z = (w-1)_{|\Omega^c}$, a direct calculation [3] leads to the minimization of $J(\Omega)$, instead of $H(w)$, where

$$J(\Omega) = \inf \int_{\Omega^c} ||\nabla z||^2 dx + \inf \int_{\Omega} ||\nabla y||^2 dx - (2|D| - |\Omega|) \int_{\Omega} y dx$$

$$- \iint_{\Omega \times \Omega} [y(x_1) - y(x_2)]^+ dx_1 dx_2 \qquad (3.8)$$

the first inf being taken over $z \in R(\Omega^c)$ and the second over $y \in G(\Omega)$.

From a practical point of view, it is possible to verify that this functional is given by

$$J(\Omega) = -2 \int_S \frac{z}{n} ds + \inf \left(- \int_{\Omega} ||\nabla y||^2 dx \right)$$

the infimum being taken over the solutions of the inner problem of Eq. 3.4.

According to the remark made at the end of Section 2 one does not try to differentiate $y(\Omega_t)$, a solution of the inner problem (which is not unique, so it may have bifurcations branches), but directly differentiates $J(\Omega_t)$, as an infimum, by the previous technique. One has

$$J(\Omega_t) = \inf_{z \in R(\Omega^c)} \int_{\Omega_t^c} ||\nabla(z \circ T_t^{-1})||^2 dx$$

$$+ \inf_{y \in G(\Omega)} \int_{\Omega_t} ||\nabla(y \circ T_t^{-1})||^2 dx$$

$$- (2|D| - |\Omega_t|) \int_{\Omega_t} y \circ T_t^{-1} dx$$

$$- \iint_{\Omega_t \times \Omega_t} [y \circ T_t^{-1}(a) - y \circ T_t^{-1}(b)]^+ da\, db$$

One now has to differentiate these four terms with respect to t and evaluate the result at $t = 0$, with fixed functions z or y. But the two first terms have already been differentiated in the previous two examples. Thus, attention turns to the two last terms.

First,

$$\frac{d}{dt}\left[(2|D|-|\Omega_t|)\int_{\Omega_t} y\circ T_t^{-1}dx\right]\Big|_{t=0} = \left(\int_\Gamma V(0)\cdot n\, ds\right)\left(\int_\Omega ydx\right)$$

$$-(2|D|-|\Omega|)\int_\Omega \nabla y\cdot V(0)\, dx - (2|D|-|\Omega|)\int_\Gamma y\, V(0)\cdot n\, ds$$

where the last term is zero, because $y = 0$ on Γ.

To proceed, one needs the two following lemmas (from Ref. 3):

Lemma 3.1. The half semi-derivative in the direction e_i is given by

$$\frac{\partial}{\partial x_i}\int_\Omega |y(x) - y(u)|\, du = \frac{\partial}{\partial x_i} y(x) \int_\Omega \text{sgn}(y(x) - y(u))\, du$$

$$+ \left|\frac{\partial}{\partial x_i} y(x)\right| \quad |P(y,x)|.$$

Now, since $|P(y,x)| = 0$ for $x \in \Omega$, one gets the two sided derivative in the direction e_i.

Lemma 3.2.

$$2\,\overline{\beta}_\Omega(y)(x) - |P(y,x)| = \int_\Omega \text{sgn}(y(x) - y(u))du + |\Omega|, \text{ a.e.x}$$

Using these results in the last term of the equation for $J(\Omega_t)$ one gets

$$\iint_{\Omega_t\times\Omega_t} [y\circ T_t^{-1}(x)-y\circ T_t^{-1}(u)]^+dxdu = \iint_{\Omega\times\Omega} [y(x)-y(u)]^+J_t(x)J_t(u)dxdu$$

The derivate at $t = 0$ of this term is equal to

$$\iint_{\Omega\times\Omega} [y(x)-y(u)]^+\ [\text{div}V(0,x)+\text{div}V(0,u)]\ dxdu$$

$$= \int_\Omega \text{div}V(0,x)\left[\int_\Omega |y(x) - y(u)|\, du\quad dx\right]$$

for $a^+ + (-a)^+ = |a|$. By Lemma 3.1, this equals

$$-\int_\Omega \left(\int_\Omega \operatorname{sgn}[y(x)-y(u)]\, du\right) \nabla y(x)\cdot V(0,x)\, dx$$

$$+\int_\Gamma \big(V(0,x)\cdot n(x)\big)\left(\int_\Omega |y(x)-y(u)|\, du\right) dx$$

By Lemma 3.2, this further reduces to

$$-\int_\Omega [2\,\overline{\beta}_\Omega(y) - |\Omega|]\big(\nabla y\cdot V(0)\big)\, dx + \int_\Gamma \big(V(0)\cdot n\big)\left(\int_\Omega y(u)du\right) ds$$

Then, since y is a solution of the inner problem of Eq. 3.4, using the green formula one gets

$$dJ(\Omega;V) = \operatorname{Inf} \int_\Gamma [(\tfrac{\partial z}{\partial n})^2 - (\tfrac{\partial y}{\partial n})^2]\,(V(0).n)\, ds \tag{3.9}$$

the infimum being taken over the solutions y of the inner problem of Eq. 3.4, which are of maximal norm,

$$-\int_\Omega |\nabla y|^2 dx = \min\{F(w) : w \in H_0^1(\Omega)\}$$

Now one is in the same situation as for the first eigenvalue of the plate displacement problem of Ref. 2, Section 5.3, i.e. the mapping $V \to dJ(\Omega;V)$ is not linear and J has no gradient at a domain Ω, even if Ω is very smooth. The difficulty here is that the solution $y(\Omega)$ is not unique (as the eigenvalue may be repeated).

Now consider the upper gradient, which is the closed convex subset in the distribution vector space defined by

$$G \in \mathcal{I}J(D) \iff dJ(\Omega;V) \le \langle G, V(0)\rangle, \text{ for all } v$$

Now one has the following theorem [3].

Theorem 3.3. Let $\mathcal{I}J(\Omega)$ be nonempty (this is a hypothesis). Then, if Ω is a local minimum for J (for a topology such that $\Omega_t(V) = T_t(V)(\Omega)$ converges to Ω when $t \downarrow 0$),

(1) the upper gradient is reduced to a single element
(2) J possesses a gradient at Ω, which is the single element of the upper gradient $\mathcal{I}J(\Omega)$

(3) the gradient is equal to zero, i.e. $IJ(\Omega)$ is reduced to zero.

Theorem 3.2 characterizes the upper gradient of J at any optimal domain. On the other hand, let w realize the minimum of H in $H_0^1(D)$, then w-1 is a solution for the global problem, w is then in $C^1(\overline{D})$, and $\Omega = \{x \in D : w(x) > 1\}$ is an open set in D with boundary Γ, $|\Gamma| = 0$, $S \cap \Gamma = \phi$, which realizes the minimum of the functional $J(\Omega)$ among all the such open sets Ω in D:

Theorem 3.4 [3]. Let w be a solution of $\text{Min}\{H(w) : w \in H_0^1(D)\}$ and $\Gamma = \{x \in D : w = 1\}$. Then for any open set Q in D, $|Q| = 0$, $Q \cap S = \phi$, one has $J(\Omega) \leq J(Q)$, i.e. Ω realizes the minimum of the domain functional J[†].

Corollary 3.5. Let Ω be a Minimum[††] solution for J, then the upper gradient is reduced to zero. If Γ is regular enough, $(\frac{\partial z}{\partial n})^2 = (\frac{\partial y}{\partial n})^2$, on Γ for any solution of the inner problem in Ω of maximal norm, z being the solution of the exterior problem in $\Omega^c = D \setminus (\Omega \cup \Gamma)$.

With the consideration of the signs of z and y, $\frac{\partial y}{\partial n} = \frac{\partial z}{\partial n}$, on Γ for any solution y of maximum norm. Then Γ is a solution for the free boundary problem.

Corollary 3.6. Let Ω be a minimum solution for J. Then if its boundary Γ is regular enough, Γ is a solution of the free boundary problem of Eqs. 3.1, 3.2, and 3.3.

[†]Let $J(Q) < \min_{w \in H_0^1(D)} H(w) = H(w^*), \Omega = \{x : w^* > 1\}$, z is the solution of the exterior problem in Q^c, y a solution in Q, extremal for F_Q, $H(z\circ + y\circ + 1) < H(w^*)$.

[††]One shows that Ω^c is a convex open set, each convex component of Ω is simply connected.

4. ON THE LEVEL CURVES OF u

4.1 Level Curves As The Boundary Of A Domain Defined By A Field V

As has been seen, many boundaries appear as level curves of a function u. In general, one can suppose $\Gamma = u^{-1}(0)$. Then, it is necessary to turn to some considerations about the level curves of a regular function u. Throughout this section, the function u defined on a regular open set Ω in $\mathbb{R}^n$, for simplicity n = 2, will always possess the following properties:

(1) $u \in C^1(\overline{\Omega})$
(2) u is strictly positive on Ω
(3) u is equal to zero on $\partial\Omega = \Gamma$, $\Gamma = u^{-1}(0)$
(4) $||\nabla u(x)|| > 0$ in Ω, except for a finite number of points of Ω.

For simplicity, suppose there is just one single exceptional point $x_u \in \Omega$, called the center of u, i.e. with

$$||\nabla u(x)|| > 0, \text{ for all } x \in D^{\bullet} = D \setminus x_u$$

necessarily, u is maximum at x_u and one may define

$$M = M(u) = u(x_u) = \text{Max } u > 0$$

Then, for $0 \leq t < M$, the level curve $u^{-1}(t)$ is a C^1 manifold. It is the boundary of a regular domain $\Omega_t = \{x \in \Omega : u(x) > t\}$, where $u^{-1}(t)$ is oriented by the normal vector field exterior to Ω_t,

$$x \in u^{-1}(t), \; n_t(x) = -||\nabla u||^{-1} \nabla u(x)$$

In fact, n_t is the restriction to $u^{-1}(t)$ of a unitary field n defined on $D^{\bullet}$ by

$$n = -||\nabla u||^{-1} \nabla u \qquad (4.1)$$

Because x_u is single, Ω_t is simply connected called a simple geometry (twice geometry for two exceptional points).

Theorem 4.1. Let u belong to $W^{2,\infty}(D^\bullet)$, then:

(1) the level curves of u are constructed by the (autonomous) vector field

$$V = |\nabla u|^{-2} \nabla u \tag{4.2}$$

which is locally Lipchitzian on $D^\bullet$, i.e.

$$u^{-1}(t+r) = T_r(||\nabla u||^{-2} \nabla u)\,(u^{-1}(t)) \tag{4.3}$$

(2) Let $\frac{\partial u}{\partial n}$ be strictly negative on Γ. Then all the level curves are constructed by the field V with Γ,

$$0 \le t < M(u)$$

$$u^{-1}(t) = T_t(V)(S)$$

i.e., following the terminology of Section 1,

$$u^{-1}(t) = \Gamma_t(V)$$

(3) The field V is proportional to the normal field n. Any field defining the level curves may be written as

$$W = V + W_T$$

where W_T is tangent to the level curves, i.e. $W_T \cdot n = 0$.

The proof of this theorem is given at the end of the section.

The level curves are just a particular case of deformation of the surface Γ by the field V, for $0 \le t < M$. One immediately gets the derivatives of all the functionals involving integration on the level curves, by giving expressions for the mean curvatures. On $u^{-1}(t)$,

$$\frac{\partial u}{\partial n_t} = n_t \cdot \nabla u = -||\nabla u|| \tag{4.4}$$

and $V = v\, n$, with

$$v = V \cdot n_t = -||\nabla u||^{-1} \tag{4.5}$$

If n is a unitary extension of n_t, then

$$\text{div } n = -H \quad (\text{the mean curvature of } u^{-1}(t))$$

$$= -\nabla(||\nabla u||^{-1}) \cdot \nabla u - ||\nabla u||^{-1} \Delta u$$

But,

$$-\nabla v = +\nabla(||\nabla u||^{-1}) = -||\nabla u||^{-3} D^2u \; \nabla u = ||\nabla u||^{-2} D^2u \; n$$

and

$$\text{div } n = -H$$

$$= ||\nabla u||^{-1} \left(\frac{\partial^2 u}{\partial n^2} - \Delta u\right) = -v\left(\frac{\partial^2 u}{\partial n^2} - \Delta u\right) \tag{4.6}$$

where, D^2u is the symmetric matrix $D^2u_{ij} = \partial^2_{ij}u$ and

$$\frac{\partial^2 u}{\partial n^2} = D^2u \; n \cdot n$$

It is also helpful to use

$$\nabla(v^m) = (-1)^m \nabla(||\nabla u||^{-m})$$

$$= (-1)^{m+1} m \; ||\nabla u||^{-m-2} D^2u \; \nabla u$$

$$= -m \; v^{m+1} \; D^2u \; n$$

for m a positive integer, and

$$\frac{\partial}{\partial n}(v^m) = -m \; v^{m+1} \; \frac{\partial^2 u}{\partial n^2} \tag{4.7}$$

The Eulerian (or material) derivative of the normal on $u^{-1}(t)$, is then

$$\dot{n} = \frac{d}{dt}(n_t \circ T_t) = \nabla n \cdot V = -\nabla(||\nabla u||^{-1} \partial_i u) \cdot (||\nabla u||^{-2} \nabla u)$$

or

$$\dot{n} = v^2 \left(\frac{\partial^2 u}{\partial n^2} I - D^2 u \right) n \tag{4.8}$$

One may calculate the derivative of a functional on a level curve. The example J_2 of Ref. 1, Section 5, gives

$$\frac{d}{dt} \int_{u^{-1}(t)} f_t \, ds = \int_{u^{-1}(t)} (\dot{f}_t + f \operatorname{div} n \, v) \, ds$$

$$= \int_{u^{-1}(t)} \left[\dot{f}_t - f \left(\frac{\partial^2 u}{\partial n^2} - \Delta u \right) v^2 \right] ds$$

If f_t is the restriction to $u^{-1}(t)$ of a function f, the material derivative of f_t (see Ref. 1, Section 2) is given by

$$\dot{f}_t = \nabla f \cdot (||\nabla u||^{-2} \nabla u) = v \frac{\partial f}{\partial n}$$

and

$$\frac{d}{dt} \int_{u^{-1}(t)} f(x) ds(x) = \int_{u^{-1}(t)} \left[\frac{\partial f}{\partial n} - f \left(\frac{\partial^2 u}{\partial n^2} - \Delta u \right) v \right] v \, ds \tag{4.9}$$

As a concrete example, consider variation of the length of the level curve

$$\frac{d}{dt} \int_{u^{-1}(t)} ds = - \int_{u^{-1}(t)} \left(\frac{\partial^2 u}{\partial n^2} - \Delta u \right) v^2 \, ds$$

Then, the level curves have a constant length when $\left(\frac{\partial^2 u}{\partial n^2} - \Delta u \right) = 0$, which implies that the mean curvature H = 0, which is impossible for a closed curve.

Proof of Theorem 4.1. For necessity, let W be a field that defines the level curves of u. For $X \in \Gamma$, $\dot{x}(t,X) = W(t,x(t,X))$, with $x(0,X) = X$ and

$$x(t,X) \in u^{-1}(t) \iff u(x(t,X)) = t$$

By differentiating with respect to t, one gets

$$\nabla u(x) \cdot W(t,x) = 1$$

Then, $W = ||\nabla u||^{-2} + W_T$

For sufficiency, let $f(t) = u(x(t,X))$. With $X \in \Gamma$ and $\dot{x}(t,X) = (||\nabla u||^{-2} \nabla u)(x(t,X))$,

$$f'(t) = \nabla u(x) \cdot (||\nabla u||^{-2} \nabla u)(x) = 1$$

and $f(0) = 0$, so $f(t) = t$, i.e. $x(t,X) \in u^{-1}(t)$. This completes the proof.

4.2 Monotone Rearrangement of u

For $0 \leq t < M$, define $\nu(u)(t)$ is the volume of the part of Ω that is under the level curve $u^{-1}(t)$, i.e.

$$s = \nu(u)(t) = \text{measure } \{x \in \Omega : u(x) < t\} = |\Omega_t^c| \tag{4.10}$$

If the function u is given, its level curves can be parametrized by the volume s (as a curve can be parametrized by its arc length), the parameter being $0 \leq s < |\Omega|$, instead of t. One has

$$s = \nu(u)(t) = \int_{\Omega_t^c} dx$$

and

$$\frac{ds}{dt} = \nu(u)'(t) = - \int_{u^{-1}(t)} v \, ds = \int_{u^{-1}(t)} ||\nabla u||^{-1} ds \tag{4.11}$$

The minus sign arises because on $u^{-1}(t)$, the normal exterior to Ω_t^c is $-n$. Thus, $\nu(u)$ is strictly increasing and one gets, with Eq. 4.9,

$$\frac{d^2s}{dt^2} = \nu(u)''(t) = -\frac{d}{dt}\int_{u^{-1}(t)} v\, ds$$

$$= \int_{u^{-1}(t)} \left[\frac{\partial^2 u}{\partial n^2}(v^3 - v) + \Delta u\, v^2\right] ds$$

The Monotone Rearrangement is the inverse of the volume,

$$u^*(s) = \nu(u)^{-1}(s) = t \tag{4.12}$$

is a strictly increasing function.

Since the level curves of u are parametrized by s, $0 \leq s < |\Omega|$, to know the function u one just has to know the value taken by u on the level curve s. This value is

$$t = u^*(s)$$

By the derivative of the inverse function, one gets

$$\frac{dt}{ds} = \frac{d}{ds} u^*(s)$$

$$= \left(\int_{u^{-1}(t)} ||\nabla u||^{-1} ds\right)^{-1} = -\left(\int_{u^{-1}(t)} v\, ds\right)^{-1} \tag{4.13}$$

and taking the derivative with respect to s as $\frac{d}{ds} = \frac{d}{dt}\frac{dt}{ds}$,

$$\frac{d^2t}{ds^2} = -\left(\int_{u^{-1}(t)} v\, ds\right)^{-3} \frac{d}{dt}\int_{u^{-1}(t)} v\, ds$$

$$= \left(\int_{u^{-1}(t)} v\, ds\right)^{-3} \int_{u^{-1}(t)} \left[\frac{\partial^2 u}{\partial n^2}(v^3 - v) - \Delta u\, v^2\right] ds \tag{4.14}$$

Classically, one considers the function $\beta(u)$ defined on Ω by

$$\beta(u) = \nu(u)\circ u \tag{4.15}$$

That is,

$$\beta(u)(x) = \text{measure } \{y \in \Omega : u(y) < u(x)\}$$

$$= \nu(u)(t), \text{ with } t = u(x)$$

$$= s, \text{ the parameter of the level curve } u^{-1}(t)$$

one thus has

$$u = u^* \circ \beta(u) \tag{4.16}$$

or in the form of a commutative mapping diagram,

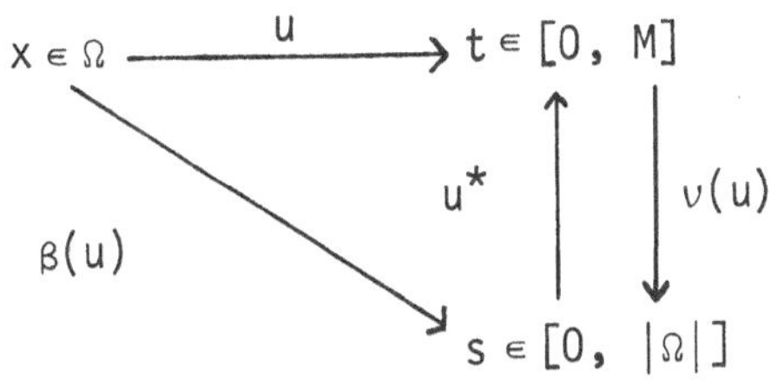

For x in Ω,

$$\beta(u)(x) = s$$

defines the (parametrized) level curve to which x belongs, and

$$t = u(x) = u^*(s)$$

defines the value taken by u on the level curve s.

One has immediately,

$$\left.\begin{aligned} \nabla\beta(u) &= \nu(u)'\circ u\ \nabla u = \frac{ds}{dt}\nabla u \\ \Delta\beta(u) &= \nu(u)''\circ u\ ||\nabla u||^2 + \nu(u)'\circ u\ \Delta u \\ &= \frac{d^2s}{dt^2}\,||\nabla u||^2 + \frac{ds}{dt}\,\Delta u \end{aligned}\right\} \tag{4.17}$$

One says that two functions (defined on Ω) have the same level curves if and only if each is constant on any level curve of the other. Thus, two proportional functions have the same level curves, but the converse is false: if one changes u^* and keeps the same β in the expression $u = u^*\circ\beta$, one gets two non-proportional functions, having the same level curves. It is just an exercise to prove the following result.

Proposition 4.2. Two regular, simple (with a center) functions u and w have the same level curves if and only if $\beta(u) = \beta(v)$. Further, u and $\beta(u)$ have the same level curves.

Thus, the data of a simple function (with a center) u is equivalent to the data of the following two functions:

$\beta = \beta(u)$, the geometry of which gives the level curves

and

u^*, which assigns to u a value on each level curve.

The search for a solution u (of a partial differential equation, for example) may be looked on as the search for two (independent) functions u^* and β, the search for β being the search for the geometry. If one has chosen the order of the geometry, for example simple (with a single center), it is the search for a deformation of $\Omega^{\cdot} = \Omega - \{x_u\}$, by a diffeomorphism of Ω. This diffeomorphism may be taken in the form $T_r(W)$, as in the previous sections (the deformation of an oepn set Ω by a regular field W). One now writes $T_r = T_r(W)$, $r \geq 0$, instead of $T_t = T_t(W)$, because now $t = u^*(s)$ would lead to confusion.

Let W be a regular field defined on $\mathbb{R}^+ \times \Omega$, with the following property: for all $r \neq 0$ and for all $x \in \Gamma$,

$$W(r,x) = 0, \text{ or } W(r,x)\cdot n(x) = 0$$

Then, the mapping

$$T_r = T_r(W): \quad \begin{array}{l} \Omega \longrightarrow \Omega \\ \Gamma \longrightarrow \Gamma \\ X \longrightarrow x(r,X) \end{array}$$

is a bi-C^k diffeomorphism of $\bar{\Omega}$, for which Γ is globally invariant, if $W \in C^{\circ}(\mathbb{R}^+, C^k(\bar{\Omega},\mathbb{R}^n))$.

Now, if u is a regular simple function on Ω, with a center x_u, then $u \circ T_r^{-1}$ is another such function with center $x_u^r = T_r(x_u)$. The range of values of $u \circ T_r^{-1}$ is then the same for u and

$$\nu(u \circ T_r^{-1})\ (t) = \int_{\Omega_t^c(u)} \det(DT_r)\ dx \tag{4.18}$$

and

$$\frac{\partial}{\partial r}(\nu(u \circ T_r^{-1})(t)) = \int_{\Omega_t^c(u)} \operatorname{div} V(r)\ dx \tag{4.19}$$

Then, suppose that

$$\left.\begin{array}{l}
\operatorname{div} W(r) = 0, \text{ on } \Omega \text{ for any } r \\
\text{then} \\
\nu(u \circ T_r^{-1}) = \nu(u) \\
\text{and} \\
(u \circ T_r^{-1})^* = u^* \\
\text{but,} \\
\beta(u \circ T_r^{-1}) = \beta(u) \circ T_r^{-1}
\end{array}\right\} \tag{4.20}$$

4.3 Approximation By Deformation Of The Geometry Of A Solution For An Equation Arising From Plasma Physics

The Equation [6,7]. One seeks a regular simple function w on Ω such that $w \in C^2(\Omega \backslash x_w)$ (the center x_w is an unknown), with

$$\left.\begin{array}{l}
-\Delta w = \dfrac{d^2}{ds^2} w^* \circ \beta(w), \text{ in } \Omega \\
w = 0, \text{ on } \Gamma
\end{array}\right\} \tag{4.21}$$

The Deformation Method. Let u, a given regular simple function, be given on Ω, with $u = 0$, on Γ. One seeks w as

$$w = u \circ T^{-1}$$

where T is a diffeomorphism of $\overline{\Omega}$, such that $\det(DT) = \det(DT^{-1}) = +1$, on Ω. The unknown of the problem is then T, which is approached by a deformation $T_r(V)$, for a field V that shall be constructed for a sufficiently large range of r: ($V(r)$ having for any r its divergence equal to zero).

From the previous properties of Eqs. 4.18, 4.19, and 4.20 of $T_r(V)$, one has

$$\frac{d^2}{ds^2}(u\circ T_r^{-1})^* \circ \beta(u\circ T_r^{-1}) = (\frac{d^2}{ds^2} u^* \circ \beta(u))\circ T_r^{-1}$$

Now, for a given function u, put

$$f = \frac{d^2}{ds^2} u^* \circ \beta(u) \tag{4.22}$$

Then if

$$e(r) = - \Delta(u\circ T_r^{-1}) - f\circ T_r^{-1} \tag{4.23}$$

the problem is to construct the field V, for r large enough, to get $e(r)$ as small as possible, since u_r is a solution if and only if $e(r) = 0$. One then introduces the cost function

$$j(r) = \int_\Omega e(r)^2 \, dx$$

and calculate the derivative $j'(r)$. It is a short exercise to show that

$$j'(r) = \int_\Omega G_r(x)\cdot V(r,x) \, dx$$

where

$$G_r(x) = Z\,(\Delta u_r - f_r) \tag{4.24}$$

is the gradient, with

$$Z = \nabla(f - \Delta u) \tag{4.25}$$

given (with the data of u) and

$$u_r = u \circ T_r^{-1}$$

$$f_r = f \circ T_r^{-1}$$

One calls G_r the gradient at the deformation u_r (for $j(r)$)

Construction Of The Field V. At any time r, one takes $V(r)$ in an opposite direction to G_r, but with div $V(r) = 0$. It is proposed that $V(r)$ be chosen as being the solution of the Stoke's System in Ω, i.e.

$$\left.\begin{aligned} &V(r,x) = (V_1(r,x), V_2(r,x)) \\ &-\Delta V(r) + G_r = \nabla p_r, \text{ in } \Omega \\ &\operatorname{div} V(r) = 0, \text{ in } \Omega \\ &V(r)\cdot n = 0, \text{ on } \Gamma \end{aligned}\right\} \tag{4.26}$$

Then,

$$j'(r) = -\int_\Omega ||\nabla V(r)||^2 \, dx$$

4.4 Poloidal Coordinates

Let Ω be a regular, simply connected, open set in $\mathbb{R}^n$ and V a regular field defined on $\mathbb{R}^+ \times U$, where U is a neighborhood either of Γ in $\Omega^c \cup \Gamma$ or of Γ in $\Omega \cup \Gamma$, as shown in Figs. 4.1(a) and 4.1(b), respectively.

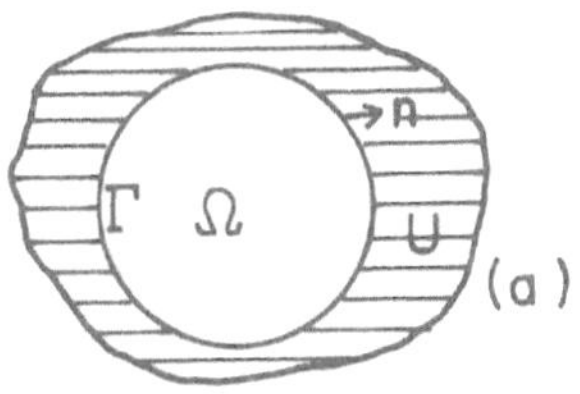

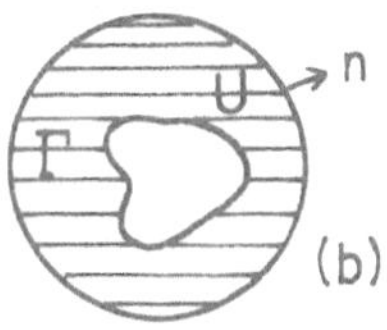

Figure 4.1 Neighborhoods of Γ

Now, for these two situations one supposes that n is the normal field on Γ exterior to Ω, $\Gamma_t = \Gamma_t(V) = T_t(V)(\Gamma)$, and on Γ,

$$V(0,x)\cdot n(x) > 0$$

or

$$V(0,x)\cdot n(x) < 0$$

Then, by continuity, for t small enough, one still has

$$\left.\begin{array}{l} V(t,x)\cdot n_t(x) > 0 \\ \text{or} \\ V(t,x)\cdot n_t(x) < 0 \end{array}\right\} \quad (4.27)$$

for all $x \in \Gamma_t(V)$, where n_t is the normal field on $\Gamma_t(V)$, exterior to $\Omega_t = T_t(V)(\Omega)$. Let

$$t^* = \sup\{t \geq 0 : t \text{ such that Eq. 4.27 is true}\}$$

and

$$Q = \bigcup_{0 \leq t < t^*} \Gamma_t(V) = \{x = x(t,X) : 0 \leq t < t^*, X \in \Gamma\}$$

as shown in Fig. 4.2.

Figure 4.2 Sets Q

Theorem 4.3. The set Q is open, with a boundary formed by two connected pieces $\Gamma \cup \Gamma_{t^*}$. The mapping

$T(V) : \Gamma \times [0, t^*[\longrightarrow Q$

that is,

$x(t,X) = T_t(V)(X)$

is a continuous, one-to-one mapping.

Proof. Let $t_2 > t_1$ and X_1, X_2 in Γ such that $x = x(t_1,X_1) = x(t_2,X_2)$. Then $x \in \Gamma_{t_1} \cap \Gamma_{t_2}$. This is impossible, since from the property of Eq. 4.27, the mapping $t \to \Omega_t(V)$ is a strictly monotone mapping. Then, $\Gamma_{t_1} \subset \Omega_{t_2}$ (strict inclusion) and Γ_{t_1} cannot encounter the boundary of Ω_{t_2}. Thus T(V) is one-to-one. Now, the continuity of T(V) follows from the regularity properties of the mapping $(X,t) \to T_t(V)(X)$, for a field that is smooth enough, say here $V \in C^0(I, C^k(U, \mathbb{R}^n))$, $k \geq 1$, which completes the proof. (In this development, minimal regularity has not always been made explicit, i.e. the minimal index k possible. Here, $k = 1$ is correct).

No confusion should be made between the two mappings

$T(V) : \Gamma \times [0, t^*[\longrightarrow Q$

and

$(X, T) \longrightarrow x(t,X)$

for a given t. For example, on $0 \leq t < t^*$

$T_t(V) : U \longrightarrow Q$

$X \longrightarrow x(t,X)$

The Poloidal coordinates consist of using T(V) to change the cartesian variables

$x = (x_1,\ldots,x_n)$, in the open set Q

to the poloidal coordinates

(X, T), $X \in \Gamma$, $0 \leq t < t^*$

Decomposition Of The Measure On Q. If T(V) is continuous, then the measure on Q can be transported by T(V) on to $\Gamma \times I$, where $I \subset [0, t^*[$. Let h be a local parametrization for the surface Γ, $h: B_0 \subset \mathbb{R}^{n-1} \longrightarrow \Gamma$ and $z = (z', z_n)$ be the variable in $\mathbb{R}^n$, so that if $z = (z', 0) \in B_0$, then $X \in \Gamma$, $X = h(z')$ and any x in $Q' = T(V)(h(B_0), I)$ may be written (in a single form) $x = x(t, h(z'))$ and the determinant of the mapping $(z', t) \to x$ is

$$d = \left|\frac{\partial x}{\partial t}, D_{z'} x \circ h\right| = |V(t,x), D_{z'} x \circ h|$$

Then, if g is a continuous function given on Q, with its support in Q', one has

$$\int_Q g(x)\, dx = \iint_{I \times B_0} g \circ (T_t \circ h)\, |d|\, dt dz'$$

Calculation Of The Determinant d. Now, Γ is the boundary of the regular open set Ω. This means that, without loss of generality, one may suppose the local parametrization h of Γ as being the restriction to B_0 of a local parametrization of Ω. Here one still denotes h, defined on the unit ball B of $\mathbb{R}^n$, z the variable in B, and e_n in B the normal vector to B_0, $e_n = (0, 0, \ldots, 1)$.

If one expands the determinant d on its first column $\frac{\partial x}{\partial t} = V(t,x)$, one gets

$$d = \sum_{i=1}^{n} V_i(t,x)(i^{th} \text{ cofactor} = \sum_{i=1}^{n} V_i(t,x)(-1)^{i+1} i^{th} \text{ minor det.})$$

Now, $M(T_t \circ h)$, being the cofactor matrix introduced in Section 2, one has

$$i^{th} \text{ cofactor} = M(T_t \circ h)\, e_n$$

Then the determinant is

$$d = (V(t,x), M(T_t \circ h)\, e_n)$$

and

$$\int_Q g(x) \ dx = \iint_{I \times B_0} g \circ (T_t \circ h) \ |(V, M(T_t \circ h) \cdot e_n)| dt \ dz'$$

But, by cofactor matrix properties seen in Section 2, one has

$$M(T_t \circ h) = (M(T_t) \circ h) \ M(h)$$

$$n = ||M(h) \ e_n||^{-1} \ M(h) \ e_n$$

Then, for any $t \in I$, one has

$$\int_{B_0} g \circ (T_t \circ h) \ |(V(t, h(z')), M(T_t \circ h) \cdot e_n)| \ dz'$$

$$= \int_{B_0} (g \circ T_t) \circ h \ |(V(t), M(T_t) \cdot n)| \circ h \ ||M(h) \ e_n|| \ dz'$$

$$= \int_\Gamma g \circ T_t \ |(V(t), M(T_t) \cdot n)| \ ds$$

$$= \int_{\Gamma_t(v)} g \ |(V(t), n_t)| \ ds_t$$

where $||M(h) \ e_n|| dz' = ds$ on S and on Γ_t,
$n_t = [||M(T_t) \cdot n||^{-1} \ M(T_t) \ n] \circ T_t^{-1}$ and $ds_t = ||M(T_t) \ n|| \circ T_t^{-1} \ ds$
have been used.

    Theorem 4.4.   Under all the hypotheses formulated in this section, for any continuous and integrable function g on Q one has

$$\int_Q g(x) \ dx = \int_0^{t^*} \left[\int_{\Gamma_t(V)} g \ |(V(t), n_t)| \ ds_t \right] dt$$

A Particular Formulation For Thm. 4.4; Federer's Theorem.

Now Ω is a simply connected, regular, open set in $\mathbb{R}^n$, Γ is its boundary, u is a simply s-regular function on Ω, as in the previous section, with

$$u \in C^1(\Omega) \ ; \ u = 0, \text{ on } \Gamma \ ; \ u > 0, \text{ on } \Omega;$$

$$||\nabla u|| > 0, \text{ in } \Omega^{\bullet} = \Omega \setminus \{x_u\} \ ; \ x_u \text{ is the center}; \tag{4.28}$$

$$\text{and } M = u(x_u) = \text{Max}\{u(x) : x \in \bar{\Omega}\}$$

The field V is $V = ||\nabla u||^{-2} \nabla u$ and $\Gamma_t = u^{-1}(t)$ is the level curve. On Γ_t, $n_t = -||\nabla u||^{-1} \nabla u$ and $v = (V,n) = -||\nabla u||^{-1} < 0$ on $\Gamma_t(V)$, for $0 < t < M$. In the notations of the previous theorem, $I =]0, M[$ and $Q = \Omega^{\bullet}$, but to avoid the singular situation: $\Gamma_{t^*} = \Gamma_M$ reduced to the single point x_u, consider for $r > 0$, r small enough

$$I_r = [r, M-r]$$

and

$$Q_r = \{x \in D \ / \ r \le u(x) \le M-r\}$$

Then, for any continuous function g on $\Omega^{\bullet}$ and for any r (small enough), the previous theorem can be written:

$$\int_{Q_r} g(x) \, dx = \int_r^{M-r} \left[\int_{u^{-1}(t)} g||\nabla u||^{-1} \, ds \right] dt$$

Then if g is integrable on Ω, the function

$$t \to \int_{u^{-1}(t)} g||\nabla u||^{-1} \, ds$$

is integrable on [0, M] and one gets the following theorem.

<u>Theorem 4.5.</u> Let u be a simple regular function (satisfying Eq. 4.28) on Ω and g be a continuous function on $\Omega^{\bullet}$ that is integrable on Ω, then

$$\int_{\Omega} g(x) \, dx = \int_0^{\text{Maxu}} \left[\int_{u^{-1}(t)} g||\nabla u||^{-1} \, ds \right] dt \tag{4.29}$$

By density, this result may be extended to $g \in W^{1,p}(\Omega)$, $1 \leq p \leq +\infty$, $1 \geq \frac{1}{p}$, in particular, $g \in H^1(\Omega)$.

This result may suppose that

$\frac{\partial u}{\partial n} = 0$, on Γ (or on a part of Γ), with $||\nabla u|| > 0$ on $\Omega^{\bullet}$.

5. DOMAIN DIFFERENTIAL EQUATION FOR THE LEVEL CURVES $u^{-1}(t)$ OF THE SOLUTION OF A PARTIAL DIFFERENTIAL EQUATION

5.1 Poloidal Weak Formulation For The Laplace Equation (P.W.F.)

Let Ω be a regular open set, Γ be its boundary, u be a regular simple function on Ω (see Eq. 4.28), $M = \text{Max } u = u(x_u)$, x_u be the center, $\Omega^{\bullet} = \Omega \setminus \{x_u\}$, and the field $V = -||\nabla u||^{-2} \nabla u$ generate $u^{-1}(t)$. One may now use test functions h on Ω, expressed in separated poloidal variables, as follows:

(1) p denotes a regular function defined on [0, M], $p \in C^1([0,M])$, with $p(0) = 1$ and support of p in [0,M[, i.e. p is equal to zero in a neighborhood of M, as shown in Fig. 5.1.
(2) g denotes a regular function defined on the boundary Γ, $g \in C^1(\Gamma)$.
(3) h denotes the function defined (and differentiable) on Ω by:

$$h(x) = [(p \circ u)(x)][g \circ T_{u(x)}(-V)(x)]$$

that is, for $x \in \Omega$, $x \in u^{-1}(t) \Longleftrightarrow u(x) = t$, $x = x(t,X) \Longleftrightarrow x = T_t(V)(X)$, and $h(x) = p(t)\ g(X)$.

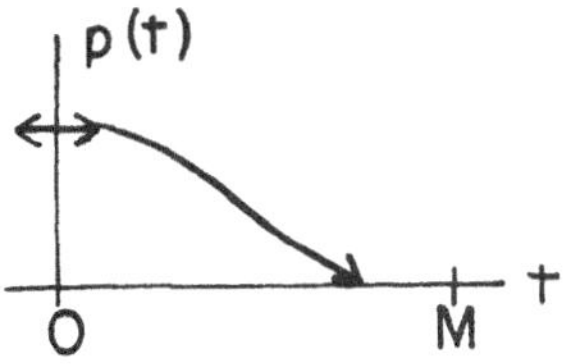

Figure 5.1 Function p

This decomposition for h is unique, for $(t,X) \to x(t,X)$ is a one-to-one mapping between $[0,M[\times\Gamma$ and $\Omega^\bullet$ and h is equal to zero in a neighborhood of x_u.

The normal derivative of the test function h on the level curve $u^{-1}(t)$, $x \in \Omega$ is

$$\nabla_x h(x) = p'(t)\ \nabla u(x)\ g(X) + p(t)\ \nabla_x(g\circ T_{u(x)}(-V)(x))$$

The function $x \to g(X) = g\circ T_{u(x)}(-V)(x)$ is constant on the integral curves of the fields $-V$, V, n, or ∇u (curves which globally are the same). In particular, one has, $n = -||\nabla u||^{-1}\ \nabla u$ and $n\cdot\nabla(g\circ T_{u(x)}(-V))(x) = 0$, in Ω, so one gets

$$\frac{\partial}{\partial n} h(x) = (n\cdot\nabla h)(x) = -\ p'(t)\ ||\nabla u(x)||\ g(X)$$

Now, suppose that the function u is the solution of the Dirichlet problem in $\Omega^\bullet$, i.e. $u \in C^2(\Omega^\bullet)$

$$\left.\begin{aligned} -\Delta u &= f, \text{ in } \Omega^\bullet \\ u &= 0, \text{ on } \Gamma,\ u(x_u) = M \end{aligned}\right\} \tag{4.30}$$

where f is continuous on $\bar{\Omega}$. Then, for any test function h, one has (for the first member of 4.30)

$$\int_\Omega -\Delta u\ h\ dx = \int_\Omega \nabla u\cdot\nabla h\ dx - \int_\Gamma \frac{\partial u}{\partial n} h\ ds$$

$$= \int_0^{Maxu} \left[\int_{u^{-1}(t)} ||\nabla u||^{-1}\ \nabla u\cdot\nabla h\ ds\right] dt - \int_\Gamma \frac{\partial u}{\partial n} h\ ds$$

$$= \int_0^{Maxu} p'(t) \left[\int_{u^{-1}(t)} ||\nabla u||(x)\ g(X)\ ds(x)\right] dt - \int_\Gamma \frac{\partial u}{\partial n} h\ ds$$

$$= -\int_0^{Maxu} p(t)\ \frac{d}{dt}\left[\int_{u^{-1}(t)} ||\nabla u||(x)\ g(X)\ ds(x)\right] dt$$

$$- \int_\Gamma [||\nabla u||g + \frac{\partial u}{\partial n} h]\ ds$$

where $||\nabla u||^{-1}\ \nabla u \cdot \nabla h = -\frac{\partial h}{\partial n}$, on $u^{-1}(t)$, integration by parts in the variable t, $p(M) = 0$, and $u^{-1}(0) = \Gamma$ have been used. On Γ, one has $h = g$ and $||\nabla u|| = -\frac{\partial u}{\partial n}$, so the last integral is zero.

The weak equality of the two members of Eq. 4.30 now gives for any h, i.e. for any functions p and g,

$$-\int_0^{Maxu} p(t)\ \frac{d}{dt}\left[\int_{u^{-1}(t)} ||\nabla u||(x)\ g(X)\ ds(x)\right] dt$$

$$= \int_0^{Maxu} p(t)\left[\int_{u^{-1}(t)} f(x)\ g(X)\ ||\nabla u(x)||^{-1}\ ds(x)\right] dt$$

Thus, u is solution of the problem

$$\frac{d}{dt}\int_{u^{-1}(t)} ||\nabla u||(x)\ g(X)\ ds(x) = \int_{u^{-1}(t)} f(x)\ g(X)\ ||\nabla u(x)||^{-1} ds(x) \tag{5.1}$$

for all $g \in C^1(\Gamma)$ where equality is between continuous functions on $]0, M[$.

The converse is true.

Theorem 5.1. Let u be a simple regular function on Ω, (see Eq. 4.28). Then $u \in C^2(\Omega^\bullet)$ is a solution of the Dirichlet problem:

$- \Delta u = f$, in $\Omega^\bullet$

$u = 0$, on Γ

$u(x_u) = M$

if and only if u is a solution for the Poloidal Weak formulation of Eq. 5.1.

Proof. One has just to prove the sufficiency. For this, one calculates the derivative with respect to t of the integral on the level curve in Eq. 5.1. Using the result of Eq. 4.9,

$$\frac{d}{dt}\int_{u^{-1}(t)} ||\nabla u||(x)\; g(X)\; ds$$

$$= \int_{u^{-1}(t)} \frac{\partial}{\partial n}\Big[||\nabla u||(x)\; g(X) + ||\nabla u||\; g(X)\Big(\frac{\partial^2 u}{\partial n^2} - \Delta u\Big) v\Big] v\; ds$$

$$= \int_{u^{-1}(t)} g(X)\Big[\frac{\partial^2 u}{\partial n^2} - \Big(\frac{\partial^2 u}{\partial n^2} - \Delta u\Big)\Big] v\; ds$$

$$= - \int_{u^{-1}(t)} g(X)\; u||\nabla u||^{-1}\; ds$$

where $\frac{\partial}{\partial n}[||\nabla u||\; g(X)] = g(X)\;\frac{\partial}{\partial n}(||\nabla u||)$ and $\frac{\partial}{\partial n}(||\nabla u||) = -\frac{\partial}{\partial n}(v^{-1}) = \frac{\partial^2 u}{\partial n^2}$ have been used. Then, Eq. 5.1 may be written

$$\int_{u^{-1}(t)} g(X)(\Delta u+f)\; ||\nabla u||^{-1}\; ds = 0$$

for all $g \in C^1(\Gamma)$, i.e. for all t,

$$- \Delta u = f, \text{ on } u^{-1}(t),\; 0 < t < M$$

This completes the proof.

Note that $V = -||\nabla u||^{-2}\;\nabla u$ is the field that constructs the level curves $u^{-1}(t)$ and $v = V{\cdot}n = -||\nabla u||^{-1}$ is the real unknown, if one wants to characterize u by its level curves $\Gamma_t(V)$, starting, e.g., from $\Gamma = \Gamma_0(V)$. If one discretizes the variable t as $t_k = t_{k-1} + q$, $q > 0$, and if v_k is the approximation of v on $\Gamma_k = u^{-1}(t_k)$, then, if $\Gamma, \Gamma_1, \ldots, \Gamma_k$ have been constructed, v_k^{-1} appears as the solution on Γ_k of the following linear problem: for all $g \in C^1(\Gamma_k)$,

$$\int_{\Gamma_k} v_k^{-1}\; g\; ds = \int_{\Gamma_{k-1}} (qf\; v_{k-1} + v_{k-1}^{-1})\, g(y+hv_{k-1}n)\, ds(y)$$

Γ_k being constructed from Γ_{k-1} and v_{k-1} as $\{x = y+hv_{k-1}n : y \in \Gamma_{k-1}\}$.

5.2 Application To An Equation Arising From Plasma Physics

Proposition 5.2. Let u be a simple regular function on Ω, $u \in C^2(\Omega^\bullet)$, $u(x_u) = M$, $u = 0$, on Γ, and $v = -||\nabla u||^{-1}$, with v finite on Γ. Then u is solution of the problem

$$- \Delta u = \frac{d^2}{ds^2} u^* \circ \beta(u), \text{ in } \Omega^\bullet$$

if and only if for all t, $0 < t < M$, and for all $g \in C^1(\Gamma)$,

$$\frac{d}{dt} \int_{u^{-1}(t)} v^{-1}(x)\, g(X)\, ds(x)$$

$$= - \frac{1}{2} \frac{d}{dt} \left[\int_{u^{-1}(t)} v\, ds \right]^{-2} \left[\int_{u^{-1}(t)} g(X)\, v(x)\, ds \right]$$

with $x = x(t,X)$, $X \in \Gamma$.

Proof. By the results of Eq. 4.14, one has, with $s = \beta(u)(x)$,

$$\frac{d^2 u^*}{ds^2}(s) = - \frac{1}{2} \frac{d}{dt} \left[\int_{u^{-1}(t)} v\, ds \right]^{-2}$$

where s is constant on the level curves (since s is the parametrization). One has only to write Eq. 5.1 to complete the proof.

6. FREE BOUNDARY PROBLEM IN THE GRAD EQUATIONS ARISING FROM PLASMA PHYSICS

A heuristical approach is now given for a free boundary problem in the Grad equation, related to the equilibrium of a confined plasma, assuming the fluid is adiabatic (See Refs. 7, 8, and 9),

$$\left.\begin{aligned} &- \Delta u(x) - \left(\frac{d^2}{ds^2} u^*\right) \circ \overline{\beta}(u)(x) = f(x), \text{ in } \Omega \\ &u = 0, \text{ on } \Gamma \end{aligned}\right\} \qquad (6.1)$$

where u^* is the monotone rearrangement for u (see Section 3) and $\overline{\beta}(u)$ has already been defined in Section 3 as

$$\overline{\beta}(u)(x) = \text{measure } (\{y \in \Omega : u(y) \leq u(x)\})$$

Now, a correct formulation and a weak existence result for its solutions has recently been given by Roger Temam [6] as follows: Let $\overline{\nu}(u)(t) = \text{measure } \{y \in \Omega : u(y) \leq t\}$, so $\overline{\beta}(u) = \overline{\nu}(u) \circ u$, when u is a regular function, say $||\nabla u(x)|| > 0$ in a neighborhood of the level curve $u^{-1}(\overline{t})$. It has been seen that in a neighborhood of $\overline{t}$, one has $\overline{\nu}(u)(t) = \nu(u)(t)$ and $\nu(u)$ is differentiable and strictly increasing. When u is a simply regular function, with $||\nabla u|| > 0$ in $\Omega^{\bullet} = \Omega \setminus x_u$, x_u is the center, $u(x_u) = M = \text{Max } u$, u being positive on Ω. It has further been seen that $\nu(u)$ is invertible, so one defines u^* as

$$u^* = \nu(u)^{-1}$$

But, in general $\overline{\nu}(u)$ is not invertible and, with Temam, define

$$u^*(s) = \inf \{t : t \in \mathbb{R}, \overline{\nu}(u)(t) > s\}$$

From Ref. 7, the mapping $u \to u^*$ is a contraction mapping from $L^2(\Omega)$ into $L^2(0, |\Omega|)$, and the set $W(\Omega) = \{u \in H_0^1(\Omega) : u^* \in H^1(0, |\Omega|)\}$ is not equal to $H_0^1(\Omega)$, for example $0 \notin W(\Omega)$. Let

$$h_\Omega(u) = \frac{1}{2}\int_\Omega ||\nabla u||^2 \, dx + \frac{1}{2}\int_0^{|\Omega|} \left(\frac{du^*}{ds}\right)^2 ds - \int_\Omega fu \, dx$$

In a very simple proof, using compactness and weak lower semi-continuity of h_Ω on $W(\Omega)$, Temam [7] gave an existence result for the optimization problem $\inf \{h_\Omega(w) : w \in W(\Omega)\}$. This problem possesses at least one solution u in $W(\Omega)$, $h_\Omega(u) = \inf \{h_\Omega(w) : w \in W(\Omega)\}$.

Now, if u is a regular simple function on Ω that realizes its maximum at its center x_u, then u is a solution of the problem

$$\left.\begin{aligned} & -\Delta u - \left(\frac{d^2}{ds^2} u^*\right)\circ\overline{\beta}(u) = f, \text{ in } \Omega \\ & u = 0, \qquad\qquad \text{on } \Gamma \\ & \frac{du^*}{ds}\,(|\Omega|) = 0 \end{aligned}\right\} \tag{6.2}$$

One now says that u is a regular, simple, variational solution of Eq. 6.2.

6.1 The Free Boundary Problem

Introduce now the Domain Functional

$$J(\Omega) = \inf\,\{h_\Omega(w) \;:\; w \in W(\Omega)\}$$

Let V be a regular field, the mapping $u \to u\circ T_r(V)^{-1}$ is one-to-one from $W(\Omega)$ onto $W(\Omega_r)$, so

$$J(\Omega_r) = \inf\,\{h_{\Omega_r}(w\circ T_r^{-1}) \;:\; w \in W(\Omega)\}$$

One may now formulate the following assumption:

(H) For any regular field V and r small enough (the bound on r depends on V), all the variational solutions of Eq. 4.27 in $\Omega_r(V)$ are simple regular and positive on Ω_r.

This is the same as assuming that any variational solution u_r of Eq. 4.27 in Ω_r satisfies the following:

u_r has a center x_u^r, $\Omega_r^{\bullet} = \Omega_r \setminus x_u^r$, $u_r \in C^2(\Omega_r^{\bullet})$, $||\nabla u|| > 0$ on $\Omega_r^{\bullet}$

$u_r > 0$ in Ω_r, $M_r = u(x_u^r)$ is the maximum of u_r on Ω_r.

Let

$$F(r,u) = h_{\Omega_r}(u\circ T_r^{-1})$$

for $u = u_r \circ T_r^{-1}$, u_r a variational solution of Eq. 4.27 in Ω_r, the mapping $r \to F(r,u)$ is differentiable.

Calculation of $\frac{\partial}{\partial r} F(r,u)\Big|_{r=0}$. Let u be a variational solution of Eq. 4.27. Now,

$$\frac{d}{dr} |\Omega_r|\Big|_{r=0} = \int_\Gamma V(0)\cdot n \, ds$$

and since

$$\frac{du^*}{ds}(|\Omega|) = 0$$

one has

$$\frac{\partial}{\partial r}F(r,u)\Big|_{r=0} = \text{the classical terms} + \int_0^{|\Omega|} \frac{du^*}{ds}\frac{d}{ds}\left(\frac{\partial}{\partial r}(u\circ T_r^{-1})^*\Big|_{r=0}\right) ds$$

Recall

$$\nu(u\circ T_r^{-1})(t) = \int_{\{x \in \Omega : u(x)<t\}} \det(DT_r)\, dx$$

Then,

$$\frac{\partial}{\partial r}\, \nu(u\circ T_r^{-1})(t)\Big|_{r=0} = \int_{\{x \in \Omega : u(x)<t\}} \operatorname{div}V(0)\, dx$$

and since

$$(u\circ T_r^{-1})^* \circ \nu(u\circ T_r^{-1})(t) = t$$

taking the derivative with respect to r, at r=0, one gets

$$\frac{\partial}{\partial r}(u\circ T_r^{-1})^*\Big|_{r=0}(s) = -\frac{du^*}{ds}(s) \int_{\{x \in \Omega : u(x)<t\}} \operatorname{div}V(0)\, dx$$

Now, following the derivative results, one gets on the level curves

$$\frac{d}{dt}\int_{\{x\in\Omega:u(x)<t\}} \mathrm{div}V(0)dx = +\int_{u^{-1}(t)} \mathrm{div}V(0)\|\nabla u\|^{-1}d\sigma$$

and

$$\frac{d}{ds}\left(\frac{\partial}{\partial r}(u\circ T_r^{-1})^*\Big|_{r=0}\right)(s)$$

$$= -\frac{d^2}{ds^2}u^*(s)\left[\int_{u^{-1}(t)} V(0)\cdot(\|\nabla u\|^{-1}\nabla u)d\sigma + \int_\Gamma V(0)\cdot n\, d\sigma\right]$$

$$-\left(\frac{du^*}{ds}(s)\right)^2\int_{u^{-1}(t)} \mathrm{div}\ V(0)\ \|\nabla u\|^{-1}\ ds$$

and

$$A = \frac{\partial}{\partial r}\left[\frac{1}{2}\int_0^{|\Omega_r|}\left(\frac{d}{ds}(u\circ T_r^{-1})^*\right)^2 ds\right]\Big|_{r=0} = A_1 + A_2 + A_3$$

where

$$A_1 = -\int_0^{|\Omega|}\frac{d^2}{ds^2}u^*(s)\left(\int_{u^{-1}(t)}\nabla u\cdot V(0)\ \|\nabla u\|^{-1}\ d\sigma\right)\left(\frac{du^*}{ds}\right)\ ds$$

$$A_2 = -\left(\int_0^{|\Omega|}\frac{d^2u^*}{ds^2}\ \frac{du^*}{ds}\ ds\right)\int_\Gamma V(0)\cdot n\ d\sigma$$

$$A_3 = -\int_0^{|\Omega|}\left(\frac{du^*}{ds}\right)^2\int_{u^{-1}(t)} \mathrm{div}V(0)\ \|\nabla u\|^{-1}\ \frac{du^*}{ds}\ ds$$

Making the change of variable $t = u^*(s)$, $0 \le t \le M$, $ds = \frac{du^*}{ds}dt$, and using the Federer's theorem (Theroem 4.5) for A_1 and A_3, one gets

$$A_1 = -\int_\Omega\left(\frac{d^2u^*}{ds^2}\right)\circ\beta(u)(\nabla u\cdot V(0))\ dx$$

$$A_2 = \frac{1}{2}\left(\frac{du^*}{ds}\right)^2(0) \int_\Gamma V(0)\cdot n \, d\sigma$$

$$A_3 = - \int_D \left(\frac{du^*}{ds}\right)^2 \circ \beta(u) \text{ div } V(0) \, dx$$

$$= \int_D \nabla\left[\left(\frac{du^*}{ds}\right)^2 \circ \beta(u)\right]\cdot V(0) \, dx$$

$$- \left(\frac{du^*}{ds}\right)^2(0) \left[\int_\Gamma V(0)\cdot n \, d\sigma\right]$$

Now, $\nabla\left[\left(\frac{du^*}{ds}\right)^2 \circ \beta(u)\right] = 2\left(\frac{du^*}{ds}\ \frac{d^2u^*}{ds^2}\right) \circ \beta(u)\ \nabla(\beta(u))$ and from Section 2, $\nabla(\beta(u)) = \left(\frac{du^*}{ds}\right)^{-1} \nabla u.$ Thus,

$$A_3 = 2\int_\Omega \left(\frac{d^2u^*}{ds^2}\right) \circ \beta(u)\ (\nabla u\cdot V(0))\, dx - \left(\frac{du^*}{ds}\right)^2(0) \int_\Gamma V(0)\cdot n \, d\sigma$$

and finally, taking the derivative with respect to r, at r=0, of

$$\frac{1}{2}\int_\Omega ||\nabla(u\circ T_r^{-1})||^2 \, dx - \int_\Omega f u\circ T_r^{-1} \, dx$$

as in Section 1, one gets (u being a solution of Eq. 6.2)

$$\frac{\partial}{\partial r}F(r,u)\Big|_{r=0} = -\frac{1}{2}\int_\Gamma\left[\left(\frac{\partial u}{\partial n}\right)^2 + \left(\frac{du^*}{ds}\right)^2\right] V(0)\cdot n \, d\sigma$$

$$+ \int_\Omega \left[\Delta u + \left(\frac{d^2u^*}{ds^2}\right)\circ \beta(u) + f\right] \nabla u\cdot V(0)\, dx$$

Now, in using Theorem 5.1 of Ref. 2, i.e. the derivative of an infimum is the infimum of the derivative, and using the assumption:

(H') with V given, for r small, say $0 \le r \le 1$, the infimum in the definition of $J(\Omega_r)$ can be taken on a bounded subset of $W(\Omega)$, which is compact for a weak topology, i.e., a topology for which $w \to F(r,w)$ is lower semi-continuous.

Thus, one gets the Eulerian semi-derivative of the domain functional J, at Ω, as

$$dJ(\Omega;V) = \inf \frac{\partial}{\partial r} F(r,u)\Big|_{r=0}$$

the infimum being taken on the variational solutions u of Eq. 6.2. One thus gets the following result:

Under the assumption (H) - (H'), the domain functional $J(\Omega)$ defined by $J(\Omega) = \inf_{w \in W(\Omega)} h_\Omega(w)$ has an upper gradient $J(\Omega)$ that is not empty, since for each variational solution u of Eq. 6.2,

$$-\frac{1}{2}\left[\left(\frac{\partial u}{\partial n}\right)^2 + \left(\frac{d}{ds}u^*(0)\right)^2\right] \in \mathcal{I}J(\Omega)$$

If Ω is a local minimum for J, then $\mathcal{I}J(\Omega)$ is reduced to zero and, for any variational solution u of Eq. 4.27, one has (Ω,u) as the solution of the problem

$$\left.\begin{aligned} &-\Delta u - \frac{d^2u^*}{ds^2} \circ \beta(u) = f, \text{ in } \Omega \\ &u = \frac{\partial u}{\partial n} = 0, \text{ on } \Gamma \\ &\frac{du^*}{ds}(0) = \frac{du^*}{ds}(|\Omega|) = 0 \end{aligned}\right\} \qquad (6.3)$$

Note that Eq. 6.3 is the Free Boundary Problem, problem which was suggested by Professor H. Grad (personal communication, June 1979).

REFERENCES

1. Zolesio, J.P., "The Material Derivative (or Speed) Method For Shape Optimization", Optimization of Distributed Parameter Structures (Eds. E.J. Haug and J. Cea), Sijthoff & Noordhoff, Alphen aan den Rijn, Netherlands, 1980.
2. Zolesio, J.P., "Semi Derivative Of Repeated Eigenvalues," Optimization of Distributed Parameter Structures (Eds. E.J. Haug and J. Cea), Sijthoff & Noordhoff, Alphen aan den Rijn, Netherlands, 1980.
3. Zolesio, J.P., Identification de Domaines par Deformation, Thesis, Nice University, 1979.

4. Mossino, J. and Zolesio, J.P., "Formulation Variationnelle de Problèmes Issus de la Physique des Plasmas", C.R. Acad. Sci., t. 285, 1977, p. 1033.
5. Zolesio, J.P., "Solution Variationnelle d'un Problème de Valeur Propre Non Linéaire et Frontière Libre en Physique des Plasmas", C.R. Acad. Sci., t. 292, 1979.
6. Temam, R., "Monotone Rearrangement Of A Function And Grad-Mercier Equation Of Plasma Physics", Proc. Intl. Meeting On Recent Methods in Non Linear Analysis, Rome, May 1978. (proceedings E. de Giorge, E. Magenes et U. Mosco, ed. Pittagora Editrice Bologna, '79).
7. Grad, H. and Hu, P.N., "Classical Diffusion", Proc. Workshop on High Plasmas, Varenna, 1977.
8. Grad, H., Hu, P.N., and Stevens, D.C., "Adiabatic Evolution Of Plasma Equilibrium", Bull. Amer. Phys. Soc., Vol. 19, 1974, p. 865.
9. Grad, H., "Alternating Dimension Plasma Transport in Three Dimensions", Proceedings of the International Symposium Held in Versailles, December 1979.

IMPLEMENTATION OF SOME METHODS OF SHAPE OPTIMAL DESIGN

Bernard Rousselet

Département de Mathematique, Université de Nice,
06034 Nice Cédex, France

ABSTRACT

This paper concerns implementation of several optimization methods based on shape design sensitivity analysis. The methods are first described on a tutorial, one dimensional example and then for higher dimensional examples.

1. INTRODUCTION

This paper concerns implementation of shape design sensitivity analysis methods derived by Cea [1] and Zolesio [2]. The starting formulas may be design derivatives of functionals with respect to shape of the boundary or distribution of material on a fixed domain.

In the first case, the iterative process from Ω_n to Ω_{n+1} is defined by a vector field G. This vector field may be estimated by solving a partial differential equation. This idea is the basis of the theoretical approach presented in Ref. 2 (see also Ref. 1).

In the second case, the boundary formulas may be used in connection with finite elements in two ways. If one is looking at boundaries that are partly fixed and partly the graph of a function, boundary formulas enable one to move the whole mesh of finite elements. Obviously, the same thing may be done if the whole moving boundary is defined in polar coordinates. Alternately, instead of moving the whole mesh, one can have it fixed and move from Ω_n to Ω_{n+1} by adding and deleting some triangles or quadrilaterals in the finite element mesh.

Whatever way is used, the implementation is not all a straight forward job, (except, of course, in dimension one), even if one is considering academic structures governed by simple equations. Large scale finite element codes are very expensive to use and are not yet written for design sensitivity analysis.

For several reasons; lack of money, lack of time, and of the small number of people working in this area; only a few methods and few examples have been really implemented to obtain numerical results.

In this paper, each method is first described for a simple one dimensional example and then for higher dimensional ones. The implementation for eigenvalue problems is examined in Ref. 3. For problems of localization of an object of given shape see Ref. 2 and for transmission problems (connected with the design of a hole in a plate) see Ref. 1. The method of dealing with a constraint on the volume of Ω is described in Ref. 3. It applies as well to the static problems.

2. A ONE DIMENSIONAL MODEL PROBLEM

2.1. The Optimization Problem

The state of the system is given as a solution of the boundary-value problem

$$\left.\begin{aligned} -y'' + y = f \quad , \qquad & \text{in } \Omega =]0,\ell[\\ y'(0) = 0 = y'(\ell) \qquad & \end{aligned}\right\} \tag{2.1}$$

where f is defined on $\mathbb{R}$ and the left boundary is chosen equal to zero, to simplify the formulas. The variational formulation of this problem is to find y so that

$$\int_0^\ell (y'v' + yv)\, dx = \int_0^\ell fv dx \tag{2.2}$$

for every smooth function v.

The functional to minimize is

$$J(\Omega) = \int_0^\ell y dx \tag{2.3}$$

Recall that if one uses vector fields $V(x)$ to perturb the domain (here it is the interval $]0,\ell[$, various formulas have been obtained [1,2] for derivatives with respect to shape. They are simply recalled here, where 0 is kept fixed and thus $V(0) = 0$.

Derivative of the Functional. The derivative of the functional of Eq. 2.3 is

$$J'(\Omega,V) = \int_0^\ell Y dx + \int_0^\ell (yV)' dx$$

$$\text{where } Y(x) = \lim_{\varepsilon\to 0} \frac{y_\varepsilon(z) - y_0(z)}{\varepsilon} \tag{2.4}$$

and y_ε is the solution of Eq. 2.1 in $(I + \varepsilon V)(\Omega) = \Omega_\varepsilon$, I being the identity mapping.

Derivative of the State Equation. The functions v that appear in Eq. 2.2, in which Ω is replaced by Ω_ε, are considered as restrictions of functions defined on all of $\mathbb{R}$, so they do not dpend explicitly on ε.* Thus, differentiation of Eq. 2.2 with respect to ε yields

$$\int_0^\ell (Y'v' + Yv)dx + \int_0^\ell \left[(y'v' + yv - fv)V\right]' dx = 0 \tag{2.5}$$

Integrating the last terms of Eqs. 2.4 and 2.5 yields

$$J'(\Omega,V) = \int_0^\ell Y dx + y(\ell)V(\ell) \tag{2.6}$$

$$\int_0^\ell (Y'v' + Yv)dx + [y(\ell)v(\ell) - f(\ell)v(\ell)]V(\ell) = 0 \tag{2.7}$$

where the conditions $V(0) = y'(\ell) = 0$ have been used.

One needs also an expression of J' explicitly in terms of V. This is obtained by using Eq.s 2.5 and 2.6. Integrating the first term of Eq. 2.5 by parts, one gets

$$\int_0^\ell (-Yv'' + Yv)dx + Yv'\Big|_0^\ell = - \int_0^\ell \left[(y'v' + yv - fv)V\right]' dx \tag{2.8}$$

*Of course this does not work for Dirichlet problems, see Ref 1 and 4.

Adjoint Equations. If one sets $-p'' + p = 1$ with boundary conditions $p'(0) = 0 = p'(\ell)$, then Eq. 2.6 may be written in the form

$$J'(\Omega,V) = \int_0^\ell Ydx + y(\ell)V(\ell) = \int_0^\ell Y(-p'' + p)dx + y(\ell)V(\ell) \tag{2.9}$$

Integrating the first term on the right by parts and using Eq. 2.5 with

$$J'(\Omega,V) = -\int_0^\ell \left[(y'p' + yp - fp)V\right]' dx + y(\ell)V(\ell) \tag{2.10}$$

or,

$$J'(\Omega,V) = -\left[y'(\ell)p'(\ell) + y(\ell)p(\ell) - f(\ell)p(\ell)\right] V(\ell) + y(\ell)V(\ell) \tag{2.11}$$

Using the boundary condition of Eq. 2.1, this gives

$$J'(\Omega,V) = - y(\ell)p(\ell) - f(\ell)p(\ell)\ V(\ell) + y(\ell)V(\ell) \tag{2.12}$$

As noticed in Refs. 1 and 2, the derivative $J'(\Omega,V)$ depends only on the value of V at the boundary. Following the idea of the steepest descent method, the challenge is now to select V so that J decreases rapidly (locally), using Eq. 2.11 or 2.12.+

In the one dimensional problem studied here, this is rather crude. However, the aim of this section is not to give efficient methods for the one dimensional case, rather to describe tools that may be used in higher dimensions. Before beginning the description of the algorithms, it may be noted that as soon as one has a code to solve the state equation, one can use it to solve the adjoint equation of Eq. 2.10. It is emphasized that the left hand side of the adjoint equation is the same as for the state equation. This is to be expected for conservative linear systems, since they can be written in self adjoint form.

+The situation is quite different of the classical optimization methods in Hilbert spaces. Here, there is not an obvious best choice of direction of decrease. The situation is similar to the Banach space context (see Refs. 1 and 2).

2.2 Use of the Boundary Formula of Eq. 2.12 (BF Methods)

Using Eq. 2.12, one may set

$$G(\ell) = \rho(\ell)[y(\ell)p(\ell) - f(\ell)p(\ell) - y(\ell)] \qquad (2.13)$$

where $\rho(\ell)$ is some positive constant. From a theoretical point of view, this formula is only meaningful if the value of the right-hand side is well defined. That may not be the case if f is irregular. With smooth data, Eq. 2.13 will at least indicate if the point ℓ has to be moved to the right or left, with a small amount of computations, except for solving Eq. 2.1. The value of $\rho(\ell)$ is evaluated with a standard one dimensional search technique (see e.g. Ref. 5). Some ways of using Eq. 2.13 in connection with finite elements are now described.

BF1 Method. If one is solving Eq. 2.1 with finite differences or finite elements, with a grid of step h (h=1/n) (see Fig. 2.1), one will add or delete a number of intervals such that the length added or deleted is equal to $G(\ell)$. They will be added or deleted according to the sign of $G(\ell)$, in general, there will not be an interger n such that $nh = |G(\ell)|$, so one will take (for example) n such that $nh \leq |G(\ell)| < (n + 1)h$. One could also use a variable step-length.

BF2 Method. There is an alternate way of using Eq. 2.13 in computations. Instead of adding or deleting intervals of length h, one can hold the number of intervals fixed and change their length to $h = (1 + G(\ell))/n$, assuming $\ell + G(\ell) > 0$. The main differences, for digital computations are the following:

(1) Except perhaps in the first few steps of the algorithm, $G(\ell)$ will be of moderate size compared to h. Thus, in BF1 only a few lines of the linear system to be solved will be added or deleted, the others being unchanged.

(2) On the other hand, all equations are changed in BF2. In this very simple example, the amount of new computations in BF2 is minor, however. The lines of the matrix will be identical, except for a multiplicative factor, so the only changes will be in the right hand side. If one is considering state equations with variable coefficients or two dimensional equations, however, the difference would not be so immaterial.

(3) The size of the arrays in which y and f are stored should be taken large enough in BF1 so that adding new intervals is possible. In BF2, the size of arrays is constant.

One can refine BF1 in the following way, which may be more crucial in higher dimensions: In the first few changes of intervals, one uses a rather large h and then decreases it as long as the interval]0,1[is approaching the optimum interval. This kind of method has never been actually implemented (to the author's knowledge). See, however, Ref. 1 on transmissions problems. Moreover, Eq. 2.12 may be be used, in an entirely different spirit, in a method that is entirely new and seems very appealing [6].

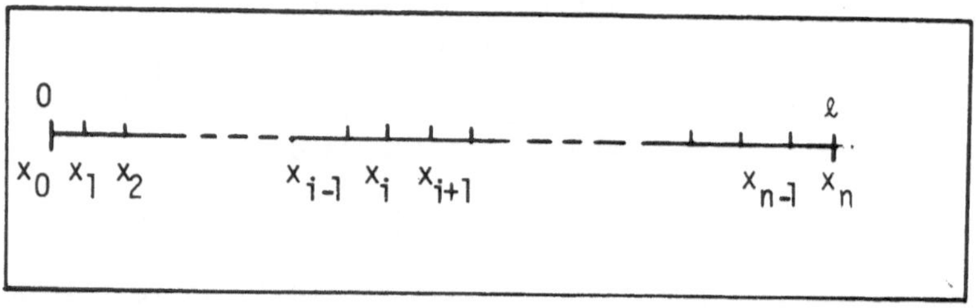

Figure 2.1 Grid on $]0,\ell[$

2.3 Boundary Formula and Green Functions (BFG Methods)

Green functions may be used in connection with finite elements to solve a large variety of problems, as follows:

(1) Fourth order problems (e.g., plates) by solving two second order problems

(2) Domains with corners

(3) Boundary value problems in unbounded domains

(4) Free boundary problems.

See Refs. 6 and 7 for the use of Green functions. Their use in connection with formulas of the type of Eq. 2.12 seems to be new. It enables one to save computing time and size of arrays in computer memory. The idea is to introduce a Green function of Eq. 2.1, which means a solution (in the distributional sense) of the equation

$$-g_k'' + g_k = \delta_k \tag{2.14}$$

where k is a real number $0 < k < 1$. In the following, the subscript k is dropped for convenience. The solution g is smooth, so using integration by parts one gets

$$\int_0^k gy''dx = -\int_0^k g'y'dx + gy'\Big|_0^h$$

But,

$$-\int_0^k g'y'dx = -\lim_{\varepsilon\to 0}\int_0^{k-\varepsilon} g'y'dx$$

$$= -\lim_{\varepsilon\to 0}\left[-\int_0^{k-\varepsilon} g''ydx + g'(k-\varepsilon)y(k-\varepsilon) - g'(0)y(0)\right]$$

$$= \int_0^{k-0} gydx - g'(k)y(k) + g'(0)y(0)$$

Hence,

$$\int_0^k gy''dx = \int_0^{k-0} gydx - g'(k-0)y(h) + g'(0)y(0)$$

$$+ g(k)y'(k) - g(0)y'(0)$$

and since $y'' = y-f$, this gives

$$-\int_0^k gfdx = g(k)y'(k) - g'(k-0)y(k) + g'(0)y(0) \tag{2.15}$$

The last term of Eq. 2.15 may be chosen to be zero, by letting $g'(0) = 0$ and one other boundary condition may be imposed on g, e.g. $g(k) = 1$. The Green function g may now be computed analytically. In $]0,k[$ and $]k,\infty[$, it satisfies $-g'' + g = 0$. Denote by g_- (resp g_+) the restriction of g to $]0,k[$ (resp $]k,\infty[$). Moreover, $g_+(k) = g_-(k)$ and $g_+'(k+0) = g_-'(k-0) - 1$. One now imposes the conditions $g'(0) = 0$ and $g_+(k) = 1 = g_-(k)$. An easy computation then yields

$$\left.\begin{aligned} g_- &= \frac{\cosh(x)}{\cosh(k)} \\ g_+ &= \tanh(k)\sinh(x-k) + e^{k-x} \end{aligned}\right\} \tag{2.16}$$

Equation 2.15 may be then written, using Eq. 2.16 as

$$y'(k) - \tanh(k)y(k) = -\int_0^k \frac{\cosh(x)f(x)dx}{\cosh(k)}$$

or

$$\cosh(k)y'(k) - \sinh(k)y(k) = -\int_0^k \cosh(x)f(x)dx \tag{2.17}$$

It is important to notice that Eq. 2.17 is valid for any solution of $-y'' + y = f$, satisfying only $y'(0) = 0$, whatever the other boundary condition may be.

In this special example, the simplest way of using Eq. 2.17 for optimization is to set $k = \ell$, hence

$$y(\ell) = \frac{\ell}{\sinh(\ell)} \int_0^\ell \cosh(x)f(x)dx \tag{2.18}$$

and use it in Eq. 2.13 to determine $G(\ell)$.

If Eq. 2.1 was an equation with variable coefficients or an equation in higher dimensions, the computation of $y(\ell)$ would not in general be possible analytically (it would be possible only with simple geometries) [6].

In the one dimensional case, one should compute g once to use in $[0,k]$. This enables one to solve, instead of Eq. 2.1, the following problem:

$$\left.\begin{aligned} &-y'' + y = f, \text{ in }]k,\ell[\\ &\cosh(k)y'(k) - \sinh(k)y(k) = -\int_0^k \cosh(x)f(x)dx \\ &y'(\ell) = 0 \end{aligned}\right\} \tag{2.19}$$

This procedure is time saving in electronic computations, since fewer nodes are needed to solve Eq. 2.19. The boundary condition at k seems more complex, but it raises no special problem here (in higher dimensions see Ref. 7).

2.4 Use of Distributed Formulas (DF Methods)

Distributed formulas are based on Eq. 2.11, which is

$$J'(\Omega,V) = \int_0^{\ell} [(y - y'p' - yp + fp)V]'dx$$

Notice of these methods was first given in Ref. 8. The basic ideas is to look for G (the "best choice" of V) in a Sobolev space $H^m(\Omega)$, where m is large enough so that G may be smooth. Here m=1 will be enough. The function G will be defined as a solution, such that G(0) = 0, of

$$\int_0^{\ell} (G'\psi' + G\psi)dx = \int_0 [(-y + y'p' + yp - fp)\psi]'dx \qquad (2.20)$$

for every smooth ψ with $\psi(0) = 0$. It is interesting, when the data are smooth enough, to find the differential equation of which G is a solution. Integrating by parts in Eq. 2.20 yields

$$\left.\begin{aligned} &-G'' + G = 0, \quad \text{in }]0,\ell[\\ &G'(\ell) = -y(\ell) + y(\ell)\ f(\ell) - f(\ell)p(\ell) \\ &G(0) = 0 \end{aligned}\right\} \qquad (2.21)$$

Equation 2.21 may be solved analytically, to obtain

$$G(x) = \frac{\cosh(x)}{\sinh(\ell)} [-y(\ell) + y(\ell)p(\ell) - f(\ell)p(\ell)]$$

and setting $x = \ell$, one has

$$G(\ell) = \tanh(\ell)\left(-y(\ell) + y(\ell)p(\ell) - f(\ell)p(\ell)\right) \qquad (2.22)$$

which differs from Eq. 2.13 only by the factor $\tanh(\ell)$. Of course the signs agree. Due to the normalizing factor $p(\ell)$, the two methods provide nearly the same algorithm, a situation that would be quite different in higher dimensions. This method may also be used with irregular data. Some ways of using Eqs. 2.11 and 2.20 in connection with finite elements are now described.

<u>DF1 Method</u>. The first DF method is as follows:

(1) Solve Eq. 2.1 in $]0,\ell[$, which yields y

(2) Solve the adjoint equation of Eq. 2.9, which yields p

(3) Solve Eq. 2.20, which yields G

(4) With a one dimensional search technique (Ref. 5), compute a normalizing factor so that the actual field used will be ρG.

(5) Move all the nodes x_i of $[0,\ell]$ to

$$x_i' = x_i + \rho G(x_i)$$

(6) After some terminal check, if necessary go on with the same procedure in $]0,\ell + \rho G(\ell)[$.

DF2 Method. As a variant of DF1, instead of solving Eq. 2.20, one could solve (for smooth data)

$$\int_0^\ell (G'\psi' + G\psi)dx = [-y(\ell) + y(\ell)p(\ell) - f(\ell)p(\ell)]\psi(\ell) \quad (2.23)$$

DF3 Method. The difference between methods DF3 and DF1 is that, instead of moving all the nodes, one uses only the boundary value $G(\ell)$ to add or delete new intervals as in BF1. The difference between methods DF3 and DF1 is that $G(\ell)$ is computed by solving Eq. 2.20, instead of direct computation of $G(\ell)$ with Eq. 2.13, which may be meaningless.

2.5 Use of Green Functions (DFG Methods)

Distributed methods using Green functions (already used in BFG methods) introduce a Green function, which provides a formula of the type,

$$g(k)y'(k) - g'(k-0)y(k) = -\int_0^k gfdx \quad (2.24)$$

The computational method is as follows:

$$\left.\begin{array}{l} \text{(1) Compute } y \text{ in }]k,\ell[\text{ by solving} \\ -y'' + y = f, \quad \text{in }]k,\ell[\\ \text{with boundary condition of Eq. 2.24 and} \\ y'(\ell) = 0 \end{array}\right\} \quad (2.25)$$

(2) Solve the adjoint equation with a similar method.

(3) Instead of looking for G in $H^1(0,\ell)$, look for it in $H^1(k,\ell)$, as a solution, such that $G(0) = 0$, of

$$\int_k^\ell (V'\psi' + V\psi)dx = \int_k^\ell [(-y + y'p' + yp - fp)\psi]'dx \qquad (2.26)$$

for every smooth ψ with $\psi(0) = 0$

(4) Normalize G with a one dimensional search technique (Ref. 5).

(5) Move the nodes of $]k,\ell[$, with the formula

$$x_i' = x_i + \rho G(x_i)$$

(6) Make a terminal check and repeat if necessary.

Being a distributed method, BFG may be used even with irregular data. On the other hand, BFG is memory and time saving, because the equations for y, p, and G are solved only in small domains.

Finally it is important to notice that the following method, which seems straightforward to give a distributed formula, amounts to a boundary formula. One may look for V in a space of shape functions $V = \sum_i V_i\phi_i$, where $\phi_i(x) = \delta_{i,j}$ (see Fig. 2.2 for example).

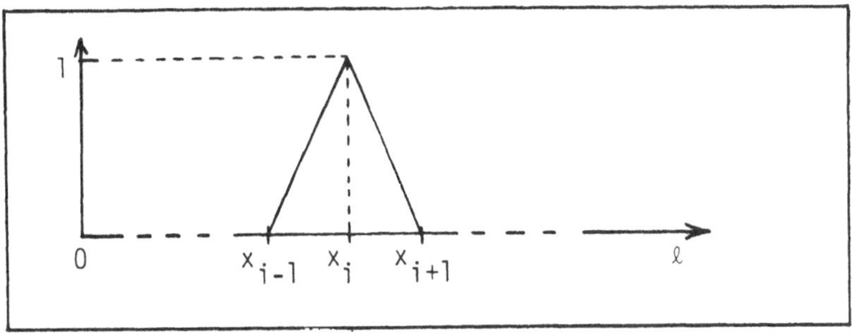

Figure 2.2 Shape Functions

Then Eq. 2.11 gives

$$J'(\Omega,V) = \sum_i V_i \int_0^{\ell} [(y - y'p' - yp + fp)\phi_i]' dx$$

and one sets $G = G_i$, $i=1,\dots,n$ where

$$G_i = -\int_0^{\ell} [(y - y'p' - yp + fp)\phi_i]' dx$$

and uses G_i to move all the nodes, as in DF1. In fact, all the inner nodes would not move, because $G_i = 0$ for the inner nodes. The integral from 0 to 1 is in fact from x_{i-1} to x_{i+1} and its value is

$$\Big[(y - y'p' - yp + fp)\phi_i\Big]_{x_{i-1}}^{x_{i+1}}$$

which is zero, so that only the boundary values of G_i are useful. In fact, a discrete version of Eq. 2.20 would be

$$(AG)_i = -\int_0^{\ell} [(y - y'p' - y' + fp)\phi_i]' dx$$

where A is a matrix.

Equation 2.20 appears not only as a way to regularize the vector field G, but also to define it throughout Ω in such a way the interior of Ω may be moved consistently. However, the use of shape functions yields an effective boundary method (see Section 2.2).

3. A TWO DIMENSIONAL MODEL PROBLEM

3.1 Design Sensitivity Equations

The state equation considered here is the Neumann problem

$$\left.\begin{aligned} -\Delta y + y &= f, \quad \text{in} \\ \frac{\partial y}{\partial n} &= 0, \quad \text{on } \Gamma \end{aligned}\right\} \qquad (3.1)$$

where Ω is a bounded open set of R_n (n = 2 or 3 in most physical situations), with boundary Γ, or in variational form

$$\int_\Omega (\nabla y \cdot \nabla v + yv)dx = \int_\Omega fvdx \tag{3.2}$$

for every smooth v.

Differentiation of Integrals. Recall the following results (well known in fluid mechanics, see Refs. 6 and 2). Let Ω be some smooth open set of R^n and V some smooth vector field (defined in a neighborhood of Ω). Define Ω_ε as the set of $x + \varepsilon V(x) \equiv T_\varepsilon(x)$, when $x \in \Omega$. For ε small enough, T_ε is a diffeomorphism (see Ref. 2). More generally, one can consider $T_t(x) = x + V(t,x)$, which may be also a diffeomorphism for large t, under suitable hypothesis (see Ref. 2).

If ϕ is a smooth function of x, which may depend also smoothly on t,

$$\frac{d}{dt}\int_{\Omega_t} \phi(t,x)dx_t = \int_{\Omega_t} \frac{\partial\phi}{\partial t}dx_t + \int_{\Omega_t} \mathrm{div}\,(gV)dx_t \tag{3.3}$$

or, applying Green's formula to the last integral,

$$\frac{d}{dt}\int_{\Omega_t} \phi(t,x)dx_t = \int_{\Omega_t} \frac{\partial\phi}{\partial t}dx_t + \int_{\Gamma_t} \phi(V\cdot n)d\sigma_t \tag{3.4}$$

It is emphasized that in these formulas the derivative of ϕ with respect to t is taken with x fixed (not depending on t). It is not a material derivative (see Refs. 2 and 4). These formulas play the role of differentiation with respect to the length in the one dimensional case.

As in Section 2, one sets

$$Y(x) = \frac{\partial}{\partial t} y(t,x) \tag{3.5}$$

Using this formula, one may proceed as in Section 2 for the one dimensional example.

The Derivative of the Functional. The derivative of the functional

$$J(\Omega) = \int_\Omega y dx \tag{3.6}$$

is

$$\frac{d}{dt} J(\Omega_t)|_{t=0} = J'(\Omega,V) = \int_\Omega Y dx + \int_\Omega \mathrm{div}(yV)dx \tag{3.7}$$

or

$$J'(\Omega,V) = \int_\Omega Y dx + \int_\Gamma y(v \cdot n)d\sigma \tag{3.8}$$

Differentiation of the State Equation. The derivative of the variational state equation of Eq. 3.2 is

$$\int_\Omega (\nabla Y \cdot \nabla v + Yv)dx + \int_\Omega \mathrm{div}[(\nabla y \cdot \nabla v + yv - fv)V]dx = 0 \tag{3.9}$$

or, applying Green's formula to the last integral,

$$\int_\Omega (\nabla Y \cdot \nabla v + Yv)dx + \int_\Gamma (\nabla y \cdot \nabla v + yv - fv)(V \cdot n)d\sigma = 0 \tag{3.10}$$

Adjoint Equation. Integrating the first integral of Eq. 3.9 by parts yields

$$\int_\Omega Y(-\Delta v + v)dx + \int_\Gamma Y \frac{\partial v}{\partial n} d\sigma = -\int_\Omega \mathrm{div}\Big([\nabla y \cdot \nabla v + yv - fv]V\Big)dx$$

which suggests that one set

$$\left.\begin{aligned} -\Delta p + p &= 1, \quad \text{in } \Omega \\ \frac{\partial p}{\partial n} &= 0, \text{ on } \Gamma \end{aligned}\right\} \tag{3.11}$$

or in variational form

$$\int_\Omega (\nabla p \cdot \nabla \phi + p\phi)dx = \int_\Omega \phi\, dx \tag{3.12}$$

for every smooth ϕ.

<u>Explicit Formula for $J'(\Omega,V)$</u>. This yields finally the formula

$$J'(\Omega,V) \doteq -\int_\Omega \mathrm{div}\Big([\nabla y \cdot \nabla p + yp - fp]V\Big)dx + \int_\Omega \mathrm{div}(yV)dx \tag{3.13}$$

or

$$J'(\Omega,V) = -\int_\Gamma (\nabla y \cdot \nabla p + yp - fp)(v \cdot n)d\sigma + \int_\Gamma y(v \cdot n)d\sigma \tag{3.14}$$

As in Section 2, the problem is now to select V as in a steepest descent method (see Ref. 6). This selection of V is still denoted G.

3.2 Use of Boundary Formula 3.14 (BF Methods)

Consider now the same BF1 and BF2 methods of Section 2. The function G is given on Γ by

$$(G \cdot n)|_\Gamma = p_\Gamma(\nabla y \cdot \nabla p + yp - fp - y)|_\Gamma \tag{3.15}$$

and the challenge is of using this formula in connection with finite elements in Ω to build a better open set, denoted by Ω, i.e. designing a process to build an iterative sequence Ω_n such that Ω_{n+1} is better than Ω_n.

Recall the arrays used to build stiffness and mass matrices. The set is Ω divided in triangles with

(1) The array of summits (TS); the coordinates of all the summits (or nodes)

(2) The array of triangles (It); the number of the summits of each triangle

(3) The array of neighbors (NV); the number of the neighboring points of each summit

(4) The stiffness matrix is denoted by A and the load vector by F.

BF1 Method. The method is described with a mesh of triangles. On each triangle, the shape functions are affine. On each side AB of a triangle that is a piece of the boundary Γ (see Fig. 3.1) (no distinction is made here between Γ and its approximation with finite elements) one computes G, given by Eq. 3.15. On Γ, $\frac{\partial y}{\partial n} = 0 = \frac{\partial p}{\partial n}$, so that $\nabla y \cdot \nabla p = \frac{\partial y}{\partial s}\frac{\partial p}{\partial s}$, where $\frac{\partial}{\partial s}$ denotes tangential derivative. But p and y are linear on each side AB so that $\frac{\partial y}{\partial s}\frac{\partial p}{\partial s}$ is a constant on AB. Finally, the restriction of $G_{|\Gamma}$ to AB is a polynomial of degree two. The same is true on BC and CD, but the polynomials are not supposed to be continuous in B and C. Hence, if one draws (see Fig 3.2) $x + (G \cdot n)(x)$, one gets pieces of quadratic curves. Thus a new boundary is defined with corners that are quite artificial. A crude way of smoothing the perturbed boundary is to use the bisection of $\widehat{B_0' \, B \, B_1'}$, $\widehat{C_0' \, C \, C_1'}$, etc., and to use a point B' such that $BB' = \frac{1}{2}(BB_0' + BB_1')$, $CC' = \frac{1}{2}(CC_0' + CC_1')$, etc., as shown in Fig. 3.2, and then to interpolate with segments to get a new boundary. One then divides these quadrilaterals to get triangles. Of course, one could also compute with quadrilaterals.

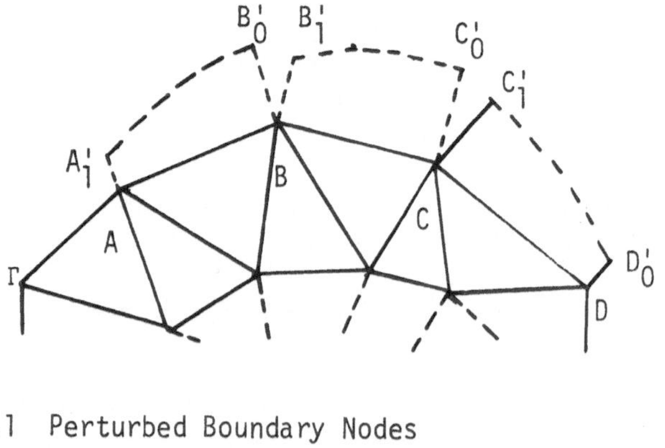

Figure 3.1 Perturbed Boundary Nodes

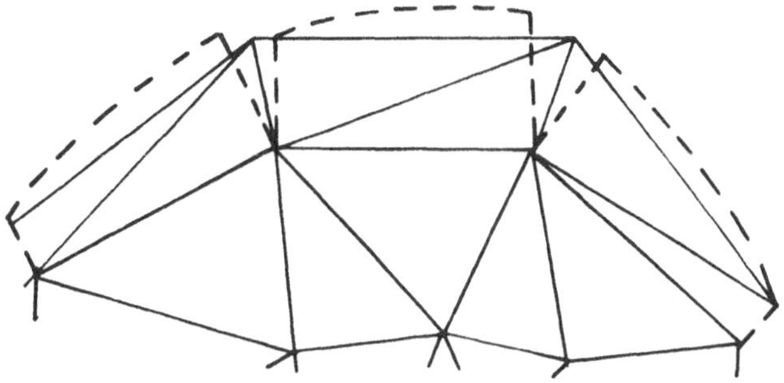

Figure 3.2 Smoothed Boundary Nodes

If one was willing to use smooth enough finite elements (which amounts selecting their degree large enough), these problems of discontinuity would disappear and the method would appear to be less "ad hoc" (with the drawback of more computing time and core storage).

So far only the addition of new elements has been described, i.e. when G•n is positive. If G•n is still positive, but small compared with the area of a quadrilateral of size h, nothing is added. The same is true if G•n is negative, but small compared to h^2. If this is not the case, triangles (or quadrilaterals) are deleted, the area of which is nearly equal to $\left| \int\int_{AB} (G \cdot n) d\sigma \right|$. In this whole process G should be normalized so that G•n is not too large, compared to h^2.

There is an alternate way of implementing the adding and deleting of new elements. It is to look for Ω as a subset of a big rectangle C and to discretize this rectangle with a fixed grid of finite elements. The elements added or deleted are taken from this fixed grid. The area of the elements added or deleted is naturally a multiple of some fixed number (the area of a standard finite element of the rectangle C), so that the change may not be exactly the one indicated by Eq. 3.15. However, the implementation is much simpler. This idea has been used for a transmission problem in a different context (see Ref. 9).

There were IM nodes in Ω and the arrays TS (summits), IT (triangles), and NV (neighbors) were used to build the stiffness matrix A and the load vector F. In Ω', there will be

$IM + I^+ - I^-$ nodes (I^+ nodes added and I^- deleted). This requires modification of the arrays, as follows:

(1) The coordinates of the nodes added (deleted) have to be brought (deleted) into (out of) TS

(2) The triangles have to be inserted in (or removed from) IT

(3) Nodes near the old boundary Γ have generally new neighbors which causes changes in NV.

In the stiffness matrix A, the rows and columns that correspond to inner nodes do not change. Some rows and columns are deleted while some are added. The same procedure is followed for redefinition of F. Of course, if one needs a matrix A with a small bandwidth the nodes should be renumbered.

BF2 Method. The BF2 method is now described in the case where the moving boundary is sought as the graph of a function. Of course, method BF1 applies to this case, but the extension of method BF2 to more general situations is less obvious (one should use for example polar coordinates).

The boundary of Ω is made up of a moving part Γ_1 and a fixed one Γ_0 (see Fig. 3.3). In design sensitivity analysis, $V \cdot n$ is taken to be zero on Γ_0 and Eq. 3.15 is used on Γ_1. As in method BF1, $G \cdot n$ is computed on each boundary side of triangles that lie along the segment KL in Fig. 3.3. It is in general discontinuous (except if one is working with smooth enough finite elements as in method BF1). One computes

$$I_{AB} = \int_{AB} (G \cdot n) d\sigma$$

on each element boundary AB on the segment KL and moves AB upward or downward*, depending on the sign of $G \cdot n$. The movement is such that A'B' is parallel to AB and the area of ABA'B' is equal to I_{AB}.

This perturbation provides a new boundary that is in general discontinuous (Fig. 3.4). It is smoothed with an averaging process, as in method BF1. These problems of discontinuities would disappear if one was using finite elements of degree large

*In fact one should move perpendicularly to the element. The process described here is more simple.

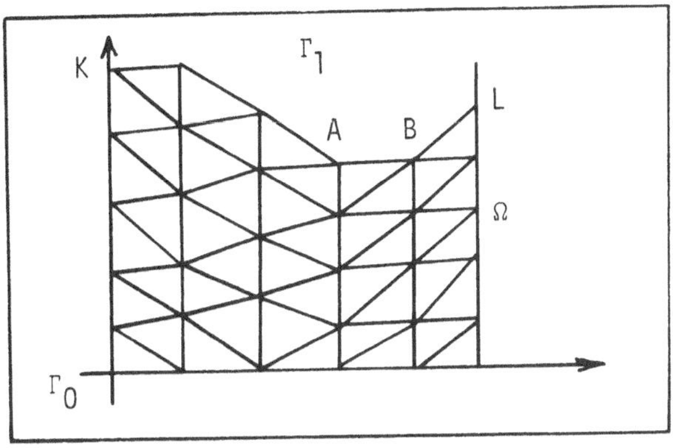

Figure 3.3 Domain with Partially Fixed Boundary

enough. This would also be the case with first degree approximation in problems for which G.n is given only in terms of y and p on the boundaries [10,11] and not their gradients.

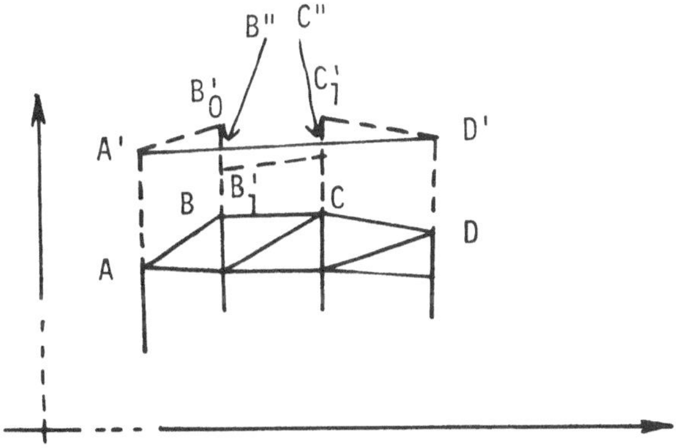

Figure 3.4 Perturbed and Smoothed Boundary

The efficiency of these algorithms has to be checked in actual computations. In particular, the decrease of computing time due to the use of finite elements of degree, as opposed to smoother ones, may have some drawback in accuracy and stability.

The boundary formula of Eq. 3.15 may also be used in connection with Green functions, as in the one dimensional case. This procedure saves a large number of nodes, so high degree finite elements may then be used with a reasonable amount of core storage and computer time. The method is not described here, for lack of space. The use of Green functions is described in connection with distributed formulas.

There is still another way of getting an improved boundary that is continuous, one may use a family of vector fields of the form

$$V = (0, \sum_i \underset{\sim}{V}_i \phi_i)$$

where ϕ_i is a (piecewise affine) shape function and $\underset{\sim}{V}_i$ are real numbers. This defines V on the boundary Γ in the following way (see Fig. 3.5):

$$V|_{\Gamma_i} = (0, \underset{\sim}{V}_i \phi_i + \underset{\sim}{V}_{i+1} \phi_{i+1})$$

so that $V(A_i) = (0, \underset{\sim}{V}_i)$ and V is affine between two summits A_i and A_{i+1} (see Figure 3.5).

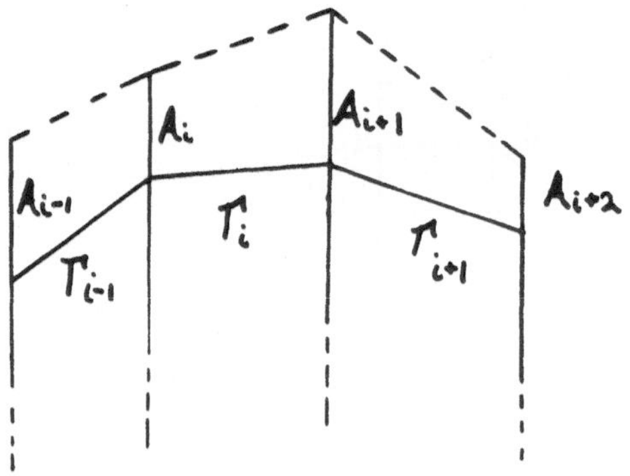

Figure 3.5 Smooth Perturbed Boundary

Equation 3.14 may be written with γ_i coordinate along Γ_i as

$$J'(\Omega,V) = -\Sigma \int_{\Gamma_i} \left(\frac{\partial y}{\partial \gamma_i}\frac{\partial p}{\partial \gamma_i} + yp - fp + y\right)$$

$$\times\left(\underset{\sim}{V}_i\phi_i + \underset{\sim}{V}_{i+1}\phi_{i+1}\right)\nu_{iy}d\gamma_i$$

where ν_{iy} is the second component of the unit outward normal vector to Γ_i. Hence, G is to be defined by the formula

$$G = (0,\Sigma \underset{\sim}{G}_i\phi_i)$$

where

$$\underset{\sim}{G}_i = \int_{\Gamma_{i-1}} \left(\frac{\partial y}{\partial \gamma_{i-1}}\frac{\partial p}{\partial \gamma_{i-1}} + yp - fp + y\right)\phi_i\nu_{(i-1)y}d\gamma_{i-1}$$

$$+ \int_{\Gamma_i} \left(\frac{\partial y}{\partial \gamma_i}\frac{\partial p}{\partial \gamma_i} + yp - fp + y\right)\phi_i\nu_{iy}d\gamma_i$$

This unambigously defines the movement of any summit A_i and thus yields a new boundary, which is also piecewise affine. This method may be used either by moving all the triangles or by adding or deleting triangles. Of course this method may be used for general domains.

3.3 USE OF DISTRIBUTED FORMULAS (D.F. METHODS)

Distributed formula methods are based on Eq. 3.13, which is

$$J'(\Omega,V) = -\int_\Omega \operatorname{div}[(\nabla y\cdot\nabla p + yp - fp)V]dx + \int_\Omega \operatorname{div}(yV)dx$$

A first way of using this formula is to search for G as a polynomial in the complex variable z (see Ref. 12 for its use in actual computations). One can also look for G in a Sobolev space $(h^m(\Omega))^n$. To have G continuous, one must take m = 2 in two dimensional problems (n = 2) (see Refs. 2 and 6). The function G may be sought, for example, as a solution of

$$\int_{\Omega} \Delta G \cdot \Delta\psi dx = \int_{\Omega} \text{div}[(\nabla y \cdot \nabla p + yp - fp)\psi]dx$$

$$- \int_{\Omega} \text{div}(y\psi)dx \tag{3.16}$$

DF1 Method. The algorithm proceeds as in the one dimensional case, using Eqs 3.1, 3.12, and 3.16. However, one should check that the triangles of the new mesh form a good triangulation.

DF2 Method. See the discussion of the one dimensional case.

DF3 Method. As in the one dimensional case, one solves Eq. 3.16 to compute G, but uses only the boundary values of G as in a BF method. This yields a field that is continuous on the boundary, even with degree one approximations for y and p. However, one should use high degree of approximation to solve Eq. 3.16.

DFG Method. One introduces a domain Ω_0, designed to be fixed in the iterations in the interior of Ω, as shown in Fig. 3.6. The boundary of Ω_0 is S. One introduces the functions g_x as solutions of

$$-\Delta g_x + g_x = \delta x \tag{3.17}$$

where δ_x is the Dirac measure about x.

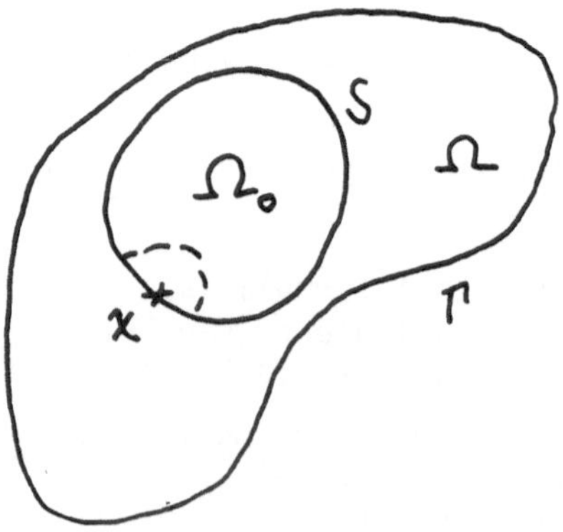

Figure 3.6 Interior Subdomain

The aim now is to derive, with these functions, a boundary condition on S that is satisfied by a solution of $-\Delta y + y = f$ (whatever the boundary condition on Γ). To derive it, as in the one dimensional case, one starts with

$$\int_{\Omega_0} g_x \Delta y d\xi = -\int_{\Omega_0} \nabla g_x \cdot \nabla y d\xi + \int_S g_x \frac{\partial y}{\partial n} d\sigma$$

If Ω_ε denotes Ω_0 minus half a circle of radius ε around x, one has

$$\int_{\Omega_\varepsilon} g_x \Delta y d\xi = -\int_{\Omega_\varepsilon} \nabla g_x \cdot \nabla y d\xi + \int_{S_\varepsilon} g_x \frac{\partial y}{\partial n} d\sigma$$

$$= \int_{\Omega_\varepsilon} (\Delta g_x) y d\xi + \int_{S_\varepsilon} g_x \frac{\partial y}{\partial n} d\sigma - \int_{S_\varepsilon} \frac{\partial g_x}{\partial n} y d\sigma$$

Then

$$\int_{\Omega_\varepsilon} g_x(\Delta y - y) d\xi = \int_{\Omega_\varepsilon} (\Delta g_x - g_x) y d\xi + \int_{S_\varepsilon} \left(g_x \frac{\partial y}{\partial n} - \frac{\partial g_x}{\partial n} y \right) d\sigma$$

or,

$$-\int_{\Omega_\varepsilon} g_x f dx = \int_{S_\varepsilon} g_x \frac{\partial y}{\partial n} d\sigma - \int_{S_\varepsilon} \frac{\partial g_x}{\partial n} y d\sigma$$

which, goes as $\varepsilon \to 0$, approaches

$$-\int_{\Omega_0} g_x f dx = \int_S g_x \frac{\partial y}{\partial n} d\Gamma - \int_S \frac{\partial g_x}{\partial n} y d\sigma + \gamma \frac{y(x)}{2} \tag{3.18}$$

where γ is a constant, which may be computed analytically [12]. Equation 3.18 is satisfied by any solution of $-\Delta y + y = f$, without taking into account the boundary condition. Hence the solution of Eq. 3.1 may be computed in $\Omega \backslash \Omega_0$ by solving the boundary-value problem

$$-\Delta y + y = f, \quad \text{in } \Omega$$

$$\int_{\Omega_0} g_x f d\xi + \int_S \left(g_x \frac{\partial y}{\partial n} - \frac{\partial g_x}{\partial n} y \right) d\sigma + \frac{\gamma y(x)}{2} = 0, \text{ on } S \tag{3.19}$$

$$\frac{\partial y}{\partial n} = 0, \text{ on } \Gamma$$

It is known that the values of the derivatives of the functionals depend only on the normal component of V on Γ, hence one can select a G that is zero in Ω_0 and a solution of the variational equation

$$\int_{\Omega \backslash \Omega_0} \Delta G \Delta \psi dx = \int_{\Omega \backslash \Omega_0} \text{div}[(\nabla y \cdot \nabla p + yp - fp)\psi]dx$$

$$- \int_{\Omega \backslash \Omega_0} \text{div}(y\psi)dx \tag{3.20}$$

where $G = 0$ and $\frac{\partial G}{\partial n} = 0$ on S. Equation 3.20 should hold for any $\psi \in (H^2(\Omega \backslash \Omega_0))^2$, which is a vector field such that $\psi = 0$ and $\frac{\partial \psi}{\partial n} = 0$ on S.

4. CONCLUSIONS

Several methods that can be used to solve shape optimal design problems have been presented. All these methods have been described only for very simple examples, but it is clear that they apply as well to more complex situations, e.g. to fourth order problems (plates) and elasticity problems in 3 dimensions.

Eigenvalue problems are described in Ref. 3 for the theory, as well as for the implementation. In that paper the design sensitivity of a fourth order vibrating plate is described. A constraint on the volume of Ω is also treated. The implementation of some special problems are also described in Refs. 2 and 6.

To date, there is no general proof of convergence of the methods described here and there is no way of telling which method to choose for solving a given problem (see Ref. 14).

The supremum norm of the solution of a static problem is a non differentiable functional. However a necessary condition involving Lagrange multipliers (one of them being a measure) has been obtained. It may be the starting point of a numerical algorithm. This is new and is sketched in Ref. 15.

Evolution problems (involving time) have not yet been studied with these methods. This remains to be done, as well as to study nonlinear problems, e.g. contact problems or the design of structures for large deflections. These seem to be exciting areas that have important applications.

REFERENCES

1. Cea, J., "Problems of Shape Optimal Design", Optimization of Distributed Parameter Structures, (Eds. E.J. Haug and J. Cea), Sijthoff and Noordhoff, Alphen aan den Rihn, Netherlands, 1980.
2. Zolesio, J-P., "The Material Derivative (or Speed) Method for Shape Optimization", Optimization of Distributed Parameter Structures , (Eds. E.J. Haug and J. Cea), Sijthoff and Noordhoff, Alphen aan den Rihn, Netherlands, 1980.
3. Rousselet, B., "Dependence of Eigenvalues with Respect to Shape", Optimization of Distributed Parameter Structures, (Eds. E.J. Haug and J. Cea), Sijthoff and Noordhoff, Alphen aan den Rihn, Netherlands, 1980.
4. Rousselet, B., "Shape Design Sensitivity Methods for Structural Mechanics", Optimization of Distributed Parameter Structures, (Eds. E.J. Haug and J. Cea), Sijthoff and Noordhoff, Alphen aan den Rihn, Netherlands, 1980.
5. Cea, J., Lectures on Optimization - Theory and Algorithms (Notes by M.K.V. Murthy), Tata Institute of Fundamental Research, Bombay, Springer-Verlag, 1978.
6. Cea, J., "Numerical Methods of Shape Optimal Design", Optimization of Distributed Parameter Structures, (Eds. E.J. Haug and J. Cea), Sijthoff and Noordhoff, Alphen aan den Rihn, Netherlands, 1980.
7. Bardos, C., Cea, J., Grisvard, P., to appear.
8. Céa, J., Une méthode numérique four la recherche d´um domaine optimal, Publication IMAN de l'Universite´de Nice, 1972.
9. Cea, J., Gioan, A., and Michel, J., "Quelques re´sultats per l´identification de domaines", Calcolo, 1943.
10. Dervieux, A., Palmerio, B., Theses de l'Université de Nice, Identification de domaines et problemes de frontiere libre.

11. Bégis, D., and Glowinski, R., Application de la méthode des éléments finis à la résolution d'un problème de domaine optimal; Com. Coll. Méth. Calcul. Scient. et Tech., 1973, I.R.I.A., Paris.
12. Rousselet, B., Identification de domaines et problemes de valeus propres., Thèse de l'Université de Nice, 1977.
13. Courant, R. and Hilbert, D., Methods of Mathematical Physics, Interscience Publishers, 1953.
14. Zolesio, J-P., Identification de domaines par Déformations, Thèse de l'Université de Nice, 1979.
15. Rousselet, B., "Multiple Eigenvalues and Supremum Norm Constraints", Optimization of Distributed Parameter Structures, (Eds. E.J. Haug and J. Cea), Sijthoff and Noordhoff, Alphen aan den Rihn, Netherlands, 1980.

DEPENDENCE OF EIGENVALUES WITH RESPECT TO SHAPE

Bernard Rousselet

Departemente de Mathematiques, Université de Nice,
06034 Nice Cédex, France

ABSTRACT

Analysis of shape design sensitivity of eigenvalues is addressed here. A tutorial one dimensional example is first discussed, before more complex examples are discussed. Implementation of the method with volume constraints is also discussed. Some numerical experiments are presented.

1. INTRODUCTION

This paper is devoted to analysis of shape design sensitivity of eigenvalues of systems occupying some open bounded subset Ω of R^n (in practical applications n = 1, 2 or 3) with respect to variations of Ω.

Section 2 is devoted to a very simple one dimensional example for which analytical computations are possible. This example is used as a check of the methods presented here for pedagogical interest. Shape sensitivity results are then derived for a vibrating beam, a second order membrane problem, and a fourth order plate problem, all with simple eigenvalues. An insight into the behavior of repeated eigenvalues is then given. See Refs. 1, 2, and 3 for other examples and methods of shape optimal design.

Implementation of the method developed here is then discussed, the highlight being the differences between the eigenvalue and static problems discussed in Ref. 4. The case of multiple eigenvalues is dealt with using necessary optimality conditions. See Ref. 5 for a treatment of the supremum norm constraint for static problems solved using a necessary optimality condition.

It should be emphasized that shape design sensitivity of the discretized problems seems an untractable challenge, except perhaps for some special problems. Shape sensitivity is an example in which the distributed parameter approach, as opposed to the discrete approach, offers substantial advantage.

2. A SIMPLE ONE DIMENSIONAL EXAMPLE

The problem of vibration of a string of density one, submitted to a tension one, is considered. It is governed by the boundary-value problem

$$\left.\begin{aligned} -y'' &= \zeta y, \text{ in }]0,\ell[, \; y \not\equiv 0 \\ y(0) &= 0 = y(\ell) \end{aligned}\right\} \qquad (2.1)$$

It is an elementary computation to find that

$$\left.\begin{aligned} \zeta &= \frac{\pi^2}{\ell^2} \\ y &= \sin \frac{\pi}{\ell} x \end{aligned}\right\} \qquad (2.2)$$

so that

$$\left.\begin{aligned} \frac{d\zeta}{d\ell} &= \frac{-2\pi^2}{\ell^3} \\ \frac{dy}{d\ell} &= \left(\cos \frac{\pi x}{\ell}\right)\left(\frac{-\pi x}{\ell^2}\right) \end{aligned}\right\} \qquad (2.3)$$

$$\frac{dy}{d\ell}(x) = -\frac{x}{\ell} y'(x) \qquad (2.4)$$

Thus,

$$\frac{dy}{d\ell}(\ell) = -y'(\ell) \qquad (2.5)$$

The most effective way of deriving $\frac{d\zeta}{d\ell}$, without using Eq. 2.2, which is in general impossible to get, is the following. One starts by recalling the variational formulation of Eq. 2.1 for every smooth v that satisfies v = 0 at 0 and ℓ,

$$\int_0^\ell y' \; v' \; dx = \zeta \int_0^\ell y \; v \; dx \tag{2.6}$$

Consider Eq. 2.6 with v = y* and take its derivative with respect to ℓ, setting $dy/d\ell \equiv Y$ leads to

$$\left.\begin{aligned} &\int_0^\ell y'^2 \; dx = \zeta \int_0^\ell y^2 \; dx \\ &2 \int_0^\ell Y' \; y' \; dx + y'^2(\ell) = \frac{d\zeta}{d\ell} \int_0^\ell y^2 \; dx \\ &\qquad\qquad + 2\zeta \int_0^\ell Y \; y \; dx + \zeta \; y^2(\ell) \end{aligned}\right\} \tag{2.7}$$

or, using $y^2(\ell) = 0$,

$$\frac{d\zeta}{d\ell}\left(\int_0^\ell y^2 \; dx\right) = y'^2(\ell) + 2\left[\int_0^\ell Y' \; y' \; dx - \zeta \int_0^\ell Y \; y \; dx\right] \tag{2.8}$$

It would be a motal error to argue that the last expression is zero by using Eq. 2.6 with v = Y, not because Y is not smooth enough but because Y does not satisfy $Y(\ell) = 0$ (see Eq. 2.5 and Ref. 6 on static Dirichlet problem shape sensitivity). In another way, one can say that

$$\int_0^\ell Y' \; y' \; dx = - \int_0^\ell Y \; y'' \; dx + Y \; y' \Big|_0^\ell \tag{2.9}$$

and Y(0) = 0, because the boundary condition y(0) = 0 is at a fixed point, whereas $y(\ell) = 0$ is at a moving boundary point and one is taking derivatives with respect to ℓ at x fixed (material derivatives are not being used here).

---

*More precisely, (since y is defined up to a multiplicative constant), consider a selection of y that is continuous with respect to ℓ.

The use of Eq. 2.9 and $-y'' = \zeta y$ in Eq. 2.8 yields

$$\left(\int_0^{\ell} y^2\, dx\right) \frac{d\zeta}{d\ell} = y'^2(\ell) + 2Y(\ell)\, y'(\ell) \tag{2.10}$$

The value of $Y(\ell)$ may be computed by the following argument*: One introduces $g(\bar{\ell}) = y_{\ell+\bar{\ell}}(\ell+\bar{\ell})$ where $y_{\ell+\bar{\ell}}$ is the solution of Eq. 2.1 in $[0,\ell+\bar{\ell}]$. One thus has $g(\bar{\ell}) = 0$ and $g'(\bar{\ell}) = 0$ (also $g'(0) = 0$), but $g'(\bar{\ell}) = \frac{\partial y_{\ell+\bar{\ell}}}{\partial \bar{\ell}}(\ell+\bar{\ell}) + \frac{\partial y_{\ell+\bar{\ell}}}{\partial x}(\ell+\bar{\ell})$ and for $\bar{\ell} = 0$, $g'(0) = Y(\ell) + y'(\ell)$ so, that $Y(\ell) = -y'(\ell)$. Hence Eq. 2.10 may be rewritten as

$$\frac{d\zeta}{d\ell} = \frac{-y'^2(\ell)}{\int_0^{\ell} y^2\, dx} \tag{2.11}$$

This method is specific to eigenvalue problems, but one can also derive Eq. 2.11 by taking the derivative of Eq. 2.6 (in a way similar to the case of static problems). In this case, one gets also the equation satisfied by $\frac{\partial y}{\partial \ell} = Y$.

Naturally instead of being just interested in the dependence with respect to ℓ, one may be interested in the dependence with respect to some vector field, as in the static case. One is thus interested in the dependence of the eigenvalue ζ_ε of Eq. 2.1 on ε, in $(I + \varepsilon V)(]0,\ell[)$. For simplicity, assume $V(0) = 0$. One gets from Eq. 2.7

$$2\int_0^{\ell} Y'\, y'\, dx + \int_0^{\ell} (y'^2 V)'\, dx$$

$$= \frac{d\zeta}{d\ell}\int_0^{\ell} y^2\, dx + 2\zeta\int_0^{\ell} Y\, y\, dx + \zeta\int_0^{\ell} (y^2 V)'\, dx$$

With an argument similar to the one used previously, one has

---

*One can also use the material derivative. (See Ref. 6 for its definition.)

$$\frac{d\zeta}{d\ell} = \frac{-\int_0^\ell (y'^2 V)' \, dx}{\int_0^\ell y^2 \, dx} \tag{2.12}$$

from which one can also deduce Eq. 2.11, be integrating the numerator and selecting $V(\ell) = 1$.

If one was considering homogeneous Neumann boundary conditions, an appropriate boundary-value problem is

$$\left.\begin{aligned} - y'' &= \zeta y \\ y'(0) &= 0 = y'(\ell) \end{aligned}\right\} \tag{2.13}$$

One would find, with the same method, that

$$\frac{d\zeta}{d\ell} = \frac{-\zeta y^2(\ell)}{\int_0^\ell y^2 \, dx} \tag{2.14}$$

3. A FOURTH ORDER, ONE DIMENSIONAL EXAMPLE

As a more realistic example, consider a clamped-clamped vibrating beam of constant cross sectional area. The governing boundary-value problem is

$$\left.\begin{aligned} EI\, y^{(4)} &= \zeta \rho h y \\ y(0) = y'(0) &= 0 = y(\ell) = y'(\ell) \end{aligned}\right\} \tag{3.1}$$

The variational formulation is

$$EI \int_0^\ell y'' \, v'' \, dx = \zeta \rho h \int_0^\ell y \, v \, dx$$

As in Section 2, one takes the derivative of

$$EI \int_0^\ell y''^2 \, dx = \zeta \rho h \int_0^\ell y^2 \, dx \tag{3.2}$$

with respect to ℓ, to obtain

$$2EI \int_0^{\ell} y'' \, Y'' \, dx + EI \, y''^2(\ell)$$

$$= \frac{d\zeta}{d\ell} \rho \, h \int_0^{\ell} y^2 \, dx + 2 \, \zeta \, \rho \, h \int_0^{\ell} y \, Y \, dx + \zeta \, \rho \, h \, y^2(\ell) \qquad (3.3)$$

the last term being equal to zero. Integrating the first term by parts gives

$$\int_0^{\ell} y'' \, Y'' \, dx = \int_0^{\ell} y^{(4)} \, Y \, dx - y''' \, Y \Big|_0^{\ell} + y'' \, Y' \Big|_0^{\ell}$$

As in Section 2, one observes that $Y(0) = Y'(0) = 0$ and

$$\left.\begin{aligned} Y(\ell) &= \frac{\partial y}{\partial \ell}(\ell) = - y'(\ell) = 0 \\ Y'(\ell) &= \frac{\partial}{\partial x}\frac{\partial y}{\partial \ell}(\ell) = \frac{\partial}{\partial \ell}\frac{\partial y}{\partial x}(\ell) = - y''(\ell) \end{aligned}\right\} \qquad (3.4)$$

Thus,

$$\int_0^{\ell} y'' \, Y'' \, dx = \int_0^{\ell} y^{(4)} \, Y \, dx - y''^2(\ell)$$

Using this result in Eq. 3.3 gives

$$- EI \, y''^2(\ell) = \frac{d\zeta}{d\ell} \rho \, h \int_0^{\ell} y^2 \, dx$$

or

$$\frac{d\zeta}{d\ell} = \frac{- EI \, y''^2(\ell)}{\rho \, h \int_0^{\ell} y^2 \, dx} \qquad (3.5)$$

4. A SIMPLE TWO DIMENSIONAL EXAMPLE

Consider a vibrating membrane, with the governing equations

$$\left.\begin{aligned} - T \, \Delta y &= \zeta \, h \, y, \quad \text{in } \Omega \\ y &= 0, \quad \text{on } \Gamma \end{aligned}\right\} \qquad (4.1)$$

where Ω is a smooth, bounded open subset of R^2, T is the linear tension in the membrane, and h is the constant area mass density. The corresponding variational equation is

$$T \int_{\Omega} \nabla y \cdot \nabla v \; dx = \zeta \, h \int_{\Omega} y \; v \; dx \tag{4.2}$$

The next step is to take the derivative of

$$T \int_{\Omega} \nabla y \cdot \nabla y \; dx = \zeta \, h \int_{\Omega} y^2 \; dx \tag{4.3}$$

using formulas of derivation of integrals with respect to shape.

Recall that if Ω is a smooth open set of R^n and V is a smooth vector field that is defined in a neighborhood of Ω, one may define Ω_ε as the set of points $x + \varepsilon V(x) = T(x)$, for $x \in \Omega$. For ε small enough, T is a diffeomorphism. (Refs. 1, 2, and 6). If ϕ is a smooth function of x and possibly of ε, then

$$\frac{d}{d\varepsilon} \int_{\Omega_\varepsilon} \phi(\varepsilon,x) \; dx_\varepsilon = \int_{\Omega_\varepsilon} \frac{\partial \phi}{\partial \varepsilon} \, dx_\varepsilon + \int_{\Omega_\varepsilon} \mathrm{div}(\phi V) \; dx_\varepsilon \tag{4.4}$$

or

$$\frac{d}{d\varepsilon} \int_{\Omega_\varepsilon} \phi(\varepsilon,x) \; dx_\varepsilon = \int_{\Omega_\varepsilon} \frac{\partial \phi}{\partial \varepsilon} \, dx_\varepsilon + \int_{\Gamma_\varepsilon} \phi(V \cdot n) \; dx_\varepsilon \tag{4.5}$$

As in the preceding, one defines

$$Y(x) = \frac{\partial}{\partial \varepsilon} y(\varepsilon,x) \tag{4.6}$$

These formulas play the role of differentiation with respect to the length in the one dimensional case. Using Eq. 4.4, one can take the derivative of Eq. 4.3 to obtain

$$\begin{aligned} &2T \int_{\Omega} \nabla y \cdot \nabla Y \; dx + T \int_{\Omega} \mathrm{div}(|\nabla y|^2 V) \; dx \\ &= \zeta'(\Omega,V) \; h \int_{\Omega} y^2 \; dx + 2 \, \zeta \, h \int_{\Omega} y \; Y \; dx \\ &\quad + \zeta \, h \int_{\Omega} \mathrm{div}(y^2 V) \; dx \end{aligned} \tag{4.7}$$

But,

$$\int_\Omega \nabla y \cdot \nabla Y \, dx = - \int_\Omega (\Delta y) \, Y \, dx + \int_\Gamma \frac{\partial y}{\partial n} Y \, d\sigma \tag{4.8}$$

and $Y\big|_\Gamma$ may be computed in the following way:

$$y_{(I+\varepsilon V)(\Omega)}(x+\varepsilon V(x)) = 0, \text{ for } x \in \Gamma$$

Hence,

$$\left.\frac{\partial y}{\partial \varepsilon}\right|_{\varepsilon=0} + \nabla y \cdot V = 0$$

See also Ref. 6. Thus, Eq. 4.8 may be written, noting that $y = 0$ on Γ implies $\nabla y \cdot V = \frac{\partial y}{\partial n} V \cdot n$,

$$\int_\Omega \nabla y \cdot \nabla Y \, dx = - \int_\Omega Y \Delta y \, dx - \int_\Gamma \left(\frac{\partial y}{\partial n}\right)^2 V \cdot n \, d\sigma$$

or

$$\int_\Omega \nabla y \cdot \nabla Y \, dx = \zeta \int_\Omega y \, Y \, dx - \int_\Gamma \operatorname{div}(|\nabla y|^2 V) \, d\sigma \tag{4.9}$$

One may now use Eq. 4.9 in Eq. 4.7 to get

$$- T \int_\Omega \operatorname{div}(|\nabla y|^2 V) \, dx = \zeta'(\Omega, V) \, h \int_\Omega y^2 \, dx$$

or,

$$\zeta'(\Omega, V) = \frac{- T \int_\Omega \operatorname{div}(|\nabla y|^2 V) \, dx}{\int_\Omega h \, y^2 \, dx} \tag{4.10}$$

which may also be written (for smooth data) as

$$\zeta'(\Omega, V) = \frac{- T \int_\Gamma \left(\frac{\partial y}{\partial n}\right)^2 (V \cdot n) \, d\sigma}{\int_\Omega h \, y^2 \, dx} \tag{4.11}$$

5. A VIBRATING PLATE

Consider now a clamped plate of constant thickness h, Young's modulus E, material density ρ (all constants), and $D = \frac{Eh^3}{12(1-\nu^2)}$ where $0 < \nu < 0.5$ is Poisson's ratio. The eigenvalue problem is

$$\left.\begin{array}{ll} D\,\Delta^2 y = \zeta\,\rho\,h\,y, & \text{in } \Omega \\ y = 0 = \frac{\partial y}{\partial n}, & \text{in } \Gamma \end{array}\right\} \tag{5.1}$$

where Ω is a smooth, bounded, open subset of R^2. The variational equation is (see Refs. 7 or 9)

$$a_\Omega(y,v) = \zeta\, b_\Omega(y,v) \tag{5.2}$$

for every smooth v satisfying $v = 0$, $\frac{\partial v}{\partial n} = 0$ on Γ, where

$$a_\Omega(y,v) = D\int_\Omega [\Delta y \Delta v + (1-\nu)(2y_{12}v_{12} - y_{11}v_{22} - y_{22}v_{11})]\,dx \tag{5.3}$$

$$b_\Omega(y,v) = \rho\, h \int_\Omega y\, v\, dx \tag{5.4}$$

One now proceeds as for the membrane by computing the derivative of

$$a_\Omega(y,y) = \zeta\, b_\Omega(y,y) \tag{5.5}$$

to obtain

$$\begin{aligned} &D\int_\Omega 2\Delta y \Delta Y dx + 2D(1-\nu)\int_\Omega (2y_{12}Y_{12} - y_{11}Y_{22} - y_{22}Y_{11})\,dx \\ &+ D\int_\Gamma [(\Delta y)^2 + 2(1-\nu)(y_{12}^2 - y_{11}y_{22})]\, V\cdot n\, d\sigma \\ &= 2\,\zeta\,\rho\,h \int_\Omega y\, Y\, dx + \zeta\,\rho\,h\int_\Gamma y^2\, V\cdot n\, d\sigma \\ &+ \zeta'(\Omega,V)\,\rho\,h\int_\Omega y^2\, dx \end{aligned} \tag{5.6}$$

Integrating the first two integrals by parts yields

$$2D \int_\Omega \Delta y \Delta Y \, dx = 2D \int_\Omega (\Delta^2 y) Y \, dx - 2D \int_\Gamma \left(\frac{\partial}{\partial n} \Delta y\right) Y \, d\sigma + 2D \int_\Gamma \Delta y \frac{\partial Y}{\partial n} \, d\sigma \qquad (5.7)$$

and

$$2D(1-\nu) \int_\Omega (2y_{12}Y_{12} - y_{11}Y_{22} - y_{22}Y_{11}) \, dx$$

$$= 2D(1-\nu) \int_\Gamma \left(- \frac{\partial^2 y}{\partial s^2} \frac{\partial Y}{\partial n} + \frac{\partial^2 y}{\partial s \partial \nu} \frac{\partial Y}{\partial s} \right) d\sigma \qquad (5.8)$$

For the last integral, $\frac{\partial}{\partial s}$ is the tangential derivative at the boundary.

Using Eq. 5.7 and 5.8 in Eq. 5.6 gives

$$2D \int_\Gamma \Delta y \frac{\partial Y}{\partial n} \, d\sigma - 2D \int_\Gamma \left(\frac{\partial}{\partial n} \Delta y\right) Y \, d\sigma$$

$$+ 2D(1-\nu) \int_\Gamma \left(- \frac{\partial^2 y}{\partial s^2} \frac{\partial Y}{\partial n} + \frac{\partial^2 y}{\partial n \partial s} \frac{\partial Y}{\partial s} \right) d\sigma + D \int_\Gamma (\Delta y)^2 V \cdot n \, d\sigma$$

$$+ D \int_\Gamma 2(1-\nu)(y_{12}^2 - y_{11}y_{22}) V \cdot n \, d\sigma = \zeta \rho h \int_\Gamma y^2 V \cdot n \, d\sigma$$

$$+ \zeta'(\Omega, V) \rho h \int_\Omega y^2 \, dx \qquad (5.9)$$

One now needs to compute ∇Y on Γ as

$$\nabla Y = \nabla \frac{\partial y}{\partial \varepsilon} = \frac{\partial}{\partial \varepsilon} \nabla y$$

but on Γ, one has $\nabla y = 0$, because $y = 0$ and $\frac{\partial y}{\partial n} = 0$. Hence $\frac{\partial y}{\partial s} = 0$ on Γ. This implies that $Y = -\nabla y \cdot V = 0$. Its material derivative* is thus zero and hence $\frac{\partial}{\partial\varepsilon}\frac{\partial y}{\partial x_i} = -\nabla\left(\frac{\partial y}{\partial x_i}\right)\cdot V$ and

$$\nabla Y = D^2 y \cdot V = -\begin{bmatrix} V\cdot n \, \frac{\partial^2 y}{\partial n^2} + V\cdot s \, \frac{\partial^2 y}{\partial n \partial s} \\ V\cdot n \, \frac{\partial^2 y}{\partial n \partial s} + V\cdot s \, \frac{\partial^2 y}{\partial s^2} \end{bmatrix}$$

But since $\frac{\partial y}{\partial n} = 0$, one has also $\frac{\partial^2 y}{\partial s \partial n} = 0$ and $\frac{\partial y}{\partial s} = 0$ implies $\frac{\partial^2 y}{\partial s^2} = 0$. Hence,

$\frac{\partial Y}{\partial n} = - V\cdot n \, \frac{\partial^2 y}{\partial n^2}$, $\frac{\partial Y}{\partial s} = 0$ and $\Delta y = \frac{\partial^2 y}{\partial n^2}$ on Γ; as well as

$$y_{12}^2 - y_{11} y_{22} = \frac{\partial^2 y}{\partial s \partial n} - \frac{\partial^2 y}{\partial s^2}\frac{\partial^2 y}{\partial n^2} = 0$$

Using these results in Eq. 5.9, one obtains

$$- 2D \int_\Gamma \left(\frac{\partial^2 y}{\partial n^2}\right)^2 V\cdot n \, d\sigma + D \int_\Gamma \left(\frac{\partial^2 y}{\partial n^2}\right)^2 V\cdot n \, d\sigma$$

$$= \zeta'(\Omega, V) \, \rho \, h \int_\Omega y^2 \, dx$$

or

$$\zeta'(\Omega, V) = \frac{- D \int_\Gamma \left(\frac{\partial^2 y}{\partial n^2}\right)^2 V\cdot n \, d\sigma}{\int_\Omega \rho \, h \, y^2 \, dx} \tag{5.10}$$

Notice that since $\frac{\partial^2 y}{\partial s^2} = 0$, the numerator is also equal to

$$- D \int_\Gamma (\Delta y)^2 (V\cdot n) \, d\sigma$$

---

*See Ref. 6 for its definition.

6. THE CASE OF MULTIPLE EIGENVALUES

The dependence of multiple eigenvalues has already been addressed in Ref. 10. One is only interested here in the dependence of multiple eigenvalues on shape variations. From Ref. 10 (see also Refs. 11, and 12), one knows that if he wants to describe the local behavior of an eigenvalue of multiplicity m, he has to consider an m x m matrix. The derivation of this matrix is presented here.

The directional derivatives of $\zeta(\Omega)$ of multiplicity m are

$$\zeta_i'(\Omega,V) = \lim_{\substack{t\to 0\\ t>0}} \frac{\zeta_i((I+tV)(\Omega)) - \zeta(\Omega)}{t} \tag{6.1}$$

They are the eigenvalues of the matrix M with entries

$$M_{i,j} = - D \int_\Gamma \frac{\partial^2 y_i}{\partial n^2} \frac{\partial^2 y_j}{\partial n^2} (V\cdot n)\, d\sigma \tag{6.2}$$

where y_i, $i = 1,\ldots,m$, is a b_Ω-orthonormal basis of the eigenspace of $\zeta(\Omega)$, i.e.

$$\rho\, h \int_\Omega y_i\, y_j\, dx = \delta_{ij}$$

Naturally as in Ref. 7, if for $t > 0$, $\zeta_i((I + tV)(\Omega))$ are all simple eigenvalues and if one uses as a basis for the eigenspace of $\zeta(\Omega)$ the limits (when $t \to 0$, $t > 0$) of an orthonormal basis of eigenvectors associated with $\zeta_i((I + tV)(\Omega))$, then

$$\zeta_i'(\Omega,V) = - D \int_\Gamma \left(\frac{\partial^2 y_i}{\partial n^2}\right)^2 (V\cdot n)\, d\sigma \tag{6.3}$$

7. IMPLEMENTATION

Consider now only the problem of maximizing the smallest eigenvalue of a vibrating membrane. The case of a constraint of constant volume will be introduced in the next section. The case of repeated eigenvalues is addressed in Ref. 5 (see also Refs. 3, 8 and 9). The development of this section refers to Ref. 4.

Among available methods BF1 and BF2 (see Ref. 4) apply quite straightforwardly. Instead of using Eq. 3.14 of Ref. 4, one uses Eq. 4.11 of this paper.

For the distributed formulation, instead of Eq. 3.16 of Ref. 4, one uses

$$\int_\Omega \Delta G \cdot \Delta\psi \, dx = -T \int_\Gamma \left(\frac{\partial y}{\partial n}\right)^2 (\psi \cdot n) \, d\sigma \tag{7.1}$$

where the eigenfunction y has been normalized by $\int_\Omega h\, y^2 \, dx = 1$ and the Green function G is found by formally solving

$$\left.\begin{aligned} &\Delta^2 G = 0, && \text{in } \Omega \\ &\Delta G = 0, && \text{on } \Gamma \\ &\left(\frac{\partial}{\partial n} \Delta G\right) \cdot n = -T \left(\frac{\partial y}{\partial n}\right)^2, && \text{on } \Gamma \end{aligned}\right\} \tag{7.2}$$

Use of Green function for the simplest example of Eq. 2.1 is now described, i.e. for

$$\left.\begin{aligned} &-y'' = \zeta y, \text{ in }]0,\ell[\\ &y(0) = 0 = y(\ell) \end{aligned}\right\} \tag{7.3}$$

One knows from Eq. 2.11 that

$$\zeta'(]0,\ell[,V) = -y'^2(\ell)\, V(\ell)$$

where y is normalized with $\int_0^\ell y^2 \, dx = 1$. Assume one wants to implement a distributed formula of the type of Eq. 7.1. The scope includes solving Eq. 7.3 and an equation analogous to Eq. 7.1, only in some neighborhood of Γ (see the same idea for static problems in Ref. 4). One introduces here a function that satisfies $-g_k'' = \delta_k$, together with some boundary conditions that will be enforced later. As in Ref. 4, one has

$$\int_0^k g\, y''\, dx = -\int_0^k g'\, y'\, dx + g\, y'\Big|_0^k$$

$$= g\, y'\Big|_0^k - \lim_{\varepsilon\to 0}\left[\int_0^{k-\varepsilon} g''\, y\, dx + g'\, y\Big|_0^{k-\varepsilon}\right]$$

and finally

$$\int_0^k g\, y''\, dx = -\zeta\int_0^k g\, y\, dx + g(k)\, y'(k) - g(0)\, y'(0)$$

$$- g'(k-0)\, y(k)$$

One enforces $g(0) = 0$ and (for example) $g(k) = k$. An elementary computation yields $g = (x-k)\, \chi_{]-\infty,k[} + k$. Hence,

$$-\zeta\int_0^k x\, y\, dx = k\, y'(k) - y(k)$$

If one would like to use boundary formulas, he would put $k = \ell$. Consider also a fixed interval $[0,k]$ and solve Eq. 7.3 in the following way:

$$-y'' = \zeta\, y, \text{ in }]k,\ell[$$

$$-\lambda\int_0^k x\, y\, dx = k\, y'(k) - y(k)$$

$$y(\ell) = 0$$

Then look for G that is zero in $[0,k]$ (one knows from the theory, see e.g. Ref. 2, that the derivative depends only on the vector field, in fact only its normal component, on the boundary) as a solution of

$$\int_k^\ell G'' \cdot \psi\, dx = -\, y'^2(\ell)\, \psi(\ell)$$

8. THE CONSTRAINT ON THE VOLUME OF Ω

In most structural optimization problems, there is a constraint on the volume of Ω (or one wants to minimize volume, with a constraint on the eigenvalue or the static deflection). Consider, for simplicity, the vibrating membrane of section 4,

$$\left.\begin{array}{ll} -T\,\Delta\, y = \zeta\, h\, y & \text{in } \Omega \\ y = 0 & \text{on } \Gamma \end{array}\right\} \tag{8.1}$$

The objective is to maximize the fundamental eigenvalues, such that $\int_\Omega 1\, dx \geq w_0 > 0$, where w_0 is some prescribed number. The derivative of ζ is, from Eq. 4.11,

$$\zeta'(\Omega, V) = -T \int_\Gamma \left(\frac{\partial y}{\partial n}\right)^2 (V\cdot n)\, d\sigma \tag{8.2}$$

where y is normalized with $\int_\Omega h\, y^2\, dx = 1$

The necessary condition of optimality is, by the usual Lagrange multiplier method (λ_0 and λ_1 being non-negative numbers not both zero),

$$-\lambda_0\, T \int_\Gamma \left(\frac{\partial y}{\partial n}\right)^2 (V\cdot n)\, d\sigma + \lambda_1 \int_\Gamma (V\cdot n)\, d\sigma = 0 \tag{8.3}$$

which should hold for every vector field V.

Since the functions $V\cdot n$ are dense in $L^2(\Gamma)$, this yields the following point-wise necessary condition:

$$-\lambda_0\, T \left(\frac{\partial y}{\partial n}\right)^2 + \lambda_1 = 0$$

from which it is impossible to have $\lambda_0 = 0$ with $(\lambda_0, \lambda_1) \neq (0,0)$. Defining $\mu = \dfrac{\lambda_1}{\lambda_0}$, one has

$$T \left(\frac{\partial y}{\partial n}\right)^2 = \mu \tag{8.4}$$

which means that at the optimum $\left(\frac{\partial y}{\partial n}\right)^2$ has to be constant on Γ. Equations 8.3 and 8.4 hint at the following algorithm:

Step 1 - Start from some estimated Ω
Step 2 - Compute y (either with finite elements or using Green functions
Step 3 - Estimate μ by the formula $\mu = \frac{T}{\text{mes } \Gamma} \int_\Gamma \left(\frac{\partial y}{\partial n}\right)^2 d\sigma$
Step 4 - Using Eq. 8.3, make a choice of G such that

$$- T \int_\Gamma \left(\frac{\partial y}{\partial n}\right)^2 (G \cdot n)\ d\sigma + \mu \int_\Gamma (G \cdot n)\ d\sigma > 0 \tag{8.5}$$

(In fact choose G such that the left hand side is as large as possible. One can use, for example, a formula of the type of Eq. 7.1, with an appropriate right hand side.)

Step 5 - Estimate a new Ω and restart the sequence, if necessary.

The choice of V such that Eq. 8.5 is satisfied may be done by any boundary or distributed method.

9. SOME NUMERICAL RESULTS

In this section some numerical results are presented from Ref. 11. They are based on the use of polynomials in the complex variable z (which works only for two dimensional problems). Use of these polynomials requires a sensitivity formula presented in a different way from Eqs. 4.10 or 4.11.

The easiest way to get this formula is to take the derivative of

$$\int_\Omega |\nabla y|^2\ dx = \zeta \int_\Omega y^2\ dx \tag{9.1}$$

using material derivatives. Instead of Eq. 4.4, use is made of

$$\frac{d}{d\varepsilon} \int_{\Omega_\varepsilon} \phi(\varepsilon, x)\ dx_\varepsilon = \int_{\Omega_\varepsilon} \frac{d\phi}{d\varepsilon}\ dx_\varepsilon + \int_{\Omega_\varepsilon} \phi\ \text{div}V\ dx_\varepsilon \tag{9.2}$$

to get

$$2 \int_\Omega \nabla y \cdot (\nabla y)^{\bullet}\, dx + \int_\Omega |\nabla y|^2 \,\mathrm{div}V\, dx$$

$$= 2\zeta \int_\Omega y\, \dot{y}\, dx + \zeta \int_\Omega y^2 \,\mathrm{div}V\, dx + \zeta'(\Omega,V) \int_\Omega y^2\, dx \qquad (9.3)$$

But,

$$\nabla \dot{y} = (\nabla y)^{\bullet} + DV\, \nabla y$$

where DV denotes the Jacobian matrix. Hence, since $y|_\Gamma = 0$, Eq. 9.3 gives

$$- 2 \int (\nabla y,\, DV\, \nabla y)\, dx + \int_\Omega |\nabla y|^2 \,\mathrm{div}V\, dx$$

$$= \zeta \int_\Omega y^2 \,\mathrm{div}V\, dx + \zeta'(\Omega,V) \int_\Omega y^2\, dx$$

Finally, for y normalized with $\int_\Omega y^2\, dx = 1$,

$$\zeta'(\Omega,V) = \int_\Omega (|\nabla y|^2 - \zeta y^2)\, \mathrm{div}V\, dx - 2 \int_\Omega (\nabla y,\, DV\, \nabla y)\, dx \qquad (9.4)$$

This result was derived by a change of variable in Ref. 11. It is easy to prove equivalence of this result with Eq. 4.11, using integration by parts.

Now suppose

$$\left.\begin{aligned} V_{1,1} &= V_{2,2} \\ V_{1,2} &= - V_{2,1} \end{aligned}\right\} \qquad (9.5)$$

which amounts to saying that $V_1(x+iy) + iV_2(x+iy) = F(x+iy)$ is a holomorphic function, by the Canchy-Riemann formula. It is known that

$$V_{1,1} = \mathrm{Re}\,(F'(x+iy)) \qquad (9.6)$$

where F' denotes the derivative in the complex sense. Hence,

$$\operatorname{div} V = 2 \operatorname{Re} (F'(x+iy)) \tag{9.7}$$

and

$$(\nabla y, DV \nabla y) = V_{1,1} |\nabla y|^2$$

$$= \operatorname{Re} (F'(x+iy)) |\nabla y|^2 \tag{9.8}$$

Using Eqs. 9.7 and 9.8 enables one to rewrite Eq. 9.4 as

$$\zeta'(\Omega,F) = - 2\zeta(\Omega) \int_\Omega y^2 \operatorname{Re} (F'(x+iy)) \, dx \, dy \tag{9.9}$$

Note that dx in Eq. 9.4 is equivalent to dx dy in Eq. 9.9.

Set

$$F(x+iy) = \Sigma \, b_p(x+iy)^p \text{ (finite sum)} \tag{9.10}$$

Then,

$$\zeta'(\Omega,F) = - 2\zeta(\Omega) \int_\Omega y^2 \operatorname{Re} (p \, b_p \, z^{p-1}) \, dx \, dy \tag{9.11}$$

If one wants to minimize

$$J(\Omega) = f(\lambda_1(\Omega), \ldots., \lambda_m(\Omega)) \tag{9.12}$$

where f is some smooth function of m variables and if one sets $b_p = \gamma_p + i \, \delta_p$, then

$$J'(\Omega,F) = \sum_{j=1}^{m} D_j \, f \, \lambda_j'(\Omega,F) \tag{9.13}$$

and the best local choice of decrease, within the family of deformation of Eq. 9.10, is given by

$$\gamma_p = - 2p \sum_{j=1}^{m} \lambda_j \frac{\partial f}{\partial \lambda_j} \int_\Omega y_j^2 \operatorname{Re} ((x+iy)^{p-1}) \, dx \, dy \tag{9.14}$$

(Equation continued on next page)

$$\delta_p = 2p \sum_{j=1}^{m} \lambda_j \frac{\partial f}{\partial \lambda_j} \int_\Omega y_j^2 \ \mathrm{Im}\ ((x+iy)^{p-1})\ dx\ dy \qquad (9.14)$$

The whole mesh is then moved according to the formula

$$(x+iy) \longrightarrow (x+iy) + \rho\ G(x+iy)$$

where

$$G + \Sigma\ (\gamma_p + i\ \delta_p)(x+iy)^p$$

and γ_p and δ_p are given by Eq. 9.14.

One of the practical problems of implementation is to get a good finite element mesh, while changing the shape of Ω. It is possible to impose constraints so that the triangles do not become too small or with too small angles. However, it is time and core storage consuming. Obviously, this should be done for any practical use of the method. However, these constraints have not been taken into account in the numerical experiments reported here.

Two experiments are presented in the following figures; they are concerned with the minimization of $J(\Omega) = \sum_{i=1}^{m} \left|\frac{1}{\lambda_i}(\Omega) - \frac{1}{\alpha_i}\right|^2$, where α_i are given numbers and where $\lambda_i(\Omega)$ are the smallest eigenvalues of the Laplacian. Considering $\left|\frac{1}{\lambda_i}(\Omega) - \frac{1}{\alpha_i}\right|^2$ instead of $|\lambda_i(\Omega) - \alpha_i|^2$ in the functional amounts to putting more emphasis on the smallest eigenvalues, which is desired. Only 37 nodes are used to have a reasonable computing time (around 1 minute of IBM 370-168).

The first experiment (Fig. 9.1) starts with a circle, the fifth lowest eigenvalues of which are 0.9441190E01, 0.2634304E02, 0.2634304E02, 0.5192597E02, and 0.5192597E02. The desired eigenvalues are 0.7735564E01, 0.1927194E02, 0.2387913E02, 0.3922726E02, and 0.4177514E02. Thus, the functional value is 0.81552×10^{-3}.

At the eighth iteration (Fig. 9.2), the domain is somewhat larger, as is to be expected. The smallest eigenvalue has to decrease, hence the area has to increase, as is well known for Dirichlet boundary conditions [13]. The first eigenvalue is

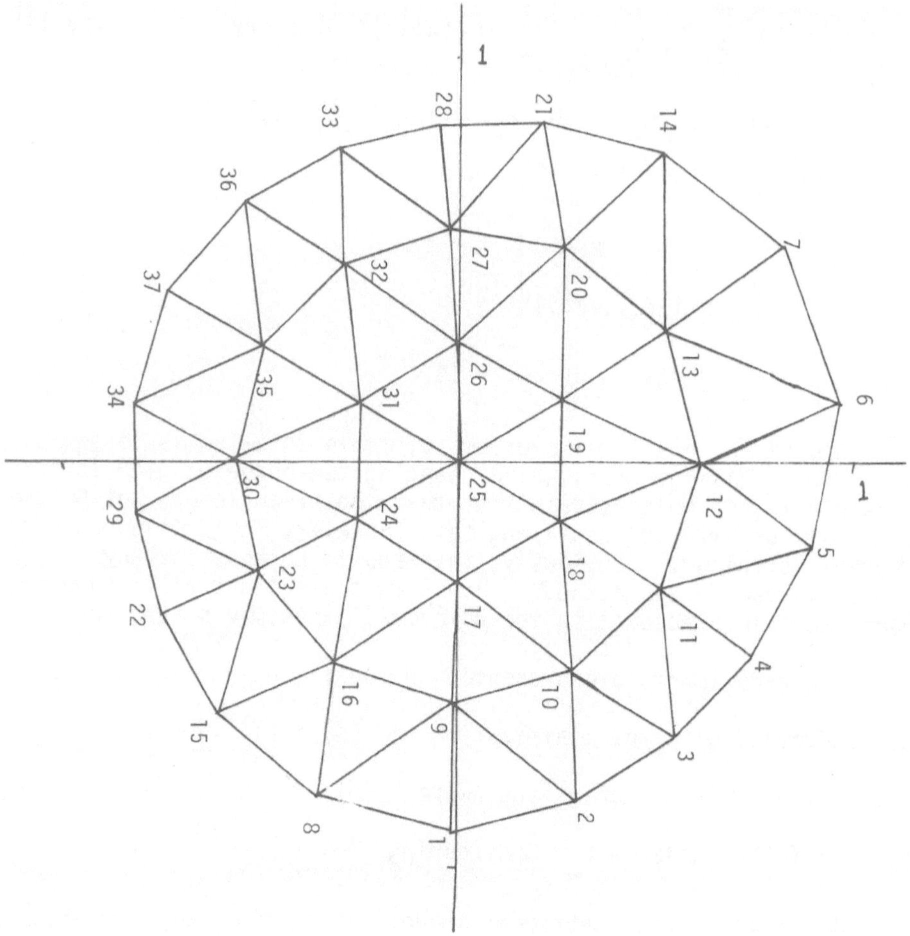

Figure 9.1 Domain At Iteration 1

already correct, up to the third digit. The first five eigenvalues are 0.7731445E01, 0.2086767E02, 0.2222555E02, 0.4197814E02, and 0.4275209E02. The other four eigenvalues have also much improved and the functional is equal to 0.15749×10^{-4}. These values will be refined in the process to the seventeenth step (Fig. 9.3), where the eigenvalues are 0.7739401E01, 0.1928198E02, 0.2351551E02, 0.3976425E02, and 0.4275909E02. No substantial improvement is obtained while running to the forty fifth iteration (Fig. 9.4), where the eigenvalues are 0.7733947E01, 0.1923563E02, 0.2359982E02, 0.4005354E02, and 0.4224717E02. This

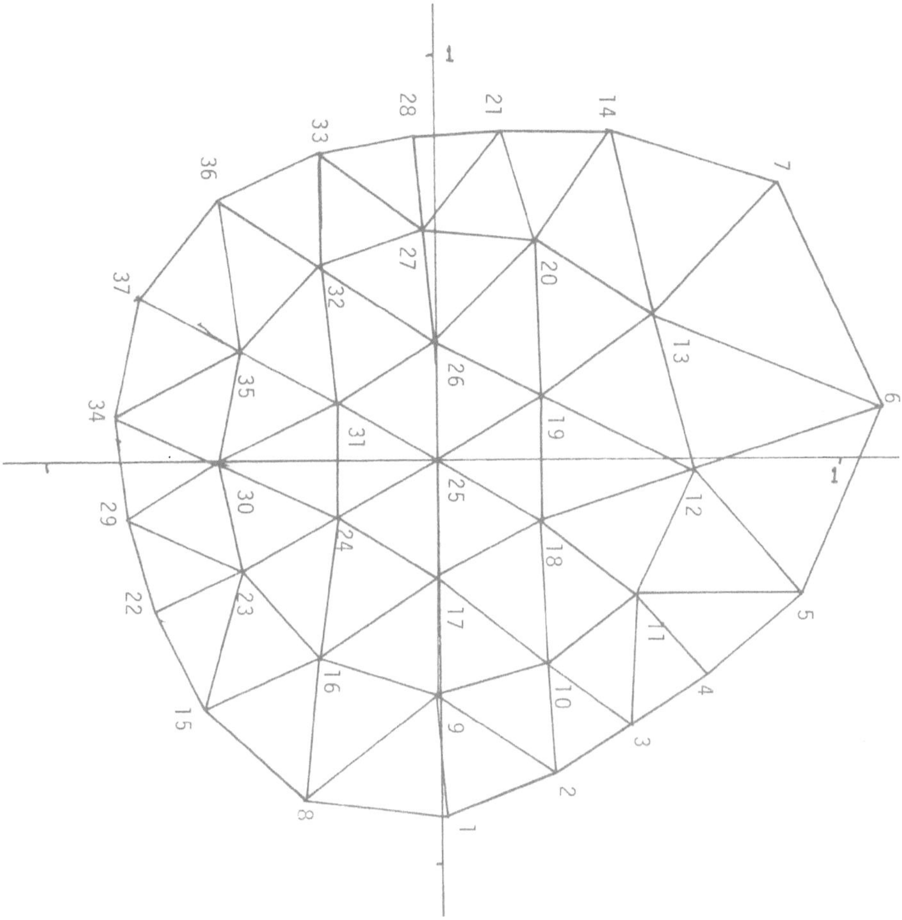

Figure 9.2 Domain At Iteration 8

behavior (no substantial improvement after a few steps) is well known for first order methods in optimization. One should notice that the eigenvalues are very close to the desired ones after very few steps.

The presentation switches now to another experiment which starts also from an alternate circle (Fig. 9.5), with eigenvalues 0.4993687E01, 0.1393351E02, 0.1393351E02, 0.2746498E02, and 0.2746498E02. This alternate domain is somewhat larger than the previous one (it appears smaller in Fig. 9.5 because the unit length has been chosen smaller than the previous experiments).

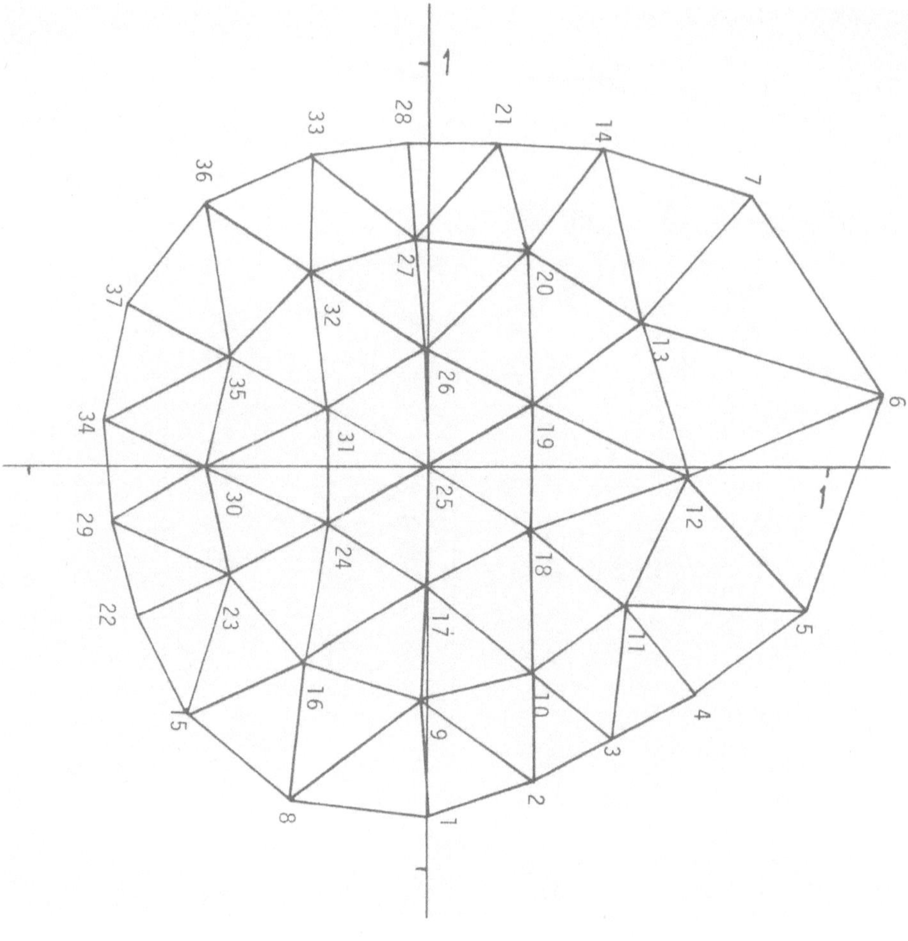

Figure 9.3 Domain At Iteration 17

The desired eigenvalues are now 0.1499723E02, 0.2943144E02, 0.5401754E02, 0.5659455E02, and 0.7782078E02, so that the functional is equal to 0.23013 x 10^{-1}. At the second iteration (Fig. 9.6) the shape is no longer convex and the eigenvalues are improved (but not very close to the desired ones). The eigenvalues are 0.6173562E01, 0.1588333E02, 0.1680725E02, 0.3101770E02, and 0.3448895E02. The repeated eigenvalues have split into separate ones. Notice one triangle (Fig. 9.7), the eigenvalues are 0.1051095E02, 0.1803863E02, 0.3506289E02, 0.4792440E02, and 0.5915188E02, which are improved, but the shape is substantially changed. Some triangles have become too small (compared with

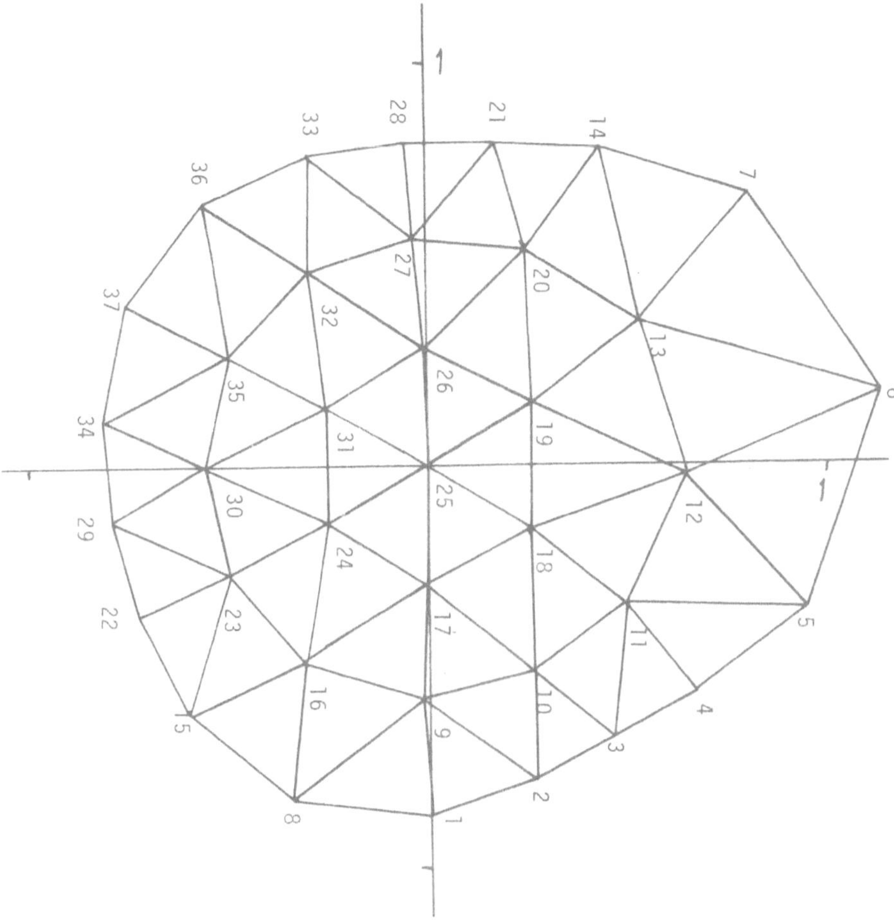

Figure 9.4 Domain At Iteration 45

the average area of the triangles) and some have small angles. Here, $J = 0.13973.10^{-2}$. However, calculations went on until the seventh iteration to get the shape displayed in Fig. 9.8. The eigenvalues, 0.1542195E02, 0.2535592E02, 0.5393070E02, 0.7191239E02, and 0.9127593E02, are (substantial improvement) but the behavior of the mesh is still worse.

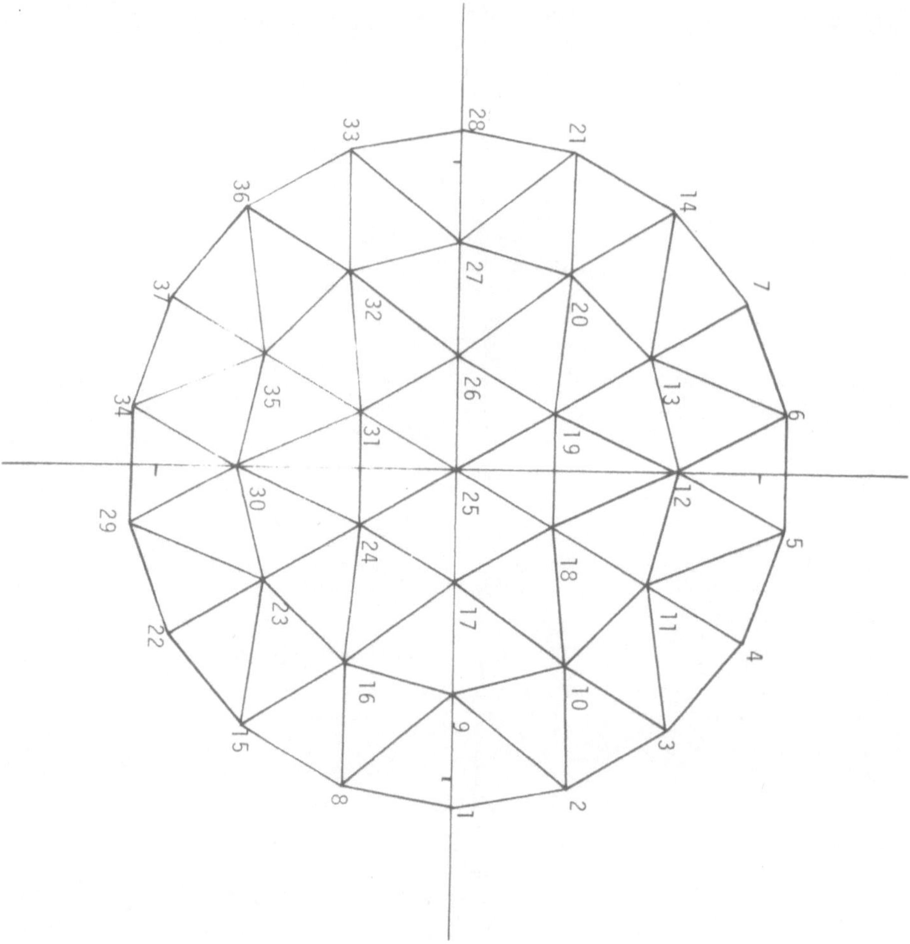

Figure 9.5 Alternate Domain At Iteration 1

Redefining the mesh of triangles would enable one to go on in the optimization process, but such algorithms were not at hand when these experiments were carried out. However, these two experiments show clearly that the implementation of general shape optimal design is quite possible and that a substantial improvement of the functional may be achieved.

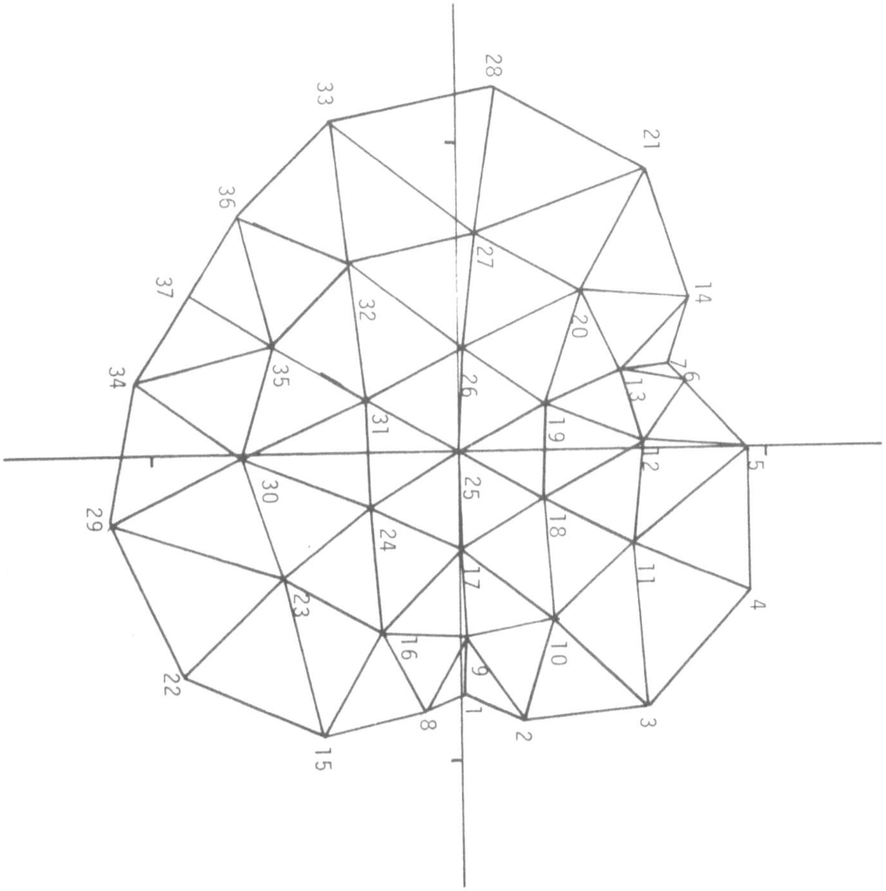

Figure 9.6 Alternate Domain At Iteration 2

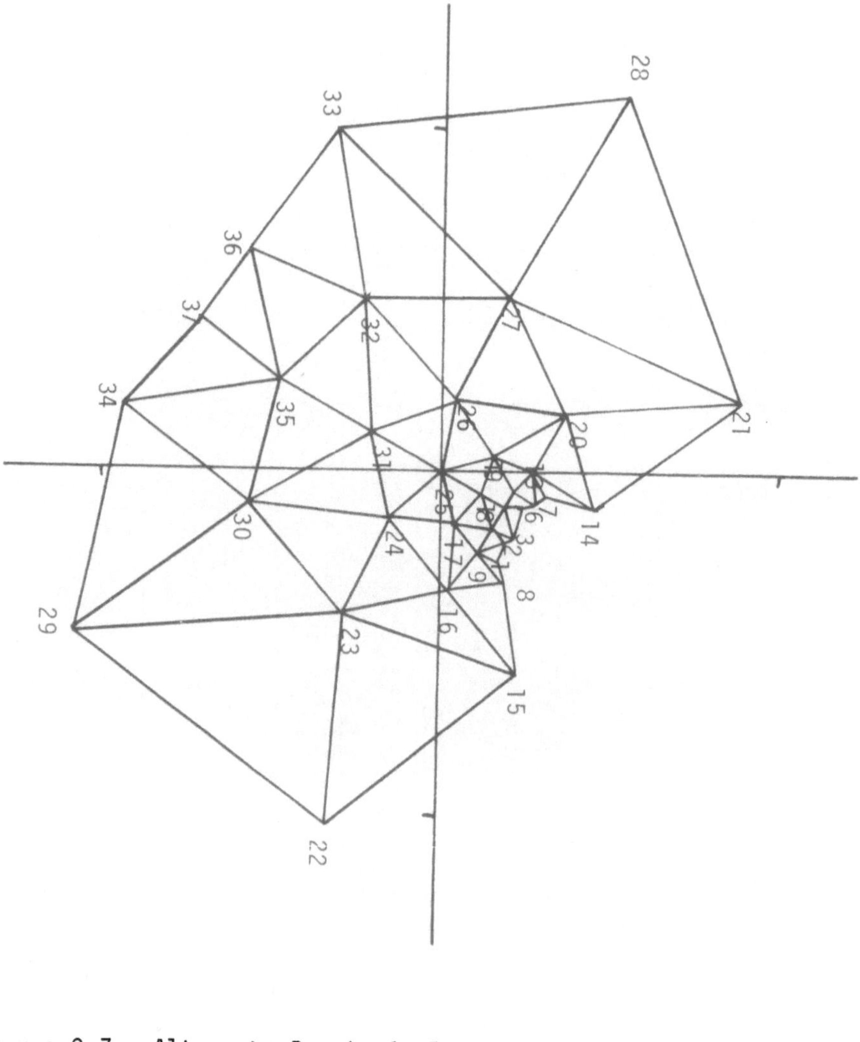

Figure 9.7 Alternate Domain At Iteration 3

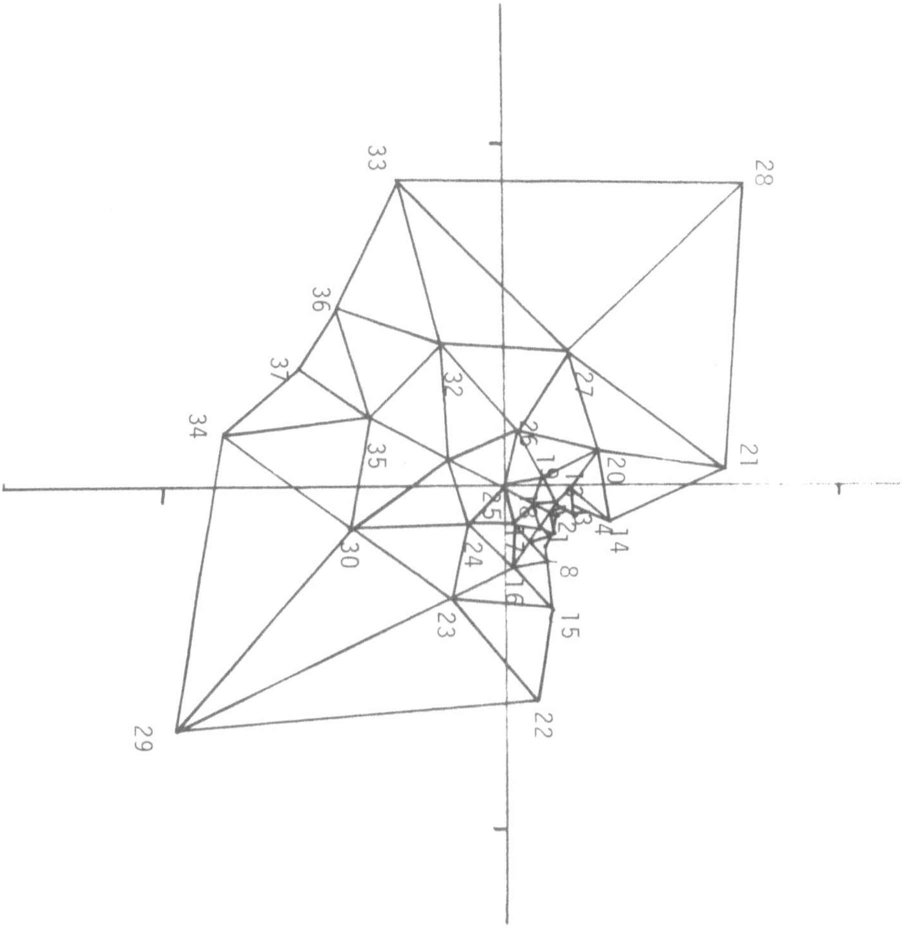

Figure 9.8 Alternate Domain At Iteration 7

10. CONCLUSIONS

Implementation has been described only for the case of simple eigenvalues. A necessary optimality condition has been obtained very recently in Ref. 8. This formula may be approximated to obtain an algorithm (see Ref. 5).

Some numerical results have also been presented: they will be connected with the problem of designing the shape of a membrane such that its first lowest frequencies be as near as possible to prescribed numbers (see Ref. 11).

Finally, although it has not been described here, implementation for the design of more complex structures may be done along the same lines. This applies, in particular to plates and two or three dimensional elastic bodies.

REFERENCES

1. Cea, J., "Numerical Methods of Shape Optimal Design," Optimization of Distributed Parameter Structures (Eds. E.J. Haug and J. Cea), Sijthoff & Noordhoff, Alphen aan den Rijn, Netherlands, 1980.
2. Zolesio, J-P., "The Material Derivative (or Speed) Method for Shape Optimization," Optimization of Distributed Parameter Structures (Eds. E.J. Haug and J. Cea), Sijthoff & Noordhoff, Alphen aan den Rijn, Netherlands, 1980.
3. Zolesio, J-P., "Semi-Derivative of Multiple Eigenvalues," Optimization of Distributed Parameter Structures (Eds. E.J. Haug and J. Cea), Sijthoff & Noordhoff, Alphen aan den Rijn, Netherlands, 1980.
4. Rousselet, B., "Implementation of Some Methods of Shape Optimal Design," Optimization of Distributed Parameter Structures (Eds. E.J. Haug and J. Cea), Sijthoff & Noordhoff, Alphen aan den Rijn, Netherlands, 1980.
5. Rousselet, B., "Multiple Eigenvalues and Supremum Norm Constraints," Optimization of Distributed Parameter Structures (Eds. E.J. Haug and J. Cea), Sijthoff & Noordhoff, Alphen aan den Rijn, Netherlands, 1980.
6. Rousselet, B., "Shape Design Sensitivity Methods for Structural Mechanics," Optimization of Distributed Parameter Structures (Eds. E.J. Haug and J. Cea), Sijthoff & Noordhoff, Alphen aan den Rijn, Netherlands, 1980.
7. Ciarlet, P.G., The Finite Element Method for Elliptic Problems, North-Holland, Amsterdam, 1978.
8. Rousselet, B., "Condition nécessaire d'optimalité en présence de valeuss propres multiples," Note C.R.A.S., Paris, to appear 1980.

9. Haug, E.J. and Rousselet, B., "Design Sensitivity Analysis of Eigenvalue Variations," Optimization of Distributed Parameter Structures (Eds. E.J. Haug and J. Cea), Sijthoff & Noordhoff, Alphen aan den Rijn, Netherlands, 1980.
10. Rousselet, B., "Singular Dependence of Repeated Eigenvalues," Optimization of Distributed Parameter Structures (Eds. E.J. Haug and J. Cea), Sijthoff & Noordhoff, Alphen aan den Rijn, Netherlands, 1980.
11. Rousselet, B., Thèse de l'Universitè de Nice, 1977.
12. Rousselet, B., "Optimal Design and Eigenvalue Problems," Proceedings 8th IFIP Conference, Lect. Notes in Cont. and Inf. Sciences, Springer Verlag, 1978.
13. Courant, R. and Hilbert, D., Methods of Mathematical Physics, Vol. I and Vol. II, Interscience Publishers, Inc., New York, 1953, 1962.

DESIGN OF A MASS-OPTIMIZED THERMAL DIFFUSER*†

M. Delfour,[1] G. Payre,[2] J.P. Zolesio[3]

ABSTRACT

In this paper, a domain optimization problem arising from space communication research is presented. It is, in fact, a nonstandard illustration of domain optimization, with elliptic equations and non-differentiable constraints. Techniques results of Refs. 1, 2 and 3 are applied in a systematic way. Numerical results are presented, using a penalty method. The example presented illustrates a cost functional f that depends on both a distributed parameter L and a domain Ω. The gradient (on the domain variations) is not a regular function defined on the boundary, but involves a Dirac measure at one point.

---

*This research was supported in part by NSERC Grant A-8730.

†This problem has been communicated to the authors by Dr. Victor Wehrle, Communication Research Center, Department of Communication, Ottawa, Canada.

(1) Centre de recherche de mathématiques appliqueés, Université de Montréal, Montréal, Qué., Canada H3C 3J7.

(2) Département de Génie Chimique, Université de Sherbrooke, Sherbrooke, Qué., Canada.

(3) Département de Mathématiques, Université de Nice, 06034 Nice, Cedex, France.

1. INTRODUCTION

Technological development of devices for the space industry often requires designing parts that can perform a given task with the least possible mass, or volume. This paper deals with an example of such a problem, which occurs when a thermal diffuser is needed between a high-power solid state device and the heat-pipes transmitting the heat flux to a radiator.

The thermal diffuser receives a constant thermal power flux from the solid state device. At its other end, its temperature is kept at a uniform value by isothermalizing heatpipes. However, the thermal power flux at the heatpipes cannot exceed a given level, which is far less than the one received by the diffuser. Hence one is faced with the problem of finding the shape of a diffuser with minimum weight, with the requirement that it must reduce the outward thermal flux to a sufficiently low level.

2. STATEMENT OF THE PHYSICAL PROBLEM

The mounting surface of the solid-state device is supposed to be circular. Therefore, the thermal diffuser will have an axisymmetric shape (Fig. 2.1); the temperature distribution $T(r,z)$ in the diffuser D satisfies the partial differential equation

$$\frac{1}{r}\frac{\partial}{\partial r}\left(r\,\frac{\partial T}{\partial r}\right)+\frac{\partial^2 T}{\partial z^2}=0 \quad \text{in} \quad D \tag{2.1}$$

and the boundary conditions

$$-k\,\frac{\partial T}{\partial n}=q_1<0\,, \quad \text{on} \quad S_1=\left\{(r,z):\ 0<r<R_1,\ z=0\right\} \tag{2.2}$$

$$-k\,\frac{\partial T}{\partial n}=0, \quad \text{on} \quad S_2=\left\{(r,z):\ r=R_1+\rho(z),\ 0<z<L\right\} \tag{2.3}$$

$$T=0\,, \quad \text{on} \quad S_3=\left\{(r,z):\ 0<r<R_2=R_1+\rho(L),\ z=L\right\} \tag{2.4}$$

$$\frac{\partial T}{\partial r}=0, \text{on}\ S_4=[(r,z):\ r=0,\ \ 0<z<L] \tag{2.5}$$

The parameters R_1, q_1, and k are given and the function $\rho(z)$ and the value of L are to be found, to define the shape of the diffuser.

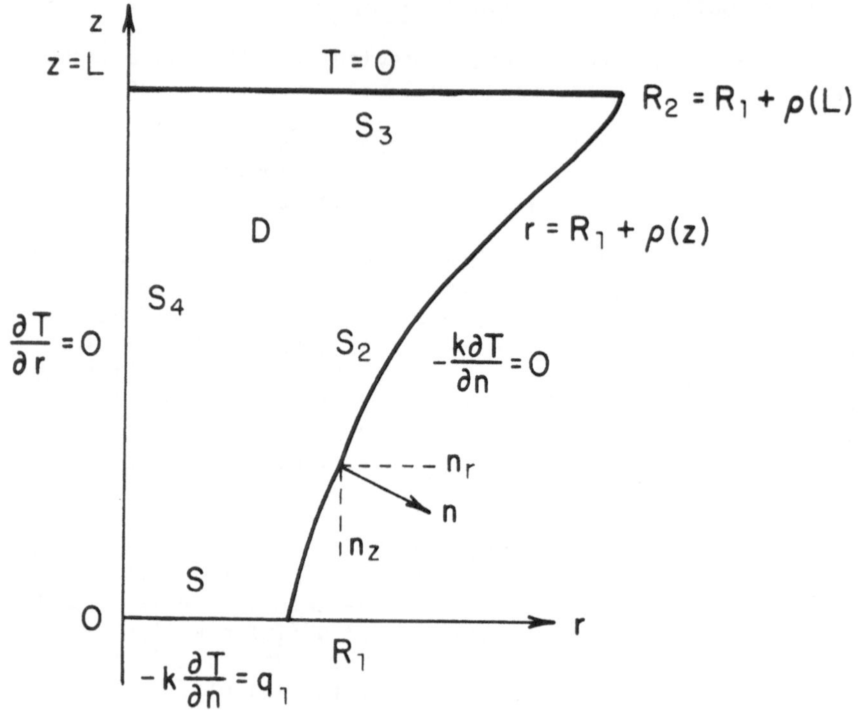

Figure 2.1 Shape of Diffuser

The non-homogeneous Neumann type boundary condition of Eq. 2.2 fixes the given heat thermal flux q_1 at the mounting surface. The boundary S_2 is adiabatic and the thermal gradient is therefore orthogonal to the outward normal, n, the components of which are $(n_r = 1/\sqrt{1 + (\frac{d\rho}{dz})^2}$ and $n_z = -\frac{d\rho}{dz}/\sqrt{1 + (\frac{d\rho}{dz})^2}$.
On S_3, the heatpipes keep the temperature at a constant value, which can be set to zero for convenience. The condition of Eq. 2.5 is the classical axisymmetrical relation on the z-axis.

The design objective is to minimize the weight of the diffuser, which is directly proportional to its volume. Hence, one can introduce the cost function

$$S(D) = 2\pi \int_0^L \int_0^{R_1 + \rho(z)} r \; dr \; dz \tag{2.6}$$

to be minimized under the constraint

$$\max_{\substack{r \\ 0 < r < R_2}} \; - k \frac{\partial T}{\partial n}(r,L) \leq q_2 > 0 \tag{2.7}$$

which forces the outward thermal flux at the boundary S_3 to be less than the maximum value q_2 that can be accepted by the heatpipes.

3. OPTIMIZATION PROBLEM

In order to obtain a formulation of the problem of Eqs. 2.1 to 2.7 that is more amenable to a mathematical treatment, a new set of variables is introduced, as follows:

$$s = \frac{z}{L}, \quad 0 \leq s \leq 1 \tag{3.1}$$

$$Y(r,s) = T(r,Ls)/L \tag{3.2}$$

$$\omega(s) = \rho(Ls)$$

where

$$\omega \in C^0(]0,1[\; , \;]-R_1 + \infty[) \quad ,$$

$$\omega(0) = 0 \tag{3.3}$$

$$\Omega = \{(r,s): \;\; 0 < s < 1, \;\; 0 < r < R_1 + \omega(s)\} \tag{3.4}$$

$$\partial\Omega = \Gamma_1 \cup \Gamma_2 \cup \Gamma_3 \cup \Gamma_4 \tag{3.5}$$

$$\Gamma_1 = \{(r,s): \;\; 0 < r < R_1, \;\; s = 0\} \tag{3.6}$$

$$\Gamma_2 = \{(r,s): \;\; r = R_1 + \omega(s), \;\; 0 < s < 1\} \tag{3.7}$$

$$\Gamma_3 = \{(r,s): \;\; 0 < r < R_1 + \omega(1), \;\; s = 1\} \tag{3.8}$$

$$\Gamma_4 = \{(r,s): \;\; r = 0, \;\; 0 < s < 1\} \tag{3.9}$$

$$q = - q_1/k > 0 \tag{3.10}$$

$$C = q_2/k > 0 \tag{3.11}$$

The original problem of Eqs. 2.1 to 2.7 then becomes (see Fig. 3.1): Find the number L and the real continuous function $\omega(s)$, $s \in [0,1]$, $\omega(0) = 0$, that minimize the cost function $J(L,\omega)$

$$J(L,\omega) = L \int_0^1 (R_1 + \omega(s))^2 \, ds \tag{3.12}$$

subject to the constraint

$$- \frac{\partial Y}{\partial n} (r,1) \leq C, \quad 0 < r < R_1 + \omega(1) \tag{3.13}$$

where Y is the solution of the boundary-value problem

$$\frac{L^2}{r} \frac{\partial}{\partial r} \left(r \frac{\partial Y}{\partial r}\right) + \frac{\partial^2 Y}{\partial s^2} = 0, \quad \text{in } \Omega \tag{3.14}$$

$$\frac{\partial Y}{\partial n} = q, \quad \text{on } \Gamma_1 \qquad \text{where } n = (0, -1) \tag{3.15}$$

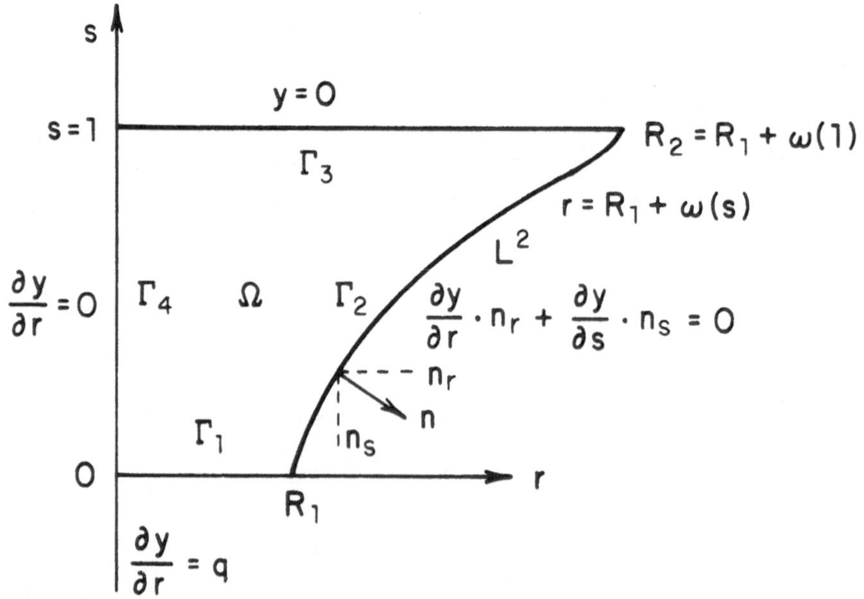

Figure 3.1 Domain of Reduced Problem

$$L^2 \frac{\partial Y}{\partial r} n_r + \frac{\partial Y}{\partial s} n_s = 0 \text{ on } \Gamma_2 \tag{3.16}$$

where

$$n_r = 1/\sqrt{1 + (\frac{d\omega}{ds})^2} \quad \text{and} \quad n_s = -\frac{d\omega}{ds}/\sqrt{1 + (\frac{d\omega}{ds})^2}$$

$$Y = 0 , \qquad \text{on } \Gamma_3 \tag{3.17}$$

$$\frac{\partial Y}{\partial r} = 0 , \qquad \text{on } \Gamma_4 \tag{3.18}$$

4. EULERIAN SEMI-DERIVATIVES FOR THE DOMAIN FUNCTIONALS

For physical reasons, the mooring boundary Γ_2 is sought as the graph of a function ω. This is a constraint on the geometry. In this case, the general material derivative (or speed) method presented in Ref. 2 is reduced to the choice of a particular vector speed $V = (0,v)$ (see Ref. 1).

This section is devoted to obtaining the gradients (see Ref. 2, Sections 1 and 2) of the cost function and the constraint, with respect to the parameter L and the function ω. The knowledge of these quantities will permit the use of well known techniques of constrained optimization.

The gradient of J is trivially given by the partial derivative with respect to scalar parameter L

$$\frac{\partial}{\partial L} J(L,\omega) = \int_0^1 (R_1 + \omega(s))^2 \, ds \tag{4.1}$$

and the Eulerian semi-derivative in the direction of the field $V = (0,v)$:

$$dJ(L,\Omega;V) = 2L \int_0^1 (R_1 + \omega(s)) \, ds \tag{4.2}$$

Note that Eq. 4.2 is not the Hadamard's formula (see Ref. 1) since v is not the normal component of the field V.

The gradient of the constraint is more difficult to obtain. First, define

$$f(L,\omega) = \int_{\Gamma_3} \left[\frac{\partial Y}{\partial n}(r,1) + C\right]^- dr \tag{4.3}$$

where Y is the solution of Eqs. 3.14 to 3.18 and $u^- = \sup(-u,0)$. An equivalent formulation of the constraint of Eq. 3.13 is then

$$f(L,\omega) \leq 0 \tag{4.4}$$

From Ref. 3, Section 3, one knows that $f(L,\Omega)$ is not differentiable on the domain Ω, but the Eulerian semi-derivative $df(L,\Omega;V)$ can be computed, as in Ref. 2, Section 9 (example 12). This semi-derivative involves the material derivative of Y. Turn now to the derivatives of Y, with respect to the parameter L and then the domain Ω.

In a second step, one can use the variational form of the boundary-value problem of Eqs. 3.14 to 3.18. To do this, introduce

$$V(\Omega) = \{\psi(r,s): \ \psi, \frac{\partial\psi}{\partial r}, \frac{\partial\psi}{\partial s} \in L_r^2(\Omega), \ \psi\big|_{\Gamma_3} = 0\} \tag{4.5}$$

where

$$L_r^2(\Omega) = \{\psi(r,s): \int_\Omega |\psi|^2 r dr ds < +\infty\} \tag{4.6}$$

and

$$a_L(y,\psi) = \int_\Omega \{L^2 \frac{\partial y}{\partial r}\frac{\partial\psi}{\partial r} + \frac{\partial y}{\partial s}\frac{\partial\psi}{\partial s}\} \, r \, dr \, ds,$$

$$\text{for all } y, \psi \in V(\Omega) \tag{4.7}$$

$$\ell(\psi) = \int_{\Gamma_1} q\,\psi\, r\, d\gamma, \quad \text{for all } \psi \in V(\Omega) \tag{4.8}$$

The solution Y of Eqs. 3.14 to 3.18 is the unique solution of the variational problem of finding $Y \in V(\Omega)$ such that

$$a_L(Y,\psi) = \ell(\psi), \quad \text{for all } \psi \in V(\Omega) \tag{4.9}$$

Conversely, a solution Y of Eq. 4.9 is a weak solution of Eqs. 3.14 to 3.18 (i.e. a solution in the distribution sense), a classical solution if it is sufficiently regular.

Define $\dot{Y}_L$ as the limit, in the Sobolev space $V(\Omega)$ defined by Eq. 4.5, of the usual differential quotient

$$\dot{Y}_1 = \lim_{t\to 0}\left[\frac{Y_{L+t} - Y_L}{t}\right] \tag{4.10}$$

where Y_{L+t} and Y_L are solutions of Eq. 4.9, with a_{L+t} and a_L, respectively. Then

$$\frac{1}{t}\left[a_{L+t}(Y_{L+t},\psi) - a_L(Y_L,\psi)\right]$$

$$= \frac{1}{t}\left[a_{L+t}(Y_{L+t},\psi) - a_{L+t}(Y_L,\psi) + a_{L+t}(Y_L,\psi) - a_L(Y_L,\psi)\right]$$

$$= \int_\Omega \left\{(L+t)^2 \frac{\partial}{\partial r}\left[\frac{Y_{L+t} - Y_L}{t}\right]\frac{\partial\psi}{\partial r} + \frac{\partial}{\partial r}\left[\frac{Y_{L+t} - Y_L}{t}\right]\frac{\partial\psi}{\partial s}\right\} rdrds$$

$$+ \int_\Omega \frac{(L+t)^2 - L^2}{t}\,\frac{\partial Y_L}{\partial r}\,\frac{\partial\psi}{\partial r}\, rdrds, \quad \text{for all } \psi \in V(\Omega)$$

Letting $t \to 0$, one obtains a characterisation of $\dot{Y}_L \in V(\Omega)$., as the solution of variational problem

$$a_L(\dot{Y}_L,\psi) = -2L \int_\Omega \frac{\partial Y_L}{\partial r}\frac{\partial\psi}{\partial r}\, rdrds \quad \text{for all } \psi \in V(\Omega) \tag{4.11}$$

Now the partial rightside derivative (on the parameter L) of the constraint function f is given by (one directly uses here the result of Ref. 3, Section 3.2)

$$\frac{\partial}{\partial L} f(L,\Omega) = \lim_{t\to 0}\left[\frac{f(L + t,\Omega) - f(L,\Omega)}{t}\right]$$

$$= \lim_{t\to 0}\frac{1}{t}\int_{\Gamma_3} -\left[\frac{\partial Y_{L+t}}{\partial n} + C\right] \chi_{\left[r:\ \frac{\partial Y_{L+t}}{\partial n} < -C\right]} dr \quad \text{(equation con't)}$$

$$- \int_{\Gamma_3} - \left[\frac{\partial Y_L}{\partial n} + C\right] \chi_{\left[r:\ \frac{\partial Y_L}{\partial n} < -C\right]} dr$$

where χ_A is the characteristic function of the set A, i.e.

$$\chi_A(r) = \begin{cases} 1, & \text{if } r \in A \\ 0, & \text{if } r \notin A \end{cases}$$

After some calculation one obtains

$$\frac{\partial}{\partial L} f(L,\Omega) = \int_{\Gamma_3} - \frac{\partial \dot{Y}_L}{\partial n} \chi_{\left[r:\ \frac{\partial Y_L}{\partial n} < -C\right]} dr$$

$$+ \int_{\Gamma_3} \left(\frac{\partial Y}{\partial n} + C\right)^- \chi_{\left[r:\ \frac{\partial \dot{Y}_L}{\partial n} = -C\right]} dr \tag{4.12}$$

Define the adjoint variable P as follows

$$\left.\begin{aligned} & L^2 \frac{1}{r}\frac{\partial}{\partial r}\left(r \frac{\partial P}{\partial r}\right) + \frac{\partial^2 P}{\partial s^2} = 0 \quad \text{in } \Omega \\ & P\big|_{\Gamma_3} = \chi_{\left[r: \frac{\partial Y_L}{\partial n} < -C\right]} \\ & L^2 \frac{\partial P}{\partial r} n_r + \frac{\partial P}{\partial s} n_s = 0 \quad \text{on } \Gamma_1 \cup \Gamma_2 \end{aligned}\right\} \tag{4.13}$$

Then by standard computations it can be shown that

$$f_L' = -2L \int_\Omega \frac{\partial Y}{\partial r} \frac{\partial P}{\partial r} r \, dr \, ds$$

So that

$$\frac{\partial f}{\partial L} = -2L \int_\Omega \frac{\partial Y_L}{\partial r} \frac{\partial P}{\partial r} rdrds \tag{4.14}$$

To obtain the gradient of f with respect to the domain Ω, one uses the domain theory results of Refs. 1 and 2. Given the domain Ω and a velocity field V such that the mapping $x \to V(t,x)$ belongs to $C^2(R^2, R^2)$, one can construct a family of domains Ω_t, $t > 0$ by means of the initial-value problem

$$x'(t) = V(t, x(t)) \qquad x(0) = X \tag{4.15}$$

which defines the mapping

$$T_t: \; X \to x(t)$$

or

$$\Omega_t = T_t(\Omega)$$

Then, if a function k(t) is given by

$$k(t) = \int_{\Omega_t} K(x,t)dx \tag{4.16}$$

the derivative of k is

$$\frac{dk}{dt}(t) = \int_{\Omega_t} \frac{\partial K}{\partial t}(x,t)dx + \int_{\Gamma_t} K(x,t)\,(V\cdot n)d\sigma \tag{4.17}$$

where $\Gamma_t = \partial\Omega_t$ and n is the outward unitary normal to Γ_t.

In order to restrict the family Ω_t to domains of admissible shapes, the velocity field V is chosen of the simple form $V = (V_r, V_s)$ where

$V_s(r,s) = 0$, $(r,s) \in \Omega_t$ (no deformation in the direction s)

$V_s(0,s) = 0$, $0 < s < 1$ (no deformation of the s-axis)

Then, denote by $Y_t \in V(\Omega_t)$ the unique solution of the variational problem

$$\int_{\Omega_t} \left\{ L^2 \frac{\partial Y_t}{\partial r} \frac{\partial \psi}{\partial r} + \frac{\partial Y_t}{\partial s} \frac{\partial \psi}{\partial s} \right\} rdrds = \int_{\Gamma_1} q\, \psi\, rdr,$$

$$\text{for all } \psi \in V(\Omega_t) \tag{4.18}$$

If $\tilde{Y}_t$ is a suitably defined extension of Y_t onto the open set $\{(r,s):\ r \in]0,+\infty[,\ s \in]0,1[\ \}$, the derivative of Y_t with respect to t can be shown to be (no confusion should be made between Y_t' and the material derivative $\dot{Y}$; see Ref. 2, Section 5.1)

$$Y_t' = \lim_{\varepsilon \to 0} \left[\frac{\tilde{Y}(t+\varepsilon) - \tilde{Y}(t)}{\varepsilon} \right] \tag{4.19}$$

(Y_t' is independent on the choice of the regular extension $\tilde{Y}$).

Taking the t-derivative of Eq. 4.18 and using Eq. 4.17, one obtains $Y_t' \in V(\Omega_t)$ as the solution of

$$\int_{\Omega_t} \left\{ L^2 \frac{\partial Y_t'}{\partial r} \frac{\partial \psi}{\partial r} + \frac{\partial Y_t'}{\partial s} \frac{\partial \psi}{\partial s} \right\} rdrds + \int_{\Gamma_t} \left\{ L^2 \frac{\partial Y_t}{\partial r} \frac{\partial \psi}{\partial r} + \frac{\partial Y_t}{\partial s} \frac{\partial \psi}{\partial s} \right\}$$

$$\times\, r(V\cdot n)d\sigma = 0\,, \qquad \text{for all } \psi \in V(\Omega) \tag{4.20}$$

The second term of Eq. 4.20 can be simplified, using the following relations:

$$\Gamma_t = \Gamma_1 \cup \Gamma_2(t) \cup \Gamma_3(t) \cup \Gamma_4$$

$$V = 0, \quad \text{on } \Gamma_4$$

$$(V\cdot n) = 0, \quad \text{on } \Gamma_1 \cup \Gamma_3(t)$$

and on $\Gamma_2(t)$, one has

$$r = R_1 + \omega_t(s)$$

$$(V\cdot n) = V_r n_r = V_r \Big/ \sqrt{1 + \left(\frac{d\omega_t}{ds}\right)^2}$$

$$d\sigma = \sqrt{1 + \left(\frac{d\omega_t}{ds}\right)^2}\; ds$$

Hence, the second term in Eq. 4.20 becomes

$$\int_{\Gamma_t} \left\{ L^2 \frac{\partial Y_t}{\partial r}\frac{\partial \psi}{\partial r} + \frac{\partial Y_t}{\partial s}\frac{\partial \psi}{\partial s} \right\} r(V\cdot n)d\sigma = \int_0^1 \left[L^2 \frac{\partial Y_t}{\partial r}\frac{\partial \psi}{\partial r} + \frac{\partial Y_t}{\partial s}\frac{\partial \psi}{\partial s} \right]_{\Gamma_2}$$

$$\times\ (R_1 + \omega(s))V_r\Big|_{\Gamma_2} ds \tag{4.21}$$

One may now differentiate the constraint f with respect to the domain with the results of Ref. 2, Section 9, and Ref. 3, Section 3.1, one gets

$$f(Y_t) = \int_{\Gamma_3} -\left[\frac{\partial Y_t}{\partial n} + C\right] \chi_{\left[r:\ \frac{\partial Y_t}{\partial n} < -C\right]} dr$$

to obtain

$$\begin{aligned} df(L,\Omega_t;V) = {} & \int_{\Gamma_3} -\frac{\partial Y_t'}{\partial n}\ \chi_{\left[r:\ \frac{\partial Y_t}{\partial n} < -C\right]} dr \\ & + \int_{\Gamma_3} \left[\frac{\partial Y_t'}{\partial n}\right] \chi_{r:\left[\frac{\partial Y_t}{\partial n} = -C\right]} dr \\ & + \left[\frac{\partial Y_t}{\partial n}(R_1 + \omega_t(1),\ 1) + C\right]^- V_r(R_1 + \omega_t(1),1) \end{aligned} \tag{4.22}$$

The second term is equal to zero, since the set $\left[r:\ \frac{\partial Y_t}{\partial n} = -C\right]$ is of measure zero. It is assumed in the following that the third term is also equal to zero, i.e. the constraint is always satisfied on the end of the boundary Γ_3,

$$\frac{\partial Y}{\partial n}(R_1 + \omega_t(1),1) > -C$$

Then, Eq. 4.22 becomes

$$df(L,\Omega;V) = \int_{\Gamma_3} - \frac{\partial Y_t'}{\partial n} \chi \left[r: \frac{\partial Y_t}{\partial n} < -C \right] dr \Big|_{t=0} \tag{4.23}$$

One now obtains the gradient of f with respect to the velocity V at the initial domain Ω, that is for t = 0. By techniques analogous to the ones used for the computation of f_L', it can be shown that

$$\dot{f}_t(Y_t)\Big|_{t=0} = - \int_0^1 \left(L^2 \frac{\partial Y}{\partial r}\frac{\partial P}{\partial r} + \frac{\partial Y}{\partial s}\frac{\partial P}{\partial s} \right)\Big|_{\Gamma_2} (R_1 + \omega(s))V_r\Big|_{\Gamma_2} ds$$

But on Γ_2, $V_r\Big|_{t=0} = \delta\omega$. So finally

$$df(L,\Omega;V) = - \int_0^1 \left(L^2 \frac{\partial Y}{\partial r}\frac{\partial P}{\partial r} + \frac{\partial Y}{\partial s}\frac{\partial P}{\partial s} \right)\Big|_{\Gamma_2} (R_1 + \omega(s))\, v(s)\, ds \tag{4.24}$$

5. PENALTY METHOD OF OPTIMIZATION

In this section, the algorithm for optimization is briefly described. This algorithm is based on the classical penalty formulation, in its simplest form. Some numerical results are presented in the following section. Let $\varepsilon > 0$ be a small number and define the penalized cost function $J_\varepsilon(L,\Omega)$ by

$$J_\varepsilon(L,\Omega) = J(L,\Omega) + \frac{1}{\varepsilon} f(L,\Omega) \tag{5.1}$$

where J is defined by Eq. 3.12 and f by Eq. 4.3.

The original problem of Eqs. 3.12 to 3.18 is then replaced by: Find the number L_ε and the function ω_ε such that

$$J_\varepsilon(L_\varepsilon,\Omega) \leq J_\varepsilon(L,\Omega) \quad \text{for all} \quad L,\Omega \tag{5.2}$$

It can be proved that any limit point of $(L_\varepsilon,\omega_\varepsilon)$ $\varepsilon \to 0$ is a local minimum solution of Eqs. 3.12 to 3.18.

The algorithm proceeds as follows. Starting with a reasonable guess $(L_\varepsilon^0,\omega_\varepsilon^0)$, a minimizing sequence $(L_\varepsilon^n,\omega_\varepsilon^n)$ is constructed by the steepest descent method, as follows:

$$\omega_\varepsilon^{n+1} = \omega_\varepsilon^n - \alpha\, J'_{\varepsilon,\omega}\,(L_\varepsilon^n,\, \omega_\varepsilon^n) \tag{5.3}$$

$$L_\varepsilon^{n+1} = L_\varepsilon^n - \beta\, J'_{\varepsilon,L}\,(L_\varepsilon^n,\, \omega_\varepsilon^{n+1}) \tag{5.4}$$

where

$$J'_{\varepsilon,\omega} = J'_\omega + \frac{1}{\varepsilon} f'_\omega \tag{5.5}$$

$$J'_{\varepsilon,L} = J'_L + \frac{1}{\varepsilon} f'_L \tag{5.6}$$

The parameters α and β are chosen in such a way that

$$J_\varepsilon(L_\varepsilon^n,\, \omega_\varepsilon^{n+1}) < J_\varepsilon(L_\varepsilon^n, \omega_\varepsilon^n)$$

$$J_\varepsilon(L_\varepsilon^{n+1}, \omega_\varepsilon^{n+1}) < J_\varepsilon(L_\varepsilon^n, \omega_\varepsilon^{n+1})$$

If $||\omega_\varepsilon^{n+1} - \omega_\varepsilon^n|| + |L_\varepsilon^{n+1} - L_\varepsilon^n| > \varepsilon_1$ (given), go to Eq. 5.3. Otherwise, stop.

6. NUMERICAL EXAMPLE

In order to illustrate the behavior of the previously described method, a concrete example is given. The numerical values of the physical parameters have been assumed to be the following:

$R_1 = 0.63$ cm (mounting surface = 1.25 cm^2)

$q_1 = 40\ W/cm^2$

$k = 2\ W/(cm \cdot s \cdot {}^\circ C)$

$q_2 = 4\ W/cm^2$

The variational problem of Eq. 4.9, which gives the state function Y, is solved by the finite element method. The triangulation of the domain Ω is obtained by dividing the quadrilaterals of a rectangular grid, stretched in the r-direction to fit the boundary Γ_2 (c.f. Fig. 6.1). The number of triangular

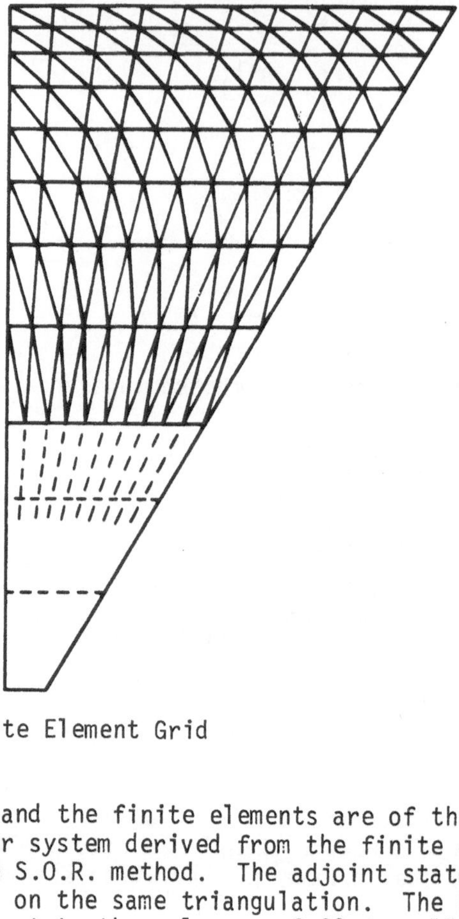

Figure 6.1 Finite Element Grid

elements is 220 and the finite elements are of the so-called P_1 type. The linear system derived from the finite element method is solved by the S.O.R. method. The adjoint state P is computed in the same way, on the same triangulation. The penalization parameter ε was set to the value $\varepsilon = 0.02$ for all the computations presented below.

The algorithm of Section 5 has been used with four different initial domains, one case for each domain (see Fig. 6.2, dashed lines). In two of these cases (case I and case II), the initial domain was such that the constraint was always satisfied at the point ($s = 1$, $r = R_2$) of Γ_3, during all iterations. However, for case III and case IV, the normal heat flux exceeds the allowable value on the whole boundary Γ_3 at the first iteration. Dropping the term in Eq. 4.22 may be the cause of the abnormal shape of the limit domain.

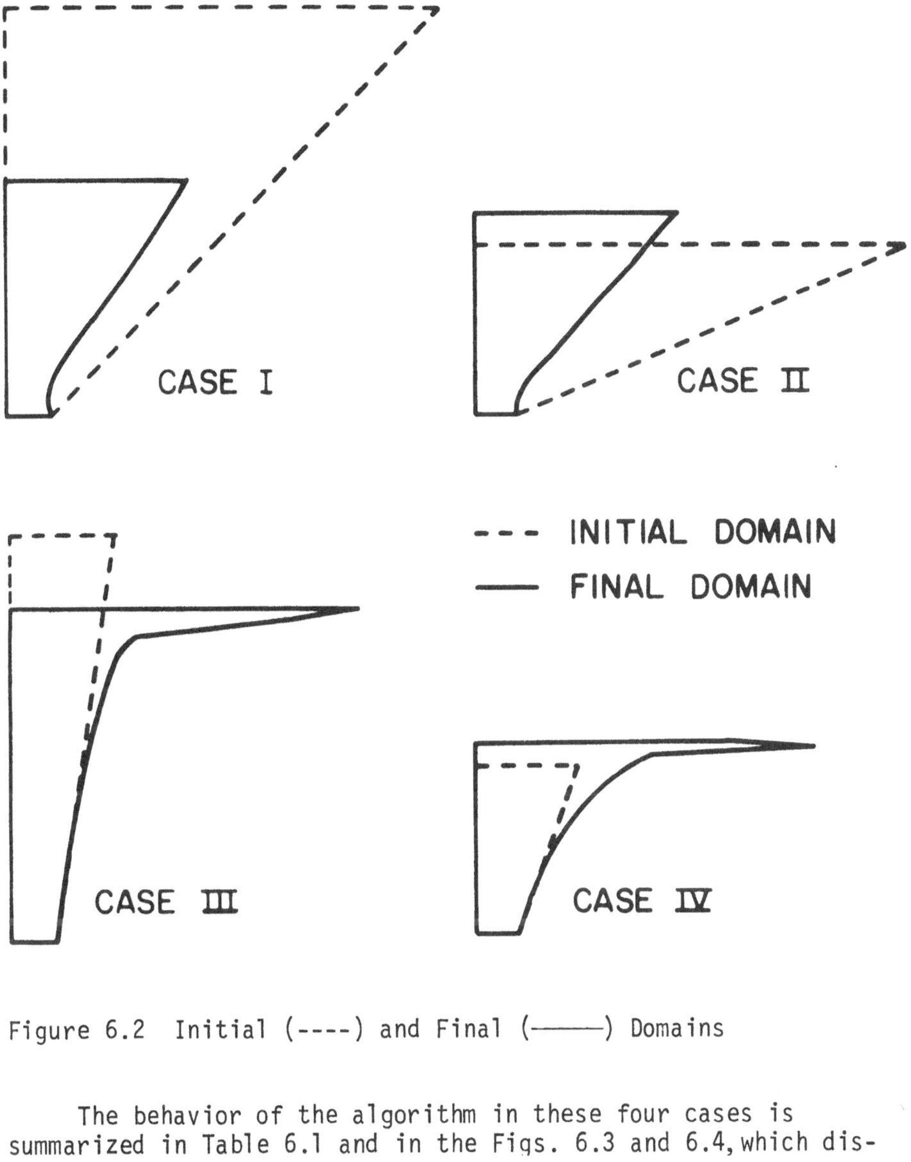

Figure 6.2 Initial (----) and Final (———) Domains

The behavior of the algorithm in these four cases is summarized in Table 6.1 and in the Figs. 6.3 and 6.4, which display the value of the cost $J(L^p,\omega^p)$ and the penalized cost $J_\varepsilon(L^p,\omega^p)$ versus the iteration index p.

TABLE 6.1 NUMERICAL RESULTS

	CASE I	CASE II	CASE III	CASE IV
J^o	97.06	40.44	7.40	3.08
J^o_ε	97.06	40.49	128.5	138.0
L^o	6.0	2.5	6.0	2.5
$\omega^o(1)$	6.0	6.0	0.9	0.9
number of iterations P	3	>18	2	3
J^P	10.00	10.57	10.55	8.63
C.P.U. time CYBER 173	29 s	81 s	24 s	28 s
L^P	3.49	2.98	4.86	2.818
$\omega^P(1)$	2.078	2.43	4.69	3.156
f^P	0.0	0.0	0.0	$0.2\ 10^{-3}$

To conclude the first stage of this study, it may be stated that the approach described herein can give an effective way to convert the optimal design technological problem into the framework of a constrained optimization problem in functional spaces. Moreover, the gradients of the cost and the constraint functionals, with respect to the minimizing parameters, have been derived and permit the use of classical method of optimization. However, the crude algorithm that was used has led to numerical difficulties and work has to be done for selecting a class of minimizing methods that are well suited to this particular problem.

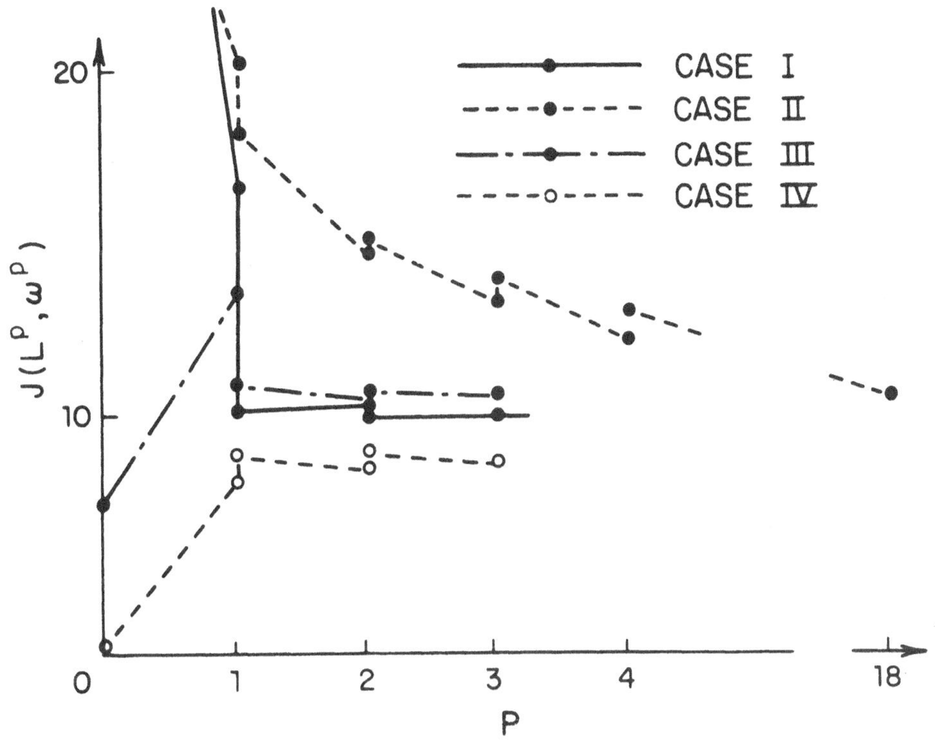

Figure 6.3 Actual Cost vs. Iteration Number

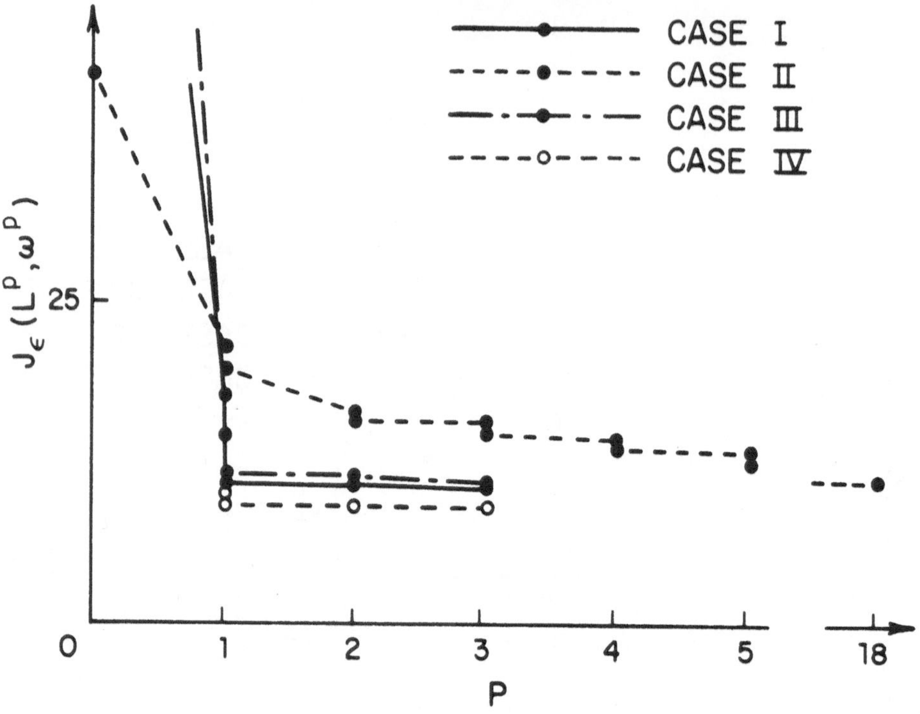

Figure 6.4 Penalized Cost vs. Iteration Number

REFERENCES

1. Cea, J., "Problems of Shape Optimal Design," Optimization of Distributed Parameter Structures (Ed. E.J. Haug and J. Cea), Sijthoff & Nordhoff, Alphen aan den Rijn, 1980.
2. Zolesio, J.P., "The Material Derivative (or Speed) Method for Shape Optimization," Optimization of Distributed Parameter Structures (Ed. E.J. Haug and J. Cea), Sijthoff & Nordhoff, Alphen aan den Rijn, 1980.
3. Zolesio, J.P., "Semi Derivative of Eigenvalues," Optimization of Distributed Parameter Structures (Ed. E.J. Haug and J. Cea), Sijthoff & Nordhoff, Alphen aan den Rijn, 1980.

A GRADIENT PROJECTION METHOD FOR OPTIMIZING SHAPE OF ELASTIC BODIES

Young W. Chun and Edward J. Haug

Department of Mechanical Engineering, Villanova University, Villanova, Pennsylvania

Department of Materials Engineering, University of Iowa, Iowa City, Iowa

ABSTRACT

Shape optimal design of an elastic solid of revolution under multiple constraints is treated. As a specific example, a device that seals a gun bore and transmits high in-bore pressure to shear loading on the projectile, is considered. The design objective is minimum weight, with constraints on stress throughout the body, tractions on one surface of the boundary, and dimensions of the body. Methods of the calculus of variations and functional analysis are used to transform the variation of a functional over a variable region as a functional over a fixed region. An adjoint variable method of operator theory is then used to reduce this variation to an explicit function of only design variations. The resulting sensitivity coefficients are used in an iterative optimization algorithm of a gradient projection method. Numerical results are presented and show that the algorithm is stable and efficient.

1. INTRODUCTION

This paper treats a problem of optimal design of the shape of a solid of revolution, under constraints on stress throughout the body, tractions on one surface of the boundary, and dimensions of the body. The specific body treated is shown in Fig. 1.1. It is a realistic model of the main load-carrying member in a modern round of anti-armor ammunition, called a sabot. A sabot is the device that seals the gun bore and transmits in-bore pressure to shear loading on the projectile, hence accelerating it down the bore. Once the projectile and sabot leave the muzzle, the sabot is separated from the low drag, fin stabilized projectile. The design objective is to minimize weight of the sabot, which must of course be accelerated down the bore with the projectile. Since pressure loading in-bore is high, strength constraints throughout the sabot must be enforced and shear-normal stress relations at the sabot-projectile boundary must be controlled. This design problem is typical of a class of shape optimal design problems encountered in pressure vessel design, dam design, structural element design, and machine element design.

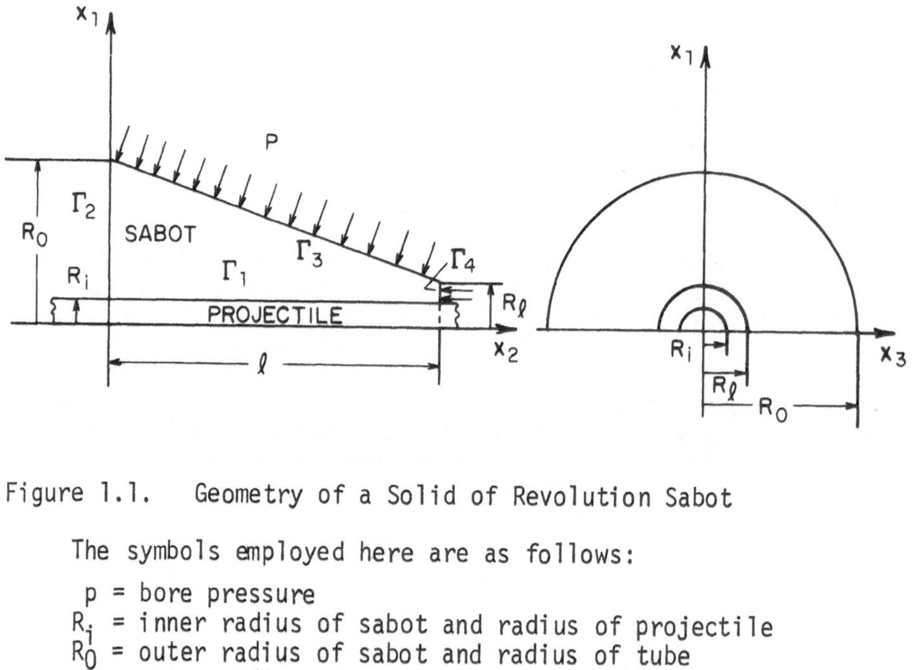

Figure 1.1. Geometry of a Solid of Revolution Sabot

The symbols employed here are as follows:

p = bore pressure
R_i = inner radius of sabot and radius of projectile
R_0 = outer radius of sabot and radius of tube
R = outer radius of sabot along Γ_3 boundary
ℓ = length of sabot.

The domain and the boundary of an axial cut through the body axis are denoted Ω and Γ, respectively. The boundary Γ is composed of four segments; Γ_3 is determined through the optimization process and Γ_1, Γ_2, and Γ_4 are straight. Sabot length ℓ and rear radius R_ℓ are to be determined as design parameters.

2. STATE EQUATIONS OF THE ELASTIC SYSTEM

The displacement field in the elastic sabot is governed by the equations of plane elasticity [1]

$$Kz \equiv - \begin{bmatrix} (\tilde{\lambda} + 2\tilde{\mu}) \left(\dfrac{\partial^2}{\partial x_1^2} + \dfrac{1}{x_1} \dfrac{\partial}{\partial x_1} - \dfrac{1}{x_1^2} \right) + \tilde{\mu} \dfrac{\partial^2}{\partial x_2^2}, & (\tilde{\lambda}+\tilde{u}) \dfrac{\partial^2}{\partial x_1 \partial x_2} \\ (\tilde{\lambda} + \tilde{\mu}) \left(\dfrac{\partial^2}{\partial x_1 \partial x_2} + \dfrac{1}{x_1} \dfrac{\partial}{\partial x_2} \right), & (\tilde{\lambda}+2\tilde{u}) \dfrac{\partial^2}{\partial x_2^2} + \tilde{\mu} \left(\dfrac{\partial^2}{\partial x_1^2} + \dfrac{1}{x_1} \dfrac{\partial}{\partial x_1} \right) \end{bmatrix} \begin{Bmatrix} z_1 \\ z_2 \end{Bmatrix} = \begin{Bmatrix} Q_1 \\ Q_2 \end{Bmatrix} \qquad (2.1)$$

where $\tilde{\lambda}$ and $\tilde{\mu}$ are Lame's constants [1], z_1 and z_2 are displacements in the x_1 and x_2 directions, and Q_1 and Q_2 are body forces in the x_1 and x_2 directions. The boundary conditions are

$$\left. \begin{array}{ll} z_1 = z_2 = 0 & \text{on } \Gamma_1 \\ T_1 = T_2 = 0 & \text{on } \Gamma_2 \\ T_1 = -pn_1, \text{ and } T_2 = -pn_2 & \text{on } \Gamma_3 \text{ and } \Gamma_4 \end{array} \right\} \qquad (2.2)$$

where T_1 and T_2 are components of surface traction T in the x_1 and x_2 directions, and n_1 and n_2 are direction cosines of the outward normal n to Γ. In operator form, Eqs. 2.1 and 2.2 are

$$Kz = Q \quad \text{in } \Omega \qquad (2.3)$$

$$Bz = q \quad \text{on } \Gamma \qquad (2.4)$$

It can be shown that the operator K with boundary operator B is symmetric [2].

Stresses and strains in the elastic body are denoted as σ and ε, each with four components, $\sigma = [\sigma_1, \sigma_2, \sigma_3, \sigma_4]^T$ and

$\varepsilon = [\varepsilon_1, \varepsilon_2, \varepsilon_3, \varepsilon_4]^T$. Stresses in terms of strains and displacements are

$$\sigma = [E]\varepsilon = [D]z \tag{2.5}$$

where

$$[E] = \begin{bmatrix} \tilde{\lambda}+2\tilde{\mu} & \tilde{\lambda} & \tilde{\lambda} & 0 \\ \tilde{\lambda} & \tilde{\lambda}+2\tilde{\mu} & \tilde{\lambda} & 0 \\ \tilde{\lambda} & \tilde{\lambda} & \tilde{\lambda}+2\tilde{\mu} & 0 \\ 0 & 0 & 0 & \tilde{\mu} \end{bmatrix} \tag{2.6}$$

$$[D] = \begin{bmatrix} (\tilde{\lambda}+2\tilde{\mu})\frac{\partial}{\partial x_1} + \tilde{\lambda}\frac{1}{x_1} & \tilde{\lambda}\frac{\partial}{\partial x_2} \\ (\tilde{\lambda}+2\tilde{\mu})\frac{1}{x_1} + \tilde{\lambda}\frac{\partial}{\partial x_1} & \tilde{\lambda}\frac{\partial}{\partial x_2} \\ \tilde{\lambda}\left(\frac{\partial}{\partial x_1} + \frac{1}{x_1}\right) & (\tilde{\lambda}+2\tilde{\mu})\frac{\partial}{\partial x_2} \\ \tilde{\mu}\frac{\partial}{\partial x_2} & \tilde{\mu}\frac{\partial}{\partial x_1} \end{bmatrix} \tag{2.7}$$

The surface traction vector $T = [T_1, T_2]^T$ can be expressed in terms of stresses and the unit normal $n = [n_1, n_2]^T$ as

$$T = [\bar{\sigma}]n \tag{2.8}$$

where $[\bar{\sigma}]$ is the matrix

$$[\bar{\sigma}] = \begin{bmatrix} \sigma_1 & \sigma_4 \\ \sigma_4 & \sigma_3 \end{bmatrix} \tag{2.9}$$

3. FORMULATION OF THE SHAPE OPTIMAL DESIGN PROBLEM

Sabot length ℓ is a design parameter and the slope v along Γ_3 is defined as a design variable over the boundary Γ_3. The surface Γ_3 is defined by the additional initial value state equations

$$\frac{dR}{dx_2} = v \quad , \qquad R(0) = R_0 \tag{3.1}$$

Let the volume (equivalently weight) of material in the sabot be the cost functional that is to be minimized,

$$\psi_0 \equiv \int_0^{\ell} \int_{R_i}^{R(x_2)} 2\pi x_1 dx_1 dx_2 = \pi \int_0^{\ell} (R^2 - R_i^2) dx_2 \tag{3.2}$$

The design problem is to find ℓ and v to minimize ψ_0, subject to the conditions:

(1) <u>Yield Stress Constraint.</u> Von-Mises yield stress criterion is employed, in the form

$$\phi_1 = \sigma_1^2 + \sigma_2^2 + \sigma_3^2 - \sigma_1\sigma_2 - \sigma_2\sigma_3 - \sigma_3\sigma_1 + 3\sigma_4^2 - \sigma_y^2$$

$$\leq 0, \; x \in \Omega \tag{3.3}$$

where σ_y is the yield stress of the material. Equation 3.3 can be written in equivalent functional form [2,3] as

$$\psi_1 \equiv 2\pi \int_0^{\ell} \int_{R_i}^{R(x_2)} \phi_1^2 (1 + \operatorname{sgn} \phi_1)^2 x_1 dx_1 dx_2 = 0 \tag{3.4}$$

where sgn ϕ = 1 if $\phi > 0$, 0 if $\phi = 0$, and -1 if $\phi < 0$.

(2) <u>Shear Force Distribution along Γ_1.</u> Along the interface Γ_1 between the sabot and projectile, shear force is required to be less than or equal to the maximum sustainable friction force. That is,

$$\phi_2 \equiv |\sigma_4| - k\,|\sigma_1| \leq 0 \quad , \qquad x \in \Gamma_1 \tag{3.5}$$

where k is the coefficient of friction along Γ_1. In functional form, Eq. 3.5 is

$$\psi_2 \equiv 2\pi R_i \int_0^{\ell} \phi_2^2(1 + \text{sgn}\ \phi_2)^2\, dx_2 = 0 \tag{3.6}$$

(3) Lower Bound on R_ℓ. From a manufacturing point of view, a lower bound R_{min} is allowed for the sabot radius at the rear, R_ℓ. Integrating Eq. 3.1, this constraint may be written as

$$\psi_3 \equiv R_{min} - R_0 - \int_0^{\ell} v dx_2 \le 0 \tag{3.7}$$

(4) Design Variable Constraint. A bound is placed on the slope of the curve Γ_3,

$$\omega \equiv v \le v_{max} \quad , \qquad x \in \Gamma_3 \tag{3.8}$$

4. DESIGN SENSITIVITY ANALYSIS OF A GENERAL SHAPE OPTIMAL DESIGN PROBLEM

The state equations 2.3 and 2.4 may be stated in the general form of

$$K(u,b)z = Q(x,u,b) \quad , \qquad x \in \Omega \tag{4.1}$$

$$B(v,b)z = q(x,v,b) \quad , \qquad x \in \Gamma \tag{4.2}$$

where x is the independent variable, which ranges over a domain Ω in R^2 with boundary Γ. The vector $u(x) = [u_1(x),\ldots,u_m(x)]^T$ is a distributed design variable over the domain Ω, $v(x) = [v_1(x),\ldots, v_s(x)]^T$ is a design variable over the boundary Γ (often determining location of the boundary), and $b = [b_1,\ldots,b_\alpha]^T$ is a vector of design parameters.

The optimal design problem of Eqs. 3.2, 3.4, 3.6, 3.7, and 3.8 may also be stated in the general state space formulation [3] as follows:

Find u, v, and b to minimize the functional

$$\psi_0 = g_0(b) + \int_\Gamma h_0(x,z,v,b)d\Gamma + \int_\Omega f_0(x,z,u,b)d\Omega \tag{4.3}$$

subject to the constraints

$$\psi_\alpha = g_\alpha(b) + \int_\Gamma h_\alpha(x,z,v,b)d\Gamma + \int_\Omega f_\alpha(x,z,u,b)d\Omega$$

$$\begin{cases} = 0 \ , & \alpha = 1,\ldots,r' \\ \leq 0 & \alpha = r'+1,\ldots,r \end{cases} \tag{4.4}$$

$$\phi_i(x,u) \begin{cases} = 0, \ i=1,\ldots,q' \\ \leq 0, \ i=q'+1,\ldots,q \end{cases} \quad x \in \Omega \tag{4.5}$$

$$\omega_j(x,v) \begin{cases} = 0, \ j=1,\ldots,s' \\ \leq 0, \ j=s'+1,\ldots,s \end{cases} \quad x \in \Gamma \tag{4.6}$$

The most important step in solving a shape optimal design problem numerically is to obtain explicit formulas for variations of the cost functional ψ_0 and constraint functionals ψ_α due to small variations in u, v, b, and the shape of Ω and Γ. Consider a general form of functional

$$\psi = g(b) + \int_\Gamma h(x,z,\sigma,v,b)d\Gamma + \int_\Omega f(x,z,\sigma,u,b)d\Omega \tag{4.7}$$

which incorporates stress constraints, as well as the cost functional and displacement constraints.

The first variation $\delta\psi$, which is derived in Ref. 4 in detail and analyzed theoretically in Refs. 5, 6, and 7, is given as

$$\delta\psi = \frac{\partial g}{\partial b}\,\delta b + \int_\Gamma \left[\left(\frac{\partial h}{\partial z} + T^*\right)\delta z + \frac{\partial h}{\partial \sigma}\,\delta\sigma + \left(\frac{\partial h}{\partial v} - \frac{\partial c^*}{\partial v}\right)\delta v \right.$$

$$\left. + \left(\frac{\partial h}{\partial b} - \frac{\partial c^*}{\partial b}\right)\delta b\right] d\Gamma + \int_\Omega \left[\left(\frac{\partial f}{\partial u} - \frac{\partial A}{\partial u}\right)\delta u \right.$$

$$\left. + \left(\frac{\partial f}{\partial b} - \frac{\partial A}{\partial b}\right)\delta b\right] d\Omega - \int_\Gamma c(\lambda,\delta z)d\Gamma$$

(Equation continued on next page)

$$- \int_{\Omega} \frac{\partial[\{A(\lambda,z)-\lambda^T Q\}\delta x_i]}{\partial x_i} d\Omega - \int_{\Gamma} \frac{\partial[c^*(\lambda,z)\delta x_i]}{\partial x_i} d\Gamma$$

$$+ \int_{\Gamma} \frac{\partial(h\delta x_i)}{\partial x_i} d\Gamma + \int_{\Omega} \frac{\partial(f\delta x_i)}{\partial x_i} d\Omega \tag{4.8}$$

where T*, c*, and A and the related operator D* are determined by the following integrations by parts:

$$\int_{\Omega} \frac{\partial f}{\partial \sigma} \delta\sigma \, d\Omega = \int_{\Omega} \frac{\partial f}{\partial \sigma} D\delta z \, d\Omega = \int_{\Omega} D^* \left(\frac{\partial f}{\partial \sigma}\right) \delta z \, d\Omega + \int_{\Gamma} T^* \delta z \, d\Gamma \tag{4.9}$$

$$\int_{\Omega} \lambda^T K \delta z \, d\Omega = \int_{\Omega} A(\lambda,\delta z) d\Omega + \int_{\Gamma} c^*(\lambda,\delta z) d\Gamma \tag{4.10}$$

where the highest order of derivatives of the adjoint variable λ and δz that appear in $A(\lambda,\delta z)$ is the same. A term $c(\lambda,\delta z)$ is further defined as

$$c(\lambda,\delta z) = c^*(\lambda,\delta z) + c_2(\lambda,\delta z) \tag{4.11}$$

where $c_2(\lambda,\delta z)$ is obtained by integrating $A(\lambda,\delta z)$ by parts once again to obtain

$$\int_{\Omega} A(\lambda,\delta z) d\Omega = \int_{\Omega} \delta z^T K \lambda \, d\Omega + \int_{\Gamma} c_2(\lambda,\delta z) d\Gamma \tag{4.12}$$

since K is symmetric.

In deriving $\delta\psi$ in Eq. 4.8, the changes in shape of Ω and Γ are described by the transformed variable x*, defined as

$$x_i^* = x_i + \delta x_i + O(\varepsilon) \tag{4.13}$$

where δx_i is the principal linear parts of a Taylor expansion of x_i^* and $O(\varepsilon)/\varepsilon \to 0$ as $\varepsilon \to 0$. The form of δx_i and its relation to design variables is selected to conform to the specific domain under consideration.

The adjoint variable λ is defined as a solution of the adjoint operator equation [2,3,4]

$$K\lambda = \frac{\partial f}{\partial z} + D^* \left(\frac{\partial f}{\partial \sigma}\right) \tag{4.14}$$

The boundary conditions for λ that are associated with Eq. 4.14 can be determined by demanding that all the coefficients of the remaining independent component of δz and $\delta\sigma$ in the boundary integral of Eq. 4.8 be zero. The boundary variations, however, must satisfy the first variation of Eq. 4.2

$$B\delta z + \frac{\partial(Bz)}{\partial v}\,\delta v + \frac{\partial(Bz)}{\partial b}\,\delta b = \frac{\partial q}{\partial v}\,\delta v + \frac{\partial q}{\partial b}\,\delta b \tag{4.15}$$

Assuming that all this calculation has been completed and $\lambda(x)$ determined for each ψ_α, Eq. 4.8 may be written for ψ_α, $\alpha = 0,1,\ldots,r$,

$$\delta\psi_\alpha = \ell^{\alpha^T}\,\delta b + \int_\Gamma \pi^{\alpha^T}\,\delta v\;d\Gamma + \int_\Omega \Lambda^{\alpha^T}\,\delta u\;d\Omega \tag{4.16}$$

where ℓ^α, π^α, and Λ^α are the coefficients of δb, δv, and δu in Eq. 4.8 after solution of Eq. 4.14 for λ^α associated with ψ_α and substitution.

Completing the perturbation calculation, one has from Eqs. 4.5 and 4.6

$$\delta\phi_i = \frac{\partial\phi_i}{\partial u}\,\delta u \tag{4.17}$$

$$\delta\omega_j = \frac{\partial\omega_j}{\partial v}\,\delta v \tag{4.18}$$

Now one can employ a numerical optimization method to iteratively solve the problem. The method employed here is the generalized gradient projection method of Ref. 3.

5. NUMERICAL RESULTS AND DISCUSSION

Numerical results presented in this section are based on the following data: $\ell = 4$ in., $R_0 = 1.5$ in., $R_i = 0.4$ in., $R_{min} = 0.6$ in., $P = 55{,}000$ psi, $k = 0.6$, and $v_{max} = 0$. The material is taken to be Aluminum, with material properties $E = 10^7$ psi, $\nu = 0.33$, $\sigma_y = 1.2 \times 10^5$ psi, and $\rho = 0.1$ lbm/in^3.

Inertial force on the accelerating sabot is included in the formulation and calculations. Acceleration input for the inertial force was calculated using approximate equations of interior ballistics, with the following data: W_p (weight of projectile excluding sabot) = 4.5 lbs., V_0 (muzzle velocity) = 5,500 ft/sec, and U_0 (tube length) = 150 in. The weight of the sabot W_s was obtained at each iteration and the sum of W_p and W_s was used in calculating the acceleration.

The transformed variables $x_i{*}$ of Eq. 4.13 selected to represent shape of the sabot are

$$x_1{*} = x_1 + \delta x_1 = x_1 + \frac{x_1 - R_i}{R - R_i}\,\delta R \tag{5.1}$$

$$x_2{*} = x_2 + \delta x_2 = x_2 + \frac{x_2}{\ell}\,\delta\ell \tag{5.2}$$

It may be noted that $\delta x_1 = 0$ when $x_1 = R_i$ (along Γ_1), $\delta x_1 = \delta R$ when $x_1 = R$ (along Γ_3), $\delta x_2 = 0$ when $x_2 = 0$ (along Γ_2), and $\delta x_2 = \delta\ell$ when $x_2 = \ell$ (along Γ_4).

The initial shape of the boundary Γ_3 was taken as a straight line with $R_\ell = 0.6$ in., as shown in Fig. 5.1. Triangular ring elements were used in the finite element analysis, with 44 grid points and 60 elements. With this initial shape, no constraints were violated. The cost function was initially $\psi_0 = 12.69$ in^3.

After 21 iterations, ψ_0 was reduced to 10.16 in^3, all constraints were active, and convergence criteria were satisfied [3]. The length ℓ was shortened to 3.84 in. and the shape of the final design is as shown by the solid curve in Fig. 5.2. The initial and final values of height and slope along Γ_3, are given in Table 5.1. The computing time was 4.4 sec. per iteration on an IBM 360/65.

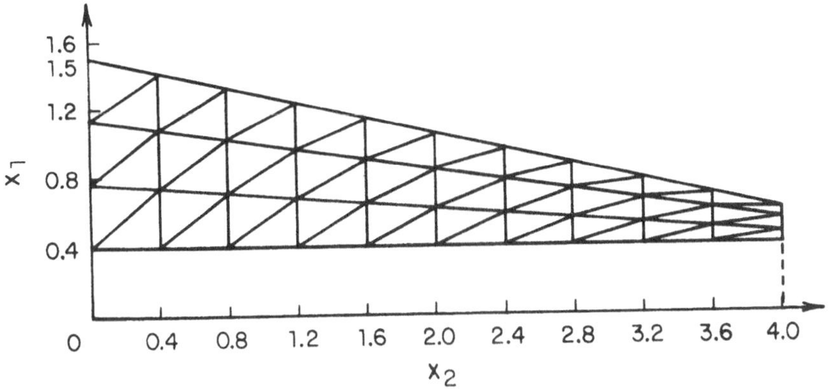

Figure 5.1. Initial Design and Descritization

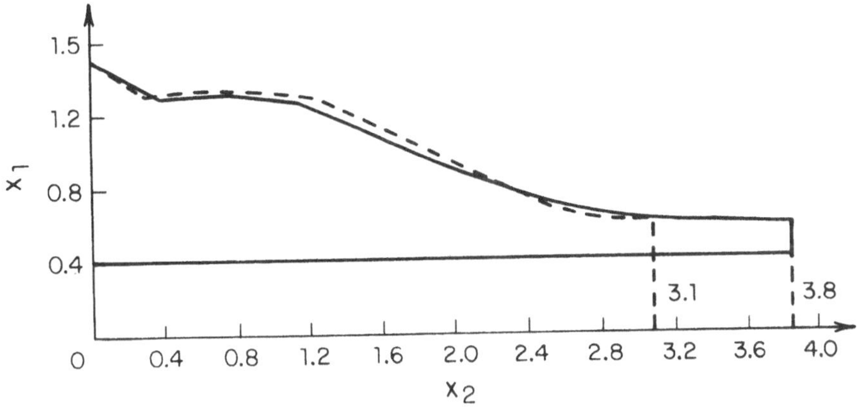

Figure 5.2. Final Designs ($v_{max} = 0$)

If one cuts off the tail portion from the optimum design, to obtain an infeasible design, and continues the calculation, the optimization algorithm converges to the design shown by the dotted curve in Fig. 5.2. The cost function for this design is $\psi_0 = 10.11 \text{ in}^3$. One may notice that material was added in the middle section of the sabot, instead of restoring it to the tail. This shows that cost is less sensitive to the length of the sabot than to the shape of the boundary Γ_3, thus enabling the designer to choose an appropriate length from other considerations.

TABLE 5.1. INITIAL AND FINAL DESIGNS, FOR INITIALLY STRAIGHT Γ_3

Initial	x_2-coord. (in.)	0.000	0.400	0.800	1.200	1.600	2.000	2.400	2.800	3.200	3.600	4.000
	height (in.)	1.500	1.410	1.320	1.230	1.140	1.050	0.960	0.870	0.780	0.690	0.600
	slope	-0.225	-0.225	-0.225	-0.225	-0.225	-0.225	-0.225	-0.225	-0.225	-0.225	-0.225
Final	x_2-coord. (in.)	0.000	0.384	0.768	1.153	1.539	1.921	2.305	2.690	3.074	3.458	3.842
	height (in.)	1.500	1.296	1.334	1.264	1.083	0.912	0.774	0.662	0.606	0.597	0.599
	slope	-1.057	-0.004	-0.008	-0.396	-0.478	-0.388	-0.343	-0.223	-0.066	0.000	0.000

TABLE 5.2. INITIAL AND FINAL DESIGNS FOR INITIALLY CONCAVE Γ_3.

Initial	x_2-coord. (in.)	0.000	0.400	0.800	1.200	1.600	2.000	2.400	2.800	3.200	3.600	4.000
	height (in.)	1.500	1.328	1.176	1.041	0.924	0.825	0.744	0.681	0.636	0.609	0.600
	slope	-0.450	-0.405	-0.360	-0.315	-0.270	-0.225	-0.180	-0.135	-0.090	-0.045	0.000
Final	x_2-coord. (in.)	0.000	0.388	0.776	1.164	1.552	1.940	2.328	2.716	3.104	3.492	3.880
	height (in.)	1.500	1.291	1.330	1.271	1.097	0.898	0.746	0.625	0.594	0.601	0.600
	slope	-1.069	-0.008	0.000	-0.334	-0.531	-0.441	-0.356	-0.200	-0.004	0.000	0.000

TABLE 5.3. INITIAL AND FINAL DESIGNS FOR INITIALLY CONVEX Γ_3

Initial	x_2-coord.(in.)	0.000	0.400	0.800	1.200	1.600	2.000	2.400	2.800	3.200	3.600	4.000
	height (in.)	1.500	1.491	1.464	1.419	1.356	1.275	1.176	1.059	0.924	0.771	0.600
	slope	0.000	-0.045	-0.090	-0.135	-0.180	-0.225	-0.270	-0.315	-0.360	-0.405	-0.450
Final	x_2-coord.(in.)	0.000	0.379	0.759	1.138	1.517	1.897	2.276	2.655	3.035	3.414	3.793
	height (in.)	1.500	1.298	1.337	1.268	1.091	0.924	0.788	0.675	0.613	0.598	0.599
	slope	-1.064	-0.002	-0.006	-0.391	-0.474	-0.386	-0.346	-0.233	-0.087	-0.004	-0.008

The problem was again solved with the two initial designs shown in Fig. 5.3. For the concave initial design, both ψ_1 and ψ_2 were violated. The optimum design was obtained in 24 iterations. For the starting design with the convex Γ_3 boundary, only ψ_2 was initially violated and it also took 24 iterations to obtain the optimum design. In both cases the optimum designs were nearly identical with the design shown by the solid curve in Fig. 5.2. Tables 5.2 and 5.3 show the initial and final designs for each of these cases.

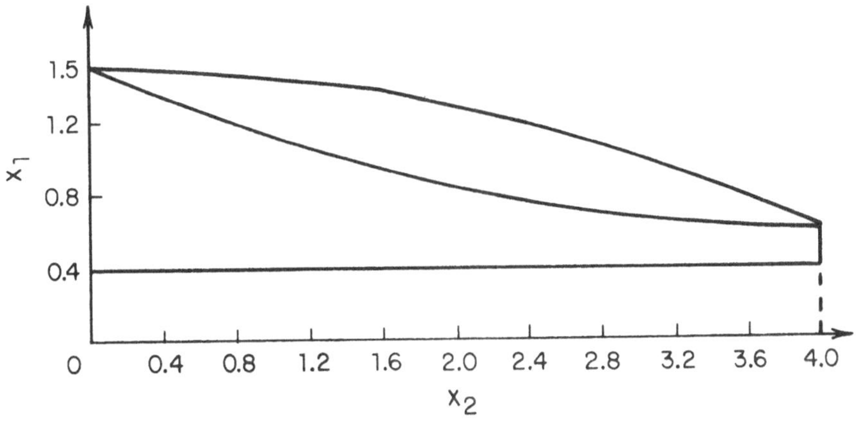

Figure 5.3. Initial Designs with Concave and Convex Γ_3

These numerical results indicate that the design sensitivity analysis is accurate and that the generalized gradient projection algorithm is stable. In view of the results presented, it is felt that there would be no fundamental conceptual or computational difficulties in applying the algorithm to other classes of shape optimal design problems that are governed by the equations of two-dimensional elasticity.

REFERENCES

1. Timoshenko, S.P. and Goodier, J.N., Theory of Elasticity, 3rd Edition, McGraw Hill, New York, 1970.
2. Chun, Y.W. and Haug, E.J., Two-Dimensional Optimal Design, Technical Report No. 38, Division of Materials Engineering, University of Iowa, Iowa City, Iowa, January 1978.

3. Haug, E.J. and Arora,J.S., Applied Optimal Design, Wiley-Interscience, 1979.
4. Chun, Y.W. and Haug, E.J., "Two-Dimensional Shape Optimal Design," International Journal for Numerical Methods in Engineering, Vol. 13, No. 2, 1978, pp. 311-336.
5. Cea, J., "Shape Optimal Design: Problems and Numerical Methods," Optimization of Distributed Parameter Structures (Ed. E.J. Haug and J. Cea), Sijthoff & Nordhoff, Alphen ann den Rihn, Netherlands,1980.
6. Rousselet, B., "Shape Design Sensitivity Methods for Structural Mechanics," Optimization of Distributed Parameter Structures (Ed. E.J. Haug and J. Cea), Sijthoff & Nordhoff, Alphen ann den Rijn, Netherlands, 1980.
7. Rousselet, B. and Haug, E.J., "Design Sensitivity Analysis of Shape Variation," Optimization of Distributed Parameter Structures (Ed. E.J. Haug and J. Cea), Sijthoff & Nordhoff, Alphen ann den Rijn, Netherlands,1980.

EXISTENCE OF OPTIMAL GEOMETRIES FOR A MODEL PROBLEM OF ANTIPLANE STRAIN

Robert V. Kohn

Courant Institute of Mathematics, 251 Mercer Street, New York, N.Y. 10012

ABSTRACT

In the context of plastic limit analysis for a model problem involving antiplane strain of an infinite cylinder, Strang has noted that the limit multiplier has a simple characterization, in terms of the geometry of the cylinder's cross-section. We use this characterization to show that there is an optimal way to remove at most k cavities from the cross-section so as to leave the limit multiplier fixed and remove the maximum possible area.

1. THE PROBLEM

The optimization of structural geometry in the context of plastic limit analysis has been studied extensively. Relatively little attention, however, has been paid to the question of existence of an optimum geometry in any general context. Some work now in progress is described here, concerning such an existence question, in the context of a model problem involving antiplane strain. In order to avoid obscuring the main points with technicalities, a certain amount of mathematical imprecision will be tolerated. Above all, an objective of this paper is to convince the reader that this area is not well understood and that it leads to interesting geometrical questions of a type not previously addressed.

The model problem to be studied is as follows: Let Ω be a domain in $\mathbb{R}^2$, with boundary $\Gamma = \Gamma_0 \ \Gamma_1$, and let $f:\Gamma_1 \to \mathbb{R}$ be a bounded function. Think of Ω as determining an infinite cylinder

$\Omega \times \mathbb{R}$, with part of its boundary $\Gamma_0 \times \mathbb{R}$ clamped and part $\Gamma_1 \times \mathbb{R}$ loaded longitudinally and uniformly by a force (0,0,f). The nonzero stresses are σ_{xz} and σ_{yz}. Suppose that the cylinder is occupied by a perfectly plastic material, with yield criterion $\sigma_{xz}^2 + \sigma_{yz}^2 = 1$. According to the duality theorems of plastic limit analysis, the limit multiplier of this structure and load is

$$\lambda(\Omega) = \inf\{ \int_\Omega |\nabla u|: \quad u:\Omega\to\mathbb{R}, \quad u|_{\Gamma_0} = 0, \quad \int_{\Gamma_1} uf = 1\} \qquad (1.1)$$

Here the functions u represent admissible velocities of failure (0,0,u). This model was introduced, and Eq. 1.1 was proven with mathematical rigor, in Ref. 1. See also Refs. 2 and 3 for related work.

If $H \subset \Omega$, consider the cylinder with section $\Omega_H = \Omega \sim H$. One can think of this as lightening the original cylinder by removing material, so one should require that f = 0 on $H \cap \Gamma_1$ (thus one is not removing any of the loaded boundary), and consider f to be defined on Ω_H by taking f = 0 on H (thus H is neither clamped nor loaded). This smaller cylinder has limit multiplier

$$\lambda(\Omega_H) = \inf\{ \int_{\Omega_H} |\nabla u|: \quad u:\Omega\to\mathbb{R}, \ u|_{\Gamma_0} = 0, \quad \int_{\Gamma_1} uf = 1\}$$

As will become clear later on, it may well happen that $\lambda(\Omega_H) = \lambda(\Omega)$, so the natural optimization question of interest here is this: Is there a set H with maximal area such that $\lambda(\Omega_H) = \lambda(\Omega)$? In other words, is there a maximum amount of material that can be removed from the original cylinder without reducing its strength? If so, of course, one would like to characterize H.

It is not hard to imagine that there might be no optimum H. It could be that perforating Ω with arbitrarily fine holes is better than removing any single set, for example. If there is to be no optimum, this is the kind of behavior that must be ruled out.

2. OPTIMALITY CRITERIA

A sufficient condition for H to be optimal in a problem like this was given by Mroz in 1963 [4]. If there exists $u:\Omega \to \mathbb{R}$ such that

$$\lambda(\Omega) = \lambda(\Omega_H) = \int_{\Omega_H} |\nabla u|$$

$$u|_{\Gamma_0} = 0$$

$$\int_{\Gamma_1} uf = 1$$

and such that for some $c > 0$

$$\left. \begin{array}{ll} |\nabla u| = c, & \text{on } H \\ |\nabla u| \geq c, & \text{on } \Omega_H \\ |\nabla u| \leq c, & \text{on } H \end{array} \right\} \tag{2.1}$$

then H is optimal. The proof is easy. Given any other set H' with $\lambda(\Omega_{H'}) = \lambda(\Omega_H)$, one has

$$\begin{aligned} \lambda(\Omega_{H'}) &\leq \int_{\Omega_{H'}} |\nabla u| \\ &= \int_{\Omega_H} |\nabla u| + \int_{H\sim H'} |\nabla u| - \int_{H'\sim H} |\nabla u| \\ &\leq \lambda(\Omega_H) + c|H\sim H'| - c|H'\sim H| \end{aligned}$$

so that $|H| \geq |H'|$.

However, the Mroz criterion doesn't help one find the optimum H, or prove that one exists. The principal result in this direction is as follows: For each $k > 0$, it can be shown that there is an optimum set H_k within the class of sets having at most k components. Moreover, H_k consists of straight line segments and curves that are, in a certain sense, graphed over Γ_1. The following conjectures concerning the behavior of H_k as $k \to \infty$ is offered:

(1) in some cases, $|H_k|$ continues to increase as $k\to\infty$, so that no optimum exists with finitely many components;
(2) the length of H_k stays bounded as $k\to\infty$; and
(3) an optimum H, possibly with infinitely many components, does indeed exist.

In a paper as short as this, one can not give much in the way of proofs, but at least the main points may be sketched. The fundamental tool is the identification of the failure velocities u that are minimizers in the definition of $\lambda(\Omega)$. As Strang pointed out [1], a minimizing function u can always be found, which is a multiple of the characteristic function of a set, $u = c\chi_E$ for some $c \in \mathbb{R}$ and $E \subset \mathbb{R}^2$, where $\int_\Omega |\nabla \chi_E|$ is interpreted as the arc length of $(\partial E \cap \overline{\Omega})$. Thus

$$\lambda(\Omega_H) = \min\left\{ \frac{\int_{\Omega_H} |\nabla \chi_E|}{\left|\int_{\Gamma_1 \cap E} f\right|} : \; E \subset \mathbb{R}^2, \;\; \Gamma_0 \subset \mathbb{R}^2 \sim E \right\} \tag{2.2}$$

If E is a minimizer for Eq. 2.2, then $\partial E \cap \Omega$ must consist of straight line segments and ∂E must hit Γ_1 in an angle θ, such that $\cos\theta = \pm\lambda f$. Intuitively, one may understand these assertions as saying that the cylinder always fails by shear across a surface, namely across the surface along which the admissible shear stresses can no longer equilibrate the applied load λf. (See Ref. 5 for an analogous treatment of a related problem.) Here, the more geometric terminology of calling E a failure mode for Ω_H, whenever E is a minimizer in Eq. 2.2, is adopted.

3. EXAMPLE

As an example, consider $\Omega = \{(x,y) : 0 < x < 1, \; 0 < y < 1\}$, with

$\Gamma_0 = \{(x,y) : y = 0\}$ and

$f = 1$ on $\partial\Omega \cap \{(x,y) : x = 0\}$

$f = -1$ on $\partial\Omega \cap \{(x,y) : x = 1\}$

$f = 0$ on $\partial\Omega \cap \{(x,y) : y = 1\}$

(see Fig. 3.1). Minimizing sets for Eq. 2.2 are by no means unique, but two examples are $E = \{(x,y) : x \leq 0\}$ and $E = \{(x,y) : x \geq 1\}$. It is clear that all failure modes for Ω lie along the loaded boundary, and that $\lambda(\Omega) = 1$.

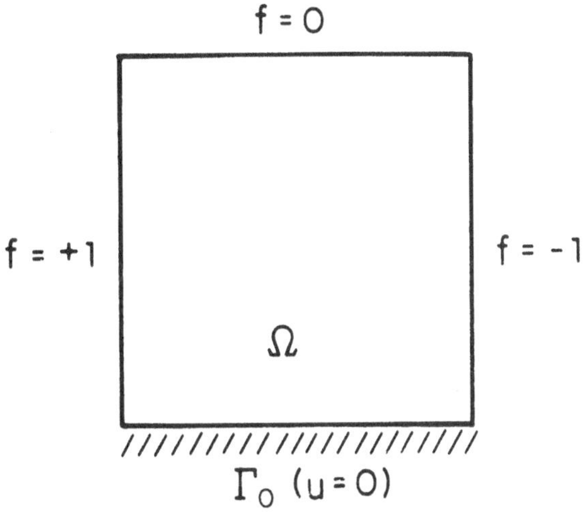

Figure 3.1 The Domain Studied.

When can one remove a set from Ω without reducing λ? The answer is, one can if and only if there are points of Ω through which no failure mode passes. In particular, a necessary condition for H to be optimum is that through each point of Ω_H there should pass at least one failure mode for Ω_H.

This condition is, however, not sufficient. Figure 3.2 gives an example of a choice of H that satisfies the criterion, but is not optimal. In it, each arc of H is on a circle and some examples of failure modes have been drawn. The choice of H in Fig. 3.2 can not be optimal, because one can improve it by filling in the tip of the cusp on the upper cavity and removing material from the edge of the lower one (see Fig. 3.3).

That is the sort of geometrical argument that leads one to conclude that in an optimum H, any two cavities that are joined by a failure mode have straight line segments as the corresponding parts of their boundaries. The passage from this idea to proving the existence of an optimum H with k components requires some argument, but no new geometrical insight.

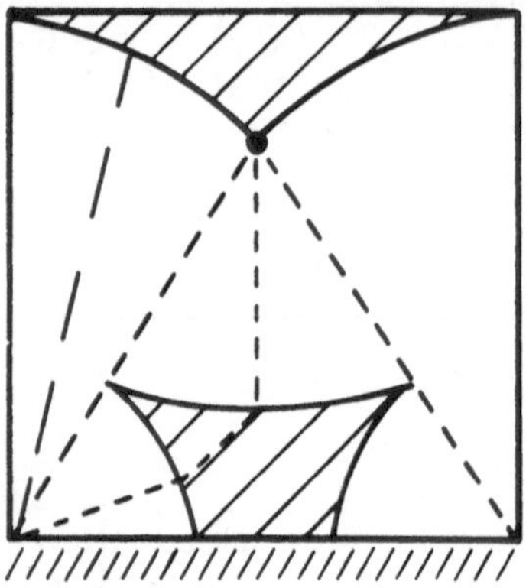

Figure 3.2 A Non-Optimum Domain.

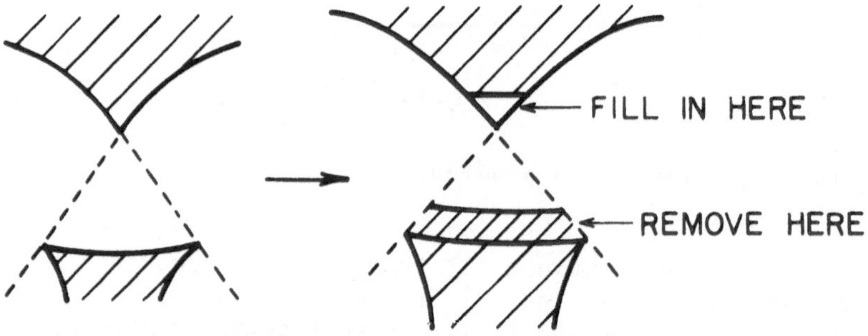

Figure 3.3 An Improved Design.

This example is concluded by showing an optimum way to choose H. It is not hard to show that there is a function u satisfying the Mroz condition for the set H pictured in Fig. 3.4.

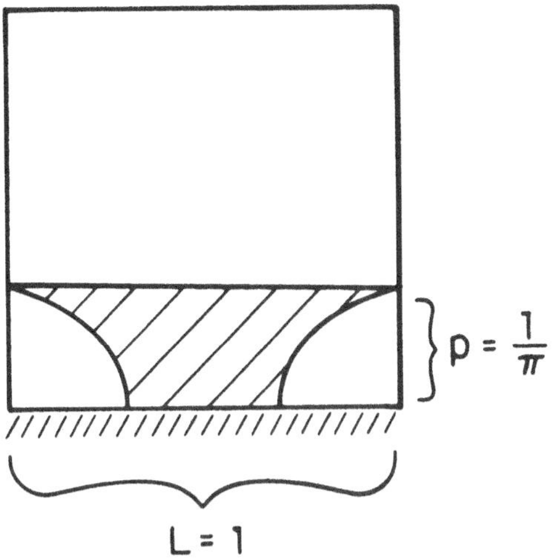

Figure 3.4 The Optimum Domain.

REFERENCES

1. Strang, G., "A Minimax Problem in Plasticity Theory," Functional Analysis Methods in Numerical Analysis (M.Z. Nashed ed.), Springer Lecture Notes in Mathematics no. 701, 1979, p. 319.
2. Matthies, H., and Strang, G., "The Saddle Point of a Differential Program," Energy Methods in Finite Element Analysis (R. Glowinski, E. Rodin, O.C. Zienkiewicz eds.), John Wiley, N.Y., 1979.
3. Teman, R., and Strang, G., "Duality and Relaxation in the Variational Problems of Plasticity," to appear in Journal de Mechanique.
4. Mroz, Z., Archiwun Mech. Stosow., 15, 63, 1963.
5. Keller, J.B., "Plate Failure Under Pressure," SIAM Review Vol. 22, 1980, p. 227.

APPLICATION OF MATHEMATICAL MODELS TO IMPROVE THE MECHANICAL BEHAVIOR OF A LARGE SUPERCONDUCTING TOROIDAL FIELD COIL UNDER MAGNETIC BODY FORCES.*

J. Erb[1] and Herbert Zehlein[2]

1) University of Karlsruhe
Institut für Experimentelle Kernphysik
Karlsruhe, FRG
2) Kernforschungszentrum Karlsruhe
Institut für Reaktorentwicklung
75 Karlsruhe 1, Postfach 3640
Karlsruhe, FRG

ABSTRACT

Some magnetoelastic problems that are relevant for the design of large, D-shaped, superconducting toroidal field magnets for future nuclear fusion reactors are discussed. The first requirement is the definition of an optimum shape of the magnet contour, leading to constant tension distribution in the structure. An iterative numerical procedure to find the ideal contour, referring to a given geometric arrangement of torus coils, is described. The real design of the EURATOM-coil, contributed to the Large Coil Task (LCT), is based on this optimum shape. Moreover structural analyses to study the actual design, regarding its optimality and safety margins, were performed. It was shown that real support conditions and manufacturing tolerances of the coil and its casing may lead to sensible deviations from the ideal stress distribution, however. In particular, the mechanical interaction between coil and casing for different basic operational

*The work reported here is based on a contract between Kernforschungszentrum Karlsruhe, and Max-Planck-Institut für Plasmaphysik, Garching, concerning the cooperation in the field of superconductivity. It was supported by EURATOM. The preparation of the results was strongly assisted by Miss Manes and Mr. Messemer.

and failure load cases were parametrically studied. The outcome of these investigations was a quantification of allowable parameter variations, covering possible design alternatives. Rather than trying a direct optimization of the actual coil, the selection of an optimum combination of design parameters is then left to the final stages of design detailing.

The shape optimization of the coil contour falls well within the topic of shape optimal design. The 3D finite element method (FEM) application, however, brought out a combination of design constraints, which cannot be idealized away. Limits must be set on gaps, contact and detachment zones, local peaks of different stress tensor components, etc., occurring under different load cases. The complexity of this situation may arouse the interest of the optimization specialist, because it is a challenge to apply new theories to a practical problem that is of paramount importance in fusion technology.

1. INTRODUCTION

The European Atomic Energy Community (EURATOM), as one of the contracting parties of the International Energy Agency (IEA) Implementing Agreement for the Development of Superconducting Magnets for Fusion Power, will contribute one test coil to the Large Coil Task (LCT), an international project currently building a large experimental facility at Oak Ridge National Laboratory (ORNL), USA. Within EURATOM, the Kernforschungszentrum Karlsruhe (KfK) has the responsibility for this coil, in contractual collaboration with the Max-Planck-Institut für Plasma-Physik (IPP), which acts as a cooperating laboratory associated with EURATOM [1, 2, 3]. The coil will be manufactured by SIEMENS, Erlangen, FRG, with KRUPP, Essen, FRG, as a subcontractor for the production of the casing. The coil design is governed by the philosophy of using, to the utmost extent, well-developed cryogenic materials and established manufacturing techniques that permit the extrapolation to even larger magnets. One of the salient features among the design objectives is the structural integrity [4 to 7] of the coil.

In designing magnets for a Tokamak-like fusion facility, it seems plausible to start structure calculations with a coil shape that provides constant mechanical tension around the circumference of the magnet, for a particular inner filament of the coil. Empirical formulae for such a contour are well known [8 to 12], but it is desirable to have a direct and more general method, when computing a particular magnet [13]. From the requirement that the product of the magnetic field and the radius of curvature, which is proportional to the tension everywhere along the filament, has a constant value along the same filament, a simple differential equation is established for the desired contour.

It is solved numerically, taking into account the actual field of the magnets considered. Since the toroidal magnet, the shape of which has to be modified, contributes to the magnetic field, the procedure is necessarily iterative. If the initial contour of the toroidal coils is suitably chosen, the procedure converges to the shape into which a toroidal system of completely flexible coils would move, under the influence of electromagnetic forces arising only from the current flowing. A FORTRAN program for solving the differential equations mentioned was written and combined with a code for field calculations [14]. Actual parameters of the coil under construction were used. Starting from the particular Princeton-D-shape for the toroidal magnets the program converges, after 6 iterations, very close to the ideal contour.

Using a simplified geometry derived from this ideal shape, preliminary reference design studies were supported by parametric finite element analyses of the coil and casing structures of the LCT-coil. In this work, emphasis was placed on the methodology applied to study the 3-dimensionality influences that are inherent to full-size structures. The effects identified by these investigations must be known for the evaluation of the margins of optimality and structural safety, under the most relevant operational and fault magnetic body forces. This paper concentrates on these magnetic body force load cases. Other design loads (quench pressure loads, combined load cases, shrinkage during cooldown, anisotropy, and prestress effects), which contribute further to the complexity of the design process, are only briefly mentioned.

For the static analyses reported in this paper the ASKA finite element code was used [15]. More detailed FE models are being applied by SIEMENS and KRUPP, utilizing the NASTRAN and ANTRAS codes, respectively.

2. THE LARGE SUPERCONDUCTING TOROIDAL FIELD COIL, UNDER MAGNETIC BODY FORCES

The design of superconducting magnets for fusion experiments requires the solution of specific problems, such as:

(1) computation of a relatively complicated distribution of body forces,
(2) finding the optimum shape of the main field coils, with respect to stress distribution,
(3) mechanical interaction of the current-carrying coil-winding - the origin of the body forces - with the casing into which the winding is embedded. Since the winding cannot be fixed within the casing by welding techniques, contact problems must be solved.

The first point does not cause major difficulties, since the treatment of body forces is straightforward in the available finite element software. The optimum shape of the magnets, however, strongly depends on the specific body force pattern. Therefore these body forces shall be described in more detail later. Point (2) is a real optimization problem with respect to a unique quantity, namely the hoop stress within the coil. The third point is a mixture of three dimensional stress analysis problems, from which a proper selection of design properties can only be derived after a considerable number of parameter variations.

An exemplary treatment of the above problems is given for the EURATOM coil contributed to the international research program, which is currently under way at ORNL, USA under the designation LCT (Large Coil Task). This project is an experiment aimed to demonstrate the feasibility of superconducting magnets for fusion devices of the Tokamak reactor type. It is planned to put 6 superconducting toroidal field coils together in the Large Coil Test Facility (LCTF) at ORNL during the next years. Three of these 6 coils will be delivered by US manufacturers, the other three will come from Japan, Switzerland, and West Germany (EURATOM Coil).

2.1 Load Cases

Torus Geometry. Figure 2.1 shows the principal layout of LCTF, which contains 6 torus coils assembled with hexagonal symmetry around a central bucking post of approximately 1 m diameter. Each coil has a roughly rectangular cross section of 0.46 m x 0.57 m and consists of approximately 600 windings, each winding carrying ca. 11,000 Amps. The coils are shaped like the capital letter D; therefore one often calls them D-coils.

Figure 2.2 shows a side view of the EURATOM coil. For manufacturing reasons, the contour of the inner filament, and consequently of any other filament, of the coil is composed of 3 arcs of different radii. By this simplification of its geometrical description, the contour differs a little from the optimum ideal shape, which will be discussed in this section. The principal designations required for the mathematical description of the coil and torus geometry are indicated in Fig. 2.3.

Magnetic Field. To give an impression of the magnetic field in a relevant toroidal arrangement, Fig. 2.4(a) shows some field lines in the plane $z = 0$. In this figure, the field lines are not symmetrical about the midplane of each magnet. Rather, they are only symmetrical about the midplane of coils 1 and 4

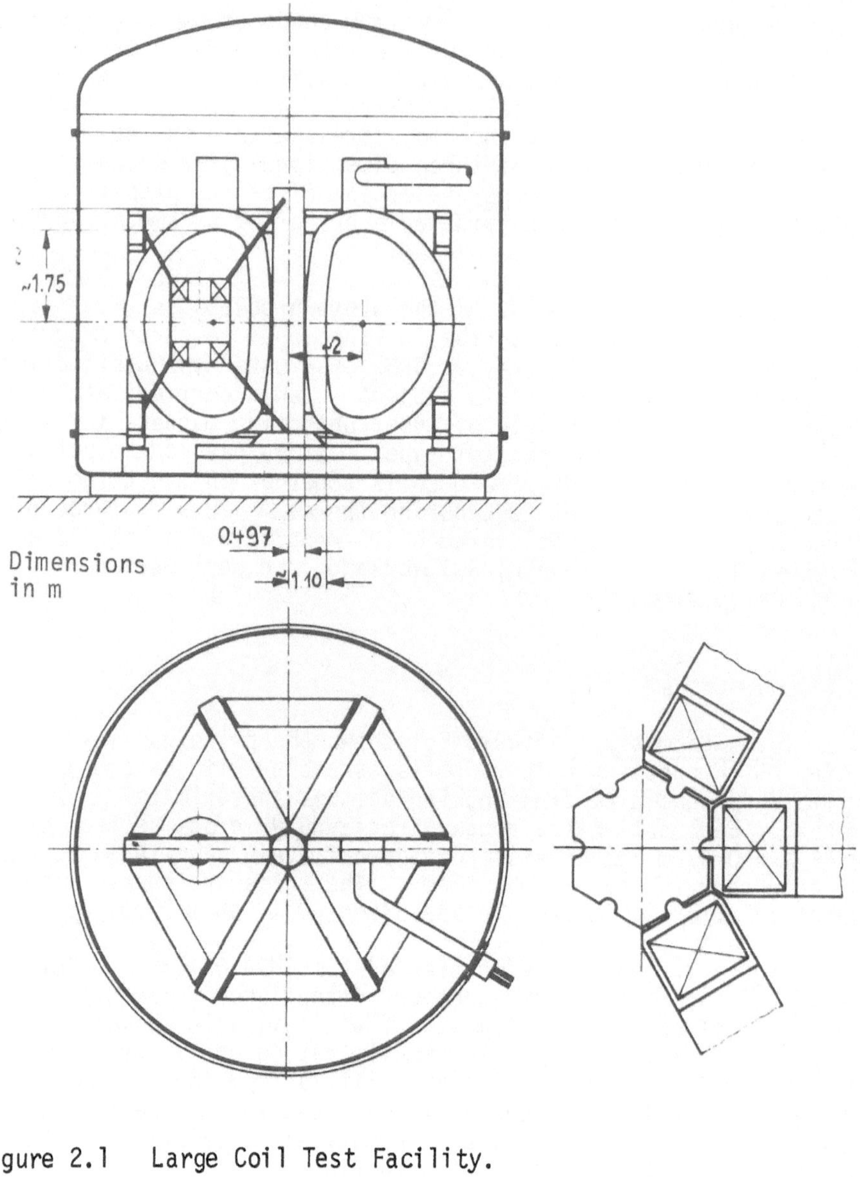

Figure 2.1 Large Coil Test Facility.

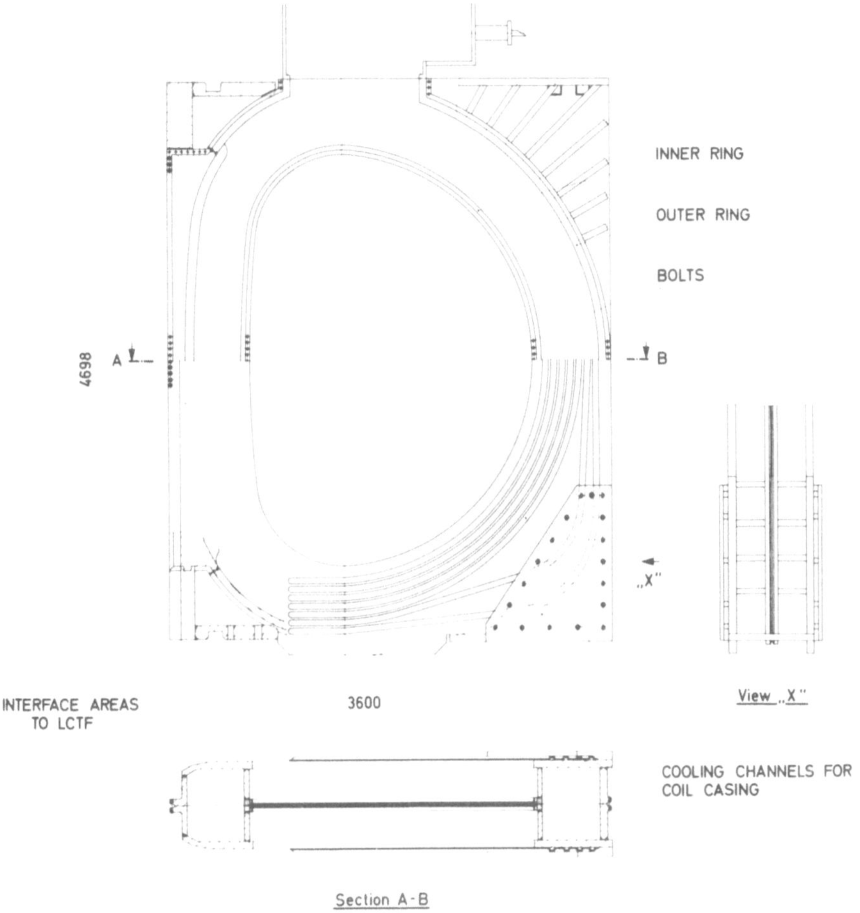

Figure 2.2 Side View Of The KfK/Euratom Reference Design.

(x - axis). This is because coil no. 1 (identified by the arrow in Fig. 2.4(a) carries (in the case presented here, which is not the normal operation mode) a higher current than the 5 other coils. As one can see, the field is weaker in the area between the coils than inside the bore region of the coils (field ripple). The field ripple must not exceed certain limits in a real fusion device. Therefore, in actual fusion reactor designs, the number of torus coils is always higher than 6 (usually 12 to 24 coils). In LCTF however, there will be no plasma, a situation which allows for a smaller number of coils.

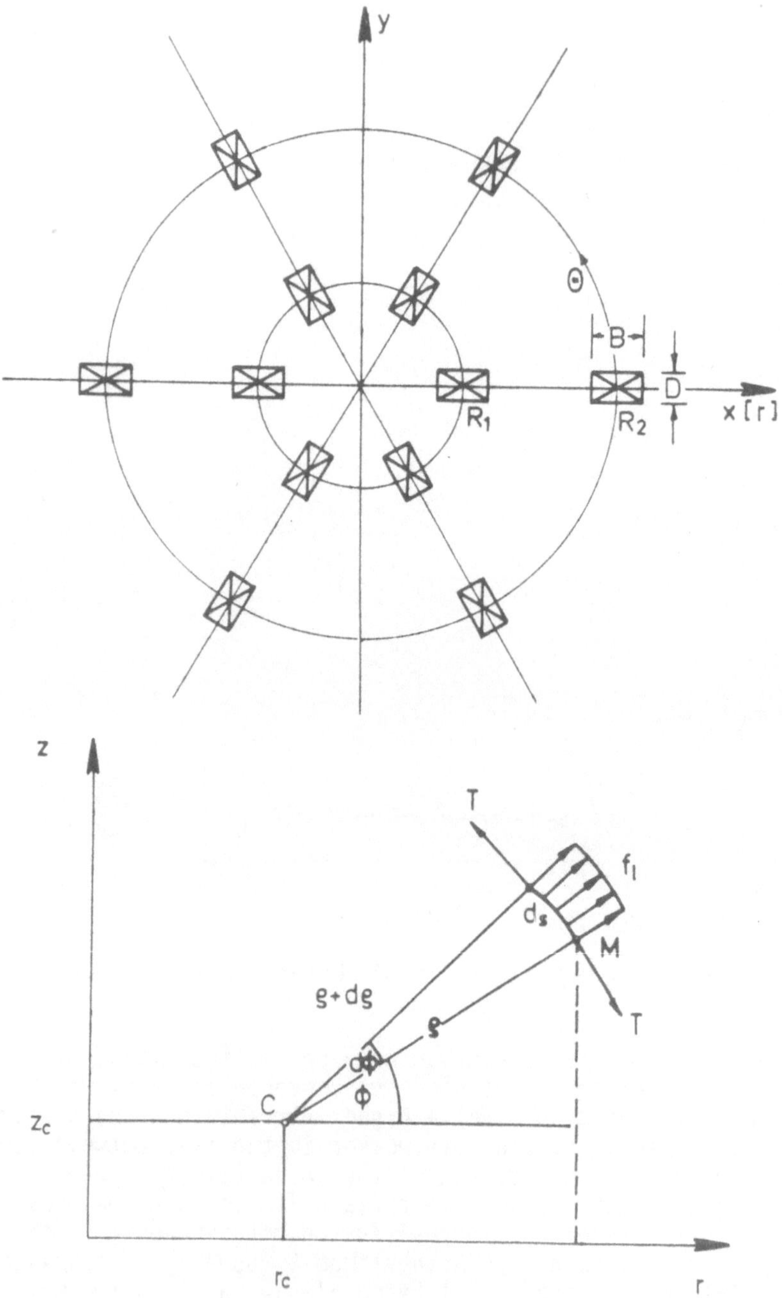

Figure 2.3 Torus Geometry And Conductor Element.

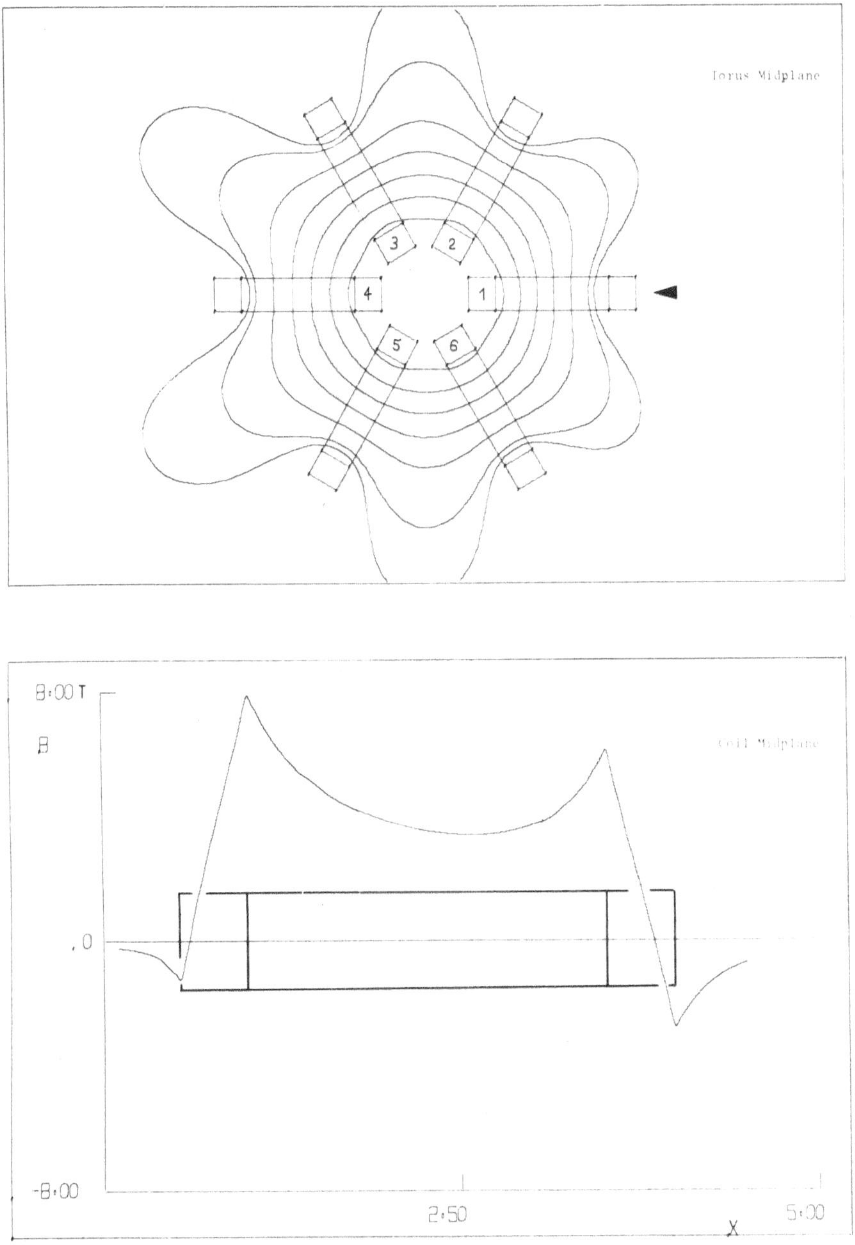

Figure 2.4 Magnetic Field Distribution.

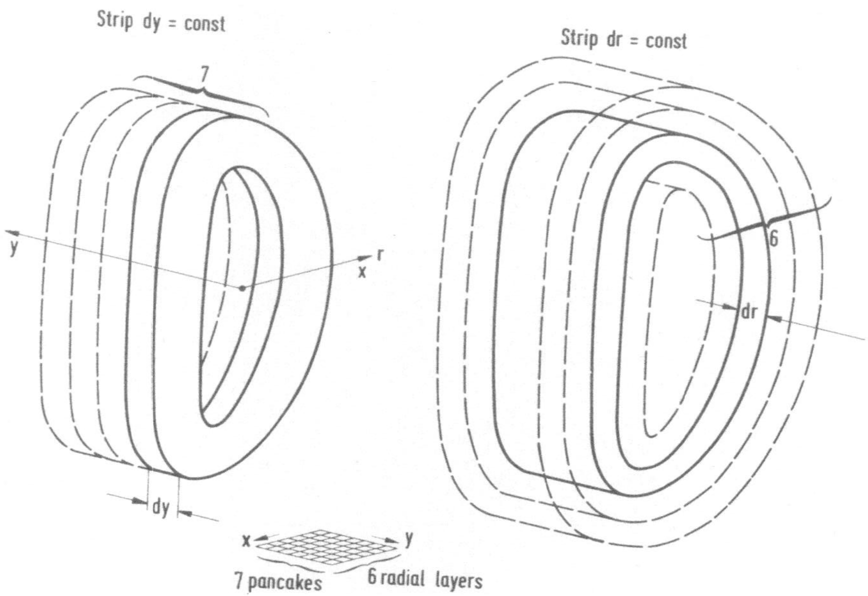

Figure 2.4 (continued)

The critical locations in a superconducting torus magnet, as far as the magnetic field is concerned, are the windings near the surface of the plasma chamber along the inner coil leg, because the magnetic field reaches its peak value there. Figure 2.4(b) shows the y-component (or θ-component, toroidally spoken) of the magnetic induction, as a function of x (torus radial direction) through the midplane of coil no. 1. Within the actual coil, the field varies linearly and reaches its peak value of 8 Tesla (80 kGauss) at the surface of the inner leg (χ = 1.02 m). If ever, things may obviously become critical first in this high field region since,

(1) superconductivity can exist only if field, temperature, and current density do not exceed certain critical values, and
(2) in order to minimize costs, one tries to raise field and current as much as possible.

"Critical" means here that superconductivity is liable to break-down locally, which could cause the whole magnet to become normal-conducting. This instability is called quenching. Of course, with magnets of such large size a quench must be avoided by all means.

Magnetic Body Forces. Magnetic forces arise from the interaction between current and magnetic field. Since the Lorentz force is the vector cross product of current and magnetic induction, the forces acting on a current-carrying filament are always perpendicular to that filament. Their magnitude is proportional to both the field value and the current. Figure 2.5 shows the in-plane force per filament length, for three different filaments. As one can see, inner radius filaments always experience a force directed outward. The force on outer radius filaments may change sign at the top and bottom of the coil (see Fig. 2.5(b)) whereas the outermost filaments always carry a force directed inward. This spatial distribution of the filament forces corresponds to the fact that the main field component, the y-component, changes sign within the coil (see Fig. 2.4(b)).

Figures 2.6 to 2.14 give a rather complete picture of the 3-dimensional distribution of the body forces, for 3 different load cases. These pertain to:

(1) normal operation (symmetric loading) shown in Figs. 2.6 to 2.8,
(2) the working conditions after the accidental fault of a neighbor coil (asymmetrix loading), which tend to turn the coil sideways around the inner leg's vertical nose touching the bucking post, shown in Figs. 2.9 to 2.11, and

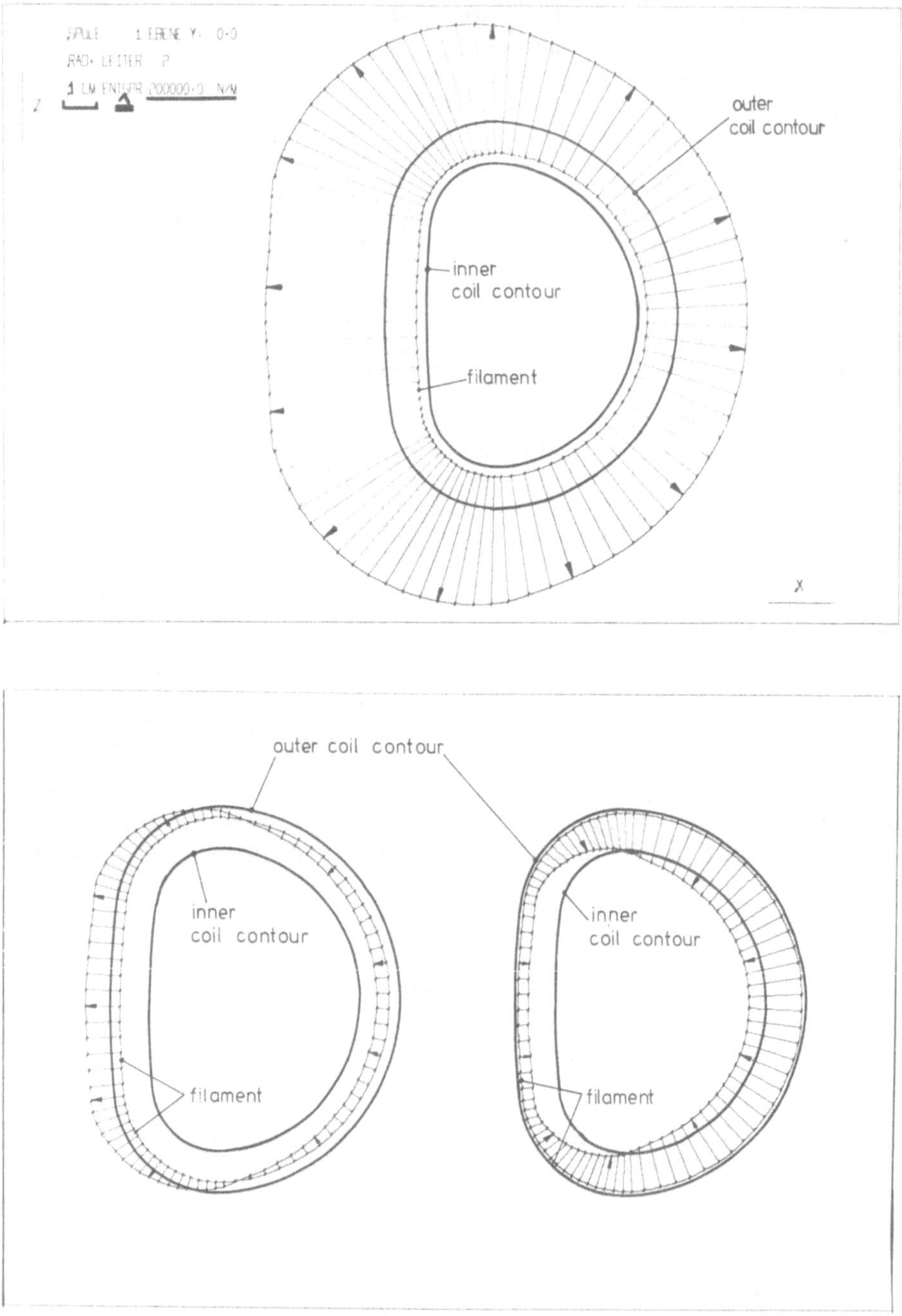

Figure 2.5 Radial Magnetic Filamentary Body Force Component.

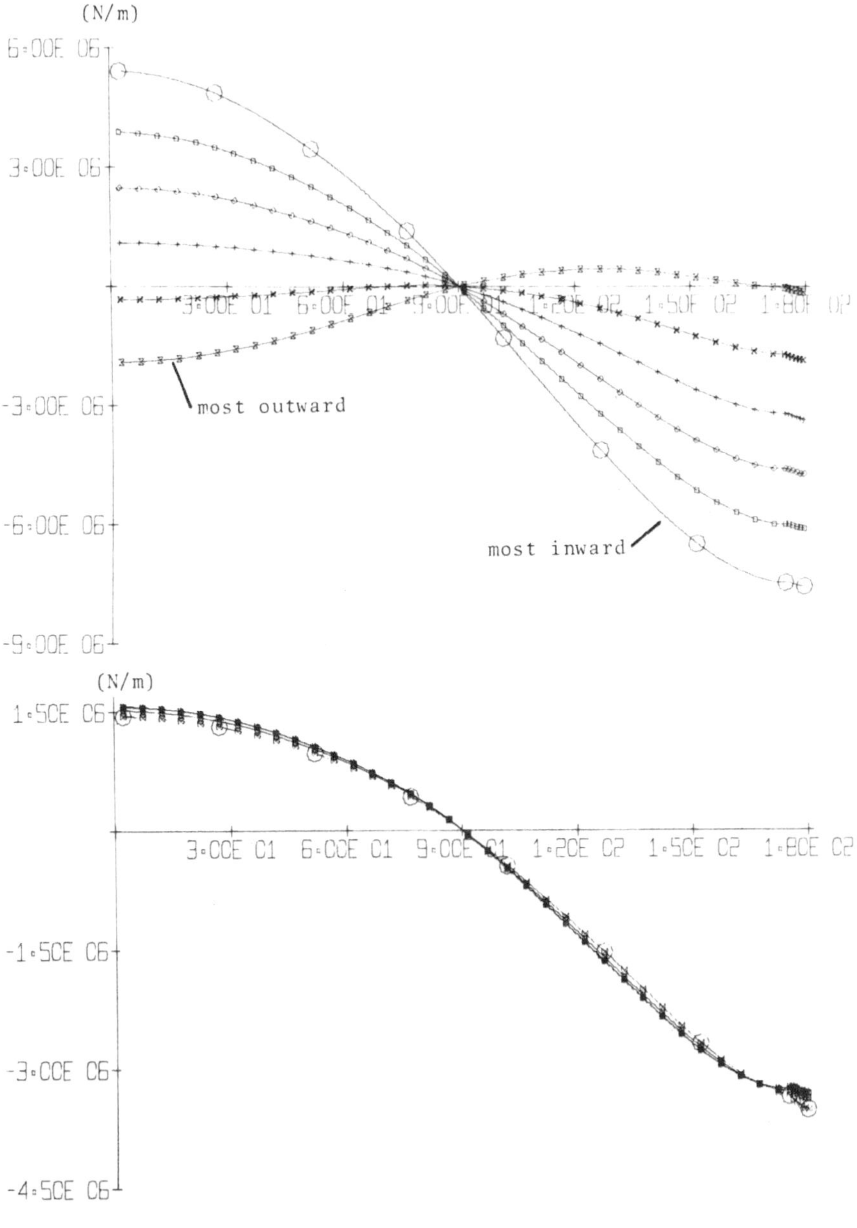

Figure 2.6 3D-Body-Force Distribution: X-Component Of The Symmetric Load Case.

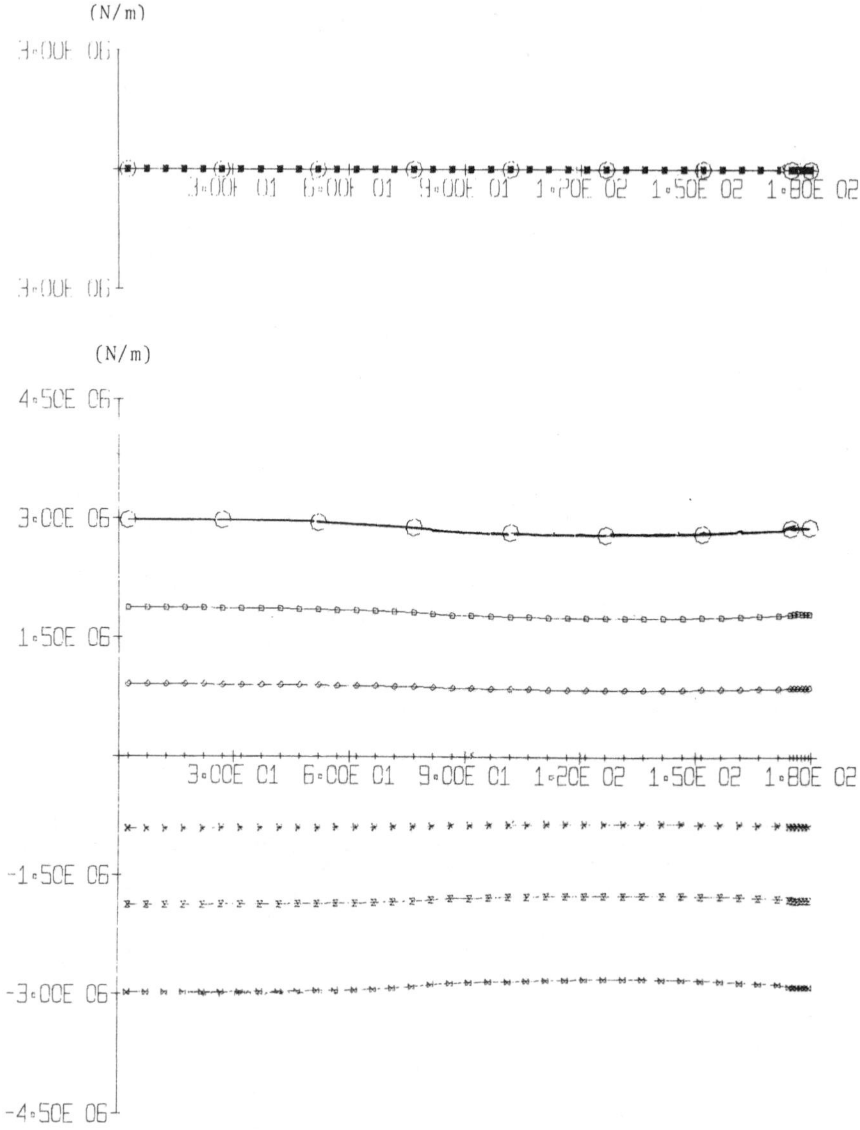

Figure 2.7 3D-Body-Force Distribution: Y-Component Of The Symmetric Load Case.

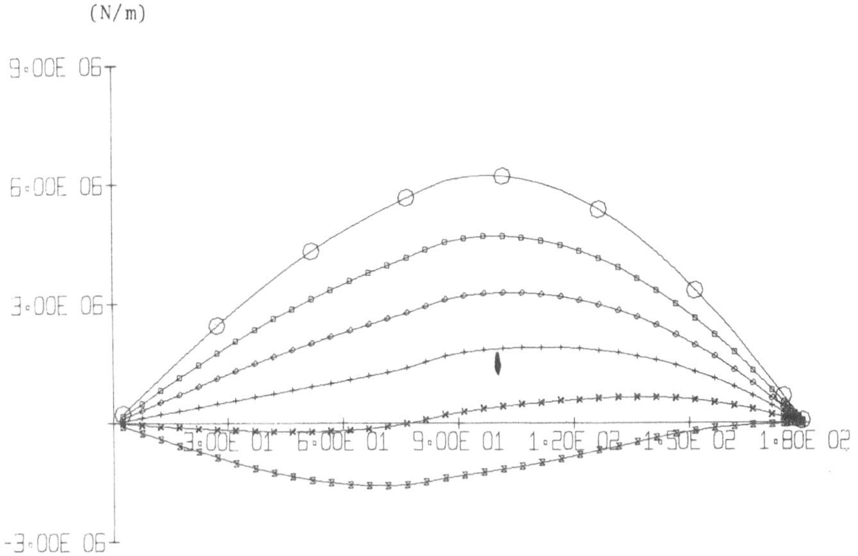

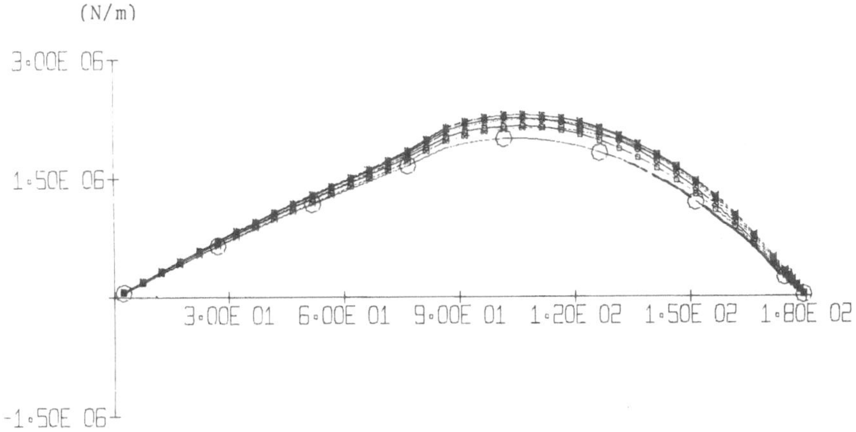

Figure 2.8 3D-Body-Force Distribution: Z-Component Of The Symmetric Load Case.

(3) the effect of the pulsed poloidal field, giving rise to Lorentz forces along the coil, which tend to impose a torque around the x-axis on it (overturn loading), shown in Figs. 2.12 to 2.14.

Each figure shows one cartesian force component in the global

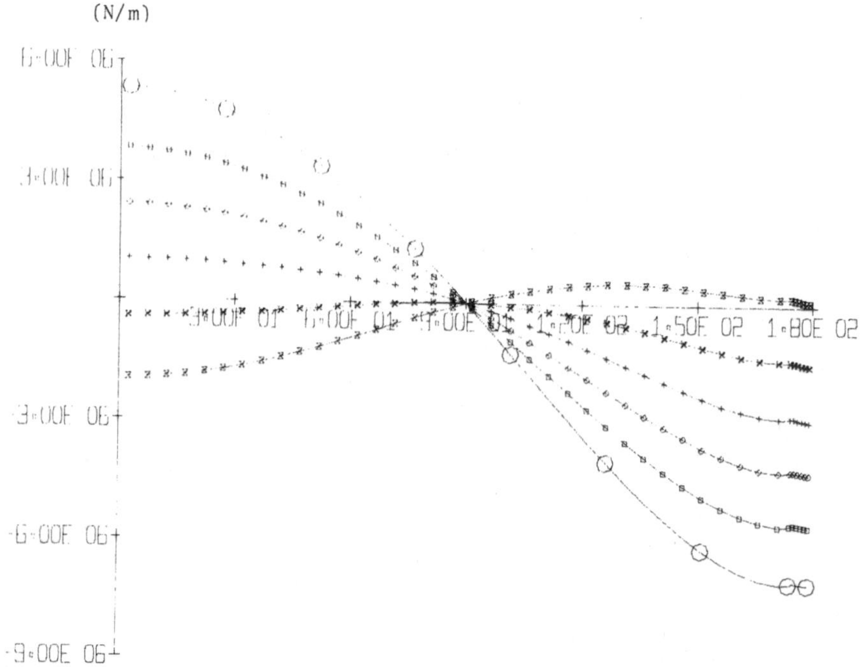

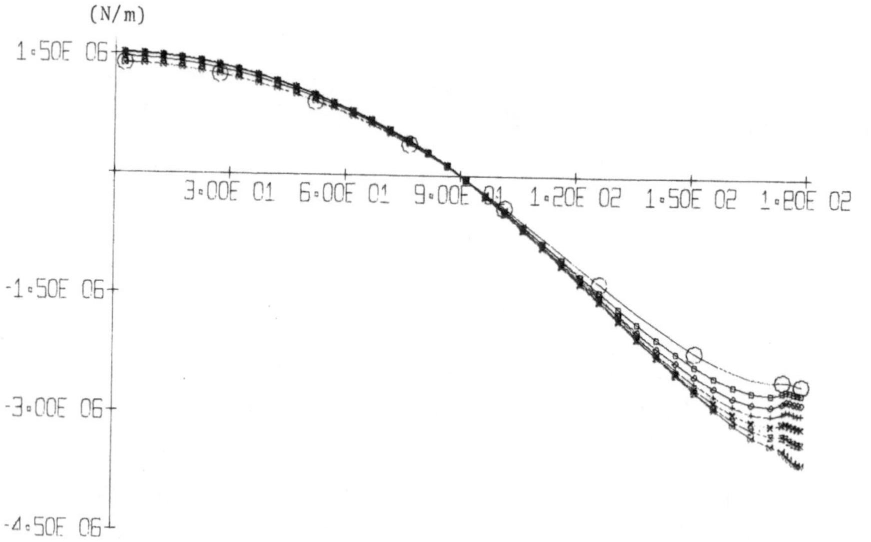

Figure 2.9 3D-Body-Force Distribution: X-Component Of The Asymmetric Load Case.

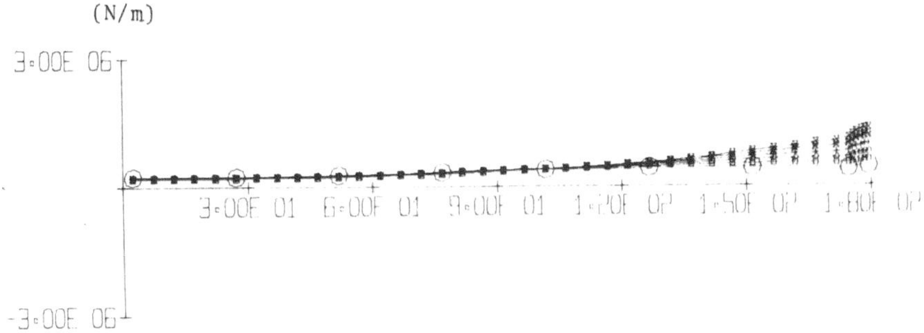

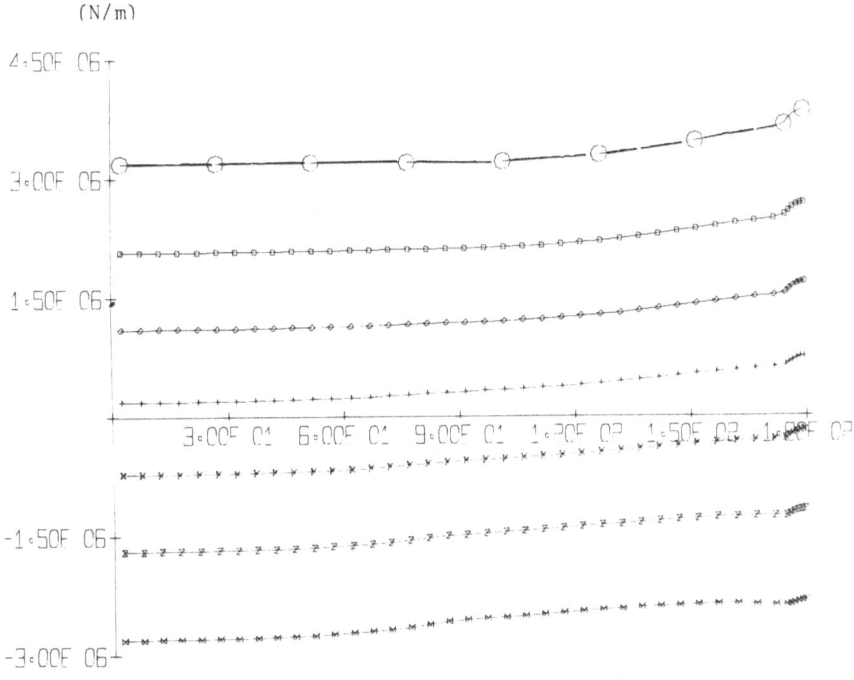

Figure 2.10 3D-Body-Force Distribution: Y-Component Of The Asymmetric Load Case.

(x, y, z)-coordinate system, plotted as a function of the azimuth angle ϕ, measured along the coil contour, as indicated in Fig. 2.3(b). The top graphs in Figs. 2.6 to 2.14 differ from the bottom graphs in the way the coil is subdivided, to demonstrate the spatial distribution of the body forces. In the diagrams on the top of these figures, the curves belong to 6 piecewise concentric axially deep rings (layers), following the direction of

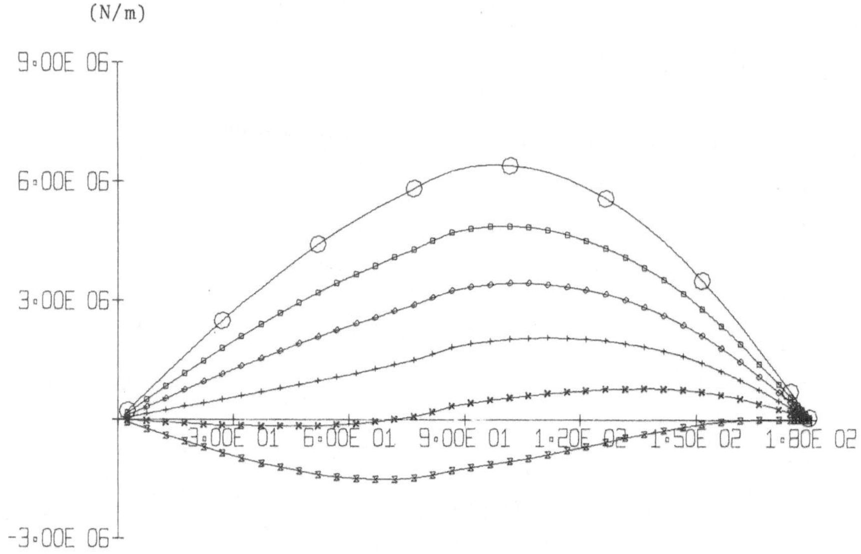

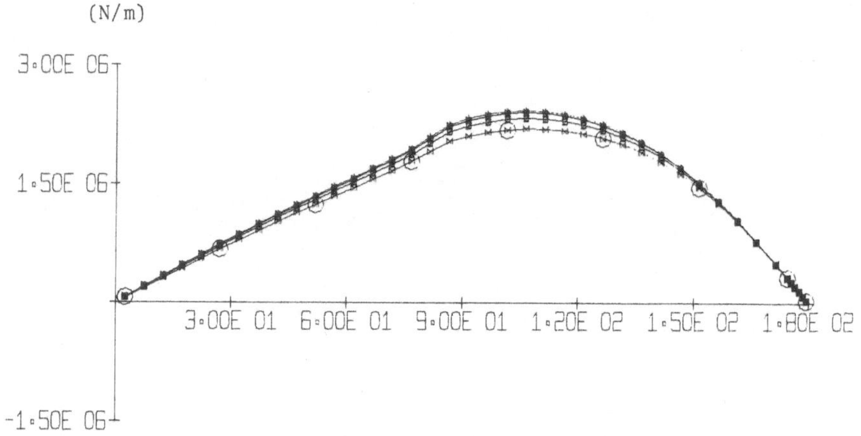

Figure 2.11 3D-Body-Force Distribution: Z-Component Of The Asymmetric Load Case.

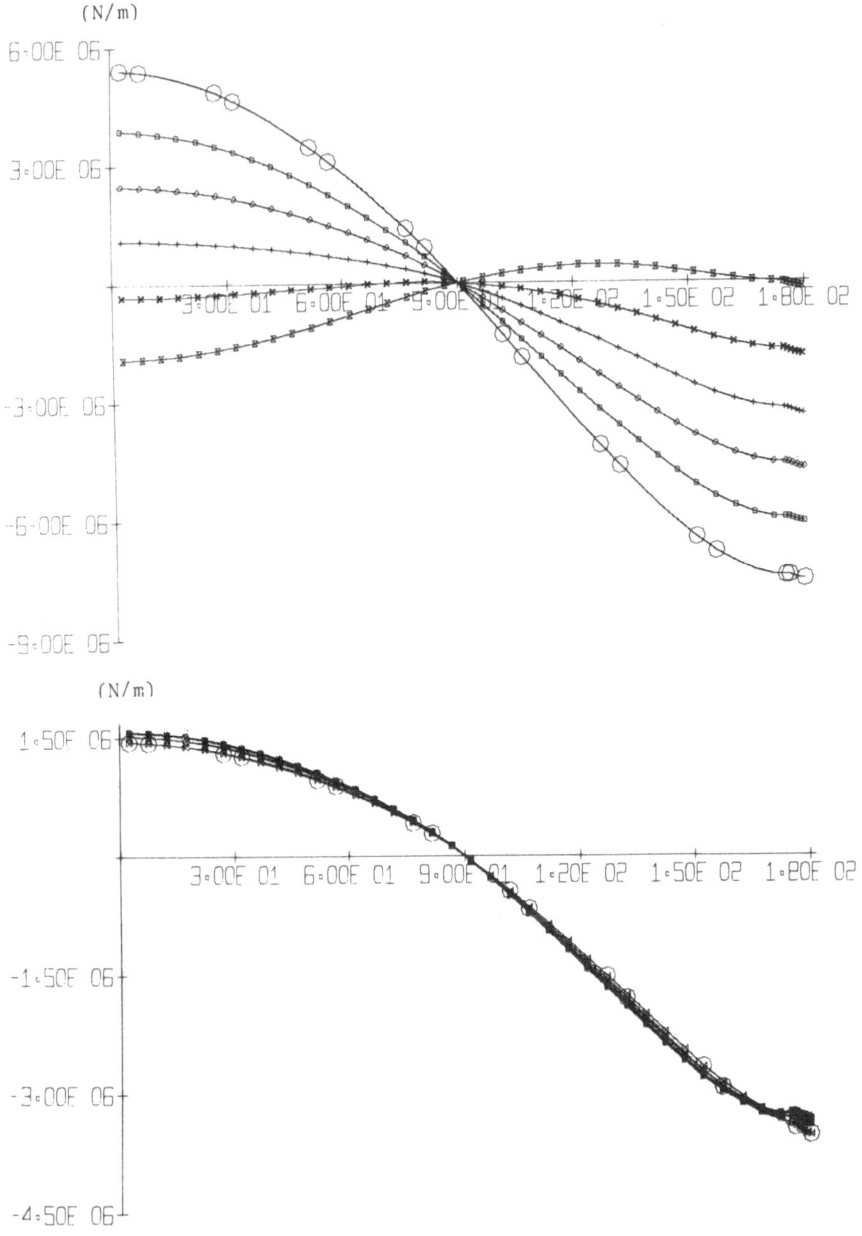

Figure 2.12 3D-Body-Force Distribution: X-Component Of The Overturn Load Case.

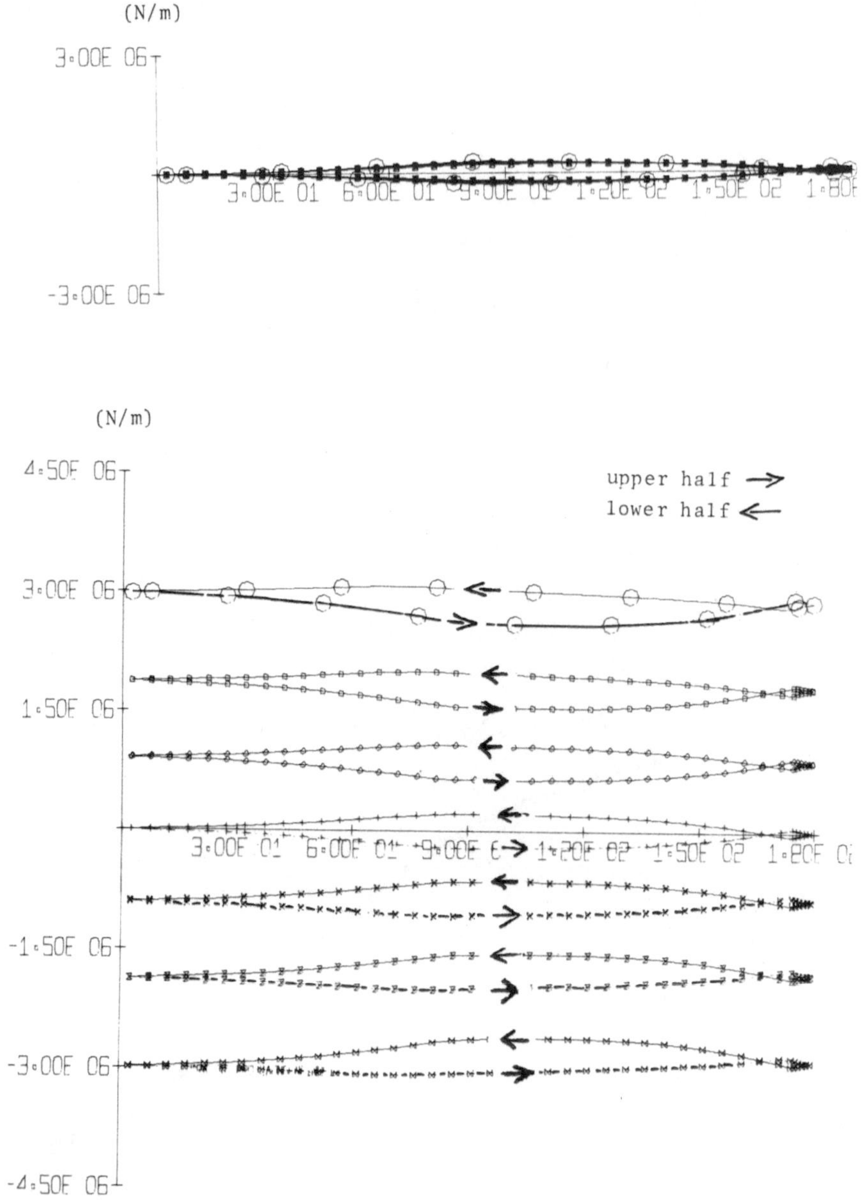

Figure 2.13 3D-Body-Force Distribution: Y-Component Of The Overturn Load Case.

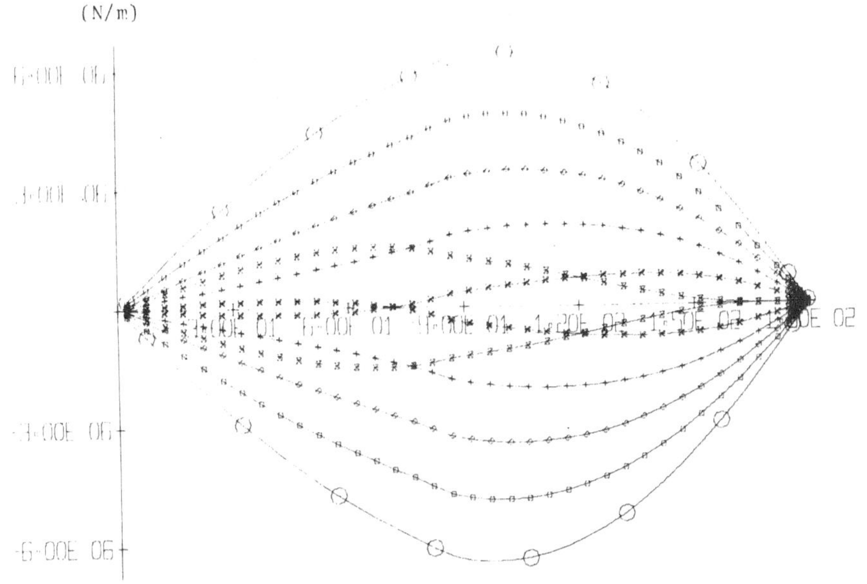

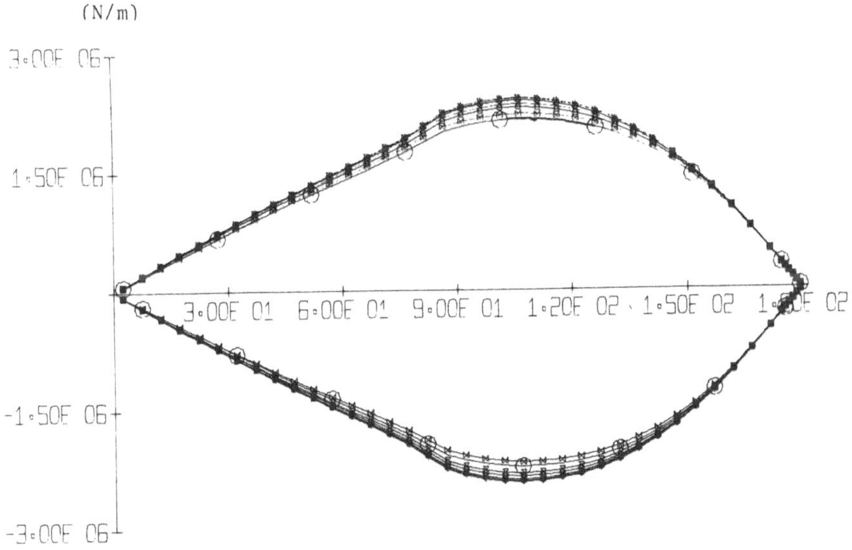

Figure 2.14 3D-Body-Force Distribution: Z-Component Of The Overturn Load Case.

the current into which the coil is thought to be cut. Each curve belongs to one imaginary layer (dr = const), as shown on the right of Fig. 2.4(c). In the diagrams at the bottom of Figs. 2.6 to 2.14, the curves belong to 7 axially cut parallel flat rings of equal size (slices, pancakes) which are perpendicular to the global y-axis. Each curve there belongs to one pancake (see the left of Fig. 2.4(c)). The azimuth angle ϕ is plotted from 0 degrees (outer leg) to 180 degrees (inner leg), covering the upper part of the coil ($z > 0$). For the lower part ($z < 0$) in the overturn load case, the angle returns from 180 degrees to 0 degrees, in the same diagram.

Perusing Figs. 2.6 to 2.14, one can see that because there is no difference in force magnitudes between the upper and lower parts of the coil, for the symmetric as well as for the asymmetric load case, plotting only the curves for $z > 0$ is sufficient for giving a complete picture. For interpretation one only has to keep in mind that the X- and Y-components coincide for $z < 0$ and $z > 0$, whereas the Z-component simply changes sign when going through $z = 0$. Accordingly, there will be no torque about the x-axis in the symmetric and in the asymmetric load cases. In the overturn load case (Figs. 2.12 to 2.14), however, only the X-component coincidence, as well as the sign-antisymmetry of the Z-component, continue to hold (the latter is demonstrated in Fig. 2.14; upper curves $Z > 0$ for $z > 0$, lower curves $Z < 0$ for $z < 0$). The Y-components, however, become unequal for the upper and lower part in the overturn load case, because the Lorentz forces due to the poloidal field are superimposed on the axial force component, due to the toroidal field in different directions for both halves of the coil. This inequality is even more pronounced by a slight ϕ-dependence of the toroidal field Y-component (compare Figs. 2.7 and 2.13) and results in a net outer torque acting on the coil, about the x-axis. In Fig. 2.13 the double curves belong to different pancakes. The lower branches pertain to one half of the coil (e.g. the upper half) and the upper branches pertain to the other half (e.g. the lower half). Both branches are produced through superimposing perturbations of equal magnitude, but different sign to the slightly ϕ-dependent $F_y(\phi)$-curves (like those in Fig. 2.7). The slight ϕ-dependency of the $F_y(\phi)$-curves is responsible for the different shapes of both branches (see Fig. 2.13).

A qualitative illustration of the overturning Y-component is given in Fig. 2.15. There, the Lorentz force T may weaken (as in regions 2 and 4) or strengthen (as in regions 1 and 3) the inward directed basic axial component F_y.

Ideal Contour (D-Shape). In Figure 2.3, a small section ds of a current carrying conductor filament is shown, in which the mechanical tension T is in balance with the magnetic pressure f_1

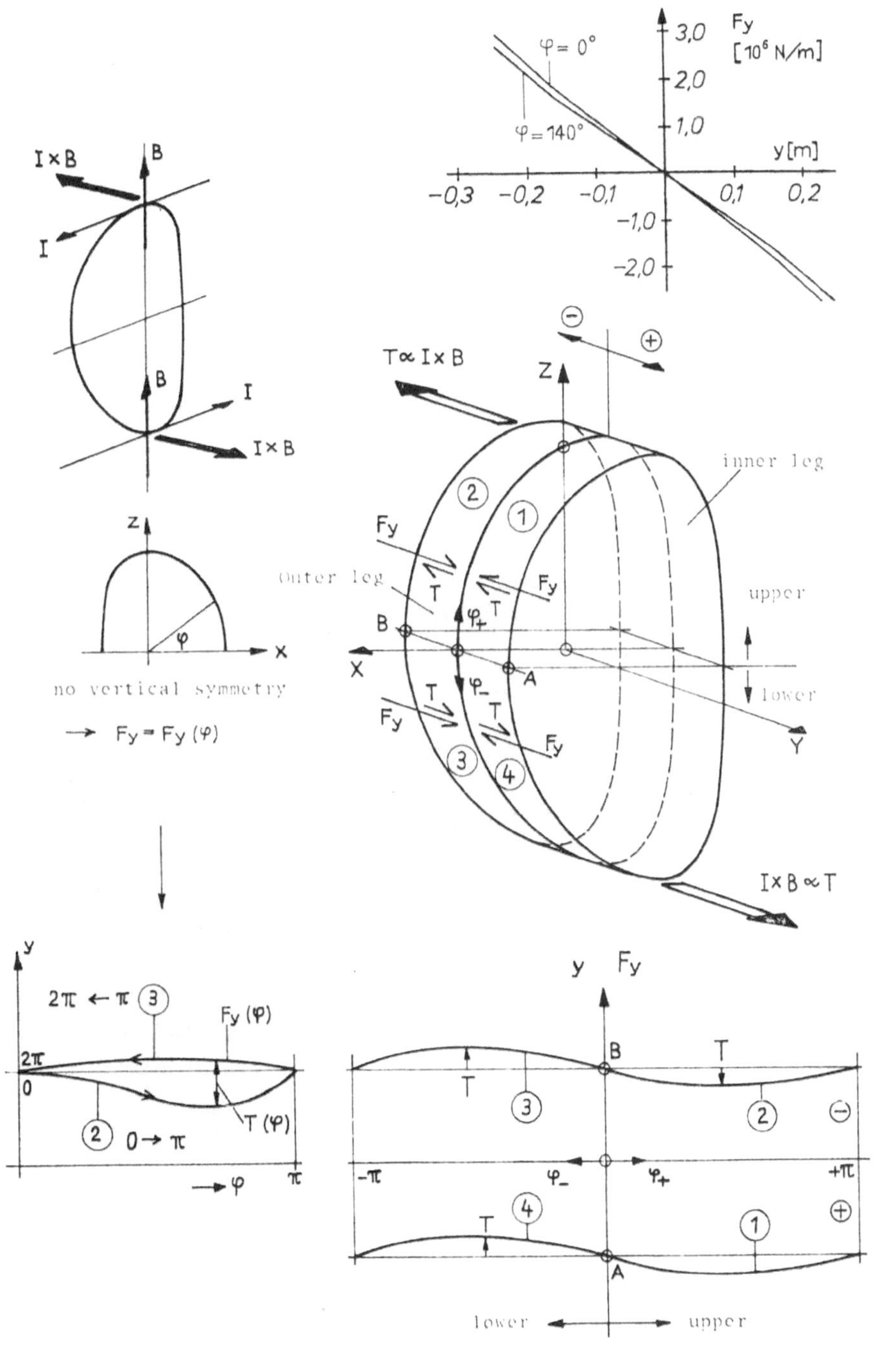

Figure 2.15 Overturning Y-Component Load Case.

(force per unit length). Under this condition, mechanical tension exists only in the tangential direction. The condition of balance is

$$T = f_1 \rho \tag{2.1}$$

where ρ is the radius of curvature, given by

$$\rho = \left[1 + \left(\frac{dz}{dr}\right)^2\right]^{3/2} / \frac{d^2z}{dr^2} \tag{2.2}$$

The magnetic pressure f_1 is proportional to the magnetic field B and the current I. For constant tension and constant current, the Eq. 2.1 is

$$B\rho = \text{const} \tag{2.3}$$

The magnetic field B_y of an ideal torus (thin layer of uniformly wound conductor) is inversely proportional to the radius r. Therefore, the radius of curvature is $\rho = k\,r$ and the differential equation of Eq. 2.2 is

$$\frac{d^2z}{dr^2} = \frac{1}{k\,r}\left[1 + \left(\frac{dz}{dr}\right)^2\right]^{3/2} \tag{2.4}$$

This equation has been investigated by many authors [8, 9, 10]. The solution z(r) of Eq. 2.4, completed by a straight piece at $r = R_1$, is called the ideal D-shape or Princeton-D. The inner and outer radii R_1 and R_2 must be specified as an input parameter, adjusting the particular shape of the Princeton-D.

In a more realistic treatment of the problem [11], the magnetic field of 2n line currents is used. Moses and Young [12] have found the empirical formula

$$\frac{\rho(r)}{r} = \left[\frac{\rho 2}{R_2}\left(1 + \frac{1}{n}\right) - \frac{1}{n}\ln\frac{r}{R_2}\right] / \left(1 + \frac{1}{n}\cos\phi\right) \tag{2.5}$$

with $\rho_2 = \rho\ (R_2, \phi = 0)$, to obtain the radius of curvature and show its dependence on r, ϕ, and the coil number n.

A direct method is now presented, based on numerical integration of the differential equation for the desired magnet contour. In general, the magnetic field is dependent on the torus coordinates r, y, and z (see Fig. 2.3, where the z-axis is perpendicular to the paper plane). Therefore, the differential equation for the constant tension contour in a plane of constant y is

$$\rho = \left[1 + \left(\frac{dz}{dr}\right)^2\right]^{3/2} \Big/ \frac{d^2z}{dr^2} = \frac{k_0}{B_y(r,y,z)} \qquad (2.6)$$

where k_0 is a parameter that in general depends on y. For technical reasons, the contour of the magnets producing the torus field in a Tokamak should be independent of y. Therefore, one specializes Eq. 2.6 for the mid-plane of the magnets (i.e. y = 0) and neglects the y-dependence of B_y, which, is relatively weak in actual magnets. To define the solution, the initial conditions are chosen as

$$\left.\begin{aligned} z(R_0) &= z_0 \\ \left.\frac{dz}{dr}\right|_{R_0} &= 0 \end{aligned}\right\} \qquad (2.7)$$

For a closed loop z(r), two values, R_1 and R_2, must exist where the first derivative $\frac{dz}{dr}$ is infinite. Here, z_0 is a boundary value that may be any arbitrary real number (open boundary). For convenience, put $z_0 = 0$. For arbitrary fields B, Eq. 2.6 can only be solved numerically.

In order to get a self consistent solution, i.e. a contour that gives rise to the B-values fed into the differential equation, a computer program was written, combining a code for solving systems of differential equations with a magnetic field code (HEDO2, see Ref. 14). The two geometrical quantities R_1 and R_2 are read in and Eq. 2.6 is solved, for the first time under the assumption that B(r) is $\sim 1/r$. The resulting curve z(r) is the ideal D-shape, or Princeton-D. This curve is taken as the geometric center line of real magnets, having a finite winding cross section (see Fig. 2.3) that produces a Tokamak-like B-field. Now the magnetic field B(r,z), along the geometric center line z(r) of one of these magnets, is calculated. This new magnetic field is used when solving Eq. 2.6 again. The procedure outlined here is more general than empirical-formula-methods for the derivation of optimum shapes, like the Moses-Young contour. To solve the differential equation (DE) usually defining the filament middle line of a rectangular cross section magnet, the present method takes into account the actual fields of a real arrangement of arbitrarily oriented planar magnets. At least one of these magnets is iteratively contour-optimized with respect to tension constancy. In each of these iterative steps, the solution of the DE for the filament is taken as a descriptor for the new magnet shape, which is then used for the 3D field computation in the next iteration.

The program starts at R_0 ($R_0^2 = R_1 R_2$), with $dz/dr = 0$, and integrates with $r < R_0$, until $dz/dr \to \infty$. This happens for some value $r = R_1'$, depending on the choice of the parameter k_0. If R_1' is different from the value R_1 that has been read in, the parameter k_0 is modified and the process is repeated until $R_1' \approx R_1$. Now, k_0 is fixed and the system is solved with $R > R_0$, resulting in $dz/dr \to \infty$ at $r = R_2$. This completes the first iteration. The procedure is repeated, taking the new curve $z(r)$ as the geometric center line of the magnets in question and computing corresponding field values. The curve emerging from that iteration is compared to the curve from the preceding iteration. If the two curves coincide (within certain limits), the procedure terminates, otherwise a new iteration is performed. In the final contour the parameter k_0 has the same value as the, now spatially independent, product $B\rho$, which is the hoop stress divided by current density, as required in the optimization criterion. This result can be interpreted as a fully stressed design where the azimuthal gradient of the hoop stress has the absolutely lowest possible value of zero, everywhere.

For the LCT-coil defined by the data given in Table 2.1, the result of the calculation is shown in Fig. 2.16. The convergence of the method is very fast. The accuracy is determined by the accuracy of the magnetic field calculation program. A comparison of the results with the contours calculated according to Eq. 2.5 shows a maximum difference of only a few mm for the LCT coil.

2.2 Real Design Effects

In a Tokamak reactor, the toroidal field coils are usually placed around the vacuum vessel and integrated into a complex mechanical load support structure that has to resist the enormous forces and torques occurring under operational and fault conditions. Therefore, the coil winding is embedded in a stiff casing, which diffuses the magnetic loads to the support structure and serves as a housing for the Helium cooling system. It is obvious that in such a real design, the stress distribution is

TABLE 2.1 DATA FOR LCT

Coil number	6
R_1 (m)	0.926
R_2 (m)	3.757
radial width B (m)	0.46
axial thickness D (m)	0.56
winding number	588
current (kA)	11

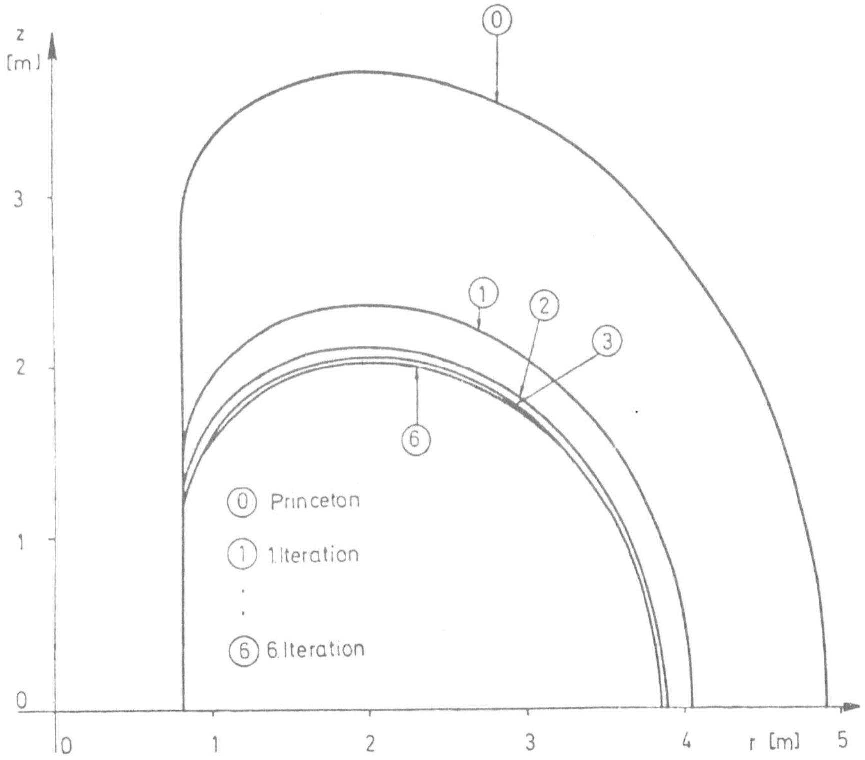

Figure 2.16 Iterative Contour Optimization.

more complex than was assumed in the contour optimization procedure. The real design; with its finite cross section, different support conditions under various load cases, and interaction between the winding and the casing; requires a 3-dimensional analysis to gather the insight that is necessary for improvement of the mechanical behavior of the toroidal field coil.

Finite Element Model. For static analysis of the LCT-coil, the ASKA code [15] was used to set up a 3-dimensional finite element model. It consists of the 10 substructures shown in Fig. 2.17. The casing is modelled by triangular plate elements (TRIB 3). The winding model consists of brick elements (HEXE 8). The choice of finite element software such as ASKA, which makes available the substructure technique together with the option to rotate local coordinates (node, as well as elementwise) seems mandatory. These modeling capabilities are attractive tools for the description of the gliding contact between coil and casing and for studies of anisotropic behavior of the composite winding [3, 4, 5, 6].

Support Conditions. In the hexagonal torous geometry (Fig. 2.3), considerable radially inward net forces require a strong structural support [16] of the coil in this direction. The influence of various simplified support conditions was previously studied by Gralnick [17]. His analytical solutions have enhanced former studies considerably [8, 9, 10]. For the present purpose, the more general finite element model was made available for a parameter study, varying the support conditions along the inner leg of the bare coil winding.

In the design process, this problem arises when elastic steel bladders, to be inserted in the gap between casing and winding, must be specified. The study varied the length α and the amplitude value of the 3 different support stiffness distributions shown in Fig. 2.18. By a suitable choice of these parameters, it is possible to control the global deformation pattern of the winding. Figure 2.18 shows (upper halves pertain to the long, lower halves to the short support length) that shape I gives a breathing mode, shape II leads to an inward inflection along the inner leg, as well as a shortening of the outer leg radius, and shape III keeps the vertical dimensions of the winding almost constant, while stretching its radial extension considerably.

These characteristics may be utilized by the designer to master the gap occurring along the inner ring under magnetic body forces (see the following subsection). The hoop stresses (Figs. 2.19(a) and 2.19(b)) confirm the typical curved beam behavior, with maximum bending stress amplitudes in the narrow curvature region (ϕ = 80°). The peaks are only slightly influenced by the

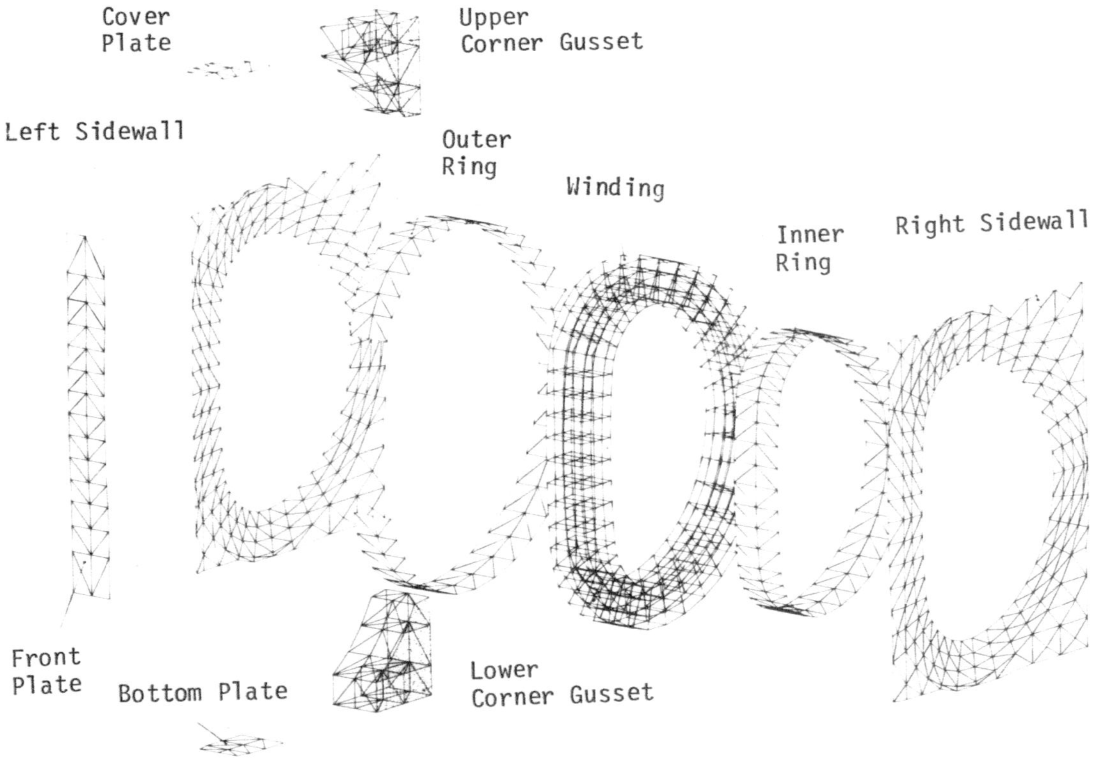

Figure 2.17 Finite Element Model Of The LCT-Coil.

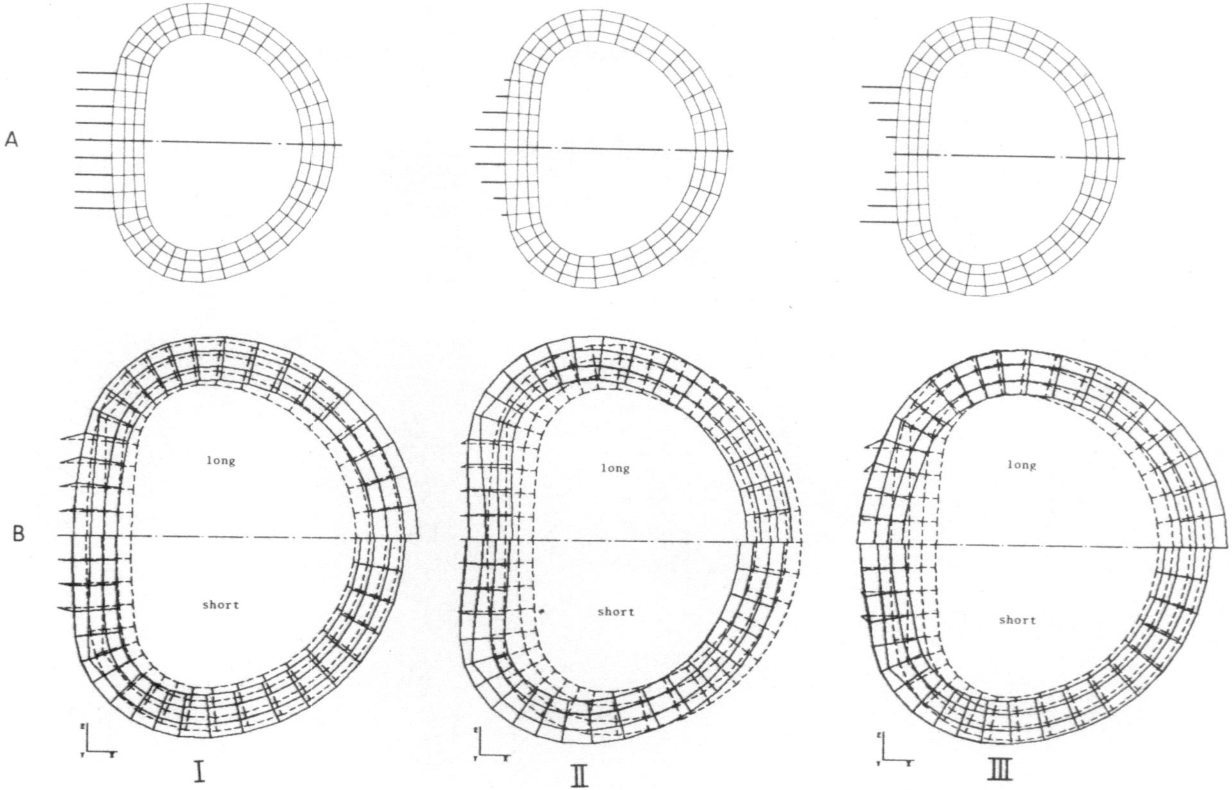

Figure 2.18 Elastic Support Parametrization For The Bare Winding.

support stiffness. But the scatter of the values at $\phi = 0$ is remarkable and the 3-dimensionality of the stress distribution is most pronounced in the support region, as expected (see Fig. 2.20(a), a = surface values, i = mid-plane values). The shear stresses are harmless, under symmetric magnetic body force loading (Fig. 2.20(b)). The softer the support, the wider the spread of the hoop stress peaks at $\phi = 0$ (see Fig. 2.21 for support III; a and i as in Fig. 2.20) and, simultaneously, the stronger the influence on the deformed shape. Practical limits on the range of the support stiffness parameter are the hoop stress peaks in the narrow curvature at $\phi = 80°$ on the soft side and the rigid support value reached asymptotically by the curves (Fig. 2.21) on the stiff side. After an appropriate perusal of the feasible parameter value ranges, optimal parameter combinations can be identified (see Table 2.2 for recommended configurations).

TABLE 2.2 MAXIMUM EQUIVALENT STRESSES

Maximale Vergleichsspannungen σ_V [N/mm^2]
Maximum Equivalent Stresses

Bettungsform		σ_V (K1)	σ_V (K2)	σ_V (K3)
I	lang/long	144,3	135,2	132,4
	ϕ	67°	67°	67°
	kurz/short	127,7	129,2	129,8
	ϕ	70°	67°	67°
II	lang/long	132,2	125,7	130,2
	ϕ	0°	70°	67°
	kurz/short	152,8	123,3	129,7
	ϕ	0°	73°	70°
	lang/long	201,3	143,6	132,7
	ϕ	0°	67°	67°
	kurz/short	147,9	136,8	130,7
	ϕ	10°, 67°	67°	67°

—— recommendable parameter combinations

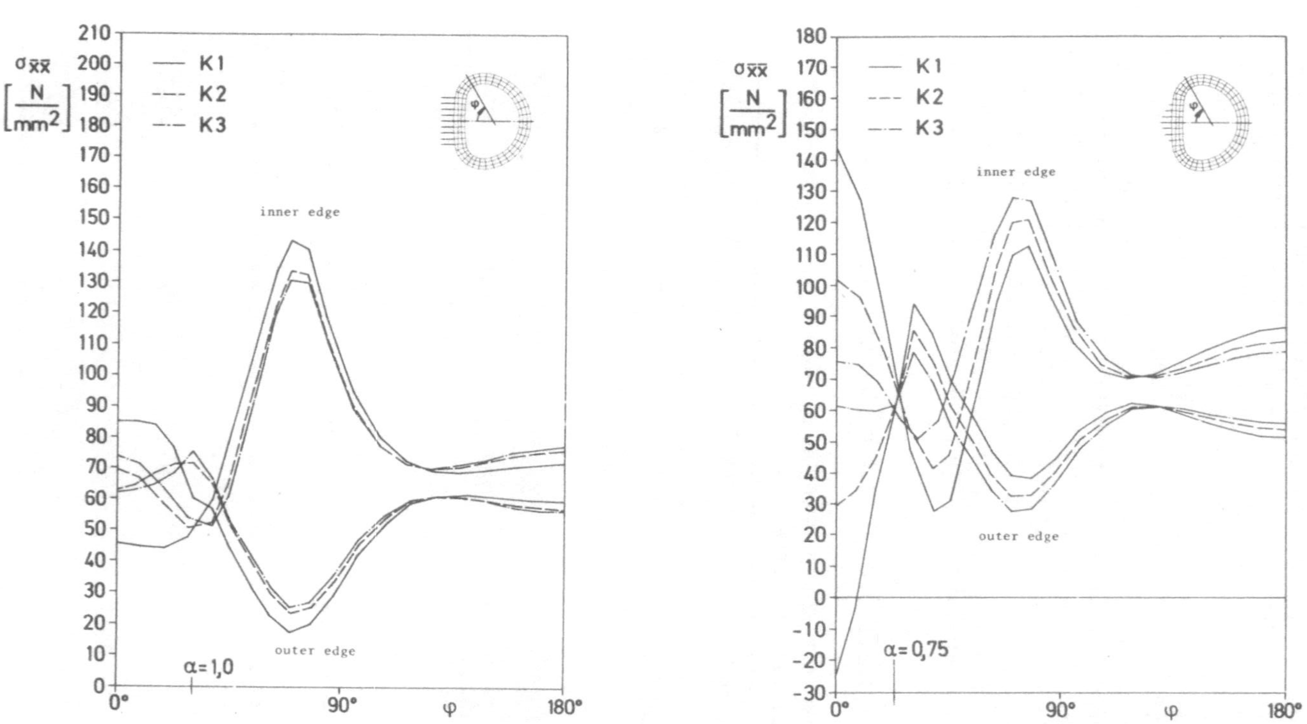

Figure 2.19 Bare Winding Hoop Stresses Under Symmetric Loading.

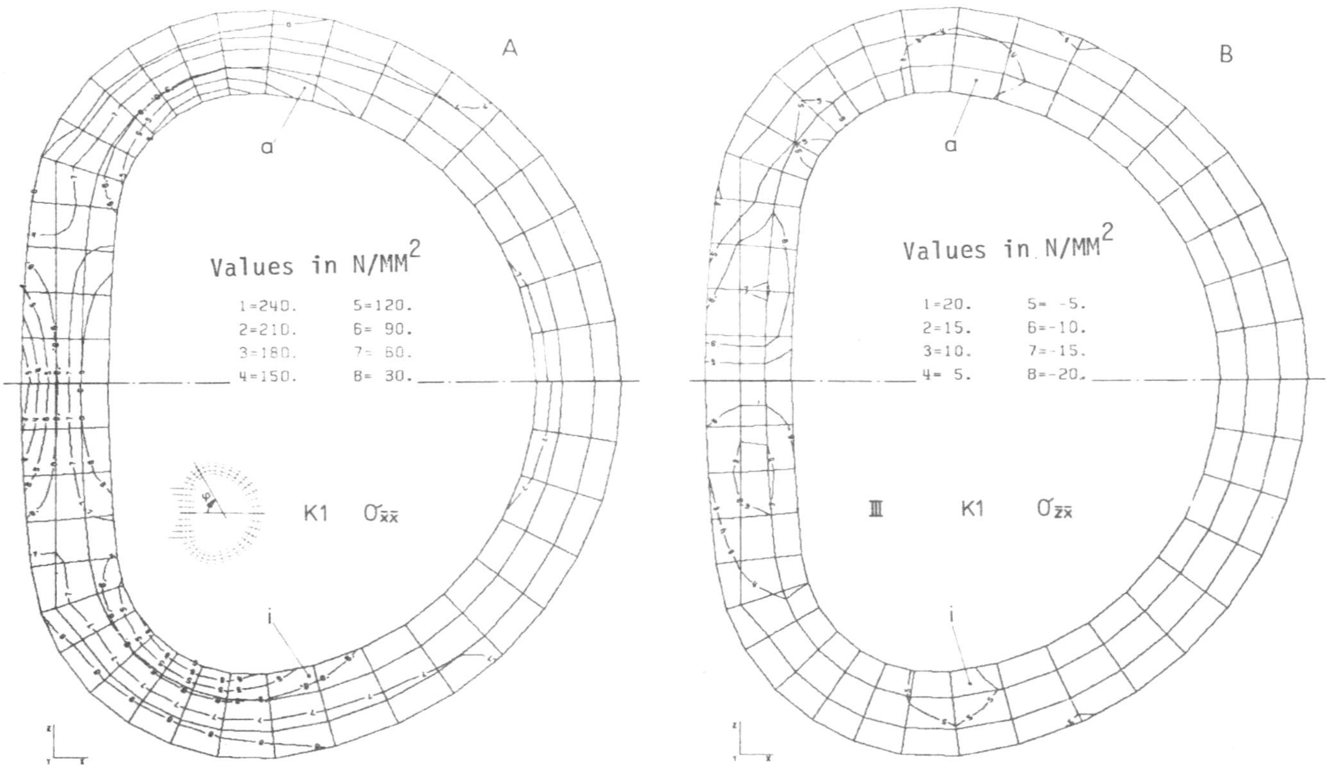

Figure 2.20 3D-Distribution Of The Hoop And Interlaminate Shear Stresses.

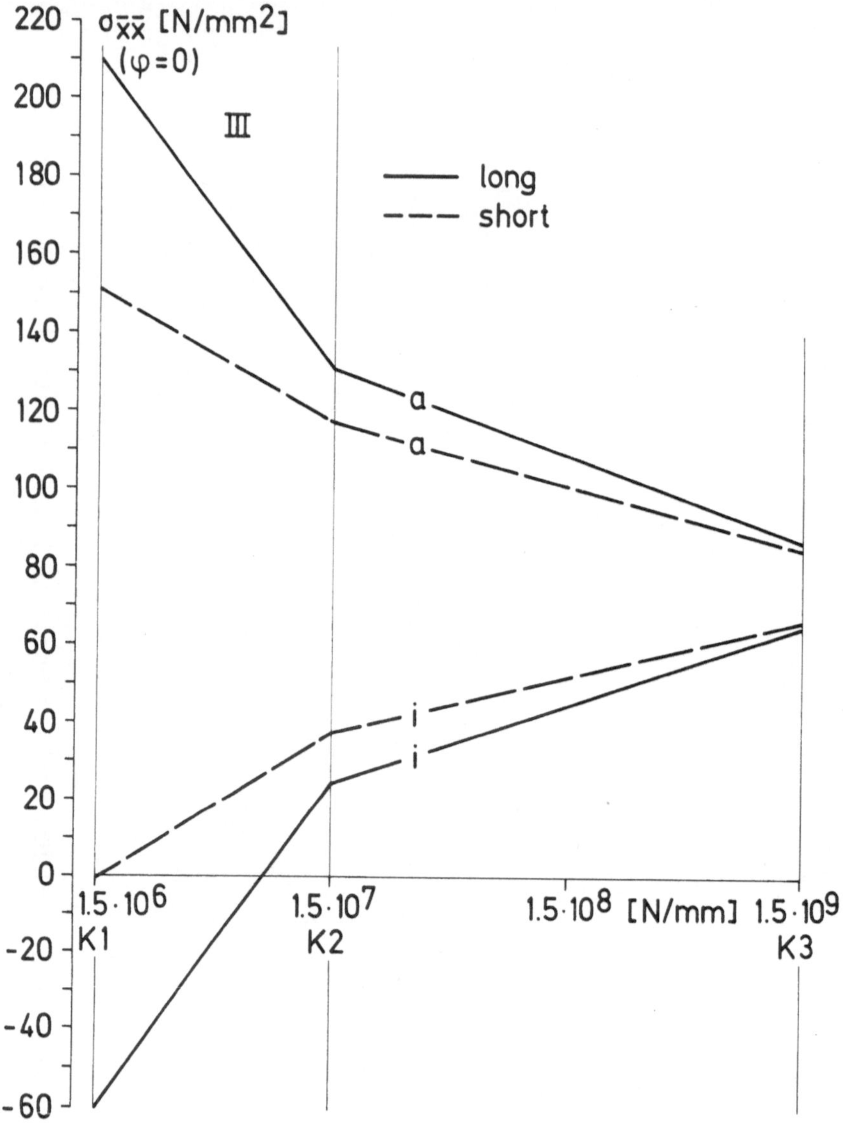

Figure 2.21 Hoop Stresses At The Middle Of The Support Region As A Function Of The Support Parameters.

Interaction Between Winding And Casing. The knowledge from the study of the mechanical behavior of the bare winding may usefully be applied to design of the complete coil, where the winding is coupled to the casing mainly through contact forces along the outer ring. Due to the breathing under magnetic body forces, the winding bends the outer ring of the casing outward (see double lines in Fig. 2.22), glides along the outer wall, as well as along the side walls, and detaches along the inner ring. This gives rise to a gap of about 0.35 mm between the coil and casing (see distance between the curves in Fig. 2.23). This gap must be limited, because of safety considerations regarding the failure of the conductor envelope duct under fault conditions, producing quench pressure; (see the following subsection). Figure 2.24 shows the healthy supporting effect that the casing has on the winding stresses. They are reduced by a factor of 2 to 3 (compare mean values of the 2 curves shown in Fig. 2.24 with those in Fig. 2.19). The bending stress peaks are considerably reduced.

A good design tends to make the winding fully stressed under normal operation, carrying equivalent stresses (Fig. 2.25) about equal to those found in the casing. This leaves enough margin of safety within the casing to stand the non-symmetric loadings. Asymmetric side forces, for example, give rise to additional gap zones over the side wall interface, on the concave side of the deformed coil (reaching 0.19 mm in some of the nodes marked in Fig. 2.26). Asymmetric loading includes the highest equivalent stress peak value encountered in the 3 magnetic body force load cases ($115 N/mm^2$, see Fig. 2.27), as well as an extraordinary peak value for the shear stress component, which tends to twist the winding cross section (see Fig. 2.28). This peak occurs near the branching point joining the outer ring to the front plate. There, the casing is too stiff to alleviate the twist of the winding cross section. The equivalent stresses in casing and winding are about 10% lower in the overturn load case (Figs. 2.29 and 2.30). There are two causes that lead to the asymmetry between the upper and the lower half of the coil in this case: the magnetic body forces the topology of the kinematic coupling between the coil and casing (Figs. 2.29 and 2.30, see marked nodes). The detachment zones appear as zero-force intervals in the contact force diagrams in Fig. 2.31.

Evaluation Of Margins Of Further Improvement. As was described in Section 2.1, the final value $B\rho = k_0$ gives a measure for the achieved constant hoop stress. It can be shown, for an ideal torus in which B is proportional to 1/r, that the hoop stress is [18]

$$T_{\text{ideal torus}} = I\,(B\rho)/2 \qquad (2.8)$$

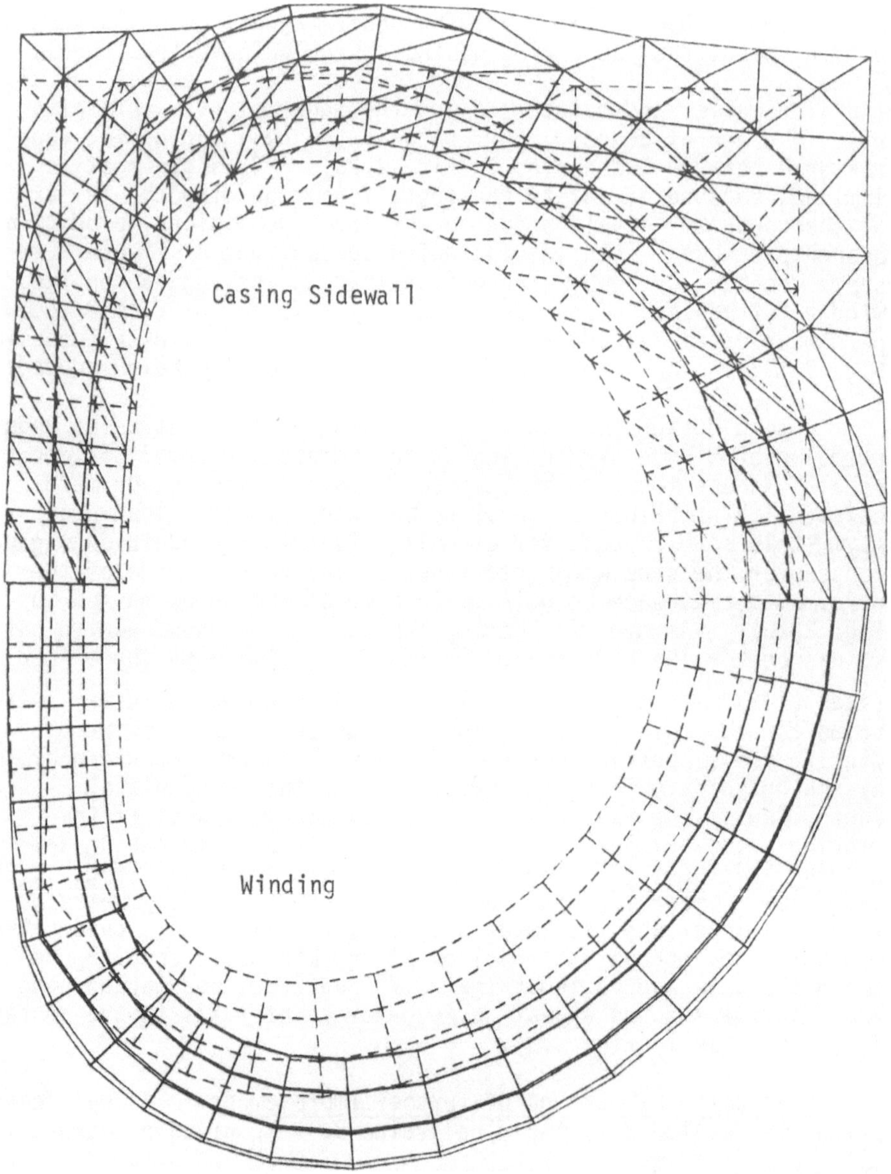

Figure 2.22 Deformation Of Winding And Casing Under Symmetric Loading.

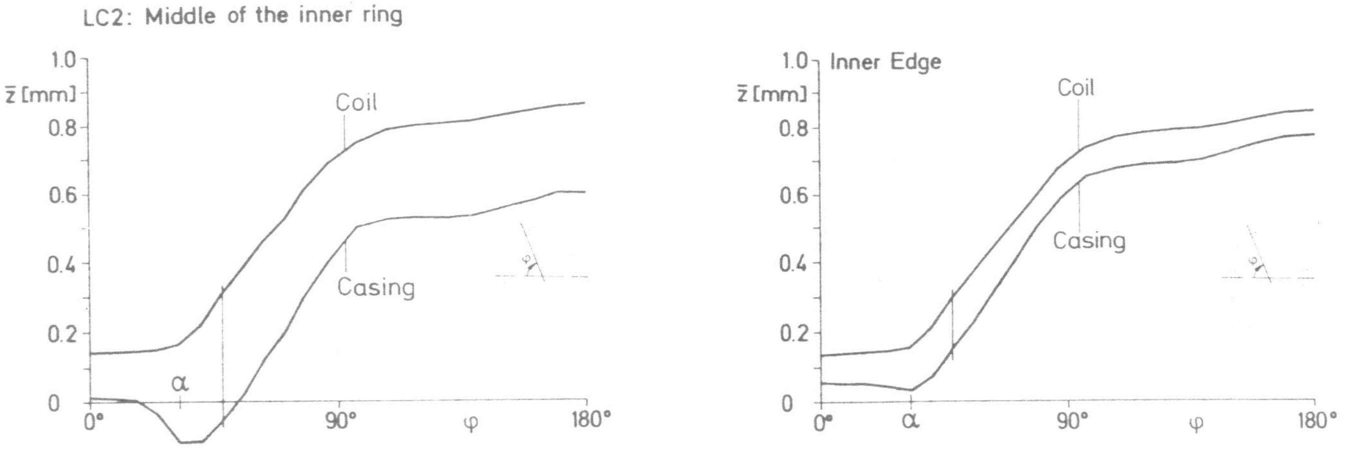

Figure 2.23 Gap Along The Inner Ring Due To Breathing Under Symmetric Loading.

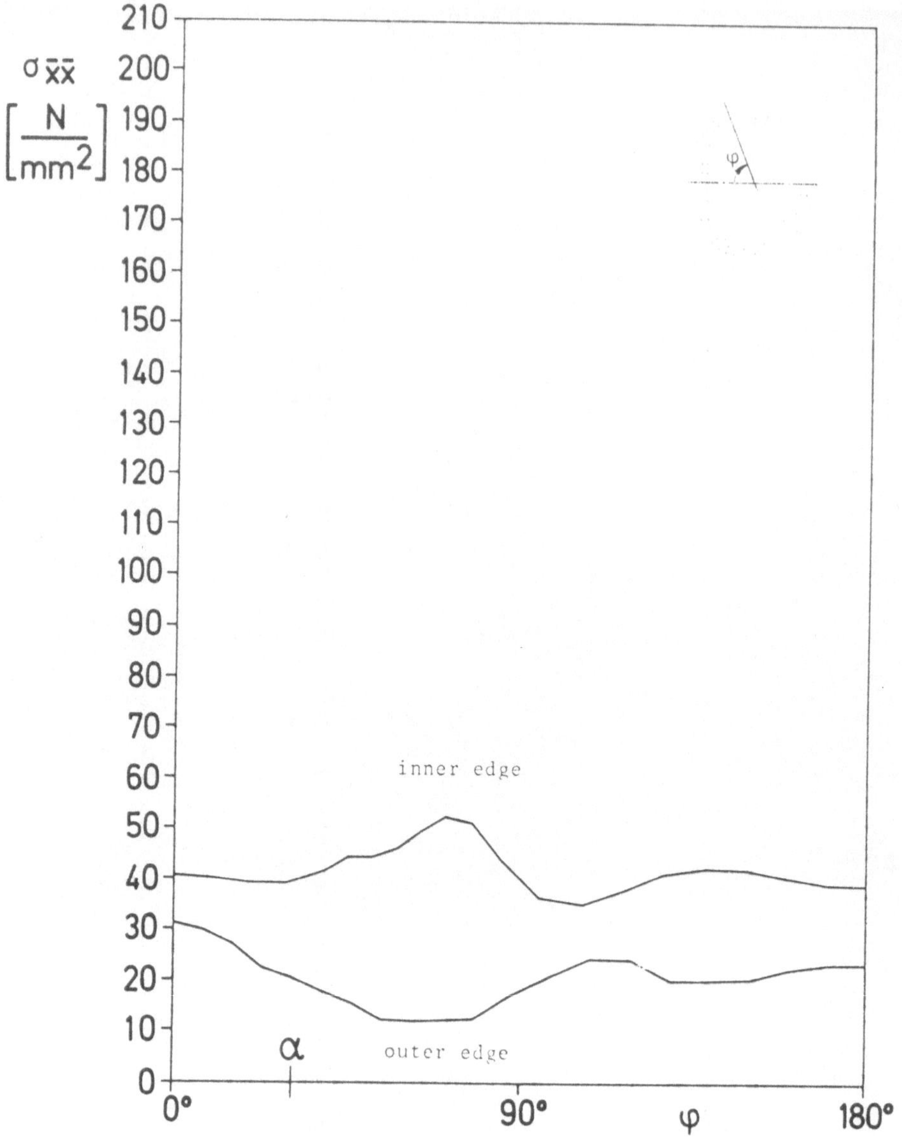

Figure 2.24 Hoop Stresses In The Winding Supported By The Casing (Symmetric Loading).

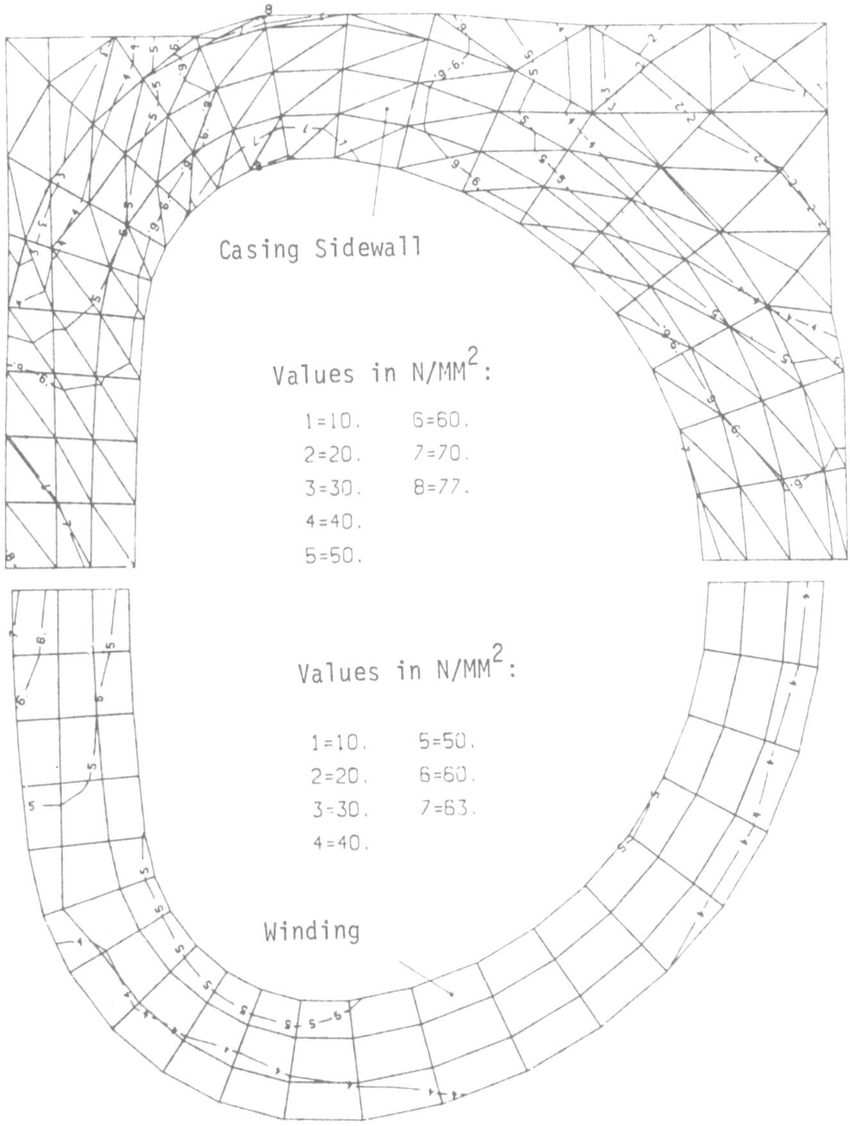

Figure 2.25 Equivalent Stresses Under Symmetric Loading.

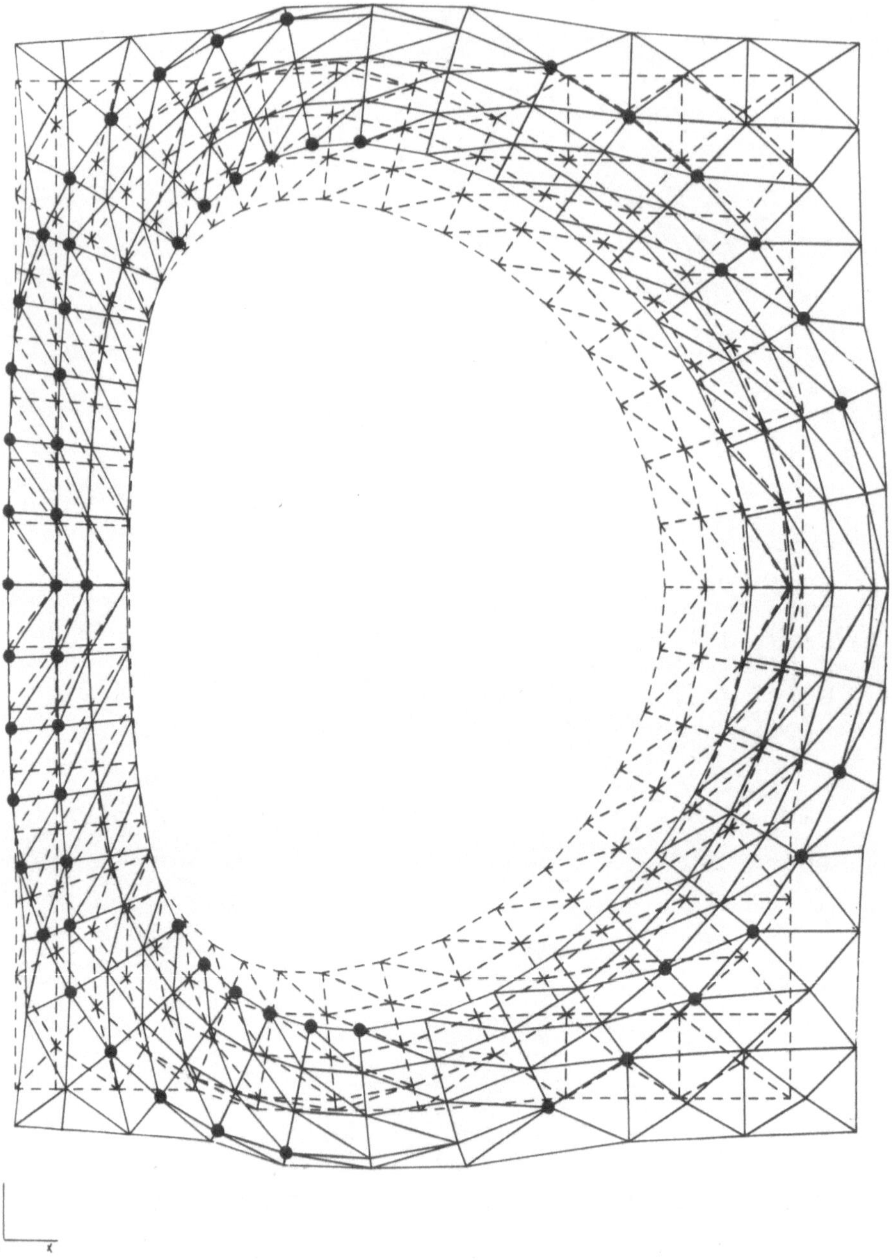

Figure 2.26 Gap Zones Occurring Between Side Wall And Winding Under Asymmetric Loading (Marked Nodes).

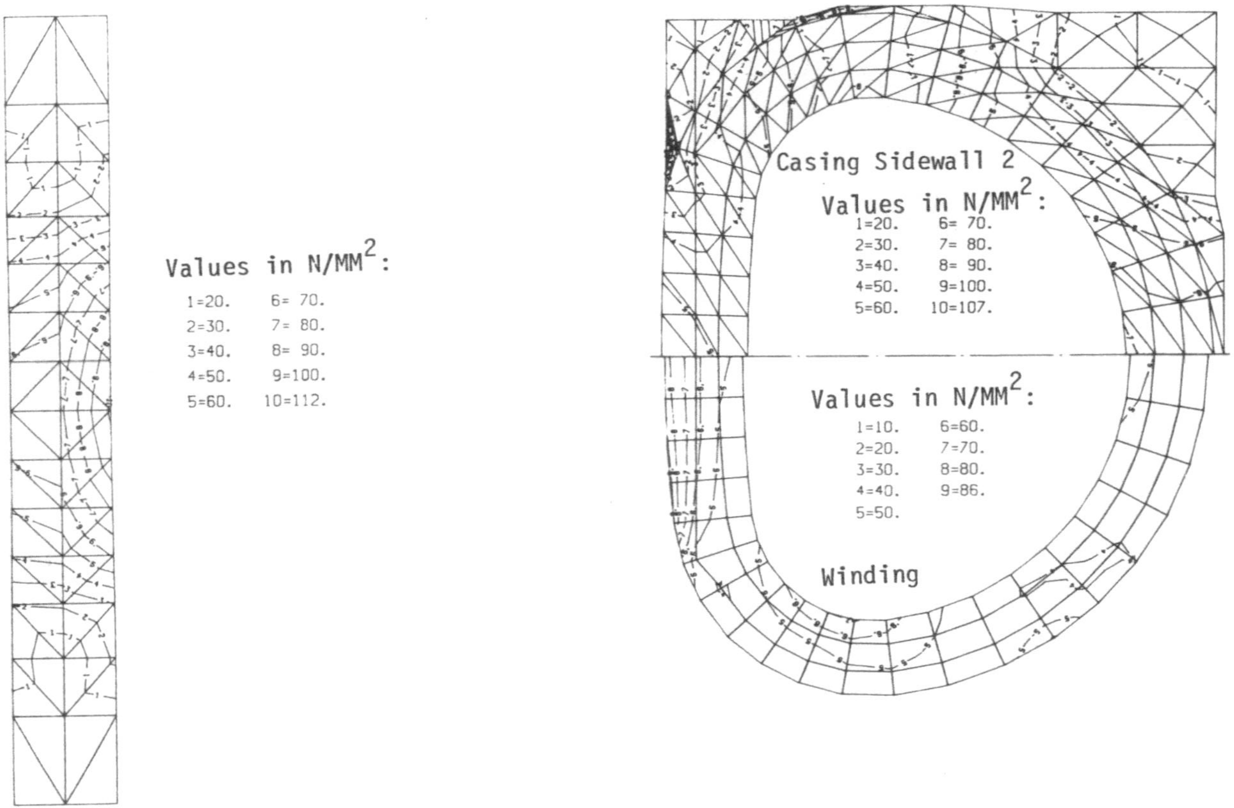

Figure 2.27 Equivalent Stresses Under Asymmetric Loading.

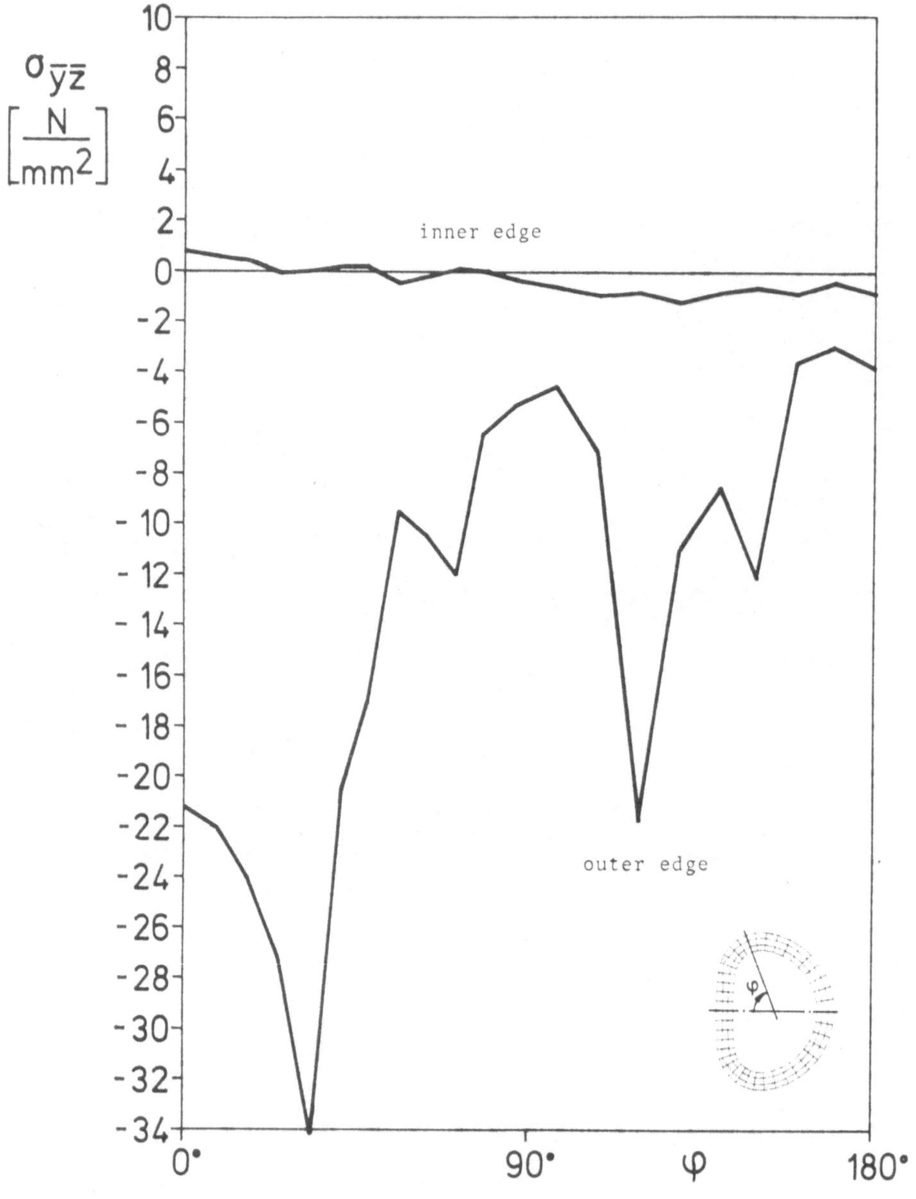

Figure 2.28 Cross-Conductor Shear Stresses Under Asymmetric Loading.

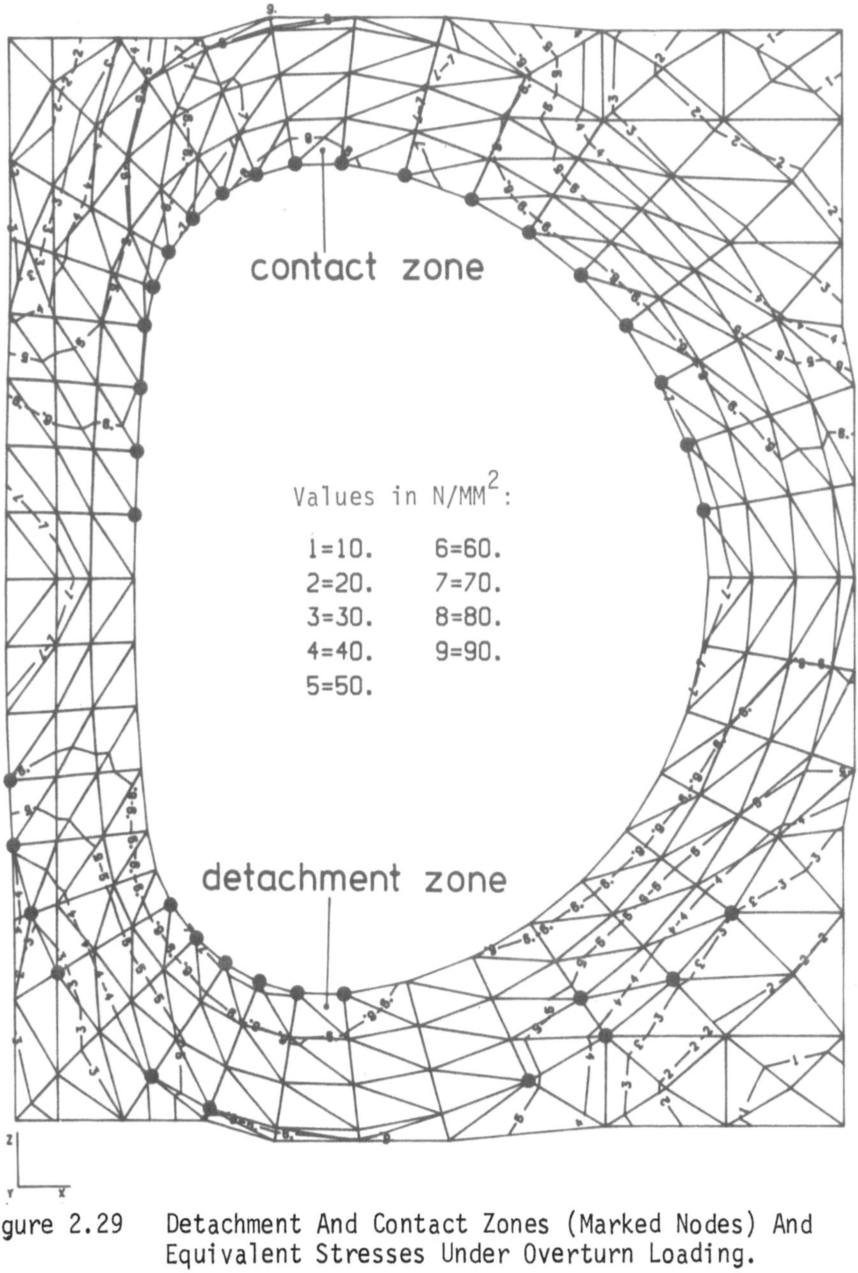

Figure 2.29 Detachment And Contact Zones (Marked Nodes) And Equivalent Stresses Under Overturn Loading.

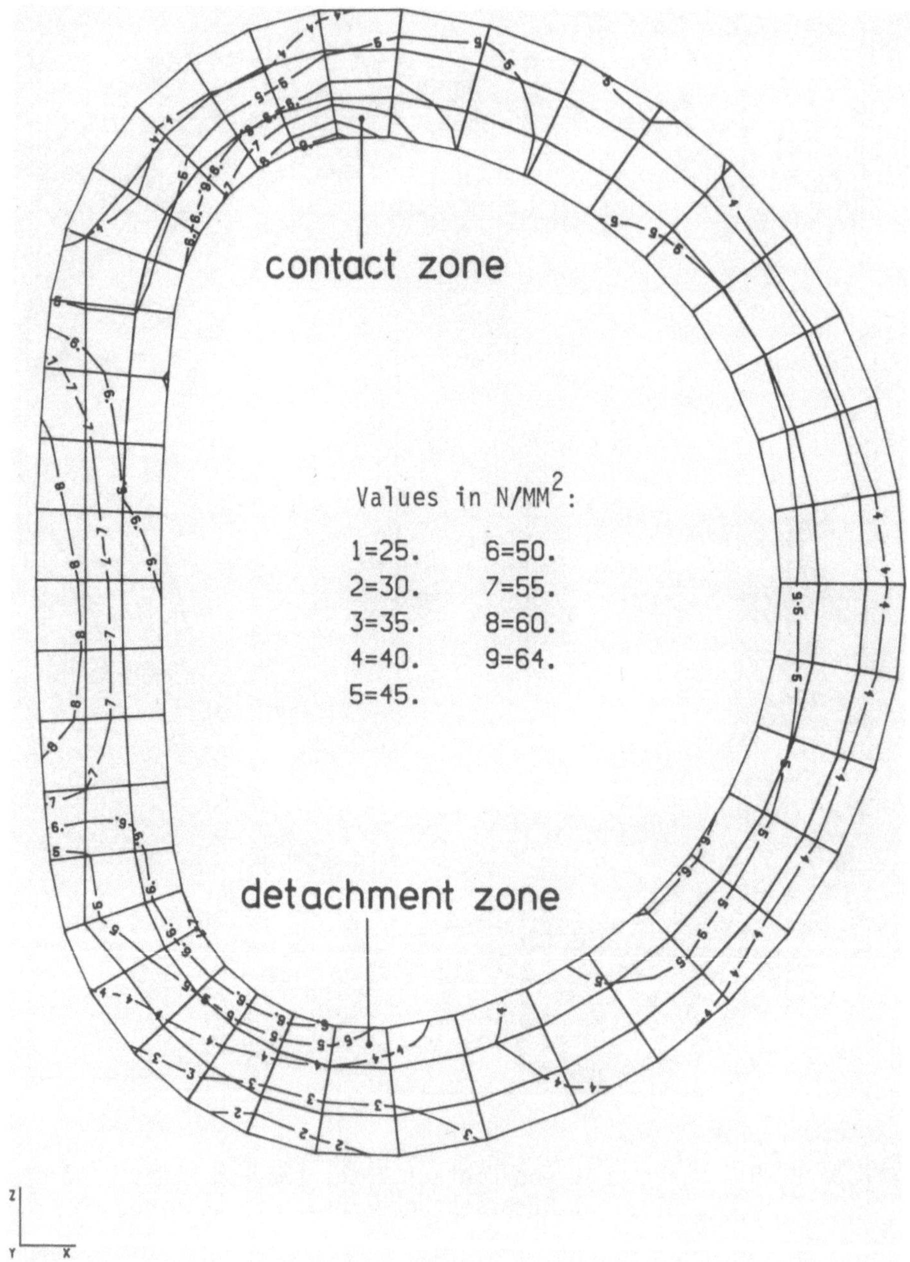

Figure 2.30 Equivalent Stresses Within The Winding Under Overturn Loading.

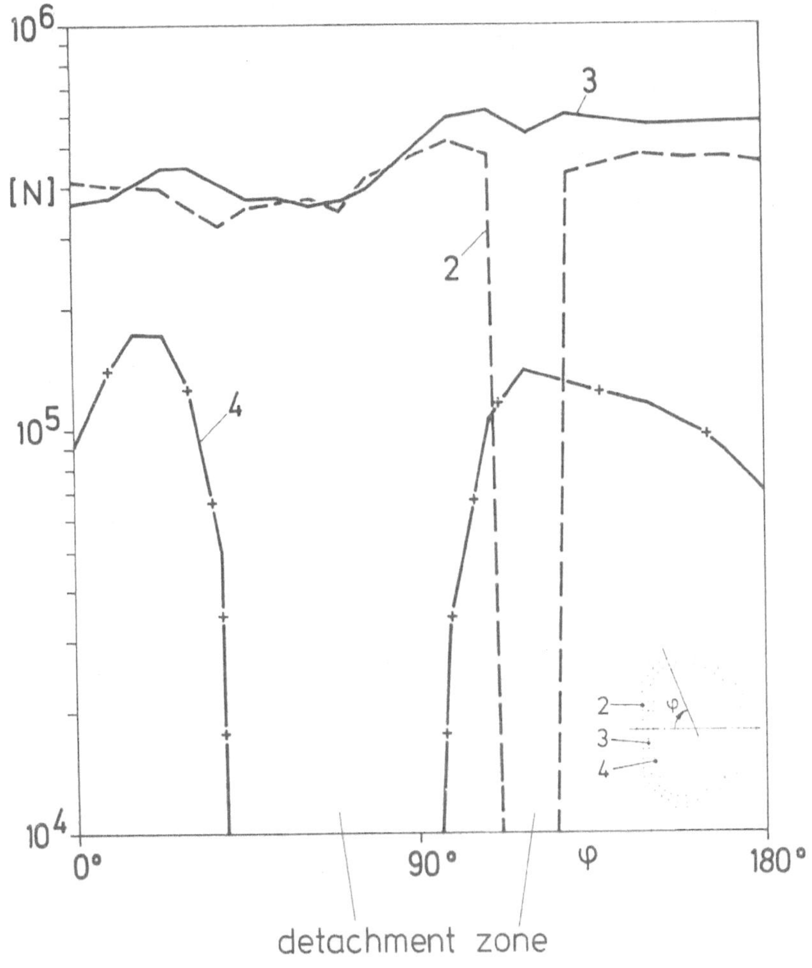

Figure 2.31 Distribution Of Contact Forces And Detachment Zones Over A Sidewall Under Overturn Loading.

The value of the product $(B\rho)I$ may be taken as a rough estimate of the hoop stress (order of magnitude) occurring along the filament. The factor 0.5 in Eq. 2.8 pertains to the ideal torus B-field and must be replaced by a more complex expression when the real magnetic field, as illustrated in Fig. 2.4 is considered. In Table 2.3, the sequence of values for $B\rho$ (in Tesla·m) over the iteration process is given, distinguishing between the maximum and minimum value along the contour. This allows one to derive the mean value, as well as the interval covered, over the circumference, as given in Column 3 and 4 of Table 2.3. Finally,

Column 5 gives the stepwise reduction of the interval width, which is a measure of hoop stress variation along the contour. The objective of the contour optimization, namely to achieve hoop stress constancy, is obviously to reduce the value in Column 5 of Table 2.3 to zero. This was practically achieved after 4 iteration steps.

TABLE 2.3 IMPROVEMENT OF HOOP STRESS CONSTANCY

Iteration step n	Min ($B\rho$)	Max ($B\rho$)	($B\rho$)m	Δ	$\Delta\Delta$
0	3.0	5.40	4.2	2.4	
1	3.6	4.1	3.85	0.5	1.9
2	3.69	3.82	3.755	0.13	0.37
3	3.70	3.75	3.725	0.05	0.08
4	3.70	3.73	3.72	0.03	0.02
⋮	⋮	⋮	⋮	⋮	0
6	3.68	3.71	3.695	0.03	

$(B\rho)m = (Min+Max)/2$ Objective: $\Delta\Delta \rightarrow 0$

$\Delta = Max-Min$

$\Delta\Delta = \Delta_{n-1}-\Delta_n$

The maximum values are more relevant in the present example, because they cover a larger part of the contour than the minimum values. Therefore it seems appropriate to take the values $B_{max} = 8T$, $\rho = 0.7$ m, $I = 11$ kA to calculate very roughly the expected hoop stress value over the conductor cross section area $A = 400$ mm^2 in the narrow curvature region.

$$\frac{B\rho I}{2A} = 75 \text{ N/mm}^2 \tag{2.9}$$

The hoop stress peak along the inner edge should reach about twice this value (see Fig. 2.19). To evaluate the change of $B\rho$, which is a measure for the hoop stress, $B\rho_1$ in Table 2.3 is taken and related to the final value, yielding

$$\frac{B\rho_1 - B\rho_6}{B\rho_6} = \frac{0.39}{3.71} = 10.6\% \qquad (2.10)$$

The result shows that the hoop stress decay during the iteration process is in the 10 to 15 percent range.

For comparison with the 3D-effects of the real design, the discussion must refer to the hoop stresses in the winding (mean and amplitude values). When the influence of the support given by the casing is evaluated, the stresses arising there are usually represented by the Von Mises equivalent stresses, as in Figs. 2.25 and 2.29. Special effects like peak shear stress components or gaps between casing and winding, in a particular load case, then require a separate treatment.

For the bare winding, the mean hoop stress is about 65 N/mm^2 which is close to Eq. 2.9. The amplitudes (see Fig. 2.19) depend on the particular support parameter combination. They vary between 40 N/mm^2 and 100 N/mm^2 [3, 6]. The comparison of these curves with the results for the complete coil (winding and casing) is interesting (see Fig. 2.24). In Table 2.4, the average values representing the neutral fiber middle curve, between the curve in Fig. 2.24, as well as the amplitudes of both curves referring to the mean values, are given for the 3 magnetic load cases in question [6].

TABLE 2.4 HOOP STRESS MEAN AND AMPLITUDE VALUES (N/mm^2) FOR CASING AND WINDING

Load Case:	Symmetric	Asymmetric	Overturn
Mean Hoop Stress	30	20 to 35* (30)	32
Amplitude Range	18 to 22	25 to 35	20 to 30

*not constant with azimuth, because of asymmetry; varies about a value of 30.

Obviously, the girth given by the outer ring of the casing does not only bring the breathing tension below the value expected, according to the contour optimization, but also decreases the bending stresses within the winding remarkably. From this, it appears that the casing reduces the 3D-bending-effects, in a way that makes the winding act more like a nearly flexible filament, as was assumed in the contour optimization.

The influence of support stiffness on the bending stresses of the bare winding are within the same range of 10 to 15% as Eq. 2.10 for the peaks in the narrow curvature part (Fig. 2.19). It is dramatic, however, for the peaks that may occur along the inner leg (Figs. 2.19, 2.20, and 2.21). Regarding this effect, recommended parameter combinations were found, after a parameter study of varying the support conditions. The findings are given in Table 2.2. Under asymmetric loadings eventually two additional design limits may exist:

(1) a very high shear stress peak, as shown in Fig. 2.28 and
(2) the most severe stress within the casing are found, among all load cases (see the value of 115 N/mm^2 reached at the middle of the front plate corner as shown in Fig. 2.27).

For all 3 load cases, the winding lifts off from the inner ring, due to the breathing behavior. This gives rise to a gap of about 0.35 mm width (see Fig. 2.23). This may be dangerous, because the safety against quenching pressure within a conductor, near that surface, might become insufficient [4]. The study of this problem has shown that the deformation of a quenching conductor within a cluster is not a safety problem. But the deformation of the conductor in the surface layer of the winding could be too large, if the gap occurring there is not limited by a proper design. Typical results for displacements under 10 bar are 1.17 mm for point A and 0.6 x 10^{-3} mm for point B, in Fig. 2.32.

Summing up the evaluation of design margins, one sees that the following influences must be quantified for design improvements:

(1) the supporting influence of the casing on the winding,
(2) the interaction between winding and casing, regarding hoop stress peaks along the inner leg support region, as well as gaps along the inner ring, and
(3) shear stress and equivalent stress peaks under asymmetric loading.

The first two influences are of the same magnitude as the margin of improvement achieved during the shape optimization. Therefore, the 3D-effects may be controlled by a proper casing design, in a global sense. The third kind of influence, however, is local and is peculiar to a specific loading, so that such effects must be mastered by local design changes.

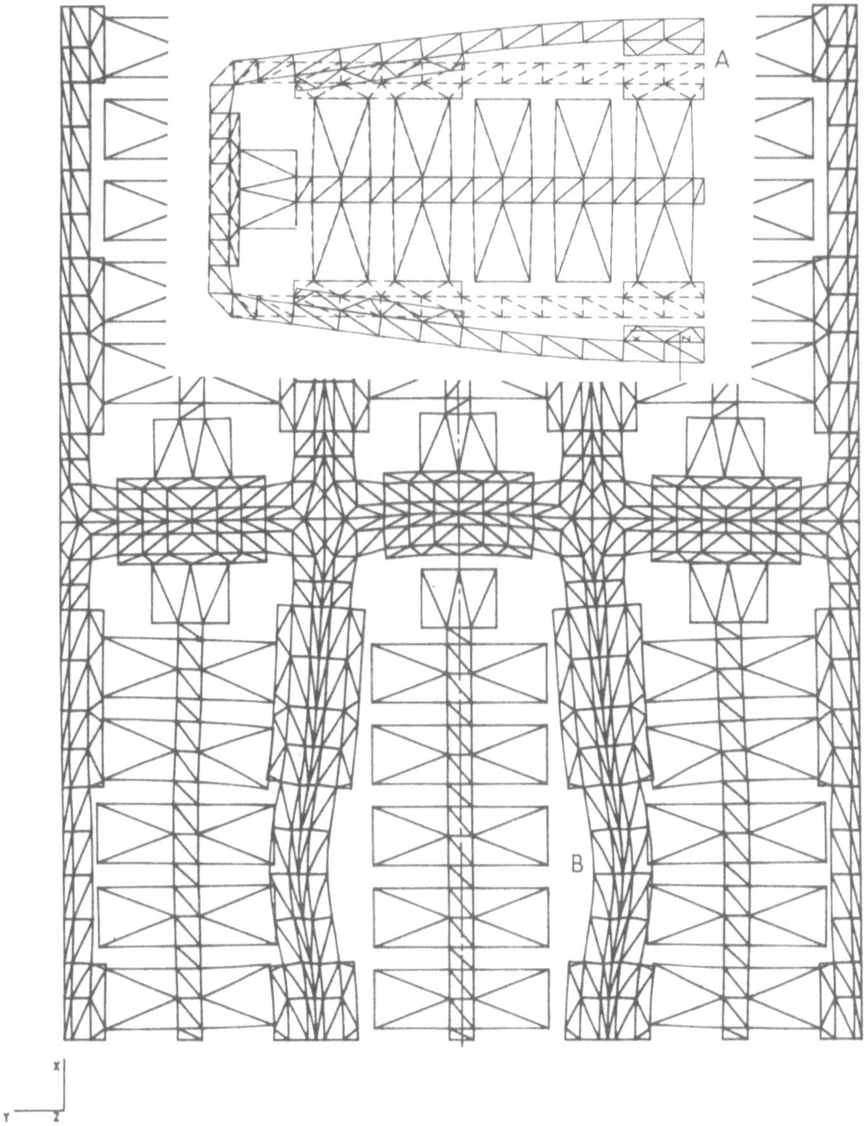

Figure 2.32 Conductor Cluster.

3. SOME ADDITIONAL INFLUENCES ON THE DESIGN PROCESS

There are other performance and safety-relevant arguments for design optimization, which can only be mentioned here:

(1) manufacture (cost, weldability, prestresses),
(2) pressure vessel safety,
(3) fracture mechanics under cryogenic conditions,
(4) cooldown stresses, sudden warm-up after cooling system failure,
(5) mechanical interactions with the outer support structure, and
(6) anisotropy of the composite coil structure (recently proven to have influence in the 10% range [5]).

The investigation of these problems is under way (see, e.g. Ref. 19).

4. CONCLUSIONS

It can be assumed that, besides localized stress peak constraints, the real design effects on the mechanical behavior of a toroidal field coil like the EURATOM LCT test coil are in the same range as the improvement achieved by the contour shape optimization. Therefore, it seems justified to improve the preliminary design in the 2 stage process, which was applied: ideal contour iteration and quantification of 3D effects. It also seems appropriate to complete the first stage, before a finite element model is derived. In the second stage, sophisticated finite element modelling is necessary to explore the range of disposable parameters that are prone to optimization. The selection of an optimum combination of them, however, can in such cases only be achieved after the choice and quantification of an optimum combination of objectives, from multitude possible ways. The extraction of a unique goal function, which may lead directly to an optimum compromise in such complex situations, without prior knowledge from analytical studies, is an unresolved problem that may arouse the interest of optimization experts.

REFERENCES

1. Komarek, P. and Krauth, H., "The 'Large Coil Task', An International Contribution To The Development Of Superconducting Magnets For Nuclear Fusion," Kerntechnik, Vol. 20, No. 6, 1978, pp. 274-281.
2. Haubenreich, P.N., et al., "Plan For The Large Coil Program," Proc. 8th Symp. Eng. Probl. Fusion Research, San Francisco, USA, 1979, Paper 27-01.

3. Krauth, H., et al., "Design Of The EURATOM Test Coil For The Large Coil Task," Proc. 8th Symp. Eng. Probl. Fusion Research, San Francisco, USA, 1979.
4. Erb, J., Grünhagen, A., Messemer, G. and Zehlein, H., "Finite Element Structural Analysis Of Coil And Casing Of A Large Superconducting Toroidal LCT-Magnet," 5th Internatl. Conf. on Structural Mechanics in Reactor Technology - SMIRT 5, Berlin, FRG, 1979, Paper N2/1-4.
5. Behrens, C.P., Krauth, H., Messemer, G. and Zehlein, H., "Festigkeitsmäßige Auslegung der LCT-Spule mit der Methode der Finiten Elemente," Proc. KTG'80 Reaktortagung, Bonn, 1980, pp. 917-926.
6. Messemer, G. and Zehlein, H., Unpublished internal reports; to appear as KfK-report, 1980.
7. Kajita, T. and Miyamoto, K., "Stress Distribution Of Coils For Toroidal Magnetic Field," Jap. J. appl. Phys., Vol. 15, No. 10, 1976, pp. 1965-1971.
8. File, J., Mills, R.G. and Sheffield, G.V., "Large Superconducting Magnet Designs For Fusion Reactors," IVth Symp. on Eng. Problems of Fusion Research, Washington, D.C., 1971.
9. Shafranov, V.D., "Optimum Shape Of A Toroidal Solenoid," Sov. Phys.-Techn. Phys., Vol. 17, 1973, p. 1433.
10. Gralnick, S.L. and Tenney, F.H., "Analytic Solutions For Constant-Tension Coil Shapes," J. of Appl. Phys., Vol. 47, 1976, p. 2710.
11. File, J. and Sheffield, G.V., "A Large Superconducting Magnet For Fusion Research," Proc. of the 4th Int. Conf. on Magnet Technology, Brookhoven, 1972.
12. Moses, R.W. Jr. and Young, W.C., "Analytic Expressions For Magnetic Forces On Sectored Toroidal Coils," VIth Symp. on Eng. Problems of Fusion Research, San Diego, 1975.
13. Erb, J. and Maurer, W., "Method For Determining The Magnet Shape In Toroidal Arrangements," Proc. 5th Internatl. Conf. on Structural Mechanics in Reactor Technology - SMIRT 5, Berlin, FRG, 1979, Paper N2/1-6.
14. Preis, H., Berechnung des Magnetischen Feldes, der Magnetischen Kräfte und des Betriebsverhaltens Großer Spulensysteme für Fusionsexperimente, Report IPP III/24, Garching, April, 1976.
15. ASKA, Part I - Linear Static Analysis, User's Reference Manual, ISD-Report No. 73, Stuttgart 1971; Rev. C., 1975.
16. Litherland, P.S., Conceptual Design Of The LCP Coil Support Structure, ORNL/TM-6/95, Oak Ridge, USA, June, 1968.
17. Gralnick, S.L., et al., "Compatibility Consideration For Zero Moment Takomak Toroidal Field Coils," Nucl. Technology, Vol. 45, 1979, pp. 233-243.
18. Dustmann, H., Erb, J., Krauth, H. and Maurer, W., Die Form von Magneten in toroidalen Anordnungen, Report KfK 2554, July, 1978.
19. Krauth, H. and Nyilas, A., Report KfK, to appear.

Part 7

DESIGN SENSITIVITY ANALYSIS

DESIGN SENSITIVITY ANALYSIS OF STATIC RESPONSE VARIATIONS*

Edward J. Haug and Bernard Rousselet

Materials Division, College of Engineering, University of Iowa, Iowa City, Iowa 52242

Département de Mathematiques, Université de Nice, 06034 Nice Cédex, France

ABSTRACT

The dependence of the solution of boundary-value problems of structural mechanics on design variables that specify material properties and distribution is central to the theory of structural optimization. Prototype problems are analyzed here using a variational method to include beams, plates, and plane elastic solids. Basic ideas and properties of Sobolev spaces that form the foundation for the variational formulation of boundary-value problems are introduced and discussed. Symmetry and positive definiteness properties of the elliptic differential operators that govern system behavior are used to show that their inverses, hence displacement fields, are Frechet differentiable with respect to design variables. Formulas for the derivatives are derived and used to obtain computable expressions for design sensitivity coefficients (first variations) of integrals that arise in optimal design formulations. These results establish an extension of the concept of "well posed" problems of structural mechanics, to include continuity (in fact differentiability) of static structural response with respect to distributed design variables and design parameters.

*Research of first author sponsored by National Science Foundation Project No. ENG77-19967. Travel support for joint efforts by the authors was provided by NATO Research Grant No. 1458.

1. INTRODUCTION

In spite of the well developed theory and computational methods for analysis of structural response to external load and determination of natural frequencies and buckling loads for a given structure, analysis of dependence of these response measures on design variations remains a relatively crude art. When the design variable is a parameter, one can calculate a difference quotient as an approximation of the derivative of structural response with respect to the design variable. If the design variable is a function, such as variable beam cross-section or variable plate thickness, this approach is not applicable. The need for a generally applicable method of design sensitivity analysis has become more acute as methods of structural design optimization have evolved.

Design sensitivity analysis methods employed in structural optimization have been heuristic in nature for the most part, assuming the problem is well posed [1] to justify formal perturbation analysis [2,3]. The idea of well posed problems of mathematical physics was introduced and studied by Hadamard in the 1920s [1]. The term "well posed" for initial- and boundary-value problems in the modern literature has come to be defined to mean that (i) the problem has a solution, (ii) the solution is unique, and (iii) the solution depends continuously on data defining the problem. The first two parts of this definition have a precise meaning, but the third is indefinite as to what constitutes "data".

For initial- and boundary-value problems, data are usually taken to be the functions appearing in the initial and boundary conditions and the nonhomogeneous terms in the differential equation. This definition of "data" is far too limited when one enters the realm of engineering design. Structural systems are defined and described by design data that may appear in unconventional and often subtle ways in the governing system of state equations. The size of beam elements and the thickness of plate elements may play the role of design variables that appear as functions in the coefficients of differential operators. This is an important "dependence on data" that goes well beyond the scope of the usual definition of well posed problems. Such problems have been treated formally in the engineering literature [2,3]. A second order nonsymmetric class of systems has recently been analyzed from a more fundamental point of view [4,5], using methods that are here combined with an adjoint equation approach [3] that has served well in numerous applications, but which relies on a stronger form of design sensitivity analysis.

Attention is restricted here to static structural systems governed by linear elliptic boundary-value problems. For such systems, the dependence of the solution of nonhomogeneous boundary-value problems on design functions and parameters is investigated. This class of problem includes deflection of beams, plates, and plane elastic solids. Design data can include material distribution along the length of beams, over the central surface of plates, and over thin elastic solids. Results obtained here for static response are used to develop comparable results in subsequent papers concerning eigenvalue dependence on design and to predict the effect of structural shape variation on response.

The results presented in this paper are developed in detail in Ref. 4. They rest on a foundation laid by Lions [6] and Fichera [7] in variational characterization of the solution of elliptic boundary-value problems and by Kato [8] in perturbation of bilinear forms and operators. In order to be concrete concerning classes of structures to which the methods apply, three examples are formulated in Section 2. These examples are then used throughout the paper to illustrate theoretical results. Frechet derivatives of bilinear forms arising in variational characterization of the equations of mechanics and Frechet differentiability of inverse state operators are proved to exist and are calculated in Section 3. An adjoint variable technique is presented in Section 4, in which the theoretical results and formulas of Section 3 are used to calculate Frechet derivatives of functionals arising in structural design.

Operator differentiability results presented here are used in Refs. 9 and 10 as tools for eigenvalue design sensitivity analysis and analysis of the effect of variation in the shape of a structural element. Results of these papers are applied for numerical optimization of structures in Refs. 11, 12 and 13. They are used for development of optimality criteria in Ref. 14.

2. PROTOTYPE PROBLEMS CONSIDERED

The purpose of this section is to formulate boundary-value problems that depend on design variables for concrete examples of structural systems.

2.1 Beam

Consider the beam of Fig. 2.1, with normalized axial coordinates, clamped supports, and variable cross-sectional area $h \in L^{\infty}(0,1)$. Here $L^{\infty}(a,b)$ is the space of Lebesque measurable functions $h(x)$ on the interval $a \le x \le b$, such that $||h||_{\infty} \equiv \inf\{M>0: |h(x)| \le M \text{ a.e., in } [a,b]\}$. For an introductory treatment of such function spaces, the reader is referred to Ref. 15. It is presumed that all dimensions of the cross-section vary with

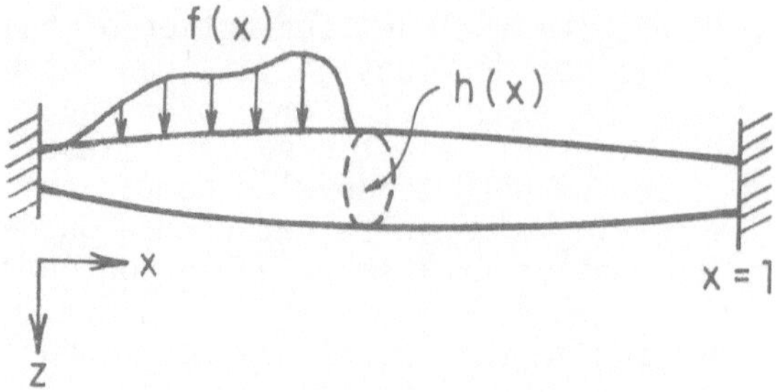

Figure 2.1 Clamped Beam of Variable Cross-Sectional Area h(x)

the same ratio, so the moment of inertia of cross-sectional area is $I(x) = \alpha h^2(x)$, where α is a positive constant depending on the shape of cross-section. The boundary-value problem for displacement z(x) is formally written as

$$\begin{aligned} &(E\alpha h^2(x) z_{xx})_{xx} = f(x) \\ &z(0) = z_x(0) = z(1) = z_x(1) = 0 \end{aligned} \tag{2.1}$$

where E is Young's modulus, f(x) is distributed load, and subscript notation is used to indicate derivatives.

The material constant E and material distribution function h(x) may be viewed as design variables, since they serve to specify the structure and may be selected by the designer. To simplify notation, they are denoted as a design vector $u = [E, h(x)]^T$ in a product space $R \times L^\infty(0,1)$ [15]. Presuming a solution z(x) of Eq. 2.1 exists, it is intuitively clear that it will depend on the design variable u. This dependence may be denoted z(x;u), which is a function defined on $0 \le x \le 1$ that depends on the design vector u. It is the nature of this dependence of solution of Eq. 2.1 on u that is to be analyzed.

In operator notation, the differential equation is

$$\bar{A}_u z \equiv (E\alpha h^2(x) z_{xx})_{xx} = f \tag{2.2}$$

where the subscript u denotes dependence of the differential operator on the design vector u. For $h \in L^\infty(0,1)$, the boundary-value problem of Eq. 2.1 is only formal. If h is twice continuously differentiable ($h \in C^2(0,1)$) and z is four times continuously differentiable, ($z \in C^4(0,1)$), then the problem of Eq. 2.1 has a classical meaning. Considering the classical case in which all functions are sufficiently smooth, one can write

$$\int_0^1 (\bar{A}_u z - f)v dx = 0 \tag{2.3}$$

for any solution of z of $\bar{A}_u z = f$, which must hold for any integrable function v. Conversely, if Eq. 2.3 holds for all twice continuously differentiable functions v that satisfy the boundary conditions of 2.1 and if $z \in C^4(0,1)$, then the differential equation of Eq. 2.1 is satisfied. This is true since the space of functions $\{v \in C^2(0,1): v(0) = v_x(0) = v(1) = v_x(1) = 0\}$ is dense in $L^2(0,1)$ [15].

One can now carry out two integrations by parts of the first term in Eq. 2.3 to obtain

$$0 = \int_0^1 E\alpha h^2 z_{xx} v_{xx}\, dx - \int_0^1 fv\, dx$$

$$+ [(E\alpha h^2 z_{xx})_x\, v - E\alpha h^2 z_{xx} v_x]\Big|_0^1$$

$$= \int_0^1 E\alpha h^2 z_{xx} v_{xx}\, dx - \int_0^1 fv\, dx \tag{2.4}$$

where the boundary terms vanish because v is required to satisfy the boundary conditions of Eq. 2.1. Defining the bilinear form

$$a_u(z,v) = \int_0^1 E\alpha h^2 z_{xx} v_{xx}\, dx \tag{2.5}$$

Eq. 2.4 is just

$$a_u(z,v) = (f,v) \tag{2.6}$$

where $(\cdot,\cdot)$ denotes the $L^2(0,1)$ scalar product;

$$(w,v) \equiv \int_0^1 w(x)v(x)\, dx.$$

It is important to note that the restriction of h to $C^2(0,1)$

and z to $C^4(0,1)$ are not only unnatural, but also unnecessary. The bilinear form $a_u(z,v)$ of Eq. 2.5 is well defined for $h \in L^\infty(0,1)$ and for any $z(x)$ and $v(x)$ that have second derivatives that are in $L^2(0,1)$. Thus, the variational equation of Eq. 2.6 may be satisfied for a function z having only one continuous derivative, with a possibly irregular second derivative that is only required to be in $L^2(0,1)$, and satisfying the boundary conditions of Eq. 2.1. Such a function is called the variational or generalized solution of the boundary-value problem of Eq. 2.1.

An alternate view of the variational formulation of the beam equation may be obtained from the minimum total potential energy characterization of beam bending. That is, the displacement $z(x)$ is to minimize

$$PE = \int_0^1 [E\alpha h^2 (z'')^2 - fz]dx$$

It is clear that the potential energy is well defined as long as $z'' \in L^2(0,1)$ and does not require z to be $C^4(0,1)$. Equating the first variation of PE to zero, with the variation $\eta(x)$ having two derivatives, $\eta'' \in L^2(0,1)$, and η satisfying the boundary conditions of Eq. 2.1, one obtains

$$\delta PE = \int_0^1 [E\alpha h^2 z''\eta'' - f\eta]dx = 0$$

But this is just Eq. 2.6, with η identified with v. Recovery of the differential equation of Eq. 2.1 is only possible if integration by parts can be justified, hence requiring either restrictive and physically unjustifiable assumptions on differentiability of z and h, or introduction of the idea of distributional derivatives [6,7,15], which in reality make the boundary-value problem of Eq. 2.1 equivalent to the variational equation of Eq. 2.6. Thus, the variational formulation is more natural from the point of view of mechanics than the fourth-order differential equation of Eq. 2.1.

The variational formulation of the problem yields a greater degree of generality if one defines solutions in Sobolev spaces [6,7,16] of functions. For functions of one variable x, the Sobolev space $H^m(0,1)$ is the collection of all functions that may be obtained as limits of $C^\infty(0,1)$ functions in the norm

$$||z||_{H^m} = \left[\sum_{i=0}^{m} \int_0^1 \left(\frac{d^i z(x)}{dx^i}\right)^2 dx\right]^{1/2} = \left[\sum_{i=0}^{m} \left\|\frac{d^i z}{dx^i}\right\|_{L^2}^2\right]^{1/2} \qquad (2.7)$$

Define a function of compact support on the open interval $]0,1[$ as a function $z(x)$ that is zero outside an interval $[\varepsilon,1-\varepsilon]$ for some $\varepsilon > 0$. The space of functions in $H^m(0,1)$ that are limits of $C^\infty(0,1)$ functions with compact support is denoted $H_0^m(0,1)$. It is known [6,7,16] that the set of all functions in $H^2(0,1)$ that satisfy the boundary conditions of Eq. 2.1 is precisely the space $H_0^2(0,1)$.

Of particular importance in analysis of beam problems by the variational method is the Sobolev Imbedding Theorem [16, Thm. 5.4]. Existence theory for variational equations of the form of Eq. 2.6 [7] guarantees that there will be a solution $z \in H^2(0,1)$. The natural question now is, "how smooth is $z \in H^2(0,1)$"? For functions of one variable, the Sobolev Imbedding Theorem [15, Thm. 5.4] asserts that $z \in C^1[0,1]$ and that there is a constant $C > 0$ such that

$$\max_{i=0,1} \max_{0 \leq x \leq 1} \left| \frac{d^i z(x)}{dx^i} \right| \leq C \, ||z||_{H^2(0,1)}$$

This result explains why principal boundary conditions are preserved when one takes H^2 limits of smooth functions that satisfy principal boundary conditions. For a compact introduction to Sobolev spaces and their application to structural mechanics, the reader is referred to the outstanding article of Fichera [7]. For a comprehensive treatment of the subject, see the book by Adams [16].

In a Sobolev space setting, the variational formulation of the boundary-value problem of Eq. 2.1 is to find a function $z \in H_0^2(0,1)$ such that Eq. 2.6 (with $h \in L^\infty(0,1)$) is satisfied for all $v \in H_0^2(0,1)$. This problem may be reformulated as an operator equation by defining the unbounded, self-adjoint linear operator A_u such that

$$(A_u z, v) = a_u(z,v) \tag{2.8}$$

for all $v \in H_0^2(0,1)$. The domain of the operator is the subspace $D(A_u)$ of $H_0^2(0,1)$ such that $A_u z \in L^2(0,1)$. It is shown in Ref. 17 that $D(A_u)$ is dense in $L^2(0,1)$ and, due to the Sobolev imbedding theorem [16], the identity map from $D(A_u) \subset H_0^2(0,1)$ to $L^2(0,1)$ is compact. It is further shown in in Refs. 7 and 17 that the operator equation

$$A_u z = f \tag{2.9}$$

has a unique solution in $D(A_u)$ for all $f \in L^2(0,1)$. Thus, the operator A_u defined by Eq. 2.8 is an extension of $\bar{A}_u$ of Eq. 2.2 and the operator equation of Eq. 2.9 is a generalization of Eq. 2.1, in the sense that: (a) any classical solution of Eq. 2.1 is a solution of Eq. 2.9, and (b) even though a classical solution of Eq. 2.1 may not exist for $h \in L^\infty(0,1)$ and $f \in L^2(0,1)$, there is a solution of Eq. 2.9 that is in fact the "natural solution" of the structural mechanics problem. For a proof of existence and uniqueness the reader is referred to the article of Fichera [7] or to Ref. 17.

In addition to the existence properties of Eq. 2.9, it is shown by Fichera [7] that the bilinear form $a_u(z,v)$ satisfies the following properties:

$$a_u(z,v) \leq K \, ||z||_{H^2} \, ||v||_{H^2} \tag{2.10}$$

$$a_u(z) \equiv a_u(z,z) \geq \gamma \, ||z||^2_{H^2} \tag{2.11}$$

where $K < \infty$ and $\gamma > 0$, provided $E \geq E_0 > 0$ and $h(x) \geq h_0 > 0$. Equation 2.10 states a form of upper bound on the bilinear form, while Eq. 2.11 is a lower bound that may also be written

$$(A_u z,z) \geq \gamma \, ||z||^2_{H^2} \tag{2.12}$$

which is a strong ellipticity property of the operator A_u.

Since Eq. 2.9 has a unique solution in $H_0^2(0,1)$, one may write

$$z(x;u) = A_u^{-1} f \tag{2.13}$$

to emphasize the dependence on u. Now, from Eq. 2.12,

$$(A_u z,z) = (f,A_u^{-1} f) \geq \gamma \, ||A_u^{-1} f||^2_{H^2} \tag{2.14}$$

By the Schwarz inequality [15],

$$||f||_{L^2} \, ||A_u^{-1} f||_{L^2} \geq |(f,A_u^{-1} f)|$$

Since $||v||_{H^2} \geq ||v||_{L^2}$ for all $v \in H^2(0,1)$, one has finally from Eq. 2.14

$$||A_u^{-1} f||_{H^2} \leq \frac{1}{\gamma} ||f||_{L^2} \tag{2.15}$$

for all $f \in L^2(0,1)$. Thus, the operator A_u^{-1} is continuous. It remains only to determine the regularity of dependence of A_u^{-1} on u.

While the foregoing analysis has been carried out with the clamped-clamped beam of Fig. 2.1, with boundary conditions of Eq. 2.1, the same results are valid for many other boundary conditions, to include the following support conditions and associated boundary conditions [7,17]:

Simply supported -

$$z(0) = z_{xx}(0) = z(1) = z_{xx}(1) = 0 \tag{2.16}$$

Cantilevered -

$$z(0) = z_x(0) = z_{xx}(1) = [E\alpha h^2(1)\ z_{xx}(1)]_x = 0 \tag{2.17}$$

Clamped-simply supported -

$$z(0) = z_x(0) = z(1) = z_{xx}(1) = 0 \tag{2.18}$$

The reader may note that since the boundary terms in Eq. 2.4 vanish if z and v satisfy these boundary conditions, the bilinear form $a_u(z,v)$ of Eq. 2.5 is valid for all boundary conditions. It is shown in Refs. 7 and 17 that the variational characterization of the solution in Eq. 2.6 is valid if z and v satisfy only boundary conditions of Eqs. 2.16, 2.17, or 2.18 that involve derivatives of order one or less. That is, boundary conditions involving derivatives of order two or three are natural and need not be satisfied by z and v in Eq. 2.6. It is further shown in Ref. 7 that Eqs. 2.10 and 2.11 are also valid for this class of functions. Hence, bounded invertibility of A_u is retained for the boundary conditions of Eqs. 2.16, 2.17, and 2.18.

2.2 Plate

Consider now the clamped plate of variable thickness $h(x) \geq h_0 > 0$, $h \in L^\infty(\Omega)$ shown in Fig. 2.2. The boundary-value problem for the displacement z is written in operator form as

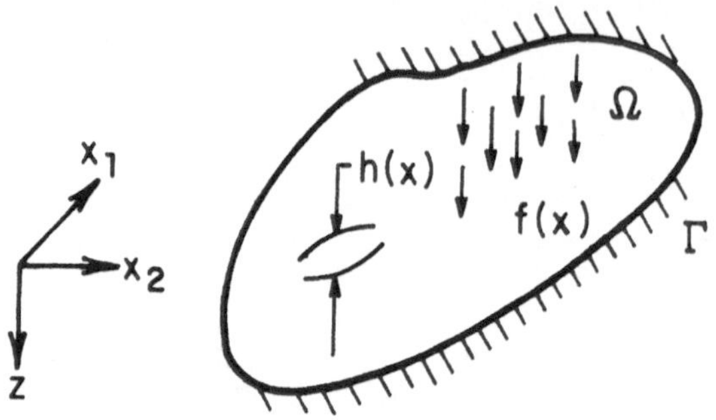

Figure 2.2 Clamped Plate of Variable Thickness h(x)

$$\left.\begin{array}{lll} \bar{A}_u z = f & & \text{in } \Omega \\ z = 0 \ , & \dfrac{\partial z}{\partial n} = 0 & \text{on } \Gamma \end{array}\right\} \tag{2.19}$$

where the operator $\bar{A}_u$ is defined as

$$\bar{A}_u z = [\hat{D}(u)(z_{11} + \nu z_{22})]_{11} + [\hat{D}(u)(z_{22} + \nu z_{11})]_{22}$$
$$+ 2(1 - \nu)[\hat{D}(u)z_{12}]_{12} \tag{2.20}$$

and a subscript i denotes the operation $\frac{\partial}{\partial x_i}$,

$$\hat{D}(u) = Eh^3/[12(1 - \nu^2)], \qquad E > E_0 > 0$$

is Young's modulus, ν is Poisson's ratio, and $u = (E,h(x))$.

The operator equation of Eq. 2.19 is equivalent to the variational formulation of Eq. 2.3, but with integration over , for functions $z \in C^4(\Omega)$, $h \in C^2(\Omega)$, and v and z satisfying boundary conditions of Eq. 2.19. Integrating by parts, one has

$$0 = \iint_\Omega (\bar{A}_u z - f)v \, d\Omega$$

$$= \iint_\Omega \hat{D}(u)[z_{11}v_{11} + \nu z_{22}v_{11} + z_{22}v_{22} + \nu z_{11}v_{22}$$

(equation con't)

$$+ 2(1-\nu)z_{12}v_{12}]d\Omega$$

$$- \iint_{\Omega} fv\, d\Omega + \int_{\Gamma} \Big\{[\hat{D}(u)(z_{11} + \nu z_{22})]_1 v n_1$$

$$- \hat{D}(u)(z_{11} + \nu z_{22})v_1 n_1$$

$$+ [\hat{D}(u)(z_{22} + \nu z_{11})]_2 v n_2 - \hat{D}(u)(z_{22} + \nu z_{11})v_2 n_2$$

$$+ 2(1-\nu)[\hat{D}(u)z_{12}]_1 v n_2 - 2(1-\nu)\hat{D}(u)z_{12}v_2 n_1\Big\} d\Gamma$$

$$\equiv a_u(z,v) - (f,v) \tag{2.21}$$

where n_1 and n_2 are components of the outward unit normal vector and the boundary terms vanish because v satisfies the boundary conditions of Eq. 2.19. For a smooth bounday $\partial v/\partial\tau = 0$, where τ is the tangential direction on Γ, since $v = 0$ on Γ. Since the normal and tangential derivatives of v are zero on Γ, both component derivatives are $v_1 = v_2 = 0$ on Γ. The bilinear form $a_u(z,v)$ is defined as the first integral over Ω in Eq. 2.21 and

$$(f,v) = \iint_{\Omega} fv\, d\Omega$$

is the $L^2(\Omega)$ scalar product on Ω.

The variational equation of Eq. 2.21 is also valid for the plate, with $h \in L^\infty(\Omega)$. Just as in the case of the beam, it is unnatural and unnecessary to restrict consideration of solutions of the variational equation to $C^4(\Omega)$. One may define admissible solutions in $H^2(\Omega)$, which is the completion of $C^\infty(\Omega)$ with the Sobolev norm

$$\|z\|_{H^2} = \left[\iint_{\Omega}\left[|z|^2 + |z_1|^2 + |z_2|^2 + |z_{11}|^2 + |z_{12}|^2 + |z_{22}|^2\right]d\Omega\right]^{1/2}$$

$$= \left[\sum_{i+j\le 2} \left\| \frac{\partial^{i+j} z}{\partial x_1^{\,i}\, \partial x_2^{\,j}} \right\|^2_{L^2(\Omega)}\right]^{1/2} \tag{2.22}$$

Further, the functions in $H^2(\Omega)$ satisfying the boundary conditions of Eq. 2.19 are limits in the norm of Eq. 2.22 of functions in $C^\infty(\Omega)$ that are zero outside compact subsets of the interior of Ω.

This space of functions is denoted as the Sobolev space $H_0^2(\Omega)$ [7,16,17].

Just as in the case of the beam, one may now define the Fredrichs extension A_u of the operator $\bar{A}_u$ for all $z \in D(A_u) \subset H_0^2(\Omega)$ that $A_u z \in L^2(\Omega)$. It is shown in Ref. 7 that this operator satisfies the bounds of Eqs. 2.10, 2.11, and 2.12. Thus, Eq. 2.13 is valid and A_u^{-1} is bounded, as in Eq. 2.15. The problem of determining the dependence of the inverse plate operator on u is of the same form as that for the beam. While the calculation is not as trivial as in the case of the beam, it is shown in Ref. 7 that the foregoing results are valid for a simply supported plate; i.e., boundary conditions

$$z = 0, \quad \frac{\partial^2 z}{\partial n^2} + \nu\left(\frac{\partial^2 z}{\partial \tau^2} - \frac{1}{s}\frac{\partial z}{\partial n}\right) = 0 \quad \text{on } \Gamma \tag{2.23}$$

where s is the radius of curvature of Γ divided by $1 - \nu$.

It should be noted that even though the algebra and calculus associated with the plate are more complex than for the beam, exactly the same operator properties hold. As will be seen in the next section, the dependence of the solution of the operator equation governing the plate is of the same form as in the case for the beam.

2.3 Plane Elasticity

Finally, consider the variable thickness, thin elastic body with in-plane loading and fixed edges shown in Fig. 2.3. While Poisson's ratio ν is constant, the variable thickness $h(x)$ may be treated by defining a variable effective Young's Modulus $E(x) = \alpha h(x) \geq \alpha h_0 > 0$. Here the design variable is just $u = h(x)$, the equations for elasticity for displacement components $z(x) = [z^1(x), z^2(x)]^T$ may be written as

$$\left.\begin{aligned} A_u z &\equiv \frac{-\alpha}{2(1+\nu)}\begin{bmatrix} [4h\nu(z_1^1 + z_2^2) + 2hz_1^1]_1 + [h(z_2^1 + z_1^2)]_2 \\ [h(z_2^1 + z_1^2)]_1 + [4h\nu(z_1^1 + z_2^2) + 2hz_2^2]_2 \end{bmatrix} \\ &= \begin{bmatrix} f^1 \\ f^2 \end{bmatrix}, \text{ in } \Omega \qquad z^1 = z^2 = 0, \quad \text{on } \Gamma \end{aligned}\right\} \tag{2.24}$$

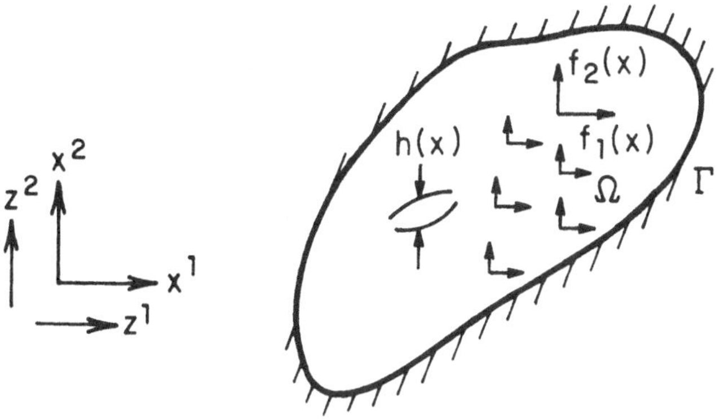

Figure 2.3 Clamped Elastic Solid of Variable Thickness h(x)

In this problem, a classical solution would require $z^i \in C^2(\Omega)$, i=1,2, and $h \in C^1(\Omega)$. Forming the scalar product of both sides of Eq. 2.24 with $v(x) = [v^1(x), v^2(x)]^T$ and integrating by parts, on obtains the variational equation

$$
\begin{aligned}
0 &= (A_u z - f, v) \\
&= -\frac{\alpha}{2(1+\nu)} \iint_\Omega \Big\{ [4h\nu(z_1^1 + z_2^2) + 2hz_1^1]_1 v^1 + [h(z_2^1 + z_1^2)]_2 v^1 \\
&\quad + [h(z_2^1 + z_1^2)]_1 v^2 + [4h\nu(z_1^1 + z_2^2) + 2hz_2^2]_2 v^2 \Big\} d\Omega \\
&\quad - \iint_\Omega (f^1 v^1 + f^2 v^2)\, d\Omega \\
&= \frac{\alpha}{2(1+\nu)} \iint_\Omega h \Big\{ [4\nu(z_1^1 + z_2^2)][v_1^1 + v_2^2] + 2(z_1^1 v_1^1 + z_2^2 v_2^2) \\
&\quad + (z_2^1 + z_1^2)(v_2^1 + v_1^2) \Big\} d\Omega - \iint_\Omega (f^1 v^1 + f^2 v^2)\, d\Omega \\
&\quad - \frac{\alpha}{2(1+\nu)} \int_\Gamma h \Big\{ [4\nu(z_1^1 + z_2^2)][v^1 n_1 + v^2 n_2]
\end{aligned}
$$

(equation cont'd)

$$+ 2(z_1^1 v^1 n_1 + z_2^2 v^2 n_2) + (z_2^1 + z_1^2)(v^1 n_2 + v^2 n_1)\Big\} d\Gamma$$

$$\equiv a_u(z,v) - (f,v) \qquad (2.25)$$

where the boundary terms vanish if v is required to satisfy the boundary conditions $v^1 = v^2 = 0$ on Γ. The variational equation of Eq. 2.25 gives a generalized solution provided only that $z \in H^1(\Omega) \times H^1(\Omega)$, $h \in L^\infty(\Omega)$, and $f \in L^2(\Omega)$. Here the Sobolev norm on $H^1(\Omega) \times H^1(\Omega)$ is

$$\|z\|_{H^1 \times H^1} = \left[\iint_\Omega (|z^1|^2 + |z^2|^2 + |z_1^1|^2 + |z_2^1|^2 + |z_1^2|^2 + |z_2^2|^2) d\Omega\right]^{1/2}$$

$$= \left[\sum_{i=1}^{2} \sum_{j+k\leq 1} \left\| \frac{\partial^{j+k} z^i}{\partial x_1^{\,j} \partial x_2^{\,k}} \right\|^2_{L^2(\Omega)}\right]^{1/2} \qquad (2.26)$$

The subspace of functions of $H^1(\Omega) \times H^1(\Omega)$ satisfying the boundary conditions of Eq. 2.24 is $H_0^1(\Omega) \times H_0^1(\Omega)$, which is the completion in the norm of Eq. 2.26 of $C^\infty(\Omega)$ functions that vanish outside compact subsets of the open set Ω.

All of the bounds and hence the bounded invertibility of the Friedrich's extension A_u of $\bar{A}_u$ follow just as for the beam and plate, where $\nu > 0$ and $h(x) \geq h_0 > 0$. This result is proved in Ref. 7 for the clamped boundary conditions and for more general boundary conditions. It is noted that for even this class of complex elastic systems, symmetry and strong ellipticity properties of the operator hold. As in the case of the preceding examples, the design variable u appears in a regular way in the coefficients of the divergence form differential operator A_u. One again wishes to determine the regularity of dependence of the state variable vector z(x;u) on the design variable u.

The example problems of this section have been selected to illustrate classes of distributed parameter structures in which design dependence arises in a consistent way. In each case, Dirichlet boundary conditions are treated in detail. Selection of these boundary conditions is a convenience, rather than a requirement. If one were to use the trace boundary operator theory in its full generality [6,7], he could also treat Neuman and a variety of mixed boundary conditions that arise naturally in applications, with only a penalty in analytic and algebraic messiness. A detailed treatment of these extensions, however, would make this paper unnecessarily complex for the intended purpose of ullustrating uniformity of form and method in problems of structural system design. For a comprehensive treatment of the more general boundary conditions, the reader is referred to Ref. 7.

3. FRECHET DIFFERENTIABILITY OF THE INVERSE STATE OPERATOR WITH RESPECT TO DESIGN

To investigate differentiability of A_u^{-1}, return now to the operator setting. In each of the problems of Section 2, a displacement state variable z(x) is determined by an operator equation of the form

$$A_u z = f \,, \qquad z \in D(A_u) \tag{3.1}$$

where $D(A_u)$ is a subspace of an appropriate Sobolev space V that is defined by boundary conditions of the problem; H_0^2 for the beam and plate with Dirichlet boundary conditions and $H_0^1 \times H_0^1$ for the fixed boundary plane elasticity problem. The forcing function f is in $L^2(\Omega)$. In each problem of Section 2, one has the following properties:

(i) The vector space $D(A_u)$ is a subspace of a Sobolev space V, inherenting with its scalar product and norm.

(ii) The vector space $D(A_u) \subset L^2(\Omega)$ and by the Sobolev Imbedding Theorem [16, Thm. 5.4] the identity operator from $D(A_u)$ into $L^2(\Omega)$ is compact.

(iii) The operator $A_u : D(A_u) \to L^2(\Omega)$ is self adjoint and strongly elliptic; i.e., there is a constant $\gamma \neq 0$ such that $(Az,z) \geq \gamma^2 ||z||_V^2$, for all $z \in D(A_u)$. This property holds for all values of design variables and domains under construction.

Under these hypotheses, it is proved in Ref. 17 that the solution of Eq. 3.1 is the unique solution of the variational equation

$$a_u(z,v) = (f,v) \,, \qquad \text{for all } v \in V \tag{3.2}$$

where $a_u(z,v)$ is the symmetric bilinear form that defines the Friedrich's extension A_u of the operators $\bar{A}_u$ of Section 2.

Theorem 3.1 *(Differentiability Theorem for Bilinear Forms):* Let $a_u(z,z) \equiv a_u(z)$ be the quadratic form associated with a_u. Each bilinear form $a_u(z,v)$ of Section 2 is Frechet differentiable with respect to u, in the sense of relatively bounded perturbations; i.e., the differential of $a_u(z,v)$ is a linear form $a^{(1)}_{u,\delta u}(z,v)$ in δu, such that with $a^{(1)}_{u,\delta u}(z) \equiv a^{(1)}_{u,\delta u}(z,z)$,

$$|a^{(1)}_{u,\ u}(z)| \leq e_1(u)||\delta u||a_u(z) \tag{3.3}$$

and with

$$a^{2}_{u,\delta u}(z) \equiv a_{u+\delta u}(z) - a_u(z) - a^{(1)}_{u,\delta u}(z) \tag{3.4}$$

the bound

$$|a^{2}_{u,\delta u}(z)| \leq e_2(u,\delta u)||\delta u||a_u(z) \tag{3.5}$$

is valid, with $e_2(u,\delta u) \to 0$ as $||\delta u|| \to 0$.

Proof: To prove the theorem, one considers each system individually.

3.1 Beam

A formal variation calculation from Eq. 2.5 yields

$$a^{(1)}_{u,\delta u}(z,v) = (\delta E)\alpha \int_0^1 h^2 z_{xx} v_{xx}\, dx + E\alpha \int_0^1 2h\delta h z_{xx} v_{xx}\, dx \tag{3.6}$$

With $||\delta u|| = |\delta E| + ||\delta h||_{L^\infty}$, one has

$$|a^{(1)}_{u,\delta u}(z)| \leq \frac{1}{E}\left[E\alpha \int_0^1 h^2 (z_{xx})^2\, dx\right] |\delta E|$$

$$+ \frac{2}{h_0}\left[E\alpha \int_0^1 h^2 (z_{xx})^2\, dx\right] ||\delta h||_{L^\infty} \leq \max\left(\frac{1}{E},\frac{2}{h_0}\right) ||\delta u|| a_u(z)$$

$$|a^2_{u,\delta u}(z)| = \left|\alpha \int_0^1 [E(\delta h)^2 + 2\delta E h \delta h + \delta E(\delta h)^2](z_{xx})^2\, dx\right|$$

$$\le \left[\frac{||(\delta h)^2||_{L^\infty}}{h_0^2} + \frac{2|\delta E|\ ||\delta h||_{L^\infty}}{E_0 h_0} + \frac{|\delta E|\ ||(\delta h)^2||_{L^\infty}}{E_0 h_0^2}\right]$$

$$\cdot \int_0^1 E\alpha h^2 (z_{xx})^2\, dx$$

$$\le ||\delta h||_{L^\infty} \left\{\frac{2||\delta h||_{L^\infty}}{h_0^2} + \frac{2|\delta E|}{E_0 h_0}\right\} a_u(z),$$

if $||\delta u|| < 1$ and $E_0 > 1$

$$\le 2||\delta h||_{L^\infty} \max\left(\frac{1}{h^2}, \frac{1}{E_0 h_0}\right) ||\delta u|| a_u(z)$$

This establishes Eqs. 3.3 and 3.5, since

$$2||\delta h||_{L^\infty} \max\left(\frac{1}{h_0^2}, \frac{1}{Eh_0}\right) \to 0 \quad \text{as} \quad ||\delta u|| \to 0.$$

3.2 Plate

A formal variational calculation from Eq. 2.21 leads to

$$a^{(1)}_{u,\delta u}(z,v) = \iint_\Omega \left[\delta E \frac{h^3}{12(1 - \nu^2)} + \delta h(x) \frac{3Eh^2}{12(1 - \nu^2)}\right]$$

$$[z_{11}v_{11} + \nu(z_{22}v_{11} + v_{22}z_{11}) + z_{22}v_{22}$$

$$+ 2(1 - \nu)z_{12}v_{12}]\, d\Omega \tag{3.7}$$

With $||\delta u|| = |\delta E| + ||\delta h||_{L^\infty}$ and since a_u is linear in E, one has

$$|a^{(1)}_{\delta u}(z)| \leq \max\left(\frac{1}{E_0}, \frac{3}{h_0}\right) a_u(z)||\delta u||$$

$$|a^2_{u,{}^2u}(z)| = \left|\iint_\Omega \frac{E\hat{h}}{12(1-\nu)^2}(\delta h)^2\left[(z_{11})^2 + 2\nu z_{22}z_{11} + (z_{22})^2 + 2(1-\nu)(z_{12})^2\right] d\Omega\right|$$

$$\leq (C||\delta h||^2_{L^\infty}) a_u(z)$$

$$\leq C||\delta h||_{L^\infty}||\delta u||a_u(z)$$

where $\hat{h}$ is evaluated near h, by Taylor's formula. Since $||\delta h||_{L^\infty} \to 0$ as $||\delta u|| \to 0$, Eqs. 3.3 and 3.5 are verified.

3.3 Plane Elasticity

A formal variational calculation from Eq. 2.25 leads to

$$a^{(1)}_{u,\delta u}(z,v) = \frac{1}{2(1+\nu)}\iint_\Omega \alpha\delta h\left\{v^1_1\left[4\nu(z^1_1 + z^2_2) + 2z^1_1\right] + v^1_2(z^1_2 + z^2_1) + v^2_1(z^1_2 + z^2_1) + v^2_2\left[4\nu(z^1_1 + z^2_2) + 2z^2_2\right]\right\} d\Omega \qquad (3.8)$$

$$|a^{(1)}_{u,\delta u}(z)| \leq \frac{1}{h_0}||\delta h||_{L^\infty} a_u(z)$$

so Eq. 3.3 holds. Also since $a_u(z)$ is linear in u, $a^2_{u,\delta u}(z) = 0$, and Eq. 3.5 holds trivially. This completes the proof.

In Section 2 and above, the consistency of operator and bilinear functional properties for a class of structural systems governed by linear elliptic boundary-value problems. In this section these properties are used to prove that the inverse operator associated with Eq. 3.1 is Frechet differentiable with respect to the design vector u.

Theorem 3.2 *(Frechet Differentiability of the Inverse State Operator):* Let the operator A_u and bilinear form $a_u(z,v)$ correspond through the identity

$$a_u(z,v) = (A_u z,v)_{L^2} \qquad (3.9)$$

for all $z \in D(A_u)$ and all $v \in V$, where the space V is dense in $L^2(\Omega)$, with compact imbedding into $L^2(\Omega)$. Further, let

$$a_u(z) \geq \gamma||z||_V^2 \tag{3.10}$$

for $\gamma > 0$ and for all $z \in V$. Let $a^{(1)}_{u,\delta u}(z)$ be the Frechet differential of $a_u(z)$ with respect to u with property

$$|a^{(1)}_{u,\delta u}(z)| \leq e_1(u)||\delta u||a_u(z) \tag{3.11}$$

and $a^2_{u,\delta u}(z)$ be the remainder term with the property

$$|a^2_{u,\delta u}(z)| = |a_{u+\delta u}(z) - a_u(z) - a^{(1)}_{u,\delta u}(z)|$$

$$\leq e_2(u,\delta u)||\delta u||a_u(z) \tag{3.12}$$

Then the Frechet derivative of A_u^{-1} with respect to u exists and is given as

$$\delta_u A_u^{-1} = -G_u^{-1} C_1(u,\delta u) G_u^{-1} \tag{3.13}$$

where the operators G_u and $C_1(u,\delta u)$ are defined as part of the proof, which follows.

Proof: As a result of Eq. 3.10 [8, p. 281], there is a non-negative, self-adjoint invertible operator G_u (the square root of A_u) such that $G_u G_u = A_u$ and

$$a_u(z,v) = (G_u z, G_u v) \tag{3.14}$$

The domain of G_u is exactly the domain of the form a_u, which contains $D(A_u)$. Further, as a result of Eqs. 3.11 and 3.12 and Lemma VI.3.1 of Ref. 8, there are continuous operators $C_1(u,\delta u)$ and $C_2(u,\delta u)$ from $L^2(\Omega)$ into $L^2(\Omega)$ such that

$$a^{(1)}_{u,\delta u}(z,v) = (C_1(u,\delta u)\; G_u z,\; G_u v) \tag{3.15}$$

$$a^2_{u,\delta u}(z,v) = (C_2(u,\delta u)\; G_u z,\; G_u v) \tag{3.16}$$

where $C_1(u,\delta u)$ is linear in u, since $a^{(1)}_{u,\delta u}(z,v)$ is linear in u, and the norms of $C_1(u,\delta u)$ and $C_2(u,\delta u)$ are bounded by

$$\left.\begin{aligned} ||C_1(u,\delta u)|| &\le e_1(u)||\delta u|| \\ ||C_2(u,\delta u)|| &\le e_2(u,\delta u)||\delta u|| \end{aligned}\right\} \tag{3.17}$$

Note that for every z in the domain of $A_{u+\delta u}$

$$\begin{aligned} (A_{u+\delta u}z,v) &= a_{u+\delta u}(z,v) = a_u(z,v) + a^{(1)}_{u,\delta u}(z,v) + a^2_{u,\delta u}(z,v) \\ &= ((G_u + C_1G_u + C_2G_u)z,G_uv) \\ &= (G_u(I + C_1 + C_2)G_uz,v) \end{aligned} \tag{3.18}$$

for all $v \in V$, where the arguments of the operators $C_1(u,\delta u)$ and $C_2(u,\delta u)$ have been suppressed for notational convenience. Since V is dense in $L^2(\Omega)$ and from the definition of V,

$$A_{u+\delta u} = G_u(I + C_1 + C_2)G_u{}^{\dagger} \tag{3.19}$$

Since $||C_1 + C_2|| \le ||C_1|| + ||C_2|| \le [e_1(u) + e_2(u,\delta u)]||\delta u||$, for sufficiently small $||\delta u||$, $I + C_1 + C_2$ has an inverse. Thus,

$$A^{-1}_{u+\delta u} = G^{-1}_u(I + C_1 + C_2)^{-1}\, G^{-1}_u \tag{3.20}$$

Note that

$$\begin{aligned} A^{-1}_{u,\delta u} - A^{-1}_u + G^{-1}_uC_1G^{-1}_u &= G^{-1}_u[(I + C_1 + C_2)^{-1} - I]G^{-1}_u \\ &\quad + G^{-1}_uC_1G^{-1}_u \end{aligned} \tag{3.21}$$

Thus,

$$||A^{-1}_{u+\delta u} - A^{-1}_u + G^{-1}_uC_1G^{-1}_u|| \le ||G^{-1}_u||^2||(I + C_1 + c^2)^{-1} - I + C_1|| \tag{3.22}$$

---

†From Eq. 3.18, $D(A_{u+\delta u}) \subset D(G_u(I + C_1 + C_2)G_u)$, but $A_{u+\delta u}$ is defined to be the maximal operator such that the first equality in Eq. 3.18 holds. Thus the domains are in fact equal.

Now, manipulating and applying the triangle inequality, one has

$$||(I+C_1+C_2)^{-1} - I + C_1|| = ||(I+C_1+C_2)^{-1} - I + C_1 + C_2 - C_2||$$

$$\leq ||(I+C_1+C_2)^{-1} - I + C_1 + C_2|| + ||C_2|| \quad (3.23)$$

For $||\delta u||$ sufficiently small, $||C_1 + C_2|| < 1/2$ and one can apply the theory of Neumann series [8, p. 30] to obtain the bound

$$||(I+C_1+C_2)^{-1} - I + C_1 + C_2|| \leq \frac{||C_1+C_2||^2}{1 - ||C_1+C_2||}$$

$$\leq 2||C_1+C_2||^2$$

The triangle inequality now yields

$$||(I+C_1+C_2)^{-1} - I + C_1 + C_2||$$

$$\leq 2||C_1||^2 + 4||C_1|| \; ||C_2|| + 2||C_2||^2 \quad (3.24)$$

Thus, Eqs. 3.11, 3.12, 3.16, 3.21, and 3.23 yield

$$||A^{-1}_{u+\delta u} - A^{-1}_u + G^{-1}_u \; C_1(u,\delta u)G^{-1}_u||$$

$$\leq ||G^{-1}_u||^2 \Big\{ 2\Big[e_1^2(u) + 2e_1^2(u)e_2(u,\delta u)$$

$$+ e_2^2(u,\delta u)\Big] \; ||\delta u||^2 + e_2(u,\delta u) \; ||\delta u|| \Big\} \quad (3.25)$$

$$= ||G^{-1}_u||^2 \Big\{ 2\Big[e_1^2(u) + 2e_1(u)e_2(u,\delta u) + e_2^2(u,\delta u)\Big] ||\delta u||$$

$$+ e_2(u,\delta u)\Big\} ||\delta u|| \equiv e_3(u,\delta u) \; ||\delta u|| \quad (3.26)$$

and $e_3(u,\delta u) \to 0$ as $||\delta u|| \to 0$. Thus, A^{-1}_u is Frechet differentiable and the proof is complete.

This result establishes the Frechet differentiability with respect to u of the solution z(x;u) of the operator equation of of Eq. 3.1. That is,

$$z(x;u) = A_u^{-1} f \tag{3.27}$$

as a mapping from $u \in U$ into V, has the Frechet differential

$$\delta_u z = (\delta_u A_u^{-1}) f = -G_u^{-1} C_1(u, \delta u)\, G_u^{-1} f \tag{3.28}$$

The importance of this result is theoretical at this point, since the explicit forms of G_u, G_u^{-1}, and $C_1(u,\delta u)$ are not known and in fact may not be readily computable. Computation of explicit design derivatives of functionals involved in a variety of structural problems is carried out using a state space adjoint computation in Ref. 3. A rigorous derivation is given in the following section of this paper.

An extension of the results presented here to forcing functions $f(x)$ that are not in $L^2(\Omega)$ is presented in Ref. 4. This is of value for problems in which applied loads are modeled as concentrated loads and moments. These forcing functions must be viewed as distributions, or bounded linear functionals, acting on displacement fields in $H_0^2(\Omega)$ or $H_0^1(\Omega)$. These spaces of distributions are denoted $H^{-2}(\Omega)$ and $H^{-1}(\Omega)$, so the operators A_u must be viewed as mappings from $H_0^i(\Omega)$ to $H^{-i}(\Omega)$, i=2 or 1. While the analysis presented in Ref. 4 for this extension is technically more complex, the same results proved here for $f \in L^2(\Omega)$ are shown to be valid.

4. CALCULATION OF DESIGN DERIVATIVES IN STRUCTURAL ENGINEERING

Rational formulations of problems of optimal structural design typically involve a cost functional that is to be minimized and inequality constraints on functionals that represent performance or failure criteria. For numerous examples, the reader is referred to Ref. 18. Consider a typical functional

$$\psi = \psi(u,z) \tag{4.1}$$

that is a differentiable mapping from $U \times V$ into the reals, where U is a Hilbert space and the L^2 Hilbert space structure is employed on V.

The total differential of ψ is

$$\delta\psi = (D_u\psi, \delta u)_U + (D_z\psi, \delta z)_{L^2} \tag{4.2}$$

Using the result of Eq. 3.28, one has

$$\delta\psi = (D_u\psi,\delta u)_U - (D_z\psi, G^{-1}C_1(u,\delta u)G_u^{-1}f)_{L^2} \tag{4.3}$$

Using self-adjointness of G_u and the fact that $G_uG_u = A_u$, one may manipulate to obtain

$$\begin{aligned}
\delta\psi &= (D_u\psi,\delta u)_U - (G_u^{-1}D_z\psi, C_1(u,\delta u)G_u^{-1}f)_{L^2} \\
&= (D_u\psi,\delta u)_U - (G_u^{-1}D_z\psi, G_u^{-1}G_uC_1(u,\delta u)G_uA_u^{-1}f)_{L^2} \\
&= (D_u\psi,\delta u)_U - (G_u^{-1}G_u^{-1}D_z\psi, G_uC_1(u,\delta u)G_uz)_{L^2} \\
&= (D_u\psi,\delta u)_U - (G_uA_u^{-1}D_z\psi, C_1(u,\delta u)G_uz)_{L^2}
\end{aligned} \tag{4.4}$$

Defining $A_u^{-1}D_z\psi = -\lambda$, or equivalently λ as the solution of the adjoint equation

$$A_u\lambda = -D_z\psi \tag{4.5}$$

and using Eq. 3.14, Eq. 4.4 may be rewritten as

$$\delta\psi = (D_u\psi,\delta u)_U + a^{(1)}_{u,\delta u}(z,\lambda) \tag{4.6}$$

which is written explicitly in terms of δu.

The form of Eq. 4.6 is the same as that obtained using heuristic arguments in the structural optimization literature [3,18]. To be more specific, consider an integral functional for the clamped beam

$$\psi = \int_0^L F(z(x),E,h(x))\,dx$$

By direct calculation and use of Eq. 3.6 in Eq. 4.6, one has

$$\delta\psi = \int_0^L \left(\frac{\partial F}{\partial E}\,\delta E + \frac{\partial F}{\partial h}\,\delta h\right) dx + \int_0^L \frac{\partial F}{\partial z}\,\delta z\; dx$$

$$= \left\{\int_0^L \left(\frac{\partial F}{\partial E} + \alpha h^2 z_{xx}\lambda_{xx}\right) dx\right\}\delta E$$

$$+ \int_0^L \left\{\left(\frac{\partial F}{\partial h} + 2E\alpha h z_{xx}\lambda_{xx}\right)\delta h\right\} dx \tag{4.7}$$

where λ satisfies the clamped boundary conditions of the beam and the differential equation of Eq. 4.5, which is

$$(E\alpha h^2 \lambda_{xx})_{xx} = -\frac{\partial F}{\partial z}, \quad \lambda \in D(A_u) \subset H_0^2(0,1) \equiv V \tag{4.8}$$

Similarly, for a functional

$$\psi = \iint_\Omega F(z(x),E,h(x))\; d\Omega$$

associated with a clamped plate, Eqs. 4.6 and 3.7 yield

$$\delta\psi = \left[\iint_\Omega \left\{\frac{\partial F}{\partial E} + \frac{h^3}{12(1-\nu^2)}\,[z_{11}\lambda_{11} + \nu(z_{22}\lambda_{11} + \lambda_{22}z_{11}) + z_{22}\lambda_{22}\right.\right.$$

$$\left.\left. + 2(1-\nu)z_{12}\lambda_{12}]\right\} d\Omega\right]\delta E + \iint_\Omega \left\{\frac{\partial F}{\partial h} + \frac{Eh^2}{4(1-\nu^2)}\,[z_{11}\lambda_{11}\right.$$

$$\left. + \nu(z_{22}\lambda_{11} + \lambda_{22}z_{11}) + z_{22}\lambda_{22} + 2(1-\nu)z_{12}\lambda_{12}]\right\}\delta u\; d\Omega \tag{4.9}$$

where λ satisfies the operator equation

$$A_u\lambda = -\frac{\partial F^T}{\partial z}, \quad \lambda \in D(A_u) \subset H_0^2(\Omega) \equiv V \tag{4.10}$$

and A_u is defined as an extension of Eq. 2.20.

While slightly more complicated from an algebraic point of view, a similar functional for the plane elasticity problem is

$$\psi = \iint_\Omega F(z(x),h(x))\; d\Omega$$

By direct calculation and use of Eq. 3.8 in Eq. 4.6, one has

$$\delta\psi = \iint_{\Omega} \left\{ \frac{\partial F}{\partial h} + \frac{\alpha}{2(1+\nu)} \left\{ \lambda_1^1 \left[4\nu(z_1^1 + z_2^2) + 2z_1^1 \right] + \lambda_2^1 (z_2^1 + z_1^2) \right.\right.$$

$$\left.\left. + \lambda_1^2 (z_2^1 + z_1^2) + \lambda_2^2 \left[4\nu(z_1^1 + z_2^2) + 2z_2^2 \right] \right\} \right\} \delta h \, d\Omega \qquad (4.11)$$

where λ satisfies the operator equation

$$A_u \lambda = - \frac{\partial F^T}{\partial z}, \; \lambda \in D(A_u) \subset H_0^1(\Omega) \times H_0^1(\Omega) \equiv V \qquad (4.12)$$

where $\frac{\partial F}{\partial z} = \left[\frac{\partial F}{\partial z^1}, \frac{\partial F}{\partial z^2} \right]$ and A_u is defined as an extension of Eq. 2.24.

For applications of these results in design sensitivity analysis and optimization, the reader is referred to Refs. 3, 11-14, and 18. The method of analysis and results of this paper demonstrate that the solution of a large class of boundary-value problems of mechanics depend differentiably on design variables. This may be stated as an extension of Hadamard's definition of well-posed boundary value problem; i.e., problems of mechanics treated here are "well-posed in design".

Of more practical value, these results allow the engineer to linearize his problem and carry out design sensitivity analysis and iterative optimization. Due to the more elementary nature of ordinary differential equations arising in modern control theory, the control engineer has always had the "mathematical right" to linearize his differential equation. The results contained herein now allow the mechanical and structural design engineer the same latitudes with their more complex partial differential equations.

REFERENCES

1. Hadamard, J., Lectures on Cauchy's Problem, Dover Publications, New York, 1952.
2. Farshad, M., "Variations in Eigenvalues and Eigenvectors in Continuum Mechanics," AIAA Journal, Vol. 12, 1974, pp. 560-561.
3. Haug, E.J. and Arora, J.S., "Design Sensitivity Analysis of Elastic Mechanical Systems," Computer Methods in Applied Mechanics and Engineering, Vol. 15, 1978, pp. 35-62.

4. Haug, E.J. and Rousselet, B., "Design Sensitivity Analysis in Structural Mechanics I, Static Response Variations", J. Structural Mechanics, Vol. 8, No. 1, 1980, pp. 17-41.
5. Rousselet, B., "Optimal Design and Eigenvalue Problems," Optimization Techniques, Proceedings of the 8th I.F.I.P. Conference, Vol. 6, Lecture Notes In Control and Information Sciences, Springer-Verlag, New York, 1978, pp. 343-352.
6. Lions, J.L. and Magenes, E., Non-Homogeneous Boundary Value Problems and Applications, Vol. I, Springer-Verlag, New York, 1972.
7. Fichera, G., "Existence Theorems in Elasticity," Handbuch Der Physik, Vol. VI a/2, 1972, pp. 347-389.
8. Kato, T., Pertubation Theory for Linear Operators, Springer-Verlag, New York, 1966.
9. Haug, E.J. and Rousselet, B., "Design Sensitivity Analysis of Eigenvalue Variations", Optimization of Distributed Parameter Structures (Eds. E.J. Haug and J. Cea), Sijthoff & Noordhoff, Alphen aan den Rijn, Netherlands, 1980.
10. Rousselet, B. and Haug, E.J., "Design Sensitivity Analysis of Shape Variations", Optimization of Distributed Parameter Structures (Eds. E.J. Haug and J. Cea), Sijthoff & Noordhoff Alphen aan den Rijn, Netherlands, 1980.
11. Haug, E.J., "A Gradient Projection Method for Structural Optimization", Optimization of Distributed Parameter Structures (Eds. E.J. Haug and J. Cea), Sijthoff & Noordhoff Alphen aan den Rijn, Netherlands, 1980.
12. Haug, E.J. and Arora, J.S., "Distributed Parameter Structural Optimization for Dynamic Response", Optimization of Distributed Parameter Structures (Eds. E.J. Haug and J. Cea), Sijthoff & Noordhoff, Alphen aan den Rijn, Netherlands, 1980.
13. Choi, K.K., "A Numerical Method for Optimizing Structures With Repeated Eigenvalues", Optimization of Distributed Parameter Structures (Eds. E.J. Haug and J. Cea), Sijthoff & Noordhoff, Alphen aan den Rijn, Netherlands, 1980.
14. Choi, K.K. and Haug, E.J., "Optimization of Structures With Repeated Eigenvalues", Optimization of Distributed Parameter Structures (Eds. E.J. Haug and J. Cea), Sijthoff & Noordhoff, Alphen aan den Rijn, Netherlands, 1980.
15. Reed, M. and Simon, B., Methods of Modern Mathematical Physics I: Functional Analysis, Academic Press, New York, 1972.
16. Adams, R.A., Sobolev Spaces, Academic Press, New York, 1975.
17. Aubin, J.P., Approximation of Elliptic Boundary-Value Problems, Wiley Interscience, New York, 1972.
18. Haug, E.J. and Arora, J.S., Applied Optimal Design, Wiley, New York, 1979.

DESIGN SENSITIVITY ANALYSIS OF EIGENVALUE VARIATIONS*

Edward J. Haug and Bernard Rousselet

Materials Division, College of Engineering, University of Iowa, Iowa City, Iowa 52242

Départemente de Mathematiques, Université de Nice, 06034 Nice Cédex, France

ABSTRACT

The dependence of eigenvalues of distributed parameter structures on design variables that specify material properties and distribution is characterized. Prototype problems considered include vibration of strings, membranes, beams, plates, and plane elastic slabs and buckling of beams. Symmetry and positive definiteness properties of the elliptic differential operators that govern system response are used to show that the eigenvalues depend continuously on design. Further, it is shown that simple eigenvalues are Frechet differentiable with respect to design, but that repeated eigenvalues can only be expected to be Gateaux (directionally) differentiable with respect to design. The latter fact is shown to have substantial consequences in classes of optimal design problems in which the fundamental eigenvalue is known to be repeated at an optimum design. Explicit and computable formulas for derivatives (first variation) of both simple and repeated eigenvalues of each of the prototype problems are presented.

---

*Research of first author sponsored by National Science Foundation Project No. ENG 77-19967. Travel support for joint efforts by the authors was provided by NATO Research Grant No. 1458.

1. INTRODUCTION

Design sensitivity analysis of static structural response with respect to design variation presented in the first paper of this sequence [1] is extended here to include eigenvalue response. The effect of design variation on natural frequencies of a variety of structural systems and on buckling of a beam is considered.

Inverse operator differentiability results of Ref. 1 are used as a starting point for investigating eigenvalue differentiability. From a mathematical point of view, however, differentiability of eigenvalues with respect to design is more difficult than differentiability of the inverse static operator. The difficulty arises primarily in the case of repeated eigenvalues. It is often argued that repeated eigenvalues only accidentally and rarely occur, so they need not be of concern. It is known [2,3], however, that some optimum designs satisfying constraints on eigenvalues must necessarily have repeated eigenvalues. Since the principal application of design sensitivity analysis is in optimal design, it is clear that the repeated eigenvalue problem must be carefully treated.

As in the first paper of this sequence, prototype structural systems and the design variables considered are defined in Section 2. Existence of derivatives of eigenvalues with respect to design variables is proved in Section 3 and general formulas for derivatives are derived. Proofs in the area of eigenvalue perturbation are technically complex and are generally no constructive. The reader is therefore referred for several proofs to Ref. 4, where technical details may be found. In Section 4, results obtained are applied to the prototype problems of Section 2.

2. PROTOTYPE PROBLEMS CONSIDERED

The conventional differential operator formulation of prototype problems is first presented. The more flexible and rigorous variational formulation (generalization) is then presented. The technical justification of the variational formulation follows in a similar way as for the static problems treated in detail in the first paper of this sequence [1].

In each problem formulated here, the formal operator eigenvalue problem is of the form

$$\bar{A}_u y = \zeta \bar{B}_u y \tag{2.1}$$

where $\bar{A}_u$ is the formal differential operator encountered in the static response problem of [1] and $\bar{B}_u$ is a much simpler continuous operator, except for buckling problems. The symbol $y \neq 0$ denotes an eigenfunction, to distinguish it from the static response z studied in Ref. 1 and ζ is the associated eigenvalue. If one assumes a high degree of differentiability of the eigenfunction y and the design variable u, he may form the L^2 scalar product of both sides of Eq. 2.1 with a smooth function v that satisfies the same boundary conditions as y, to obtain the variational equation

$$a_u(y,v) \equiv (\bar{A}_u y,v) = \zeta(\bar{B}_u y,v) \equiv \zeta b_u(y,v) \tag{2.2}$$

Conversely, if Eq. 2.2 holds for all v in a smooth class of functions and if y and u are sufficiently regular, they y and ζ constitute the solution of the eigenvalue problem of Eq. 2.1 [5-8].

As in the first paper [1], one now defines a subspace V of an appropriate Sobolev space H^m as generalized candidate solutions of Eq. 2.1. The generalized solution $y \in V$, $y \neq 0$, is then characterized by the variational equation

$$a_u(y,v) = b_u(y,v) \ , \qquad \text{for all } v \in V \tag{2.3}$$

where the design variable u is now only required to be in L^∞. As in Ref. 1, one now defines extensions A_u and B_u of the formal operators $\bar{A}_u$ and $\bar{B}_u$ such that for $y \in D(A_u) \subset V$

$$\left.\begin{aligned} (A_u y,v) &= a_u(y,v) \\ (B_u y,v) &= b_u(y,v) \end{aligned}\right\} \quad \text{for all } v \in V \tag{2.4}$$

The domain of B_u is such that $D(A_u) \subset D(B_u)$. Thus, one has the generalized operator eigenvalue problem

$$A_u y = \zeta B_u y \ , \qquad y \in D(A_u) \ , \qquad y \neq 0 \tag{2.5}$$

which is valid for physically meaningful designs $u \in L^\infty$. The regularity conditions associated with functions in $D(A_u)$ are as in Ref. 1, which are also more meaningful physically than the extreme smoothness conditions associated with the formal operators of Eq. 2.1.

Since the extension of candidate solutions to the space $D(A_u)$ is dictated completely by the operator A_u, technical definition of generalized solutions is exactly as in Ref. 1, where the reader is referred for details. In this section, the operator eigenvalue equation will be stated for each problem studied, the bilinear forms $a_u(y,v)$ and $b_u(y,v)$ defined, and the space V identified.

2.1 Vibration of a String

A perfectly flexible string of variable linear density, $h \in L^\infty(0,1)$, $h(x) \geq h_0 > 0$, and tension $T \geq T_0 > 0$, is shown in Fig. 2.1. The operator eigenvalue equation is

$$\bar{A}_u y \equiv -Ty_{xx} = \zeta hy \equiv \zeta \bar{B}_u y \tag{2.6}$$

where $\zeta = \omega^2$, ω being the natural frequency. The boundary conditions are

$$y(0) = y(1) = 0 \tag{2.7}$$

Here, $u = (h(x),T)$ and the bilinear forms of Eq. 2.2 are obtained by integration by parts as

$$\left.\begin{aligned} (\bar{A}_u y,v) &= a_u(y,v) = T\int_0^1 y_x v_x dx \\ (\bar{B}_u y,v) &= b_u(y,v) = \int_0^1 hyv\, dx \end{aligned}\right\} \tag{2.8}$$

where y and v satisfy boundary conditions of Eq. 2.7. Since only first order derivatives appear in the formula for $a_u(y,v)$, it is logical to select $V \subset H^1(0,1)$. The boundary conditions of Eq. 2.7 are satisfied in a generalized sense [5,6] if V is further restricted to $V = H_0^1(0,1)$. It is readily verified [6,7,8] that the form $a_u(y,y)$ is V-elliptic, so all the theory of Ref. 1 concerning A_u holds for this problem.

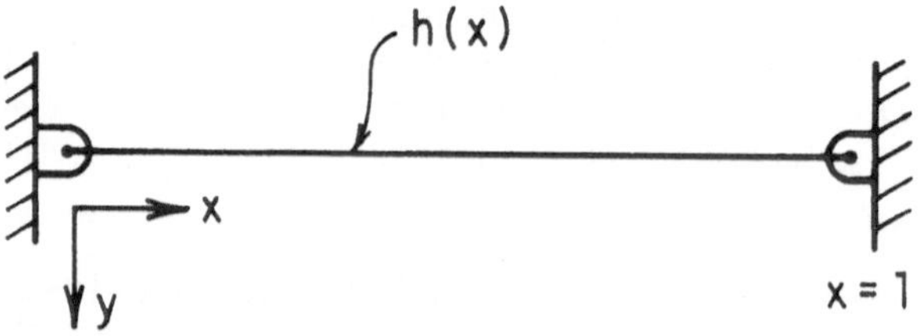

Figure 2.1 Vibrating String with Linear Mass Density h(x)

2.2 Vibration of a Beam

For a beam of variable cross-sectional area, h(x), let $h \in L^\infty(0,1)$, $h(x) \geq h_0 > 0$ (such that the second moment of the cross-sectional area is $I(x) = \alpha h^2(x)$), Young's Modulus $E \geq E_0 > 0$, and material density $\rho \geq \rho_0 > 0$, as shown with clamped-clamped supports in Fig. 2.2. The formal operator eigenvalue equation is

$$\bar{A}_u y \equiv (E\alpha h^2 y_{xx})_{xx} = \zeta\rho h y \equiv \zeta \bar{B}_u y \tag{2.9}$$

where $\zeta = \omega^2$, ω being the natural frequency. Boundary conditions for the clamped-clamped beam are

$$y(0) = y_x(0) = y(1) = y_x(1) = 0 \tag{2.10}$$

Here $u = (h(x), E, \rho)$ and the bilinear forms of Eqs. 2.2 are obtained by integration by parts as

$$\left.\begin{aligned} (\bar{A}_u y, v) &= a_u(y,v) = E\alpha \int_0^1 h^2 y_{xx} v_{xx} dx \\ (\bar{B}_u y, v) &= b_u(y,v) = \rho \int_0^1 hyv\, dx \end{aligned}\right\} \tag{2.11}$$

where y and v satisfy boundary conditions of Eq. 2.10. Since only second derivatives arise in $a_u(y,v)$, it is logical to select $V \subset H^2(0,1)$. The boundary conditions of Eq. 2.10 are satisfied in a generalized sense [5,6] if V is selected to be $V = H_0^2(0,1)$. All properties of $a_u(y,v)$ that are of interest here are demonstrated in Ref. 1.

2.3 Buckling of a Beam

If a beam is subjected to an axial load P, as shown in Fig. 2.3, then buckling can occur if P is larger than a critical load ζ.

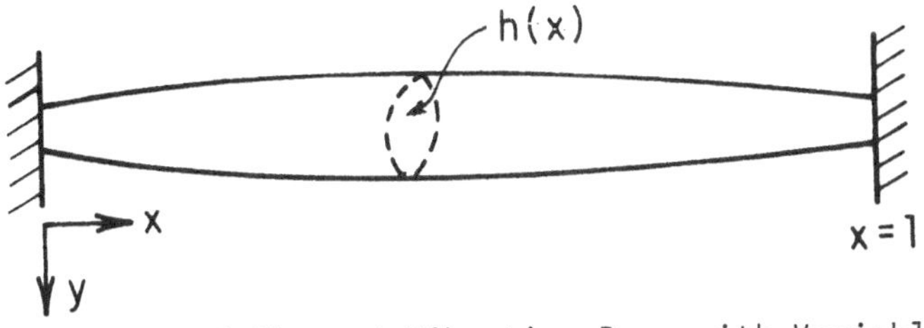

Figure 2.2 Clamped-Clamped Vibrating Beam with Variable Cross-Sectional Area h(x)

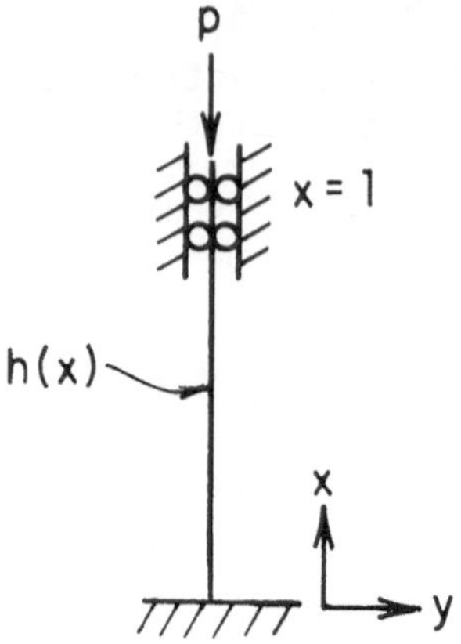

Figure 2.3 Clamped-Clamped Column with Variable Cross-Sectional Area h(x)

With the same design variables as in beam vibration, the formal operator eigenvalue equations is

$$\bar{A}_u y \equiv (E\alpha h^2 y_{xx})_{xx} = -\zeta y_{xx} \equiv \zeta \bar{B}_u y \tag{2.12}$$

with boundary conditions as in Eq. 2.10. Since material density does not arise in beam buckling, the design variable is $u = (h(x),E)$. Integration by parts with these operators in Eq. 2.2 yields the bilinear forms

$$\left.\begin{aligned} (\bar{A}_u y,v) &= a_u(y,v) = E\alpha \int_0^1 h^2 y_{xx} v_{xx} dx \\ (\bar{B}_u y,v) &= b_u(y,v) = \int_0^1 y_x v_x dx \end{aligned}\right\} \tag{2.13}$$

where y and v satisfy the boundary conditions of Eq. 2.10. Since $a_u(y,v)$ and $b_u(y,v)$ involve derivatives of order no higher than second order and the boundary conditions are the same as in the case of the vibrating beam, one may again select $V = H_0^2(0,1)$.

2.4 Vibration of a Membrane

Consider a vibrating membrane with variable mass density $h(x)$ per unit area, $h \in L^\infty(\Omega)$, $h(x) \geq h_0 > 0$, and membrane tension T (force per unit length), as shown in Fig. 2.4. The formal operator eigenvalue problem is

$$\bar{A}_u y \equiv -T\nabla^2 y = \zeta h(x) y \equiv \zeta \bar{B}_u y \tag{2.14}$$

where $\zeta = \omega^2$, ω being natural frequency, and the boundary condition is

$$y = 0 \quad \text{on } \Gamma \tag{2.15}$$

Here $u = (h(x),T)$ is the design variable and the bilinear forms of Eq. 2.2 are

$$\left.\begin{aligned} (\bar{A}_u y,v) &= a_u(y,v) = T \iint_\Omega (y_1 v_1 + y_2 v_2)\, d\Omega \\ (\bar{B}_u y,v) &= b_u(y,v) = \iint_\Omega hyv\, d\Omega \end{aligned}\right\} \tag{2.16}$$

where a subscript i denotes $\partial/\partial x_i$, $i = 1,2$, and y and v satisfy the boundary condition of Eq. 2.15. As in the case of the vibrating string, $V = H_0^1(\Omega)$, the bilinear form $a_u(y,v)$ is V-elliptic [7,8], and all properties needed for the differentiability theory of [1] follow.

2.5 Vibration of a Plate

Consider a clamped vibrating plate of variable thickness $h \in L^\infty(\Omega)$, $h(x) \geq h_0 > 0$, Young's modulus $E \geq E_0 > 0$, and material density $\rho \geq \rho_0 > 0$, as shown in Fig. 2.5. The formal operator eigenvalue problem is

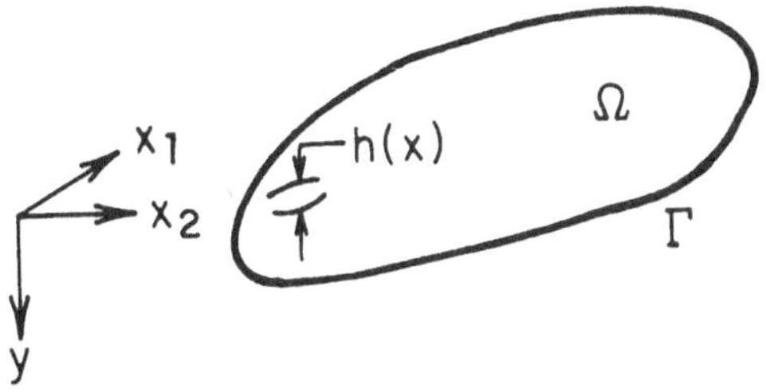

Figure 2.4 Membrane of Variable Area Density h(x)

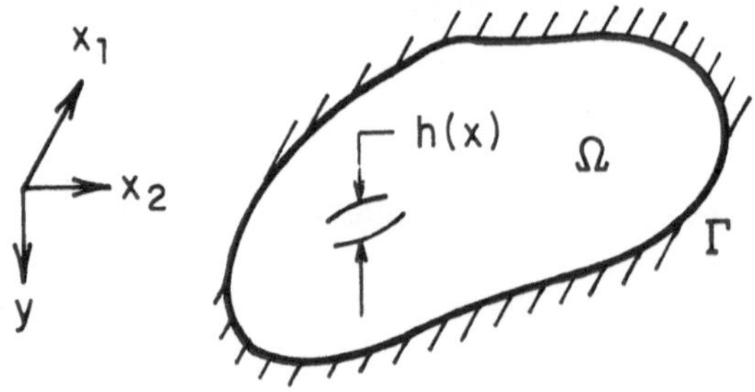

Figure 2.5 Clamped Plate of Variable Thickness h(x)

$$\bar{A}_u y \equiv [\hat{D}(u)(y_{11} + \nu y_{22})]_{11} + [\hat{D}(u)(y_{22} + \nu y_{11})]_{22} + 2(1 - \nu)[\hat{D}(u)y_{12}]_{12} = \zeta\rho h y \equiv \zeta\bar{B}_u y \tag{2.17}$$

where

$$y_{ij} \equiv \frac{\partial^2 y}{\partial x_i \partial y_j}, \quad \hat{D}(u) = \frac{Eh^3}{12(1 - \nu^2)} \quad \zeta = \omega^2,$$

ω is a natural frequency, $0 < \nu < 0.5$ is Poissons ratio, and the boundary conditions for a clamped plate are

$$y = \frac{\partial y}{\partial n} = 0 \quad \text{on } \Gamma \tag{2.18}$$

where $\partial y/\partial n$ is the normal derivative of y on Γ. Here the design variable is $u = (h(x), E, \rho)$. Integration by parts yields the bilinear forms of Eq. 2.2 as

$$(\bar{A}_u y, v) = a_u(y,v) = \iint_\Omega \hat{D}(u)[y_{11}v_{11} + \nu(y_{22}v_{11} + y_{11}v_{22}) + y_{22}v_{22} + 2(1 - \nu)y_{12}v_{12}]d\Omega \tag{2.19}$$

$$(\bar{B}_u y, v) = b_u(y,v) = \rho \iint_\Omega hyv \, d\Omega$$

where y and v satisfy the boundary conditions of Eq. 2.18. As in the case of the vibrating beam, the natural domain of the bilinear form $a_u(y,v)$ is $V = H_0^2(\Omega)$.

2.6 In-Plane Vibration of a Variable Thickness Planar Elastic Slab

Finally, consider a clamped edge, planar elastic slab of variable thickness $h \in L^\infty(\Omega)$, $h(x) \geq h_0 > 0$, Young's modulus $E(x) = \alpha h(x)$, and density $\rho \geq \rho_0 > 0$, as shown in Fig. 2.6. The formal operator eigenvalue problem for in-plane vibration is

$$\bar{A}_u y \equiv -\frac{\alpha}{2(1+\nu)} \begin{bmatrix} [4h\nu(y_1^1 + y_2^2) + 2hy_1^1]_1 + [h(y_2^1 + y_1^2)]_2 \\ \\ [h(y_2^1 + y_1^2)]_1 + [4h\nu(y_1^1 + y_2^2) + 2hy_2^2]_2 \end{bmatrix}$$

$$= \zeta\rho h \begin{bmatrix} y^1 \\ \\ y^2 \end{bmatrix} \equiv \zeta\bar{B}_u y \qquad (2.20)$$

where $y = [y^1, y^2]^T$ is the vector, in-plane displacement, $\zeta = \omega^2$, ω is natural frequency, $0 < \nu < 0.5$, and the boundary conditions are

$$y^1 = y^2 = 0 \quad \text{on} \quad \Gamma \qquad (2.21)$$

The design variable in this problem is $u = (h(x), \rho)$. Integration by parts yields the bilinear forms of Eq. 2.2 as

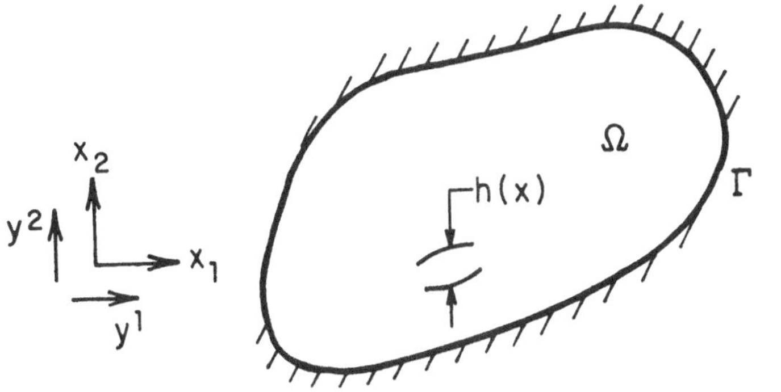

Figure 2.6 Planar Elastic Slab of Variable Thickness h(x)

$$(\bar{A}_u y,v) = a_u(y,v) = \frac{\alpha}{2(1+\nu)} \iint_\Omega h\left\{\left[4\nu(y_1^1 + y_2^2) + 2y_1^1\right] v_1^1 \right.$$

$$\left. + (y_2^1 + y_1^2)(v_2^1 + v_1^2) + \left[4\nu(y_1^1 + y_2^2) + 2y_2^2\right] v_2^2\right\} d\Omega$$

$$(\bar{B}_u y,v) = b_u(y,v) = \rho \iint_\Omega h(y^1 v^1 + y^2 v^2) d\Omega$$

where y and v satisfy the boundary conditions of Eq. 2.21. As shown in Ref. 1, the natural domain of the bilinear form $a_u(y,v)$ is $V = H_0^1(\Omega) \times H_0^1(\Omega) \equiv H_0^1(\Omega)^2$.

With these bilinear forms, the variation formulation presented at the beginning of this section now characterize the eigenvalue behavior of each of the six problems discussed. Since they all have the same basic variational form and all the bilinear forms share the regularity of design dependence demonstrated in Ref. 1, one is now in a position to derive eigenvalue-design relationships that apply equally well to all these problems. This is the objective of the following section. Following this general derivation, design sensitivity formulas specific to each of the six examples are presented.

Due to the Sobolev imbedding lemma, the identity map from $D(A_u) \subset H_0^2(\Omega)$ into $L^2(\Omega)$ is compact. This fact, the V-ellipticity and the self adjointness of the bilinear forms a_u and b_u allows one to show that the operator equation $A_u y = \zeta B_u y$, in all the previous examples, has a purely discrete and real positive spectrum, thus in each case there is a denumerable set of isolated eigenvalues of finite multiplicity (which are real positive numbers). This result rests on F. Riesz spectral theory [6].

To apply this result, one introduces the operator $C_u = A_u^{-1} B_u$ and shows that the eigenvalues of the origianl problem are the inverses of the eigenvalues of C_u. It is easy to see that $A_u^{-1} B_u$ is a self-adjoint operator, with respect to the scalar product defined by b_u. Since C_u is also compact from H_0^2 into L_2, the F. Riesz spectral theory [6] applies. It is, in particular, deduced that all the eigenvalues are semi-simple , a result that will be of uppermost importance for perturbation studies. This

†The restriction of C_u to any eigenspace can be transformed to diagonal form.

information and methods of functional analysis are used in Ref. 4 to prove the following results.

Proposition 2.1: Let X be a Banach space, A an unbounded operator with a discrete spectrum, B a bounded operator in X, and zero is supposed not to be an eigenvalue of A or of B. Then the eigenvalues of $A^{-1}B$ are exactly the inverse of numbers ζ such that there exists $y \in D(A)$, $y \neq 0$ satisfying $Ay - \zeta By = 0$.

In the case of the bukling problem, B_u is an unbounded operator in a Hilbert space $L^2(0,1)$, with domain $D(A_u) \subset D(B_u)$. In fact B_u does not depend on u (formally $B_u y = y''$) and may be viewed as a bounded mapping from $H_0^2(0,1)$ into $L^2(0,1)$. Since $\underline{A}_u^{-1}$ (the operator $\underline{A}_u$ is defined in the Appendix of Ref 4) is a bounded mapping from $H^{-2}(0,1)$ onto $H_0^2(0,1)$, one can consider $\underline{A}_u^{-1}B^*$, where B^* is the adjoint of B (mapping $L^2(0,1)$ into $H^{-2}(0,1)$), which is a $L^2(0,1)$ bounded operator (composition of two bounded operators). Moreover it is even compact: $\underline{A}_u^{-1}$ is bounded from $H^{-2}(0,1)$ onto $H^2(0,1)$ and the injection of $H^2(0,1)$ into $L^2(0,1)$ is compact (see Ref. 5). Its spectrum is thus discrete, as for the other examples. The proof of Proposition 2.1 for the column then follows exactly the same pattern as for the other examples and is given in Ref. 4.

3. DIFFERENTIATION OF EIGENVALUES WITH RESPECT TO DESIGN

3.1 A Formal Perturbation Analysis

If one views Eq. 2.5 as a relationship among y, ζ, and u, it is clear that the eigenvalue ζ will in general depend on the design u. In analogy with calculations that may be carried out in finite dimensional equations, formal perturbation analysis has been used in the engineering literature [9,10,11] to obtain formulas for a variation $\delta\zeta$ in the eigenvalue due to a variation δu in design. A heuristic derivation follows by making a linear approximation of Eq. 2.3,

$$a_{u,\delta u}^{(1)}(y,y) + 2a_u(y,\delta y) = \delta\zeta\, b_u(y,y) + \zeta b_{u,\delta u}^{(1)}(y,y) + \zeta 2b_u(y,\delta y) \tag{3.1}$$

where $b_{u,\delta u}^{(1)}$ is the first variation (or Frechet differential) of the bilinear form b_u with respect to u, defined in exactly the same way as $a^{(1)}_{u,\delta u}$ in Ref. 1, and symmetry of the forms a_u and b_u has been used. Since y is an eigenfunction and $\delta y \in V$, Eq. 2.3 implies that $a_u(y,\delta y) = \zeta b_u(y,\delta y)$, thus eliminating two terms in Eq. 3.1. Finally, if the eigenfunction y is normalized by $b_u(y,y) = 1$, then Eq. 3.1 yields simply

$$\delta\zeta = a^{(1)}_{u,\delta u}(y,y) - \zeta b^{(1)}_{u,\delta u}(y,y) \tag{3.2}$$

This derivative is valid under the assumption that the eigenvalue and eigenvector are differentiable with respect to design. This is exactly the result to be proved in this section. Such a result has been presented in Ref. 12 for a related class of operators. Differentiability for a related problem has also been proved for simple eigenvalues, through the use of the implicit function theorem in Ref. 13.

3.2 Differentiability of Bilinear Forms

As the heuristic introduction showed, the derivation of the eigenvalues involves the derivative of the associated bilinear forms. In particular, the differentiability of the eigenfunction is not needed and in fact can only be proved for simple eigenvalues (see examples 1.12 and 5.3 of Chapter II of Ref. 14).

Differentiability of the bilinear form $a_u(y,v)$ for the beam, plate, and plane elasticity operators is demonstrated in Ref. 1 and the formulas for the differential $a^{(1)}_{u,\delta u}(y,v)$ are given. These derivations may be easily repeated for a_u associated with the string and membrane and for b_u in each of the examples discussed in Section 2. Formulas for $a^{(1)}_{u,\delta u}(y,v)$ and $b^{(1)}_{u,\delta u}(y,v)$ are given here as:

<u>Vibrating string</u>, $u = (h(x),T)$:

$$a^{(1)}_{u,\delta u}(y,v) = \delta T \int_0^1 y_x v_x dx \tag{3.3}$$

$$b^{(1)}_{u,\delta u}(y,v) = \int_0^1 (\delta h) y v dx \tag{3.4}$$

Vibrating beam, $u = (h(x), E, \rho)$:

$$a^{(1)}_{u,\delta u}(y,v) = (\delta E)\alpha \int_0^1 h^2 y_{xx} v_{xx} dx + E\alpha \int_0^1 2h(\delta h) y_{xx} v_{xx} dx \quad (3.5)$$

$$b^{(1)}_{u,\delta u}(y,v) = \delta\rho \int_0^1 hyvdx + \rho \int_0^1 (\delta h) yvdx \quad (3.6)$$

Buckling beam, $u = (h(x), E)$:

$$a^{(1)}_{u,\delta u}(y,v) = (\delta E)\alpha \int_0^1 h^2 y_{xx} v_{xx} dx + E\alpha \int_0^1 2h(\delta h) y_{xx} v_{xx} dx \quad (3.7)$$

$$b^{(1)}_{u,\delta u}(y,v) = 0 \quad (3.8)$$

Vibrating membrane, $u = (h(x), T)$:

$$a^{(1)}_{u,\delta u}(y,v) = \delta T \iint_\Omega (y_1 v_1 + y_2 v_2) d\Omega \quad (3.9)$$

$$b^{(1)}_{u,\delta u}(y,v) = \iint_\Omega (\delta h) yvd\Omega \quad (3.10)$$

Vibrating plate, $u = (h(x), E, \rho)$:

$$a^{(1)}_{u,\delta u}(y,v) = \iint_\Omega \left[\frac{h^3 \delta E}{12(1-\nu^2)} + \frac{3Eh^2 \delta h}{12(1-\nu^2)}\right] \times \left[y_{11}v_{11} + \nu(y_{22}v_{11} + y_{11}v_{22}) + y_{22}v_{22} + 2(1-\nu) y_{12} v_{12}\right] d\Omega \quad (3.11)$$

$$b^{(1)}_{u,\delta u}(y,v) = \delta\rho \iint_\Omega hyvd\Omega + \rho \iint (\delta h) yvd\Omega \quad (3.12)$$

Vibrating plane elastic slab, $u = (h(x), \rho)$:

$$a^{(1)}_{u,\delta u}(y,v) = \frac{\alpha}{2(1+\nu)} \iint_\Omega (\delta h) \left\{ \left[4\nu(y_1^1 + y_2^2) + 2y_1^1\right] v_1^1 \right.$$

$$+ (y_2^1 + y_1^2)(v_2^1 + v_1^2) + \left[4\nu(y_1^1 + y_2^2) + 2y_2^2\right]v_2^2\Big\}d\Omega \quad (3.13)$$

$$b^{(1)}_{u,\Omega u}(y,v) = \delta\rho \iint_\Omega h(y^1v^1 + y^2v^2)d\Omega$$

$$+ \rho \iint_\Omega (\delta h)(y^1v^1 + y^2v^2)d\Omega \quad (3.14)$$

Apart from algebraic form of the bilinear forms and their differentials, the mathematical structure of each of the six examples is the same. Thus, the eigenvalue differentiability is proved using only the common set of operator theoretic properties. The result is applicable to each of the six examples discussed here and to any other problem that can be put in the same variational form. As shown by Fichera [7], virtually all problems of linear elasticity fall into this category, as do many other partial differential equations arising in mathematical physics. Using the foregoing formulas, it is shown in Ref. 1 that the inverse A_u^{-1} of the operator A_u defined in Eq. 2.4, as a mapping from u to an element of $B(L^2)$†, is Frechet differentiable with respect to design. In all examples except the buckling column, the operator B_u defined in Eq. 2.4 is in $B(L^2)$ and trivially Frechet differentiable with respect to design, so that $A_u^{-1}\ B_u$ is Frechet differentiable. The operator B_u associated with the column requires a separate analysis, which may be found in Ref. 4.

3.3 Regularity of the Operator Versus Regularity of the Eigenvalues

The following matrix example, borrowed from Ref. 14, shows that the regularity of the eigenvalues needs some care and precaution. Let

$$A(x_1,x_2) = \begin{pmatrix} x_1 & x_2 \\ x_2 & -x_1 \end{pmatrix}$$

---

† $B(L^2)$ denotes the space of bounded operators from L^2 into L^2.

where $(x_1,x_2) \in R^2$. The matrix A is analytic with respect to the couple (x_1,x_2). However, the eigenvalues $d_{\pm} = \pm\sqrt{x_1^2 + x_2^2}$ are not Frechet differentiable at (0,0), even though A is self-adjoint. The difficulty arises because A depends on two real parameters. Put $x_2 = 0$, $d_1 = x_1$ and $d_2 = -x_1$, which are analytic functions of x_1 and are equal to the two eigenvalues of $A(x_1,0)$. The partial derivative with respect to x_1 of these eigenvalues is (1,-1). Note however, that if one sets $\underline{d}_1 = |x_1|$ and $\underline{d}_2 = -|x_1|$, $(\underline{d}_1,\underline{d}_2)$ are still equal to the eigenvalues of $A(x_1,0)$, but they are no longer differentiable at (0,0).

This example shows that one cannot expect more than directional differentiability of the eigenvalues, which means that one can select differentiable functions of a real parameter that are equal to the eigenvalues of the operator, but one cannot expect the eigenvalues of a self-adjoint operator (which are real) are differentiable if they are ordered according to increasing magnitude. If one sets

$$\hat{d} = \frac{d_+ + d_-}{2} = 0$$

(as introduced in Ref. 14), he has the weighted mean of a clustered group of eigenvalues, which is differentiable at zero (the result is general). One may consider this latter property, together with the fact that $\hat{d}$ is an approximation of the two eigenvalues (for small values of the parameter), to be of numerical interest.

The foregoing considerations lead one to first show the continuity and then the directional differentiability of the eigenvalues.

3.4 Continuity of Eigenvalues

Theorem 3.1: Let $\{\zeta_1,\ldots,\zeta_k\}$ be m generalized eigenvalues (counted with multiplicity such that $m \geq k$) of the generalized eigenvalue problem of Eq. 2.5 with operators A_u and B_u of Section 2. For every neighborhood W of $\{\zeta_1,\ldots,\zeta_k\}$ in the real line, with no other eigenvalue in it, there exists a neighborhood S of u (in the space U of the design variable) such that for every $u + \delta u \in S$ there are exactly m generalized eigenvalues of $(A_{u+\delta u},B_{u+\delta u})$ (counted with their multiplicity) in W.

Proof: Since $u \to A_u^{-1}B_u$ is Frechet differentiable, it is also continuous in the sense of bounded operators and thus in the sense of generalized convergence of closed operators (Theorem IV.2.23 of Ref. 14). Continuity of eigenvalues of $A_u^{-1}B_u$, as stated in the theorem, then follows from IV.3.5 of Ref. 14. One proves the continuity of the generalized eigenvalues of (A_u,B_u) by noting that they are reciprocals of the eigenvalues of $A_u^{-1}B_u$, which are never zero (Proposition 2.1).

3.5 Differentiability of Eigenvalues

Recall that the eigenvalues are not generally Frechet-differentiable (see example 3.1).

Theorem 3.2: Let the operator A_u of Eq. 2.4 have a bounded inverse that is Frechet differentiable with respect to $u \in U$ and let the operator B_u of Eq. 2.4 either be bounded and Frechet differentiable with respect to $u \in U$, or be unbounded with $B = B_u$ independent of u, but A_u^{-1} B be bounded. Let $\zeta(u)$ be a generalized eigenvalue of (A_u,B_u) of Eq. 2.5 of multiplicity m_u. Then the group of m_u eigenvalues of Eq. 2.5 associated with operators $(A_{u+\delta u},B_{u+\delta u})$ for $||\delta u||$ small enough (as stated in Theorem 3.1) is differentiable at u in the direction δu (for any δu); i.e. there exist representations of the eigenvalues $(\zeta_j(u+t\delta u)$, $j=1,\ldots,m_u)$ such that

$$\zeta_j(u + t\delta u) = \zeta(u) + t\zeta_j^{(1)}(u,\delta u) + o(t),$$

$$j = 1,\ldots,m_u \tag{3.19}$$

where $\zeta_j^{(1)}(u,\delta u) = -(\zeta(u))^2\, \alpha_j^{(1)}(u,\delta u)$ and $\alpha_j^{(1)}(u,\delta u)$ are the eigenvalues of

$$P_u\left(\left(\frac{d}{ds}\,\hat{C}_{u+s\delta u}\right)_{s=0}\right)P_u$$

restricted to the subspace of V spanned by the eigenfunctions of

$\hat{C}_u$; where $\hat{C}_u = A_u^{-1}B_u$ and P_u is its spectral projector[†] associated with $\alpha(u) = \frac{1}{\zeta(u)}$. If $\zeta_j(u + t\delta u)), j = 1,\ldots,m_u$, are the eigenvalues written in increasing order of magnitude, then Eq. 3.19 is only valid for $t \geq 0$. Moreover, the directional derivative of the smallest eigenvalue is the smallest of all the directional derivatives of the group of m_u eigenvalues.

The proof of this theorem is rather technical and involves the use of the resolvent $R(\zeta,\hat{C}) = (\hat{C} - \zeta)^{-1}$ for ζ in the complex plane and the Dunford integral representation. It is given in Ref. 4.

An explicit computation of $\zeta_j^{(1)}(u,\delta u)$ is presented in the two following corollaries. In the first corollary the differentiability result formally computed at the beginning of this section is proved for the case of simple and then for multiple eigenvalues.

Corollary 3.3: Under the hypothesis of Theorem 3.2, if (u) is a simple generalized eigenvalue, it remains simple for t small enough (by Theorem 3.1) and its derivative at t = 0 is given by

$$\zeta^{(1)}_{u,\delta u} = a^{(1)}_{u,\delta u}(y_u,y_u) - \zeta(u)b^{(1)}_{u,\delta u}(y_u,y_u) \tag{3.20}$$

where y_u satisfies

$$\left.\begin{aligned} &(i)\quad A_u y_u = \zeta(u)B_u y_u \\ &(ii)\quad (B_u y_u, y_u) = 1 \end{aligned}\right\} \tag{3.21}$$

Proof: One has

$$A_{u+t\delta u}\, y_{u+t\delta u} = \zeta(u + t\delta u)\, B_{u+t\delta u}\, y_{u+t\delta u}$$

if and only if (Proposition 2.1)

$$A^{-1}_{u+t\delta u}\, B_{u+t\delta u}\, y_{u+t\delta u} = \left[\frac{1}{\zeta(u + t\delta u)}\right] y_{u+t\delta u} \; .$$

[†]The operator projecting V onto the subspace of V spanned by eigenfunctions of $\hat{C}_u$ associated with the eigenvalue $\alpha(u)$. Since $\hat{C}_u$ is self-adjoint (see the end of Section 2) for the scalar product b_u, this projection is indeed orthogonal with respect to b_u.

Since A_u and B_u are self-adjoint, $A_u^{-1} B_u$ is self-adjoint for the scalar product $(B_u y,y)$. Thus, for a simple eigenvalue, P_u is the orthogonal projector for the scalar product $(B_u y,y)$ on the line spanned by y_u; i.e.,

$$P_u y = (B_u y,y_u)y_u$$

Since the range of P_u is a scalar multiple of y_u, it is clear that y_u is an eigenvector of $P_u\left(\left(\frac{d}{ds}\hat{C}_{u+s\delta u}\right)_{s=0}\right)P_u$; i.e., $P_u\left(\left(\frac{d}{ds}\hat{C}_{u+s\delta u}\right)_{s=0}\right)P_u y_u = \lambda y_u$. Thus, since $(B_u y_u,y_u) = 1$, the eigenvalue of $P_u\left(\left(\frac{d}{ds}\hat{C}_{u+s\delta u}\right)_{s=0}\right)P_u$, which by Theorem 3.2 is $\alpha^{(1)}_{u,\delta u}$, is equal to

$$\alpha^{(1)}_{u,\delta u} = \left(B_u P_u\left(\left(\frac{d}{ds}\hat{C}_{u+s\delta u}\right)_{s=0}\right)P_u y_u, y_u\right)$$

Now, Eq. 3.13 of Ref. 1 for the derivative of A_u^{-1}; i.e., $\delta_u A_u^{-1} = -G_u^{-1}C_1G_u^{-1}$, allows one to write,† noting that $A_u^{-1}G_u = G_u^{-1}$ and

$$\left(\frac{d}{ds}\hat{C}_{u+s\delta u}\right)_{s=0} = (\delta_u A_u^{-1})B_u + (A_u^{-1} B^{(1)}_{u,\delta u}), \text{ that}$$

$$\alpha^{(1)}_{u,\delta u} = \left(P_u\left[-A_u^{-1}G_uC_1(u,\delta u)G_uA_u^{-1}B_u + A_u^{-1}B^{(1)}_{u,\delta u}\right]y_u, B_u y_u\right) \tag{3.22}$$

Now for all $v \in V$

$$(P_u^* B_u y_u, v) \equiv (B_u y_u, P_u v) = (B_u v, y_u)(B_u y_u, y_u)$$

$$= (B_u v, y_u) = (B_u y_u, v)$$

and since V is dense in L^2, one has

†For the buckling problem a generalization presented in Ref. 4 is employed, but the same result holds.

$$P_u^* B_u y_u = B_u y_u$$

Thus Eq. 3.22 may be written in the form

$$\alpha_{u,\delta u}^{(1)} = \left(\left[-A_u^{-1} G_u C_1(u,\delta u) G_u A_u^{-1} B_u + A_u^{-1} B_{u,\delta u}^{(1)} \right] y_u, B_u y_u \right)$$

$$= \left(-G_u C_1(u,\delta u) G_u A_u^{-1} B_u y_u, A_u^{-1} B_u y_u \right)$$

$$+ \left(B_{u,\delta u}^{(1)} y_u, A_u^{-1} B_u y_u \right)$$

Now one can use $A_u^{-1} B_u y_u = \frac{1}{\zeta(u)} y_u$ to get

$$\alpha_{u,\delta u}^{(1)} = \frac{1}{(\zeta(u))^2} \left(-G_u C_1(u,\delta u) G_u y_u, y_u \right) + \frac{1}{\zeta(u)} \left(B_{u,\delta u}^{(1)} y_u, y_u \right)$$

or in terms of design derivatives of the bilinear forms,

$$\alpha_{u,\delta u}^{(1)} = - \frac{1}{(\zeta(u))^2} a_{u,\delta u}^{(1)}(y_u, y_u) + \frac{1}{\zeta(u)} b_{u,\delta u}^{(1)}(y_u, y_u)$$

and Eq. 3.20 follows from this last formula, when one notes that

$$\zeta_{u,\delta u}^{(1)} = -(\zeta(u))^2 \alpha_{u,\delta u}^{(1)}$$

Corollary 3.4: Under the hypothesis of theorem 3.2, if $\zeta(u)$ is an m_u-fold eigenvalue of (A_u, B_u), its derivatives are the m_u eigenvalues of the matrix M of general term:

$$M_{ij} = a_{u,\delta u}^{(1)}(y_i, y_j) - \zeta(u)\, b_{u,\delta u}^{(1)}(y_i, y_j),$$

$$i = 1,\ldots,m, \quad j=1,\ldots,m_u. \tag{3.23}$$

where y_i (short for y_{ui}) satisfy

$$\left. \begin{aligned} &\text{(i)} \quad A_u y_i = \zeta(u) B_u y_i \\ &\text{(ii)} \quad (B_u y_i, y_j) = \delta_{ij} \end{aligned} \right\} \tag{3.24}$$

where δ_{ij} is one if $i = j$ and otherwise is zero.

Proof: As for simple eigenvalues, P_u is the orthogonal projector for the scalar product $(B_u y,y)$ on the eigensapce associated with $\zeta(u)$ of multiplicity m_u, where y_i, $i = 1,\ldots,m_u$ denotes a basis of this eigenspace that is orthonormal with respect to the $(B_u y,v)$ scalar product. Thus, for any $y \in V$,

$$P_u y = \sum_{i=1}^{m} (B_u y, y_i) y_i$$

Now an explicit formula for $N_{ij} \equiv \left[P_u \left(\left(\frac{d}{ds} C_{u+s\delta u} \right)_{s=0} \right) P_u \right]_{i,j}$ has to be written, in order to use Theorem 3.2. As for the simple eigenvalue case,

$$\left(\frac{d}{ds} \hat{C}_{u+s\delta u} \right)_{s=0} = -G_u^{-1} C_1(u,\delta u) G_u^{-1} B_u + A_u^{-1} B_{u,\delta u}^{(1)}$$

so that

$$N_{ij} = \left(B_u P_u \left[-G_u^{-1} C_1(u,\delta u) G_u^{-1} B_u + A_u^{-1} B_{u,\delta u}^{(1)} \right] P_u y_j, y_i \right)$$

Now,

$$(P_u^* B_u y_i, v) = (B_u y_i, P_u v)$$

$$= \sum_{j=1}^{m} (B_u y_i, y_j)(B_u v, y_j) = (B_u v, y_i) = (v, B_u y_i)$$

so that $P_u^* B_u y_i = B_u y_i$, which is clear since P_u is self-adjoint with respect to the $(B_u y,v)$ scalar product. Thus,

$$N_{ij} = \left(\left[-G_u^{-1} C_1(u,\delta u) G_u^{-1} B_u + A_u^{-1} B_{u,\delta u}^{(1)} \right] y_j, B_u y_i \right)$$

$$= \left(-A_u^{-1} G_u C_1(u,\delta u) G_u A_u^{-1} B_u y_j, B_u y_i \right) + \left(A_u^{-1} B_{u,\delta u}^{(1)} y_j, B_u y_i \right)$$

Now, using the self-adjointness of A_u^{-1} and

$$A_u^{-1} B_u y_j = \frac{1}{\zeta(u)} y_j$$

one gets

$$N_{ij} = \frac{1}{\zeta(u)^2}\left(-G_u C_1(u,\delta u) G_u y_j, y_i\right) + \frac{1}{\zeta(u)}\left(B^{(1)}_{u,\delta u} y_j, y_i\right)$$

One concludes, by noting $\zeta^{(1)}_{u,\delta u} = -\zeta(u)^2 \alpha^{(1)}_{u,\delta u}$, that $\zeta^{(1)}_{u,\delta u}$ are the eigenvalues of $M_{ij} = -\zeta(u)^2 N_{ij}$, which gives Eq. 3.24.

3.6 Calculation of Explicit Directional Derivatives of Repeated Eigenvalues ($m_u = 2$)

For any pair of b_u-orthonormal eigenvectors y_1 and y_2 corresponding to the twice repeated eigenvalue $\zeta = \zeta(u)$, let

$$\left.\begin{aligned} \bar{y}_1 &= y_1 \cos\phi + y_2 \sin\phi \\ \bar{y}_2 &= -y_1 \sin\phi + y_2 \cos\phi \end{aligned}\right\} \qquad (3.25)$$

Note that

$$\begin{aligned} b_u(\bar{y}_1,\bar{y}_2) &= -\cos\phi \sin\phi\, b_u(y_1,y_1) + \cos^2\phi\, b_u(y_1,y_2) \\ &\quad - \sin^2\phi\, b_u(y_2,y_1) + \sin\phi\cos\phi\, b_u(y_2,y_2) \\ &= 0 \quad \text{for all } \phi \end{aligned}$$

Similarly, $b_u(\bar{y}_1,y_1) = b_u(\bar{y}_2,y_2) = 1$ for all ϕ, so $\bar{y}_1$ and $\bar{y}_2$ are b_u-orthonormal for all ϕ.

Since the directional derivatives of $\zeta(u)$ are the eigenvalues of the matrix $\bar{M}$, where over-bar denotes M evaluated with $\bar{y}_1$ and $\bar{y}_2$, if ϕ is chosen so that $\bar{M}_{12} = 0$, then the directional derivatives of $\zeta(u)$ are $\bar{M}_{11}$ and $\bar{M}_{22}$. Thus, put

$$\begin{aligned} 0 = \bar{M}_{12} &= a^{(1)}_{u,\delta u}(\bar{y}_1,\bar{y}_2) - \zeta(u)\, b^{(1)}_{u,\delta u}(\bar{y}_1,\bar{y}_2) \\ &= -\cos\phi \sin\phi\, M_{11} + (\cos^2\phi - \cos^2\phi) M_{12} + \sin\phi\cos\phi\, M_{22} \\ &= \frac{1}{2}\sin 2\phi\,(M_{22} - M_{11}) + \cos 2\phi\, M_{12} \end{aligned}$$

Thus,

$$\tan 2\phi = \frac{2M_{12}}{M_{11} - M_{12}}$$

or

$$\phi(u,\delta u) = \frac{1}{2} \text{Arctan}\left[\frac{2M_{12}(u,\delta u)}{M_{11}(u,\delta u) - M_{22}(u,\delta u)}\right] \tag{3.26}$$

Note that:

(1) $\phi(u,\delta u)$ is not linear in δu.

(2) since $M_{ij}(u,\delta u)$ are linear in δu, $\phi(u,\alpha\delta u) = \phi(u,\delta u)$ for any real α, so $\phi(u,\delta u)$ is homogeneous of degree zero in δu.

(3) since only principal values of Arctan are needed, $|\phi| \leq \pi/4$.

With ϕ chosen so $\bar{M}_{12} = 0$,

$$\zeta^{(1)}_{u,\delta u,1} = \bar{M}_{11} = a^{(1)}_{u,\delta u}(\bar{y}_1,\bar{y}_1) - \zeta(u)\, b^{(1)}(\bar{y}_1,\bar{y}_1)$$

$$= \frac{1}{2}\Big(M_{11}(u,\delta u) + M_{22}(u,\delta u)\Big) + \sin 2\phi(u,\delta u)\, M_{12}(u,\delta u)$$

$$+ \frac{1}{2}\cos 2\phi(u,\delta u)\Big(M_{11}(u,\delta u) - M_{22}(u,\delta u)\Big) \tag{3.27}$$

Similarly,

$$\zeta^{(2)}_{u,\delta u,2} = \frac{1}{2}\Big(M_{11}(u,\delta u) + M_{22}(u,\delta u)\Big) - \sin 2\phi\,(u,\delta u)\, M_{12}(u,\delta u)$$

$$- \frac{1}{2}\cos 2\phi\,(u,\delta u)\Big(M_{11}(u,\delta u) - M_{22}(u,\delta u)\Big) \tag{3.28}$$

Since $M_{ij}(u,\delta u)$ are linear in δu and $\phi(u,\delta u)$ is homogeneous of degree zero in δu, the directional derivatives for Eqs. 3.27 and 3.28 of the repeated eigenvalue $\zeta(u)$ are homogeneous of degree one in δu. They are not, however, linear in δu. Note that the ordering of eigenvales as u is perturbed is defined by the choice of eigenfunctions in Eq. 3.25. If ordering of the eigenvalues by magnitude is desired, then Eqs. 3.27 and 3.28 will not give the directional derivatives, which are only positvely homogeneous in δu.

4. EXAMPLES

In this section, formulas are first presented for the derivative of a simple eigenvalue for each example previously introduced, making direct use of Corollary 3.3. Here one writes y instead of y_u, for notational simplicity.

Vibrating String, $u \equiv (h(x), T)$:

$$\zeta^{(1)}_{u,\delta u} = (\delta T) \int_0^1 y_x^2 \, dx - \zeta(u) \int_0^1 (\delta h) y^2 dx \tag{4.1}$$

Vibrating Beam, $u = (h(x), E, \rho)$:

$$\zeta^{(1)}_{u,\delta u} = \alpha(\delta E) \int_0^1 h^2 y_{xx}^2 \, dx + \alpha E \int_0^1 2h(\delta h) \, y_{xx}^2 \, dx$$
$$- \zeta(u) \left[(\delta\rho) \int_0^1 hy^2 \, dx + \rho \int_0^1 (\delta h) y^2 dx \right] \tag{4.2}$$

Buckling Beam, $u = (h(x), E)$:

$$\zeta^{(1)}_{u,\delta u} = \alpha(\delta E) \int_0^1 h^2 y_{xx}^2 \, dx + E\alpha \int_0^1 2h(\delta h) \, y_{xx}^2 dx \tag{4.3}$$

Vibrating Membrane, $u = (h(x), T)$:

$$\zeta^{(1)}_{u,\delta u} = (\delta T) \iint_\Omega \left(y_1^2 + y_2^2 \right) d\Omega - \zeta(u) \iint_\Omega (\delta h) y^2 d\Omega \tag{4.4}$$

Vibrating Plate, $u = (h(x), E, \rho)$:

$$\zeta^{(1)}_{u,\delta u} = \iint_\Omega \left[\frac{h^3(\delta E) + 3Eh^2(\delta h)}{12(1 - \nu^2)} \right] \left[y_{11}^2 + \nu(y_{22}y_{11} + y_{11}y_{22}) \right.$$
$$\left. + \, y_{22}^2 + 2(1 - \nu) y_{12}^2 \right] d\Omega$$
$$- \zeta(u) \left[(\delta\rho) \iint_\Omega hy^2 \, d\Omega + \rho \iint_\Omega (\delta h) y^2 \, d\Omega \right] \tag{4.5}$$

Vibrating Plane Elastic Slab, $u = (h(x), \rho)$:

$$\zeta^{(1)}_{u,\delta u} = \frac{\alpha}{2(1+\nu)} \iint_\Omega (\delta h) \Big\{ \Big[\nu\Big(y^1_1 + y^2_2\Big) + 2y^1_1\Big] y^1_1 + \Big(y^1_2 + y^2_1\Big)^2$$

$$+ \Big[\nu\Big(y^1_1 + y^2_2\Big) + 2y^2_2\Big] y^2_2 \Big\} d\Omega$$

$$- \zeta(u) \Big[(\delta\rho) \iint_\Omega h\Big((y^1)^2 + (y^2)^2\Big) \, d\Omega$$

$$+ \rho \iint_\Omega (\delta h) \Big((y^1)^2 + (y^2)^2\Big) \Big] \, d\Omega \qquad (4.6)$$

The formulas presented in this section for simple eigenvalues, as direct applications of the general perturbation theory of Section 3 to problems of Section 2, verify formulas used in the engineering literature [2,3,9,10,11]. This result clears up some controversy over the validity of such formulas for structural systems with continuously distributed design. Engineering application of perturbation formulas of Corollary 3.4 for repeated eigenvalues has only recently been made [16,17]. As shown in Refs. 2, 3, and 17, optimization of structural designs will often lead to repeated eigenvalues. In this class of applications, the theory presented here provides a new engineering capability.

In addition to providing relatively simple formulas for derivatives of eigenvalues with respect to design variables, the coefficients of design variations in these formulas are easily computed. These coefficients may be viewed as design sensitivity coefficients, which tell the designer what effect design variations will have on eigenvalues of his system. He may then use this information as an aid in design of mechanical systems. The data may also be used in iterative optimization of mechanical systems, as illustrated for structures in Refs. 11 and 15. Computational aspects of eigenvalue perturbation, particularly as regards repeated eigenvalues, are treated in Ref. 16.

As a final note, consider the directional derivatives of a repeated eigenvalue that are ordered by magnitude. The eigenvalues of the matrix M of Eq. 3.23 are given by the characteristic equation

$$\begin{vmatrix} M_{11}-\lambda & M_{12} \\ M_{12} & M_{22}-\lambda \end{vmatrix} = M_{11}M_{22} - M_{12}^2 + (M_{11} + M_{22})\lambda + \lambda^2 = 0$$

which has the solutions

$$\lambda = \frac{1}{2}\left\{(M_{11} + M_{22}) \pm \sqrt{(M_{11} + M_{22})^2 - 4(M_{11}M_{22} - M_{12}^2)}\right\} \quad (4.8)$$

Each M_{ij} is itself a linear functions of δu. Thus, Eq. 4.8 defines $\lambda = \lambda(u,\delta u)$, which is not linear for $\delta u \neq 0$, since it involves the square root of a sum of squares, which is not linear. For fixed δu, one may put $\delta u = \varepsilon\eta(x)$ and note that λ is positively homogeneous in ε, but for fixed $\eta(x)$. This shows that the derivative of the repeated eigenvalue $\zeta(u)$ is only directional. This result is consistent with the theory in Section 3 on differentiability of eigenvalues and the counterexample that shows that all one can expect in general for the repeated eigenvalue case is directional derivatives. This result, taken with recent results [2,3,17] showing that one should expect repeated eigenvalues when he optimizes a structure with bounds on eigenvalues, may be stated as a warning; "structural optimizers beware". More work is required to resolve this apparent technical difficulty, but now one has Corollary 3.4 as a tool with which to begin a study of this problem of structural optimization. Preliminary results in this direction may be found in Refs. 16 and 17.

REFERENCES

1. Haug, E.J. and Rousselet, B., "Design Sensitivity Analysis of Static Response Variations," Optimization of Distributed Parameter Structures (Ed. E.J. Haug and J. Cea), Sijthoff & Nordhoff, Alphen aan den Rijn, Netherlands, 1980.
2. Masur, E.F. and Mroz, Z., "On Singular Optimality Conditions in Structural Design," Proceedings IUTAM Symposium on Variational Methods in the Mechanics of Solids, to appear 1980.
3. Olhoff, N. and Rasmussen, S.H., "On Single and Bimodal Optimal Buckling Modes of Clamped Columns," International Journal of Solids and Structures, Vol. 13, 1977, pp. 605-614.
4. Haug, E.J. and Rousselet, E., "Design Sensitivity Analysis in Structural Mechanics II, Eigenvalue Variations," Journal Structural Mechanics, Vol. 8, No. 2, 1980, pp. 161-186.
5. Lions, J.L. and Magenes, E., Non-homogeneous Boundary Value Problems and Applications, Vol. 1, Springer-Verlag, Berlin and New York, 1972.

6. Treves, F., Basic Linear Partial Differential Equations, Academic Press, New York, 1975.
7. Fichera, G., "Existence Theorems in Elasticity," Handbuch Der Physik, Vol. VI a/2, 1972, pp. 347-389.
8. Aubin, J.P., Approximation of Elliptic Boundary-Value Problems, Wiley Interscience, New York, 1972.
9. Haug, E.J. and Komkov, V., "Sensitivity Analysis in Distributed Parameter Mechanical System Optimization," Journal of Optimization Theory and Application, Vol. 23, No. 3, 1977, pp. 445-464.
10. Farshad, M., "Variations of Eigenvalues and Eigenfunctions in Continuum Mechanics," AIAA Journal, Vol. 12, 1974, pp. 560-561.
11. Haug, E.J., Pan, U.C., and Streeter, T.D., "A Computational Method for Optimal Structural Design, Part II," International Journal of Numerical Methods in Engineering, Vol. 9, 1975, pp. 649-667.
12. Rousselet, B., "Etude de la Régularité des Valeurs Propres par Rapport à des Déformations Bilipschitziennes du Domaine," Comptes Rendus de l'Académie des Sciences, t.283, série A, 1976, p. 507.
13. Jouron, C., "Etude des Conditions Nécessaires d'Optimalité pour un Problème d'Optimisation non convex," Comptes Rendus de l'Académie des Sciences, t. 281, série A, 1975, p. 1031.
14. Kato, T., Perturbation Theory for Linear Operators, Springer-Verlag, New York, 1966.
15. Haug, E.J. and Arora, J.S., Applied Optimal Design, Wiley, New York, 1979.
16. Choi, K.K., "A Numerical Method for Optimizing Structures with Repeated Eigenvalues," Optimization of Distributed Parameter Structures (Eds. E.J. Haug and J. Cea), Sijthoff & Nordhoff, Alphen aan den Rijn, Netherlands, 1980.
17. Choi, K.K. and Haug, E.J., "Optimization of Structures with Repeated Eigenvalues," Optimization of Distributed Parameter Structures (Eds. E.J. Haug and J. Cea), Sijthoff & Nordhoff, Alphen aan den Rijn, Netherlands, 1980.

DESIGN SENSITIVITY ANALYSIS OF SHAPE VARIATION*

Bernard Rousselet and Edward J. Haug

Département de Mathematiques, Université de Nice
06034 Nice Cédex, France

Materials Division, College of Engineering,
The University of Iowa, Iowa City, Iowa 52242

ABSTRACT

The dependence of the static response and eigenvalues of distributed parameter structures on the shape of the bodies making up the structure is characterized. Prototype problems treated include membranes, shafts, plates, and plane elastic solids. A transformation function is defined that uniquely determines the shape of a body. Differential operator properties and transformation techniques of integral calculus are employed to show that static response and eigenvalues of the system depend in a continuous (in fact differentiable) way on the shape of the body. Explicit and computable formulas are presented for the derivative (first variation) of structural response and eigenvalues with respect to shape.

1. INTRODUCTION

This paper completes a sequence [1,2] on differentiability of the solution of operator equations of mechanics with respect to design, addressing the problem of domain variation. That is, a structure is to occupy a domain Ω in physical space ($\mathbb{R}^2$ or $\mathbb{R}^3$),

---

*Research of the second author sponsored by National Science Foundation Project No. ENG 77-19967. Travel support for joint efforts by the authors was provided by NATO Research Grant No. 1458.

sustain loads applied over Ω, and obey governing partial differential field equations over Ω and boundary conditions on the boundary Γ of Ω. Such field equations will be of the form

$$\bar{A}z = f \qquad x \in \Omega \quad , \qquad z \in D(\bar{A}) \tag{1.1}$$

for static response, where f is associated with applied load and the boundary conditions are satisfied by admissible displacement fields in $D(\bar{A})$. For vibration response, the governing equations are of the form

$$\bar{A}y = \zeta\bar{B}y \qquad x \in \Omega \quad , \qquad y \in D(\bar{A}) \subset D(\bar{B}) \tag{1.2}$$

where ζ is an eigenvalue that is proportional to the square of natural frequency of vibration and y is an associated eigenfunction.

It is intuitively clear that the functions z and y and the eigenvalue ζ depend on the shape and size of Ω; i.e., $z = z(x;\Omega)$, $y = y(x;\Omega)$, and $\zeta = \zeta(\Omega)$. The task addressed in this paper is to quantify these relationships and to demonstrate not only that the dependence on Ω is regular, but to calculate explicit derivatives of z and ζ with respect to Ω. These derivatives will subsequently be used for shape optimal design of structures.

The reader may be concerned with the question, "how does one analytically characterize shape?" This question is addressed in Section 3 of this paper, following definition of example problems in Section 2. In Section 4, the bilinear and linear forms appearing in the variational formulation of problems of mechanics are reduced to bilinear forms on fixed domains, which depend explicitly on the function that defines shape of the domain. Results of the first two papers are then applied in Sections 5 and 6 to derive explicit derivatives of the static response $z(x;\Omega)$ and the eigenvalue $\zeta(\Omega)$ with respect to the function that defines Ω.

The reader is forwarned that the developments and proofs presented in this paper are more tedious and technical than those in the first two papers [1,2]. Some of the more technical calculations are relegated to appendices, but even with this, calculations found in Sections 3 through 5 are likely to try the patience of even the most dedicated reader. One hopes that as advances are made in the somewhat new field of shape optimal design, more transparent arguments will be found. For advances in shape optimization using other technical approaches, the reader is referred to the papers by Cea, Zolesio, and Rousselet in Part 6 of these proceedings.

To alleviate the technical complexities noted above, the conventional design variables treated in the first two papers (denoted by a vector u in [1,2]) are suppressed here. The task of coupling pure shape optimization and design variable selection is left for specific applications, where it can best be treated in a tractable fashion.

2. PROTOTPYE PROBLEMS CONSIDERED

Except for torsion of a shaft, all the examples studied in this paper have been treated in the two previous papers [1,2], so only the variational equations are given, in the form $a(z,v) = \ell(v)$ or $a(y,v) = \zeta b(y,v)$. Here z denotes static response, y denotes an eigenfunction, and ζ denotes an eigenvalue. Left members of the variational equations are identical in both static and vibration problems and have the same domain, which is a subset of the domain of $b(y,v)$.

2.1 Membrane (Static and Vibrations)

The left member of the static variational equation is [1]

$$a(z,v) = T \iint_{\Omega} (\nabla z, \nabla v) d\Omega \tag{2.1}$$

defined on $H_0^1(\Omega)$. The right member for static response is [1]

$$\ell(v) = \iint_{\Omega} fv d\Omega \tag{2.2}$$

where $f \in C^1(\Omega)$ is required for technical reasons and for the eigenvalue problem the right member is $\zeta b(y,v)$, where [2]

$$b(y,v) = \iint_{\Omega} hyv d\Omega \tag{2.3}$$

and the density distribution $h(x)$ satisfies $h \in C^1(\Omega)$. It is possible to carry through all calculations with h piecewise smooth, but this extension will not be carried out in this writing.

2.2 Torsion of an Elastic Shaft (Static)

The prismatic shaft shown in Fig. 2.1 is twisted by a torque T, so that it undergoes an angular deformation of θ radians per unit length of shaft. The elastic deformation of the system is governed by the boundary-value problem [3]

$$\left.\begin{aligned} \bar{A}z \equiv -\nabla^2 z = 2 \quad & \text{in } \Omega \\ z = 0 \quad & \text{on } \Gamma \end{aligned}\right\} \tag{2.4}$$

where z is the Prandlt stress function. The torque-angular deflection relation is then given by

$$T = GJ\theta \tag{2.5}$$

where G is the shear modulus of the shaft material and J is the torsional constant that is given by [3]

$$J = 2 \iint_\Omega z d\Omega \tag{2.6}$$

The shear stress on the boundary is given by

$$\tau_0 = \theta G \frac{\partial z}{\partial n} \tag{2.7}$$

where n is the normal to the boundary.

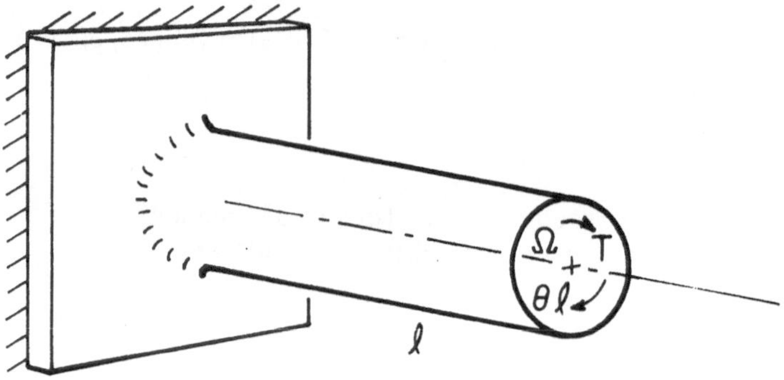

Figure 2.1 Torsion of a Prismatic Shaft.

The variational formulation for this problem is particularly simple. Static behavior is prescribed by

$$a(z,v) \equiv \iint_\Omega (\nabla z, \nabla v) d\Omega = 2 \iint_\Omega v d\Omega \equiv \ell(v) \tag{2.8}$$

for all $v \in H_0^1(\Omega)$. Note that this is just a special case of the variational equation for static deflection of a membrane, in which f = 2 and T = 1. Thus, analysis of the effect of domain change on z is just a special case of that for the static membrane problem.

2.3 Plate (Static and Vibration)

The left member of the static variational equation is [1]

$$a(z,v) = \iint_\Omega \hat{D}(h)[z_{11}v_{11} + \nu(z_{22}v_{11} + z_{11}v_{22}) + z_{22}v_{22} + 2(1 - \nu)z_{12}v_{12}]\, d\Omega \qquad (2.9)$$

defined on $H_0^2(\Omega)$, where $\hat{D} \in C^1(\mathbb{R})$ and the plate thickness is $h \in C^1(\Omega)$. The right member of the variational equation for static response is [1]

$$\ell(v) = \iint_\Omega fv d\Omega \qquad (2.10)$$

where $f \in C^1(\Omega)$. The right member of the variational equation for eigenvalue response is $\zeta b(y,v)$, where [2]

$$b(y,v) = \rho \iint_\Omega hyv d\Omega \qquad (2.11)$$

2.4 Plane Elasticity (Static and Vibration)

The displacement variables are vectors; $z = (z^1,z^2)$, $v = (v^1,v^2)$, and $y = (y^1,y^2)$. The left member of the static variational equation is [1]

$$a(y,v) = \iint_\Omega h[4\nu(z_1^1 + z_2^2)(v_1^1 + v_2^2) + 2(z_1^1v_1^1 + z_2^2v_{22}) + (z_2^1+z_1^2)(v_2^1 + v_1^2)]\, d\Omega \qquad (2.12)$$

defined on $(H_0^1(\Omega))^2$, where the slab thickness if $h \in C^1(\Omega)$. The right member for static response is [1]

$$\ell(v) = \iint_\Omega (f_1v^1 + f_2v^2)d\Omega \qquad (2.13)$$

where $f \in (C^1(\Omega))^2$. The right-hand member of the variational equation for eigenvalue response is $\zeta b(y,z)$, where [2]

$$b(y,v) = \rho \iint_\Omega h(y^1v^1 + y^2v^2)d\Omega \quad (2.14)$$

3. CHARACTERIZATION OF SHAPE

The objective is to determine the dependence of the static response and the eigenvalues of problems of Section 2 on the shape of Ω. If one is only concerned with continuity properties, it would be possible to consider the coefficients involved in the bilinear forms to be $L^\infty(\Omega)$, as in the previous papers [1,2]. Although this would be interesting from a theoretical point of view, it is more important to obtain differentiability with respect to domain shape, in order to work out an iterative type optimization method. The approach to the problem presented here is applicable only for coefficients and right-hand sides of differential equations for static response that are smooth (see Ref. 7 for the piecewise smooth case). One must first state precisely how the operators are connected to the domain Ω, which amounts to stating the dependence of the coefficients of the differential operators on the geometric domain Ω. The following hypotheses are made:

<u>Hypothesis H_1</u>:

(i) the physical domain Ω is a bounded open set of $\mathbb{R}^n$ ($n = 2$ or 3).

(ii) for every $x \in \Gamma$ (boundary of Ω) there exists a system of local coordinates and a cube $Q = (]-a,a[)^n$ (neighborhood of x) such that the points in Ω and Q satisfy $x_n \leq \phi(x_1,\ldots,x_{n-1})$ for $(x_1,\ldots,x_{n-1}) \in (]-a,a])^{n-1}$ where $\phi \in C^1(Q)$ for second order problems or $\phi \in C^2(Q)$ for fourth order problems as shown in Fig. 3.1.

<u>Hypothesis H_2</u>: All the coefficients involved in the operators defined in Section 2 are assumed to be $C^1(\Omega)$. Moreover, it is assumed that there exists $h_{min} > 0$ such that $h(x) \geq h_{min} > 0$ in Ω.

Another difficulty is that "shapes of geometrical domain Ω" are not usually considered as a vector space, so the question arises; how shall one define differentiability relative to the shape of Ω? Some previous work has been done on this subject. In Ref. 5 it is proved that the eigenvalues of the Laplacian are continuous when one considers two open sets Ω and Ω^* of $\mathbb{R}^n$ as neighbors, if and only if there exists a C^1 function F such that $\Omega = (I + F)(\Omega^*)$. This point of view has been systemized in

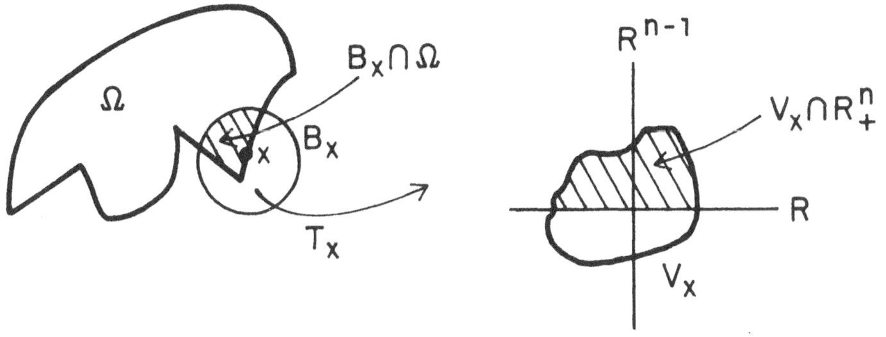

Figure 3.1 Boundary Regularity

Ref. 6 for regular domains under the so-called "Courant topology." In fact, it turns out that this is sufficient to define derivatives relative to F [7]. See also Ref. 8 for a detailed treatment. In Refs. 9-14 and in Part 6 of these proceedings, alternate approaches to optimal design problems and numerical results for domain selection are presented.

A domain is considered in Section 4 that satisfies hypothesis H_1 for every $F \in C^1$ or C^2, as appropriate, such that $||F|| \leq C < 1$ (in the corresponding norm). The mapping $(I + F)$ is a homeomorphism (with the required regularity) of a neighborhood of $\Omega_F = (I + F)(\Omega)$. Let a_F, b_F (respectively A_F and B_F) be the forms (respectively the closed differential operators) associated with Ω_F for each of the problems of Section . Then a_F and b_F satisfy hypothesis H_2. It is proved in Section 5 that the forms a_F and b_F are Frechet differentiable, in the sense of relatively bounded perturbations. Differentiability of the static response and of the eigenvalues are then proved in Section 6, through use of results proved in the first two papers [1,2]. Applications of results to prototype problems are given in Section 7.

4. REDUCTION OF VARIATIONAL EQUATIONS TO FIXED DOMAINS

In order to bring the perturbation methods and results of Refs. 1 and 2 to bear on the problem of domain variation, it is convenient to transform variational equations for the problem on the perturbed domain Ω_F to variational equations on the domain Ω. The linear and bilinear forms that result will depend on the transformation function F that defines the perturbed domain Ω_F. The modified domain Ω_F is prescribed by the transformation $\Omega \to \Omega_F$

given by $x \to x + F(x)$; or $x \to \phi(x)$, where $\phi \equiv \phi(x) = Ix + F(x)$. A function f defined on Ω_F can thus be written as a function on Ω, $\tilde{f} = f(x + F(x))$; or, $\tilde{f} = f \circ \phi$ and $f = \tilde{f} \circ \phi^{-1}$, since ϕ is a homeomorphism from Ω to Ω_F. Since the static response and eigenvalues of a system defined on the perturbed domain Ω_F are solutions of variational equations, it is first necessary to transform the variational equation to the fixed domain Ω. In general, one must transform linear and bilinear forms $\ell_F(v)$, $a_F(z,v)$, and $b_F(y,v)$ defined on $H_0^1(\Omega_F)$ or $H_0^2(\Omega_F)$. These transformations will be carried out for each problem separately.

4.1 Membrane

The static and eigenvalue behavior of a membrane on Ω_F are governed by the linear and bilinear forms of Eqs. 2.1 to 2.3, defined on Ω_F. The following lemma provides a redefinition of those forms and the associated equations on Ω.

<u>Lemma 4.1</u>: Consider the bilinear and linear forms with domains $H_0^1(\Omega)$, given by

$$\tilde{a}_F(\tilde{z},v) = T \iint_\Omega \left((D\phi^{-1})^T \nabla\tilde{z} \ , \ (D\phi^{-1})^T \nabla v\right) |D\phi| \, d\Omega \tag{4.1}$$

$$\tilde{b}_F(\tilde{y},v) = \iint_\Omega \tilde{h}\, \tilde{y}\, v |D\phi| \, d\Omega \tag{4.2}$$

$$\tilde{\ell}_F(v) = \iint_\Omega \tilde{f} v \, |D\phi| \, d\Omega \tag{4.3}$$

where $\phi \equiv \phi(x) = Ix + F(x)$ is assumed to be $C^1(\Omega)$, $D\phi$ is the Jacobian matrix of ϕ, and $|D\phi| = |\det(D\phi)|$. Then, the real numbers $\zeta = \zeta(\Omega_F)$ such that there exists $\tilde{y} \not\equiv 0$ so that

$$\tilde{a}_F(\tilde{y},v) = \zeta \tilde{b}_F(\tilde{y},v)$$

for all $v \in H_0^1(\Omega)$ are the eigenvalues of the membrane over the domain Ω_F and $y = \tilde{y} \circ \phi^{-1}$ are the associated eigenfunctions. Furthermore, the solution $\tilde{z} \in H_0^1(\Omega)$ of

$$\tilde{a}_F(\tilde{z},v) = \tilde{\ell}_F(v)$$

for all $v \in H_0^1(\Omega)$ is such that $z = \tilde{z} \circ \phi^{-1}$ is the static solution of the membrane equation on Ω_F.

Proof: The solution of the eigenvalue and static problems on Ω_F are given by the variational equations

$$a_F(y,v) = \zeta b_F(y,v) \ , \quad y \in H_0^1(\Omega_F) \text{ for all } v \in H_0^1(\Omega_F)$$

and

$$a_F(z,v) = \ell_F(v) \ , \quad z \in H_0^1(\Omega_F) \text{ for all } v \in H_0^1(\Omega_F) \ .$$

Transforming from Ω_F to Ω and noting that $D_{\tilde{X}} v = D_X \tilde{v}(D_{\tilde{X}}\phi^{-1})$ and $\nabla_{\tilde{X}} v = (D_{\tilde{X}}\phi^{-1})^T \ \nabla_X \tilde{v}$, one has

$$a_F(y,v) = \iint_{\Omega_F} (\nabla y, \nabla v) d\Omega$$

$$= \iint_{\Omega} ((D\phi^{-1})^T \ \nabla\tilde{y}, (D\phi^{-1})^T \ \nabla\tilde{v}) \ |D\phi| \ d\Omega = \tilde{a}_F(\tilde{y},\tilde{v})$$

$$b_F(y,v) = \iint_{\Omega_F} hyv d\Omega = \iint_{\Omega} \tilde{h}\tilde{y}\tilde{v} \ |D\phi| \ d\Omega = \tilde{b}_F(\tilde{y},\tilde{v})$$

$$\ell_F(v) = \iint_{\Omega_F} fv d\Omega = \iint_{\Omega} \tilde{f}\tilde{v} \ |D\phi| \ d\Omega = \tilde{\ell}_F(\tilde{v})$$

Since ϕ is a $C^1(\Omega)$ homeomorphism, there is a one-to-one correspondence between functions in $H_0^1(\Omega_F)$ and $H_0^1(\Omega)$. Thus,

$$a_F(y,v) = \tilde{a}_F(\tilde{y},\tilde{v}) = \zeta\tilde{b}_F(\tilde{y},\tilde{v}) = \zeta b_F(y,v)$$

for all $v \in H_0^1(\Omega_F)$ is the same as for all $\tilde{v} \in H_0^1(\Omega)$, hence the ~ may be dropped from the test function v and the first result of the Lemma follows. The second result follows in exactly the same way. Thus, the Lemma is proved.

It is interesting to note that the eigenspaces and static response spaces for the perturbed and unperturbed domains correspond through the change of variable; i.e., $D(\tilde{a}_F) = D(a_0)$ and $D(\tilde{b}_F) = D(b_0)$. This result does not follow for the domains of the operators.

It should also be noted that the change of variable holds not only if the homeomorphism ϕ and its inverse are $C^1(\Omega)$, but also providing ϕ and its inverse are Lipchitzian (see Lemma 3.2 of chapter 2 of Ref. 4).

4.2 Torsion of an Elastic Shaft

As shown in Section 2, the shaft problem is simply a special case of the membrane. Thus, results of Lemma 4.1 apply directly to the shaft and do not need to be restated.

4.3 Plate

The coordinate transformation for the plate is developed in the same way as for the membrane, but with more algebraic complexity.

Lemma 4.2: Consider the bilinear and linear forms of $L^2(\Omega)$, with domain $H_0^2(\Omega)$

$$\tilde{a}_F(\tilde{z},v) = \iint_\Omega \hat{D}(\tilde{h}) \Big[\langle\langle \tilde{z}_{11} \rangle\rangle \langle\langle v_{11} \rangle\rangle + \nu\left(\langle\langle \tilde{z}_{22} \rangle\rangle \langle\langle v_{11} \rangle\rangle + \langle\langle \tilde{z}_{11} \rangle\rangle \langle\langle v_{22} \rangle\rangle\right) + \langle\langle \tilde{z}_{22} \rangle\rangle \langle\langle v_{22} \rangle\rangle + 2(1-\nu) \langle\langle \tilde{z}_{12} \rangle\rangle \langle\langle v_{12} \rangle\rangle \Big] |D\phi| \, d\Omega \tag{4.4}$$

$$\tilde{b}_F(\tilde{y},v) = \rho \iint_\Omega \tilde{h}\tilde{y}v \, |D\phi| \, d\Omega$$

$$\tilde{\ell}_F(v) = \iint_\Omega \tilde{f}v \, |D\phi| \, d\Omega \tag{4.6}$$

where $\phi \equiv \phi(x) = Ix + F(x)$ is assumed to be $C^2(\Omega)$ and

$$\langle\langle \tilde{z}_{ij} \rangle\rangle \equiv X_{1,i} X_{1,j} \tilde{z}_{11} + X_{2,i} X_{2,j} \tilde{z}_{22} + (X_{1,i} X_{2,j} + X_{1,j} X_{2,i})\tilde{z}_{12} + X_{1,ij} \tilde{z}_i + X_{2,ij} \tilde{z}_2 \tag{4.7}$$

where $X_{k,i} \equiv \frac{\partial x_k}{\partial \tilde{x}_i}$ and $X_{k,ij} \equiv \frac{\partial^2 x_k}{\partial \tilde{x}_i \partial \tilde{x}_j}$. As for the membrane, the forms $\tilde{a}_F$ and $\tilde{b}_F$ define the eigenvalues of the plate on the domain Ω_F and the solution of $\tilde{a}_F(\tilde{z},v) = \tilde{\ell}_F(v)$ for all $v \in H_0^2(\Omega)$ is such that $z = \tilde{z} \circ \phi^{-1}$ is the static response of the plate in Ω_F.

Proof: The method of proof is the same as for the membrane, but the computations are more difficult since the bilinear form involves second-order derivatives. The basic formulas needed, from chain rule differentiation, are

$$\frac{\partial^2}{\partial \tilde{x}_1^2} = (X_{1,1})^2 \frac{\partial^2}{\partial x_1^2} + 2X_{1,1} X_{2,1} \frac{\partial^2}{\partial x_1 \partial x_2} + (X_{2,1})^2 \frac{\partial^2}{\partial x_2^2} + X_{1,11} \frac{\partial}{\partial x_1} + X_{2,11} \frac{\partial}{\partial x_2} \tag{4.8}$$

$$\frac{\partial^2}{\partial \tilde{x}_2^2} = (X_{1,2})^2 \frac{\partial^2}{\partial x_1^2} + 2X_{1,2} X_{2,2} \frac{\partial^2}{\partial x_1 \partial x_2} + (X_{2,2})^2 \frac{\partial^2}{\partial x_2^2} + X_{1,22} \frac{\partial}{\partial x_1} + X_{2,22} \frac{\partial}{\partial x_2} \tag{4.9}$$

$$\frac{\partial^2}{\partial \tilde{x}_1 \partial \tilde{x}_2} = X_{1,2} X_{1,1} \frac{\partial^2}{\partial x_1^2} + X_{2,1} X_{2,2} \frac{\partial^2}{\partial x_2^2} + (X_{1,2} X_{2,1} + X_{2,2} X_{1,1}) \times \frac{\partial^2}{\partial x_1 \partial x_2} + X_{1,21} \frac{\partial}{\partial x_1} + X_{2,21} \frac{\partial}{\partial x_2} \tag{4.10}$$

Beginning with Eqs. 2.9 and 2.11 and transforming, one has

$$a_F(z,v) = \iint_{\Omega_F} \hat{D}(h)[z_{\tilde{1}\tilde{1}} v_{\tilde{1}\tilde{1}} + \nu(z_{\tilde{2}\tilde{2}} v_{\tilde{1}\tilde{1}} + z_{\tilde{1}\tilde{1}} v_{\tilde{2}\tilde{2}}) + z_{\tilde{2}\tilde{2}} v_{\tilde{2}\tilde{2}} + 2(1-\nu) z_{\tilde{1}\tilde{2}} v_{\tilde{1}\tilde{2}}] d\Omega$$

where $z_{\tilde{i}\tilde{j}} \equiv \frac{\partial^2 z}{\partial \tilde{x}_i \partial \tilde{x}_j}$. By the definition of the $\langle\!\langle \cdot \rangle\!\rangle$ operation and the chain rule of differentiation in Eqs. 4.8 to 4.10, one has immediately that the transformation of $z_{\tilde{i}\tilde{j}}$, as one changes coordinates from Ω_F to Ω is just $\langle\langle \tilde{z}_{ij} \rangle\rangle$. Thus,

$$a_F(z,v) = \iint_\Omega \hat{D}(\tilde{h})\Big[\langle\langle \tilde{z}_{11}\rangle\rangle \langle\langle \tilde{v}_{11}\rangle\rangle + \nu\Big(\langle\langle \tilde{z}_{22}\rangle\rangle \langle\langle \tilde{v}_{11}\rangle\rangle$$

$$+ \langle\langle \tilde{z}_{11}\rangle\rangle \langle\langle \tilde{v}_{22}\rangle\rangle\Big) + \langle\langle z_{22}\rangle\rangle \langle\langle v_{22}\rangle\rangle$$

$$+ 2(1-\nu) \langle\langle \tilde{z}_{12}\rangle\rangle \langle\langle \tilde{v}_{12}\rangle\rangle\Big] \, |D(\phi)| \, d\Omega = \tilde{a}_F(\tilde{z},\tilde{v})$$

which is exactly Eq. 4.4. Equations 4.5 and 4.6 follow directly by change of variable of integration.

Now, the theorem follows since $a_F(z,v) = \ell_F(v)$ for all v in $H_0^2(\Omega_F)$ is the same as $\tilde{a}_F(\tilde{z},\tilde{v}) = \tilde{\ell}_F(v)$ for all $\tilde{v}$ in $H_0^2(\Omega)$ (since ϕ is a C^2 homeomorphism). Thus, one can drop the ~ over the v and obtain the solution $z = \tilde{z}\circ\phi^{-1} \in H_0^2(\Omega_F)$, ζ, and $y = \tilde{y}\circ\phi^{-1} \in H_0^2(\Omega_F)$ from

$$\tilde{a}_F(\tilde{z},v) = \tilde{\ell}_F(v) \quad , \quad \text{for } \tilde{z} \in H_0^2(\Omega) \text{ and all } v \in H_0^2(\Omega)$$

and

$$\tilde{a}_F(\tilde{y},v) = \zeta\tilde{b}_F(y,v) \ , \quad \text{for } \tilde{y} \in H_0^2(\Omega) \text{ and all } v \in H_0^2(\Omega)$$

This completes the proof of the Lemma.

4.4 Plane Elasticity

Transformation of the plane elasticity problem again resembles that for the membrane problem, but with vector variables.

Lemma 4.3: Consider the following bilinear and linear forms on $L^2(\Omega)^2$ with domain $H_0^1(\Omega)^2$

$$\tilde{a}_F(\tilde{z},v) = \iint_\Omega \tilde{h}\{4\nu([[\tilde{z}_1^1]] + [[\tilde{z}_2^2]])\,([[v_1^1]] + [[v_2^2]])$$

$$+ 2([[\tilde{z}_1^1]][[v_1^1]] + [[\tilde{z}_2^2]][[v_2^2]])$$

(cont'd.)

$$+ ([[\tilde{z}_2^1]] + [[\tilde{z}_1^2]])([[v_2^1]] + [[v_1^2]])\} \ |D\phi| \ d\Omega \tag{4.11}$$

$$\tilde{b}_F(\tilde{y},v) = \rho \iint_\Omega \tilde{h}\tilde{y}v \ |D\phi| \ d\Omega \tag{4.12}$$

$$\tilde{\ell}_F(v) = \iint_\Omega \tilde{f}v \ |D\phi| \ d\Omega \tag{4.13}$$

where

$\phi(x) \equiv Ix + F(x)$ is $C^1(\Omega)$, and $[[\tilde{z}_j^i]] \equiv X_{1,j}\,\tilde{z}_1^i + X_{2,j}\,\tilde{z}_2^i$,

and

$[[v_j^i]] = X_{1,j}\,v_1^i + X_{2,j}\,v_2^i$

As for the membrane and the plate, the forms $\tilde{a}_F$ and $\tilde{b}_F$ define the eigenvalues of the plane elasticity problem in Ω_F. The static response is $z = \tilde{z}\circ\phi^{-1}$, where $\tilde{z}$ is the solution of $\tilde{a}_F(\tilde{z},v) = \tilde{\ell}_F(v)$, for all $v \in H_0^1(\Omega)^2$.

Proof: The proof is exactly the same as for the membrane or the plate, except that here only first order derivatives are involved and the solution has two components.

5. DIFFERENTIABILITY OF BILINEAR FORMS

In this section the differentiability (in the sense of bounded perturbations) of the bilinear forms of Section 2 with respect to the shape variation function F is calculated. The development uses the expressions for the bilinear forms $\tilde{a}_F(v,w)$ on fixed domains Ω, $\tilde{a}_F(\cdot,\cdot)$ considered as a mapping from $H_0^i(\Omega) \times H_0^i(\Omega)$ (i = 1 or 2, depending on the problem) that depends on the function F. The bilinear form $\tilde{a}_F(\cdot,\cdot)$ depends on F and its derivatives, much as the bilinear forms $a_u(\cdot,\cdot)$ of Ref. 1 depended on the design variable u. The objective here is to analytically characterize the dependence of $\tilde{a}_F(\cdot,\cdot)$ on F, as the dependence of $a_u(\cdot,\cdot)$ on u was characterized in Ref. 1.

In this section (z,v) denote arbitrary functions in $H_0^i(\Omega)$ (i = 1 or 2 depending on the problem); there is no connection between this z and the solution of the static problems of Section 2. In each problem considered, a formal calculation is used to obtain the desired derivative, much as was done for the static problem in the first paper [1] and for the eigenvalue problem in the second [2]. Rigorous proofs of the validity of these formulas are then given. Due to the technical complexity of these proofs, many of the details are given in appendices.

5.1 Membrane

From Eq. 4.1 one has the bilinear form $\tilde{a}_F(z,v)$ explicitly in terms of ϕ, where $\tilde{x} = \phi(x) = x + F(x)$. One may now formally calculate the derivative of $\tilde{a}_F(z,v)$ for v and $z \in H_0^1(\Omega)$, with respect to F and then prove validity of the formula. The formal calculation is illustrated here in detail, to serve as a guide to a procedure that can be used in other problems.

Since $x = \phi^{-1}(\tilde{x})$,

$$(D\phi^{-1})_{ij} \equiv \frac{\partial x_i}{\partial \tilde{x}_j} = ([D\phi]^{-1})_{ij} \tag{5.1}$$

and $\tilde{x} = \phi(x) = x + F(x)$, so (5.2)

$$(D\phi)_{ji} \equiv \frac{\partial \tilde{x}_j}{\partial x_i} = \frac{\partial x_j}{\partial x_i} + \frac{\partial F_j}{\partial x_i} = \delta_{ij} + \frac{\partial F_j}{\partial x_i} \tag{5.3}$$

Thus, $D\phi = I + DF$, which with

$$||F(x)||_2^2 = \sum_i |F_i(x)|^2 + \sum_{i,j} |F_{i,j}(x)|^2 \tag{5.4}$$

yields

$$\frac{\partial x_j}{\partial \tilde{x}_i} = ([D\phi]^{-1})_{ji} = \delta_{ij} - \frac{\partial F_j}{\partial x_i} + o(||F(x)||_2) \tag{5.5}$$

Hence, with $z_j \equiv \frac{\partial z}{\partial x_j}$,

$$((D\phi^{-1})^T \nabla z)_i = \sum_j ([D\phi]^{-1})_{ji} z_j$$

$$= \sum_j \frac{\partial x_j}{\partial \tilde{x}_i} z_j = z_i - \sum_j \frac{\partial F_j}{\partial x_i} z_j \tag{5.6}$$

$$\det(D\phi) = \begin{vmatrix} \frac{\partial \tilde{x}_1}{\partial x_1} & \frac{\partial \tilde{x}_1}{\partial x_2} \\ \frac{\partial \tilde{x}_2}{\partial x_1} & \frac{\partial \tilde{x}_2}{\partial x_2} \end{vmatrix} = \begin{vmatrix} 1 + \frac{\partial F_1}{\partial x_1} + o(||F(x)||_2) & \frac{\partial F_1}{\partial x_2} + o(||F(x)||_2) \\ \frac{\partial F_2}{\partial x_1} + o(||F(x)||_2) & 1 + \frac{\partial F_2}{\partial x_2} + o(||F(x)||_2) \end{vmatrix}$$

$$= \begin{vmatrix} 1 + o(||F(x)||_2) & \frac{\partial F_1}{\partial x_2} + o(||F(x)||_2) \\ o(||F(x)||_2) & 1 + \frac{\partial F_2}{\partial x_2} + o(||F(x)||_2) \end{vmatrix} + \begin{vmatrix} \frac{\partial F_1}{\partial x_1} + o(||F(x)||_2) & \frac{\partial F_1}{\partial x_2} + o(||F(x)||_2) \\ \frac{\partial F_2}{\partial x_1} + o(||F(x)||_2) & 1 + \frac{\partial F_2}{\partial x_2} + o(||F(x)||_2) \end{vmatrix}$$

$$= \begin{vmatrix} 1 + o(||F(x)||_2) & 0 \\ o(||F(x)||_2) & 1 + o(||F(x)||_2) \end{vmatrix} + \begin{vmatrix} 1 + o(||F(x)||_2) & \frac{\partial F_1}{\partial x_2} + o(||F(x)||_2) \\ 0 & \frac{\partial F_2}{\partial x_2} + o(||F(x)||_2) \end{vmatrix}$$

$$+ \begin{vmatrix} \frac{\partial F_1}{\partial x_1} + o(||F(x)||_2) & 0 \\ \frac{\partial F_2}{\partial x_1} + o(||F(x)||_2) & 1 \end{vmatrix} + \begin{vmatrix} \frac{\partial F_1}{\partial x_1} + o(||F(x)||_2) & \frac{\partial F_1}{\partial x_2} + o(||F(x)||_2) \\ \frac{\partial F_2}{\partial x_1} + o(||F(x)||_2) & \frac{\partial F_2}{\partial x_1} + o(||F(x)||_2) \end{vmatrix}$$

so

$$\det(D\phi) = 1 + \frac{\partial F_2}{\partial x_2} + \frac{\partial F_1}{\partial x_1} + o(||F(x)||_2)$$

$$= 1 + \text{div } F + o(||F(x)||_2) \tag{5.7}$$

If one applies these computations to Eq. 4.1 for $\tilde{a}_F(z,v)$, he gets

$$\tilde{a}_F(z,v) = T \iint_\Omega \left\{ \sum_i \left[z_i - \sum_j \frac{\partial F_j}{\partial x_i} z_k + o(||F(x)||_2) \right] \right.$$

$$\left. \times \left[v_i - \sum_j \frac{\partial F_j}{\partial x_i} v_j + o(||F(x)||_2) \right] \right\}$$

$$\times \left| 1 + \text{div } F + o(||F(x)||_2) \right| d\Omega + \cdots \tag{5.8}$$

since for $||F||_2$ small enough, $|1 + \text{div } F + o(||F(x)||_2)|$ $= 1 + \text{div } F + o(||F(x)||_2)$,

$$\tilde{a}_F(z,v) = T \iint_\Omega \sum_i z_i v_i \, d\Omega - T \iint_\Omega \sum_{i,j} \left[\frac{\partial F_j}{\partial x_i} + \frac{\partial F_i}{\partial x_j} \right] z_i \, v_j \, d\Omega$$

$$+ T \iint_\Omega \sum_i z_i v_i \, \text{div } F \, d\Omega + \iint_\Omega o(||F(x)||_2) \, d\Omega$$

This formula can be written to first order as

$$\tilde{a}_F(z,v) = a_0(z,v) + a^{(1)}_{\Omega,F}(z,v) + \iint_\Omega o(||F(x)||_2) \, d\Omega \tag{5.9}$$

where

$$a^{(1)}_{\Omega,F}(z,v) = -T \iint_\Omega \sum_{i,j} \left[\frac{\partial F_i}{\partial x_j} + \frac{\partial F_i}{\partial x_j} \right] z_i v_j \, d\Omega$$

$$+ T \iint_\Omega \sum_i z_i v_i \, \text{div } F \, d\Omega \tag{5.10}$$

Thus, $a^{(1)}_{\Omega,F}(z,v)$ is the derivative of $\tilde{a}_F(z,v)$ with respect to F, if the above calculations are accurate.

These computations are formal, so one has to show that the foregoing pointwise developments are valid in the appropriate function space norms when one has integrals. This is a technical task that is outlined in the following, with detailed proofs given in Appendix 1.

Lemma 5.1: For the membrane, let

$$a^{(1)}_{\Omega,F}(z,v) = -2T \iint_\Omega (DF^T \nabla z, \nabla v) d\Omega + T \iint_\Omega (\nabla z, \nabla v) \operatorname{div} F \, d\Omega \tag{5.11}$$

Then, $a^{(1)}_{\Omega,F}$ is linear in F and for $||F||$ small enough, the form

$$a^2_{\Omega,F} = \tilde{a}_F - a_0 - a^{(1)}_{\Omega,F}$$

satisfies

$$|a^2_{\Omega,F}(z,z)| \leq C_2(F) \, ||z||_{H^1_0(\Omega)} \tag{5.12}$$

where

$$C_2(F) > 0 \qquad \text{and} \qquad C_2(F) = o(||F||)$$

Thus, $a^{(1)}_{\Omega,F}(z,z)$ is the Frechet derivative of $\tilde{a}_F$ with respect to F, at F = 0, Moreover,

$$|a^{(1)}_{\Omega,F}(z,z)| \leq C_1 \, ||z||^2_{H^1_0(\Omega)} \, ||F|| \tag{5.13}$$

where $C_1 > 0$. In this Lemma $||F||$ denotes the $C^1(R^n,R^n)$ norm of F.

Proof: By definition and manipulation, one has

$$\tilde{a}_F(z,z) - a_0[z,z] - a^{(1)}_{\Omega,F}(z,z)$$

$$= T \iint_\Omega \left\{ (D\phi^{-1^T} \nabla z, D\phi^{-1^T} \nabla z) - (\nabla z, \nabla z) + 2(DF^T \nabla z, \nabla z) \right\} d\Omega$$

(con't.)

$$+ T \iint_{\Omega} (\nabla z, \nabla z) \, [|D\phi| - 1 - \text{div } F] d\Omega$$

$$+ T \iint_{\Omega} \left\{ \left(D\phi^{-1^T} \nabla z, D\phi^{-1^T} \nabla z \right) - (\nabla z, \nabla z) \right\} [|D\phi| - 1] d\Omega$$

which may also be written as

$$\tilde{a}_F - a_0 - a^{(1)}_{\Omega,F} = T \iint_{\Omega} \left(\left(D\phi^{-1^T} - I + DF^T \right) \nabla z, \, D\phi^{-1^T} \nabla z \right) d\Omega$$

$$+ T \iint_{\Omega} \left(\nabla z, \left(D\phi^{-1^T} - I + DF^T \right) \nabla z \right) d\Omega$$

$$+ T \iint_{\Omega} \left(DF^T \nabla z, \, \left(I - D\phi^{-1^T} \right) \nabla z \right) d\Omega$$

$$+ T \iint_{\Omega} (\nabla z, \nabla z) \, [|D\phi| - 1 - \text{div } F] \, d\Omega$$

$$+ T \iint_{\Omega} \left(\left(D\phi^{-1^T} - I \right) \nabla z, \, D\phi^{-1^T} \nabla z \right) [|D\phi| - 1] d\Omega$$

$$+ T \iint_{\Omega} \left(\nabla z, \left(D\phi^{-1^T} - I \right) \nabla z \right) [|D\phi| - 1] d\Omega$$

Now taking absolute values and bounds yields

$$\left| \tilde{a}_F(z,z) - a_0(z,z) - a^{(1)}_{\Omega,F}(z,z) \right|$$

$$\leq T \sup_{x \in \Omega} \left| D\phi^{-1^T}(\phi(x)) - I + DF^T(x) \right| \left| D\phi^{-1}(\phi(x)) \right| \int_{\Omega} |\nabla z|^2 \, d\Omega$$

$$+ T \sup_{x \in \Omega} \left| D\phi^{-1^T}(\phi(x)) - I + DF^T(x) \right| \int_{\Omega} |\nabla z|^2 \, d\Omega$$

$$+ T \sup_{x \in \Omega} \left| DF^T(x) \right| \left| I - D\phi^{-1^T}(\phi(x)) \right| \int_{\Omega} |\nabla z|^2 \, d\Omega$$

(con'td.)

$$+ T \sup_{x\in\Omega} \left| |D\phi(x)| - 1 - \operatorname{div} F(x) \right| \int_\Omega |\nabla z|^2 \, d\Omega$$

$$+ \sup_{x\in\Omega} \left| (D\phi^{-1^T}(\phi(x)) - I) \right| \left| D\phi^{-1^T}(\phi(x)) \right| \left| |D\phi(x)| - 1 \right| \int_\Omega |\nabla z|^2 \, d\Omega$$

$$+ T \sup_{x\in\Omega} \left| D\phi^{-1^T}(\phi(x)) - I \right| \left| |D\phi(x)| - 1 \right| \int_\Omega |\nabla z|^2 \, d\Omega \tag{5.14}$$

This bound yields Eq. 5.12, if it is shown that every term involving a supremum over $x \in \Omega$ is small of the order $o(||F||_{C^1})$. Note that $||F||_{C^1} = \sup_{x\in\Omega} (|F(x)| + |DF(x)|)$, where $|F(x)|$ denotes the Euclidean norm of the vector $F(x)$ and $|DF(x)|$ denotes the associated matrix norm of the matrix $DF(x)$. These results are shown in Appendix 1, hence completing the proof.

Lemma 5.2: Let

$$b^{(1)}_{\Omega,F}(y,v) = \iint_\Omega (h'(x)F(x))yv \, d\Omega + \iint_\Omega hyv \operatorname{div} F \, d\Omega \tag{5.15}$$

Then $b^{(1)}_{\Omega,F}(y,v)$ is linear in F and for $||F||_{C^1}$ small enough,

$$b^2_{\Omega,F}(y,y) = b_F(y,y) - b_0(y,y) - b^{(1)}_{\Omega,F}(y,y) \tag{5.16}$$

satisfies the inequality

$$|b^2_{\Omega,F}(y,y)| \le e_2(F) \, ||y||^2_{L^2(\Omega)}$$

where $e_2(F) = o(||F||_{C^1(\mathbb{R}^n)})$ and $e_2 > 0$. Moreover there is an $e_1 > 0$ such that

$$|b^{(1)}_{\Omega,F}(y,y)| \le e_1 \, ||y||^2_{L^2(\Omega)} ||F||_{C^1}$$

Proof: Details of the proof are omitted, since it is much simpler than the proof of Lemma 5.1, as only bounded operators are involved. Note, however, that the regularity assumptions on h are used in this Lemma, as was pointed out at the beginning of Section 3.

The bound on $b^{(1)}_{\Omega,F}$ is straightforward. Since

$$b^2_{\Omega,F}(y,y) = \iint_\Omega \tilde{h}(x)y^2 \, |D\phi| \, d\Omega - \iint_\Omega h(x)y^2 \, d\Omega$$

$$- \left[\iint_\Omega (h'(x)F(x))y^2 \, d\Omega + \iint_\Omega hy^2 \operatorname{div} F \, d\Omega\right]$$

which may be written

$$b^2_{\Omega,F}(y,y) = \iint_\Omega (\tilde{h}(x) - h(x) - h'(x)F(x))y^2 \, |D\phi| \, d\Omega$$

$$+ \iint_\Omega h(x)y^2 \left(|D\phi| - 1 - \operatorname{div} F\right) d\Omega$$

$$+ \iint_\Omega (h'(x)F(x))y^2 \left(|D\phi| - 1\right) d\Omega$$

The second term is of the same kind as one of the terms encountered in the proof of Lemma 5.1. With this observation, the detailed proof is written at the end of Appendix 1 (Eqs. A.9 and A.10).

5.2 Plate

The method of treating the plate is analogous to the membrane. Some technical results of the membrane are used, but the computations are much more intricate. Recall first that

$$\frac{\partial x_i}{\partial \tilde{x}_k} = \left(D\phi^{-1}(\phi(x))\right)_i e_k \quad \text{and} \quad \frac{\partial^2 x_i}{\partial \tilde{x}_k \partial x_\ell} = \left(D^2\phi^{-1}(\phi(x))\right)_i (e_k, e_\ell)$$

A formal computation to derive the derivative (in the sense of bounded perturbations) of $\tilde{a}_F$ is now to be written. Recall Eq. 4.4

$$\tilde{a}_F(z,v) = \iint_\Omega \hat{D}(\tilde{h}) \Big[\langle\langle z_{11}\rangle\rangle \langle\langle v_{11}\rangle\rangle + \nu\Big(\langle\langle z_{22}\rangle\rangle \langle\langle v_{11}\rangle\rangle$$

$$+ \langle\langle z_{11}\rangle\rangle \langle\langle v_{22}\rangle\rangle\Big) + \langle\langle z_{22}\rangle\rangle \langle\langle v_{22}\rangle\rangle$$

$$+ 2(1-\nu) \langle\langle z_{12}\rangle\rangle \langle\langle v_{12}\rangle\rangle\Big] |D\phi| \, d\Omega$$

Since the quadratic form $\tilde{a}_F(z,z)$ is composed with analogous terms, computations are carried through only for the term

$$\tilde{a}_{1,F}(z,v) \equiv \iint_\Omega \hat{D}(\tilde{h}) \langle\langle z_{11}\rangle\rangle \langle\langle v_{11}\rangle\rangle |D\phi| \, d\Omega$$

Equation 4.7 gives

$$\langle\langle z_{11}\rangle\rangle = \left(\frac{\partial x_1}{\partial \tilde{x}_1}\right)^2 z_{11} + \left(\frac{\partial x_2}{\partial \tilde{x}_1}\right)^2 z_{22} + 2\left(\frac{\partial x_1}{\partial \tilde{x}_1} \frac{\partial x_2}{\partial \tilde{x}_1}\right) z_{12}$$

$$+ \frac{\partial^2 x_1}{\partial \tilde{x}_1^2} z_1 + \frac{\partial^2 x_2}{\partial \tilde{x}_1^2} z_2$$

Recall Eq. 5.5

$$\frac{\partial x_i}{\partial \tilde{x}_j} = \delta_{ij} - \frac{\partial F_i}{\partial x_j} + o(||F(x)||_2)$$

where

$$||F(x)||_2 = \left[\sum_i |F_i(x)|^2 + \sum_{i,j} |F_{i,j}(x)|^2\right]^{1/2}$$

By taking derivatives formally (rigorous derivation in Appendix 2),

$$\frac{\partial^2 x_i}{\partial \tilde{x}_j^2} = - \frac{\partial^2 F_i}{\partial \tilde{x}_j^2} + o(||F(x)||_3) \qquad (5.17)$$

where

$$||F(x)||_3 = \left[\sum_i |F_i(x)|^2 + \sum_j |F_{i,j}(x)|^2 + \sum_{i,j,k} |F_{i,jk}(x)|^2\right]^{1/2}$$

Now,

$$\left(\frac{\partial x_i}{\partial \tilde{x}_i}\right)^2 = 1 - 2\frac{\partial F_i}{\partial x_i} + o(||F(x)||_2)$$

$$\frac{\partial x_i}{\partial \tilde{x}_i}\ \frac{\partial x_j}{\partial \tilde{x}_i} = \left(1 - \frac{\partial F_i}{\partial x_i}\right)\left(-\frac{\partial F_j}{\partial x_i}\right) + o(||F(x)||_2)$$

$$= -\frac{\partial F_j}{\partial x_i} + o(||F(x)||_2)$$

$$\left(\frac{\partial x_j}{\partial \tilde{x}_i}\right)^2 = o(||F(x)||_2)$$

On the other hand, $\tilde{h}(x) = h(x + F(x)) = h(x) + h'(x)F(x) + o(||F(x)||_2)$. As in Eq. 5.8, these computations yield (also using Eq. 5.7),

$$\tilde{a}_{1,F}(z,v) = \iint_\Omega [\hat{D}(h) + \hat{D}'(h(x))\ h'(x)F(x) + o(||F(x)||_2)]$$

$$\times \left[\left(1 - 2\frac{\partial F_1}{\partial x_1}\right) z_{11} - 2\frac{\partial F_2}{\partial x_1} z_{12} - \frac{\partial^2 F_1}{\partial x_1^2} z_1 - \frac{\partial^2 F_2}{\partial x_1^2} z_2 + o(||F(x)||_2)\right]$$

$$\times \left[\left(1 - 2\frac{\partial F_1}{\partial x_1}\right) v_{11} - 2\frac{\partial F_2}{\partial x_1} v_{12} - \frac{\partial^2 F_1}{\partial x_1^2} v_1 - \frac{\partial^2 F_2}{\partial x_1^2} v_2 + o(||F(x)||_2)\right]$$

$$\times [1 + \text{div } F + o(||F(x)||_2)\quad d\Omega$$

Thus,

$$\tilde{a}_{1,F}(z,v) = \tilde{a}_{1,0}(z,v) + \tilde{a}^{(1)}_{1,\Omega,F}(z,v) + o(||F(x)||_2)$$

where

$$\tilde{a}_{1,0}(z,v) = \iint_\Omega \hat{D}(h) z_{11} v_{11}\, dx$$

is the unperturbed form and

$$\tilde{a}^{(1)}_{1,\Omega,F}(z,v) = \iint_\Omega (\hat{D}'(h)h'F(x) + \hat{D}(h)\ \mathrm{div}\ F)\ z_{11}v_{11}\, d\Omega$$

$$+ \iint_\Omega \hat{D}(h) \left(- 4 \frac{\partial F_1}{\partial x_1} z_{11}v_{11} - 2 \frac{\partial F_2}{\partial x_1} (z_{12}v_{11} + z_{11} v_{12}) - \frac{\partial^2 F_1}{\partial x_1^2} (z_1 v_{11} + z_{11}v_1) - \frac{\partial^2 F_2}{\partial x_1^2} (z_2 v_{11} + z_{11} v_2) \right) d\Omega$$

A complete statement of the Frechet derivative of the whole left-hand bilinear form is given in the following lemma.

Lemma 5.3: The following bilinear form with domain $H_0^2(\Omega)$ is the Frechet derivative of $\tilde{a}_F(z,v)$, for the plate, with respect to the domain:

$$\begin{aligned} a^{(1)}_{\Omega,F}(z,v) = \iint_\Omega \Big\{ & \hat{D}'(h)h'\ F[z_{11}v_{11} + \nu(z_{11}v_{22} + z_{22}v_{11}) + z_{22}v_{22} + 2(1-\nu)z_{12}v_{12}] \\ & + \hat{D}(h)\ [Z_{11}v_{11} + V_{11}z_{11} + \nu(Z_{11}v_{22} + V_{22}z_{11}) + \nu(Z_{22}v_{11} + V_{11}z_{22}) + Z_{22}v_{22} + V_{22}z_{22} \\ & + 2(1-\nu)(Z_{12}v_{12} + V_{12}z_{12})] \\ & + \hat{D}(h)\ [z_{11}v_{11} + \nu(z_{11}v_{22} + z_{22}v_{11}) + z_{22}v_{22} + 2(1-\nu)z_{12}v_{12}]\ (\mathrm{div}\ F) \Big\}\, d\Omega \end{aligned} \qquad (5.18)$$

where

$$Z_{11} = -(2F_{1,1}z_{11} + 2F_{2,1}z_{12} + F_{1,11}z_1 + F_{2,11}z_2)$$

$$Z_{22} = -(2F_{1,2}z_{12} + 2F_{2,2}z_{22} + F_{1,22}z_1 + F_{2,22}z_2)$$

$$Z_{12} = -(F_{1,2}z_{11} + F_{2,1}z_{22} + (F_{1,1} + F_{2,2})z_{12} + F_{1,21}z_1 + F_{2,12}z_2)$$

$$V_{11} = -(2F_{1,1}v_{11} + 2F_{2,1}v_{12} + F_{1,11}v_1 + F_{2,11}v_2)$$

$$V_{22} = -(2F_{1,2}v_{12} + 2F_{2,2}v_{22} + F_{1,22}v_1 + F_{2,22}v_2)$$

$$V_{12} = -(F_{1,2}v_{11} + F_{2,1}v_{22} + F_{1,1}v_{12} + F_{2,2}v_{12} + F_{1,21}v_1 + F_{2,12}v_2) \tag{5.19}$$

$a^{(1)}_{\Omega,F}$ is linear in F and, for $||F||_{C^2}$ small, the form $a^2_{\Omega,F} = a_F - a_0 - a^{(1)}_{\Omega,F}$ satisfies

$$|a^2_{\Omega,F}(z,z)| \leq C_2(F) \quad ||z||_{H^2_0(\Omega)} \tag{5.20}$$

where $C_2(F) > 0$ and $C_2(F) = o(||F||_{C^2})$. Moreover

$$|a^{(1)}_{\Omega,F}(z,z)| \leq C_1 \quad ||z||^2_{H^2_0(\Omega)} \quad ||F||_{C^2} \tag{5.21}$$

Proof: As in the formal computations, the lemma will be proved only for the term

$$\iint_\Omega \hat{D}(\tilde{h}) \left\langle\!\left\langle z_{11}\right\rangle\!\right\rangle^2 \quad |D\phi| \; d\Omega$$

For this term, the appropriate part of Eq. 5.20 is

$$\left|\iint_\Omega \Big[\hat{D}(h) \left\langle\!\left\langle z_{11}\right\rangle\!\right\rangle^2 |D\phi| - \hat{D}(h)\, z^2_{11} - \hat{D}'(h)h'\, Fz^2_{11} - 2\hat{D}(h)Z_{11}z_{11} - \hat{D}(h)z^2_{11} \operatorname{div} F\Big] d\Omega\right| \leq C_2(F) \quad ||z||^2_{H^2_0(\Omega)} \tag{5.22}$$

with $C_2(F) > 0$, $C_2(F) = o(||F||_{C^2})$.

Introducing new terms and manipulating shows that Eq. 5.22 is satisfied if the following inequality holds.

$$\left|\iint_{\Omega} \left\{(\hat{D}(\tilde{h}) - \hat{D}(h) - \hat{D}'(h)h'F) \left\langle\!\left\langle z_{11}\right\rangle\!\right\rangle^2 |D\phi| \right.\right.$$

$$+ \hat{D}(h) \left(\left\langle\!\left\langle z_{11}\right\rangle\!\right\rangle^2 - z_{11}^2 - 2Z_{11}z_{11}\right) |D\phi|$$

$$+ \hat{D}(h)z_{11}^2 \ (|D\phi| - 1 - \text{div } F) + 2\hat{D}(h)Z_{11}z_{11}(|D\phi| - 1)$$

$$+ \hat{D}'(h)h'F\left(\left\langle\!\left\langle z_{11}\right\rangle\!\right\rangle^2 - z_{11}^2\right) |D\phi|$$

$$\left.\left. + \hat{D}'(h)h'F \ z_{11}^2(|D\phi| - 1)\right\} d\Omega\right| \leq C_2(F) \ ||z^2||_{H_0^2(\Omega)} \tag{5.23}$$

This inequality is also implied by Eq. 5.24, where one has taken bounds and applied the triangle inequality, to obtain

$$\iint_{\Omega} |\hat{D}(\tilde{h}) - \hat{D}(h) - \hat{D}'(h)h'F| \left\langle\!\left\langle z_{11}\right\rangle\!\right\rangle^2 |D\phi| \ d\Omega$$

$$+ \iint_{\Omega} \hat{D}(h) \left|\left\langle\!\left\langle z_{11}\right\rangle\!\right\rangle^2 - z_{11}^2 - 2Z_{11}z_{11}\right| |D\phi| \ d\Omega$$

$$+ \iint_{\Omega} \hat{D}(h)z_{11}^2 \ ||D\phi| - 1 - \text{div } F| \ d\Omega$$

$$+ \iint_{\Omega} \left\{2\hat{D}(h) \ |Z_{11}| \ |z_{11}| \ ||D\phi| - 1|\right.$$

$$+ |\hat{D}'(h)h'F| \ \left|\left\langle\!\left\langle z_{11}\right\rangle\!\right\rangle^2 - z_1^2\right| \ |D\phi|$$

$$\left. + |\hat{D}'(h)h'F \ |z_{11}^2 \ ||D\phi| - 1|\right\} d\Omega \leq C_2(F) \ ||z||_{H_0^2(\Omega)}^2 \tag{5.24}$$

which is implied by

$$\sup_{x\in\Omega} |\hat{D}(\tilde{h}(x)) - \hat{D}(h(x)) - \hat{D}'(h(x))h'(x)F(x)|\ |D\phi(x)|$$

$$\times \iint_\Omega \langle\langle z_{11}\rangle\rangle^2\ d\Omega$$

$$+ \sup_{x\in\Omega} \hat{D}(h(x))\ \ |D\phi(x)| \iint_\Omega \left|\langle\langle z_{11}\rangle\rangle^2 - z_{11}^2 - 2Z_{11}z_{11}\right|\ d\Omega$$

$$+ \sup_{x\in\Omega} \hat{D}(h(x))\ \ ||D\phi(x)| - 1 - \operatorname{div} F(x)| \iint_\Omega z_{11}^2\ d\Omega$$

$$+ \sup_{x\in\Omega} \hat{D}(h(x))\ ||D\phi(x)| - 1| \iint_\Omega |Z_{11}|\ |z_{11}|\ d\Omega$$

$$+ \sup_{x\in\Omega} |\hat{D}'(h(x))h'(x)F(x)|\ |D\phi(x)| \iint_\Omega \left|\langle\langle z_{11}\rangle\rangle^2 - z_{11}^2\right|\ d\Omega$$

$$+ \sup_{x\in\Omega} |\hat{D}'(h(x))h'(x)F(x)|\ ||D\phi(x)| - 1| \iint_\Omega z_{11}^2\ d\Omega$$

$$\leq C_2(F)\ ||z||^2_{H_0^2(\Omega)} \tag{5.25}$$

If one compares this with the analogous inequality for the membrane, the only new thing to show is that

$$\iint_\Omega \langle\langle z_{11}\rangle\rangle^2\ dx$$

is uniformly bounded in F for bounded F. This is an easy consequence of the following inequality:

$$\iint_\Omega \left|\langle\langle z_{11}\rangle\rangle^2 - z_{11}^2 - 2Z_{11}z_{11}\right| d\Omega \leq C_2(F)\ ||z||^2_{H_0^2}$$

with

$$C_2(F) = o(||F||_{C^2}) \tag{5.26}$$

$$\iint_\Omega |Z_{11}|\ |z_{11}||\ d\Omega \le C_2(F)\ ||z||^2_{H^2_0}\ ,$$

$$C_2(F) \to 0 \quad \text{as} \quad ||F||_{C^2} \to 0 \tag{5.27}$$

$$\iint_\Omega \left| \langle\langle z_{11} \rangle\rangle^2 - z^2_{11} \right| d\Omega \le C_2(F)\ \ ||z||^2_{H^2_0} \tag{5.28}$$

These results are proved in Appendix 2, which then completes the proof of the lemma.

The right-hand bilinear form and the right-hand linear form are exactly the same as for the membrane, so that one may refer to Lemma 5.2 for their derivatives.

5.3 Plane Elasticity

The derivation of the derivative of the bilinear form is similar to the membrane and the plate. To formally calculate the derivative, one has simply to recall that

$$[[\tilde{z}^i_j]] = \sum_{k=1}^{2} X_{k,j}\ z^i_k$$

where $X_{k,j} = \dfrac{\partial x_k}{\partial \tilde{x}_j}$. It has been shown in Eq. 5.5 that

$$X_{k,j} = \delta_{kj} - F_{k,j} + o(||F(x)||_2)$$

so that

$$[[\tilde{z}^i_j]] = z^i_j - \sum_{k=1}^{2} F_{k,j}\ z^i_k$$

In the same way,

$$[[\tilde{v}^i_j]] = v^i_j - \sum_{k=1}^{2} F_{k,j}\ v^i_k$$

If one sets (as for the plate):

$$Z_j^i = \sum_{k=1}^{2} F_{k,j}\, z_k^i \qquad \text{and} \qquad V_j^i = - \sum_{k=1}^{2} F_{k,j}\, v_k^i$$

and uses

$$\tilde{h}(x) = h(x) + h'(x) \cdot F(x) + o(||F(x)||_2)$$

$$|D\phi| = 1 + \operatorname{div} F + o(||F(x)||_2)$$

one gets the first term of the bilinear form $\tilde{a}_F$ pf Eq. 4.11 the following way:

$$\iint_\Omega 4\nu\tilde{h}([[\tilde{z}_1^1]] + [[\tilde{z}_2^2]])([[v_1^1]] + [[v_2^2]])\, |D\phi|\, d\Omega$$

$$= \iint_\Omega 4\nu h(z_1^1 + z_2^2)(v_1^1 + v_2^2)\, d\Omega$$

$$+ \iint_\Omega \Big\{ 4\nu h'(x) \cdot F(x)(z_1^1 + z_2^2)(v_1^1 + v_2^2)$$

$$+ 4\nu h[(Z_1^1 + Z_2^2)(v_1^1 + v_2^2) + (z_1^1 + z_2^2)(V_1^1 + V_2^2)]\Big\} d\Omega + o(||F||_2)$$

A complete statement of the Frechet derivative of the whole left-hand bilinear form is given in the following lemma.

<u>Lemma 5.4</u>. The following bilinear for with domain $(H_0^1(\Omega))^2$ is the Frechet derivative of $\tilde{a}_F(z,v)$ for plane elasticity, with respect to the domain

$$a^{(1)}_{\Omega,F}(z,v) = \iint_\Omega \Big\{ h' \cdot F[4\nu(z_1^1 + z_2^2)(v_1^1 + v_2^2) + 2z_1^1 v_1^1 + 2z_2^2 v_2^2$$

$$+ (z_2^1 + z_1^2)(v_2^1 + v_1^2)] + h[4\nu(Z_1^1 + Z_2^2)(v_1^1 + v_2^2)$$

$$+ 4\nu(z_1^1 + z_2^2)(V_1^1 + V_2^2) + 2(Z_1^1 v_1^1 + z_1^1 V_1^1)$$

$$+ 2(Z_2^2 v_2^2 + z_2^2 V_2^2)$$

$$+ (Z_2^1 + Z_1^2)(v_2^1 + v_1^2) + (z_2^1 + z_1^2)(V_2^1 + V_1^2)]$$

$$+ h[4\nu(z_1^1 + z_2^2)(v_1^1 + v_2^2) + 2z_1^1\ v_1^1 + 2z_2^2\ v_2^2$$

$$+ (z_2^1 + z_1^2)(v_2^1 + v_1^2)]\ \mathrm{div}\ F\Big\}\ d\Omega$$

where

$$Z_j^i = -F_{1,j}\ z_1^i - F_{2,j}\ z_2^i \quad \text{and} \quad V_j^i = -F_{1,j}\ v_1^i - F_{2,j}\ v_2^i$$

The form $a^{(1)}_{\Omega,F}$ is linear in F and, for $||F||_{C^1}$ small enough, the form

$$a^2_{\Omega,F} = \tilde{a}_F - a_0 - a^{(1)}_{\Omega,F}$$

has associated with it $C_2(F) > 0$, $C_2(F) = o(||F||_{C^1})$, such that

$$|a^2_{\Omega,F}(z,z)| \leq C_2(F)\ \ ||z||_{H_0^1(\Omega)}{}^2$$

Moreover, there is a $C_1 > 0$ such that

$$|a^{(1)}_{\Omega,F}(z,z)| \leq C_1\ \ ||z||_{H_0^1(\Omega)}{}^2\ \ ||F||_{C^1}$$

The proof of this lemma follows from Lemmas 5.1 and 5.3. The right-hand bilinear and linear forms have again the same form as those of the membrane, so that one may refer to Lemma 5.3 for their derivatives.

With the foregoing results, one can now investigate differentiability of the forms in the sense of bounded perturbations.

<u>Proposition 5.5</u>: All the forms of the examples studied in this paper are Fréchet differentiable, in the sense of relatively bounded perturbations. That is,

$$|a^{(1)}_{\Omega,F}(z,z| \leq C_1\ \ ||F||\ a_0(z,z)\ \ , \qquad C_1 > 0$$

$$|a^2_{\Omega,F}(z,z)| \leq C_2(F)\ \ ||F||\ a_0(z,z)\ \ , \quad C_2(F) > 0$$

$$C_2(F) = o(||F||)$$

with $||F||$ denoting norm either in C^1, or in C^2.

For proof, one has only to note that these inequalities follow from the inequalities obtained for each example and then to use the uniform ellipticity of the form $a_0(z,z)$.

6. DIFFERENTIATION OF STATIC RESPONSE AND EIGENVALUES

In this section the shape derivatives of bilinear forms found in Section 5 are used to extend the operator derivative formulas found in the first and second papers [1,2] to derivatives with respect to shape.

6.1 Static Response

The main difference between differentiability here and in Ref. 1 is the right-hand side of the state equation, which now depends on the design variable F. In every case treated in Section 4, the right-hand side is of the form $\tilde{f}|D\phi|$, where $\tilde{f}$ is a $\mathbb{R}^2$ valued function for the planar elasticity problem and a real valued function in the other examples. Thus, the static response is $\tilde{z}_F = \tilde{A}_F^{-1}\,\tilde{f}|D\phi|$.

Following Theorem 3.2 of Ref. 1, the differentiability of $F \to \tilde{A}_F^{-1}$ follows from the differentiability, in the sense of bounded perturbations in Proposition 5.1, for $F \in C^1(\mathbb{R}^n, \mathbb{R}^n)$ for second-order problems, and $F \in C^2(\mathbb{R}^n, \mathbb{R}^n)$ for the plate problem. It remains only to show that $F \to \tilde{f}|D\phi|$ is Frechet differentiable from $C^1(\mathbb{R}^n)$ or $L^2(\mathbb{R}^n)$, which amounts to proving that

$$\left[\iint_\Omega (\tilde{f}|D\phi| - f - f'F - f\,\mathrm{div}\,F)^2\,d\Omega\right]^{1/2} = o(||F||_{C^i})$$

$i = 1$ or 2, which is in turn implied by

$$\sup_{x\in\Omega} |\tilde{f}|D\phi| - f - f'F - f\,\mathrm{div}\,F| = o(||F||_{C^i})$$

$i = 1$ or 2. Manipulating, one obtains

$$\sup_{x\in\Omega} |(\tilde{f} - f - f'F)\,|D\phi|| + \sup_{x\in\Omega} |f(|D\phi| - 1 - \mathrm{div}\,F)|$$

$$+ \sup_{x\in\Omega} |f'F(|D\phi| - 1)| = o(||F||_{C^i})$$

i = 1 or 2. Then the first term has the proper estimate by using Eq. A.9 of Appendix 1 and the second term by using Eq. A.8 of Appendix 1. Equation A.6 of Appendix 1 implies also that

$$\sup_{x \in \Omega} \left| |D\phi| - 1 \right| = o(\|F\|_{C^i})$$

i = 1 or 2 which yields the estimate for the third term. This completes the proof of Frechet differentiability of the static response.

One may now write the derivative. Denoting by $C_1(\Omega,F)$ the $L^2(\Omega)$ bounded operator such that $a^{(1)}_{\Omega,F}(z,v) = (C_1(\Omega,F)\tilde{G}_F z, \tilde{G}_F v)$ as in the proof of Theorem 3.2, one gets the Frechet derivative of $F \to \tilde{A}_F^{-1}$ as being $F \to G_0^1\, C_1(\Omega,F) G_0^{-1}$. Then the derivative of the static response is given by*

$$z^{(1)}_{\Omega,F} = -G_0^{-1}\, C_1(\Omega,F)\, G_0^{-1}\, f + A_0^{-1}(f'F + f \operatorname{div} F) \tag{6.1}$$

To be of practical use, this formula may be used with an adjoint equation as in the fourth section of Ref. 1.

6.2 Eigenvalues

Derivatives of eigenvalues with respect to shape variation may now also be derived. Here the situation is exactly the same as in Ref. 2. The eigenvalue problem is defined by the bilinear form

$$\tilde{a}_F(y_F, v) = \zeta(\Omega_F)\, \tilde{b}_F(y,v) \quad , \qquad \text{for all} \quad v \in V \tag{6.2}$$

both being Frechet differentiable with respect to F (in the sense of bounded perturbations for $\tilde{a}_F$). As in section 3 of Ref. 2, one deduces that $F \to \tilde{A}_F^{-1}\, \tilde{B}_F$ is a $L^2(\Omega)$ bounded operator that is Frechet differentiable. Further, the eigenvalues of $\tilde{A}_F^{-1}\, \tilde{B}_F$ are

---

*It should be emphasized that $z(x + F(x)) = \tilde{z}_F(x)$, so

$$z^{(1)}_{\Omega,F}(x) = \lim_{\varepsilon \to 0} \frac{\tilde{z}_{\varepsilon F}(x) - z(x)}{\varepsilon} = \lim_{\varepsilon \to 0} \frac{z_{I+\varepsilon F}(x + \varepsilon F(x)) - z(x)}{\varepsilon}$$

which is the "material derivative" of continuum mechanics.

the inverse of ζ of Eq. 6.2. Thus, the continuity and differentiability are exactly as stated in section 3 of Ref. 2, where one should replace u by Ω and δu by F in the formulas. Finally, explicit formulas involving the bilinear forms are, from Eq. 3.2 of Ref. 2 for a simple eigenvalue

$$\zeta^{(1)}_{\Omega,F} = a^{(1)}_{\Omega,F}(y_0,y_0) - \zeta(\Omega)\, b^{(1)}_{\Omega,F}(y_0,y_0) \tag{6.3}$$

where y_0 and $\zeta(\Omega)$ are the eigenfunction and smallest eigenvalue of Eq. 6.2, with F = 0, $b_0(y_0,y_0) = 1$. For an m-times repeated eigenvalue Eq. 3.23 of Ref. 2 defines the m derivatives of $\zeta(\Omega_F)$, $\zeta^{(1)}_{\Omega,F,i}$, $i = 1,\dots,m$, as the eigenvalues of the matrix M of general term

$$M_{ij} = a^{(1)}_{\Omega,F}(y_i,y_j) - \zeta(\Omega)\, b^{(1)}_{\Omega,F}(y_i,y_j) \tag{6.4}$$

where y_i (short for $y_{\Omega i}$) are the m eigenfunctions of Eq. 6.2, with F = 0 and $b_0(y_i,y_j) = \delta_{ij}$, $i,j = 1,\dots,m$.

7. EXAMPLES

To illustrate use of the expressions developed in section 6, the effect of domain variation on an integral functional of the static response is first treated. This result and the equations for eigenvalue variation are then illustrated for a simple example of variation of a rectangular domain. Each of the problems of section 2 are analyzed.

Consider a general integral functional

$$\psi = \iint_\Omega L(z)\, d\Omega \tag{7.1}$$

where z is the solution of a variational equation (one of those treated in section 2)

$$a_F(z,v) = \ell_F(v) \quad , \qquad \text{for all} \quad v \in V \tag{7.2}$$

where V is a Hilbert space of Sobolev type associated with the problem. Initially treating the effect of variation F in shape and variation in state z separately, one has a differential of Eq. 7.1 that to first order is

$$\delta_F\psi = \iint_\Omega \frac{\partial L}{\partial z} z^{(1)}_{\Omega,F}\, d\Omega + \iint_\Omega L(\text{div } F)d\Omega \tag{7.3}$$

using $L_2(\Omega)$ scalar product notation and Eq. 6.1,

$$\delta_F\psi = -\left(\frac{\partial L}{\partial z},\ G_0^{-1}\, C_1(\Omega,F)G_0^{-1}f\right)_{L_2}$$

$$+\left(\frac{\partial L}{\partial z},\ A_0^{-1}(f'F + f\text{ div } F)\right)_{L_2} + (L,\text{ div } F)_{L_2} \tag{7.4}$$

Since G_0, G_0^{-1}, $C_1(\Omega,F)$, and A_0^{-1} are all $L_2(\Omega)$-symmetric operators and $G_0^{-1}G_0^{-1} = A_0^{-1}$,

$$\delta_F\psi = -\left(G_0A_0^{-1}\frac{\partial L}{\partial z},\ C_1(\Omega,F)G_0A_0^{-1}f\right)_{L_2}$$

$$+\left(A_0^{-1}\frac{\partial L}{\partial z},\ (f'F + f\text{ div } F)\right)_{L_2} + (L,\text{ div } F)_{L_2}$$

Defining $\lambda = A_0^{-1}\frac{\partial L}{\partial z}$, or equivalently λ as the solution of the operator equation

$$A_0\lambda = \frac{\partial L}{\partial z} \tag{7.5}$$

and noting from Eq. 1.1 that $z = A_0^{-1}f$,

$$\delta_F\psi = -a^{(1)}_{\Omega,F}(\lambda,z) + (\lambda,f'F)_{L_2} + (\lambda f + L,\text{ div } F)_{L_2} \tag{7.6}$$

Note that since the operator equation of Eq. 7.5 for λ is the same as Eq. 1.1 for z, with only different right-hand sides, the same numerical calculations can be used to obtain both z and λ. Thus calculations leading to Eq. 7.6 are efficient. Note also that the basic form of Eq. 7.6 is the same for all problems treated in section 2, hence for other problems of structural mechanics. The main calculation involved in using Eq. 7.6 is development of the expression for derivative of the energy bilinear form $a_F(\lambda,z)$.

As a simple illustration, consider each of the problems of section 2 defined on the domain Ω_F of Fig. 7.1, with only the movable boundary described by the function $g(x_1)$. An acceptable (but not unique transformation function F is thus

$$\begin{bmatrix} x_1 \\ x_2 \end{bmatrix} \to \begin{bmatrix} x_1 \\ x_2 \end{bmatrix} + \begin{bmatrix} 0 \\ x_2 g(x_1) \end{bmatrix} \tag{7.7}$$

Thus

$$F(x) = \begin{bmatrix} 0 \\ x_2 g(x_1) \end{bmatrix} \tag{7.8}$$

and

$$\left.\begin{aligned} \operatorname{div} F &= g(x_1) \\ DF &= \begin{bmatrix} 0 & 0 \\ x_2 g'(x_1) & g(x_1) \end{bmatrix} \end{aligned}\right\} \tag{7.9}$$

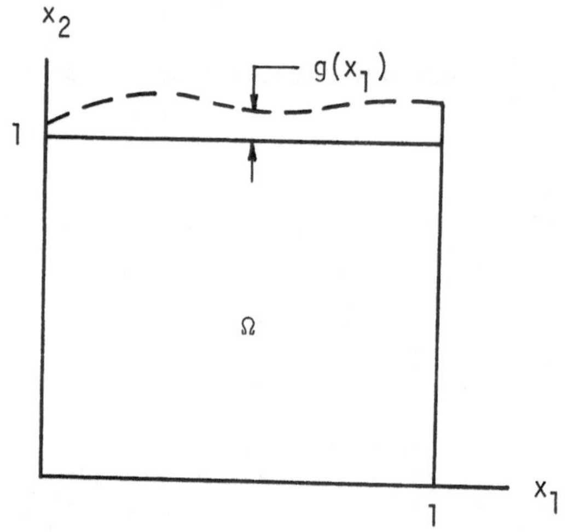

Figure 7.1 Variation of an Initially Square Domain

7.1 Membrane

For the membrane, let $L = z^2$ in Eq. 7.1 and $f(x) = 1$ in Eq. 1.1. Then from Eqs. 7.6, 7.9, and 5.10;

$$\delta_g\psi = \iint_\Omega [T(z_1\lambda_2\ x_2g'(x_1) + z_2\lambda_1\ x_2g'(x_1) + 2z_2\lambda_2g(x_1)$$

$$- T(z_1\lambda_1 + z_2\lambda_2)g(x_1) + (\lambda + z^2)g(x_1)]d\Omega$$

$$= \iint_\Omega [(-T\ z_1\lambda_1 - T\ z_2\lambda_2 + \lambda + z^2)g(x_1)$$

$$+ T(z_1\lambda_2 + z_2\lambda_1)x_2g'(x_1)]d\Omega \qquad (7.10)$$

where λ is the solution of the boundary-value problem

$$\left.\begin{array}{ll} -\nabla^2\lambda = 2z \ , & x \in \Omega \\ \lambda = 0 \ , & x \in \Gamma \end{array}\right\} \qquad (7.11)$$

For a simple eigenvalue ζ of Eq. 1.2 with $My = y$, i.e. $h = 1$, Eqs. 6.3, 7.9, 5.11, and 5.15 yield

$$\zeta^{(1)}_{\Omega,g} = \iint_\Omega [-2T\ y_1y_2x_2g'(x_1) - 2Ty_2^2g(x_1)$$

$$+ T(y_1^2 + y_2^2)g(x_1) - \zeta y^2g(x_1)]d\Omega$$

$$= \iint_\Omega [(T\ y_1^2 + T\ y_2^2 - \zeta y^2)g(x_1) - (2T\ x_2y_1y_2)g'(x_1)]d\Omega$$

$$(7.12)$$

7.2 Shaft

Let $L = 2z$ in Eq. 7.1, corresponding to the problem of shaft design for maximum torsional stiffness in Eq. 2.6. Then, with $f = 2$ and $T = 1$, the equation for stiffness sensitivity to shape variation is similar to the result for the membrane

$$\delta_gJ = \iint_\Omega [-z_1\lambda_1 + z_2\lambda_2 + 2\lambda + z)g(x_1)$$

$$+ (z_1\lambda_2 + z_2\lambda_1)x_2g'(x_1)]d\Omega \qquad (7.13)$$

where λ is the solution of the boundary-value problem

$$\left.\begin{array}{ll} -\nabla^2\lambda = 2 \ , & x \in \Omega \\ \lambda = 0 \ , & x \in \Gamma \end{array}\right\} \qquad (7.14)$$

But Eq. 7.14 for λ is the same as the equation for z, hence $\lambda = z$. Thus, Eq. 7.13 can be written in the simplified form

$$\delta_g J = \iint_\Omega [(-z_1^2 + z_2^2 + 3z)g(x_1) + (2z_1 z_2)x_2 g'(x_1)]d\Omega \qquad (7.15)$$

where no additional adjoint problem must be solved.

7.3 Plate

Let $L(z) = z^2$ in Eq. 7.1, h = 1 in Eqs. 2.9 and 2.11, and f = 1 in Eq. 2.10. Then from Eqs. 7.6, 7.9, 5.18, and 5.19,

$$\delta_g \psi = - \iint_\Omega \Big\{ \hat{D}[Z_{11}\lambda_{11} + \Lambda_{11}z_{11} + \nu(Z_{11}\lambda_{22} + \Lambda_{22}z_{11})$$

$$+ \nu(Z_{22}\lambda_{11} + \Lambda_{11}z_{22}) + Z_{22}\lambda_{22} + \Lambda_{22}z_{22}$$

$$+ 2(1 - \nu)(Z_{12}\lambda_{12} + \Lambda_{12}z_{12}) + (z_{11}\lambda_{11} + \nu(z_{11}\lambda_{22} + z_{22}\lambda_{11})$$

$$+ z_{22}\lambda_{22} + 2(1 - \nu)z_{12}\lambda_{12})g] - (\lambda + z^2)g \Big\} d\Omega$$

where

$$Z_{11} = -(2x_2 g' z_{12} + x_2 g'' z_2)$$

$$Z_{22} = -(2g z_{22})$$

$$Z_{12} = -(x_2 g' z_{22} + g z_{12} + g' z_2)$$

$$\Lambda_{11} = -(2x_2 g' \lambda_{12} + x_2 g'' \lambda_2)$$

$$\Lambda_{22} = -(2g\lambda_{22})$$

$$\Lambda_{12} = -(x_2 g' \lambda_{22} + g\lambda_{12} + g'\lambda_2)$$

Thus,

$$\delta_g \psi = \iint_\Omega \Big[\Big\{ \hat{D}[3z_{22}\lambda_{22} + 2(1 - \nu)z_{12}\lambda_{12} - z_{11}\lambda_{11}$$

$$+ \nu(\lambda_{22}z_{11} + z_{22}\lambda_{11})] + \lambda + z^2 \Big\} g$$

$$+ 2\hat{D}\, x_2(z_{12}\lambda_{11} + \lambda_{12}z_{11}) + \nu x_2(z_{12}\lambda_{22} + \lambda_{12}z_{22})$$

$$+ (1 - \nu)(x_2 z_{22}\lambda_{12} + z_2\lambda_{12} + x_2\lambda_{22}z_{12} + \lambda_2 z_{12}) \quad g'$$

$$+ \hat{D}\left\{x_2(z_2\lambda_{11} + \lambda_2 z_{11}) + \nu x_2(z_2\lambda_{22} + \lambda_2 z_{22})\right\} g'' \Big]\!\Big] d\Omega \tag{7.16}$$

The adjoint variable λ in Eq. 7.16 satisfies the plate equation with right side $\partial L/\partial x = 2z$, or equivalently the variational equation

$$a(\lambda, v) = \iint_{\Omega} 2z\lambda d\Omega \tag{7.17}$$

for all $v \in H_0^2(\Omega)$ and $a(\cdot,\cdot)$ of Eq. 2.9.

For a simple eigenvalue ζ of Eq. 5.2 with $My = \rho y$; i.e. $h = 1$ and ρ constant in Eq. 2.11; Eqs. 6.3, 7.9, 5.18, and 5.15 yield

$$\zeta_{\Omega,g}^{(1)} = \iint_{\Omega} \Big[\!\Big[\left\{\hat{D}[y_{11}^2 - 3y_{22}^2 - 2(1 - \nu)y_{12}^2 - 2\nu y_{11}y_{22}] - \rho\zeta y^2\right\} g(x_1)$$

$$- \left\{4\hat{D}[x_2 y_{11}y_{12} - x_2 y_{22}y_{12} + (1 - \nu)y_2 y_{12}]\right\} g(x_1)$$

$$- \left\{2\hat{D}x_2 y_2(y_{11} + \nu y_{22})\right\} g''(x_1) \Big]\!\Big] d\Omega \tag{7.18}$$

7.4 Plane Elasticity

The plane elasticity problem for a rectangular domain similar to the membrane, but with a vector variable $z = [z^1, z^2]^T$ for displacement. Consider the rectangular domain of Fig. 7. . Let $L = (z^1)^2 + (z^2)^2$ in Eq. 7.1, $h = 1$ in Eqs. 2.12 and 2.14, and $f = [1,1]^T$ in Eq. 2.13. Then from Eqs. 7.6, 7.9, and 5.29.

$$\delta_g\psi = \iint_{\Omega} \Big[\!\Big[\left\{2(1 + 2\nu)(z_2^2\lambda_2^2 - z_1^1\lambda_1^1) + z_2^1\lambda_2^1\right.$$

$$\left. - z_1^2\lambda_1^2 + \lambda^1 + \lambda^2 + (z^1)^2 + (z^2)^2\right\} g(x_1)$$

$$+ \left\{2(2\nu + 1)(z_1^1\lambda_2^1 + z_2^1\lambda_1^1)\right.$$

$$+ (4\nu + 1)(z^1_2\lambda^2_2 + z^2_2\lambda^1_2)$$

$$+ (z^2_2\lambda^2_1 + z^2_1\lambda^2_2)\Big\} x_2 g'(x_1) \Big]\!\Big] d\Omega \tag{7.19}$$

The adjoint variable $\lambda = [\lambda^1, \lambda^2]^T$ in Eq. 7.19 satisfies the plane elasticity equations with right side $\partial L/\partial z = [2z_1, 2z_2]^T$, or equivalently the variational equation

$$a(\lambda, v) = \iint_\Omega (2z^1\lambda^1 + 2z^2\lambda^2) d\Omega \tag{7.20}$$

for all $v \in (H^1_0(\Omega))^2$ and $a(\cdot,\cdot)$ of Eq. 2.12.

For a simple eigenvalue of Eq. 1.2 with $My = \rho y$; i.e., $h = 1$ and ρ constant in Eq. 2.14; Eqs. 6.3, 7.9, 5.29, and 5.15 applied to $b(\cdot,\cdot)$ of Eq. 2.14 yield

$$\zeta^{(1)}_{\Omega,g} = \iint_\Omega \Big[\!\Big[\Big\{ 2(1 + 2\nu)[(y^2_2)^2 - (y^1_1)^2] + (y^1_2)^2 - (y^2_1)^2$$

$$- \rho\zeta[(y^1)^2 + (y^2)^2] \Big\} g(x_1)$$

$$+ \Big\{ 4(1 + 2\nu)(y^1_1 y^1_2) + 2(1 + 4\nu)(y^1_2\, y^2_2)$$

$$+ 2y^2_2\, y^2_1 \Big\} x_2\, g'(x_1) \Big]\!\Big] d\Omega \tag{7.21}$$

APPENDIX 1

It is shown in this appendix that the pointwise developments carried out in the introduction to the differentiability of bilinear forms in section 5 are in fact uniform.

The expression

$$D\phi^{-1}(\phi(x))^T - I + DF^T(x) \tag{A1.1}$$

is first considered. Recall that

$$D\phi^{-1}(\phi(x)) = (D\phi(x))^{-1} = (I + DF(x))^{-1} \tag{A1.2}$$

It is a standard result of the algebra of linear operators or matrices (recalled in Ref. 1 section 3) that if $w \in X$, a normed algebra of linear operators, then

$$|(1 + w)^{-1} - 1 + w|_X \leq 2|w|_X^2 \quad , \quad \text{for} \quad |w|_X < \frac{1}{2} \tag{A.3}$$

Set $w = DF^T(x)$, then Eq. A.3 yields a bound for Eq. A.1, using Eq. A.2

$$|D\phi^{-1}(\phi(x)) - I + DF^{-1}(x)|_2 \leq 2\ |DF(x)|_2^2 \tag{A.4}$$

where $|\cdot|_2$ denotes the matrix norm associated with the usual Euclidean norm for vectors. Taking supremum over $x \in \Omega$ of both sides yields the required bound for this term.

The same kind of bound is now to be shown for

$$|D\phi(x)| - 1 + \text{div } F \quad ,$$

The following result is first established: If $w \in L(R^n)$

$$|\det(I + w) - 1 - \text{tr}w| \leq \sum_{k=2}^{n} \binom{n}{k} |\omega|_2^k \tag{A.5}$$

where $\text{tr}w$ is the trace of w and $\binom{n}{k}$ is the binomial coefficient. After choice of a basis, always denoting by w the matrix formed from the column vectors w_i in this basis, and $e_i = (0,\ldots,1,\ldots,0)^T$, the 1 being in the ith position, one has

$$\det(I + w) = \det(e_1 + w_1, e_2 + w_2, \ldots, e_i + w_i, \ldots, e_n + w_n)$$

Using the multilinear property of the determinant, this may be expanded as

$$\begin{aligned} \det(I + w) = {} & \det(e_1,\ldots,e_n) + \det(w_1,e_2,\ldots,e_n) + \ldots \\ & + \det(e_1,\ldots,w_n) + \det(w_1,w_2,e_3,\ldots,e_n) + \ldots \\ & + \det(w_1,\ldots,w_n) \end{aligned} \tag{A.6}$$

To get the bound of Eq. A.5, note that there are $\binom{n}{k}$ terms in Eq. A.6 involving k factors w_i (and $(n-k)$ factors e_j). Since $|w_i|_2 \leq |w|_2$, this yields Eq. A.5.

Equation A.5 yields also

$$|\det(I + w) - 1 - \operatorname{tr} w| \le |w|_2^2 \left(\sum_{k=2}^{n} \binom{n}{k} |w|_2^k \right)$$

and for $|w|_2 \le 1$, one has

$$|\det(I + w) - 1 - \operatorname{tr} w| \le |w|_2^2 \left[\sum_{k=2}^{n} \binom{n}{k} \right] = |w|_2^2 (2^n - n - 1) \tag{A.7}$$

Now set $w = DF(x)$ and note that $\operatorname{tr} DF(x) = \operatorname{div} F(x)$ so Eq. A.7 yields

$$||\det(I + DF(x)) - 1 - \operatorname{div} F(x)| \le C|DF(x)|_2^2 \tag{A.8}$$

(for DF small enough $|\det(I + DF(x))| = \det(I + DF(x))$. Taking the supremum of both sides of Eqs. A.1, A.4, and A.8 for $x \in \Omega$ yields the desired bounds in Eq. 5.14. Equations A.4 and A.8 are the two uniform bounds needed to complete the proof of Lemma 5.1.

Two results needed for the proof of Lemma 5.2 are now to be shown; first, if $h \in C^1(R^n)$

$$\sup_{x \in \Omega} |\phi(x + F(x)) - \phi(x) - \phi'(x)F(x)| = o(||F||_{C^1}) \tag{A.9}$$

In fact,

$$h(x + F(x)) - h(x) = \int_0^1 h'(x + tF(x))\, F(x)dt$$

so that

$$\sup_{x \in \Omega} |h(x + F(x)) - h(x) - h'(x)F(x)|$$

$$\le \sup_{t \in [0,1]} \sup_{x \in \Omega} |h'(x + tF(x)) - h'(x)| \; ||F||_{C^1}$$

If $||F||_{C^1} \le 1$, $x + tF(x)$ belongs, for almost all $x \in \Omega$ and every $t \in [0,1]$, to a compact neighborhood of Ω. On this compact neighborhood h', which is continuous, is uniformly continuous. Therefore for every $\varepsilon > 0$, there exists $\alpha \in [0,1]$ such that if $||F||_{C^1} \le \alpha$, then

$$\sup_{t\in[0,1]} \sup_{x\in\Omega} |h'(x + tF(x)) - h'(x)| \leq \varepsilon$$

which shows Eq. A.9. The second estimate needed in the proof of Lemma 5.2 is more straightforward ($h \in C^1(R^n)$),

$$\left|\iint (h'(x)F(x))y^2(|D\phi| - 1)d\Omega\right| = o(||F||_{C^1}) \tag{A.10}$$

Since h' is continuous,

$$\sup_{x\in\Omega} |h'(x)\ F(x)| \leq \sup_{x\in\Omega} |h'(x)|\ ||F||_{C^1}$$

From Eq. A.8, one deduces $\sup_{x\in\Omega} ||D\phi| - 1| \leq C||F||_{C^1}$. Equation A.10 is easily deduced from these two estimates.

APPENDIX 2

The three bounds of Eqs. 5.26 to 5.28 are derived in this appendix, as part of the proof of differentiability of the bilinear form of the plate.

The bound of Eqs. 5.26 to be shown is:

$$I \equiv \iint_\Omega \left|\langle\langle z_{11}\rangle\rangle^2 - z_{11}^2 - 2Z_{11}z_{11}\right|\ d\Omega \leq C_2(F)\ ||z||^2_{H_0^2} \tag{A.11}$$

with $C_2(F) = o(||F||_{C^2})$,

$$\langle\langle z_{11}\rangle\rangle = \left(\frac{\partial x_1}{\partial\tilde{x}_1}\right)^2 z_{11} + \left(\frac{\partial x_2}{\partial\tilde{x}_1}\right)^2 z_{22} + 2\left(\frac{\partial x_1}{\partial\tilde{x}_1}\frac{\partial x_2}{\partial\tilde{x}_1}\right) z_{12}$$

$$+ \frac{\partial^2 x_1}{\partial\tilde{x}_1^2} z_1 + \frac{\partial^2 x_2}{\partial\tilde{x}_1^2} z_2$$

and

$$Z_{11} = -\left(2F_{1,1}z_{11} + 2F_{2,1}z_{12} + F_{1,11}z_1 + F_{2,11}z_2\right)$$

As suggested by the formal expansions of $\left(\frac{\partial x_i}{\partial \tilde{x}_j}\right)$ in the treatment of the plate of section 5, the left-side of Eq. A.11 may be bounded by

$$I \leq I_1 + I_2 + I_3 + I_4 + I_5 \tag{A.12}$$

where

$$I_1 = \iint_\Omega \left|\left[\left(\frac{\partial x_1}{\partial \tilde{x}_1}\right)^4 - 1 + 4F_{1,1}\right]\right| \, |z_{11}|^2 \, d\Omega \leq C_4(F) \, ||z||^2_{H_0^2(\Omega)} \tag{A.13}$$

$$I_2 = \iint_\Omega 4\left|\left[\left(\frac{\partial x_1}{\partial \tilde{x}_1}\right)^2 \left(\frac{\partial x_1}{\partial \tilde{x}_1}\,\frac{\partial x_2}{\partial \tilde{x}_1}\right) + F_{2,1}\right]\right| \, |z_{12}| \, |z_{11}| \, d\Omega$$

$$\leq C_5(F) \, ||z||^2_{H_0^2(\Omega)} \tag{A.14}$$

$$I_3 = \iint_\Omega 2\left|\left[\left(\frac{\partial x_1}{\partial \tilde{x}_1}\right)^2 \frac{\partial^2 x_1}{\partial \tilde{x}_1^2} + F_{1,11}\right]\right| \, |z_{11}| \, |z_1| \, d\Omega < C_6(F) \, ||z||^2_{H_0^2(\Omega)} \tag{A.15}$$

$$I_4 = \iint_\Omega 2\left|\left[\left(\frac{\partial x_1}{\partial \tilde{x}_1}\right)^2 \frac{\partial^2 x_2}{\partial \tilde{x}_1^2} + F_{2,11}\right]\right| \, |z_{11}| \, |z_2| \, d\Omega \leq C_7(F) \, ||z||^2_{H_0^2(\Omega)} \tag{A.16}$$

$$I_5 = \iint_\Omega \left[\left(\frac{\partial x_2}{\partial \tilde{x}_1}\right)^4 z_{22}^2 + 4\left(\frac{\partial x_1}{\partial \tilde{x}_1}\,\frac{\partial x_2}{\partial \tilde{x}_1}\right)^2 z_{12}^2 + \left(\frac{\partial^2 x_1}{\partial \tilde{x}_1}\right)^2 z_1^2\right.$$

$$+ \left(\frac{\partial^2 x_2}{\partial \tilde{x}_1}\right)^2 z_2^2 + 2\left(\frac{\partial x_1}{\partial \tilde{x}_1}\right)^2\left(\frac{\partial x_1}{\partial \tilde{x}_1}\right)^2 |z_{11}| \, |z_{22}|$$

$$+ 4\left(\frac{\partial x_2}{\partial \tilde{x}_1}\right)^2 \left|\frac{\partial x_1}{\partial \tilde{x}_1}\,\frac{\partial x_2}{\partial \tilde{x}_1}\right| |z_{22}| \, |z_{12}| + 2\left(\frac{\partial \dot{x}_2}{\partial \tilde{x}_1}\right)^2 \left|\frac{\partial^2 x_1}{\partial \tilde{x}_1^2}\right| |z_{22}||z_1|$$

$$+ 2\left(\frac{\partial x_2}{\partial \tilde{x}_1}\right)^2 \left|\frac{\partial^2 x_2}{\partial \tilde{x}_1^2}\right| |z_{22}|\ |z_2| + 4\left|\frac{\partial x_1}{\partial \tilde{x}_1}\frac{\partial x_2}{\partial \tilde{x}_1^2}\right| \left(\frac{\partial^2 x_1}{\partial \tilde{x}_1^2}\right)^2 |z_{12}|\ |z_1|$$

$$+ 4\left|\frac{\partial x_1}{\partial \tilde{x}_1}\frac{\partial x_2}{\partial \tilde{x}_2}\right| \frac{\partial^2 x^2}{\partial \tilde{x}_1^2} |z_{12}|\ |z_2| + 2\left|\frac{\partial^2 x_1}{\partial \tilde{x}_1}\frac{\partial^2 x_2}{\partial \tilde{x}_1^2}\right| |z_1|\ |z_2|\Bigg]\, d\Omega$$

$$\leq C_8(F)\ ||z||^2_{H_0^2(\Omega)} \qquad \text{(A.17)}$$

where $C_i(F) = o(||F||_{C^2})$ for $i = 4,\ldots,8$. Recall from Eq. A.4

$$\sup_{x\in\Omega} |D\phi^{-1}(x) - I + DF^T(x)| \leq 2\,||F||^2_{C^1}$$

Now, since $\dfrac{\partial x_i}{\partial \tilde{x}_j} = \left((D\phi^{-1})(e_j)\right)_i$

$$\sup_{x\in\Omega}\left|\frac{\partial x_i}{\partial \tilde{x}_j} - \delta_{ij} + \frac{\partial F_i}{\partial \tilde{x}_j}\right| \leq 2\ ||F||^2_{C^1(\mathbb{R}^n,\mathbb{R}^n)} \qquad \text{(A.18)}$$

from which one can deduce by elementary manipulations that

$$\sup_{x\in\Omega}\left|\left(\frac{\partial x_1}{\partial x_1}\right)^4 - 1 + 4\,F_{1,1}\right| = o(||F||_{C^1})$$

which yields the estimate of Eq. A.13.

In the same way Eq. A.18 implies

$$\sup_{x\in\Omega}\left|\left(\frac{\partial x_1}{\partial \tilde{x}_1}\right)^2\left(\frac{\partial x_1}{\partial \tilde{x}_1}\frac{\partial x_2}{\partial \tilde{x}_1}\right) + F_{2,1}\right| = o(||F||_{C^1})$$

which yields the bound of Eq. A.14.

Since Eqs. A.15, A.16, and A.17 involve second-order derivatives of F, the following estimate is needed:

$$\sup_{x\in\Omega} \left|\left|\frac{\partial}{\partial\tilde{x}_1}(D\phi)^{-1} + \frac{\partial}{\partial x_1} DF\right|\right| = o(||F||_{C^2}) \tag{A.19}$$

The proof of this starts by computing $\frac{\partial}{\partial\tilde{x}_1}(D\phi)^{-1}$, which means in fact $\frac{\partial}{\partial\tilde{x}_1}(D\phi(\phi^{-1}(\tilde{x}))^{-1}$, but

$$\frac{\partial}{\partial\tilde{x}_1}[D\phi(\phi^{-1}(\tilde{x}))] = \frac{\partial}{\partial\tilde{x}_1}[I + DF(\phi^{-1}(\tilde{x}))]$$

$$= \frac{\partial}{\partial\tilde{x}_1} DF(\phi^{-1}(\tilde{x})) = \sum_k \left(\frac{\partial}{\partial x_k} DF\right)\frac{\partial x_k}{\partial\tilde{x}_1}$$

and (derivative of an inverse)

$$\frac{\partial}{\partial\tilde{x}_1}[D\phi(\phi^{-1}(\tilde{x}))]^{-1} = -[D\phi(\phi^{-1}(\tilde{x}))]^{-1}\left[\frac{\partial}{\partial\tilde{x}_1} D\phi(\phi^{-1}(\tilde{x}))\right]$$

$$\times [D\phi(\phi^{-1}(\tilde{x}))]$$

Thus, Eq. A.19 is implied by

$$\sup_{x\in\Omega}\left|\left|-(D\phi)^{-1}\left(\sum_k \frac{\partial x_k}{\partial\tilde{x}_1}\frac{\partial}{\partial x_k} DF\right)(D\phi)^{-1} + \frac{\partial}{\partial x_1} DF\right|\right| = o(||F||_{C^1})$$

which is in turn implied by

$$\sup_{x\in\Omega} \; ||(D\phi)^{-1}||^2 \left|\sum_k \frac{\partial x_k}{\partial\tilde{x}_1}\frac{\partial}{\partial x_k} DF - \frac{\partial}{\partial x_1} DF\right|$$

$$+ ||(D\phi)^{-1} - I|| \left|\left|\frac{\partial}{\partial x_1} DF\right|\right| \; ||(D\phi)^{-1}||$$

$$+ \left|\left|\frac{\partial}{\partial x_1} DF\right|\right| \; ||(D\phi)^{-1} - I|| = o(||F||_{C^2})$$

Using Eq. A.4 and manipulating, one easily gets the estimate for the last two terms. For the first of these, one has to use Eq. A.18 and manipulate. This completes the proof of Eq. A.19. From Eqs. A.19 and A.18 one gets Eq. A.15.

The same technique yields the estimate for Eq. A.16. Finally, one should break Eq. A.17 into pieces and apply to each piece the same kind of techniques. This completes the proof of Eq. A.11.

Equations 5.17 and 5.18 are much easier to prove. For Eq. 5.17, one has just to write explicitly Z_{11}, using Eq. 5.9 and then take bounds. One deduces from Eq. 5.16 that for F small enough

$$\iint_{\Omega} |\langle\langle z_{11}\rangle\rangle^2 - z_{11}^2| \, d\Omega \leq 2 \iint_{\Omega} |Z_{11}| \, |z_{11}| \, d\Omega + e_1(F) \, ||z||^2_{H_0^2(\Omega)}$$

which yields Eq. 5.18.

REFERENCES

1. Haug, E.J. and Rousselet, B., "Design Sensitivity Analysis of Static Response Variations," Optimization of Distributed Parameter Structures (Eds. E.J. Haug and J. Cea), Sijthoff and Noordhoff, Alphen aan den Rijn, Netherlands, 1980.
2. Haug, E.J. and Rousselet, B., "Design Sensitivity Analysis of Eigenvalue Variations," Optimization of Distributed Parameter Structures (Eds. E.J. Haug and J. Cea), Sijthoff and Noordhoff, Alphen aan den Rijn, Netherlands, 1980.
3. Wang, C.T., Applied Elasticity, McGraw-Hill, New York, 1953.
4. Necas, J., "Les Méthodes Directes en Théorie Des Equations Elliptiques," Masson, 1967.
5. Courant, R.S. and Hilbert, D., Methods of Mathematical Physics, Vol. I, Wiley-Interscience, New York, 1953.
6. Micheletti, A.M., "Perturbazione dello Spettro di un Operatore Ellitico di Tipo Variazionale, in Relazione ad una Variazione del Campo," Annali di Matematica pura ed. applicata, Vol. SCVII, p. 267-282, 1973.
7. Rousselet, B., "Identification de Domaines et Problèmes de Valeurs Propres," thèse (3ème cycle), Université de Nice, France, 1977.
8. Murat, F. and Simon, J., "Sur le Controle par un Domaine Géométrique," Publications du Laboratoire d'Analyse Numérique (L.A. 189), Université de Paris VI, 1976.
9. Ben, Dali, "Existence et Regularisation dans un Probleme d'Identification de Domaine: Application a l'Analyse Numerique d'un Cas Modele," These de l'Universite d'Alger, 1975.
10. Cea, J., "Une Méthode Numérique pour la Recherche d'un Domaine Optimal," Publications de l'I.M.A.N., Université de Nice, France, 1977.

11. Cea, J., Gioan, A. and et Michel, J., "Adaptation de la Méthode de Gradient à un Problème d'Identification de Domaine," in Computing Methods in Applied Sciences and Engineering, Part 2 (Eds. R. Glowinski and J.L. Lions), Springer-Verlag, Berlin, pp. 391-402, 1974.
12. Dervieux, A. and Palmerio, B., "Une Formula de Hadamard dans des Problèmes d'Identification de Domaines I et II," Comptes Rendus de l'Académie des Sciences, t.280, série A, 1975, p. 1697, and t.280, série A, 1975,p. 1781.
13. Djadane, A., "Regularité d'une Functionelle de Domaines et Application à l'analyse Numérique d'un Problème Modèle," thèse de l'Université d'Alger, 1975.
14. Zolesio, J.P., "Localisation de Support d'un Contrôle Optimal," Comptes Rendus de l'Académie des Sciences, série A, t.284, p. 191, 1977.

SINGULAR DEPENDENCE OF REPEATED EIGENVALUES

Bernard Rousselet

Département de Mathématiques, Université de Nice,
06034 Nice Cédex, France

ABSTRACT

This lecture is introductory to the mathematical study of design sensitivity analysis of eigenvalue variations. This paper begins with finite dimensional systems, considering first the design sensitivity of simple eigenvalues. Making some regularity assumptions, a very simple derivation of eigenvalue design sensitivity is given. The behavior of repeated eigenvalues is shown to be nearly as difficult in the discrete as in the distributed parameter case, provided one is familiar with some ideas of functional analysis. Making some regularity assumptions on the behavior of eigenvalues and eigenvectors in the neighborhood of a repeated eigenvalue, a simple derivation of the directional derivative of eigenvalues with respect to design is given. Two expressions for these derivatives are given.

1. INTRODUCTION TO THE PROBLEM

1.1 Background Studies

If one considers a matrix typical eigenvalue problem

$$A_u x = \zeta x \tag{1.1}$$

where A_u depends on a vector parameter $u = [u_1,\ldots,u_p]^T$, the following result has been extensively used in structural optimization [1,2] as well as in other fields [3]. Assuming that A_u is a self-adjoint matrix ($A_u^T = A_u$) and that $\zeta(u)$ is a simple eigenvalue of A_u, then

$$\frac{\partial \zeta}{\partial u_i} = \left(\frac{\partial A}{\partial u_i} x_u, x_u\right) \tag{1.2}$$

where $(x_u, x_u) = 1$ and x_u is an eigenvector of A_u associated with the eigenvalue $\zeta(u)$ and A_u is assumed to be differentiable with respect to u.

This result may be extended to the generalized eigenvalue problem that is encountered when one approximates vibrating structures using finite elements. Let

$$A_u x = \zeta B_u x \tag{1.3}$$

where B_u is also self adjoint, differentiable with respect to u, and positive definite. Then

$$\frac{\partial \zeta}{\partial u_i} = \left(\frac{\partial A}{\partial u_i} x_u, x_u\right) - \zeta(u)\left(\frac{\partial B}{\partial u_i} x_u, x_u\right) \tag{1.4}$$

where x_u is an eigenvector of Eq. 1.3 associated with the eigenvalue $\zeta(u)$ and is normalized by

$$(B_u x_u, x_u) = 1 \tag{1.5}$$

Assuming that ζ and x are differentiable* functions of u, it is easy to obtain Eq. 1.4 by differentiating Eq. 1.3, yielding

$$\frac{\partial A}{\partial u_i} x + A \frac{\partial x}{\partial u_i} = \frac{\partial \zeta}{\partial u_i} Bx + \zeta \frac{\partial B}{\partial u_i} + \zeta B \frac{\partial x}{\partial u_i}$$

then taking the scalar product of both sides with x

$$\left(\frac{\partial A}{\partial u_i} x, x\right) + \left(A \frac{\partial x}{\partial u_i}, x\right) = \frac{\partial \zeta}{\partial u_i} (Bx, x) + \zeta\left(\frac{\partial B}{\partial u_i} x, x\right) + \zeta\, B\left(\frac{\partial x}{\partial u_i}, x\right)$$

*This result may be proved for simple eigenvalues by using the implicit function theorem. This method has been used also by several authors in the distributed case. For an account of these results see Ref. 4.

and using the self-adjointness of A(u) and B(u), as well as Eq. 1.3. Equation 1.5 now yields

$$\frac{\partial \zeta}{\partial u_i} = \left(\frac{\partial A}{\partial u_i} x, x\right) - \zeta\left(\frac{\partial B}{\partial u_i} x, x\right)$$

which is exactly Eq. 1.4.

The aim of this paper is to derive a formula similar to Eq. 1.4 in the distributed case for simple as well as for multiple eigenvalues. A short idea of the proofs is only given (see Ref. 6 for a detailed account). This extended formula is a basis for a rigorous derivation of necessary optimality condition (using Lagrange multipliers) for weight minimization with eigenvalue constraints [7,8].

It has been recently proved by Olhoff and Rasmussen [9] and Mazur and Mroz [10] that in a certain clamped column that is optimized for maximum buckling load, a repeated eigenvalue may occur. Several other examples are provided in Refs. 7 and 11. Hence, dealing with multiple eigenvalues is not only an interesting mathematical problem, but it is also an important practical one.

1.2 A Distributed Example

The extension of the previous paragraph to the distributed case is straightforward if one assumes differentiability. Of course, the proofs do involve some functional analysis. The method is now to be sketched for vibration of a beam. The beam to be considered is of variable cross-sectional area $h \in L^\infty(0,1)$, such that $h(x) \geq h_0 > 0$. The second moment of the cross sectional area is taken as $I(x) = \alpha h^2(x)$, where α depends of the shape of the cross section. If E denotes Young's modulus and ρ material density, the equation providing the eigenvalue $\zeta = \omega^2$ (ω being the natural frequency) of a clamped-clamped beam is

$$(E\alpha h^2 y_{xx})_{xx} = \zeta\rho\, hy, \text{ in } [0,1] \qquad (1.6)$$

with boundary conditions

$$y(0) = y_x(0) = y(1) = y_x(1) = 0 \qquad (1.7)$$

The problem to be addressed is the dependence of ζ with respect to h. It is not possible to differentiate the operators with respect to h, as was done in Section 1.1 for the finite dimensional case. This is due to the lack of regularity of h, which is not differentiable. Moreover, the process of linearizing Eq. 1.6 in the neighborhood of a smooth h (with smooth variations) is not straightforward, since the operator is unbounded in $L^2(0,1)$ in the sense of functional analysis because it involves derivatives. However, this may be done formally.

A way to deal with this problem is to use the bilinear forms involved in the variational formulation of the boundary value problem of Eqs. 1.6 and 1.7.

Indeed, if y is the solution of Eqs. 1.6 and 1.7, then multiplying Eq. 1.6 by a smooth function v that satisfies the boundary conditions of Eq. 1.7 (because these are essential boundary conditions), integrating twice by parts, and using Eq. 1.7 yields

$$a_h(y,v) = \zeta b_h(y,v) \tag{1.8}$$

where

$$a_h(y,v) = E\alpha \int_0^\ell h^2 y_{xx} v_{xx} dx \tag{1.9}$$

$$b_h(y,v) = \rho \int_0^\ell hyvdx \tag{1.10}$$

On the other hand, it is classical (see e.g. Ref. 12) that if y satisfies Eq. 1.8 for every smooth v fullfilling the boundary conditions of Eq. 1.7 and if y is smooth enough, then y is a classical solution of Eqs. 1.6 and 1.7. Otherwise, it is called a weak solution.

One is then led to define the derivative of these bilinear forms as being

$$a'_{h,\bar{h}}(y,v) = 2E\alpha \int_0^\ell h\bar{h} y_{xx} v_{xx} dx \tag{1.11}$$

$$b'_{h,\bar{h}}(y,v) = \rho \int_0^\ell hyvdx \tag{1.12}$$

This intuitive manipulation rests on a sound functional analysis foundation. The bilinear forms are Frechet derivable in the sense of relatively bounded perturbations (see Ref. 13 for the background and Ref. 5 for its use in structural optimization).

Then if one assumes as in Section 1.1 that ζ and y are differentiable with respect to h, one can easily deduce a formula that provides $\zeta'(h,\bar{h})$, denoted here $\zeta'_{h,\bar{h}}$. Taking the derivative of Eq. 1.8 yields

$$a'_{h,\bar{h}}(y,v) + a_h(y'_{h,\bar{h}},v) = \zeta'_{h,\bar{h}}b_h(y,v) + \zeta b'_{h,\bar{h}}(y,v) + \zeta b_h(y'_{h,\bar{h}},v) \qquad (1.13)$$

If one now sets v=y, uses the self-adjointness of a_h, b_h, and uses the fact that $y'_{h,\bar{h}}$ may be considered as a test function v (this is indeed a smoothness assumption for $y'_{h,\bar{h}}$), then*

$$\zeta'_{h,\bar{h}} = a'_{h,\bar{h}}(y,y) - b'_{h,\bar{h}}(y,y) \qquad (1.14)$$

where y has been normalized with $b_h(y,y) = 1$. If a_h and b_h were finite dimensional bilinear forms, Eq. 1.14 would be the same as Ea. 1.4.

For a simple eigenvalue, a formula of the type of Eq. 1.14 has been derived in related problems by several authors (see e.g. Ref. 4).

1.3 The Case of Multiple Eigenvalues

The extension of Eq. 1.4 to the case of multiple eigenvalues is a more subtle challenge. Continuity results are more difficult in the distributed case than in the discrete case. In fact, continuity of the whole spectrum is in general not valid (see Ref 6). However, if $\zeta(h)$ is an m-fold eigenvalue, then for every neighborhood of $\zeta(h)$ (containing no other eigenvalues), for $\bar{h}$ small enough there are exactly m eigenvalues $\zeta_1(h+\bar{h}),\ldots, {}_m(h+\bar{h})$, some of which may be equal (see Ref. 6).

Apart from continuity considerations (which will not be discussed further here, but which is true in many structural examples), the behavior of multiple eigenvalues is nearly as subtle in the discrete case as in the distributed case. Hence Section 2 analyzes a two dimensional example.

*In these thickness design problems one can also derive $\zeta'_{h,\bar{h}}$ by taking the derivative of $a_h(y,y) = b_h(y,y)$ ((1.8) with v=y).

2. A TWO DIMENSIONAL EXAMPLE

The spring mass system shown in Fig. 2.1 is presented in Refs. 7 and 11.

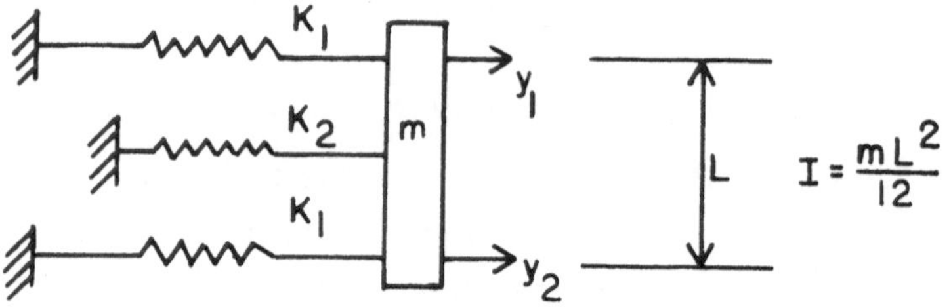

Figure 2.1 Two-degree of Freedom Spring-mass System

The eigenvalue problem for vibration of this system is

$$\begin{bmatrix} 4k_1 + k_2 & k_2 \\ k_2 & 4k_1 + k_2 \end{bmatrix} Y = \zeta \begin{bmatrix} 2 & 1 \\ 1 & 2 \end{bmatrix} Y \tag{2.1}$$

where horizontal motion of the bar is ignored.

The weight to be minimized is presumed to be of the form $J(k) = c_1k_1 + c_2k_2$, where c_1, c_2 are known constants. The constraints to be satisfied are $\lambda_0 \leq \lambda_1$, and $\lambda_0 \leq \lambda_2$, where λ_0 is a given constant, The design variables are k_1 and k_2.

This problem is solved in Refs. 7 and 11, where it is noticed that, depending on the values of the parameters c_1 and c_2, the eigenvalue at the optimum may be simple or double.

It is an easy computation to solve Eq. 2.1 for the eigenvalues

$$\left.\begin{aligned} \zeta_1 &= 4k_1 \\ \zeta_2 &= \frac{4k_1 + 2k_2}{3} \end{aligned}\right\} \tag{2.2}$$

If $k_2 \neq 4k_1$, the eigenvalues are simple and the eigenvectors are

$Y_1 = p_1[1,-1]^T$ and $Y_2 = p_2[1,1]^T$, where p_1 and p_2 are normalizing constants.

If $k_2 = 4k_1$ then $\lambda_1 = \lambda_2 = k_2 = 4k_1$ (conversely if $\lambda_1 = \lambda_2$ then $k_2 = 4k_1$) and every nonzero vector is an eigenvector.

Equation 2.2 provides (whether $k_2 = 4k_1$ or not)

$$\frac{\partial\zeta_1}{\partial k_1} = 4, \; \frac{\partial\zeta_1}{\partial k_2} = 0, \; \frac{\partial\zeta_2}{\partial k_1} = \frac{4}{3}, \; \frac{\partial\zeta_2}{\partial k_2} = \frac{2}{3} \tag{2.3}$$

and if $K = (x_1, x_2)$

$$\left.\begin{aligned} \zeta_1'\;(k,K) &= 4x_1 \\ \zeta_2'\;(k,K) &= \frac{4}{3}x_1 + \frac{2}{3}x_2 \end{aligned}\right\} \tag{2.4}$$

On the other hand, one can compute the differential of the matrix A as

$$A'(k,K) = \begin{bmatrix} 4 & 0 \\ 0 & 4 \end{bmatrix} x_1 + \begin{bmatrix} 1 & 1 \\ 1 & 1 \end{bmatrix} x_2$$

and then try to use Eq. 1.4, which is indeed valid, not only for partial derivatives, but also for the derivative computed in any direction. One gets

$$\left(A'(k,K)Y,Y\right) = 4\left(y_1^2 + y_2^2\right)x_1 + \left(y_1^2 + 2y_1y_2 + y_2^2\right)x_2 \tag{2.5}$$

with

$$2y_1^2 + 2y_1y_2 + 2y_2^2 = 1 \tag{2.6}$$

and Y is any eigenvector of Eq. 2.1.

In the case of simple eigenvalues (i.e. $k_2 \neq 4k_1$), one gets the normalized eigenvectors $Y_1 = \frac{1}{\sqrt{2}}[1,-1]^T$, $Y_2 = \frac{1}{\sqrt{6}}[1,1]^T$, such that

$$\left.\begin{aligned}\left(A'(k,K)Y_1,Y_1\right) &= 4x_1\\ \left(A'(k,K)Y_2,Y_2\right) &= \frac{4}{3}x_1 + \frac{2}{3}x_2\end{aligned}\right\}\qquad (2.7)$$

which is of course consistant with Eqs. 2.1 and 2.4.

In the case of multiple eigenvalues, it has been noted previously that any nonzero vector is an eigenvector. If one uses Y_1 and Y_2 in Eq. 2.5, one gets Eq. 2.7, which provides the right answer.

However, at a point where $k_2 = 4k_1$, there is no reason to select these eigenvectors (in actual practice, this is impossible). Any vector $Y = (y_1,y_2)$ that satisfies Eq. 2.6 is a valid eigenvector. For arbitrary θ, take

$$y_1 = \frac{1}{\sqrt{2}}\cos\theta - \frac{1}{\sqrt{6}}\sin\theta$$

$$y_2 = \sqrt{\frac{2}{3}}\sin\theta = \frac{2}{\sqrt{6}}\sin\theta$$

which satisfy Eq. 2.6, but for which

$$(A'(k,K)Y,Y) = 4\left(\frac{1}{2} - \frac{1}{2\sqrt{3}}\sin 2\theta + \frac{1}{3}\sin^2\theta\right)x_1 + \frac{1}{2}\left(\cos\theta + \frac{1}{\sqrt{3}}\sin\theta\right)^2 x_2$$

where it is clear that $(A'(k,K)Y,Y)$ may take nearly arbitrary values, with no connection with the derivative of ζ_1 and ζ_2.

Nevertheless, the situation is not so desparate, as will be seen in the next section.

3. DEPENDENCE OF MULTIPLE EIGENVALUES (MATRIX CASE)

Consider here the general eigenvalue problem of Section 1.1

$$A(u)X = \zeta B(u)X \qquad (3.1)$$

where $A(u)$ and $B(u)$ are symmetric $n \times n$ matrices with real entries,

B is positive definite, ζ are thus real numbers, and A(u) and B(u) depend smoothly on u.

The problem under study is the dependence of $\zeta(u)$ in the neighborhood of u_0 where $\zeta(u_0)$ is an m-fold eigenvalue. From Section 2 it is clear that one cannot expect the derivative of the eigenvalues to be given by a formula of the type of Eq. 2.7.

Some assumptions are now made, the proofs of which are very technical. They are partly proved in Ref. 6. The method consists of using perturbation theory, as stated in Ref. 13, based on the Dunford integral. The following three assumptions are made:

(1) The eigenvalues are continuous functions of u, i.e. if $\zeta(u_0)$ is an m-fold eigenvalue, then for any neighborhood of $\zeta(u_0)$ containing no other eigenvalue there are, for $\bar{u}$ small enough, exactly m eigenvalues $\zeta_1(y_0 + \bar{u}),\ldots,\zeta_m(u_0 + \bar{u})$ in this neighborhood.

(2) These m eigenvalues have directional derivatives, i.e.

$$\lim_{\substack{t\to 0\\ t>0}} \frac{\zeta_i(u_0 + t\bar{u}) - \zeta(u_0)}{t} \quad \text{exists.}$$

(3) If all the eigenvalues $\zeta_1(u_0 + t\bar{u}),\ldots, \zeta_m(u_0 + t\bar{u})$ are simple for $t > 0$, one can associate eigenvectors $X_1(u_0 + t\bar{u}),\ldots,X_m(u_0 + t\bar{u})$ such that $X_i(u_0 + t\bar{u})$ goes to some limit denoted $X_i(u_0 + 0^+\bar{u})$ when $t \to 0$, $t > 0$; moreover

$$\frac{X_i(u_0 + t\bar{u}) - X_i(u_0 + 0^+\bar{u})}{t} \quad \text{goes to some limit } X_i'(u_0,\bar{u}) \text{ when}$$

$t \to 0$, $t > 0$.

In the example of Section 2, these assumptions were clearly satisfied. However, two features were very special to this example. First, one was able to select two eigenvalues in such a way that they were both Frechet-differentiable. Secondly, the eigenvectors associated to the two simple eigenvalues do not depend on the value of u. See Ref. 7 for simple examples showing that this will not happen in general.

Even in the example of the previous paragraph, if one is interested in the smallest eigenvalue $\zeta(k) = \mathrm{Min}\left(4k_1, \dfrac{4k_1 + 2k_2}{3}\right)$

$\zeta(k)$ is not Frechet differentiable at points (k_1, k_2) where $k_2 = 4k_1$.

Two lemmas are now proved that will enable one to compute the directional derivatives of the eigenvalues.

Lemma 3.1 If one considers the eigenvalue problem of Eq. 3.1, if $\zeta(u_0)$ is an m-fold eigenvalue, and if assumptions (1), (2), and (3) are satisfied, then

$$\left(A'(u_0,\bar{u})X_i(u_0 + 0^+\bar{u}),\ X_j(u_0 + 0^+\bar{u})\right)$$

$$- \zeta(u_0)\left(B'\ u_0,\bar{u})X_i(u_0 + 0^+\bar{u}),\ X_j(u_0 + 0^+\bar{u})\right)$$

$$= \delta_{ij}\zeta'(u_0,\bar{u}) \tag{3.2}$$

where $X_i(u_0 + 0^+\bar{u})$ $(i=1,\ldots,m)$ are supposed to be B orthonormal.

The proof starts, as for simple eigenvalues, by considering Eq. 3.1 written at $u_0 + t\bar{u}$, with $X_i(u_0 + 0^+\bar{u})(i=1,\ldots,m)$ being B orthonormal, as

$$A_{u_0+t\bar{u}}\ X_i(u_0 + t\bar{u}) = \zeta(u_0 + t\bar{u})B_{u_0+t\bar{u}}\ X_i(u_0 + t\bar{u}) \tag{3.3}$$

All the functions appearing in Eq. 3.3 are supposed to have limits as $t \to 0$, $t > 0$ and to have right side derivatives. Hence,

$$A'(u_0,\bar{u})X_i(u_0 + 0^+\bar{u}) + A(u_0)X_i'(u_0,\bar{u})$$

$$= \zeta'(u_0,\bar{u})B_{u_0}X_i(u_0 + 0^+\bar{u}) + \zeta(u_0)B'(u_0,\bar{u})X_i(u_0 + 0^+\bar{u})$$

$$+ \zeta(u_0)B_{u_0}X_i'(u_0,\bar{u}) \tag{3.4}$$

If one multiplies Eq. 3.4 by $X_i(u_0 + 0^+\bar{u})$, one gets Eq. 3.2 for $i=j$. This is a formula of the same type as for simple eigenvalues. On the other hand, if one multiplies by $X_j(u_0 + 0^+\bar{u})$ with $j \neq i$, then

$$\left(A'(u_0,\bar{u})X_i(u_0 + 0^+\bar{u}),\ X_j(u_0 + 0^+\bar{u})\right)$$
$$+ \left(A_{u_0}X_i'(u_0,\bar{u}),\ X_j(u_0,\bar{u})\right)$$
$$= \zeta'(u_0,\bar{u})\left(B_{u_0}X_i(u_0 + 0^+\bar{u}),\ X_j(u_0 + 0^+\bar{u})\right)$$
$$+ \zeta(u_0)\left(B'(u_0,\bar{u})X_i(u_0 + 0^+\bar{u}),\ X_j(u_0 + 0^+\bar{u})\right)$$
$$+ \zeta(u_0)\left(B(u_0)X_i'(u_0,\bar{u}),\ X_j(u_0 + 0^+\bar{u})\right) \tag{3.5}$$

Using the symmetry of A and B and Eq. 3.1, one gets

$$\left(A'(u_0,\bar{u})X_i(u_0 + 0^+\bar{u}),\ X_j(u_0 + 0^+\bar{u})\right)$$
$$- \zeta(u_0)\left(B'(u_0,\bar{u})X_i(u_0 + 0^+\bar{u}),\ X_j(u_0 + 0^+\bar{u})\right)$$
$$= -\zeta'(u_0,\bar{u})\left(B(u_0)X_i(u_0 + 0^+\bar{u}),\ X_j(u_0 + 0^+\bar{u})\right) \tag{3.6}$$

For $t > 0$, $X_i(u_0 + t\bar{u})$, $i=1,\ldots,m$, are orthogonal with respect to the scalar product (BX,Y) (eigenvectors of simple eigenvalues), hence so are $X_i(u_0 + 0^+\bar{u})$, $i=1,\ldots,m$, and Eq. 3.6 is then identical to Eq. 3.2. This completes the proof.

The interest of Eq. 3.2 is more of theoretical than practical interest, because one does not know in general the limits of the eigenvectors $X_i(u_0 + 0^+\bar{u})$. Moreover even if one knows these limits, one expects that in numerical computations they would not be very accurate. The following lemma enables one to construct a more practical formula, which is true even if the assumption (3) is not satisfied.

Lemma 3.2 Under the assumptions of Lemma 3.1, if $Y_i(u_0)$, $i=1,\ldots,m$, is a B-orthonormal basis of the eigenspace of $\zeta(u_0)$, then $\zeta_i'(u_0,\bar{u})$ are the eigenvalues of the matrix M of entries

$$M_{i,j} = \left(A'(u_0,\bar{u})Y_i(u_0),\ Y_j(u_0)\right)$$
$$-\zeta(u_0)\left(B'(u_0,\bar{u})Y_i(u_0),\ Y_j(u_0)\right) \tag{3.7}$$

To prove Eq. 3.7 one first notices that Eq. 3.2 may be written in the form $X^T(A' - \zeta B')X = Z'$, where $X = [X_1,\ldots X_m]$ and $Z' = \text{diag}\left(\zeta_i'(u_0,\bar{u})\right)$. Then, since the X_i, $i=1,\ldots,m$ are B-orthonormal, $X^TBX = XBX^T = I$, where I denotes the identity matrix. Thus, $A' - \zeta B' = XBZ'BX^T$, hence

$$Y^T(A' - \zeta B')Y = Y^TXBZ'BX^TY \tag{3.8}$$

Moreover

$$\begin{aligned} Y^TXBBX^TY &= Y^TXB(X^TBX)BX^TY \\ &= Y^T(XBX^T)B(XBX^T)Y = Y^TBY = I \end{aligned}$$

Thus BX^TY is an orthogonal matrix, which proves that $(Y^TXB)Z'(BX^TY)$ has the same eigenvalues as Z', which are $\zeta_i'(u_0,\bar{u})$. Finally, the eigenvalues of $Y^T(A' - \zeta B')Y$, which is exactly the matrix M, are $\zeta_i'(u_0,\bar{u})$. This proves Lemma 3.2.

4. CONCLUSIONS

4.1 Distributed Case and Multiple Eigenvalues

The results of Section 3 may be extended to the distributed case. If one considers a system governed by the eigenvalue problem, in the variational form

$$a_u(y,v) = b_u(y,v) \tag{4.1}$$

for every $v \in V$, where V is some Hilbert space of Sobolev type, e.g. $V = H_0^2(0,1)$ for the vibrating clamped beam of Section 1. Buckling of bars, vibration of plates, and vibration of planar elastic solids also fall in this general framework [6].

Theorem If $\zeta(u_0)$ is an m-fold eigenvalue, then the directional derivatives are the eigenvalues of the matrix M of entries

$$M_{i,j} = a'_{u_0,\bar{u}}(y_i,y_j) - \zeta(u_0)b'_{u_0,\bar{u}}(y_i,y_j)$$

where y_i, $i=1,\ldots,m$, is a b-orthonormal basis of the eigenspace of $\zeta(u_0)$ and $a'_{u_0,\bar{u}}$, $b'_{u_0,\bar{u}}$ are the derivatives of the bilinear forms a_u and b_u.

For the vibration of a clamped beam, $u = h$, $\bar{u} = \bar{h}$, and

$$a'_{h,\bar{h}}(y,v) = 2E\alpha \int_0^{\ell} h\bar{h}y_{xx}v_{xx}dx$$

$$b'_{h,\bar{h}}(y,v) = \rho \int_0^{\ell} \bar{h}yvdx$$

A rigorous proof of this result is technical and involves some functional analysis [6] but all this mathematical theory is not needed to use the result.

4.2 Concluding Remarks

A heuristic approach has been given for design sensitivity analysis for multiple eigenvalues. As long as one is only interested in the results, the distributed case is very similar to the discrete one.

These formulas are a basis for derivation of necessary conditions of optimality for weight minimization with eigenvalue constraints [7,8,11].

REFERENCES

1. Farshad, M., "Variations in Eigenvalues and Eigenvectors in Continuum Mechanics", AIAA J., Vol. 12, 1974, pp. 560-561.
2. Haug, E.J., and Arora, J.S., "Design Sensitivity Analysis of Elastic Mechancial Systems", Computer Methods in Applied Mechanics and Engineering, Vol. 15, 1978, pp. 35-62.
3. Joseph, D.D., "Parameter and Domain Dependence of Eigenvalues of Elliptic Partial Differential Equations", Arch. Rat. Mech. Anal., Vol. 35, 1967, pp. 169-177.
4. Lions, J.L., "Remarks on the Theory of Optimal Control of Distributed Systems", Communication at the Conference on Control Theory of Systems Governed by Partial Differential Equations, (Ed. Aziz, Wingate, Balas), Academic Press, New York, 1977.

5. Haug, E.J. and Rousselet, B., "Design Sensitivity Analysis of Static Response Variations", Optimization of Distributed Parameter Structures, (Eds. E.J. Haug and J. Cea),Sijthoff and Noordhoff, Alphen aan den Rihn, Netherlands, 1980.
6. Haug, E.J. and Rousselet, B., "Design Sensitivity Analysis of Eigenvalue Variations", Optimization of Distributed Parameter Structures, (Eds. E.J. Haug and J. Cea), Sijthoff and Noorhoff, Alphen aan den Rihn, Netherlands, 1980.
7. Choi, K.K., Haug, E.J., and Rousselet, B., Problems of Structural Optimization Involving Multiple Eigenvalues, Technical Report, Materials Division, College of Engineering, The University of Iowa, November, 1979.
8. Rousselet, B., "Condition Necessaire d'optimalite en Presence de Valeurs Propres Multiples, Comptes Rendus de l'Academie des Sciences, Paris, 1980, to appear.
9. Olhoff, N. and Rasmussen, S.H., "On Single and Biomodal Optimum Buckling Loads of Clamped Columns", International Journal Solids and Structures, Vol. 13, 1977, pp. 605-614.
10. Masur, E.F. and Mroz, Z., "Non Stationary Optimality Conditions in Structural Design", International J. Solids and Structures, Vol. 15, 1979, pp. 503-512.
11. Choi, K.K. and Haug, E.J., "Optimization of Structures with Repeated Eigenvalues", Optimization of Distributed Parameter Structures, (Eds. E.J. Haug and J. Cea), Sijthoff and Noordhoff, Alphen aan den Rihn, Netherlands, 1980.
12. Courant, R. and Hilbert, D., Methods of Mathematical Physics, Interscience Publishers, New York, 1953.
13. Kato, T., Perturbation Theory for Linear Operators, Springer-Verlag, New York, 1966.

SEMI DERIVATIVES OF REPEATED EIGENVALUES

Jean-Paul Zolésio

Departement de Mathematiques, Universite de Nice,
06034 Nice Cedex, France

ABSTRACT

Using the well known Rayleigh quotient, the first eigenvalue of a symmetric matrix or of an elliptic, symmetric operator may be written as an infimum. Two basic theorems are developed in this paper that give a result about the Hadamard semi-derivative of an infimum. This result is an extension of a first result given in Ref. 1, which has been used in Ref. 2 for studying non-differentiable functionals. Basically, this result says that the derivative of an infimum is the infimum of the derivative.

When using this result, one must be very careful about the hypotheses, which are of a topological (smoothness) nature. An interesting aspect of this result is the simplicity of the arithmetic and that most of the old results about the regular dependence of the eigenvalue or the coefficients of a matrix, for example, are immediate and are now extended to the non-differentiable situation. Analogous results for the nth eigenvalue (n = 2,3,...) are given in Ref. 3, but attention is restricted here to the first eigenvalue.

Before giving the results about the eigenvalue, it is helpful to first give basic results as examples for the semi-derivative of non-differentiable functions. Consider first the Hadamard semi-derivative (see Ref. 4 for more details) since it is the best property that a non-differentiable function may have (when Frechet differentiability fails). It is, in particular, stronger than semi-Gateaux differentiability and it contains, for the situation considered here, the Clarke semi-derivative properties (see Ref. 5).

1. SEMI-DERIVATIVE OF AN INFIMUM

Let F be a function defined from $E \times W$ into $\mathbb{R}$, where E is a Banach space and W a topological compact set. Let f be the scalar function defined on E by

$$f(u) = \operatorname*{INF}_{w \in W} F(u,w) \tag{1.1}$$

If F is lower semi-continuous on W, this infimum is realized for some w. Denote by $W^*(u)$ the set of the minimizing elements, i.e. $W^*(u) = \text{Arg min } \{F(u,w) : w \in W\} \equiv \{w \in W : F(u,w) = \inf_{\alpha \in W} F(u,\alpha)\}$
The set $W^*(u)$ is a closed set in W, hence a compact set.

Let the Hadamard semi-derivative of f(u) in the direction v be defined as

$$d_H f(u;v) = \lim_{\substack{t \overset{>}{\to} 0 \\ z \to v \text{ in } E}} [f(u + tz) - f(u)]/t \tag{1.2}$$

and of F(u,w) with respect to u in the direction v be defined as

$$F'_u(u,w;v) = \lim_{\substack{t \overset{>}{\to} 0 \\ z \to v \text{ in } E}} [F(u + t_z, w) - F(u,w)]/t \tag{1.3}$$

Assuming that F'_u exists and is positively homogeneous in the direction v, one may prove that "the derivative of the minimum is the minimum of the derivative", more precisely,

$$d_H f(u;v) = \operatorname*{INF}_{w \in W^*(u)} F'_u(u,w;v) \tag{1.4}$$

Theorem 1.1: Let E be a Banach space and W be a topological compact set. Assume that:

(H1) $(u,w) \longrightarrow F(u,w)$ is lower semi-continous on $E \times W$

(H2) For any u and v in E and w in W, the Hadamard semi-derivative $F'_u(u,w;v)$ exists

(H3) For any w in W, $u \longrightarrow F(u,w)$ is continuous

(H4) For any u and z in E and w in W and for all $\rho > 0$, $F'_u(u,w;\rho u) = \rho F'_u(u,w;v)$

(H5) $(u,w,z) \to F_u'(u,w;z)$ is lower semi-continuous on $E \times E \times W$.

Then the function f defined by Eq. 1.1 possesses a Hadamard semi-derivative at each $u \in E$ and in all direction $v \in E$. Furthermore, $d_H f(u;v)$ is given by Eq. 1.4.

Proof: The proof is carried out in two steps.

Step 1. For t and z fixed, there is a $w_t^* \in W^*(u + tz)$ such that

$$f(u+tz) - f(u) = F(u+tz,\ w_t^*) - F(u,w^*)$$

$$\leq F(u+tz,\ w^*) - F(u,w^*)$$

for all $w^* \in W^*(u)$. Thus

$$\limsup_{\substack{t \overset{>}{\to} 0 \\ z\to v \text{ in } E}} (f(u+tz)-f(u))/t \leq \lim_{\substack{t \overset{>}{\to} 0 \\ z\to v}} (F(u+tz,w^*)-f(u,w^*))/t$$

for all $w^* \in W^*(u)$. From hypothesis H2, this limit exists and is equal to $F_u'(u,w^*,v)$. Thus,

$$\bar{d}_H f(u;v) \overset{\Delta}{=} \limsup_{\substack{t \overset{>}{\to} 0 \\ z\to v \text{ in } E}} (f(u+tz) - f(u))/t$$

$$\leq \inf_{w^* \in W^*(u)} F_u'(u,w^*;v)$$

Step 2. Let

$$\underline{d}_H f(u;v) \overset{\Delta}{=} \liminf_{\substack{t \overset{>}{\to} 0 \\ z\to v \text{ in } E}} (f(u+tz) - f(u))/t$$

If $\underline{d}_H f(u;v)$ is the smallest accumulation point, then there exists a sequence (t_n, z_n), with $t_n > 0$, $t_n \to 0$, and $z_n \to v$ in E as $n \to +\infty$, such that $\underline{d}_H f(u;v) = \lim_{n\to\infty} (f(u+t_n z_n) - f(u))/t_n$

now,

$$f(u+t_n z_n) - f(j) = F(u+t_n z_n, w_{t_n}^*) - F(u,w^*)$$

$$\geq F(u+t_n z_n, w_{t_n}^*) - F(u, w_{t_n}^*)$$

for all $w^*_{t_n} \in W^*(u + t_n z_n)$.

Now, let

$$g(s) = F(u + st_n z_n, w^*_{t_n})$$

Then,

$$f(u + t_n z_n) - f(u) \geq g(1) - g(0)$$

But, since by hypothesis H3 g is continuous, there exists θ, $0 \leq \theta \leq 1$, such that

$$g(1) = g(0) \geq dg(\theta;1)$$

Now, by hypothesis H2,

$$dg(\theta;1) = F'_u(u + \theta t_n z_n, w^*_{t_n}; t_n z_n)$$

then by hypothesis H4,

$$(f(u + t_n z_n) - f(u))/t_n \geq F'_u(u + \theta_n t_n z_n, w^*_{t_n}; z_n)$$

Since W is a compact topological set, the sequence $w^*_{t_n}$ may be chosen such that $w^*_{t_n} \to \bar{w}$ in W, $n \to +\infty$. Then by hypothesis H5, one gets

$$\underline{d}_H f(u;v) \geq F'_u(u,\bar{w};v)$$

Now,

$$F(u + t_n z_n, w^*_{t_n}) \leq F(u + t_n z_n, w)$$

for any w in W. Taking the limit as $n \to \infty$ (by hypothesis H3) in the right member and the lim inf in the left member, by hypothesis H1 one gets

$$F(u,\bar{w}) \leq F(u,w)$$

for all $w \in W$. This proves that $\bar{w} \in W^*(u)$. Thus,

$$F'_u(u,\bar{w};v) \leq \underline{d}_H f(u;v) \leq \bar{d}_H f(u;v)$$

$$\leq \operatorname{Inf}_{w^* \in W^*(u)} F'_u(u,w^*;v)$$

where the last inequality follows from step 1 of the proof. Since $\bar{w} \in W^*(u)$, these four terms are equal and the limit exists and the proof is complete.

Note that if F possesses a gradient in the variable u, i.e. $F'_u(u,w;v)$ is linear and continuous in v, then df(u;v) is concave and upper semi-continuous. It is linear in the direction v if and only if $W^*(u)$ is reduced to a single element or if $F'_u(u,w;v)$ takes the same value for each element of $W^*(u)$.

Note also that even if $F'_u(u,w;v)$ is linear on v, when $W^*(u)$ is not reduced to a single element, f is not Hadamard differentiable. Neither is it Gateaux differentiable at u, since (as in general $w \to F'_u(u,w;v)$ is not constant on $W^*(u)$)

$$- d_H f(u; -v) = \max_{w \in W^*(u)} F'_u(u,w;v)$$

$$> d_H f(u;v) = \min_{w \in W^*(u)} F'_u(u,w;v)$$

2. HADAMARD'S SEMI-DERIVATIVE OF THE SUPREMUM NORM

Let Ω be a regular bounded domain in $\mathbb{R}^n$ and for u in $C^0(\bar{\Omega})$ (i.e. u is the restriction to $\bar{\Omega}$ of a continuous function in $\mathbb{R}^n$) let

$$f(u) = \max_{x \in \bar{\Omega}} u(x)$$

Since u is continuous and $\bar{\Omega}$ is a compact set, this maximum is achieved on a subset K(u) of $\bar{\Omega}$, K(u) is itself a compact subset of $\bar{\Omega}$ (it is enough in fact that u is lower semi-continuous). Then, Theorem 1.1 may be written

$$d_H f(u;v) = \max_{x \in K(u)} v(x)$$

Since in this case, $F(u,x) = u(x)$ and $F'_u(u,x;v) = v(x)$, so hypotheses H1 to H5 are realized.

Thus, one may write

$$d_H f(u;v) \geq v(x) = \langle \delta_x, v \rangle$$

for any v in $C^\circ(\bar{\Omega})$ and any x in $K(u)$ = Arg Max u. Here, δ_x is the Dirac measure at the point x.

Let $M(\bar{\Omega})$ be the space of continuous measures on $\bar{\Omega}$. This discussion concluded by noting that the Hadamard subgradient

$$\partial_H f(u) \equiv \{\mu \in M(\bar{\Omega}) : d_H f(u;v) \geq \langle \mu, v \rangle$$
$$\text{for all } v \in C^\circ(\bar{\Omega})\}$$

contains the closure of the convex set generated by the Dirac measure δ_x, $x \in K(u)$.

3. HADAMARD SEMI-DERIVATIVE OF THE L^1 NORM

3.1 A First Example

For $u \in L^1(\Omega)$, put

$$f(u) = \int_\Omega |u| du = \max_{w \in W} \int_\Omega wu \, dx$$

where

$$W = \{w \in L^\infty(\Omega) : |w(u)| \leq 1 \quad \text{a.e. } x \text{ in } \Omega\}$$

Then

$$F(u,w) = \int_\Omega uw \, du$$

satisfies hypotheses H1 to H5 of Thm 1.1, with W taken equipped with the weak star topology (see Ref. 6), for which it is compact.

Theorem 1.1 thus gives

$$d_H f(u;v) = \max_{w \in W^*(u)} \int_\Omega w \, v \, dx$$

now, $w \in W^*(u)$ if and only if $w \in W$ and

$$w(x) = \begin{cases} +1, & \text{for a.e. } x \text{ such that } u(x) > 0 \\ -1, & \text{for a.e. } x \text{ such that } u(x) < 0 \end{cases}$$

Then, if one puts

$$\operatorname{sgn}(u)(x) = \begin{cases} +1, & \text{if } u(x) > 0 \\ 0, & \text{if } u(x) = 0 \\ -1, & \text{if } u(x) < 0 \end{cases}$$

and

$$u^{-1}(0) = \{x \in \Omega : u(x) = 0\}$$

one has

$$d_H f(u;v) = \int_\Omega \operatorname{sgn}(u)v \, dx + \int_{u^{-1}(0)} |v| \, dx$$

Note that this derivative is not linear in the direction v, but it satisfies the hypothesis H4 of Thm 1.1, namely for $\rho > 0$,

$$d_H f(u;\rho v) = \rho d_H f(u;v)$$

3.2 A Second Example

The constraint $u \leq \phi$ in Ω (ϕ a given function) may be written

$$(u - \phi)^+ = 0$$

where

$$a^+ = \begin{cases} a, & \text{if } a \geq 0 \\ 0, & \text{if } a < 0 \end{cases}$$

Then, the constraint $u \leq \phi$ may be written

$$f(u) = \int_\Omega (u - \phi)^+ dx = 0$$

But,

$$f(u) = \operatorname*{Max}_{w \in T} \int_\Omega w(u - \phi) \, dx$$

where

$$T = \{w : 0 \leq w(x) \leq 1, \text{ a.e. } x \text{ in } \Omega\}$$

Then, as W, T is weak-star compact and with

$$F(u,w) = \int_{\Omega} w(u - \phi)dx$$

Thm. 1.1 may be applied as

$$F'_u(u,w;v) = \int_{\{x:u(x) > \phi\}} v\ dx + \int_{\{x:u(x) = \phi(x)\}} w\ v\ dx$$

and one gets

$$d_H f(u;v) = \int_{\{x:u(x) > \phi\}} v\ dx + \int_{\{x:u(x) = \phi(x)\}} v^+\ dx$$

4. THE HADAMARD SEMI-DERIVATIVE OF THE FIRST EIGENVALUE (POSSIBLY REPEATED) OF A MATRIX

Let $A = (a_{ij})$ and $B = (b_{ij})$ be two $n \times n$ real symmetric matrices, with $<Bx,x> \geq \alpha \, ||x||^2$ for some $\alpha > 0$ and for all $x \in \mathbb{R}^n$. The first eigenvalue λ of the equation $Ax = \lambda Bx$ is given by

$$\lambda = \inf_{w \in W} <Ax,x>$$

where

$$W = \{u \in \mathbb{R}^n : <Bx,x> = 1\}$$

is a compact set. This infimum is realized on a compact subset W^* of W, where W^* is just the set of the B-normalized eigen-vectors corresponding to λ. The eigenvalue λ is repeated if and only if W^* is not reduced to two opposite elements.

If one denotes by $d^+_{ij}\lambda$ the right side derivative of λ with respect to the a_{ij} coefficient, one gets

$$d^+_{ij}\lambda = \inf_{w \in W^*} x_i x_j$$

and it is clear that the mapping $a_{ij} \to \lambda$ is differentiable if and only if λ is not repeated, in which case

$$\frac{\partial\lambda}{\partial a_{ij}} = x_i x_j$$

for any B-normalized eigenvector x.

If, for example, λ is twice repeated, let x^1 and x^2 be two linearly independent eigenvectors and $d_{ij}^{-}\lambda$ be the left side derivative of with respect to the coefficient a_{ij}, then

$$d_{ij}^{+}\lambda - d_{ij}^{-}\lambda = |x_i^1 x_j^1 - x_i^2 x_j^2|$$

which is strictly positive, then λ is not differentiable.

5. SEMI-DERIVATIVE OF THE FIRST EIGENVALUE (POSSIBLY REPEATED) FOR CONTINUOUS PROBLEMS

By a continuous problem is meant partial differential equations in infinite dimensional vector spaces. In fact, when one looks for solutions in Sobolev spaces, different topologies occur, so that while the eigenfunctions y are normalized, they are not in a bounded set of the variational Sobolev space of the problem. Then, one cannot directly use Thm 1.1 as in the finite dimensional case, since the set W is not compact. An adaptation of Thm 1.1 that alleviates this situation is now given.

5.1 Relaxation of the Compactness Hypothesis on W

The hypothesis that W is a compact set in Thm 1.1 may be avoided in the following way:

Theorem 5.1. Let E be a Banach space, G a Hilbert space, ε an open set in E, W a weakly closed subset of G, and F a mapping, $F : \varepsilon \times W \to \mathbb{R}$ such that:

(H_0) For any $\bar{u}$ in ε, $F(u,w) \to +\infty$ when $||w||_G \to +\infty$, $w \in W$ this convergence being uniform for u in some neighborhood of $\bar{w}$.

(H_w) W being equipped with the weak topology of G, assume the hypotheses H1 to H4 of Thm 1.1.

(H_s) W being equipped with the strong topology of G, assume the hypothesis H5 of Thm. 1.1.

Then,

$$f(u) = \inf_{w \in W} F(u,w) = \operatorname{Min}_{w \in W} F(u,w) \tag{5.1}$$

Let $W^*(u)$ be the subset of W where the infimum in Eq. 5.1 is achieved. Assume

(H_6) Their exists a compact subset K of W (compact for the strong topology of G) such that $W^*(u) \subset K$, for all u,

Then,

$$d_H f(u;v) = \operatorname{Min}_{w \in W^*(u)} F'_u(u,w;v) \tag{5.2}$$

Proof: The hypothesis H_0 may be written in the following way:

For all $\bar{u} \in \varepsilon$, there exists $N_{\bar{u}}$ (a neighborhood of $\bar{u}$ in ε) such that for all $A \in \mathbb{R}$, there exists such that $||w|| > r \Rightarrow F(u,w) > A$, for all $u \in N_{\bar{u}}$.

Let $B = B(r) = \{w \in G : ||w|| \leq r\}$ with $r = F(u,w_0)$, for some $w_0 \in W$. Then, $f(u) = \inf_{w \in W} F(u,w) = \inf_{w \in W \cap B} F(u,w)$. The set $W \cap B$ is weakly compact and $w \to F(u,w)$ is weakly lower semi-continuous. Then, the infimum is achieved.

At this point one has only to repeat the proof of Thm. 1.1, while the subsequence converges strongly in G, from hypothesis H_6. This completes the proof.

5.2 Hadamard Semi-Derivative With Respect to the Design Variable of a Repeated Eigenvalue

Let Ω be a regular domain in $\mathbb{R}^n$, n = 2 or 3, with boundary Γ and h a regular, strictly postive and bounded function given in Ω. Consider the eigenvalue problem

$$\left.\begin{aligned} \Delta(h\Delta y) &= -\lambda \Delta y, \text{ in } \Omega \\ y = \frac{\partial y}{\partial n} &= 0, \quad \text{on } \Gamma \end{aligned}\right\} \tag{5.3}$$

where $\lambda = \lambda(h)$ is the first eigenvalue and y is any eigenfunction.

It is well known that (Rayleigh quotient)

$$\lambda = \inf_{y \in W} \int_\Omega h(\Delta y)^2 \, dx$$

where

$$W = \{y \in H_0^2(\Omega) : \int_\Omega ||\nabla y||^2 \, dx = 1\}$$

and H_0^2 is the second order Sobolev space on Ω, satisfying the boundary conditions.

Taking

$$F(h,y) = \int_\Omega h(\Delta y)^2 \, dx$$

$$E = C^k(\bar{\Omega})$$

$$\varepsilon = \{h \in E : h(x) > 0, \quad \text{for all } x \in \bar{\Omega}\}$$

$$G = H_0^2(\Omega)$$

the hypotheses H_0, H_W, and H_S of Thm. 5.1 are satisfied. One now has only to verify hypothesis H_6.

The set $W^*(h)$ is the set of normalized eigenfunctions y. If y belongs to $W^*(h)$, then y is a solution of Eq. 5.3, i.e. y is a solution of a partial differential equation with regular second member $-\lambda\Delta y$. Then, if h is regular enough, say $k \geq 3$, one knows (see Ref. 7) that y is in a bounded set of $H^3(\Omega)$. Then, from the Rellich Theorem (see Ref. 8) one knows that y is in a compact subset K of $H_0^2(\Omega)$. Then, Thm. 5.1 gives, for any direction v in $C^k(\Omega)$,

$$d_H\lambda(h;v) = \min_{y \in W^*(h)} \int_\Omega v(\Delta y)^2 \, dx \tag{5.4}$$

It is clear that λ is differentiable, i.e.

$$d_H\lambda(h;v) = - d_H\lambda(h; -v)$$

if and only if λ is not repeated. Thus, the eigenvalue possesses a gradient

$$d_H \lambda (h;v) = \int_\Omega v(\Delta y)^2 \, dx$$

that is linear and continuous in the direction v, the gradient being $(\Delta y)^2$ (for either of the two normalized eigenfunctions y).

If, on the other hand, λ is twice repeated, let y_1 and y_2 be two (normalized) eigenfunctions. Then any normalized eigenfunction y (in W*) may be written $y = y_\theta = \cos\theta \, y_1 + \sin\theta \, y_2$. Thus,

$$d_H\lambda(h;v) = \underset{0 \leq \theta \leq 2\pi}{\text{Min}} \left[\cos^2\theta \int_\Omega v(\Delta y_1)^2 \, dx \right.$$

$$\left. + 2 \cos\theta \sin\theta \int_\Omega v \, \Delta y_1 \, \Delta y_2 \, dx + \sin^2\theta \int_\Omega v \, (\Delta y_2)^2 \, dx \right]$$

This minimum is reached for same θ_v and one has

$$d_H\lambda(h;v) = \int_\Omega v \, (\Delta y_{\theta_v})^2 \, dx$$

But, the function $(\Delta y_{\theta_v})^2$ is not a gradient, as in the non-repeated case, since it depends on the direction v.

5.3 Eulerian Semi-Derivative With Respect to Domain of the First Eigenvalue

Consider the problem $y(\Omega_t) \in H_0^2(\Omega_t)$ and $\lambda(\Omega_t) > 0$, such that

$$\left.\begin{aligned} \Delta^2 y(\Omega_t) &= \lambda(\Omega_t) \, y(\Omega_t), && \text{in } \Omega_t \\ y(\Omega_t) &= \frac{\partial}{\partial n_t} y(\Omega_t) = 0, && \text{on } \Gamma_t \end{aligned}\right\} \tag{5.4}$$

where Ω_t is a regular domain in R^n, built from Ω by a field V (see Ref. 2), Γ_t is its boundary and n_t is the normal exterior to Ω_t. The eigenvalue $\lambda(\Omega_t)$ is given by the Rayleigh quotient

$$\lambda(\Omega_t) = \inf\left\{ \int_{\Omega_t} (\Delta y)^2 dx \; : \; \int_{\Omega_t} y^2 dx = 1 \right\}$$

Let $J_t = \det(DT_t)$ (see Ref. 2 for notation), then one has

$$\lambda(\Omega_t) = \inf\left\{\int_{\Omega_t} (\Delta[J_t^{-1/2}\, z_o\, T_t^{-1}])^2 dx \,:\, z\in H_0^2(\Omega),\right.$$

$$\left.\int_\Omega z^2\, dx = 1\right\}$$

(simply because $y = J_t^{-1/2}\, z_o T_t^{-1}$ verifies $\int_{\Omega_t} dx = 1$, and any such y may be written in this form)

Then, let

$$F(t,z) = \int_{\Omega_t} (\Delta[J_t^{-1/2} z_o T_t^{-1}])^2 dx$$

and

$$W = \left\{z \in H_0^2(\Omega) \,:\, \int_\Omega z^2 dx = 1\right\}$$

with

$\varepsilon = E = \mathbb{R}$ and $G = H_0^2(\Omega)$. One can now apply Thm. 5.1 and a simple calculation gives

$$\frac{\partial}{\partial t} F(\cdot,z) = \int_\Omega 2\ \Delta z\ \Delta(-\nabla z \cdot V - \frac{1}{2}\, z \text{ div } V)\ dx$$

$$+ \int_\Gamma (\Delta z)^2\ (V\cdot n)\ ds$$

Then W^*, being the set of the $L^2(\Omega)$-normalized eigenfunctions, one gets (by Thm. 5.1):

$$d\lambda(\Omega;V) = \inf_{z\in W^*}\left\{-2\int_\Omega \Delta z \Delta\ [\nabla z\cdot V(0)]dx\right.$$

$$\left.+ \int_\Gamma (\Delta z)^2\ (V(0)\cdot n)\ ds - \int_\Omega \Delta z\ \Delta[z \text{ div } V\ (0)]dx\right\} \quad (5.5)$$

Using the Green formula for the bi-Laplace operator,

$$\int_\Omega \Delta f\ \Delta g dx = \int_\Gamma \left[\frac{\partial g}{\partial n}\ \Delta f - \frac{\partial}{\partial n}\ (\Delta f) g\right]\ ds + \int_\Omega g\ \Delta^2 f\ dx$$

one obtains, for the two integrals on Ω:

$$2\int_\Omega \Delta z\ \Delta[\nabla z\cdot V(0)]dx = 2\int_\Omega \nabla z\cdot V(0)(-\lambda z)dx + \int_\Gamma \cdots$$

$$= -\lambda\int_\Omega \nabla(z^2)\cdot V\ dx + \int_\Gamma \cdots \qquad (5.6)$$

and

$$\int_\Omega \Delta z\ \Delta[z \operatorname{div} V(0)]dx = -\lambda\int_\Omega z^2\ \operatorname{div} V(0)\ dx + \int_\Gamma \cdots \qquad (5.7)$$

and the two integrals on Ω in Eq. 5.5 give (with Eqs. 5.6 and 5.7 and Stoke's Formula)

$$\lambda\int_\Omega \operatorname{div}(z^2 V(0))dx = \lambda\int_\Gamma z^2(V(0)\cdot n)\ ds = 0$$

for $z = 0$ on Γ. Thus, only the integral terms on Γ remain and

$$d\lambda(\Omega;V) = \inf -2\int_\Gamma \frac{\partial}{\partial n}\ (\nabla z\cdot V)\Delta z - \frac{\partial}{\partial n}\ (\Delta z)\nabla z\cdot V\ ds$$

$$-\int_\Gamma \frac{\partial}{\partial n}\ (z \operatorname{div} V)\Delta z - \frac{\partial}{\partial n}\ (\Delta z) z \operatorname{div} V\ ds$$

$$+\int_\Gamma (\Delta z)^2\ (V(0)\cdot n)\ ds$$

Now, $z=0$ on Γ, so $\nabla z = \frac{\partial z}{\partial n}\ n$ on Γ. But $\frac{\partial z}{\partial n} = 0$ on Γ, so $\nabla z = 0$ on Γ and

$$\frac{\partial}{\partial n}\ (z \operatorname{div} V) = \frac{\partial z}{\partial n} \operatorname{div} V + z\ \frac{\partial}{\partial n} \operatorname{div} V = 0 \quad \text{on} \quad \Gamma$$

Thus,

$$d\lambda(\Omega;V) = \inf -2\int_\Gamma \frac{\partial}{\partial n}\ (\nabla z\cdot V)\Delta z\ ds + \int_\Gamma (\Delta z)^2(V(0)\cdot n)\ ds$$

Now, $\nabla z = 0$ on Γ, i.e. $\partial_i z = 0$ on Γ for any i, so

$$\nabla(\partial_i z) = \frac{\partial}{\partial n}(\partial_i z)\, n \text{ on } \Gamma, \text{ i.e.}$$

$$\partial_j(\partial_i z) = \frac{\partial}{\partial n}(\partial_i z) n_j$$

Now,

$$\partial_j(\nabla z \cdot V) = \partial_j(\Sigma_i \partial_i z V_i) = \Sigma_i \partial^2_{ij} z\, V_i + \Sigma_i \partial_i z \partial_j V_i$$

and

$$\frac{\partial}{\partial n}(\nabla z \cdot V) = \langle D^2 z V, n\rangle$$

where $D^2 z_{ij} = \partial^2_{ij}\, z$ on Γ and

$$\frac{\partial}{\partial n}(\nabla z \cdot V) = \Sigma_{ij} \frac{\partial}{\partial n}(\partial_i z)\, n_j V_j n_i$$

$$= (n \cdot V(0))\, \Sigma_i \frac{\partial}{\partial n}(\partial_i z) n_i$$

Note that

$$\Sigma_i \frac{\partial}{\partial n}(\partial_i z) n_i = \Sigma_{ik}\, \partial_k(\partial_i z) n_k n_i = \langle d^2 z n, n\rangle = \frac{\partial^2 z}{\partial n^2}$$

But on any level curve of z, one has (see Ref. 9)

$$\Delta z = \frac{\partial^2 z}{\partial n^2} + H \frac{\partial z}{\partial n} \tag{5.8}$$

where H is the mean curvature of Γ. Since here $\frac{\partial z}{\partial n} = 0$, one has simply

$$\frac{\partial^2 z}{\partial n^2} = \Delta z$$

and finally

$$\frac{\partial}{\partial n}(\nabla z \cdot V) = \Delta z$$

so

$$d\lambda(\Omega;V) = - \text{Max} \int_\Gamma (\Delta z)^2 \, (V\cdot n) \, ds \tag{5.9}$$

the Maximum being taken on the eigenfunctions z, which are L^2 normalized. It is thus clear that the mapping $V \to d\lambda(\Omega;V)$ is concave and that $d\lambda(\Omega;V)$ just depends on $v = (V\cdot n)$ on Γ. Also, λ is differentiable if and only if λ is not repeated.

In general, λ is of multiplicity greater than 1, W^* is not reduced to two opposite elements, and this infimum is not linear in V(0). Thus, λ does not possess a gradient at Ω.

Nevertheless, it is shown in Ref. 5 that one can use the expression for $d\lambda(\Omega;V)$ to get:

1. necessary optimality conditions

2. a steepest descent method for λ using a continuous selection of the upper gradient $I\lambda(\Omega)$, which is the closed convex set of all the distributions G having the same structure as the gradient, but with only:

$$G \in I\lambda(\Omega) \iff d\lambda(\Omega;V) \le \langle G, V(0)\rangle, \text{ for all } V$$

Here it has been verified that $I\lambda(\Omega)$ is the closed convex cone generated by the distributions obtained as

$$G = {}^t\gamma_\Gamma(g_n \, n)$$

where

$$g_n = - (\Delta z)^2$$

for any $z \in W^*$, i.e. z is a $L^2(\Omega)$-normalized eigenfunction of Δ^2, relative to $\lambda(\Omega)$.

REFERENCES

1. Lemaire, B., Jeux Differentiel, Thesis, University of Paris, 1971.
2. Zolesio, J.P., "The Material Derivative (or Speed) Method for Shape Optimization," Optimization of Distributed Parameter Structures (Ed. E.J. Haug and J. Cea), Sijthoff & Nordhoff, Alphen aan den Rijn), 1980.
3. Zolesio, J.P., Identification de Domaine Par Deformation, Thesis, University of Nice, 1979.

4. Penot, J.P., "Calcul Sous Differential," J. Funct. Anal., Vol. 27, No. 2, 1977, p. 248-276.
5. Aubin, J.P., Micro Cours Sur la Derived de F. Clarke, Université de Montreal, 1976.
6. Céa, J., Optimization, Thesis et Algorithme, Dunod, Paris, 1971.
7. Lions, J.L., and Magenes, E., Problems aux Limites Non Homogenes, Dunod, Paris, 1968.
8. Adams, ., Sobolev Spaces, Academic Press, New York, 1975.
9. Zolesio, J.P., "Domain Variational Formulation for Free Boundary Problem," Optimization of Distributed Parameter Structures (Ed. E.J. Haug and J. Cea), Sijthoff & Nordhoff, Alphen aan den Rijn, 1980.

SHAPE DESIGN SENSITIVITY METHODS FOR STRUCTURAL MECHANICS

Bernard Rousselet

Département de Mathématiques, Université de Nice,
06034 Nice Cédex, France

ABSTRACT

A derivation of shape design sensitivity is provided, including several approaches to derive shape design sensitivity formulas. Several alternate formulas are obtained, which look quite different but are in fact equivalent if the data are smooth enough.

1. INTRODUCTION

The objective of this paper is to develop usable shape design sensitivity methods for structural optimization. No attempt is made to use the minimal smoothness assumptions for the equations or the geometric domain Ω (see Refs. 1 and 2 for weak smoothness assumption). Rigorous proof of design sensitivity is provided, for smooth Ω, in Ref. 3, so here it is shown that deriving shape design sensitivity is not as difficult as one may have imagined from glancing at Ref. 3.

The first six sections of the paper are devoted to different approaches applied to a very simple second-order problem, while Section 7 is devoted to the clamped plate.

2. SHAPE DESIGN SENSITIVITY USING EXPANSIONS

A clamped membrane loaded with a unit distributed load is considered. It is governed by the boundary value problem

$$\left.\begin{aligned} -T\,\Delta z &= 1 \quad \text{in } \Omega \\ z &= 0 \quad \text{in } \Gamma \end{aligned}\right\} \tag{2.1}$$

or in variational form

$$T \int_\Omega (\nabla z \cdot \nabla v)\,dx = \int_\Omega v\,dx \tag{2.2}$$

which must hold for every smooth v such that $v = 0$ on Γ.

Following Ref. 3, the perturbation of the geometrical domain is defined with a one-to-one mapping of the form $I + F$ where I is the identity mapping and F is any function from R^2 to R^2, which is small enough so that $I + F$ is one-to-one. Moreover, it is supposed to be Fréchet differentiable.

Now, if one denotes $\Omega_F = (I+F)(\Omega)$, one would like to characterize the dependence of z with respect to F. Set

$$\tilde{z}_{\Omega_F}(X) = z_{\Omega_F}(X+F(X)) = z_{\Omega_F}(x) \tag{2.3}$$

where the subscript Ω_F emphasizes the dependence of z with respect to Ω_F, which follows from Eq. 2.1. Note that z_{Ω_F} is defined in Ω_F, whereas $\tilde{z}_{\Omega_F}$ is defined in Ω_0. Indeed, the derivative of $\tilde{z}$ with respect to F is going to be computed. Thus, it will be equal to the total derivative of z_F with respect to F. Dependence through Eq. 2.1 and through the movement of the points where z_F must therefore be computed. If one is only interested in directional derivatives, this is the material derivative that is well known in continuum mechanics, especially in fluid mechanics (see for example Ref. 4).

From the chain rule,

$$\tilde{z}'_{\Omega_F}(X) = z'_{\Omega_F}(X+F(X))(I+F'(X))$$

where for a scalar function y,

$$y' \equiv (\nabla y)^T$$

and for a vector function F,

$$F' = \left[\frac{\partial F_i}{\partial X_j}\right]$$

Thus

$$\nabla_x z_{\Omega_F} = (I+F'(x))^{-1^T} \nabla_X \tilde{z}_{\Omega_F} \tag{2.4}$$

Now starting from Eq. 2.2, written in Ω_F, and using the change of variable $x = X + F(X)$ an equation in Ω is obtained. It is

$$T \int_\Omega \left((I+F'(X))^{-1^T} \nabla \tilde{z}_{\Omega_F} \, (I+F'(X))^{-1^T} \nabla \tilde{v}\right) |\det(I+F'(X))| dX$$

$$= \int_\Omega \tilde{v} \, |\det(I+F')| dX \tag{2.5}$$

which should hold for any smooth $\tilde{v}$ that is zero on Γ and where $\tilde{z}_{\Omega_F} = 0$ on Γ. First-order expansions with respect to F are now used. The well known formula $(1+\varepsilon)^{-1} = 1 - \varepsilon + o(\varepsilon)$ is valid in the algebra of matrices (see Ref. 3), so that $(I+F(x))^{-1} = 1 - F(x) + o(F)$. Moreover, in R^2

$$\det(I+F') = \begin{vmatrix} 1 + F_{1,1} & F_{1,2} \\ F_{2,1} & 1 + F_{2,2} \end{vmatrix} = 1 + F_{1,1} + F_{2,2}$$

$$+ F_{1,1}F_{2,2} - F_{1,2}F_{2,1} \tag{2.6}$$

so that

$$\det(I+F') = 1 + \operatorname{div} F + o(F) \tag{2.7}$$

For F small enough,

$$|\det(I+F')| = 1 + \operatorname{div} F + o(F)$$

and one has also that

$$\tilde{z}_{\Omega_F} = \tilde{z}_0 + \tilde{z}^{(1)}_{\Omega,F} + o(F) \tag{2.8}$$

In fact, $\tilde{z}_0$ is the solution of Eq. 2.2 in $\Omega_0 = \Omega$, so that $\tilde{z}_0 = z$. There is no need in expanding $\tilde{v}$ with respect to F. It depends on F only through the change of variable, so it is any smooth function in Ω that is zero on Γ (as v was any smooth function in Ω_F that was zero on Γ_F). Thus, the functions v are chosen in the following way:

$$v_F(x) = v_F(X+F(X)) = \tilde{v}(X)$$

For a rigorous proof, several technical points should be proved (see Ref. 3). The expansions of Eqs. 2.6 and 2.7 should hold uniformly for $X \in \bar{\Omega}$, not only pointwise. The expansion of Eq. 2.8 has to be proved. This is not at all a trivial task and may be wrong for some rather complex equations. If all this is done (see Ref. 3 for details), Eq. 2.5 yields

$$T\int_\Omega (\nabla z, \nabla \tilde{v})dX - T\int_\Omega (F'^T \nabla z, \nabla \tilde{v})dX$$

$$- T\int_\Omega (\nabla z, F'^T \nabla \tilde{v})dX + T\int_\Omega (\nabla \tilde{z}^{(1)}_{\Omega,F}, \nabla \tilde{v})dX$$

$$+ T\int_\Omega (\nabla z, \nabla \tilde{v}) \operatorname{div} F \, dX$$

$$= \int_\Omega \tilde{v} \, dX + \int_\Omega \tilde{v} \operatorname{div} F \, dX + O(F) + O(F')$$

Since z is the solution of Eq. 2.2, the first terms of left and right-hand sides, cancel eath other. Hence

$$T\int_\Omega (\nabla \tilde{z}^{(1)}_{\Omega,F}, \nabla \tilde{v})dX$$

$$= \int_\Omega \tilde{v} \operatorname{div} F \, dX - T\int_\Omega (\nabla z, \nabla \tilde{v}) \operatorname{div} F \, dX$$

$$+ T\int_\Omega ((F'+F'^T) \nabla z, \nabla \tilde{v}) \, dX \tag{2.9}$$

From the discussion following Eq. 2.3, $\tilde{z}^{(1)}_{\Omega,F}$ is the total derivative of z_F, with respect to F. It is denoted here as

$$\dot{z}_{\Omega,F} = z^{(1)}_{\Omega,F} \tag{2.10}$$

when it is referred to this interpretation. Its use in optimization is now discussed. Assume the cost function is

$$\psi(\Omega) = \int_\Omega z_\Omega^2 \, dx \tag{2.11}$$

where z_Ω is the solution of Eq. 2.2.

The expansion of $\psi(\Omega)$ with respect to changes in Ω is done as for z_Ω. Consider Eq. 2.11 in Ω_F and make the change of variable $x = X + F(X)$. Now,

$$\psi(\Omega_F) = \int_\Omega z^2_{\Omega_F} \, |\det(I+F(X))| \, dX$$

hence

$$\psi(\Omega_F) = \int_\Omega z_\Omega^2 \, dX + 2\int_\Omega z_\Omega \, \dot{z}_{\Omega,F} \, dX + \int_\Omega z_\Omega^2 \operatorname{div} F \, dX + o(F) + o(F')$$

which yields

$$\psi'(\Omega,F) = 2\int_\Omega z_\Omega \, \dot{z}_{\Omega,F} \, dX + \int_\Omega z_\Omega^2 \operatorname{div} F \, dX \tag{2.12}$$

To find the explicit dependence on F, Eq. 2.9 is used. One may integrate the first term of Eq. 2.9 by parts to obtain

$$- T\int_\Omega \dot{z}_{\Omega,F} \, \Delta\tilde{v} \, dX + T\int_\Omega \dot{z}_{\Omega,F} \frac{\partial v}{\partial n} \, d\sigma$$

$$= \int_\Omega \tilde{v} \operatorname{div} F \, dX - T\int_\Omega (\nabla z, \nabla \tilde{v}) \operatorname{div} F \, dX$$

$$+ T\int_\Omega ((F'+F'^T) \nabla z, \nabla \tilde{v}) dX \tag{2.13}$$

Now, since $\tilde{z}_{\Omega_F} = 0$ on Γ_F and $\tilde{z}_{\Omega_0} = 0$ on Γ, $\dot{z}_{\Omega,F} = 0$ on Γ. This is the case because the total derivative are used. See Sections 3 and 4 also. Hence, the second term of the left-hand side of Eq. 13 vanishes.

Now, Eq. 2.13 holds for any $\tilde{v}$, such that $\tilde{v} = 0$ on Γ. If one defines an adjoint variable p_Ω as the solution of

$$\left.\begin{aligned} - T\,\Delta p_\Omega &= 2\,z_\Omega, \quad && \text{in } \Omega \\ p_\Omega &= 0, && \text{on } \Gamma \end{aligned}\right\} \tag{2.14}$$

and sets $\tilde{v} = p_\Omega$ in Eq. 2.13, he obtains

$$2\int_\Omega z_\Omega\,\dot{z}_{\Omega,F}\,dX = \int_\Omega p_\Omega\,\mathrm{div}\,F\,dX - T\int_\Omega (\nabla z_\Omega, \nabla p_\Omega)\,\mathrm{div}\,F\,dX$$
$$+\,T\int_\Omega ((F'+F'^T)\,\nabla z_\Omega, \nabla p_\Omega)\,dX \tag{2.15}$$

Hence, an explicit formula for Eq. 2.12 is

$$\psi'(\Omega,F) = \int_\Omega p_\Omega\,\mathrm{div}\,F\,dX - T\int_\Omega (\nabla z_\Omega, \nabla p_\Omega)\,\mathrm{div}\,F\,dX$$
$$+\,T\int_\Omega ((F'+F'^T)\,\nabla z_\Omega, \nabla p_\Omega)\,dX$$
$$+\int_\Omega z_\Omega^2\,\mathrm{div}\,F\,dX \tag{2.16}$$

3. SHAPE DESIGN SENSITIVITY USING THE MATERIAL DERIVATIVE

Equation 2.16 may be derived through a different approach, using formulas for material derivatives that are in common use in fluid mechanics. The function F may be viewed as a vector field in Ω, defining the one parameter family of domains $\Omega_\varepsilon = (I+\varepsilon F)(\Omega_0)$, when F is fixed. The solution of Eq. 2.2 in Ω_ε depends on ε in the following two ways:

(1) It is evaluated as a solution in a domain Ω_ε which depends on ε;
(2) It is evaluated at a point that may move with ε: $x_\varepsilon = X + \varepsilon F(X)$.

In a compact subset of Ω_0, for ε small enough,

$$\frac{d}{d\varepsilon} z_{\Omega_\varepsilon}(x_\varepsilon) = \frac{\partial z_{\Omega_\varepsilon}}{\partial \varepsilon}(x_\varepsilon) + \nabla z_{\Omega_\varepsilon}(x_\varepsilon)\, F(x_\varepsilon)$$

where $\frac{\partial}{\partial \varepsilon}$ denotes the derivative of $\varepsilon \to z_{\Omega_\varepsilon}$, evaluated at a fixed point that does not depend on ε. The derivative of z_Ω in the direction F is

$$\dot{z}_{\Omega,F} = \frac{d}{d\varepsilon} z_{\Omega_\varepsilon}(x_\varepsilon)\Big|_{\varepsilon=0}$$

and one may define

$$z'_{\Omega,F} = \frac{\partial z_{\Omega_\varepsilon}}{\partial \varepsilon}\Big|_{\varepsilon=0}$$

so that

$$\dot{z}_{\Omega,F} = z'_{\Omega,F} + \nabla z_\Omega\, F \tag{3.1}$$

The derivatives $\frac{\partial}{\partial \varepsilon}$ and $\frac{\partial}{\partial x_i}$ commute (with smoothness assumptions) since they are derivatives with respect to independent variables, i.e.

$$\left(\frac{\partial y}{\partial x_i}\right)'_{\Omega,F} = \frac{\partial}{\partial x_i} y'_{\Omega,F} \tag{3.2}$$

On the other hand, using Eq. 3.1, one has

$$\left(\frac{\partial y}{\partial x_i}\right)^{\cdot}_{\Omega,F} = \left(\frac{\partial y}{\partial x_i}\right)'_{\Omega,F} + \frac{\partial^2 y}{\partial x_j \partial x_i} F_j$$

(Equation continued on next page)

$$= \frac{\partial}{\partial x_i} y'_{\Omega,F} + \frac{\partial^2 y}{\partial x_j \partial x_i} F_j$$

$$= \frac{\partial}{\partial x_i}\left(\dot{y}_{\Omega,F} - \frac{\partial y_\Omega}{\partial x_j} F_j\right) - \frac{\partial^2 y}{\partial x_j \partial x_i} F_j$$

Hence,

$$\left(\frac{\partial y}{\partial x_i}\right)^{\bullet}_{\Omega,F} = \frac{\partial}{\partial x_i} \dot{y}_{\Omega,F} - \frac{\partial y_\Omega}{\partial x_j} \frac{\partial F_j}{\partial x_i} \tag{3.3}$$

or

$$(\nabla y)^{\bullet}_{\Omega,F} = \nabla \dot{y}_{\Omega,F} - F'^T \nabla y \tag{3.4}$$

Now using (see Refs. 5-7), one has

$$\left(\int_\Omega f_\Omega(x)dx\right)^{\bullet}_{\Omega,F} = \int_\Omega f^{\bullet}_{\Omega,F}(x)dx + \int_\Omega f \operatorname{div} F \, dx$$

and the derivative of Eq. 2.2 is obtained as

$$T \int_\Omega (\nabla \dot{z}_{\Omega,F} - F'^T \nabla z, \nabla v)dx - T \int_\Omega (\nabla z, F'^T \nabla v)dx$$

$$+ T \int_\Omega (\nabla z, \nabla v) \operatorname{div} F \, dx$$

$$= \int_\Omega v \operatorname{div} F \, dx \tag{3.5}$$

In this derivation, $\dot{v}_{\Omega,F} = 0$ has been used. One can choose v as constant on the line $x_\varepsilon = X + \varepsilon F(X)$, i.e. $v_\varepsilon(x_\varepsilon) = \tilde{v}(x)$ where $\tilde{v}$ is a fixed function (it does not depend on ε). If $\tilde{v}$ is any smooth function in Ω such that $\tilde{v} = 0$ on Γ, one gets any smooth function in Ω_ε such that $v_\varepsilon = 0$ on Γ_ε.

Note that Eq. 3.5 is the same as Eq. 2.9, obtained by a different method.

There is no need to use the variational equation to derive the equation satisfied by $\dot{z}_{\Omega,F}$, one can also use Eq. 2.1 and Eq. 3.3 to compute

$$\left(\frac{\partial^2 z}{\partial x_i^2}\right)^{\bullet}_{\Omega,F} = \frac{\partial}{\partial x_i}\left(\frac{\partial z}{\partial x_i}\right)^{\bullet}_{\Omega,F} - \frac{\partial}{\partial x_j}\left(\frac{\partial z}{\partial x_i}\right)\frac{\partial F_j}{\partial x_i}$$

Thus,

$$\left(\frac{\partial^2 z}{\partial x_i^2}\right)^{\bullet}_{\Omega,F} = \frac{\partial^2}{\partial x_i^2}\dot{z}_{\Omega,F} - 2\,\frac{\partial^2 z}{\partial x_j \partial x_i}\frac{\partial F_j}{\partial x_i} - \frac{\partial z}{\partial x_j}\frac{\partial^2 F_j}{\partial x_i^2} \tag{3.6}$$

Finally, since $(\Delta z)^{\bullet}_{\Omega,F} = \dot{1}_{\Omega,F} = 0$, one has

$$\left.\begin{aligned} &\Delta\dot{z}_{\Omega,F} = 2\sum_{i,j}\frac{\partial^2 z}{\partial x_j \partial x_i}\frac{\partial F_j}{\partial x_i} + \sum_{i,j}\frac{\partial z}{\partial x_j}\frac{\partial^2 F_j}{\partial x_i^2}\ , \text{ in } \Omega \\ &\dot{z}_{\Omega,F} = 0\ , \text{ on } \Gamma \end{aligned}\right\} \tag{3.7}$$

Note that Eq. 3.7 may be written in compact form as

$$\Delta\dot{z}_{\Omega,F} = 2(F'+F'^T, z'') + (\nabla z, \Delta F)$$

where $z'' = (z'^T)'$ and $(,)$ means the contraction of tensors with respect to all their indices.

To compare this with previous results, recall Eq. 2.9, which holds for any smooth v which is zero on Γ, is

$$\begin{aligned} T\int_\Omega (\nabla \dot{z}_{\Omega,F}, \nabla v)dx &= \int_\Omega v \text{ div } F\, dx - T\int_\Omega (\nabla z, \nabla v) \text{ div } F\, dx \\ &\quad + T\int_\Omega ((F'+F'^T)\nabla z, \nabla v)dx \end{aligned} \tag{3.8}$$

Integrate the left-hand side of Eq. 3.8 by parts, to obtain

$$- T \int_\Omega \Delta \dot{z}_{\Omega,F} \; v \; dx = \int_\Omega v \; \mathrm{div}\; F \; dx - T \int_\Omega (\nabla z, \nabla v) \; \mathrm{div}\; F \; dx$$

$$+ T \int_\Omega ((F'+F'^T) \nabla z, \nabla v) dx \qquad (3.9)$$

To obtain an operator equation, eliminate the derivatives acting on v on the right-hand side to get

$$- T \sum_{i,k} \int_\Omega z_i v_i \; F_{k,k} \; dx = T \sum_{i,k} \int_\Omega (z_{ii} F_{k,k} + z_i F_{k,ki}) v \; dx$$

$$- T \sum_{i,k} \int_\Gamma z_i n_i v \; F_{k,k} \; d\sigma$$

or

$$- T \sum_{i,k} \int_\Omega (\nabla z, \nabla v) \; \mathrm{div}\; F \; dx$$

$$= T \int_\Omega [\Delta z \; \mathrm{div}\; F + (\nabla z, \nabla \; \mathrm{div}\; F)] v \; dx$$

$$(3.10)$$

In the same way,

$$T \sum_{i,j} \int_\Omega (F_{i,j} + F_{j,i}) z_j \; v_i \; dx$$

$$= - T \sum_{i,j} \int_\Omega [(F_{i,ji} + F_{j,ii}) z_j + (F_{i,j} + F_{j,i}) z_{ij}] v \; dx$$

$$+ T \sum_{i,j} \int_\Gamma (F_{i,j} + F_{j,i}) z_j \; n_i \; v \; d\sigma$$

or,

$$T \int_\Omega ((F'+F'^T) \nabla z, \nabla v) dx =$$

$$- T \int_\Omega [(\nabla z, \nabla \; \mathrm{div}\; F) + (\nabla z, \Delta F)$$

(Equation continued on next page)

$$+ (F'+F'^T, z'')]v\, dx \qquad (3.11)$$

Finally, Eqs. 3.9, 3.10, and 3.11 yield

$$- T \int_\Omega \Delta\dot{z}_{\Omega,F}\, v\, dx = \int_\Omega v \,\mathrm{div}\, F\, dx$$

$$+ T \int_\Omega (\Delta z) v \,\mathrm{div}\, F\, dx + T \int_\Omega (\nabla z, \nabla \,\mathrm{div}\, F)\, v\, dx$$

$$- T \int_\Omega (\nabla z, \nabla \,\mathrm{div}\, F) v\, dx - T \int_\Omega (\nabla z, \Delta F)\, v\, dx$$

$$- T \int_\Omega (F'+F'^T, z'')\, v\, dx \qquad (3.12)$$

or,

$$- T \int_\Omega \Delta\dot{z}_{\Omega,F}\, v\, dx = - T \int_\Omega (\nabla z, \Delta F)\, v\, dx$$

$$- T \int_\Omega (F'+F'^T, z'')\, v\, dx$$

which yields

$$\Delta\dot{z}_{\Omega,F} = (F'+F'^T, z'') + (\Delta F, \nabla z)$$

as in Eq. 3.7. Thus, again consistent results are obtained by the two methods.

4. BOUNDARY EXPRESSION OF THE FUNCTIONAL DERIVATIVE

It is proved in this section that for smooth enough data, Eq. 2.16 for the derivative of the functional,

$$\psi'(\Omega,F) = \int_\Omega p_\Omega \,\mathrm{div}\, F\, dX - T \int_\Omega (\nabla z_\Omega, \nabla p_\Omega)\, \mathrm{div}\, F\, dX$$

$$+ T \int_\Omega ((F'+F'^T)\, \nabla z_\Omega, \nabla p_\Omega) dX + \int_\Omega z_\Omega^2 \,\mathrm{div}\, F\, dX \qquad (4.1)$$

which is referred to as a distributed expression of the derivative, may be expressed with integrals over Γ, instead of integrals over Ω, and that it depends only on the normal component of F on Γ. Pioneering works in this area are Refs. 8 and 9.

One first integrates terms in Eq. 4.1 by parts, so that no derivative is acting on F, to obtain

$$\int_\Omega p_\Omega \,\text{div}\, F \,dX = - \int_\Omega \nabla p_\Omega \, F \, dX \tag{4.2}$$

$$- T \int_\Omega (\nabla z_\Omega, \nabla p_\Omega) \,\text{div}\, F \, dX$$

$$= T \int_\Omega [\nabla(\nabla z_\Omega, \nabla p_\Omega)] \, F \, dX$$

$$- T \int_\Gamma (\nabla z_\Omega, \nabla p_\Omega)(F \cdot n) d\sigma$$

$$T \int_\Omega (F' \nabla z_\Omega, \nabla p_\Omega) dX$$

$$= T \sum_{i,j} \int_\Omega F_{i,j} z_j p_i \, dX = - T \sum_{i,j} \int_\Omega F_i (z_{jj} p_i + z_j p_{ij}) dX$$

$$+ T \sum_{i,j} \int_\Gamma F_i z_j p_i n_j d\sigma \tag{4.3}$$

$$T \int_\Omega ((F' + F'^T) \nabla z_\Omega, \nabla p_\Omega) dX = - T \int_\Omega [\Delta z_\Omega (F, \nabla p_\Omega)$$

$$+ (p'' \nabla z_\Omega, F)] dX + T \int_\Gamma (F, \nabla p_\Omega) \frac{\partial z_\Omega}{\partial n} d\sigma$$

$$- T \int_\Omega [(F, \nabla z_\Omega) \Delta p_\Omega + (z''_\Omega \nabla p_\Omega, F)] dX$$

$$+ T \int_\Gamma (F, \nabla z_\Omega) \frac{\partial p_\Omega}{\partial n} d\sigma \tag{4.4}$$

and

$$\int_\Omega z_\Omega^2 \operatorname{div} F \, dX = -2 \int_\Omega z_\Omega \nabla z_\Omega F \, dX \tag{4.5}$$

where $p_\Omega = 0$ on Γ has been used to eliminate boundary terms that arise through integration by parts in Eqs. 4.2 and 4.5.

Using these equations and the facts that $z = 0$ on Γ, $T\Delta z_\Omega = 1$, and $-T\,\Delta p_\Omega = 2z_\Omega$ in Eq. 4.1, one has

$$\begin{aligned}\psi'(\Omega,F) = T \int_\Omega [\nabla(\nabla z_\Omega, \nabla p_\Omega)] F \, dX \\ - T \int_\Gamma \frac{\partial z_\Omega}{\partial n} \frac{\partial p_\Omega}{\partial n} (F\cdot n) d\sigma - T \int_\Omega (p'' \nabla z_\Omega, F) dX \\ + T \int_\Gamma \frac{\partial p_\Omega}{\partial n} \frac{\partial z_\Omega}{\partial n} (F\cdot n) d\sigma - T \int_\Omega (z'' \nabla p_\Omega, F) dX \\ + T \int_\Gamma \frac{\partial z_\Omega}{\partial n} \frac{\partial p_\Omega}{\partial n} (F\cdot n) d\sigma\end{aligned}$$

which reduces simply to

$$\psi'(\Omega,F) = T \int_\Gamma \frac{\partial z_\Omega}{\partial n} \frac{\partial p_\Omega}{\partial n} (F\cdot n) d\sigma \tag{4.6}$$

which depends only on the normal component $(F\cdot n)$ of F on Γ.

5. THE PARTIAL DERIVATIVE

The aim is now to derive the derivative of ψ by using the derivative of $z'_{\Omega,F}$, instead of the material derivative $\dot{z}_{\Omega,F}$ used in Section 3.

As in Section 3, the partial derivative with respect to F commutes with derivatives with respect to x, hence

$$\Delta z'_{\Omega,F} = 0, \text{ in } \Omega \tag{5.1}$$

To find the boundary condition satisfied by $z'_{\Omega,F}$, the simplest way is to start from $\dot{z}_{\Omega,F} = 0$ on Γ and recall that $\dot{z}_{\Omega,F} = z'_{\Omega,F} + \nabla z_\Omega \cdot F$. Hence, $z'_{\Omega,F} = - \nabla z_\Omega \cdot F$ on Γ, which is also equal to

$$z'_{\Omega,F} = - \frac{\partial z_\Omega}{\partial n} (F \cdot n) \tag{5.2}$$

Thus, $z'_{\Omega,F}$ is the solution of

$$\left.\begin{aligned} \Delta z'_{\Omega,F} &= 0, \text{ in } \Omega \\ z'_{\Omega,F} &= - \frac{\partial z_\Omega}{\partial n} (F \cdot n), \text{ on } \Gamma \end{aligned}\right\} \tag{5.3}$$

This is a nonhomogeneous Dirichlet problem. The variational formulation is standard (see for example Ref. 10). One multiplies Eq. 5.3 by a smooth function v, with $v = 0$ on Γ, and integrates over Ω. One then integrates twice by parts to get

$$\int_\Omega z'_{\Omega,F} \, \Delta v \, dx - \int_\Gamma z'_{\Omega,F} \frac{\partial v}{\partial n} d\sigma = 0$$

Using the boundary condition for $z'_{\Omega,F}$, this becomes

$$\int_\Omega z'_{\Omega,F} \, \Delta v \, dx + \int_\Gamma \frac{\partial z_\Omega}{\partial n} \frac{\partial v}{\partial n} (F \cdot n) d\sigma = 0 \tag{5.4}$$

Conversely, in general, the solution of Eq. 5.4 is a solution of Eq. 5.3 only for smooth enough data.

As in Section 4, one now defines the adjoint variable p_Ω as the solution of the boundary-value problem,

$$\left.\begin{aligned} - T \, \Delta p_\Omega &= 2 z_\Omega, \text{ in } \Omega \\ p_\Omega &= 0, \quad \text{on } \Gamma \end{aligned}\right\} \tag{5.5}$$

then with $v = p_\Omega$ in Eq. 5.4, one has

$$\int_\Omega 2 z_\Omega z'_{\Omega,F} = T \int_\Gamma \frac{\partial z_\Omega}{\partial n} \frac{\partial p_\Omega}{\partial n} (F \cdot n) d\sigma \tag{5.6}$$

From Eq. 2.12

$$\psi'(\Omega,F) = 2\int_\Omega z_\Omega \dot{z}_{\Omega,F}\, dX + \int_\Omega z_\Omega^2 \operatorname{div} F\, dX$$

$$= 2\int_\Omega z_\Omega z'_{\Omega,F}\, dX + 2\int_\Omega z_\Omega(\nabla z_\Omega, F) dX + \int_\Omega z_\Omega^2 \operatorname{div} F\, dX$$

$$= 2\int_\Omega z_\Omega z'_{\Omega,F}\, dX + \int_\Omega \operatorname{div}(z_\Omega^2 F) dX$$

Hence, using Eq. 5.6 and Green's formula

$$\psi'(\Omega,F) = T\int_\Gamma \frac{\partial z_\Omega}{\partial n}\frac{\partial p_\Omega}{\partial n}(F\cdot n)d\sigma + \int_\Gamma z_\Omega^2 (F\cdot n)d\sigma$$

or, since $z_\Omega = 0$ on Γ,

$$\psi'(\Omega,F) = T\int_\Gamma \frac{\partial z_\Omega}{\partial n}\frac{\partial p_\Omega}{\partial n}(F\cdot n)d\sigma \tag{5.7}$$

which is the same as Eq. 4.6. Note that the derivative of ψ is the same, whether one uses material or partial derivative for z and for this functional ψ, $\psi'(\Omega,F) = \dot{\psi}_{\Omega,F}$.

As a final check of these computations, compare Eq. 2.9, which gives $\tilde{z}^{(1)}_{\Omega,F} = \dot{z}_{\Omega,F}$, and Eq. 5.4, which gives $z'_{\Omega,F}$. For this, one starts from Eq. 2.9,

$$T\int_\Omega (\nabla \dot{z}_{\Omega,F}, \nabla \tilde{v}) dX = \int_\Omega \tilde{v} \operatorname{div} F\, dX - T\int_\Omega (\nabla z_\Omega, \nabla \tilde{v}) \operatorname{div} F\, dX$$

$$+ T\int_\Omega ((F'+F'^T)\nabla z_\Omega, \nabla \tilde{v}) dX$$

Integrating the left-hand side by parts,

$$- T\int_\Omega \dot{z}_{\Omega,F}\, \Delta\tilde{v}\, dX = \int_\Omega \tilde{v} \operatorname{div} F\, dX - T\int_\Omega (\nabla z_\Omega, \nabla \tilde{v}) \operatorname{div} F\, dX$$

$$+ T\int_\Omega ((F'+F'^T)\nabla z_\Omega, \nabla \tilde{v}) dX$$

One can now use $\dot{z}_{\Omega,F} = z'_{\Omega,F} + \nabla z_\Omega \cdot F$ to get

$$- T \int_\Omega z'_{\Omega,F} \Delta\tilde{v} \, dX = T \int_\Omega (\nabla z_\Omega \cdot F)\Delta\tilde{v} \, dX + \int_\Omega \tilde{v} \operatorname{div} F \, dX$$

$$- T \int_\Omega (\nabla z_\Omega, \nabla \tilde{v}) \operatorname{div} F \, dX + T \int_\Omega ((F'+F'^T)\nabla z_\Omega, \nabla \tilde{v}) dX$$

Integrating the last three terms by parts, as in Section 4, one gets

$$- T \int_\Omega z'_{\Omega,F} \Delta\tilde{v} \, dX = T \int_\Omega (\nabla z_\Omega, F) \Delta\tilde{v} \, dX - \int_\Omega (\nabla \tilde{v}, F) dX$$

$$+ T \int_\Omega [\nabla(\nabla z_\Omega, \nabla \tilde{v})] \cdot F \, dX - T \int_\Gamma \frac{\partial z_\Omega}{\partial n} \frac{\partial \tilde{v}}{\partial n} (F \cdot n) d\sigma$$

$$- T \int_\Omega [\Delta z_\Omega (F, \nabla \tilde{v}) + (\tilde{v}'' \nabla z_\Omega, F)] dX$$

$$+ T \int_\Gamma \frac{\partial \tilde{v}}{\partial n} \frac{\partial z_\Omega}{\partial n} (F \cdot n) d\sigma$$

$$- T \int_\Omega [(F, \nabla z_\Omega)\Delta\tilde{v} + (z''_\Omega \nabla \tilde{v}, F)] dX$$

$$+ T \int_\Gamma (F, \nabla z_\Omega) \frac{\partial \tilde{v}}{\partial n} d\sigma$$

Using Eq. 2.1, one gets, after manipulation,

$$- T \int_\Omega z'_{\Omega,F} \Delta\tilde{v} \, dX = + T \int_\Gamma \frac{\partial z_\Omega}{\partial n} \frac{\partial \tilde{v}}{\partial n} (F \cdot n) d\sigma$$

which is the same as Eq. 5.4.

6. EXPANSIONS FOR THE OPERATOR EQUATION

Although it may seem to be redundant, the derivative of z is going to be computed, using expansions as in Section 2, but here these expansions will be carried out for the operator equation of Eq. 2.1,

$$\left.\begin{aligned} -T\,\Delta z &= 1, \quad \text{in } \Omega \\ z &= 0, \quad \text{on } \Gamma \end{aligned}\right\} \tag{6.1}$$

The aim of the section is twofold; firstly to give one more derivation for the derivative of the solution of Eq. 6.1, with respect to shape variations, and secondly to be induced to do some computations that will be needed for shape sensitivity analysis of the clamped plate. The advantage is that these computations, needed for the plate, will thus be checked with other computations of the membrane. Due to the complexity of the computations involved in shape sensitivity, it is a good point to have many checks.

6.1 Change of Variable for the Laplacian

The first step is to make a change of variable for the Laplacian. It is known that $\Delta v = \operatorname{div} \nabla v$, and from Section 2

$$(\nabla_x v)(\phi(X)) = (\phi')^{-1^T} \nabla_X (v\circ\phi) \tag{6.2}$$

To compute the effect of the change of variable on the divergence operator, the following identity may be used:

$$\int_{\Omega'} (\operatorname{div} F)v \, dx = - \int_{\Omega'} (F, \nabla v) dx$$

where F is any smooth vector field. Making the change of variable $x = \phi(X)$ in both integrals,

$$\int_{\Omega} [(\operatorname{div}_x F)\circ\phi] (v\circ\phi) \, |\det \phi'| \, dX$$

$$= - \int_{\Omega} \left(F\circ\phi, \phi'^{-1^T} \nabla_X (v\circ\phi)\right) |\det \phi'| \, dX$$

$$= - \int_{\Omega} \left(|\det \phi'| \, \phi'^{-1} F\circ\phi, \nabla_X (v\circ\phi)\right) dX$$

Integrating by parts, the last integral yields

$$\int_\Omega |\det \phi'| \, [(\mathrm{div}_X F)\phi] \, (v_\circ\phi) dX$$

$$= \int_\Omega \mathrm{div}_X[|\det \phi'| \, \phi'^{-1} F_\circ\phi] \, v_\circ\phi \, dX$$

Since this equality is satisfied for any $v_\circ\phi$,

$$|\det \phi'| \, [(\mathrm{div}_X F)_\circ\phi] = \mathrm{div}_X \, [|\det \phi'| \, \phi'^{-1} F_\circ\phi]$$

is obtained, or since $|\det \phi'|$ is different from zero,

$$(\mathrm{div}_X F)_\circ\phi = \frac{1}{|\det \phi'|} \mathrm{div}_X[|\det \phi'| \, \phi'^{-1} F_\circ\phi] \tag{6.3}$$

By setting $F = \nabla v$, one can now get the change of variable for the Laplacian as

$$(\Delta_X v)_\circ\phi = \mathrm{div}_X(\nabla_X v)_\circ\phi$$

$$= \frac{1}{|\det \phi'|} \mathrm{div}_X[|\det \phi'| \, \phi'^{-1} \, \phi'^{-1^T} \nabla_X(v_\circ\phi)] \tag{6.4}$$

6.2 Expansions

Set $\phi = I + F$, where F is designed to be small, and recall from Eqs. 2.6 and 2.9 of Section 2 that

$$\det(I+F') = 1 + \mathrm{div}\, F + 0(F) \tag{6.5}$$

$$(I+F'(x))^{-1} = 1 - F'(x) + 0(F) \tag{6.6}$$

and hence

$$(I+F'(x))^{-1^T} = 1 - F'^T(x) + 0(F) \tag{6.7}$$

In Eq. 6.4, set $v = z$ and $\tilde{z}_{\Omega_F} = z_{\Omega_F}\circ\phi$ and expand $\tilde{z}$ as $\tilde{z}_{\Omega,F} = z_\Omega + \dot{z}_{\Omega,F} + 0(F)$ (see Section 2), to obtain:

$$- T(\Delta z_\Omega)_\circ\phi(X) = \left[\frac{-T}{1+\mathrm{div}\, F+0(F)}\right] \mathrm{div}_X \, [(1+\mathrm{div}\, F)(1-F'))(1-F'^T)\nabla z_\Omega$$

(Equation continued on next page)

$$+ \nabla \dot{z}_{\Omega,F}) + O(F)]$$

Hence,

$$- T (\Delta z)\circ\phi(X) = - T(1 - \text{div } F) \text{ div}_X (\nabla z_\Omega + \text{div } F \nabla z_\Omega$$

$$- F' \nabla z_\Omega - F'^T \nabla z_\Omega + \nabla \dot{z}_{\Omega,F} + O(F))$$

$$- T (\Delta z_\Omega)\circ\phi = - T \Delta z_\Omega - T \text{ div}(\text{div } F \nabla z_\Omega)$$

$$+ T \text{ div}((F'+F'^T)\nabla z_\Omega) - T \Delta\dot{z}_{\Omega,F}$$

$$+ T \text{ div } F \Delta z_\Omega + O(F) \tag{6.8}$$

and using Eq. 6.1, one obtains

$$- T \Delta\dot{z}_{\Omega,F} = T \text{ div}(\text{div } F \nabla z_\Omega) - T \text{ div}((F'+F'^T) \nabla z_\Omega) + \text{div } F \tag{6.9}$$

6.3 Computational Checks

The objective is now to compare this with Eq. 3.7,

$$- T \Delta\dot{z}_{\Omega,F} = - T (F'+F'^T, z'') - T (\nabla z, \Delta F) \tag{6.10}$$

this is done by using the following partial derivative calculations:

$$T \text{ div}(\text{div } F \nabla z) = T[(F_{1,1}+F_{2,2})z_1]_1 + T[(F_{1,1}+F_{2,2})z_2]_2$$

$$= T(F_{1,11}+F_{2,21})z_1 + T(F_{1,1}+F_{2,2})z_{11}$$

$$+ T(F_{1,12}+F_{2,22})z_2 + T(F_{1,1}+F_{2,2})z_{22}$$

$$= T((\nabla \text{ div } F), \nabla z) + T \text{ div } F \Delta z \tag{6.11}$$

The equation $\text{div}(fV) = (\nabla f, V) + f \text{ div } V$ could also have been used.

$$- T \operatorname{div}((F'+F'^T) \nabla z) = - T[2F_{1,1}z_1+(F_{1,2}+F_{2,1})z_2]_1$$

$$- T[(F_{2,1}+F_{1,2}) z_1+2F_{2,2}z_2]_2$$

$$= - T[2F_{1,11}z_1+2F_{1,1}z_{11}+(F_{1,21}+F_{2,11})z_2+(F_{1,2}+F_{2,1})z_{21}]$$

$$- T[(F_{2,12}+F_{1,22})z_1+(F_{2,1}+F_{1,2})z_{12}+2F_{2,22}z_2+2F_{2,2}z_{22}]$$

$$= - T[(F_{1,11}+F_{1,22})z_1+(F_{2,11}+F_{2,22})z_2]$$

$$- T[(F_{2,12}+F_{1,11})z_1+(F_{1,21}+F_{2,22})z_2]$$

$$- T[2F_{1,1}z_{11}+(F_{1,2}+F_{2,1})z_{21}+(F_{2,1}+F_{1,2})z_{12}+2F_{2,2}z_{22}]$$

$$= - T(\Delta F, \nabla z) - T(\nabla \operatorname{div} F, \nabla z) - T(F'+F'^T, z'') \qquad (6.12)$$

Finally, using Eqs. 6.11 and 6.12, as well as Eq. 6.1 in 6.9, Eq. 1.10 is obtained.

Another check is possible (and easier), to compare with Eq. 2.9. It is straightforward to use integration by parts in Eq. 2.9 and to use Dirichlet boundary condition.

7. SHAPE DESIGN SENSITIVITY FOR THE PLATE

7.1 Introduction

In this section, a clamped plate of constant thickness h is considered, with a static distributed load f. The boundary-value problem for displacement z_Ω is

$$\left.\begin{aligned} &\frac{Eh^3}{12(1-\nu^2)} \Delta^2 z_\Omega = f , && \text{in } \Omega \\ &z_\Omega = 0 = \frac{\partial z_\Omega}{\partial n} , && \text{on } \Gamma \end{aligned}\right\} \qquad (7.1)$$

where E is Young's modulus and ν is Poisson's ratio. As for the membrane of the previous sections, the derivative of z_Ω with respect to changes of Ω is to be computed.

As for the membrane, no claim is made here for rigorous proofs (they may be found in Ref. 3), but rather to describe how to derive the formulas. The use of partial derivative for the plate is not as straightforward as for the membrane, since the boundary condition involves the normal n to Γ.

The derivation is carried out using the material derivative introduced in Section 3, the expansion method introduced in Sections 2 and 6, and the partial derivative method are presented.

7.2 Use of Material Derivative (Operator Form)

Using Eq. 6.8 (divided by -T), one obtains that for any function ψ,

$$(\Delta\psi)^{\bullet}_{\Omega,F} = \text{div}(\text{div } F \,\nabla \psi) - \text{div}((F'+F'^T)\nabla \psi) + \Delta\dot{\psi}_{\Omega,F} - \text{div } F\, \Delta\psi$$

Setting $\psi = \Delta z$, one then has

$$(\Delta^2 z)^{\bullet}_{\Omega,F} = \text{div}(\text{div } F \,\nabla \Delta z) - \text{div}((F'+F'^T)\nabla \Delta z) + \Delta((\Delta z)^{\bullet}_{\Omega,F}) - \text{div } F\, \Delta^2 z$$

Hence,

$$(\Delta^2 z)^{\bullet}_{\Omega,F} = \text{div}(\text{div } F \,\nabla \Delta z) - \text{div}((F'+F'^T)\nabla \Delta z) + \Delta(\text{div}(\text{div } F\, \nabla z) - \text{div}((F'+F'^T)\nabla z) + (\Delta\dot{z}_{\Omega,F} - \text{div } F\, \Delta z) - \text{div } F\, \Delta^2 z \tag{7.2}$$

which, using Eq. 7.1, yields

$$\bar{D}(h)\Delta^2\dot{z}_{\Omega,F} = -\bar{D}(h)\,\text{div}(\text{div } F\, \nabla \Delta z) + \bar{D}(h)\,\text{div}((F'+F'^T)\nabla \Delta z) - \bar{D}(h)\,\Delta(\text{div}(\text{div } F\, \nabla z) - \text{div}((F'+F'^T)\nabla z) - \text{div } F\, \Delta z) + f'F + f\,\text{div } F \tag{7.3}$$

The boundary conditions for z on Γ are $z = 0$ and $\frac{\partial z}{\partial n} = 0$. But

$z = 0$ on Γ implies that the tangential derivative of z is zero, so that $\frac{\partial z}{\partial n} = 0$ is in fact equivalent to $\nabla z = 0$ on Γ. From Eq. 3.6,

$$0 = (\nabla z)^{\bullet}_{\Omega,F} = \nabla \dot{z}_{\Omega,F} - F'^{T} \nabla z$$

Since $\nabla z_{\Omega} = 0$ on Γ, this gives

$$\nabla \dot{z}_{\Omega,F} = 0$$

Now, $z = 0$ implies $\dot{z}_{\Omega,F} = 0$ (see Section 3), hence the boundary conditions for $\dot{z}_{\Omega,F}$ are simply

$$\left.\begin{aligned} &\dot{z}_{\Omega,F} = 0 \\ &\frac{\partial \dot{z}_{\Omega,F}}{\partial n} = 0 \end{aligned}\right\} \qquad (7.4)$$

If one wants to minimize

$$\psi(\Omega) = \int_{\Omega} z^2 \, dx$$

he first calculates

$$\psi^{\bullet}_{\Omega,F} = 2 \int_{\Omega} z \, \dot{z}_{\Omega,F} \, dx + \int_{\Omega} z^2 \, \mathrm{div}\, F \, dx$$

Multiplying both sides of Eq. 7.3 by a function v, integrating over Ω, and using integration by parts, one has

$$\overline{D}(h) \int_{\Omega} \dot{z}_{\Omega,F} \, \Delta^2 v \, dx = - \overline{D}(h) \int_{\Omega} [\mathrm{div}(\mathrm{div}\, F \, \nabla \Delta z)$$

$$+ \mathrm{div}((F'+F'^{T}) \, \nabla \Delta z)]v \, dx - \overline{D}(h) \int_{\Omega} \Delta(\mathrm{div}(\mathrm{div}\, F \, \nabla z$$

$$- \mathrm{div}((F'+F'^{T}) \, \nabla z) - (\mathrm{div}\, F) \, \Delta z)v \, dx$$

$$+ \int_{\Omega} (f'F + f \, \mathrm{div}\, F)v \, dx$$

where v satisfies the boundary conditions of Eq. 7.1.

If one defines an adjoint variable p to satisfy

$$\left.\begin{aligned} &D(h)\ \Delta^2 p = 2z, \text{ in } \Omega \\ &p = 0 = \frac{\partial p}{\partial n}, \quad \text{on } \Gamma \end{aligned}\right\}$$

he obtains

$$\dot{\psi}_{\Omega,F} = -\ D(h) \int_{\Omega} [\mathrm{div}(\mathrm{div}\ F\ \nabla\ \Delta z) + \mathrm{div}((F'+F'^T)\ \nabla\ \Delta z)]p\ dx$$

$$-\ D(h) \int_{\Omega} \Delta(\mathrm{div}(\mathrm{div}\ F\ \nabla\ z) - \mathrm{div}((F'+F'^T)\ \nabla\ z)$$

$$-\ (\mathrm{div}\ F)\ \Delta z)p\ dx + \int_{\Omega} (f'F+f\ \mathrm{div}\ F) + \int_{\Omega} z^2\ \mathrm{div}\ F\ dx$$

This last formula may be used in actual computations (perhaps with some manipulations).

7.3 Use of Expansions on the Variational Equation

The variational equation is given in Ref. 3 as

$$D(h) \int_{\Omega} [\Delta z\ \Delta v + \nu(z_{22}v_{11} + z_{11}v_{22} - 2z_{12}v_{12})]dx = \int_{\Omega} f\ v\ dx \tag{7.5}$$

where $D(h) = \dfrac{Eh^3}{12(1-\nu^2)}$

As long as one is only interested in clamped plates with $z = 0 = \frac{\partial z}{\partial n}$ and $v = 0 = \frac{\partial v}{\partial n}$ on Γ, this bilinear form may be simplified. One has (Ref. 11, I.1.2)

$$\int_{\Omega} (z_{22}v_{11} + z_{11}v_{22} - 2z_{12}v_{12})dx = \int_{\Gamma} (z_{\tau\tau}v_{\nu} - u_{\nu\tau}v_{\tau})d\sigma \tag{7.6}$$

where subscripts $\tau(\nu)$ denote tangential (normal) derivative. This boundary integral is zero, hence, one may use a simplified bilinear form and write the variational equation for the problem as

$$D(h) \int_\Omega \Delta z_\Omega \, \Delta v \, dx = \int_\Omega fv \, dx \tag{7.7}$$

One may use Eq. 6.8 to make an expansion of Δz_Ω in Eq. 7.7 (divided by -T), where Δv is not expanded (see Sections 2 and 3). This yields

$$D(h) \int_\Omega [\Delta z + \mathrm{div}(\mathrm{div}\, F \, \nabla z) - \mathrm{div}((F' + F'^T) \nabla z)$$

$$+ \Delta \dot{z}_{\Omega,F} - (\mathrm{div}\, F) \, \Delta z][\Delta v + \mathrm{div}(\mathrm{div}\, F \, \nabla v)$$

$$- \mathrm{div}((F' + F'^T) \, \nabla v) - (\mathrm{div}\, F) \, \Delta v](1 + \mathrm{div}\, F) dx$$

$$= \int_\Omega (f + f' \cdot F) \, v(1 + \mathrm{div}\, F) dx + O(F) \tag{7.8}$$

Expanding Eq. 7.8 and neglecting terms of order higher than one in F, one gets

$$D(h) \int_\Omega \{\Delta z \Delta v + \Delta z[\mathrm{div}(\mathrm{div}\, F \, \nabla v) - \mathrm{div}((F' + F'^T) \, \nabla v)$$

$$- (\mathrm{div}\, F) \, \Delta v] + [\mathrm{div}(\mathrm{div}\, F \, \nabla z) - \mathrm{div}((F' + F'^T) \, \nabla z)$$

$$+ \Delta \dot{z}_{\Omega,F} - (\mathrm{div}\, F) \Delta z] + \Delta z \, \Delta v \, \mathrm{div}\, F\} dx$$

$$= \int_\Omega (fv + (f'F)v + fv \, \mathrm{div}\, F) dx \tag{7.9}$$

Using Eq. 7.6 for simplification, this is

$$D(h) \int_\Omega \Delta \dot{z}_{\Omega,F} \, \Delta v \, dx = - D(h) \int_\Omega \Delta z[\mathrm{div}(\mathrm{div}\, F \, \nabla v)$$

$$- \mathrm{div}((F' + F'^T) \nabla v) - \mathrm{div}\, F \, \Delta v] dx - D(h) \int_\Omega [\mathrm{div}(\mathrm{div}\, F \, \nabla z)$$

$$- \mathrm{div}((F' + F'^T) \nabla z) - \mathrm{div}\, F \, \Delta z] \Delta v \, dx - D(h) \int_\Omega \Delta z \Delta v \, \mathrm{div}\, F \, dx$$

$$+ \int_\Omega [(f'F)v + fv \, \mathrm{div}\, F] dx \tag{7.10}$$

By using Eqs. 6.11 and 6.12, one can also write

$$- D(h) \int_\Omega \Delta\dot{z}_{\Omega,F}\, \Delta v \, dx = - D(h) \int_\Omega \Delta z[(\Delta F, \nabla v) - (F'+F'^T, v'')]dx$$

$$- D(h) \int_\Omega [(\Delta F, \nabla z) - (F'+F'^T, z'')]\Delta v \, dx$$

$$- D(h) \int_\Omega \Delta z \, \Delta v \, \mathrm{div}\, F \, dx + \int_\Omega [(f'F)v + fv \, \mathrm{div}\, F]dx \qquad (7.11)$$

Using partial derivatives, this last equation may be written in a similar way to that used in Ref. 3, by setting

$$Z_{11} = - (2 F_{1,1}z_{11}+2F_{2,1}z_{12}+F_{1,11}z_1+F_{2,11}z_2)$$

$$Z_{22} = - (2 F_{1,2}z_{12}+2F_{2,2}z_{22}+F_{1,22}z_1+F_{2,22}z_2)$$

$$V_{11} = - (2 F_{1,1}v_{11}+2F_{2,1}v_{12}+F_{1,11}v_1+F_{2,11}v_2)$$

$$V_{22} = - (2 F_{1,2}v_{12}+2F_{2,2}v_{22}+F_{1,22}v_1+F_{2,22}v_2)$$

Equation 7.11 is thus

$$- D(h) \int_\Omega \Delta\dot{z}_{\Omega,F}\, \Delta v \, dx = - D(h) \int_\Omega (z_{11}+z_{22})(V_{11}+V_{22})dx$$

$$- D(h) \int_\Omega (Z_{11}+Z_{22})(v_{11}+v_{22})dx$$

$$- D(h) \int_\Omega (z_{11}+z_{22})(v_{11}+v_{22})(F_{1,1}+F_{2,2})dx$$

$$+ \int_\Omega [(f_1F_1+f_2F_2)v + fv(F_{1,1}+F_{2,2})]dx \qquad (7.12)$$

As in Section 7.2, one can use Eqs. 7.10, 7.11, or 7.12 and an adjoint state to compute the derivative of any functional.

7.4 Use of Partial Derivative (Operator Form)

Consider the plate equation in operator form

$$\left.\begin{aligned} &\frac{Eh^3}{12(1-\nu^2)}\,\Delta^2 z_\Omega = f, \text{ in } \Omega \\ &\left.\begin{aligned} z &= 0 \\ \frac{\partial z}{\partial n} &= 0 \end{aligned}\right\}, \text{ on } \Gamma \end{aligned}\right\} \tag{7.13}$$

As in Section 5, one deduces that in Ω

$$\frac{Eh^3}{12(1-\nu^2)}\,\Delta^2 z'_{\Omega,F} = f'F$$

As in Section 7.1, from the boundary conditions one deduces that $\nabla z = 0$ on Γ. On the other hand, $z = 0$ on Γ implies $\dot{z}_{\Omega,F} = 0$ on Γ. Hence $z'_{\Omega,F} = \dot{z}_{\Omega,F} - \nabla z \cdot F$ which yields $z'_{\Omega,F} = 0$ on Γ. Moreover, from Section 7.2 $\nabla\, \dot{z}_{\Omega,F} = 0$ on Γ, hence on Γ

$$\frac{\partial z'_{\Omega,F}}{\partial n}\, n = \nabla z'_{\Omega,F} = -\nabla\ (\nabla z \cdot F)$$

which may be expanded, to obtain

$$\nabla\ (\nabla z \cdot F) = -\ (D^2 z,\ F) + (\nabla z,\ \nabla F) \tag{7.14}$$

Since $\nabla z = 0$,

$$\frac{\partial z'_{\Omega,F}}{\partial n}\, n = -\ (D^2 z,\ F) = -\ \frac{\partial^2 z}{\partial n^2}\,(F\cdot n)n$$

or,

$$\frac{\partial z'_{\Omega,F}}{\partial n} = -\ \frac{\partial^2 z}{\partial n^2}\,(F\cdot n)$$

Finally, $z'_{\Omega,F}$ is the solution of

$$\left.\begin{aligned} &\frac{Eh^3}{12(1-\nu^2)} \; \Delta^2 z'_{\Omega,F} = f'F, \text{ in } \Omega \\ &z'_{\Omega,F} = 0, \text{ on } \Gamma \\ &\frac{\partial z'_{\Omega,F}}{\partial n} = - \frac{\partial^2 z}{\partial n^2} (F \cdot n), \text{ on } \Gamma \end{aligned}\right\} \qquad (7.15)$$

8. CONCLUSION

In this paper, several methods have been presented for computing the derivative of a functional involving the solution of a second-order Dirichlet problem, with respect to domain. The aim was to provide several ways to derive the same equations, as a check for computations. This is necessary because of the complexity of the computations involved. However, one should notice that some ways may get the result much quicker than others.

Shape design sensitivity for a clamped plate has been derived, where the computation is much more difficult. If one uses some of the results derived for second-order problems, the computations are feasible. The hope is to have convinced the reader that shape design sensitivity is feasible, even for complex structural examples.

REFERENCES

1. Chenais, D., "Identification d'ouverts lipschitziens et de variótes à bord lipschitziennes dans des equations aux dérivées partielles," Thèse d'Etat, Nice, France, 1977.
2. Rousselet, B., "Etude de la Régularité de Valeurs Propres par Rapport a des Déformation Bilipschitziennes du Domaine Geometrique," Comptes-Rendus de l'Académie des Sciences, 283 Série A, 1976, p. 507.
3. Rousselet, B., and Haug, E.J., "Design Sensitivity Analysis of Shape Variations," Optimization of Distributed Parameter Structures (Eds. E.J. Haug and J. Cea), Sijthoff & Noordhoff, Alphen aan den Rihn, Netherlands, 1980.
4. Germain, P., Cours de Mécanique des Milieus Continus, Masson, Paris, 1973.
5. Rousselet, B., "Implementation of Some Methods of Shape Optimal Design," Optimization of Distributed Parameter Structures (Eds. E.J. Haug and J. Cea), Sijthoff & Noordhoff, Alphen aan den Rijn, Netherlands, 1980.

6. Cea, J., "Problems of Shape Optimal Design," Optimization of Distributed Parameter Structures (Eds. E.J. Haug and J. Cea, Sijthoff & Noordhoff, Alphen aan den Rihn, Netherlands, 1980.
7. Zolesio, J-P., "The Material Derivative (or Speed) Method for Shape Optimal Design," Optimization of Distributed Parameter Structures (Eds. E.J. Haug and J. Cea), Sijthoff & Noordhoff, Alphen aan den Rihn, Netherlands, 1980.
8. Dervieux, A., and Palmeria, B., "Une Formule de Hadamard dans Des Problémes d'Identification de Domaines (I) et (II)," C.R. Acad. Sci., 280 Série A, et. 1761, 1975, p. 1697.
9. Dervieux, A., and Palmeria, B., "Une Formula de Hadamard dans Des Problèmes d'Identification de Domaines I et II," Computes Rendus de l'Académis des Sciences, 280 Série A, 1975, p. 1697 and 280 Série A, 1975, p. 1781.
10. Lions, J.L., Contrôle Optimal des Systèmes Gouvernés par des Equations aux Dérivées Partielles, Dunod-Gauthier-Villars, Paris, 1968.
11. Ciarlet, P., The Finite Element Method for Elliptical Problems, North-Holland, Amsterdam, 1978.

COMPUTING EIGENVECTOR DERIVATIVES WITH GENERALIZED INVERSES

Joseph E. Whitesell, Jr.

Mechanical Engineering Department
Michigan State University
E. Lansing, MI 48824

ABSTRACT

In this paper, a method for computing the derivatives of an eigenvector of a symmetric matrix was described. These methods are most efficient when the technique used for finding the eigenvectors involves an LU factorization of L-λI (or K-λM). The improved efficiency of the method depends, in part, on reusing the LU decomposition to compute the eigenvector derivatives. Another source of efficiency in the method is the simplification in the eigenvector formulation which results when a 'natural' generalized inverse is used.

1. INTRODUCTION

While methods for computing the derivative of an eigenvalue with respect to some system parameter have been known since the nineteenth century work of Jacobi [1], eigenvector derivatives have a much shorter history. The first work was done by Fox and Kapoor [2]. They presented two methods to find eigenvector derivatives for the symmetric eigenvalue problem $(K-\lambda M)u=0$, but unfortunately both are of limited value for determining numerical values. The first method, while it requires only knowledge of the specified eigenvalue and eigenvector, requires the multiplication of an (n-1)xn matrix by its transpose and the subsequent solution of a fully populated symmetric system. The matrix multiplication is a lengthy computation and usually leads to a loss of any sparseness possessed by the factor matrices. The second method avoids these problems, but it expresses the derivative of an eigenvector in terms of a complete set of eigenvalues and eigenvectors. This formulation has only theoretical value when the eigensystem is large, since in that case, it is difficult to extract a full set of eigenvectors. In Refs. 3 to 5, the work of Fox and Kapoor was extended to non-symmetric systems, but no improvement in computational efficiency was made.

These computational difficulties were first discussed by Nelson [6]. As a remedy, he proposed a technique by which the rank (n-1) matrix $L-\lambda I$ is modified by zeroing certain of its entries. The modified matrix, which describes a system of equations that must be solved to determine the derivative is well-conditioned and retains the sparseness of the original system. This is a significant improvement, since problems of mathematical physics and engineering often involve sparse matrices.

In this paper an application of generalized inverse theory is described that leads to further improvements in computing the derivative of an eigenvector. To simplify the presentation, only eigensystems involving real symmetric matrices are considered. More general cases are considered in Ref. 7. The method exploits most of the effort spent in extracting the eigenvector itself to obtain its derivative. Each eigenvector derivative is then determined as the solution of a sparse triangular system and involves no more than $O(n^2)$ multiplications.

2. STANDARD EIGENVALUE PROBLEMS

Consider an algebraic eigenvalue problem in standard form, $Lu=\lambda u$, where L is an nxn real symmetric matrix. Any non-zero solution u is an eigenvector of L and the corresponding scalar λ is an eigenvalue of L. Since L is symmetric it can be shown that the eigenvalues λ_i of L lie on the real axis and that a set U of n orthonormal eigenvectors u_i exist, such that $u_i^T u_j=\delta_{ij}$ [8].

Suppose that the matrix L is a matrix-valued function of $\vec{x}\in D^k \subseteq R^k$ and consider the variable eigenvalue problem

$$L(\vec{x})u(\vec{x}) = \lambda(\vec{x})u(\vec{x}), \qquad L: \; D^k \rightarrow R^{nxn} \tag{2.1}$$

It is simplest to consider only variations in a single components $\vec{x}$, say $x_j \in D \subseteq R$. Dropping the subscript j, Eq. 2.1 becomes

$$L(x)u(x) = \lambda(x)u(x), \qquad L: \; D \rightarrow R^{nxn}, \; x \in D \subseteq R. \tag{2.2}$$

Here $\lambda_i:D\to R$ is a scalar-valued function, $u:D\to R^n$ is a vector-valued function and $L:D\to R^{nxn}$ is an analytic matrix-valued function. Furthermore, $L(x)$ is a linear operator on the linear space $V(x)$ of vector-valued functions $v:D\to R^n$ into itself, $L:V(x)\to V(x)$.

Kato [9] has shown that if the mapping $L(x):D\to R^{nxn}$ is analytic and if $L^T(x)=L(x)$ for $x\in D\subseteq R$, then the mapping $\lambda:D\to R$ is analytic and there exists a set $\overline{U(x)}\subset V(x)$ of n orthonormal and analytic eigenvectors such that $u_i^T(x)u_j(x)=\delta_{ij}$. Thus,

$$L(x)\frac{\partial u(x)}{\partial x} + \frac{\partial L(x)}{\partial x}u(x) = \lambda(x)\frac{\partial u(x)}{\partial x} + \frac{\partial \lambda(x)}{\partial x}u(x), \qquad u\in U(x) \tag{2.3}$$

and

$$[L(x)-\lambda(x)I]\frac{\partial u(x)}{\partial x} = -[\frac{\partial L(x)}{\partial x} - \frac{\partial \lambda(x)}{\partial x}I]u(x) \tag{2.4}$$

follow from the product rule of differentiation.

The existence of solutions of Eq. 2.4 may be verified by applying the theorem of alternatives [10]. Equation 2.4 has solutions if and only if for any $v(x)$ such that

$$v^T(x)[L(x)-\lambda(x)I] = 0 \tag{2.5}$$

then

$$v^T(x)[\frac{\partial L(x)}{\partial x} - \frac{\partial \lambda(x)}{\partial x}I]u(x) = 0 \tag{2.6}$$

Since $L(x)$ is self adjoint, Eq. 2.5 may be written as

$$[L(x)-\lambda(x)I]v(x) = 0 \tag{2.7}$$

which is satisfied only by the eigenvectors $\overline{u}(x)\in V(x)$ of $L(x)$, corresponding to $\lambda(x)$ and the zero vector. Thus,

$$\overline{u}^T(x)[\frac{\partial L(x)}{\partial x} - \frac{\partial \lambda(x)}{\partial x}I]u(x) = 0 \tag{2.8}$$

is necessary and sufficient for the existence of solutions of Eq. 2.4. Solving Eq. 2.8, one obtains

$$\left.\begin{aligned}&\frac{\partial\lambda}{\partial x} = \bar{u}^T(x)\,\frac{\partial L(x)}{\partial x}\,u(x)\\ &\text{where}\\ &\bar{u}^T(x)u(x) = 1\end{aligned}\right\} \qquad (2.9)$$

If $\lambda(x)$ is not repeated at x_0, then the normalized eigenvector $\bar{u}(x_0)$ is unique and one has the formula [1 to 6,9]

$$\frac{\partial\lambda}{\partial x} = u^T(x_0)\,\frac{\partial L(x_0)}{\partial x}\,u(x_0) \qquad (2.10)$$

If $\lambda(x_0)$ is repeated, the assumption that eq. 2.4 is normally solvable leads to complications in solving Eq. 2.9 for the eigenvalue derivative, since now $\bar{u}(x_0)$ may be chosen from a multidimensional subspace. Since this case is discussed elsewhere [9], for simplicity it is assumed here that $\lambda(x)$ is not repeated for $x \in D$.

3. SOLUTION BY GENERALIZED INVERSE

One may attempt a direct solution of Eq. 2.9, assuming that $\partial\lambda/\partial x$ is defined by Eq. 2.10. To do this, one may introduce a generalized inverse matrix [11] for the matrix $L_\lambda(x) = L(x)-\lambda(x)I$.

In general, if the matrix equation

$$Ax = b \qquad (3.1)$$

has solutions x, they may be expressed in terms of a generalized inverse matrix A^I as

$$x = A^I b + z_0 \qquad (3.2)$$

where A^I satisfies the relation

$$AA^I A = A \qquad (3.3)$$

and z_0 is any solution of the homogeneous problem [11]

$$Az_0 = 0. \qquad (3.4)$$

It is easy to show that if A^{-1} exists, such that $AA^{-1}=A^{-1}=I$ then Eq. 3.3 implies that $A^I=A^{-1}$ and Eq. 3.4 implies $z_0=0$.

From Eqs. 3.2, 3.3, and 3.4, the solutions of Eq. 2.4 may be written as

$$\frac{\partial u}{\partial x} = -L_\lambda^I[\frac{\partial L}{\partial x} - \frac{\partial \lambda}{\partial x}I]u + v \qquad (3.5)$$

where L_λ^I is any matrix satisfying

$$L_\lambda L_\lambda^I L_\lambda = L_\lambda \qquad (3.6)$$

and

$$L_\lambda v = 0 \qquad (3.7)$$

Since it has been assumed that the eigenvalue is simple, it follows from Eq. 2.2 that v is a multiple of the unit eigenvector u. The matrix L_λ^I is not necessarily unique as may be seen by the following lemma.

<u>Lemma 3.1</u>. Let A_R and A_L satisfy $AA_R=0$ and $A_LA=0$. Then if A^I is any generalized inverse of A,

$$A^{\#}=(I-A_R)A^I(I-A_L) \qquad (3.8)$$

is also a generalized inverse of A.

<u>Proof</u>. Multiplying Eq. 3.8 on both sides by A results in

$$AA^{\#}A = A(I-A_R)A^I(I-A_L)A = AA^IA = A$$

which completes the proof.

Since L_λ^I is not unique in Eq. 3.5, one is able to choose a particular generalized inverse, say $L_\lambda^\dagger$, that will make the numerical evaluation of Eq. 3.5 convenient. If one requires that the following conditions hold for $L_\lambda^\dagger$:

$$L_\lambda^\dagger u = 0 \tag{3.9}$$

$$u^T L_\lambda^\dagger = 0 \tag{3.10}$$

then substituting $L_\lambda^\dagger$ into Eq. 3.5 and using $u^T \frac{\partial u}{\partial x} = 0$ [6], one arrives at a simpler form for the eigenvector derivative,

$$\frac{\partial u}{\partial x} = -L_\lambda^\dagger \frac{\partial L}{\partial x} u \tag{3.11}$$

where $u^T \frac{\partial u}{\partial x} = 0$ is used to eliminate v from Eq. 3.5 [7]. Equations 3.9 and 3.10 hold if one selects

$$L_\lambda^\dagger = (I-uu^T)L_\lambda^I(I-uu^T) \tag{3.12}$$

where L_λ^I is any generalized inverse of L_λ. That $L_\lambda^\dagger$ is a generalized inverse of L_λ follows from Lemma 3.1, since the idempotent matrix uu^T satisfies $L_\lambda uu^T = uu^T L_\lambda = 0$.

4. NUMERICAL APPROACH

In numerical work, the algorithm (inverse iteration),

$$(L-\tilde{\lambda}I)u_{i+1} = v_i \tag{4.1}$$

$$v_i = u_i/||u_i||_\infty \tag{4.2}$$

is often used to determine an eigenvector u corresponding to the eigenvalue λ, where $\tilde{\lambda} = \lambda + \varepsilon$. Solving Eq. 4.1, one has

$$u_{i+1} = (L-\tilde{\lambda}I)^{-1}v_i \tag{4.3}$$

In practice, the matrix $(L-\tilde{\lambda}I)$ is given an LU factorization and the iteration 4.1 is carried out as a series of back substitution stages. Since each back substitution stage only involves a triangular linear system, the LU decomposition stage is the major effort in the computation.

These practical considerations provide a strong incentive to study the structure of $(L-\lambda I)^{-1}$, with the hope that it can be utilized in constructing $L_\lambda^\dagger$. In Ref. 7, it is shown that there is a relationship between $L_\lambda^\dagger$ and $(L-\tilde{\lambda}I)^{-1}$ that can be exploited in numerical work. Specifically,

$$L_\lambda^\dagger = (I-uu^T)(L-\tilde{\lambda}I)^{-1}(I-uu^T) \tag{4.4}$$

This form of $L_\lambda^\dagger$ is especially convenient, since $(L-\tilde{\lambda}I)^{-1}$ may be computed previously to determine the eigenvector u and since the idempotent matrix $I-uu^T$ is easily multiplied by a vector, i.e.,

$$(I-uu^T)b = b-(u^Tb)u \tag{4.5}$$

or

$$b^T(I-uu^T) = b^T-(b^Tu)u^T \tag{4.6}$$

In practice, instead of directly inverting $(L-\tilde{\lambda}I)^{-1}$, an LU decomposition of $L-\tilde{\lambda}I$ is formed. The following has proved to be an efficient procedure:

(1) Form an LU decomposition of $L-\tilde{\lambda}I$.
(2) Use the LU decomposition in an inverse iteration scheme to find an eigenvector u.
(3) Solve for z in

$$L_L z = (I-uu^T)\frac{\partial L}{\partial x}u$$

(4) Solve for y in

$$L_u y = z$$

(5) $\frac{\partial u}{\partial x} = (I-uu^T)y = y-(u^Ty)u$

where L_u and L_L are upper and lower triangular matrices computed in the LU factorization of $L-\tilde{\lambda}I$.

5. DIFFERENTIATION WITH RESPECT TO SEVERAL PARAMETERS

The preceding analysis concerned differentiation of eigen-systems with respect to a single variable. The situation is more complicated when several variables are involved. Consider, for example,

$$L(x_1,x_2) = \begin{bmatrix} x_1 & x_2 \\ x_2 & -x_1 \end{bmatrix} \tag{5.1}$$

where $(x_1,x_2) \in R^2$. The eigenvalues $\lambda_{12} = \pm(x_1^2 + x_2^2)^{1/2}$ of L are repeated only for $x_1=x_2=0$ [9]. Since the eigenvalues are simple in any domain of (x_1,x_2) not including the origin, one can construct an expansion of $L(x_1,x_2)$ in terms of its constituent idempotent matrices $L_i(x_1,x_2)$ [14],

$$L(x_1,x_2) = \lambda_1(x_1,x_2)\, L_1(x_1,x_2) + \lambda_2(x_1,x_2)\, L_2(x_1,x_2) \tag{5.2}$$

where

$$\left.\begin{aligned} L_1 &= \frac{L-\lambda_2 I}{(\lambda_1-\lambda_2)} \\ L_2 &= \frac{L-\lambda_1 I}{(\lambda_2-\lambda_1)} \end{aligned}\right\} \tag{5.3}$$

or,

$$L_1(x_1,x_2) = \frac{1}{2(x_1^2+x_2^2)^{1/2}} \begin{bmatrix} x_1+(x_1^2+x_2^2)^{1/2} & x_2 \\ x_2 & -x_1+(x_1^2+x_2^2)^{1/2} \end{bmatrix} \tag{5.4}$$

$$L_2(x_1,x_2) = -\frac{1}{2(x_1^2+x_2^2)^{1/2}}$$

$$\begin{bmatrix} x_1-(x_1^2+x_2^2)^{1/2} & x_2 \\ x_2 & -x_1-(x_1^2+x_2^2)^{1/2} \end{bmatrix} \qquad (5.5)$$

It can be verified that any non-zero column of $L_i(x_1,x_2)$ is an eigenvector of $L(x_1,x_2)$ corresponding to $\lambda_i(x_1,x_2)$.

However, the idempotents $L_i(x_1,x_2)$ possess no limit as $(x_1,x_2)\to(0,0)$ since

$$\lim_{x_2\to 0} L_1(0,x_2) = \frac{1}{2}\begin{bmatrix} 1 & 1 \\ 1 & 1 \end{bmatrix} \qquad (5.6)$$

whereas

$$\lim_{x_1\to 0} L_1(x_1,0) = \begin{bmatrix} 1 & 0 \\ 0 & 0 \end{bmatrix} \qquad (5.7)$$

Similarly,

$$\lim_{x_2\to 0} L_2(0,x_2) = \frac{1}{2}\begin{bmatrix} 1 & -1 \\ -1 & 1 \end{bmatrix} \qquad (5.8)$$

whereas

$$\lim_{x_1\to 0} L_2(x_1,0) = \begin{bmatrix} 0 & 0 \\ 0 & 1 \end{bmatrix} \qquad (5.9)$$

This illustrates that although the idempotents, and therefore the eigenvectors, can be continued smoothly through the origin along the x_1 or x_2 axis, their total limits do not exist at the origin. Consequently, the constituent idempotent matrices and eigenvectors are not even continuous in this example. For further discussion of this question see Refs. 7, 9, and 15.

6. GENERALIZED EIGENVALUE PROBLEMS

Other variants of the eigenvalue problem are possible. For example, in structural design work it is common to encounter eigenvalue problems of the form

$$Ku = \lambda Mu \tag{6.1}$$

where K is an nxn positive (semi)definite matrix ($K \geq 0$), and M is an nxn positive definite matrix ($M>0$), or of the form

$$(K + C\lambda + \lambda^2 M)u = 0 \; , \qquad K, C \geq 0, \; M>0 \tag{6.2}$$

The matrices K, M, and C commonly account for the stiffness, mass, and viscous damping of an underlying structural system.

Consider first the variable eigenvalue problem corresponding to Eq. 6.1,

$$K(x)u(x) = \lambda(x)M(x)u(x) \tag{6.3}$$

where $K(x) \geq 0$, $M(x)>0$, and the mappings $K:D \to R^{nxn}$ and $M:D \to R^{nxn}$ are analytic in $x \in D \subseteq R$. This problem can be related to the symmetric eigenvalue problem in standard form, through the use of the following transformation [8], where dependence on x is understood:

$$u = M^{-1/2}v$$

$$M^{-1/2}>0 \tag{6.4}$$

so

$$M^{-1/2}KM^{-1/2}v = \lambda v \tag{6.5}$$

Since $M^{-1/2}$ is positive definite and since K and M are analytic functions of $X \in D$, then $M^{-1/2}$ and therefore $M^{-1/2}KM^{-1/2}$ are analytic on D. Thus Eq. 6.3 is equivalent to the variable eigenvalue problem in standard form, treated in Section 5 and the mappings $\lambda:D \to R$ and $v:D \to R^n$ are analytic for the eigenvalues $\lambda_i(x)$ and for particular corresponding eigenvectors $v_i(x)$, $i = 1,\cdots,n$, $x \in D$.

It is not convenient to determine the derivatives of $\lambda_i(x)$ and $v_i(x)$ from Eq. 6.5, since the transformation $M^{-1/2}(x)$ may be difficult to compute and the transformed matrix of Eq. 6.5 is usually difficult to differentiate. Instead, Eq. 6.3 is used. After differentiating and rearranging Eq. 6.3, one arrives at

$$[K - \lambda M] \frac{\partial u}{\partial x} = -[\frac{\partial K}{\partial x} - \lambda\frac{\partial M}{\partial x} - \frac{\partial \lambda}{\partial x} M]u \tag{6.6}$$

Solving for $\frac{\partial u(x)}{\partial x}$ with a generalized inverse L_λ^I, one obtains

$$\frac{\partial u}{\partial x} = -L_\lambda^I[\frac{\partial K}{\partial x} - \lambda\frac{\partial M}{\partial x} - \frac{\partial \lambda}{\partial x} M]u + cu \tag{6.7}$$

where

$$L_\lambda^I = [K(x) - \lambda(x)M(x)]^I \tag{6.8}$$

and $cu(x)$ is an element in the null space of $[K(x)-\lambda(x)M(x)]$. One may assume that analytic eigenvectors exist such that

$$u_i^T(x)M(x)u_j(x) = \delta_{ij} \tag{6.9}$$

This may be seen by substituting Eq. 6.4 into

$$v_i^T(x)v_j(x) = \delta_{ij} \tag{6.10}$$

$$u_i^T M^{1/2} M^{1/2} u_j = u_i^T M u_j \tag{6.11}$$

Differentiating Eq. 6.9 gives [6]

$$\frac{\partial u_i^T}{\partial x} Mu_j + u_i^T \frac{\partial M}{\partial x} u_j + u_i^T M \frac{\partial u_j}{\partial x} = 0 \tag{6.12}$$

Suppose now that u satisfies Eq. 6.9. Then from Lemma 3.1,

$$L_\lambda^\dagger = (I-uu^TM)L_\lambda^I(I-Muu^T) \tag{6.13}$$

is also a generalized inverse of $(K-\lambda M)$, since

$$(K-\lambda M)uu^TM = Muu^T(K-\lambda M) = 0 \tag{6.14}$$

Substituting $L_\lambda^\dagger$ in Eq. 6.7 gives

$$\frac{\partial u}{\partial x} = -L_\lambda^\dagger \left[\frac{\partial K}{\partial x} - \lambda\frac{\partial M}{\partial x} - \frac{\partial \lambda}{\partial x} M\right]u + cu \tag{6.15}$$

$$= -L_\lambda^\dagger \left[\frac{\partial K}{\partial x} - \lambda\frac{\partial M}{\partial x}\right]u + cu. \tag{6.16}$$

Multiplying Eq. 6.16 by $u^T M$ and applying Eq. 6.12 shows that

$$c = -\frac{1}{2} u^T \frac{\partial M}{\partial x} u \tag{6.17}$$

Again, one must determine $L_\lambda^\dagger$ to complete the formulation. The approach is similar to the one used in the standard eigenvalue problem. It is shown in Ref. 7 that

$$L_\lambda^\dagger = (I-uu^T M)(K-\tilde{\lambda}M)^{-1}(I-Muu^T) \tag{6.18}$$

may be used to solve Eq. 6.6 and one has

$$\frac{\partial u}{\partial x} = -L_\lambda^\dagger \left(\frac{\partial K}{\partial x} - \lambda\frac{\partial M}{\partial x}\right)u + cu \tag{6.19}$$

If u is determined through the inverse iteration algorithm

$$\begin{aligned} (K-\tilde{\lambda}M)u_{i+1} &= Mv_i \\ v_i &= u_i/||u_i||_\infty \end{aligned} \tag{6.20}$$

then the algorithm given previously is essentially unchanged.

The eigenvalue derivative may be found by multiplying Eq. 6.6 by u^T and solving for $\partial\lambda/\partial x$, to obtain

$$\frac{\partial \lambda}{\partial x} = u^T\left(\frac{\partial K}{\partial x} - \lambda\frac{\partial M}{\partial x}\right) u \tag{6.21}$$

where $u^T Mu = 1$.

The damped eigensystem of Eq. 6.2 is usually more difficult to solve than the corresponding system of Eq. 6.11, without damping. However, under special circumstances, the eigenvectors of the damped system are also eigenvectors of the corresponding undamped system. This is true, for example, if C is some linear combination of K and M, i.e.

$$C = \alpha K + \beta M \qquad \alpha, \beta \in \mathcal{R} \tag{6.22}$$

It has been shown that

$$KM^{-1}C = CM^{-1}K \tag{6.23}$$

is a necessary and sufficient condition for the eigensystem of Eq. 6.2 to be uncoupled when it is transformed to the modal coordinates of the corresponding eigensystem of Eq. 6.1 [17]. If Eq. 6.23 holds for $x \in D \subseteq R$, then the eigenvectors of the damped system do not depend on C(x), so their derivatives may be computed directly from Eq. 6.19.

The eigenvalue derivatives may be determined by differentiating Eq. 6.2 and solving for $\partial\lambda/\partial x$ to obtain

$$\frac{\partial\lambda}{\partial x} = \frac{u^T[\lambda^2 \frac{\partial M}{\partial x} + \lambda\frac{\partial C}{\partial x} + \frac{\partial K}{\partial x}]u}{u^T Cu + 2\lambda u^T Mu} \tag{6.24}$$

Some caution must be taken when using this formula, since the eigenvalues are not differentiable for particular values of the matrix C. For example, suppose u is an eigenvector of Eq. 6.2, then

$$\lambda^2 u^T Mu + \lambda u^T Cu + u^T Ku = 0 \tag{6.25}$$

is satisfied by λ, so

$$\lambda = \frac{-u^T Cu \pm [(u^T Cu)^2 - 4u^T Mu\ u^T Ku]^{1/2}}{2u^T Mu} . \tag{6.26}$$

If

$$(u^T Cu)^2 = 4u^T Mu\ u^T Ku$$

then the mode corresponding to u is said to be critically damped [18] and the eigenvalue λ has no derivative, as can be seen by attempting to differentiate Eq. 6.26 [7].

REFERENCES

1. Jacobi, C.G.J., Crelle's Journal FürDie Reine Und Angewand Mathematik, Vol. 30, Berlin: De Gruyter, 1846, pp. 51-95.
2. Fox, R.L., and Kapoor, M.P., "Rates Of Change Of Eigenvalues And Eigenvectors," AIAA Journal, Vol. 6, 1968, pp. 2426-2429.

3. Plaut, R.H., and Huseyin, K., "Derivatives Of Eigenvalues And Eigenvectors In Non-Self Adjoint Systems," AIAA Journal, Vol. 11, 1973, pp. 250-251.
4. Rogers, L.C., "Derivatives Of Eigenvalues And Eigenvectors," AIAA Journal, Vol. 8, 1970, pp. 943-944.
5. Rudisill, C.S., "Derivatives Of Eigenvalues And Eigenvectors For A General Matrix," AIAA Journal, Vol. 12, 1974, pp. 721-722.
6. Nelson, R.B., "Simplified Calculation Of Eigenvector Derivatives," AIAA Journal, Vol. 14, 1976, pp. 1201-1205.
7. Whitesel, J.E., Design Sensitivity In Dynamical Structures, Ph.D. Thesis, Michigan State University, East Lansing, 1980.
8. Wilkinson, J.H., The Algebraic Eigenvalue Problem, Clarendon Press, London, 1965.
9. Kato, T., Perturbation Theory For Linear Operators, Springer-Verlag, New York, 1976, 2nd ed.
10. Bachman, G., and Narici, N., Functional Analysis, Academic Press, New York, 1966.
11. Frame, J.S., "Matrix Functions and Applications, Part I," IEEE Spectrum, March 1964.
12. Wilkinson, J.H., and Reinsch, C., Handbook For Automatic Computation, Vol. II, Linear Algebra, Springer-Verlag, New York, 1971.
13. Bathe, K.J., and Wilson, E.L., Numerical Methods In Finite Element Analysis, Prentice-Hall, Englewood Cliffs, NJ, 1976.
14. Frame, J.S., "Matrix Functions and Applications, Part II," IEEE Spectrum, April, 1964.
15. Haug, E.J., and Rousselet, B., "Design Sensitivity Analysis In Structural Mechanics II, Eigenvalue Variations," Journal of Structural Mechanics, to appear, 1980.
16. Meirovitch, L., Analytic Methods In Vibrations, MacMillan, New York, 1967.
17. Caughey, T.K., and O'Kelley, M.F.S., "Classical Normal Modes In Damped Linear Dynamic Systems," Journal of Applied Mechanics, Vol. 32, 1965, pp. 583-588.
18. Inman, D.J., and Andry, A.N., "Some Results On The Nature Of Eigenvalues Of Discrete Damped Linear Systems," Journal of Applied Mechanics, ASME, to appear.

DESIGN SENSITIVITY ANALYSIS FOR DISTRIBUTED PARAMETER STRUCTURAL SYSTEMS GOVERNED BY DOUBLE EIGENVALUE PROBLEMS*†

Jean-Louis Claudon, Megumi Sunakawa

Department of Aeronautics, University of Tokyo, Bunkyo-ku, Tokyo, Japan

Institute of Space and Aeronautical Science, University of Tokyo, Meguro-ku, Tokyo, Japan

ABSTRACT

Distributed-parameter structural systems governed by equations of the two types $[K + pE]u = 0$ and $[K + pE - \omega^2 M]u = 0$ are considered. The first equation is a simple eigenvalue problem that applies to buckling or free vibration systems. The second equation is a double eigenvalue problem that applies to free vibration systems exhibiting buckling or flutter instability. Design sensitivity analysis is considered in both cases, for an arbitrary variation of a single design function. There is more to the sensitivity analysis of the second equation than just two separate applications of results obtained for the first equation. Taking full advantage of a double eigenvalue problem formulation, two alternative sets of series expressions for up to second variations of any point on the eigenvalue curves in $p \times \omega^2$ are developed. This includes, in particular, the second variation of buckling and flutter loads. Curvature information on the eigenvalue curves is also obtained and analytical properties at

*This work was supported by the Japanese Ministry of Education.

†The authors are grateful to Professor R.H. Plaut for reading an early draft of the manuscript and for indicating a few related references. The late Professor William Prager showed a keen interest in this work. We deeply regret the recent loss of a great and respected teacher.

singular points of contact between two distinct curves are derived. A condition for such a contact to be preserved in a variation of design is presented.

1. INTRODUCTION

Given a physical system with known behavior, sensitivity analysis studies the effect produced on that behavior by changes in some governing parameters. In structural systems these parameters are typically related to design properties, such as stiffness and mass, and to loading conditions. A major application is in the field of optimal design, where the role of sensitivity analysis may be described as an analytical simulation of local changes in design space, making it possible to determine in an analytical way a strategy of best improvement of a current design. The optimum design can then be obtained numerically through iteration. Sensitivity analysis and related design optimization methods are discussed in phase setting and applied to elementary distributed parameter systems in the recent textbook by Haug and Arora [1], which also carries an extensive list of references.

The particular concern here is with distributed parameter structural systems that are governed by eigenvalue problems of the following two types:

$$[K + pE]u = 0 \tag{1.1}$$

or

$$[K - \omega^2 M]u = 0 \tag{1.2}$$

and

$$[K + pE - \omega^2 M]u = 0 \tag{1.3}$$

where K, E, and M are real linear homogeneous differential operators, defined in terms of a single continuous design function, with E and M regular; and p and ω^2 are eigenvalues with corresponding right and left eigenfunctions u and v. Such systems are in general non-self-adjoint (nonconservative), and some eigenvalues and eigenfunctions may be complex. Self-adjoint systems are conservative.

Equation 1.1 is a linear eigenvalue problem in p only, which is termed simple. It applies to buckling and to possibly nonconservative free vibration problems. Design sensitivity analysis here consists in analyzing the changes in the eigenelements p, u, and v, caused by changes in K and E, and is well documented in its discrete-system version (see for example, Refs. 2-5). Farshad [6] has extended the most elementary results to distributed parameter systems. In Section 2, Farshad's work is first followed and then expanded using the simple problem of Eq. 1.1 as an introduction to the distributed case, including notations and the basic methodology to be used throughout the sequel. Normalization of the eigenfunctions, an operation not otherwise assumed in this paper, is also discussed.

Equation 1.3 applies to non-dissipative, circulatory or non-circulatory autonomous systems, for which buckling or flutter instability may occur. The books in Refs. 8-10 are respresentative of the increasing attention paid to such systems in engineering research. Equation 1.3 is a linear eigenvalue problem in ω^2 and it is also a linear eigenvalue problem in p; thus it is termed a double eigenvalue problem. Design sensitivity analysis now consists in analyzing the evolutions of the eigenvalue curves in $p \times \omega^2$ (the characteristic curves), caused by design variations, as well as the corresponding changes in the eigenfunctions. In Section 3, design sensitivity analysis for this problem is carried out up to the second variation of the characteristic curves. Various analytical information is also obtained on the curves themselves and on the singular case in which two curves are in contact.

The emphasis in this paper is on the consequences of the fact that Eq. 1.3 is a double eigenvalue problem (let there be no confusion of terminology between double eigenvalue problems and simple eigenvalue problems with repeated eigenvalues). It will be seen that many more sensitivity analysis results may be obtained for Eq. 1.3 than can be obtained by separately analyzing two equations of the form of Eqs. 1.1 and 1.2.

2. SIMPLE EIGENVALUE PROBLEMS

2.1 First Variation of Eigenvalues and Eigenfunctions

Suppose Eq. 1.1 is solved and has distinct eigenvalues. For some eigenvalue p_m, and corresponding right eigenfunction u_m, consider the inner product

$$(v_i,[K + p_m E]u_m) = 0 \tag{2.1}$$

where v_i is any one of the left eigenfunctions. Taking the first variation of Eq. 2.1 and applying Eq. 1.1, one obtains

$$(v_i,[K+p_m E]\delta u_m) + (v_i,[\delta K + p_m \delta E]u_m) + (v_i,Eu_m)\delta p_m = 0 \tag{2.2}$$

which is the starting point for the first-order sensitivity analysis of Eq. 1.1.

First let i=m, so the first term in Eq. 2.2 vanishes, by definition of a left eigenfunction. The first variation of the eigenvalue p_m is then

$$\delta p_m = - \frac{(v_m, [\delta K + p_m \delta E]u_m)}{(v_m, Eu_m)} \tag{2.3}$$

where the denominator is equal to unity if the eigenfunctions are so normalized. The equations will be kept completely general, however, by not assuming any kind of normalization.

Now let i=n, with n≠m. By the biorthogonality property of eigenfunctions, the third term in Eq. 2.2 vanishes, so

$$(v_n, [K + p_m E]\delta u_m) + (v_n, [\delta K + p_m \delta E]u_m) = 0 \tag{2.4}$$

Due to completeness of the eigenfunctions, one can write

$$\delta u_m = \sum_j \alpha_{mj} u_j \tag{2.5}$$

where the coefficients α_{mj} are sought. Substitution into Eq. 2.4 and use of Eq. 2.1 and the biorthogonality property, results in

$$\alpha_{mn}(p_m - p_n)(v_n, Eu_n) + (v_n, [\delta K + p_m \delta E]u_m) = 0, \quad n \neq m \tag{2.6}$$

yielding the coefficients α_{mj} in Eq. 2.5. The first variation of right eigenfunctions is thus obtained as

$$\delta u_m = - \sum_{\substack{j \\ j \neq m}} \frac{(v_j, [\delta K + p_m \delta E]u_m)}{(p_m - p_j)(v_j, Eu_j)} u_j \tag{2.7}$$

Similarly the first variation of left eigenfunctions is found to be

$$\delta v_m = - \sum_{\substack{j \\ j \neq m}} \frac{(v_m, [\delta K + p_m \delta E]u_j)}{(p_m - p_j)(v_j, Eu_j)} v_j \tag{2.8}$$

Note that any additional term $\alpha_{mm} u_m$ in Eq. 2.7, or any additional term $\beta_{mm}v_m$ in Eq. 2.8, would only produce a second-order eigendirection change. Equations 2.7 and 2.8 are therefore satisfactory when no particular normalization is required. Such additional terms, on the other hand, may be used when some normalization requirement is to be passed on from the initial eigenfunctions to the varied eigenfunctions. For example, let the customary normalization $(v_m, Eu_m) = 1$ be required to be preserved, in the form

$$(v_m + \delta v_m, [E + \delta E](u_m + \delta u_m)) = 1 \tag{2.9}$$

Deleting higher-order terms, the first-order condition is

$$(\delta v_m, Eu_m) + (v_m, E\delta u_m) + (v_m, \delta E u_m) = 0 \tag{2.10}$$

where the variational delta δ implies variation for the first symbol following it only. In order to satisfy this condition, express the normalized eigenfunctions as

$$\left.\begin{aligned} \delta u_m &= \delta^* u_m + \alpha_{mm} u_m \\ \delta v_m &= \delta^* v_m + \beta_{mm} v_m \end{aligned}\right\} \tag{2.11}$$

where $\delta^* u_m$ and $\delta^* v_m$ are given by Eqs. 2.7 and 2.8, where (v_j, Eu_j) has been set equal to unity. Substituting into Eq. 2.10 and letting $\alpha_{mm} = \beta_{mm}$ (a natural choice), one obtains

$$\alpha_{mm} = \beta_{mm} = -\frac{1}{2}(v_m, \delta E u_m) \tag{2.12}$$

The matter of normalization is not further pursued in this paper. No normalization of the original eigenfunctions will be assumed, as non-normalized equations are more illuminating and can be subsequently normalized, if necessary. Normalization of varied eigenfunctions can, in all cases, be carried out using the technique described above.

2.2 Second Variation of Eigenvalues

At this point, it is helpful to introduce the following additional notation for operators:

$$\left.\begin{aligned} \underset{m}{\Delta} &= \delta K + p_m \delta E \\ \underset{m}{\Delta}^2 &= \delta^2 K + p_m \delta^2 E \end{aligned}\right\} \tag{2.13}$$

A similar notation is used in Section 3.

Taking the second variation of Eq. 2.1, with i=m, and using the biorthogonality property with Eqs. 2.7 and 2.8 results in

$$(\delta v_m, \underset{m}{\Delta} u_m) + (v_m, \underset{m}{\Delta}\delta u_m) + (v_m, \underset{m}{\Delta}^2 u_m) + 2(v_m, \delta E u_m)\delta p_m$$

$$+ (v_m, Eu_m)\delta^2 p_m = 0 \qquad (2.14)$$

Using the previous results of Eqs. 2.3, 2.7, and 2.8, one finally obtains the second variation of the eigenvalue p_m as

$$\delta^2 p_m = \frac{1}{(v_m, Eu_m)} \left\{ -(v_m, \underset{m}{\Delta}^2 u_m) + 2\frac{(v_m, \delta E u_m)}{(v_m, Eu_m)} (v_m, \underset{m}{\Delta} u_m) \right.$$

$$\left. + 2 \sum_{\substack{j \\ j \neq m}} \frac{(v_m, \underset{m}{\Delta} u_j)(v_j, \underset{m}{\Delta} u_m)}{(p_m - p_j)(v_j, Eu_j)} \right\} \qquad (2.15)$$

Similar results can be found in the literature (see for example, Ref. 2). Such results essentially concern discrete systems, where elements of governing matrices are functions of a finite number of parameters. Cross-derivatives are encountered. In distributed design-sensitivity analysis, as presented in this paper, cross-derivatives are at this point irrelevant, since design has been assumed to be completely determined by a single design function, defined on the whole (fixed) structural domain. Successive variations of the governing operators are taken with respect to arbitrary variations of that single function, resulting in expressions that are simpler than those obtained if design were discretized at an early stage. In the distributed context, cross-differentiation would occur only if design were described by several independent design functions. This case is not considered in the present work.

Note that no assumption of self-adjointness or definiteness of any operator has been made in the above calculations. The formulas derived in this section are thus applicable to any linear, non-singular eigenvalue problem of the same nature that may be encountered.

3. DOUBLE EIGENVALUE PROBLEMS

3.1 The Double Eigenvalue Problem

Equation 1.3 is solved when all pairs (p,ω^2) satisfying it have been determined, along with the corresponding right and left eigenfunctions. Special attention is given here to solutions for which p and ω^2 are both real. The set of all such pairs can be visualized in the plane $p \times \omega^2$, where it constitutes the so-called characteristic curves. Figure 3.1 shows typical characteristic curves for a system governed by an equation of the form of Eq. 1.3. Points A correspond to buckling loads and may be absent, as in the case of tangential-force systems. Points B correspond to unloaded free vibration frequencies. Points C has a horizontal tangent and is a critical flutter point. Point D has a vertical tangent.

Points belonging to the characteristic curves may be obtained either by fixing $\omega^2 = \omega_0^2$ and solving the linear eigenvalue problem for p (see points A' in Fig. 3.1), or by fixing $p = p_0$ and solving the linear eigenvalue problem for ω^2 (see points B'). Either problem has the form of Eq. 1.1, so that the results of Section 2 remain valid here, under the same assumptions. To illustrate this, a portion of a varied characteristic curve, represented by a dotted line, corresponding to some variation of the design

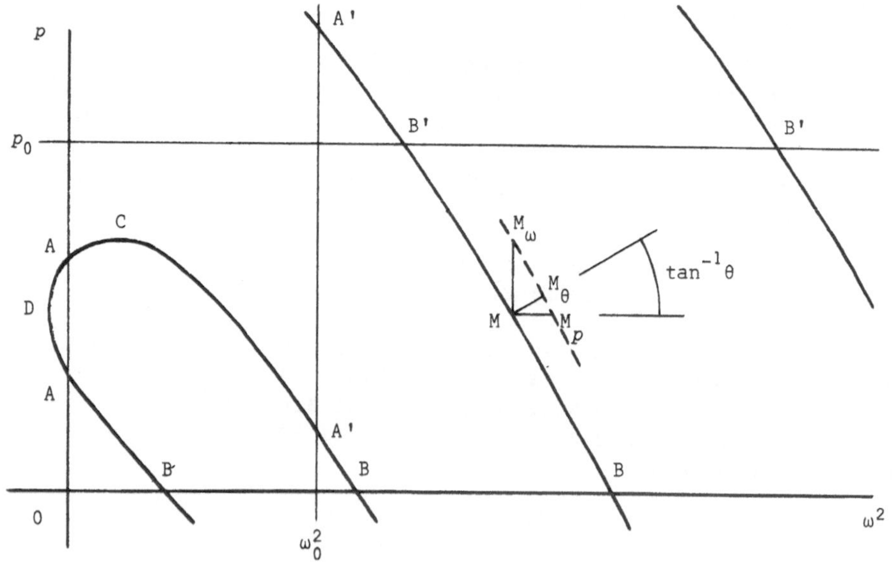

Figure 3.1 Typical Characteristic Curves in the Plane $p \times \omega^2$

function is shown in Fig. 3.1. Points such as M_ω may be studied using Eqs. 2.3 and 2.15, where K must be replaced by $K-\omega^2M$ and where the eigenfunctions correspond to the various real and complex eigenvalues p_j. The point M_p is studied in a completely similar manner, but this time p is fixed and the eigenfunctions are for the modal frequencies ω_j^2.

This formal symmetry of double eigenvalue problems leads to an observation that will play a central role in this paper. It is undoubtedly more natural to construct characteristic curves through the solution of a number of eigenproblems in ω^2, for various fixed loads p of physical interest. In doing so one obtains real and complex frequencies and modes that enjoy clear physical meanings, and the critical load is readily located. When it comes to sensitivity analysis, however, the less usual eigenproblem in p, for fixed (real) ω^2, is on a par with the more natural one. Although its solutions, involving complex loads, preclude any straightforward physical interpretation, it provides a perfectly valid solution set. Real eigenvalues are on the characteristic curves and their variations can obviously be calculated, regardless of the fact that most other eigenelements may be complex. Moreover, the less usual solution set may turn out to be free of deficiencies the physical eigensolutions may display. At a flutter point, for example, coalescence of two modes of the customary eigenproblem runs against the assumptions of distinctness and completion made Section 2. Such a point, however, is a simple eigenvalue for the eigenproblem in terms of p and does not involve any degeneracy of its eigenfunctions.

This particularity of double eigenvalue problems has two main consequences, one of which is trivial. It is always possible, using the results of Section 2 in one or the other of the two ways above, to calculate successive variations in the characteristic curves for any variation of design. The second consequence is very powerful, but has not yet been exploited as it should. Two convenient solution sets are available to express variations of eigenfunctions, in the form of Eq. 2.5. In the following, the sensitivity analysis of Eq. 1.3 is carried out in the plane $p\times\omega^2$ (not only in the directions of the coordinate axes p and ω^2). The two available eigensolution sets will make up for each other's deficiencies, in degenerate cases.

3.2 First Variation of Eigenvalues; Envelope Condition

Consider a point M on the characteristic curves obtained from Eq. 1.3, and the corresponding inner product

$$(v_m, [K + p_m E - \omega_m^2 M] u_m) = 0 \tag{3.1}$$

where all eigenelements belong to the point M and are real. Taking the first variation of this equation and using Eq. 1.3 and the definition of a left eigenfunction, one obtains the sensitivity equation

$$(v_m, \Delta_m u_m) + (v_m, E u_m)\delta p_m - (v_m, M u_m)\delta\omega_m^2 = 0 \tag{3.2}$$

where the notation $\Delta_m = \delta K + p_m \delta E - \omega_m^2 \delta M$ has been used.

Equation 3.2 is reminiscent of Eq. 2.2 of Section 2, when i=m, but it involves an extra variation term. It displays the way in which the variations of p_m and ω_m^2 are related to the given variation of the operators and to each other. First, letting $\Delta_m = 0$, one obtains the slope of the characteristic curve at point M as the limiting quotient

$$\frac{dp_m}{d\omega_m^2} = \frac{(v_m, M u_m)}{(v_m, E u_m)} \tag{3.3}$$

At a flutter point, the slope is zero and a very general form of Plaut's flutter condition [11] is obtained as

$$(v_m, M u_m) = 0 \tag{3.4}$$

Similarly, at a point with vertical tangent one has the condition

$$(v_m, E u_m) = 0 \tag{3.5}$$

implying $(v_m, K u_m) = 0$, from Eq. 3.1, if this point has zero frequency [13].

Now consider a non-zero variation of the operators. Letting $\delta\omega_m^2 = 0$ in Eq. 3.2 gives the vertical variation of point M (point M_ω in Fig. 3.1),

$$\delta p_m = -\frac{(v_m, \Delta_m u_m)}{(v_m, E u_m)} \tag{3.6}$$

This result, anticipated in the preceding section, is the same as Eq. 2.3. It gives, in particular, a formula for the first variation of the buckling points A in Fig. 3.1. In the same way, one obtains the horizontal first variation of point M (see point M_p in Fig. 3.1) as

$$\delta\omega_m^2 = \frac{(v_m, \underset{m}{\Delta} u_m)}{(v_m, M u_m)} \tag{3.7}$$

The oblique variation along a line with slope θ (see point M_θ in Fig. 3.1) will also prove useful. One now has $\delta p_m = \theta\delta\omega_m^2$, giving

$$\left.\begin{aligned} \delta p_m &= \frac{(v_m, \underset{m}{\Delta} u_m)}{(v_m, M u_m)/\theta - (v_m, E u_m)} \\ \delta\omega_m^2 &= \frac{(v_m, \underset{m}{\Delta} u_m)}{(v_m, M u_m) - \theta(v_m, E u_m)} \end{aligned}\right\} \tag{3.8}$$

Note that Eqs. 3.8 exhibit singularities when the slope θ is equal to that of the characteristic curve at the same point. An important illustration is provided by the case of characteristic curves that are tangent to the loading axis, a special case of points A in Fig. 3.1. It is easy to see that, following a variation of the governing operators, such a double buckling point will either split into two simple buckling points of disappear altogether. Only for special operator variations, satisfying the following envelope condition,

$$(v_m, \underset{m}{\Delta} u_m) = 0 \tag{3.9}$$

can Eqs. 3.6 to 3.8 be indeterminate. This indeterminacy will be resolved later.

3.3 First Variation of Eigenfunctions

In contrast to Eq. 3.1, consider at point M the inner product

$$(v_n, [K + p_m E - \omega_m^2 M] u_m) = 0 \tag{3.10}$$

with $n \neq m$, so that the left eigenfunction v_n is not the one at point M. Taking the first variation of this equation now gives

$$(v_n, [K + p_m E - \omega_m^2 M]\delta u_m) + (v_n, \Delta_m u_m) + (v_n, Eu_m)\delta p_m$$

$$- (v_n, Mu_m)\delta\omega_m^2 = 0 \qquad (3.11)$$

which is similar to Eq. 2.4. In order to express the variation δu_m in the form of Eq. 2.5, it now becomes necessary to remove the ambiguity of the subscript m by stating which one of the two available sets of eigenfunctions is to be used. If one chooses the eigenproblem in p by fixing ω^2 (accordingly, remove the subscript from ω_m^2), the biorthogonality property $(v_n, Eu_m) = 0$ reduces Eq. 3.11 to

$$(v_n, [K + p_m E - \omega^2 M]\delta u_m) + (v_n, \Delta_m u_m) - (v_n, Mu_m)\delta\omega^2 = 0 \qquad (3.12)$$

Alternatively, to choose the eigenproblem in ω^2 will bring about the biorthogonality property $(v_n, Mu_m) = 0$ and result in the equation

$$(v_n, [K + pE - \omega_m^2 M]\delta u_m) + (v_n, \Delta_m u_m) + (v_n, Eu_m)\delta p = 0 \qquad (3.13)$$

Equations 3.12 and 3.13 are of the same general form as Eq. 2.4. Equations 2.7 and 2.8 can then be extended in a straightforward manner to give similar general formulas for the eigenfunction variations δu_m and δv_m in both cases. These formulas are now written in the forms under which they will put to use in the following. Two separate sets of formulas to choose from will thus be obtained, each corresponding to one of the two available sets of eigensolutions.

<u>Variation Along Original Characteristic Curves</u>. Starting from Eq. 3.12 with $\Delta_m = 0$ and proceeding along the lines of Eqs. 2.4 to 2.8, one readily obtains

$$\delta u_m = \delta\omega^2 \times \sum_{\substack{j \\ j \neq m}} \frac{1}{p_m - p_j} \frac{(v_j, Mu_m)}{(v_j, Eu_j)} u_j \qquad (3.14)$$

$$\delta v_m = \delta\omega^2 \times \sum_{\substack{j \\ j \neq m}} \frac{1}{p_m - p_j} \frac{(v_m, Mu_j)}{(v_j, Eu_j)} v_j \qquad (3.15)$$

where all eigenfunctions u_j and v_j belong to the eigenproblem in terms of p through point M. These formulas are clearly invalid at a point with a vertical tangent, since the corresponding solution set is incomplete. One may also use Eq. 3.3 to replace $\delta\omega^2$ by δp in the right-hand side of Eqs. 3.14 and 3.15. The new expressions are then degenerate at flutter points.

Alternatively, treating Eq. 3.13 in a similar manner gives

$$\delta u_m = \delta p \times \sum_{\substack{j \\ j\neq m}} \frac{1}{\omega_m^2-\omega_j^2} \frac{(v_j,Eu_m)}{(v_j,Mu_j)} u_j \tag{3.16}$$

$$\delta v_m = \delta p \times \sum_{\substack{j \\ j\neq m}} \frac{1}{\omega_m^2-\omega_j^2} \frac{(v_m,Eu_j)}{(v_j,Mu_j)} v_j \tag{3.17}$$

where the u_j and v_j are now from the eigenproblem in terms of ω^2 at M. In the same way as above, these equations are invalid at a flutter point and degenerate at a point with vertical tangent, if $\delta\omega^2$ replaces δp.

<u>Variation on Oblique Line $\delta p = \theta\delta\omega^2$</u>. Letting $\delta p = \theta\delta\omega^2$ in Eq. 3.12 and proceeding in the same way, the first variation of eigenfunctions along a line with slope θ, when the governing operators undergo a variation, is

$$\delta u_m = -\sum_{\substack{j \\ j\neq m}} \frac{1}{p_m-p_j} \left\{ \frac{(v_j,\underset{m}{\Delta}u_m)}{(v_j,Eu_j)} - \frac{(v_m,\underset{m}{\Delta}u_m)}{(v_m,Mu_m) - \theta(v_m,Eu_m)} \frac{(v_j,Mu_m)}{(v_j,Eu_j)} \right\} u_j \tag{3.18}$$

$$\delta v_m = -\sum_{\substack{j \\ j\neq m}} \frac{1}{p_m-p_j} \left\{ \frac{(v_m,\underset{m}{\Delta}u_j)}{(v_j,Eu_j)} - \frac{(v_m,\underset{m}{\Delta}u_m)}{(v_m,Mu_m) - \theta(v_m,Eu_m)} \frac{(v_m,Mu_j)}{(v_j,Eu_j)} \right\} v_j \tag{3.19}$$

where the eigenproblem is in terms of p. As before, these formulas are invalid at a point with a vertical tangent.

Alternatively, from Eq. 3.13,

$$\delta u_m = \sum_{\substack{j \\ j\neq m}} \frac{1}{\omega_m^2-\omega_j^2} \left\{ \frac{(v_j,\underset{m}{\Delta} u_m)}{(v_j,Mu_j)} \right.$$

$$\left. + \frac{(v_m,\underset{m}{\Delta} u_m)}{(v_m,Mu_m)/\theta - (v_m,Eu_m)} \frac{(v_j,Eu_m)}{(v_j,Mu_j)} \right\} u_j \tag{3.20}$$

$$\delta v_m = \sum_{\substack{j \\ j\neq m}} \frac{1}{\omega_m^2-\omega_j^2} \left\{ \frac{(v_m,\underset{m}{\Delta} u_j)}{(v_j,Mu_j)} \right.$$

$$\left. + \frac{(v_m,\underset{m}{\Delta} u_m)}{(v_m,Mu_m)/\theta - (v_m,Eu_m)} \frac{(v_m,Eu_j)}{(v_j,Mu_j)} \right\} v_j \tag{3.21}$$

where the eigenproblem is in terms of ω^2. These two formulas are invalid at a flutter point. Note also that, consistent with the corresponding formulas for variations of the eigenvalues, Eqs. 3.18 to 3.21 exhibit singularities when the slope θ is equal to that of the characteristic curve at the same point.

Variations on Lines $\delta p = 0$ and $\delta\omega^2 = 0$. Formulas for the first variations of eigenfunctions along horizontal or vertical lines of loading-frequency plane can be obtained, as special cases of Eqs. 3.18 to 3.21, by setting the slope θ equal to zero or to infinity.

3.4 Variation of Flutter Point and Modes

At a flutter point, one has $(v_m,Mu_m) = 0$. From Eq. 3.2, the first variation of flutter load corresponding to a variation of the governing operators is then readily obtained as

$$\delta p_{flutter} = - \frac{(v_m,\underset{m}{\Delta} u_m)}{(v_m,Eu_m)} \tag{3.22}$$

The derivation of $\delta\omega^2_{flutter}$ is more tedious and makes use of the varied flutter condition $(v_m+\delta v_m,[M+\delta M](u_m+\delta u_m)) = 0$, implying

$$(\delta v_m,Mu_m) + (v_m,M\delta u_m) + (v_m,\delta Mu_m) = 0 \tag{3.23}$$

The slope of the line joining the original and the varied flutter points is, of course, unknown. The variations δu_m and δv_m in Eq. (3.23) are given by Eqs. 3.18 and 3.19, where (v_m,Mu_m) is made zero and where θ is sought. Carrying out the indicated substitution in Eq. 3.23, after some manipulation, one finally obtains

$$\theta = \frac{2\,\dfrac{(v_m,\underset{m}{\Delta}u_m)}{(v_m,Eu_m)}\displaystyle\sum_{\substack{j\\ j\neq m}}\frac{(v_m,Mu_j)(v_j,Mu_m)}{(p_m-p_j)(v_j,Eu_j)}}{(v_m,\delta Mu_m)-\displaystyle\sum_{\substack{j\\ j\neq m}}\frac{(v_m,Mu_j)(v_j,\underset{m}{\Delta}u_m)+(v_j,Mu_m)(v_m,\underset{m}{\Delta}u_j)}{(p_m-p_j)(v_j,Eu_j)}} \tag{3.24}$$

Using Eq. 3.22, the frequency variation sought is therefore

$$\delta\omega^2_{flutter} = -\frac{(v_m,\delta Mu_m)-\displaystyle\sum_{\substack{j\\ j\neq m}}\frac{(v_m,Mu_j)(v_j,\underset{m}{\Delta}u_m)+(v_j,Mu_m)(v_m,\underset{m}{\Delta}u_j)}{(p_m-p_j)(v_j,Eu_j)}}{2\displaystyle\sum_{\substack{j\\ j\neq m}}\frac{(v_m,Mu_j)(v_j,Mu_m)}{(p_m-p_j)(v_j,Eu_j)}} \tag{3.25}$$

The corresponding first variation of flutter modes is given by Eqs. 3.18 and 3.19, with the slope θ given by Eq. 3.24.

Following the same pattern, one can derive similar formulas for the first variation of a point with vertical tangent. Its component $\delta\omega^2$ is $(v_m,\underset{m}{\Delta}u_m)/(v_m,Mu_m)$ and the corresponding δp, δu_m, and δv_m can be derived using Eqs. 3.20 and 3.21. One can also readily generalize this development to the variation of a point with an arbitrary tangent.

The reader will have noticed that this also resolves the indeterminacy problem associated with the envelope condition of Eq. 3.9.

3.5 Second Variation of Eigenvalues

One may now take a further variation of Eq. 3.2, the first-order eigenvalue sensitivity equation. With the notation $\Delta_m^2 = \delta^2 K + p_m \delta^2 E - \omega_m^2 \delta^2 M$, this is

$$
\begin{aligned}
&(v_m, \Delta_m^2 u_m) + (\delta v_m, \Delta_m u_m) + (v_m, \Delta_m \delta u_m) \\
&\quad + \{2(v_m, \delta E u_m) + (\delta v_m, E u_m) + (v_m, E \delta u_m)\}\, \delta p_m \\
&\quad + (v_m, E u_m)\, \delta^2 p_m \\
&\quad - \{2(v_m, \delta M u_m) + (\delta v_m, M u_m) + (v_m, M \delta u_m)\}\, \delta \omega_m^2 \\
&\quad - (v_m, M u_m)\, \delta^2 \omega_m^2 = 0
\end{aligned}
\qquad (3.26)
$$

<u>Curvature of Characteristic Curves</u>. Equation 3.2, with $\Delta_m = 0$ yielded the slope at some point M of the original characteristic curves. Equation 3.26, with $\Delta_m^2 = \Delta_m = 0$ will yield the curvature. To derive the second variation $\delta^2 p_m$, corresponding to a $\delta \omega_m^2$, let $\delta^2 \omega_m^2 = 0$ in Eq. 3.26 and use the slope of Eq. 3.3, to obtain

$$
\begin{aligned}
\frac{\delta^2 p_m}{\delta\, \omega_m^2} &= \frac{1}{(v_m, E u_m)} \{(\delta v_m, M u_m) + (v_m, M \delta u_m)\} \\
&\quad - \frac{(v_m, M u_m)}{(v_m, E u_m)^2} \{(\delta v_m, E u_m) + (v_m, E \delta u_m)\}
\end{aligned}
\qquad (3.27)
$$

The variations δu_m and δv_m needed are given by Eqs. 3.14 and 3.15, or alternatively by Eqs. 3.16 and 3.17. If. Eqs. 3.14 and 3.15

are used, the limiting result is the following second derivative:

$$\frac{d^2 p_m}{d(\omega^2)^2} = \frac{2}{(v_m, Eu_m)} \sum_{\substack{j \\ j \neq m}} \frac{(v_m, Mu_j)(v_j, Mu_m)}{(p_m - p_j)(v_j, Eu_j)} \tag{3.28}$$

where the eigenproblem is in terms of p. This is invalid at a point with a vertical tangent. (In Ref. 10, p. 178, Huseyin presents a similar result as being the second derivative at a flutter point. The development above shows that Eq. 3.28 is valid at any point that does not have a vertical tangent.) Alternatively, if Eqs. 3.16 and 3.17 are used in Eq. 3.27, together with Eq. 3.3, the result obtained is invalid at a flutter point and is singular at a point with a vertical tangent.

Use of the standard formula for the curvature $1/\rho$ of a plane curve gives the expression

$$\frac{1}{\rho} = \frac{2(v_m, Eu_m)^2}{[(v_m, Eu_m)^2 + (v_m, Mu_m)^2]^{3/2}} \times \left| \sum_{\substack{j \\ j \neq m}} \frac{(v_m, Mu_j)(v_j, Mu_m)}{(p_m - p_j)(v_j, Eu_j)} \right| \tag{3.29}$$

where the eigenproblem is in terms of p. This expression is invalid only at points with a vertical tangent. In the special case of a flutter point, Eq. 3.29 reduces to Eq. 3.28.

If the curvature at a point with a vertical tangent is sought, or if for any reason one prefers to use the eigenproblem in terms of ω^2, calculations completely similar to those in Eqs. 3.27 to 3.29 can be carried out in the alternative way. First, setting $\delta^2 p_m = 0$ in Eq. 3.16 and using Eq. 3.3, one obtains an expression for $\delta^2\omega^2/\delta p_m$, similar to Eq. 3.27. Substitution of Eqs. 3.16 and 3.17 gives the second derivative as

$$\frac{d^2\omega_m^2}{dp_m^2} = \frac{2}{(v_m, Mu_m)} \sum_{\substack{j \\ j \neq m}} \frac{(v_m, Eu_j)(v_j, Eu_m)}{(\omega_m^2 - \omega_j^2)(v_j, Mu_j)} \tag{3.30}$$

The corresponding expression for the curvature, invalid only at flutter points, follows from this in the same way as above. At a point with a vertical tangent, it reduces to Eq. 3.30.

Second Variation on Oblique Line $\delta p = \theta\delta\omega^2$. In Eq. 3.26, the first variations δp_m and $\delta\omega_m^2$ are now given by Eqs. 3.8. As the second varied point sought is on the same oblique line, one also has $\delta^2 p_m = \theta\delta^2\omega^2_m$. Substituting this into Eq. 3.28 and using Eqs. 3.18 and 3.19, one obtains

$$\delta^2 p_m = \frac{1}{(v_m, Mu_m)/\theta - (v_m, Eu_m)}$$

$$\times \{(v_m, \underset{m}{\Delta}^2 u_m) - 2 \sum_{\substack{j \\ j\neq m}} \frac{(v_m, \underset{m}{\Delta} u_j)(v_j, \underset{m}{\Delta} u_m)}{(p_m - p_j)(v_j, Eu_j)}\}$$

$$+ \frac{2(v_m, \underset{m}{\Delta} u_m)}{[(v_m, Mu_m)/\theta - (v_m, Eu_m)]^2} \{(v_m, \delta[E - \frac{M}{\theta}]u_m)$$

$$+ \frac{1}{\theta} \sum_{\substack{j \\ j\neq m}} \frac{(v_m, Mu_j)(v_j, \underset{m}{\Delta} u_m) + (v_j, Mu_m)(v_m, \underset{m}{\Delta} u_j)}{(p_m - p_j)(v_j, Eu_j)}\}$$

$$- \frac{2(v_m, \underset{m}{\Delta} u_m)^2}{[(v_m, Mu_m)/\theta - (v_m, Eu_m)]^3} \frac{1}{\theta} \sum_{\substack{j \\ j\neq m}} \frac{(v_m, Mu_j)(v_j, Mu_m)}{(p_m - p_j)(v_j, Eu_j)} \tag{3.31}$$

where the eigenproblem is in terms of p. This formula is invalid at a point with vertical tangent, due to incompleteness of the corresponding solution set. It will also be, in general, singular when the slope θ is equal to that of the characteristic curve at the same point. The corresponding $\delta^2\omega_m^2$ follows immediately through division by θ. Alternative expressions involving the eigenvalue problem in terms of ω^2 may be obtained by using Eqs. 3.20 and 3.21, instead of Eqs. 3.18 and 3.19.

Second Variations on Lines $\delta p = 0$ and $\delta\omega^2 = 0$. Two alternative formulas for each of the second variations of eigenvalues along horizontal or vertical lines of the loading-frequency plane can be obtained as special cases of Eq. 3.31 and the three related formulas mentioned above, by setting the slope θ equal to zero or to infinity.

3.6 Second Variation of Flutter Load

Referring to first variations and to the first of Eqs. 3.8, one notices that, in the case of a flutter point, the general variation δp_m reduces to Eq. 3.6 and also to Eq. 3.22, for any θ. In other words, at a flutter point, the first variation δp_m is a constant. Consequently, any formula for a first variation δp_m at a flutter point gives, in effect, the first variation δp_m of the flutter point.

Consider now the second variation. If θ is taken as that in Eq. 3.24, then Eq. 3.31, with $(v_m, Mu_m) = 0$, gives the second variation $\delta^2 p_m$ along a line that runs through the original flutter point and the first-varied flutter point. Although that line does not in general run through the second-varied flutter point, it should be clear that the second variation $\delta^2 p_m$ is a constant at the first-varied flutter point. Therefore Eq. 3.31, with $(v_m, Mu_m) = 0$ and Eq. 3.24, give the second variation of the flutter load. To determine the corresponding second variation $\delta^2 \omega^2_{flutter}$, one needs the second variation of eigenfunctions. The calculation is straightforward, but extremely tedious, so it is not carried out here.

One can readily extend the foregoing discussion to the second variation of a point with a vertical tangent and, in general, to the second variation of a point with an arbitrary tangent.

3.7 Second-Order Envelope Condition

Equation 3.9 gives the first-order condition under which the characteristic curve at a point varies along its own tangent. The corresponding second-order envelope condition is now derived, by setting $(v_m, \Delta u_m) = 0$ in Eq. 3.31 and requiring $\delta^2 p_m$ to be indeterminate when the slope θ is equal to that of the curve at the same point. This condition is

$$(v_m,\underset{m}{\Delta}^2 u_m) - 2 \sum_{\substack{j \\ j \neq m}} \frac{(v_m,\underset{m}{\Delta} u_j)(v_j,\underset{m}{\Delta} u_m)}{(p_m - p_j)(v_j, Eu_j)} = 0 \tag{3.32}$$

where the eigenproblem is in terms of p. Another version of the same condition may be derived for the eigenproblem in terms of ω^2. The result is

$$(v_m,\underset{m}{\Delta}^2 u_m) + 2 \sum_{\substack{j \\ j \neq m}} \frac{(v_m,\underset{m}{\Delta} u_j)(v_j,\underset{m}{\Delta} u_m)}{(\omega_m^2 - \omega_j^2)(v_j, Mu_j)} = 0 \tag{3.33}$$

Either version of this condition may be used, together with Eq. 3.9. A method has been indicated to resolve the indeterminacy associated with Eq. 3.9. The indeterminacy associated with Eqs. 3.32 and 3.33 is resolved in the same way, through the extensions outlined in the preceding.

3.8 Properties at Singular Contact Point

The occurrence in optimization of contacts and switches between characteristic curves (as in Fig. 3.2) was first reported in Ref. 12. For an up-to-date introduction, the reader is referred to Ref. 13. It may be useful to emphasize that in the vicinity of the contact point of Fig. 3.2, the two branches on the left belong together and the two branches on the right belong together. Alternatively, the two upper branches belong together and the two lower branches belong together. The curves in Fig. 3.2 do not cross in the usual sense, but rather behave in a way similar to a degenerate hyperbola. A general variation of parameters has the effect of breaking the contact and defining two separate pairs of branches in one of the two ways indicated above.

The second derivatives in Eqs. 3.28 and 3.30 imply a remarkable property of such a contact point. Since there is a double root for both eigenproblems, in terms of p and in terms of ω^2, one of the denominators in each of Eqs. 3.28 and 3.30 vanishes. The second derivatives are then singular and the characteristic curves are not smooth at the contact point. A consequence is that the slopes at that point are degenerate. Referring to Eq. 3.3, one then has for a contact point

$$(v_m, Mu_m) = (v_m, Eu_m) = 0 \tag{3.34}$$

Referring to Eq. 3.1, this implies the following contact condition:

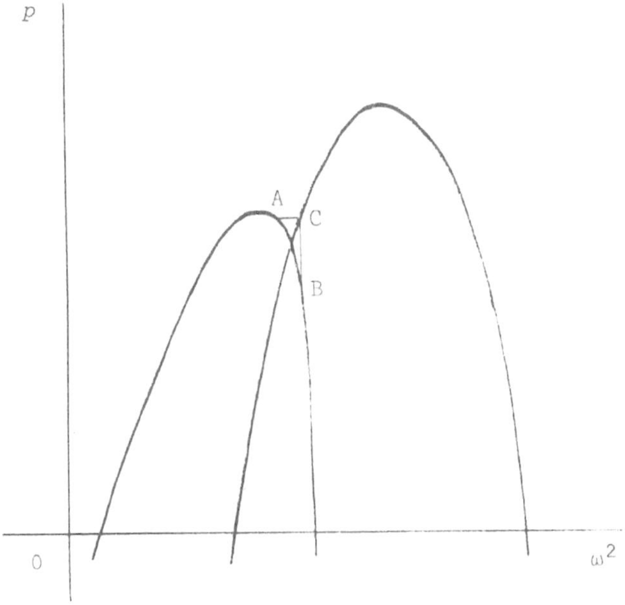

Figure 3.2 Characteristic Curves with Contact Point

$$(v_m,Ku_m) = (v_m,Mu_m) = (v_m,Eu_m) = 0 \tag{3.35}$$

It is now proved that characteristic curves will always seem to cross at a contact point. This amounts to showing that two branches not belonging together have the same slopes and curvatures (and presumably higher derivatives) in the vicinity of the contact point. Consider the point C in Fig. 3.2. With obvious notations, the first variation of its eigenfunction u_c along the corresponding branch is found from Eq. 3.14 as

$$\delta u_c = \frac{\delta\omega^2}{p_c-p_b}\,\frac{(v_b,Mu_c)}{(v_b,Eu_b)}\,u_b + \delta\omega^2 \sum_{\substack{j\neq c\\ j\neq b}} \frac{1}{p_c-p_j}\,\frac{(v_j,Mu_c)}{(v_j,Eu_j)}\,u_j \tag{3.36}$$

When C tends to the contact point, the fraction $(v_b,Mu_c)/(v_b,Eu_b)$ in the first term does not tend to zero, but p_c-p_b does, so the second term in Eq. 3.36 becomes negligible and δu_c tends to be proportional to u_b. Next, using Eq. 3.16 instead of Eq. 3.14, one finds that δu_c tends to be proportional to u_a, with the conclusion that when points A and B tend to the contact point, their right eigenfunctions become equal. A similar argument holds for left eigenfunctions. From Eq. 3.3, it is now clear that slopes of opposite braches are identical on both sides of a contact point.

It is helpful to digress shortly to discuss eigenfunctions at the contact point. Such a point is the limit of sequences of flutter points (with one right and one left eigenfunction each). It is also the limit of similar sequences of points with a vertical tangent. The smooth variation of eigenfunctions proved above shows that there is a unique eigenfunction pair at such a contact point.

Consider now the second derivatives at points A and B, when these points are equidistant from the contact point, as in Fig. 3.3. From Eq. 3.28,

$$\left.\frac{d^2p}{d(\omega^2)^2}\right|_A = \frac{2}{(v_a,Eu_a)}\left\{\frac{(v_a,Mu_e)(v_e,Mu_a)}{(p_a-p_e)(v_e,Eu_e)} + \sum_{\substack{j\neq a\\ j\neq e}}\frac{(v_a,Mu_j)(v_j,Mu_a)}{(p_a-p_j)(v_j,Eu_j)}\right\} \tag{3.37}$$

$$\left.\frac{d^2p}{d(\omega^2)^2}\right|_B = \frac{2}{(v_b,Eu_b)}\left\{\frac{(v_b,Mu_f)(v_f,Mu_b)}{(p_b-p_f)(v_f,Eu_f)} + \sum_{\substack{j\neq b\\ j\neq f}}\frac{(v_b,Mu_j)(v_j,Mu_b)}{(p_b-p_j)(v_j,Eu_j)}\right\} \tag{3.38}$$

When A and B tend to the contact point, while remaining equidistant from it, one can see, from the limiting equality of eigenvectors at A and B and at E and F, that the quantities in Eqs. 3.37 and 3.38 tend to each other (only at the contact point do Eqs. 3.37 and 3.38 become singular). This shows that curvatures are identical on both sides of a contact point.

Note that the second derivatives at E and F do not tend to those at A and B, since the first terms in the brackets of Eqs. 3.37 and 3.38 change signs for E and F. This nevertheless gives still another property at a contact point, namely a relationship

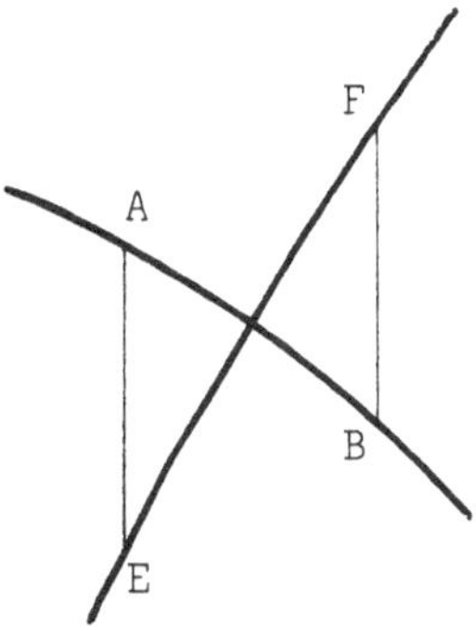

Figure 3.3 Geometry Near Contact Point

between the two different curvatures involved in its vicinity. This relationship is not written here.

3.9 Variations That Preserve Contact

In the preceding section, the fact that a general variation of the design function has the effect of breaking the contact in Fig. 3.2 has been noted. Special design variations, however, may preserve it. If Eq. 3.2 is written at the contact point and contact is broken, Eq. 3.2 becomes singular and nothing can be concluded. But if contact is preserved, one can see that

$$(v_m, \Delta_m u_m) = 0 \tag{3.39}$$

and this is a limiting case of the envelope condition, Eq. 3.9.

4. CONCLUDING REMARKS

The most up-to-date introduction to the problem of structural optimization under nonconservative loading is Ref. 13. In the present paper, a simple analytical basis for design sensitivity analysis and optimization of distributed-parameter structural systems governed by double eigenvalue problems has been developed, with no reference to control theory. As should be obvious from the preceding sections, such sensitivity analysis greatly benefits from a fully double eigenvalue problem formulation.

As any practical, iterative optimization method of the gradient type proceeds with definitely finite step sizes toward the optimum, the second variations may be used to improve redesign efficiency at each iteration. Also, the step size for each iteration may be chosen such that the second variation of the objective function (say, of the critical flutter load) remains within some fraction of the first variation. It is undoubtedly worth evaluating an extra formula leading to an automatic choice of step size.

REFERENCES

1. Haug, E.J. and Arora, J.S., Applied Optimal Design, Wiley, 1979.
2. Plaut, R.H. and Huseyin, K., "Derivatives of Eigenvalues and Eigenvectors in Non-Self-Adjoint Systems," AIAA Journal, Vol. 11, 1973, pp. 250-251.
3. Romstad, K.M., Hutchinson, J.R. and Runge, K.H., "Design Parameter Variation and Structural Response," International Journal for Numerical Methods in Engineering, Vol. 5, 1973, pp. 337-349.
4. Taylor, D.L. and Kane, T.R., "Multiparameter Quadratic Eigenproblems," Journal of Applied Mechanics, Vol. 42, 1975, pp. 478-483.
5. Cardani, C. and Mantegazza, P., "Calculation of Eigenvalue and Eigenvector Derivatives for Algebraic Flutter and Divergence Eigenproblems," AIAA Journal, Vol. 17, 1979, pp. 408-412.
6. Farshad, M., "Variations of Eigenvalues and Eigenfunctions in Continuum Mechanics," AIAA Journal, Vol. 12, 1974, pp. 560-561.
7. Haug, E.J. and Rousselet, B., "Design Sensitivity Analysis in Structural Mechanics II: Eigenvalue Variations," J. Structural Mechanics, Vol. 8, 1980, to appear.
8. Ziegler, H., Principles of Structural Stability, Blaisdell, 1968.
9. Leipholz, H., Direct Variational Methods and Eigenvalue Problems in Engineering, Noordhoff, 1977.
10. Huseyin, K., Vibrations and Stability of Multiple Parameter Systems, Sijthoff & Noordhoff, 1978.
11. Plaut, R.H., "Determining the Nature of Instability in Nonconservative Problems," AIAA Journal, Vol. 10, 1972, pp. 967-968.
12. Claudon, J.-L., "Characteristic Curves and Optimum Design of Two Structures Subjected to Circulatory Loads," Journal de Mécanique, Vol. 14, 1975, pp. 531-543.
13. Weisshaar, T.A. and Plaut, R.H., "Structural Optimization Under Nonconservative Loading," Optimization of Distributed Parameter Structures (Ed. E.J. Haug and J. Cea), Sijthoff & Noordhoff, Alphen aan den Rijn, Netherland, 1980.

INVERSE PERTURBATION METHODS FOR VIBRATION ANALYSIS

Robert E. Sandstrom

Department of Naval Architecture and Marine Engineering
University of Michigan, Ann Arbor, Michigan

ABSTRACT

A linear perturbation method is presented for modification of mode shapes and natural frequencies of structures and machines. Linearization of vibration equations is used to obtain first-order relations between design and response variations. Quadratic programming is used to find the smallest design change that yields the desired response variation.

1. INTRODUCTION

Procedures for determining the vibratory characteristics (natural frequencies and mode shapes) of a structure are well established. Typically, a vibratory analysis is carried out to meet some design specifications. To satisfy these specifications, changes in the vibratory characteristics of the structure are studied by a technique of reanalysis. Normally, changes in the stiffness and mass of the structure are specified and new vibratory characteristics are computed. This can be an expensive procedure. A more desirable approach to this problem of satisfying design specifications would be to specify the required vibratory characteristics and to solve directly for the changes in the stiffness and mass of the structure.

A first order inverse perturbation method is investigated as a possible means of solving this problem, using the finite element method as the basic analysis tool. This method, as proposed by Stetson [1,2] and extended in this paper, produces a linearized relationship between structural changes (stiffness and mass) and modal changes (natural frequency and mode shape). So called inverse perturbations involve the specification of modal changes and the calculation of required structural changes, using this linear relationship.

The idea of using perturbation methods for structural redesign is not considered to be the major contribution of this paper. Stetson has applied this method using his solution scheme for the redesign of a turbine blade [3 to 5]. The goal of his investigation was to adjust the thickness distribution of a turbine blade so that it would meet specified vibratory characteristics, using the NASTRAN finite element program.

The principal contribution of this paper comes from the solution procedure that has been developed for the perturbation equations. Stetson's solution procedure involves substantial computational effort, as explained in Refs. 3 and 5. The solution procedure presented in this paper provides a solution technique that is computationally much easier. This scheme is so simple to use that the application of perturbation methods is now possible with hand calculations, for small problems. Larger problems are easily solved using a computer to assist in the required matrix multiplications. A second benefit of the solution procedure is that it provides a direct indication as to which structural changes would be the most effective in controlling modal changes. It is felt that engineers find it particularly appealing to use such methods, which expose the physical characteristics of the structural system.

In the remaining sections of this paper, the mathematical formulation of the first order perturbation method, as proposed by Stetson, is described. Solution procedures for these equations are developed, followed by an example problem that demonstrates the potential of the method and the proposed solution procedures as an effective engineering tool.

2. MATHEMATICAL DEVELOPMENT

Typically, an eigenvalue analysis is performed on a structural system defined by

$$[m]_{n\times n}\{\ddot{\psi}\}_{n\times 1} + [k]_{n\times n}\{\psi\}_{n\times 1} = \{0\}$$

to produce the eigenvectors

$$[\phi] = [\{\psi_1\},\{\psi_2\}, \ldots,\{\psi_N\}]$$

and the natural frequencies

$$[\omega^2] = \begin{bmatrix} \omega_1^2 & & & 0 \\ & \omega_2^2 & & \\ & & \ddots & \\ 0 & & & \omega_N^2 \end{bmatrix}$$

The equations of motion can be uncoupled using the calculated eigenvectors and natural frequencies

$$[K] = [M]\,[\omega^2]$$

where

$$\text{Generalized Stiffness} = [K] = [\phi]^T[k][\phi]$$

$$\text{Generalized Mass} = [M] = [\phi]^T[m][\phi]$$

Introducing small changes in the system,

$$\text{New Mass} = [m'] = [m] + [\Delta m]$$

$$\text{New Stiffness} = [k'] = [k] + [\Delta k]$$

$$\text{New Frequencies} = [\omega'^2] = [\omega^2] + [\Delta\omega^2]$$

$$\text{New Eigenvectors} = [\phi'] = [\phi] + [\Delta\omega] = [\phi]([I] + [C])^T$$

where $C_{ii} = 0$ and C_{ij} are small. The expression describing the new mode shapes $[\phi']$ is a bit puzzling at first glance. However, by expanding $[\phi']$ one can see the logic behind this expression,

$$[\phi'] = [\{\psi_1'\},\{\psi_2'\},\ldots,\{\psi_N'\}] = [\{\psi_1\},\{\psi_2\},\ldots,\{\psi_N\}] \begin{bmatrix} 1 & C_{21}\cdots & C_{N1} \\ C_{12} & 1 & \vdots \\ \vdots & & \\ C_{1N}\cdots & & 1 \end{bmatrix}$$

where

$$\{\psi_1'\} = \{\psi_1\} + C_{12}\{\psi_2\} + \ldots C_{1N}\{\psi_N\}$$

$$\{\psi_2'\} = C_{21}\{\psi_1\} + \{\psi_2\} + \ldots C_{2N}\{\psi_N\}$$

The new eigenvectors are thus expressed in terms of the original eigenvectors. The C_{ij} terms relate the participation of the jth mode to changes in the ith mode.

The perturbed quantities also satisfy the equations of motion,

$$[K'] = [M'][\omega^{2\prime}]$$

where

$$[K'] = [\phi']^T[k'][\phi']$$

$$[M'] = [\phi']^T[m'][\phi']$$

In this procedure one considers only small perturbations that are of order Δ. Consequently, terms involving Δ^2 and higher are neglected. After a bit of manipulation and neglecting terms of order Δ^2 and higher, one obtains the general form of the first order perturbation equations as

$$[\phi]^T[\Delta k][\phi] - [\phi]^T[\Delta m][\phi][\omega^2] = [\Delta] \qquad (2.1)$$

where

$$[\Delta] \equiv [M][\Delta\omega^2] + [M]\left[[C]^T[\omega^2] - [\omega^2][C]^T\right]$$

This is essentially the same result derived by Stetson [1,2,3,4]. The left hand side of Eq. 2.1 involves only structural changes, while the right hand side involves only frequency and mode shape changes. Solution of Eq. 2.1 is discussed in the next section.

3. SOLUTION OF THE PERTURBATION EQUATIONS

Two uses of the perturbation equations defined by Eq. 2.1 are apparent. First, one could use these equations to solve directly for the changes in natural frequencies and mode shapes, as a result of specifying small changes in the mass and stiffness of the structural system. Second, one could use these equations to determine changes in stiffness and mass that would be required to satisfy specified small changes in the natural frequencies and mode shapes. The first approach is referred to as the forward solution and the second approach is referred to as the inverse solution. This paper addresses the inverse solution problem.

To facilitate the anticipated solution procedures, one expresses the first order perturbation equations of Eq. 2.1 in an alternate form. Note that each Δ_{NK} term is equal to

$$\Delta_{NK} = \{\psi_N\}^T[\Delta k]\{\psi_K\} - \omega_K^2\{\psi_N\}^T[\Delta m]\{\psi_K\} \tag{3.1}$$

where

$$\Delta_{NK} \equiv \begin{cases} M_K \Delta\omega_K^2 & N=K \\ M_N\ C_{KN}(\omega_K^2 - \omega_N^2) & N\neq K \end{cases}$$

The perturbation equations in this form exhibit some interesting characteristics. First, the structural changes $[\Delta k]$ and $[\Delta m]$ are separated from the modal changes $\Delta\omega_K^2$ and C_{KN}. Furthermore, equations corresponding to N=K relate the structural changes to the Kth natural frequency, without involving the mode shape changes. Similarly, equations corresponding to $N\neq K$ relate the structural changes to mode shape changes through the C_{KN} terms, without involving the natural frequency changes. This characteristic is attractive since this formulation permits decomposition of the equations simply in terms of Δk, Δm, $\Delta\omega^2$, and C.

To deal effectively with the inverse solution, one must give the structural changes $[\Delta k]$ and $[\Delta m]$ a practical interpretation. This can be accomplished by decomposing the system changes in L element changes. For the stiffness changes, one obtains

$$[\Delta k]_{SYSTEM} = \sum_{E=1}^{L} [\Delta k_E]$$

Furthermore, each element change can be expressed as a fractional change from the original stiffness,

$$[\Delta k_E] = [k_E]\alpha_E^k$$

where α_E^k represents the fractional change in the stiffness of element E. The superscript k refers to stiffness changes, whereas superscript m refers to mass changes. One now writes

$$[\Delta k]_{SYSTEM} = \sum_{E=1}^{L} [k_E]\alpha_E^k \tag{3.2}$$

and similarly for the masses

$$[\Delta m]_{SYSTEM} = \sum_{E=1}^{L} [m_E]\alpha_E^m \quad (3.3)$$

The system structural changes can therefore be expressed in terms of elemental structural changes. In the implementation of this procedure at an industrial level, it will be necessary to define element changes in a more realistic sense. For example, a general three dimensional beam element contains several stiffness effects.

$$[k_{BEAM}] = [k_{AXIAL}] + [k_{BENDING}] + [k_{TORSIONAL}]$$

Axial effects are controlled by cross sectional area changes, bending effects are controlled by moment of inertia changes, and torsional effects are controlled by polar moment of inertia changes. Furthermore, changes in elemental mass are directly related to cross sectional area changes. With this in mind, one can express beam element changes as

$$[\Delta k_{BEAM}] - \omega^2[\Delta m_{BEAM}] = [k_{AXIAL} - \omega^2 m_{BEAM}]\alpha_{BEAM}^{AREA}$$

$$+ [k_{BENDING}]\,\alpha_{BEAM}^{I} + [k_{TORSION}]\alpha_{BEAM}^{J}$$

Returning to the perturbation equations, one can now use Eqs. 3.2 and 3.3 to relate changes in the mode shapes and natural frequencies to changes in the stiffness and mass of each element of the system,

$$\Delta_{NK} = \{P^k\}^T\{\alpha^k\} + \{P^m\}^T\{\alpha^m\}$$

where

$$P_E^k = \{\psi_N\}^T[k_E]\{\psi_K\}$$

$$P_E^m = -\omega_K^2\{\psi_N\}^T[m_E]\{\psi_K\}$$

The P terms are referred to as perturbation influence terms. These terms provide important information for vibration design. They indicate which structural elements will be most effective in controlling modal changes defined by the Δ_{NK} terms. Large values for P suggest that small structural changes are required to meet design specification. Small values for P, on the other hand, suggest that large structural changes would be required.

Obviously, small structural changes produce the least alteration to the structural system. Small structural changes also allow solutions of higher accuracy, since the development of this method is based on small changes. The perturbation influence terms can also be used to identify structural elements that have no effect on the modal changes. These elements would be characterized by P values close to zero.

In the application of this method one specifies constraints on the modal characteristics of the system through the Δ_{NK} terms. In addition, the user selects a set of possible structure changes α_E. Suppose L structural changes are to be used to satisfy M modal changes. The perturbation equations now become a set of linear algebraic equations that can be expressed in matrix form as

$$\{\Delta_{NK}\}_{M\times 1} = [P^K]_{M\times L}\{\alpha^K\}_{L\times 1} + [P^M]_{M\times L}\{\alpha^M\}_{L\times 1}$$

$$= [P]_{M\times 2L}\{\alpha\}_{2L\times 1} \tag{3.4}$$

The solution of these perturbation equations falls into three categories. They include:

(1) Overconstrained Case - $M>2L$

(2) Unique Solution Case - $M=2L$

(3) Underconstrained Case - $M<2L$

The unique solution case simply involves inverting the perturbation matrix, provided it is not singular. The remaining cases are solved using a quadratic programming algorithm [6], which produces a least structural change solution by minimizing

$$\{\alpha\}^T\{\alpha\}$$

Additional constraints can also be imposed on this system of equations. From a practical standpoint, it may be required that certain elements grow or change at the same rate. This constraint is imposed by modifying the system of equations given above. Suppose, for example, that four elements are involved in a change to satisfy one modal constraint. The perturbation equations can be written as follows:

$$\Delta_{11} = P_1^k\,\alpha_1^k + P_2^k\,\alpha_2^k + P_3^k\,\alpha_3^k + P_4^k\,\alpha_4^k$$
$$+ P_1^m\,\alpha_1^m + P_2^m\,\alpha_2^m + P_3^m\,\alpha_3^m + P_4^m\,\alpha_4^m$$

where

$$P_E^k = \{\psi_1\}^T[k_E]\{\psi_1\}$$

$$P_E^m = -\omega_1^2\{\psi_1\}^T[m_E]\{\psi_1\}$$

If one requires that elements 2 and 3 grow together, i.e.

$$\alpha_2^k = \alpha_3^k$$

$$\alpha_2^m = \alpha_2^m$$

then this equation can be written as

$$\Delta_{11} = P_1^k\ \alpha_1^k + (P_2^k + P_3^k)\alpha_2^k + P_4^k\ \alpha_4^k$$

$$+ P_1^m\ \alpha_1^m + (P_2^m + P_3^m)\alpha_2^m + P_4^m\ \alpha_4^m$$

More generally, this involves adding together the columns in each of the matrices $[P^k]$ and $[P^m]$.

A second constraint that could be imposed on the system involves specifying the magnitude of some structural changes, while the remaining element changes satisfy the conditions of a least structural change. This type of constraint is useful when positive control of the structural changes on a particular element is required. Suppose that the solution from the previous example suggests that a 15% reduction in stiffness for element 4 is required (α_4^k = - .15). Furthermore, suppose that only a 10% reduction can be tolerated. One then fixes the value of α_4^k = - .10 and modifies the perturbation equations as follows:

$$\Delta_{11} - P_4^K\alpha_4^K = P_1^K\alpha_1^K + (P_2^K + P_3^K)\alpha_2^K$$

$$+ P_1^M\alpha_1^M + (P_2^M + P_3^M)\alpha_2^M + P_4\alpha_4^M$$

The least change solution is then used to find the remaining structural changes α_1^k, α_2^k, α_1^M, α_2^M, and α_4^M.

4. EXAMPLE PROBLEM

Consider the torsional analysis of a marine diesel propulsion unit, modeled as shown in Fig. 4.1. This unit exhibited excessive vibration at 150 Hz, which corresponded to the fourth order engine excitation at full power. A torsional analysis indicated a second natural frequency at 154.4 Hz. To eliminate this apparent resonate condition, it was decided to adjust the stiffness of this system so that the natural frequency would be increased to 20% above the full power excitation frequency.

A finite element analysis, using SAPIV, was performed on the original unit to obtain the natural frequencies and mode shapes for the perturbation analysis. Only the results corresponding to the second torsional mode were required for this problem. To apply the perturbation analysis for the second mode frequency change, one only needs to compute the Δ_{22} constraint and the perturbation influence terms corresponding to N = K = 2. The resulting perturbation equation is

$$\Delta_{22} = \sum_{E=1}^{11} P_E^{TORSION} \alpha_E$$

where α_E represents the percentage change in the torsional stiffness of element E and

$$\begin{aligned}
\Delta_{22} &= 338{,}170.0 \\
P_1 &= 243{,}929.0 \\
P_2 &= 161{,}215.0 \\
P_3 &= 14{,}113.0 \\
P_4 &= 513{,}384.0 \\
P_5 &= 72.1 \\
P_6 &= 449.5 \\
P_7 &= 694.4 \\
P_8 &= 525.7 \\
P_9 &= 343.7 \\
P_{10} &= 80.3 \\
P_{11} &= 6{,}936.0
\end{aligned}$$

Computation of the P_E terms is straightforward. For example, P_1 is computed from

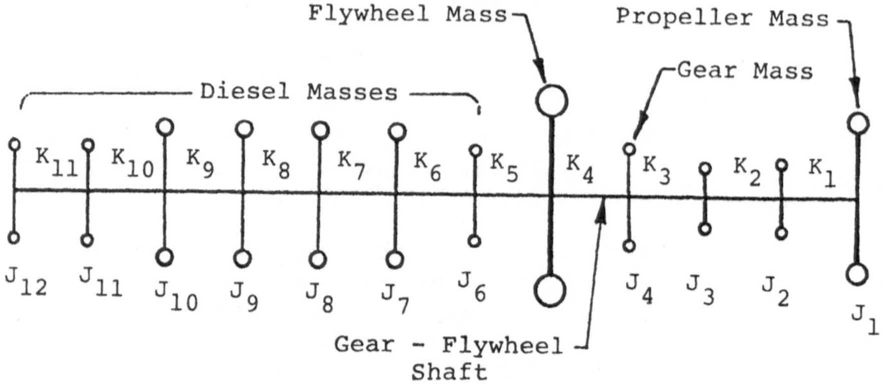

Torsional Mass Elastic Model
for a
Marine Diesel Propulsion Unit

E	J_E Rotational Mass in-lbs/sec^2	K_E Rotational Stiffness in-lbs/RAD
1	7.392	0.319×10^6
2	0.058	0.333×10^6
3	0.058	1.50×10^6
4	0.295	0.192×10^6
5	22.28	215.0×10^6
6	0.42	32.4×10^6
7	0.923	17.16×10^6
8	0.773	17.16×10^6
9	0.773	17.16×10^6
10	0.923	32.40×10^6
11	0.269	0.270×10^6
12	0.241	

Figure 4.1 Marine Diesel Propulsion Unit

$$P_1 = \{\psi_2\}^T [k_1] \{\psi_2\}$$

where

$$[k_1] = K_1 \begin{bmatrix} 1 & -1 \\ -1 & 1 \end{bmatrix}$$

and

$$\{\psi_2\} = \begin{Bmatrix} \text{Rotation at } J_1 = -.0401 \\ \text{Rotation at } J_2 = .8341 \end{Bmatrix}$$

These perturbation terms can be used to determine which parts of the structure have the most influence in satisfying a particular modal constraint. From the list of computed terms, it is evident that elements 1, 2, and 4 have the most influence on Δ_{22}. Substantial changes to the remaining elements would not have a significant effect in satisfying the design constraint Δ_{22}. Consequently elements 3 and 5 through 11 need not be considered ($\alpha_E = 0$ where E = 3,5-11).

Using the quadratic programming procedure for the under-constrained case, as described in Ref. 6, one obtains the following least structural changes:

Element	%Stiffness Increase	New K_E	%Diameter Increase
1	23.6	$.394 \times 10^6$	5.4
2	15.6	$.385 \times 10^6$	3.7
4	49.7	$.287 \times 10^6$	10.6

Since shaft diameter changes are expensive, a decision was made to investigate the effects of increasing the stiffness of only element 4, the gear-flywheel shaft. The perturbation terms clearly show that this element has the greatest influence on Δ_{22}.

The perturbation equations are easily modified to include this single element change, obtaining

$$\Delta_{22} = P_4^{TORSION} \alpha_4 \tag{4.1}$$

which gives

$$\alpha_4 = .66$$

$$K_4^{NEW} = K_4^{ORIGINAL}(1 + \alpha_4)$$

This 66% stiffness increase is accomplished by increasing the diameter of the solid shaft 14%.

The perturbation equation of Eq. 4.1 can be used to illustrate the relationship between structural stiffness and natural frequency, as shown in Fig. 4.2.

The slight curvature shown in this plot indicates that structural changes are actually linearly related to frequency change squared. It should be noted that changes in the system mass associated with small shaft diameter changes were considered negligible in this analysis.

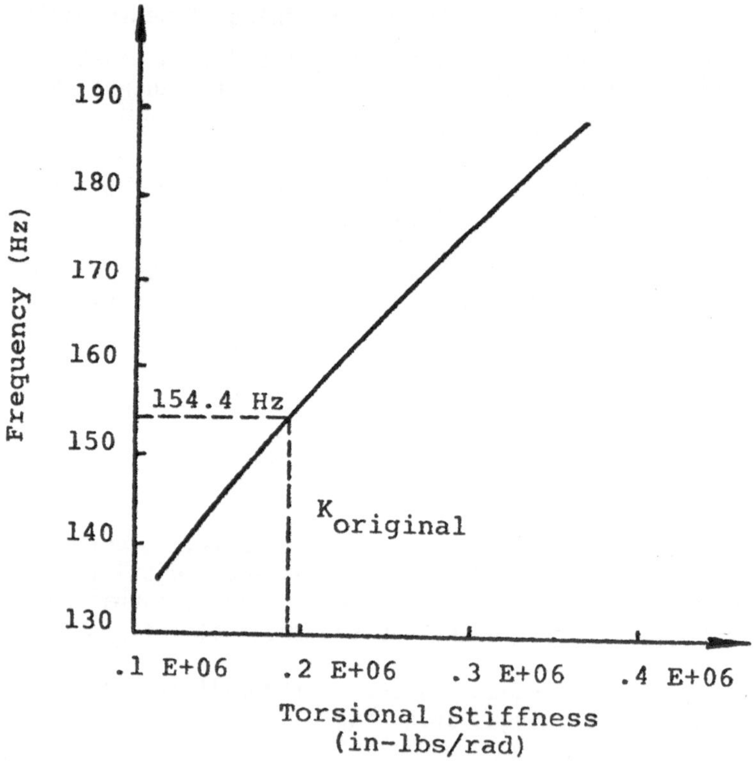

Figure 4.2 Effect on the 2nd Torsional Mode (154.4 Hz) of the Gear - Flywheel Shaft Stiffness

Accuracy of the results is subject to question, since the formulation of this method is based on the assumption of small changes. The least structural change is defined by $\min\{\alpha_E\}^T\{\alpha_E\}$. This definition indicates that alterations to elements 1, 2, and 4 produce smaller structural changes than alterations to element 4 alone. Consequently, this single element change should produce a larger error. This error can be quantified by comparing results obtained from the perturbation analysis with results obtained from reanalysis using the SAPIV finite element program. The results, shown in Table 4.1, show excellent agreement for this simple problem.

TABLE 4.1 COMPARISON BETWEEN REANALYSIS AND PERTURBATION METHODS

Shaft Stiffness (in.-lbs/rad)	Perturbation Method	Reanalysis Method	Difference
.12 E+06	137.8 Hz	137.6 Hz	+0.1 %
.18 E+06	151.8	151.8	small
.24 E+06	164.6	165.2	-0.3
.30 E+06	176.5	176.9	-0.2
.36 E+06	187.7 Hz	188.1 Hz	-0.2 %
Computer Cost	abt. $3.00	abt. $10.00	

The need exists to assess the potential and accuracy of this method for more complex problems. Preliminary investigations have indicated that the solution of some overconstrained problems has led to ridiculous results. Changes requiring mass and stiffness reductions greater than 100% are obviously impossible. Solution procedures that step or iterate toward a design goal appear to be the way to handle these overconstrained problems. This area requires further work. The most attractive results have been obtained for underconstrained problems.

REFERENCES

1. Stetson, K.A., "Perturbation Method of Structural Design Relevant to Holographic Vibration Analysis," AIAA Journal, Vol. 13, No. 4, 1975, pp. 457-459.
2. Stetson, K.A. and Palma, G.E., "Inversion of First Order Perturbation Theory and Its Application to Structural Design," AIAA Journal, Vol. 14, No. 4, 1976.
3. Stetson, K.A., Harrison, I.R., and Cassenti, B.N., Redesign of Structural Vibration Modes by Finite Element Inverse Perturbation, United Technologies Research Center Final Report R78-992945, February, 1978.
4. Stetson, K.A., Harrison, I.R., and Palma, G.E., "Redesigning Structural Vibration Modes by Inverse Perturbation Subject to Minimal Change Theory," Computer Methods in Applied Mechanics and Engineering, 1978, pp. 151-175.
5. Stetson, K.A. and Harrison, I.R., "Redesign of Structural Vibration Modes by Finite Element Inverse Perturbation," ASME Gas Turbine Conference and Product Show, ASME Paper #80-GT-167, New Orleans, March, 1980.
6. Noble, B., Applied Linear Algebra, Prentice-Hall, Englewood Cliffs, NJ, 1969, pp. 142-146.
7. Bathe, K.J. and Wilson, E.L., Numerical Methods in Finite Element Analysis, Prentice-Hall, Englewood Cliffs, NJ, 1976.

OPTIMAL DESIGN FOR ELASTIC BODIES IN CONTACT*

Robert L. Benedict* and John E. Taylor

College of Engineering, The University of Iowa,
Iowa City, Iowa 52240

John E. Taylor, Department of Aerospace Engineering,
College of Engineering, University of Michigan,
Ann Arbor, Michigan

ABSTRACT

It is shown that a Lagrangian multiplier technique can be used to transform the inequality equilibrium relations found for contact problems to equality relations. This result is used to extend Clapeyron's Theorem to contact problem cases. Using the extended Clapeyron's Theorem it was shown that receding contact problems have a linear load-response relation, a result shown previously by other means by Dundurs, and that advancing problems have a stiffening nature, a new result. An efficient numerical contact problem solution technique, based on a direct minimization of the potential energy function, by a constrained quadratic programming algorithm is developed and demonstrated on several examples. The program is capable of solving contact problems involving bodies of arbitrary shape.

A relation between the minimum stiffness and the minimum-maximum contact stress initial separation designs is developed. A numerical technique for problem solution is formulated and several solutions displayed. Use of the volume of separation in controlling designs is demonstrated. Although the technique is

*Based upon a dissertation submitted in partial fulfillment of the requirements of the Ph.D. degree. Work supported in part by the National Science Foundation.

derived assuming smooth surfaces, a non-smooth initial surface is found to cause no difficulties. The high sensitivity of design to variations in load is noted and a recommendation is made for alleviating this problem.

1. INTRODUCTION

On division of mechanics problems is into a class that has unilateral (one-sided) boundary conditions and a class with the more common bilateral (two-sided) boundary conditions. Optimal design theory for problems with bilateral boundary conditions has been well developed in the last thirty years. The two main approaches to solutions, a direct minimization of the functional and the development of optimality criteria, have both received wide attention. In contrast, it is only recently that structural design problems with unilateral constraints have been attempted by the direct method [1,2] and apparently the optimality criteria approach has been undeveloped.

The purpose of this investigation is to develop optimality criteria for a class of optimal design problems for systems governed by unilateral constraints. Problems involving two elastic bodies in unbonded contact are considered. The nonpenetration constraint that is central to the nature of these problems imposes a unilateral constraint on the displacement fields.

A contact problem formulation that is well suited to optimal design techniques is developed. This formulation is used to derive optimality criteria for design problems involving body contour design. A numerical technique for problem solution is developed and demonstrated with several computational examples.

In order to develop general optimality criteria for bodies in contact, one needs a contact problem formulation that provides information about the nature of the solution. To enhance physical insight, the problem is formulated in the continuum mechanics realm and an energy principle approach is used. After development of the optimality criteria, numerical solutions are sought to problems posed in the continuum setting. The use of energy principles allows for the rigorous derivation of finite element solution techniques.

Neither functional analysis nor variational calculus has produced a minimum principle to cover the Couloumb friction case, in which the surface tangential stress bound is a function of the normal stress. All approaches to data have been limited to requiring that the tangential stress bound is provided a priori. Problems have been solved numerically by iteration, in which the tangential stress bound is related to the normal stress result from the previous iteration [3,4].

Realizing this limitation, the rest of this work is concerned with frictionless contact. Consideration is then restricted to linearly elastic bodies that undergo small deformations under quasi-static loading. Referring to Fig. 1.1, surface tractions are limited to Body 1 and are not allowed in the contact region. With these restrictions, one may express the potential energy of the system as

$$\pi(u) \equiv \int_{\Omega^1} \frac{1}{2} \tau^1_{ij} \varepsilon^1_{ij} \, dv - \int_{\Omega^1} X^1_i u^1_i \, dv - \int_{\Gamma^1_p} u^1_i p_i \, ds$$

$$+ \int_{\Omega^2} \frac{1}{2} \tau^2_{ij} \varepsilon^2_{ij} \, dv - \int_{\Omega^2} X^2_i u^2_i \, dv \, .$$

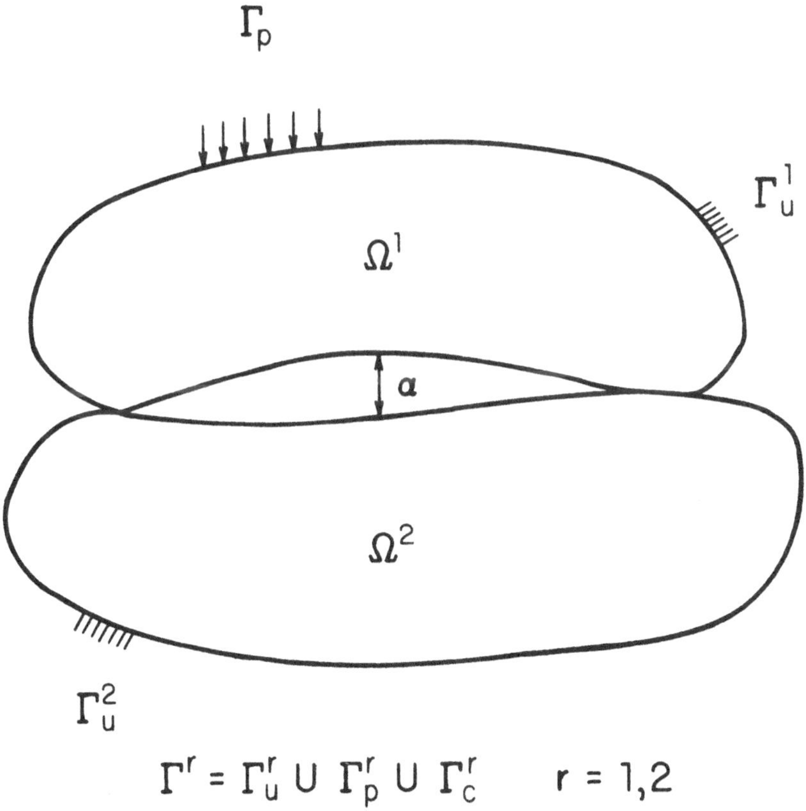

Figure 1.1 Two Elastic Bodies in Contact.

where u_i^k, τ_{ij}^k, ε_{ij}^k, and X_i^k are displacement, stress, strain, and body force in body $k(k = 1,2)$ and p_i is surface traction applied to body 1.

One may identify Γ_c^1 and Γ_c^2 as the regions on the surface of Body 1 and Body 2 where contact may occur. These are chosen to be sufficiently large to contain the actual contact regions in the problem solution. Consistent with the small deformations assumption, the initial gap α between the bodies in the unloaded state is considered to be sufficiently small that the regions Γ_c^1 and Γ_c^2 are indistinguishable. Using this pointwise equivalence, one need only refer to a single potential contact region Γ_c. One may now impose the nonpenetration constraint by defining a unilateral constraint function ϕ on Γ_c as

$$\phi(u) \equiv u_n^1 + u_n^2 - \alpha \leq 0$$

From the principle of minimum potential energy, the solution displacement field u of the contact problem is the field that solves the following problem (called P1):

$$\min_{u \in K} \pi(u) \qquad \text{subject to} \qquad \phi(u) \leq 0 \qquad \text{P1} \qquad (1.1)$$

where K is defined as the convex set of continuous functions u_i^1 and u_i^2 that satisfy bilateral kinematic boundary conditions. Since the constraint function is linear in the displacement fields, the subset $\hat{K}$ of K defined as

$$\hat{K} \equiv \{u \mid \phi(u) \leq 0 \;\text{ and }\; u \in K\}$$

is also a convex set. Then, $\hat{K}$ comprises the set of admissible displacement fields for the problem P1 of Eq. 1.1.

Problem formulation P1 is the most common variational inequality form for the contact problem statement. Conditions for the existence and uniqueness of solutions have been developed by Boucher [5]. When one body is considered rigid, existence and uniqueness proofs have been developed by Fichera [6] and uniqueness proofs by Kalker [7] and Panagiotopolous [8]. As a result of the inequality constraint on the kinematic admissiblity of the displacement field, the necessary conditions (equilibrium relations) of problem P1 include inequalities. In order to obtain equality relations in the displacement relations, an alternative formulation is used in the development that follows. A Lagrange multiplier is introduced to append the nonpenetration constraint to the potential energy functional. Kinematic admissibility is then defined by the set K, i.e. in terms of the usual requirements for problems without unilateral constraints. The variations in the displacement quantities are than free and lead to equality relations. The appended constraint and the Lagrangian functional are

$$\langle \lambda, \phi \rangle \equiv \int_{\Gamma_c} \lambda\phi \, ds \tag{1.2}$$

and

$$L(u,\lambda) \equiv \pi + \langle \lambda, \phi \rangle \tag{1.3}$$

Using a generalized Lagrange multiplier relation [9], one may relate a problem formulated with L to problem P1 of Eq. 1.1 as follows:

$$\left.\begin{array}{l} \text{if} \quad \lambda^* \geq 0, \quad u^* \in K \\ \text{and for all } \lambda \geq 0 \quad \text{and} \quad u \in K, \\ \qquad L(u^*,\lambda) \leq L(u^*,\lambda^*) \leq L(u,\lambda^*) \end{array}\right\} \quad P2 \tag{1.4}$$

then u* solves

$$\min \pi(u) \quad \text{subject to}$$

$$\phi(u) \leq 0$$

$$\text{for } u \in K$$

The Lagrangian problem formulation P2 has a saddle point at the solution of problem P1. One may identify λ as the surface contact stress, i.e. the force that holds the constraint [10].

So far, no smoothness assumptions have been made about the contact surfaces. Experience with contact problem solutions indicates that one may expect singularities in λ and in body stress and strain solution quantities. To avoid these singularities it is now required that the contact surfaces be smooth. That this smoothness is sufficient to prevent the occurrence of singularities is not proven here. However, for the Signorini contact problem, Fichera [6] has given a proof. The body of existing contact problem solutions also supports this assumption.

With ϕ continuous and differentiable and λ, τ_{ij}, and ε_{ij} nonsingular, the integrals arising in the problem P2 are all proper.

Necessary conditions for a solution of the contact problem are thus

if $u^* \in K$ minimizes $\pi(u)$

subject to

$$\phi(u) \leq 0$$

then

1. $\lambda^* \geq 0$ exists
2. $L(u,\lambda^*)$ is stationary at u^*
3. $\int_{\Gamma_c} \lambda^* \phi \, ds = 0$.

Stationary conditions for L may be derived explicitly by noting that for linear strain-displacement relations, $\delta\varepsilon_{ij}$ is given by

$$\delta\varepsilon_{ij} = \left(\frac{u_{i,j} + u_{j,i}}{2}\right) = \frac{1}{2}\left((\delta u_i)_{,j} + (\delta u_j)_{,i}\right)$$

Using symmetry of the stress tensor and the divergence theorem yields

$$\delta L = \int_{\Omega^1} (-\tau^1_{ij,j} - X^1_i)\, \delta u^1_i \, dv + \int_{\Omega^2} (-\tau^2_{ij,j} - X^2_i)\, \delta u^2_i \, dv$$

$$+ \int_{\Gamma^1_p} (\tau^1_{ij}\, \nu^1_j - p_i)\, \delta u^1_i \, ds + \int_{\Gamma^1_c} (\tau^1_{ij}\, \nu^1_j + \lambda \nu^1_i) \delta u^1_i \, ds$$

$$+ \int_{\Gamma^2_c} (\tau^2_{ij}\, \nu^2_j + \delta \nu^2_i)\, \delta u^2_i \, ds + \int_{\Gamma^1_u} \tau^1_{ij}\, \nu^1_j\, \delta u^1_i \, ds$$

$$+ \int_{\Gamma^2_u} \tau^2_{ij}\, \nu^2_j\, \delta^1_i \, ds .$$

Since the variations in u are all free, except over Γ_u, where they are zero, setting $\delta L = 0$ implies

$$\tau^1_{ij,j} + X^1_i = 0 \quad \text{in} \quad \Omega^1$$

$$\tau^2_{ij,j} + X^2_i = 0 \quad \text{in} \quad \Omega^2$$

$$\tau^1_{ij}\, \nu^1_j - p_i = 0 \text{ on } \Gamma^1_p \tag{1.5}$$

$$\tau^2_{ij}\, \nu^2_j = 0 \quad \text{on} \quad \Gamma^2_p$$

$$\tau^1_{ij}\, \nu^1_j + \lambda\nu^1_i = 0 \tag{1.6}$$

$$\tau^2_{ij}\, \nu^2_j + \lambda\nu^2_1 = 0$$

Also, since $\lambda \geq 0$ and $\phi \leq 0$, the relation

$$\int_{\Gamma_c} \lambda\phi \; ds = 0$$

implies

$$\lambda\phi = 0 \tag{1.7}$$

Equations 1.5 and 1.6 are the usual equilibrium equations and boundary conditions with the addition of λ as a normal surface traction. The condition of Eq. 1.7 shows that whenever the contact stress λ is nonzero the constraint must be tight, which indicates that the bodies are in contact. Whenever ϕ is non-zero, the bodies do not touch and λ must be zero.

An extension of Clapeyron's theorem to frictionless contact problems is to be derived. Multiplying Eqs. 1.5 by their respective displacement fields and integrating over the appropriate regions yields

$$\int_{\Omega^1} (\tau^1_{ij,j} + X^1_i)\, u^1_i \; dv + \int_{\Omega^2} (\tau^2_{ij,j} + X^2_i)\, u^2_i \; dv = 0$$

Expanding and using symmetry of the stress tensor and the divergence theorem, this may be written in the form

$$-\int_{\Omega^1} \tau^1_{ij}\, \varepsilon^1_{ij}\, dv - \int_{\Omega} \tau^2_{ij}\, \varepsilon^2_{ij}\, dv + \int_{\Gamma^1} \tau^1_{ij}\, \nu^1_j\, u^1_i\, ds$$

$$+ \int_{\Gamma^2} \tau^2_{ij}\, \nu^2_j\, u^2_i\, ds + \int_{\Omega^1} X^1_i\, u^1_i\, dv + \int_{\Omega^2} X^2_i\, u^2_i\, dv = 0 \,.$$

Applying boundary conditions of Eq. 1.6 to the boundary integral terms yields

$$-\int_{\Omega^1} \tau^1_{ij}\, \varepsilon^1_{ij}\, dv - \int_{\Omega^2} \tau^2_{ij}\, \varepsilon^2_{ij}\, dv + \int_{\Gamma_p} p_i\, u^1_i\, ds$$

$$+ \int_{\Omega^1} X^1_i\, u^1_i\, dv + \int_{\Omega^2} X^2_i\, u^2_i\, dv - \int_{\Gamma_c} \lambda(u^1_i\, \nu^1_i + u^2_i\, \nu^2_i)\, ds = 0 \qquad (1.8)$$

Defining

$$\tilde{W} \equiv \int_{\Gamma_c} \lambda(u^1_n + u^2_n)\, ds$$

one may write Eq. 1.8 as

$$2U - W + \tilde{W} = 0 \qquad (1.9)$$

which is the desired extension of Clapeyron's Theorem to the contact problem.

Recall that when λ is nonzero, the constraint must be tight, i.e.

$$u^1_n + u^2_n - \alpha = 0$$

One may then express $\tilde{W}$ as

$$\tilde{W} = \int_{\Gamma_c} \alpha\lambda\, ds \qquad (1.10)$$

which shows that

(i) $\tilde{W} \geq 0$

(ii) if $\alpha = 0$ whenever $\lambda > 0$ then $\tilde{W} = 0$

Dundurs [11] has divided contact problems into two categories. Those in which the loaded contact region lies within the unloaded contact region are called "receding" and those in which the loaded contact region anywhere exceeds the unloaded contact region are called "advancing". Condition (ii) then is met for all receding contact problems. That is, if α is zero everywhere that λ is strictly positive, then the loaded contact region must lie within the unloaded contact region and the problem is "receding". A nonzero $\tilde{W}$ then corresponds to an "advancing" contact problem.

From the definition of the compliance term W, one recalls that for linear response of a conservative system

$$\text{Work} = W/2 \tag{1.11}$$

For a single point load p acting through a displacement u, this takes the form

$$\int_0^{u_0} p(u)\, du = \frac{1}{2}\, p(u_0)u_0$$

The conservation of energy and Eq. 1.11 together imply Clapeyron's theorem.

However, for the contact problem Eq. 1.9 is

$$2U - W + \tilde{W} = 0$$

so

$$2U = W - \tilde{W}$$

and one has

$$\text{Work} + \tilde{W}/2 = W/2$$

For that point load case,

$$\int_0^{u_0} p(u)\, du + \frac{1}{2}\tilde{W} = \frac{1}{2}\, p(u_0)u_0$$

$$= \int_0^{u_0} \frac{p(u_0)}{u_0}\, u\, du$$

For receding contact problems, $\tilde{W} = 0$ implies Eq. 1.11 and that the load-response curve is linear. For advancing contact problems, one sees that $\tilde{W} > 0$ implies that

$$\int_0^{u_0} p(u)\, du < \int_0^{u_0} \frac{p(u_0)}{u_0}\, u\, du$$

so that $p(u) < \frac{p(u_0)}{u_0} u$ over some region. Thus, the actual load-response curve must at some point lie below a linear curve connecting the initial and final load states. This is the definition of strictly convex function, i.e. $p''(u) > 0$ for some region around a point where $p(u) < \frac{p(u_0)}{u_0} u$. The actual load-response then shows a stiffening nature in some region. This is shown graphically in Fig. 1.2, which is the plot of a hypothetical load-response curve for a system with a single point load.

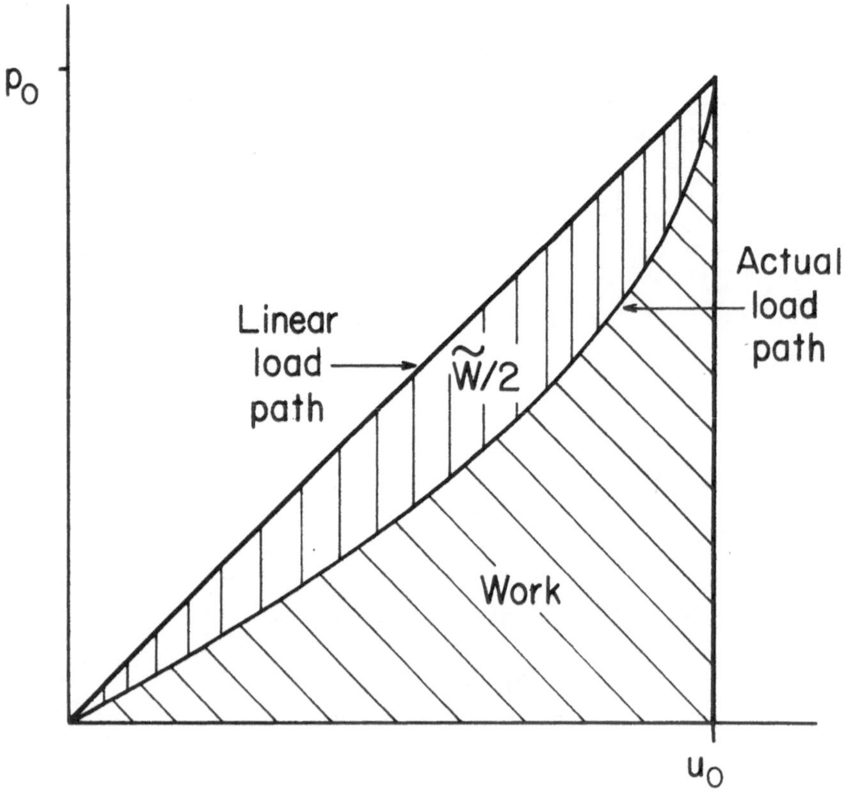

Fig. 1.2 $W/2 = \tilde{W}/2 + \text{Work}$.

That the load-response curve is linear for receding contact has already been shown by Dundurs [11], using other means. The stiffening nature of the load-response curve for the advancing case is believed to be a new result. It is further conjectured that the ratio $\tilde{W}/W$ may be a useful measure of the nonlinearity of the system. This conjecture will be discussed again in the results section.

Taking a functional analysis approach, Fichera [7] proved the existence of a very general measure for the Signorini problem, which yields information about the global nature of the reaction force the contact region exerts on the body. The $\tilde{W}$ term derived here by a Lagrangian approach may be considered as an explicit representation of such a measure for the contact problem involving two elastic bodies. The two analyses differ, in that Fichera's measure is not expressed directly in terms of the contact stress and he does not appear to have considered his measure in light of the energies associated with problem solutions.

2. INITIAL SEPARATION DESIGN

The contour design problem involves designing the initial separation over the possible contact region, with the objective of minimizing the maximum contact stress on the loaded bodies. The nature of the solution of this problem is discussed. A formulation is presented for an associated design problem whose development provides the basis for the numerical techniques developed in Section 3. To be consistent with the problem formulation, the initial separation design function α is restricted to be small and smooth. As a consequence of the restriction to small separations one may consider the stiffness of the bodies to be unchanged by the design process.

The problem may be stated as

$$\min_{\alpha} \ \sup_{\underline{x}} \ \lambda^*(\underline{x}) \qquad \text{P3} \qquad (2.1)$$

where λ^* is the contact problem solution stress for design α. Using this formulation Conry [2] and Haug and Kwak [1] studied the discretized problem by a direct programming attack.

Constraints on the size of the separation and the region over which it could exist were included in Ref. 1. In each case considered, the unconstrained problem had as a solution a design that caused a constant contact stress everywhere. Problems with a constraint limiting the region of design had as solutions designs that caused a constant contact stress over the allowed design region.

In analyzing problem P3 in the continuum setting here, an alternative approach is used. Rather than attempting to establish necessary conditions for the solution design of P3, the solution contact stress will be derived. An associated problem, whose necessary conditions require that its solution contact stress be identical to that of P3 is developed. A design that solves the associated problem will then also solve P3.

Insight into the problem can be gained by considering the following simpler problem:

$$\min_{y} \left[I \equiv \sup y(x) \right] \quad , \quad x \in (0,\ell), \quad y \in L^1$$

$$\text{subject to} \int_0^\ell y(x)\, dx = 1$$

As pointed out by Masur and Mroz [17] the obvious solution is

$$y_0 = 1/\ell$$

This solution y_0 does not make I stationary, since I is not differentiable at the solution. The unilateral nature of the supremum operator causes the first variation of I to be positive, for all variations about y_0.

The problem P3 has a similar nature. Global equilibrium requires that, for any contact problem solution, the contact stress must equilibriate the applied load, i.e.

$$\int_{\Gamma_c} \lambda \nu_i^2 b_i \, ds = \int_{\Gamma_p} p_i b_i \, ds + \int_{\Omega_1} X_i b_i \, dv = C_0 \qquad (2.2)$$

Then, the solution contact stress for problem P3 is

$$\lambda = \frac{C_0}{\text{measure } \Gamma_c} \qquad (2.3)$$

Note that this is the result found numerically by the discretized approaches mentioned. The assumption that contact stess of Eq. 2.3 is the solution stress distribution for P3 was also used by Lukasiewicz [18] in producing closed form analytical solutions for problems with simple geometries.

As an alternative approach to problem P3, consider the problem

$$\left.\begin{aligned} &\min_{\alpha} \pi(u(\alpha)) \\ &\text{subject to} \int_{\Gamma_C} \alpha \, d\underline{x} = V \end{aligned}\right\} \quad P4 \qquad (2.4)$$

where $u(\alpha)$ is the contact problem solution displacement field for a given design α. The interpretation of P4 is to minimize the stiffness of the system by choosing an initial separation from an isoperimetrically constrained set. The isoperimetric constraint on the resource here requires a constant volume between the bodies. By specifying this volume the separation between the bodies can be controlled in a global way. We will show that an optimality criteria for the problem P4 is that the contact stress satisfying Eq. 2.3 for a particular value of V. The solution design of problem P4 is then also the solution of problem P3.

<u>Solution of Problem P4</u>. Without the isoperimetric constraint on α, problem P4 is not well posed. Design solutions would admit the addition of an arbitrary constant to α, which corresponds to a rigid body displacement of Body 1. The adjointment of the isoperimetric constraint casts the problem as one of optimal remodel designs. Remodel design theory [14] is an approach to optimization problems that predicts modification, by a globally prescribed amount, to a specified reference design. The more restrictive nature of remodel design insures that the problem will remain well posed.

Consider a reference system of two bodies that are initially in contact everywhere in Γ_C. A remodel means in this case, a system that differs from the reference system by the addition of subregions in Γ_C over which some initial separation exists. The volume of initial separation is limited by the isoperimetric constraint. To be physically meaningful, the initial separation is restricted to be nonnegative. Design constraints are added to the contact problem functional L by again using Lagrangian multipliers to produce a few functional

$$F = L + \Lambda \left(\int_{\Gamma_C} \alpha \, ds - V \right) + \int_{\Gamma_C} - \gamma\alpha \, ds \qquad (2.5)$$

Problem P4 can now be expressed as

$$\min_{\alpha} \max_{\lambda} \min_{u} F \tag{2.6}$$

Necessary conditions for the problem

$$\max_{\lambda} \min_{u} F$$

are identical to those for

$$\max_{\lambda} \min_{u} L$$

and provide equilibrium equations for the contact problem for any given α. Necessary conditions for

$$\min_{\alpha} F$$

are:

$$\lambda(\underline{x}) = \Lambda - \gamma(\underline{x}) \tag{2.7}$$

$$\int_{\Gamma_c} \alpha \, ds = V \tag{2.8}$$

$$\alpha \geq 0 \tag{2.9}$$

$$\gamma\alpha = 0 \tag{2.10}$$

$$\gamma \geq 0 \tag{2.11}$$

At points where the constraint of Eq. 2.9 is not active, the value of γ must be 0 and λ equals the constant Λ. That is, wherever initial separation exists in the optimum design the contact stress is uniform. Where the constraint of Eq. 2.9 is active, γ is then less than Λ. Thus where there is no initial separation, the contact stress must be less than or equal to the contact stress in the remodeled regions. Λ then bounds the contact stress.

To see that these conditions are sufficient to minimize the stiffness of the system, the solution for design α_0, which satisfies Eqs. 2.7 through 2.11, is compared with the solution for any other design that satisfies Eqs. 2.8 through 2.11 only. Let u again denote the complete solution displacement field u_i^1, u_i^2. This solution displacement field minimizes L:

$$L(\alpha_0, \lambda_0, u_0) \le L(\alpha_0, \Lambda_0, u) \tag{2.12}$$

$$\pi(u_0) + \tilde{W}(\lambda_0, u_0) - \int_{\Gamma_c} \lambda_0\, \alpha_0\, ds$$

$$\le \pi(u) + \tilde{W}(\lambda_0, u) - \int_{\Gamma_c} \lambda_0\, \alpha_0\, ds$$

$$\pi(u) - \pi(u_0) \ge \tilde{W}(\lambda_0, u_0) - \tilde{W}(\lambda_0, u) \tag{2.13}$$

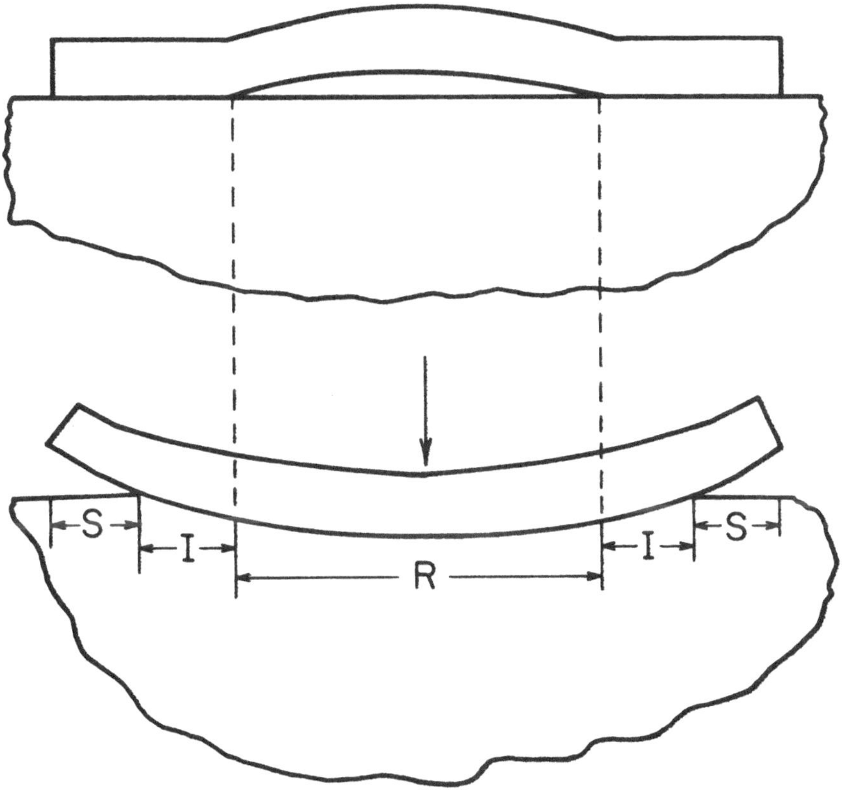

R ≡ region of initial separation in final contact
I ≡ region of no initial separation in final contact
S ≡ region of no initial separation not in final contact

Figure 2.1 Γ_c Subregions.

Thy necessary conditions of Eqs. 2.7 through 2.11 are then sufficient, if it can be shown that the right hand side of Eq. 2.13 is nonnegative for all possible choices of α that meet the design constraints. In order words, the objectives is to show that

$$\int_{\Gamma_c} \lambda_0(u_{n0}^1 + u_{n0}^2)\, ds - \int_{\Gamma_c} \lambda_0(u_n^1 + u_n^2)\, ds \geq 0 \tag{2.14}$$

One can divide Γ_c into contact and noncontact subregions and further divide the contact subregion into remodeled and reference design subregions. This is shown in Fig. 2.1 for a simple beam problem. The integrals in Eq. 2.14 can then be broken into sums of integrals over the intersections of the various subregions. The necessary conditions are used to evaluate the integrals in these subregions. Table 2.1 gives values for variables in the various subregions.

TABLE 2.1 VALUES OF VARIABLES ON SUBREGIONS

Region	λ_0	α_0	$u_{0n}^1 + u_{0n}^2$
R_0	Γ	≥ 0	α_0
I_0	$\Gamma - \gamma$	0	0
S_0	0	0	≤ 0
Region	λ	α	$u_n^1 + u_n^2$
R	λ	≥ 0	α
I	λ	0	0
S	0	0	≤ 0

Both integrals in Eq. 2.14 will vanish over the noncontact regions S and S_0. Considering the remaining 6 subregions, Eq. 2.14 is expanded as:

$$\int_{II_0} \lambda_0(u_{n0}^1 + u_{n0}^2)\, ds + \int_{IR_0} \lambda_0(u_{n0}^1 + u_{n0}^2)\, ds$$

$$+ \int_{RI_0} \lambda_0(u_{n0}^1 + u_{n0}^2)\, ds + \int_{RR_0} \lambda_0(u_{n0}^1 + u_{n0}^2)\, ds$$

$$+ \int_{SI_0} \lambda_0(u_{n0}^1 + u_{n0}^2)\, ds + \int_{SR_0} \lambda_0(u_{n0}^1 + u_{n0}^2)\, ds$$

$$- \Bigg[\int_{II_0} \lambda(u_n^1 + u_n^2)\, ds + \int_{IR_0} \lambda(u_n^1 + u_n^2)\, ds$$

$$+ \int_{RI_0} \lambda(u_n^1 + u_n^2)\, ds + \int_{RR_0} \lambda(u_n^1 + u_n^2)\, ds$$

$$+ \int_{SI_0} \lambda(u_n^1 + u_n^2)\, ds + \int_{SR_0} \lambda(u_n^1 + u_n^2)\, ds \Bigg]$$

$$= \int_{R_0} \Gamma\, \alpha_0\, ds - \Bigg(\int_{RI_0} (\Gamma-\gamma)\, \alpha\, ds + \int_{RR_0} \Gamma\, \alpha\, ds$$

$$+ \int_{SI_0} (\Gamma-\gamma)\, (u_n^1 + u_n^2)\, ds + \int_{SR_0} (u_n^1 + u_n^2)\, ds \Bigg)$$

$$= \int_{R_0} \Gamma\, \alpha_0\, ds - \Bigg(\int_{R} \Gamma\, \alpha\, ds - \int_{RI_0} \gamma\, \alpha\, ds$$

$$+ \int_{SI_0} (\Gamma-\gamma)\, (u_n^1 + u_n^2)\, ds + \int_{SR_0} \Gamma(u_n^1 + u_n^2)\, ds \Bigg)$$

$$= \Gamma \left(\int_{R_0} \alpha_0\, ds \right) - \int_{R} \alpha\, ds + \int_{RI_0} \gamma\, \alpha\, ds$$

$$- \int_{SI_0} (\Gamma-\gamma)\, (u_n^1 + u_n^2)\, ds - \int_{SR_0} \Gamma(u_n^1 + u_n^2)\, ds$$

From the isoperimetric constraint, the terms in parentheses cancel. Of the remaining three terms, the first is always nonnegative and the remaining two nonpositive, since γ, α, and λ are nonnegative and $u_n^1 + u_n^2 \leq 0$ in region S. Thus their sum is nonnegative, whereby the result π the same notation as

$$\pi(u) \geq \Pi(u_0)$$

is proved.

For the case in which $\alpha > 0$ over the interior of Γ_C, optimality criteria of Eq. 2.7 implies that

$$\lambda = \Lambda$$

This is the solution contact stress distribution for problem P3. Then for values of V sufficiently large for the entire contact region to be remodeled, the solutions for problems P3 and P4 are identical. For smaller values of V, for which the remodeled region is less than all of Γ_C, Eq. 2.7 shows that the contact stress is constant over the remodeled region and smaller than this constant elsewhere. The isoperimetric constraint then limits the design to one that lowers the maximum contact stress, but does not produce an absolute minimum. It should be noted that the numerical results of Haug and Kwak [2] for problem P3, with limits on the region of initial separation, produced contact stress distributions that meet all of the optimality criteria derived here for problem P4.

A relation has now been established between a point-wise objective functional and a global objective functional. Taylor [15] has shown that, for the usual bilateral boundary condition material distribution problem, there is an equivalence of this sort between the pointwise objective of minimizing the upper bound on specific strain energy and the global objective of maximizing the stiffness of the structure.

In addition to shedding light on the mechanics of contact problems, there is a computational interest in the relation between problems P3 and P4. Having in mind the objective of reducing the maximum contact stress one may produce designs by an efficient numerical procedure that directly minimizes the stiffness of the system. A computational program based on this approach is presented in Section 3.

3. NUMERICAL FORMULATIONS

3.1 Contact Problem Solution

A numerical method for solution of frictionless contact problems is first developed. A Finite Element approximation for the potential energy of both bodies is formed, using standard techniques. Recall that the solution displacement field is the solution of

$$\min_{u} \pi(u), \quad u \in \hat{K} \tag{3.1}$$

It is usual in the bilateral boundary case to find u by solving the necessary conditions of Eq. 3.1. Since the necessary conditions involve inequalities for contact problems, one may directly minimize the potential energy on the constrained set.

Discretization of the bodies is done on much the same basis as in the usual Finite Element solution techniques, with one exception. In the contact region it is necessary to make a one to one correspondence between nodes in the bodies, so that the nonpenetration constraint may be applied. In these problems, it is assumed that the resultant of the applied loads is normal to the contact surface and is at the same time a coordinate direction. Equation 3.1 may be expressed in discretized form as

$$\left.\begin{aligned} &\min_{u^1,u^2} \; \frac{1}{2} \underline{u}^{1^T} K^1 \underline{u}^1 - \underline{p}^T \underline{u}^1 + \frac{1}{2} \underline{u}^{2^T} K^2 \underline{u}^2 \\ &\text{subject to} \quad u_i^1 + u_i^2 \le \alpha_i \qquad i \in \Gamma_c \end{aligned}\right\} \tag{3.2}$$

where Γ_c is now the set of indices that correspond to normal displacements in the possible contact region. Kinematic admissibility of $\underline{u}^1$, $\underline{u}^2$ is assured by the usual technique of modifying the stiffness matrix. Simplifying notation by imbedding yields

$$\underline{u} \equiv \begin{Bmatrix} \underline{u}^1 \\ \underline{u}^2 \end{Bmatrix} \qquad K \equiv \begin{bmatrix} K^1 & 0 \\ 0 & K^2 \end{bmatrix}$$

The following linear transformation of coordinates proves to be useful:

$$z_i \equiv u_i^1 + u_i^2 \quad i \in \Gamma_C$$
$$z_j \equiv u_i^1 - u_i^2 \quad j \equiv \text{index of } u_i^2 \text{ in } \underline{u} \tag{3.3}$$

Let B be the matrix of this transformation and A its inverse. Transforming from $\underline{u}$ to $\underline{z}$,

$$B\underline{u} = \underline{z}$$

with

$$A = B^{-1}$$

gives

$$\underline{u} = A\underline{z} \tag{3.4}$$

In the new variable,

$$\underline{u}^T K \underline{u} = (A\underline{z})^T K A \underline{z} = \underline{z}^T A^T K A \underline{z} = \underline{z}^T \hat{K} \underline{z}$$

$$\underline{p}^T \underline{u} = \underline{p}^T A \underline{z} = \hat{\underline{p}}^T \underline{z}$$

Now, Eq. 3.2 may be written as

$$\min_{\underline{z}} \frac{1}{2} \underline{z}^T \hat{K} \underline{z} - \hat{\underline{p}}^T \underline{z}$$
$$\text{subject to} \quad z_i \leq \alpha_i \quad i \in \Gamma_C \tag{3.5}$$

This is a constrained quadratic programming problem. It differs from the formulation of Conry [3] and others, in that their formulation is written in terms of stress quantities in the contact region. Conry's formulation also requires a presolving for the global flexibility matrices of the bodies. The transformation used here affects only the constrained displacement variables. The transformation matrices then are sparse and the transformations are inexpensive.

O'Leary [16] has developed an effective algorithm for solution of the problem expressed by Eq. 3.5. This conjugate gradient technique has been used by O'Leary and Yang [17] to efficiently minimize the constrained function, arising from variational inequalities associated with the elastic-plastic torsion problem. It has the capability of constraining variables between a pair of constant bounds. The program requires an initial feasible starting point and generates a set of improved feasible iterations, until all components of the gradient are smaller than a specified error tolerance.

An outline of the overall solution technique is:

(1) Form K, $\underline{p}$

(2) Transform K, $\underline{p}$ to $\hat{K}$, $\hat{\underline{p}}$

(3) $\min_{\underline{z}} \frac{1}{2} \underline{z} \hat{K} \underline{z} - \hat{\underline{p}}^T \underline{z}$

subject to $z_i \leq \alpha_i, \quad i \in \Gamma_c$

(4) Transform $\underline{z}$ back to $\underline{u}$

The minimization in step (3) is carried out by the gradient projection quadratic programming technique mentioned. Better results are obtained if the minimizing algorithm is run twice successively, with a decrease in the error tolerance between the runs.

Two programs were written for two separate types of problems. The simplest problem that demonstrates the features of contact problems is a beam on a Winkler type foundation. The matrix K^1 was assembled by summing a supplied quadratic beam element stiffness matrix. Since a Winkler material supports no shear, K^2 is simply a diagonal matrix of stiffness coefficients. The set Γ_c contains the beam deflections and their corresponding foundation deflection variables. Other problems considered are those of plane stress. Linear triangular elements are used to discretize both bodies. The program generated K^1 and K^2 for given data that specifies nodal coordinates and connectivity. Both programs are written for loads whose resultant lies along one axis, in a two-dimensional cartesian coordinate system.

3.2 Initial Separation

For the problem with initial separation, one seeks a solution to the contact stress distribution problem, formulated as a stiffness minimization problem. In the discretization of the contact problem one now requires Γ_c to be divided into equal areas. The isoperimetric constraint may be approximated by

$$\sum_i A \alpha_i = V \tag{3.7}$$

where A is the contact surface area of an element. Equation 3.7 may be expressed as

$$\sum_i \alpha_i = \bar{V} \equiv V/A \tag{3.8}$$

This constraint is added to the discretized potential energy, using a penalty method with a weighting factor w

$$g \equiv \pi + w\,(\sum_i \alpha_i - \bar{V})^2 \tag{3.9}$$

In vector notation

$$(\sum_i \alpha_i - \bar{V})^2 = (\underline{d}^T \underline{\alpha} - \bar{V})^2$$

$$= \underline{\alpha}^T D\, \underline{\alpha} - 2\, \underline{d}^T \underline{\alpha}\, \bar{V} + \bar{V}^2$$

where d is a column vector of ones and $D \equiv dd^T$. Note that dropping the $\bar{V}^2$ term does not affect the solution variable values. The α_i can also be eliminated from the function, by noting that in the final contact region

$$u_i^1 + u_i^2 - \alpha_i = 0$$

so that

$$z_i = \alpha_i$$

One can now imbed wD in $\hat{K}$ and $w\underline{d}^T\bar{V}$ in $\hat{\underline{p}}$. The final result is

$$\left.\begin{aligned} &\min_{\underline{z}} \tfrac{1}{2}\, \underline{z}^T \bar{\bar{K}} \underline{z} - \bar{\bar{\underline{p}}}^T \underline{z} \\ &\text{subject to } z_i \geq 0 \quad \text{in } \Gamma_1 \\ &\qquad\qquad z_i \leq 0 \quad \text{in } \Gamma_2 \equiv \Gamma_c - \Gamma_1 \end{aligned}\right\} \tag{3.10}$$

It is assumed in this development that the extent of the contact region is known. Although in general one will not know the final contact region, he need not guess its extent precisely. As in Section 2, one may divide Γ_c into three regions at a solution:

R the region remodeled when $z_i > 0$.

I the region in contact but not remodeled where $z_i = 0$.

S the region not in contact where $z_i < 0$.

In choosing the boundary of Γ_1 one need only insure that Γ_1 includes R but excludes S. Any boundary falling in I is equally acceptable. If an incorrect estimate is made, the analysis reveals this immediately and indicates how to modify Γ_1. Specifically, if Γ_1 is too large a tensile contact force is found. If Γ_1 is too small the contact stress in R is not less than that in I.

An outline of the program used is then:

(1) Form K, $\underline{p}$

(2) Transform K, and $\underline{p}$ into $\hat{K}$ and $\hat{\underline{p}}$

(3) Pick Γ_1

(4) Add penalty constraint, forming $\bar{\bar{K}}$, $\bar{\bar{p}}$

(5) $\min\limits_{\underline{z}} \frac{1}{2}\ \underline{z}^T \bar{\bar{K}}\, \underline{z} - \bar{\bar{\underline{p}}}\, \underline{z}$

subject to $z_i \geq 0$ in Γ_1

$z_i \leq 0$ in Γ_2

(6) Transform $\underline{z}$ back to $\underline{u}$

If all of the optimality criteria are met, the problem is solved. If not, Γ_1 is adjusted and the program rerun. The constrained quadratic programming algorithm of O'Leary is again used successively in step [5]. This program was applied to a beam-on-a-foundation and plane stress problems, as formulated in the first part of this section.

4. RESULTS

The algorithm developed in Section 3 (Eq. 3.10) is used to solve several initial separation design problems. Problem types include a beam on a Winkler foundation, two elastic layers on a rigid half plane, and a beam on a support. An investigation of the sensitivity of the design to variations in load is also performed.

4.1 Beam Problems

Two cases of point-loaded beams on a Winkler foundation are studied. Center and end loads are considered for a constant stiffness beam. In each case, symmetry is used and therefore only the half problem is solved numerically (see Figure 4.1a and 4.1b). The beam is discretized by ten unit length quadratic beam elements and the foundation is discretized by eleven springs for a total of 33 displacement components (see Figure 4.2). The design vector $\underset{\sim}{\alpha}$ contains eleven components, which represent the initial separation at each node pair. After nondimensionalizing by the element length and stiffness, the relevant quantities are

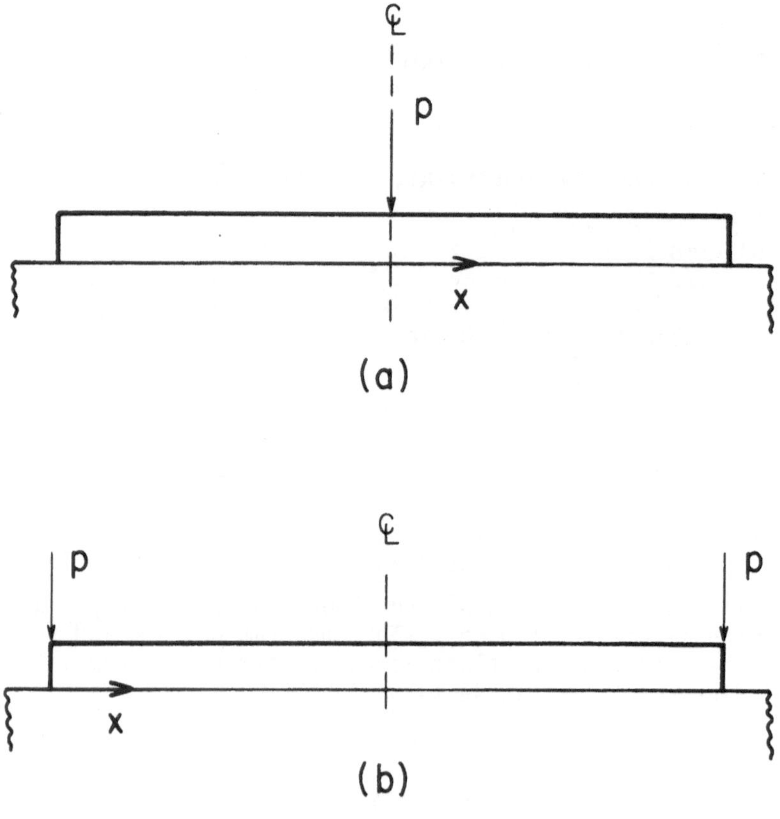

Figure 4.1 Beam Problems and Coordinate Systems.

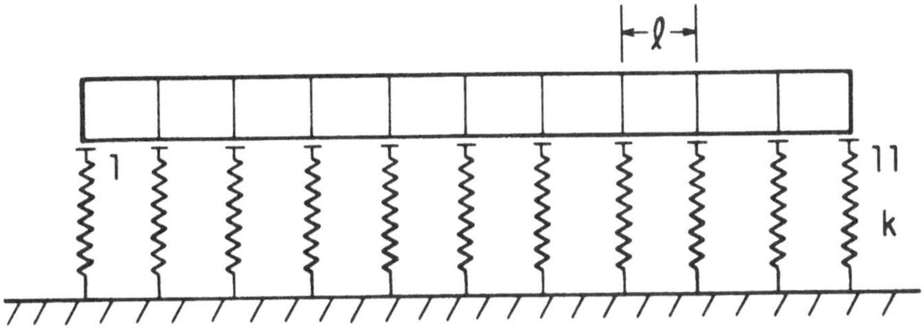

Figure 4.2 Beam Discretization.

$\bar{k} = (k/EI)\ell^4 = 1.0 \times 10^{-2}$

$\bar{p} = (p/EI)\ell^2 = 1.73 \times 10^{-3}$

$\bar{\lambda} = \lambda/EI\ \ell^3$

$\bar{\alpha} = \alpha/\ell$

A weighting factor w of 100 was found to be sufficient to enforce the isoperimetric constraint to four significant places. Error tolerance EPS1 was set to 10^{-4} and EPS2 to 10^{-8}. Decreasing EPS2 to 10^{-12} did not change solution quantities in the first five significant places. Solutions required 7 to 35 iterations of the minimizer, taking a total of 1 to 2 seconds of CPU time on the MTS Amdhal 470V6.

For each problem, V was varied from 0 to a value sufficiently large to produce a constant contact stress at the optimal design. At each step, Γ_1 was estimated from the contact region of the previous solution. Figures 4.3 and 4.6 plot solution quantities against $\bar{V}$. Figures 4.4 and 4.7 plot $\bar{\alpha}$ against $\bar{x}$ and Figures 4.5 and 4.8 plot $\bar{\lambda}$ against $\bar{x}$. Table 4.1 contains a solution summary for fully remodeled cases. The maximum contact stress is observed to have been reduced by 67% to 75%.

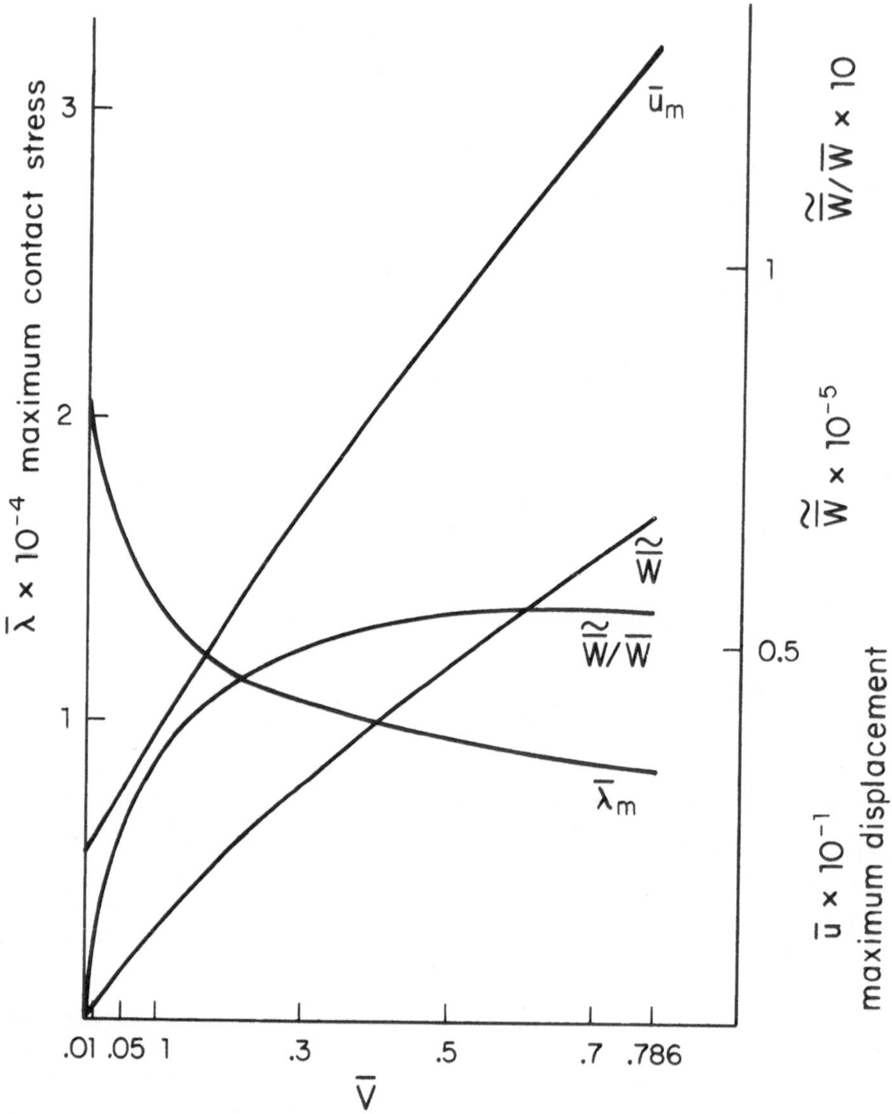

Figure 4.3 Beam-Center Load. Solution maximum contact stress ($\bar{\lambda}_m$), maximum displacement ($\bar{u}_m$), W, and ratio of $\tilde{W}/W$ as functions of value of isoperimetric constraint on volume of separation ($\bar{V}$) for a center-loaded beam.

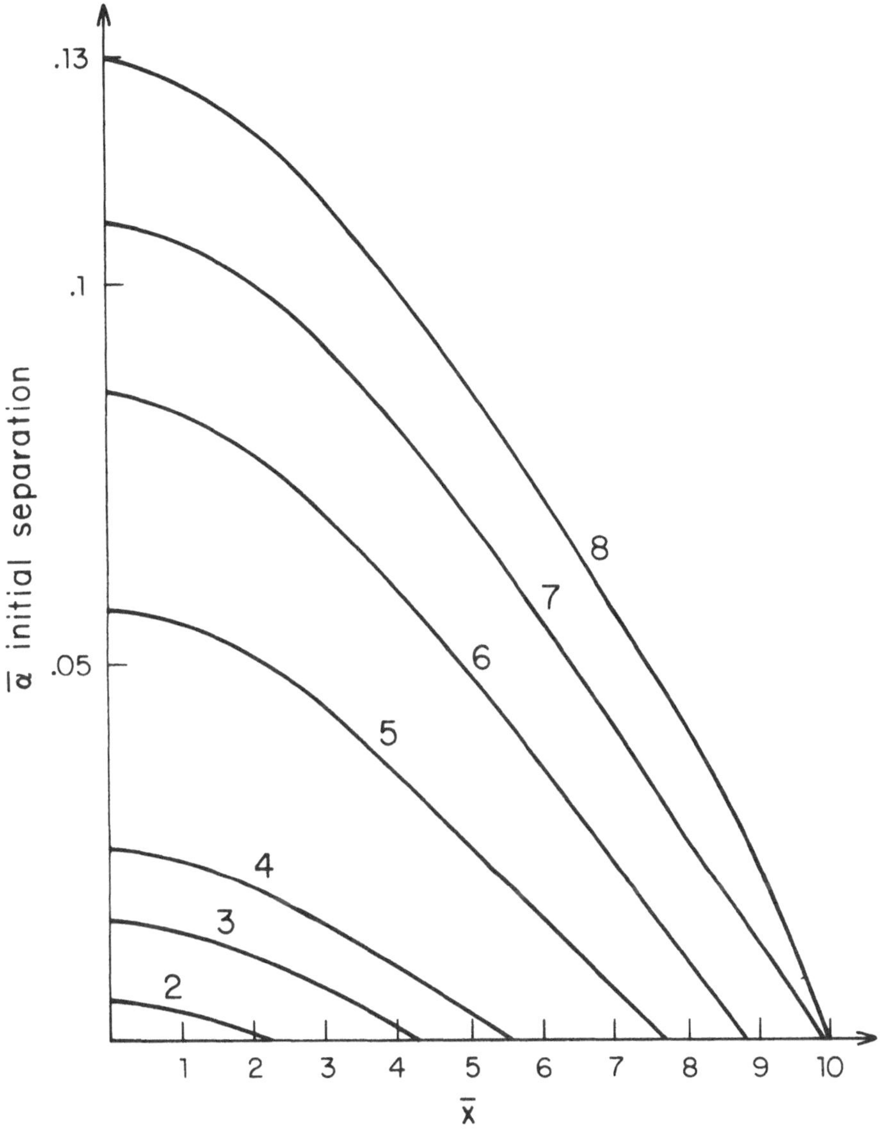

Figure 4.4 Beam-Center Load $\bar{\alpha}$ vs $\bar{x}$. Optimal initial separation designs ($\bar{\alpha}(\bar{x})$) for various values of $\bar{V}$ for a center loaded beam.

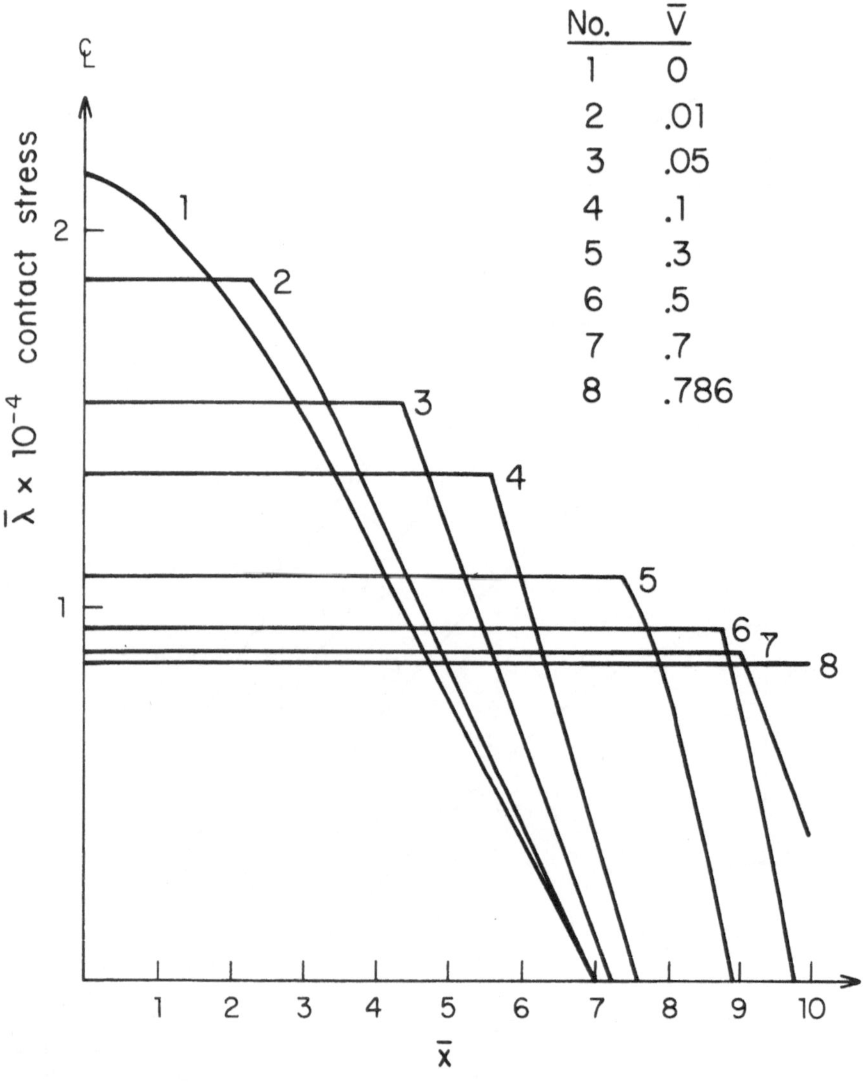

Figure 4.5 Beam-Center Load $\bar{\lambda}$ vs $\bar{x}$. Contact stress ($\bar{\lambda}(\bar{x})$) at optimal initial separation design for various values of $\bar{V}$ for a center loaded beam.

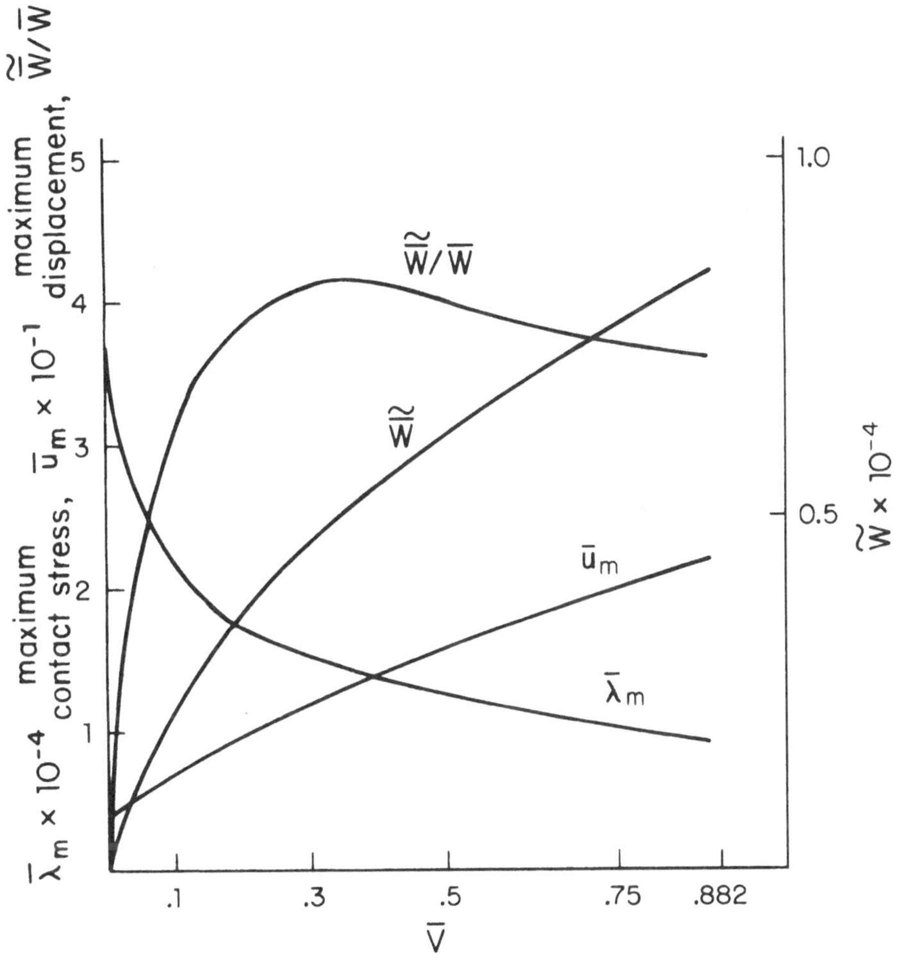

Figure 4.6 Beam-End Load End Rotations. Solution maximum contact stress ($\bar{\lambda}_m$), maximum displacement ($\bar{u}_m$), $\tilde{W}$, and ratio of $\tilde{W}/W$ as functions of value of isoperimetric constraint on volume of separation ($\bar{V}$) for an end loaded beam with end rotations allowed.

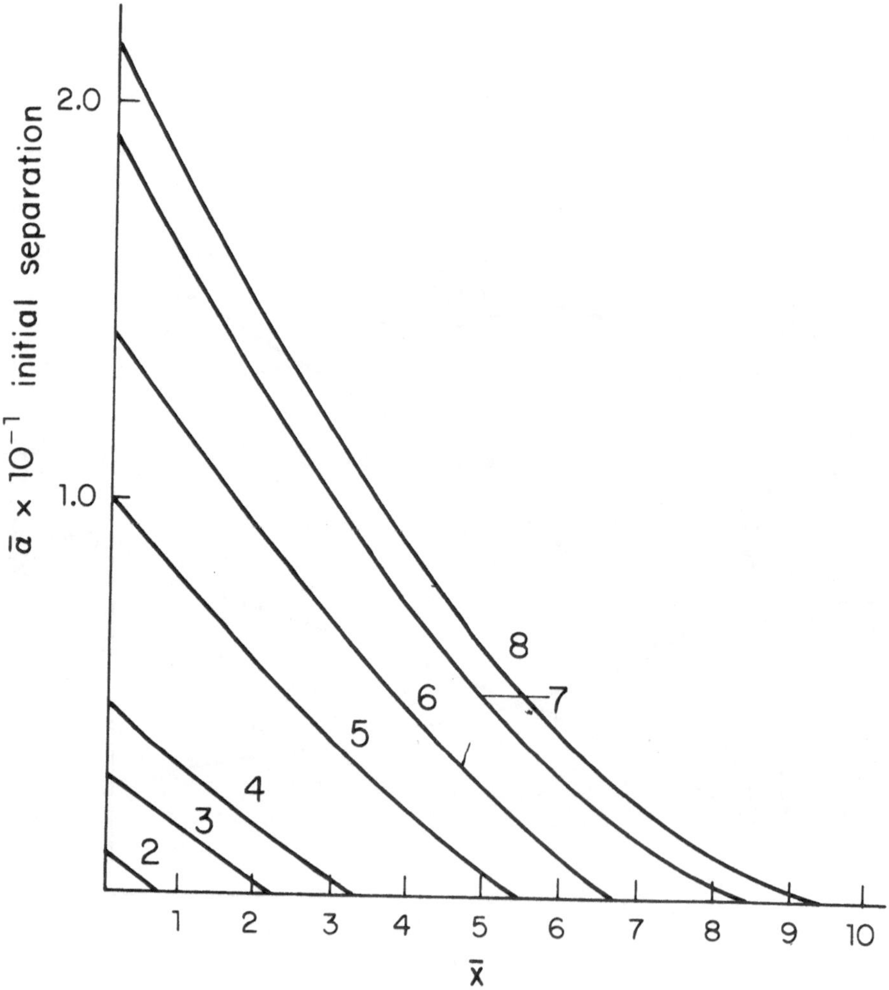

Figure 4.7 Beam End Load End Rotations $\bar{\alpha}$ vs $\bar{x}$. Optimal initial separation design ($\bar{\alpha}(x)$) for various values of $\bar{V}$ for an end loaded beam with end rotations allowed.

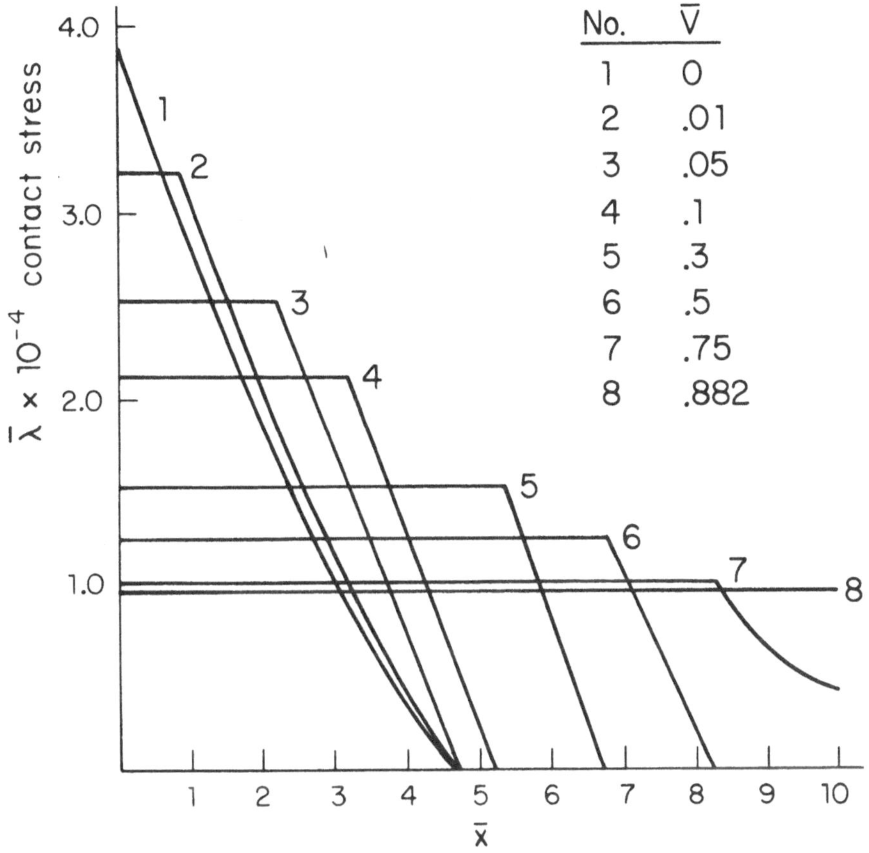

Figure 4.8 Beam-End Load End Rotation $\bar{\lambda}$ vs $\bar{x}$. Contact stress ($\bar{\lambda}(\bar{x})$) at optimal initial separation design for various values of $\bar{V}$ for an end loaded beam with end rotations allowed.

TABLE 4.1 RESULTS SUMMARY

	$\frac{\bar{\lambda}_{max}}{\bar{\lambda}_{o_{max}}}$	$\frac{\bar{\pi}}{\bar{\pi}_o}$	$\frac{\bar{w}}{\bar{w}_o}$
Beam Center Load	2.56	11.6	5.54
Beam End Load	4.02	8.30	6.11
Layers	2.88	6.64	4.67
Support	3.45	10.7	5.73

$\bar{\lambda}_{max}$ = maximum contact stress for initial design

$\bar{\lambda}_{o_{max}}$ = maximum contact stress for optimal design

$\bar{\pi}_{max}$ = potential energy of initial design

$\bar{\pi}_{o_{max}}$ = potential energy of optimal design

$\bar{w}$ = compliance of initial design

$\bar{w}_o$ = compliance of optimal design

Ratios of values of initial design to optimal design for selected quantities

Note that solutions produced by a minimization scheme do meet all of the optimality criteria (Eqs. 2.7 through 2.11 Section 2). That is, the $\bar{\lambda}$ curve is flat wherever $\bar{\alpha}$ is nonzero and $\bar{\lambda}$ is smaller outside this region. As $\bar{V}$ increases the contact region and the remodeled region increase monotonically, while $\bar{\lambda}_{max}$ and $\bar{\pi}$ decrease monotonically. Although $\bar{\pi}$ decreases steadily with increasing $\bar{V}$, a strong "diminishing return" effect is noted in $\bar{\lambda}_{max}$. Since the compliance and non-linearity of response increase with $\bar{V}$, a full remodel may not be desirable. This question is addressed in the discussion of sensitivity that follows.

TABLE 4.2 $\bar{\alpha}$ FOR CENTER LOADED BEAM

Node	$\bar{V}$ = .875	$\bar{V}$ = .786
1	.12865	.12057
2	.12646	.11838
3	.12044	.11235
4	.11134	.10326
5	.99856 E-1	.91774 E-1
6	.86575 E-1	.78493 E-1
7	.72007 E-1	.63925 E-1
8	.56576 E-1	.48494 E-1
9	.40621 E-1	.32540 E-1
10	.24398 E-1	.16318 E-1
11	.80762 E-2	.82718 E-24

For all nodes

$$\bar{\alpha}\Big|_{\bar{V} = 8.75} - \bar{\alpha}\Big|_{\bar{V} = .786} = .808$$

Initial separation ($\bar{\alpha}$) for two values of volume ($\bar{V}$) for a center loaded beam.

Table 4.2 shows the effects of specifying a value for $\bar{V}$ greater than that needed for full remodel. All $\bar{\alpha}$ are simply increased by an equal increment, leading in effect to a rigid body translation of the beam. Other solution quantities remained unchanged. This feature allows constant contact stress designs to be immediately calculated, without the use of a parametric analysis in $\bar{V}$. One need only set Γ_1 equal to all of Γ_c and choose an arbitrarily large value for $\bar{V}$. The rigid body displacement is easily subtracted from $\bar{\alpha}_i$, to yield the desired design contour.

4.2 Elastic Strips

A periodically loaded plane stress problem, consisting of an infinite elastic strip resting on another infinite elastic strip that is bonded to a rigid half-plane (see Fig. 4.9), is now considered. Since the system and its response are periodic in the direction of the length of the strip, a single half-cell 9" in length was analyzed. The strips were discretized as shown in Fig. 4.10, resulting in 128 problem variables with contact boundary conditions applied at 9 node pairs. Table 4.3 lists other program data.

The volume V was varied from 0 to an amount sufficient for a fully remodeled case. Again Γ_1 was estimated from the previous solution. Taking the previous displacement solution as an initial trial reduced CPU time required from 30 seconds to 12 seconds per solution. Figures 4.11, 4.12 and 4.13 plot solution quantities against V and X. The solutions behave much as the previous beam problem solutions. Again necessary conditions are met. Contact area, remodel area and maximum contact stress are monotonic

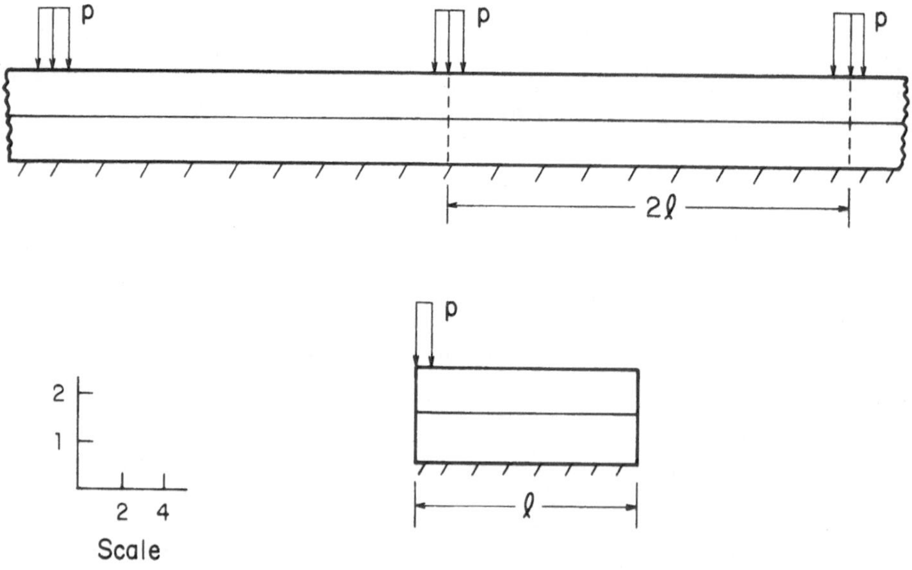

Figure 4.9 Two Elastic Strips.

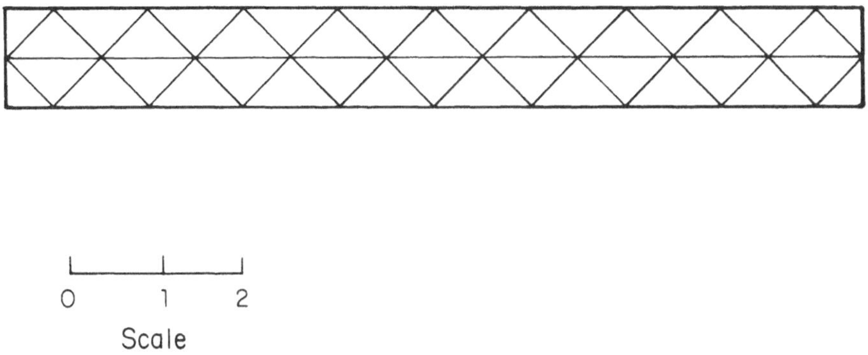

Figure 4.10 Elastic Strip Discretization.

TABLE 4.3 PROGRAM VARIABLES

	Strips	Support
E_1	3×10^7 PSI	3×10^7 PSI
E_2	3×10^6 PSI	3×10^7 PSI
ν_1	.3	.3
ν_2	.3	.3
EPS1	10^{-3}	10^{-3}
EPS2	10^{-6}	10^{-6}
W	10^8	10^8

Material constants and program variable for plane stress problems.

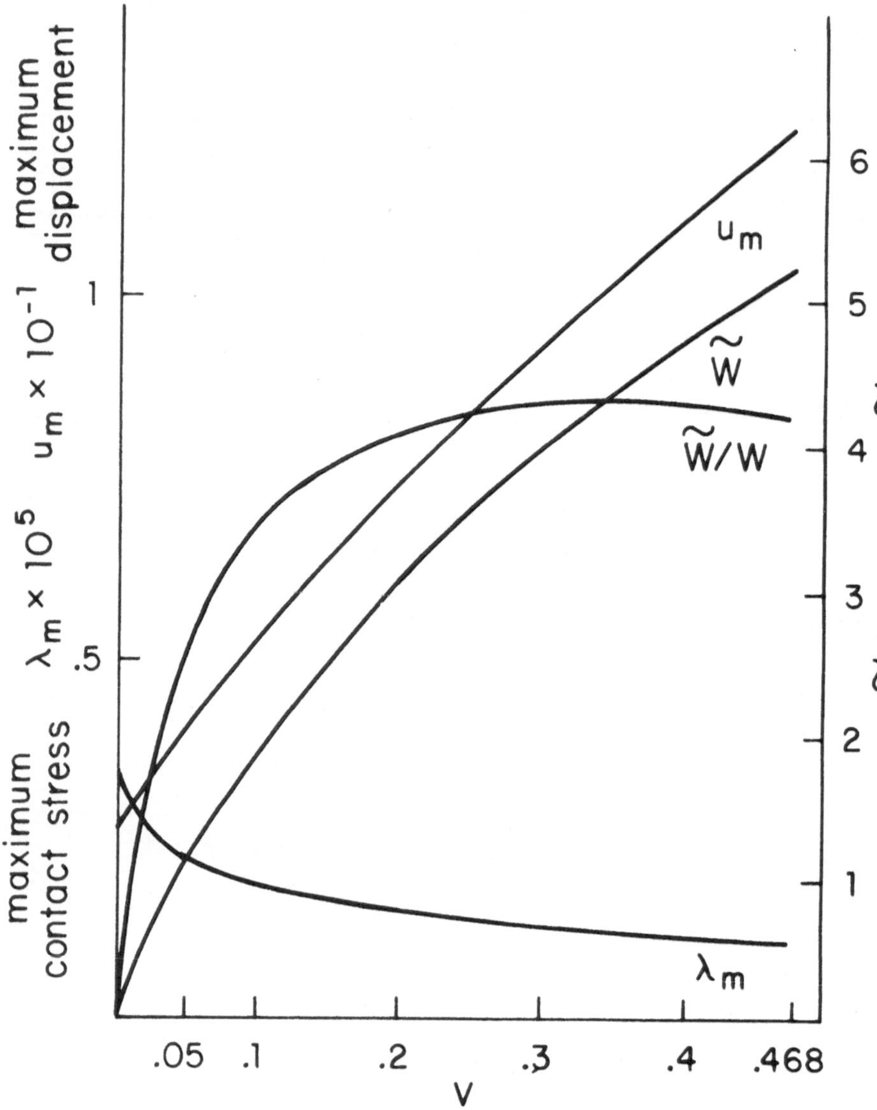

Figure 4.11 Two Elastic Layers. Solution maximum contact stress (λ_m), maximum displacement (u_m), $\tilde{W}$, and ratio of $\tilde{W}/W$ as functions of value of isoperimetric constraint on volume of separation (V) for two elastic strips.

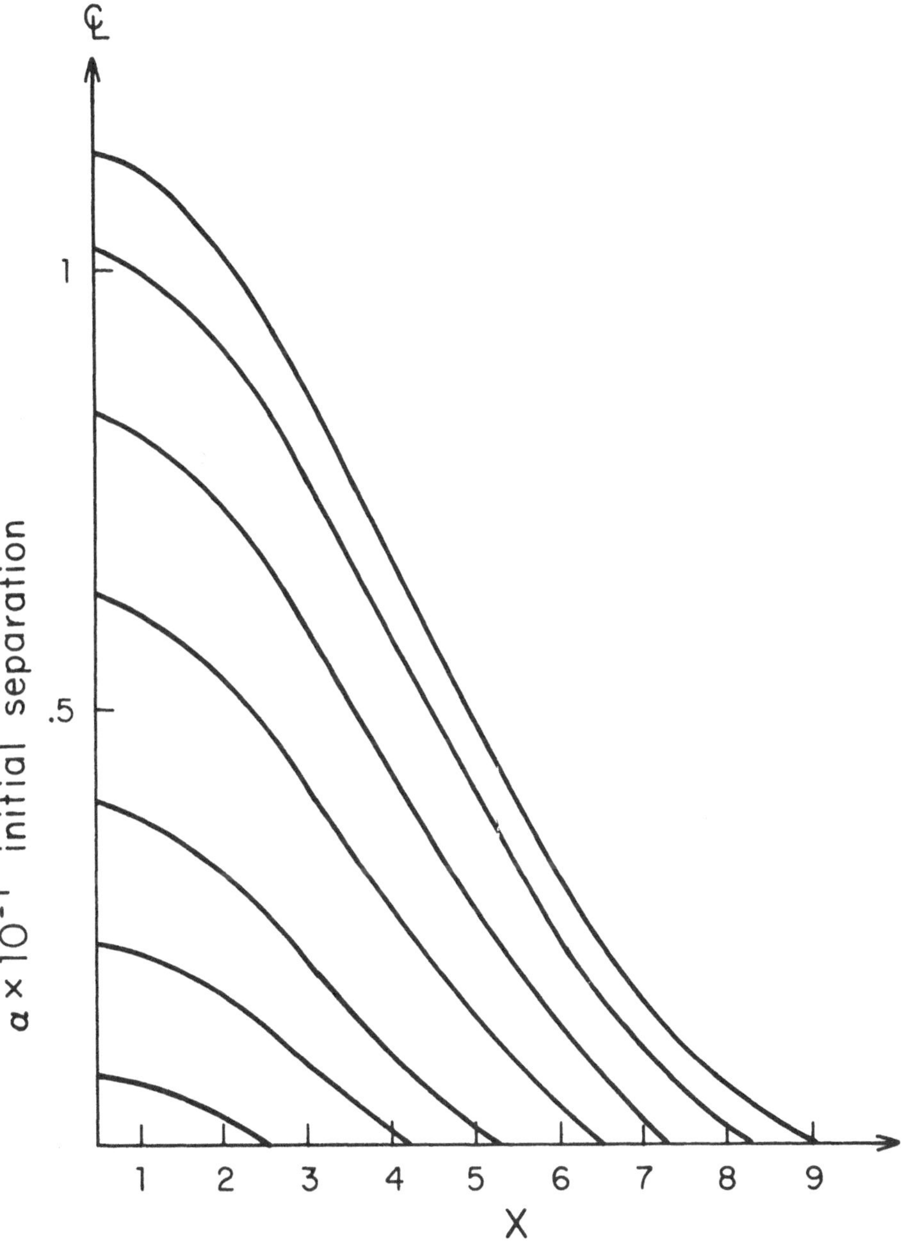

Figure 4.12 Layers α vs x. Optimal initial separation designs (α(x)) for various values of V for two elastic strips.

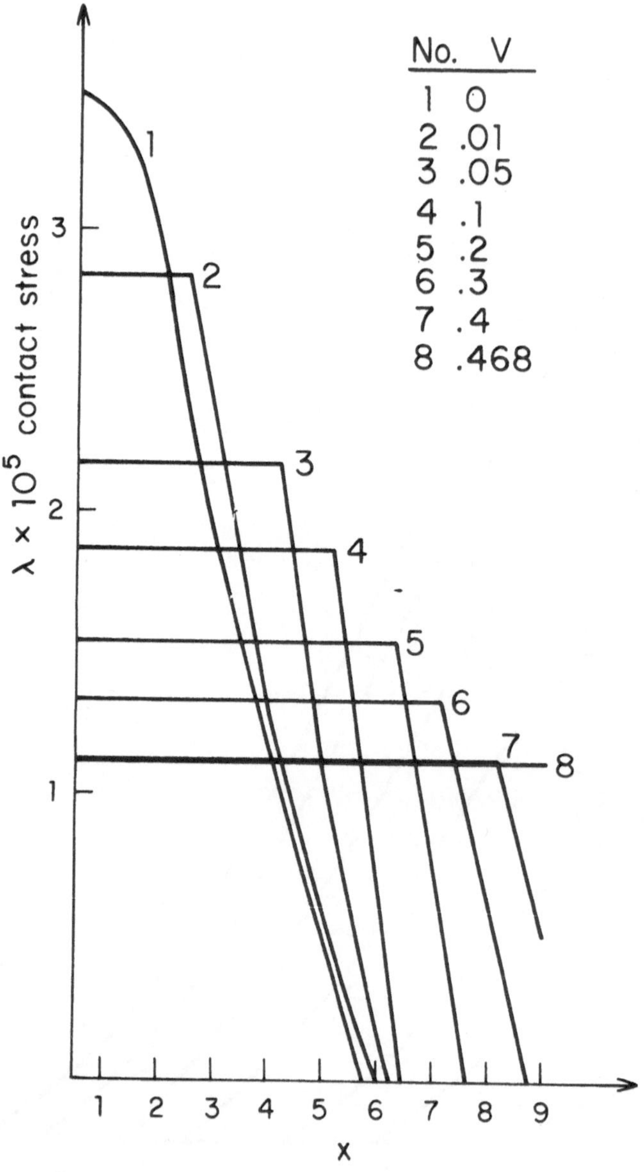

Figure 4.13 Two Elastic Layers λ vs x. Contact stress (λ(x)) at optimal initial separation design for various values of V for two elastic strips.

functions of V. Increasing w and decreasing EPS2 had no effect on the solutions to four significant places. The "diminishing return" effect noted for the beam problem is even more pronounced in this case.

4.3 Beam on a Support

A plane stress problem of a beam on a support under a distributed load, is now considered. Figure 4.14, depicts the problem and Fig. 4.15 shows the discretization. The problem contains 128 variables, with contact boundary conditions applied at 6 node pairs. Table 4.3 lists other program data.

In the initial derivation of the remodel design optimality criteria, it was assumed that α was smooth, in order to prevent singularities. In the unremodeled case it is clear that a singularity will exist in the contact stress at the edge of the support. The contact problem solution technique is based on the minimum potential energy principle, which is valid for non-smooth contact surfaces. If solution quantities change rapidly, as in this case, a much finer discretization is useful, if one is interested in studying the singularity. As design is more properly concerned with avoiding singularities than studying them, the mesh used is felt to be adequate.

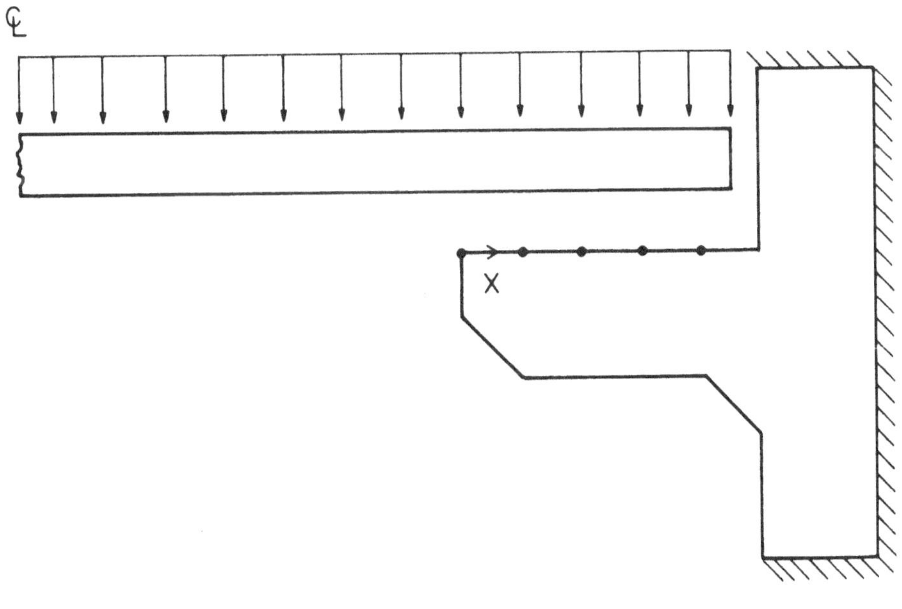

Figure 4.14 Beam on a Support.

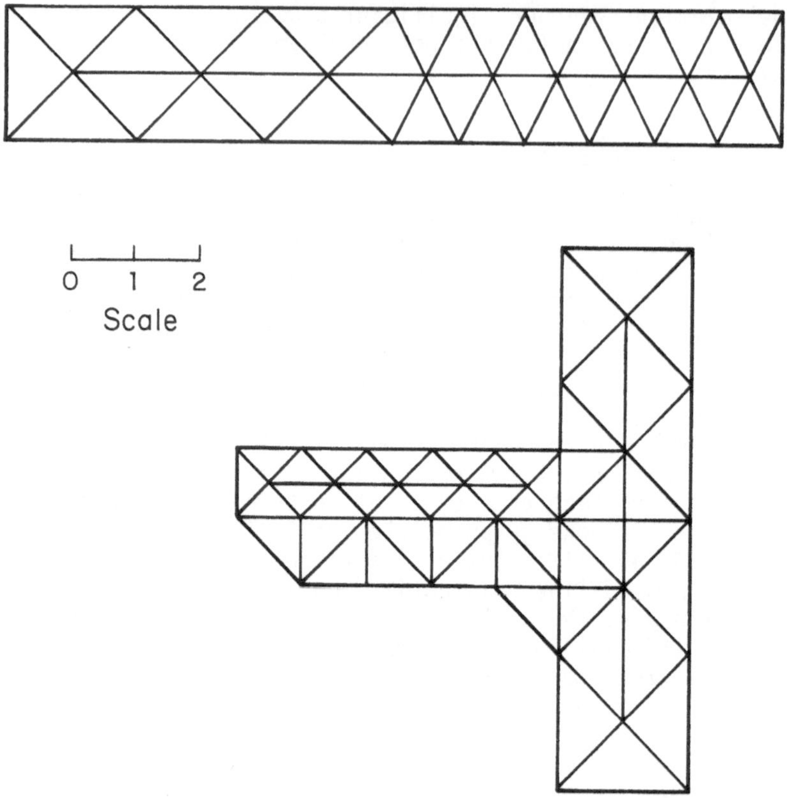

Figure 4.15 Beam on a Support Discretization.

Although the optimization technique was derived for smooth surfaces, the program encountered no difficulty in producing design solutions for an initially non-smooth surface. Any amount of remodel relieves the singularity. Increasing V then produces a family of solutions similar to the one seen in the previous problem.

V was varied from 0 to an amount sufficient for a fully remodeled case. Using Γ_1 and displacements from the previous solution yielded CPU times of about 12 seconds for a solution. The initial problem with V = 0 required 36 seconds of CPU time.

Figures 4.16, 4.17, and 4.18 plot solution quantities against V and x. Most solution quantities behave as in the layer and beam cases, with one important difference. Although the remodeled region increases monotonically with V, the contact region initially shrinks and then grows. Any initial separation relieves the singularity and drops the contact stress dramatically. Specifically, a volume V of only 2% of that needed for a full remodel drops the maximum contact stress by 42%.

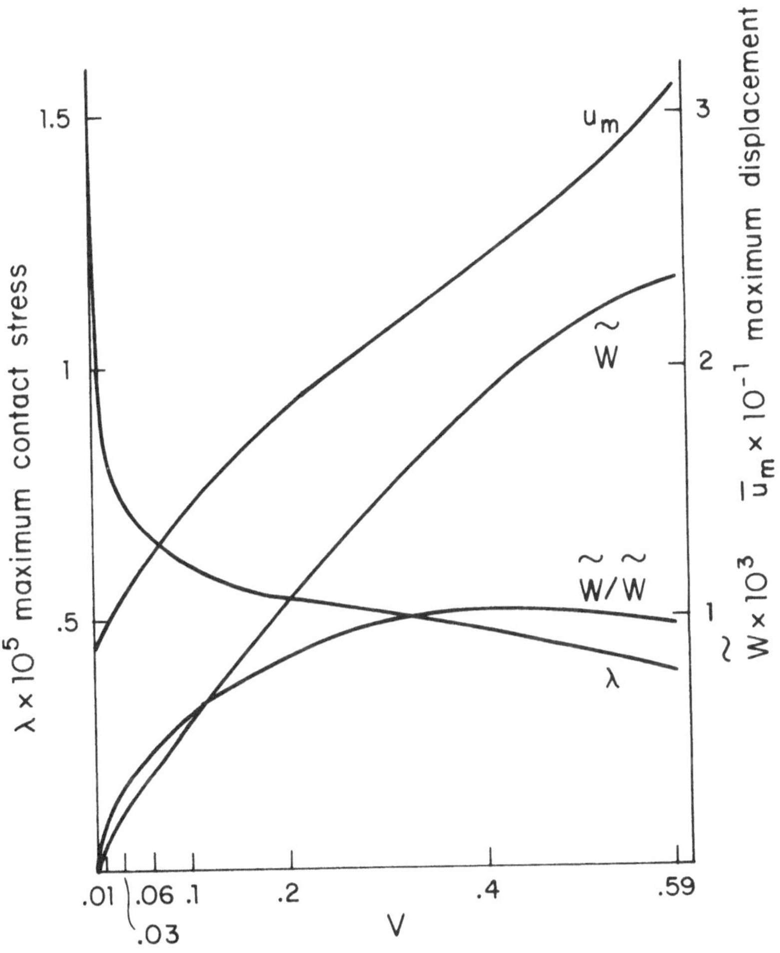

Figure 4.16 Support. Solution maximum contact stress (λ_m), maximum displacement (u_m), $\tilde{W}$, and ratio of $\tilde{W}/W$ as functions of value of isoperimetric constraint on volume of separation (V) for a beam on a support.

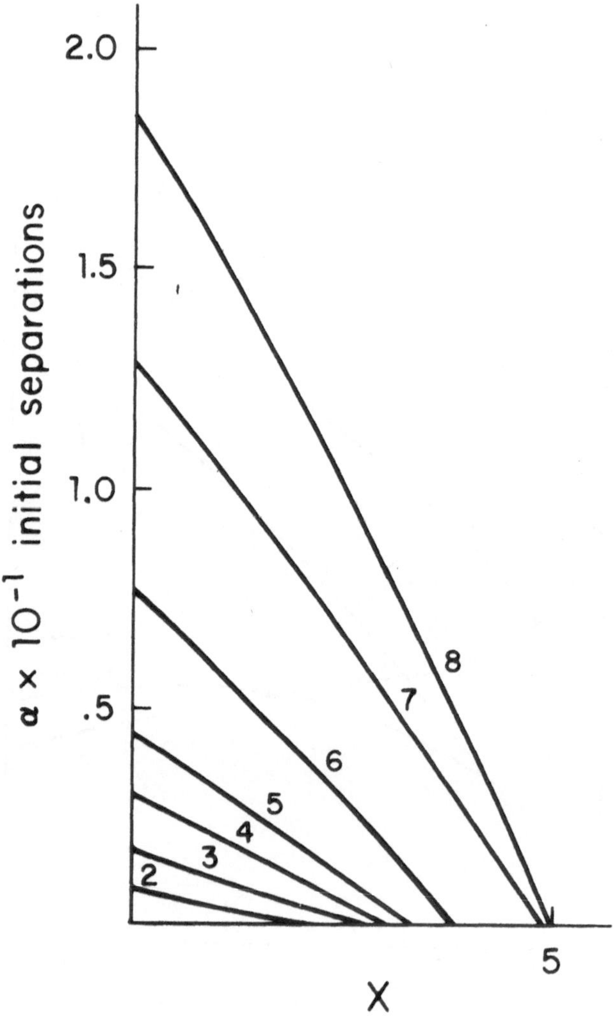

Figure 4.17 Support α vs x. Optimal initial separation designs ($\alpha(x)$) for various values of V for a beam on a support.

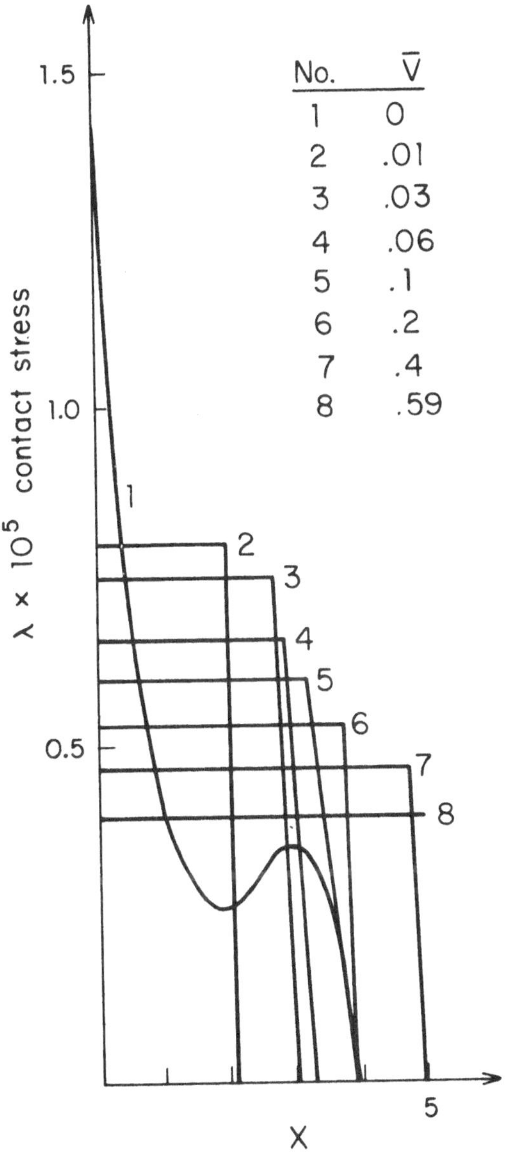

Figure 4.18 Support λ vs x. Contact stress (λ(x)) at optimal initial separation design for various values of V for a beam on a support.

4.4 Sensitivity

A numerical analysis is now conducted to check the sensitivity of the contact stress for a given design to variations in load. A center loaded beam with no remodel, partial remodel, and full remodel is subjected to load variations of ±5%. Figure 4.19 plots $\bar{\lambda}_{max}$ against $\tilde{W}/\bar{W}$ for the three load cases.

For the unremodeled case, the problem is receding and therefore $\bar{\lambda}_{max}$ varies in proportion to the load. In the partial remodel case, a load increase of 5% caused a $\bar{\lambda}_{max}$ increase of 8%, but a load decrease of 5% only caused a $\bar{\lambda}_{max}$ decrease of 2.9%. In the fully remodeled case, the 5% load increase leads to a 12.4% increase in $\bar{\lambda}_{max}$ and the 5% load decrease leads to a 2.2% increase in $\bar{\lambda}_{max}$.

This suggests that fully remodeled designs are relatively more sensitive to load variations. Recall also the strong "diminishing returns" effect noted in these problems. In practical problems of contact stress reduction, the minimum possible contact stress, achieved with total remodel, may be undesirably load sensitive. Using V to control the amount of remodel allows the production of designs that greatly reduce the maximum contact stress, but which are not unduly load sensitive.

Figure 4.20 plots the change in response, measured by the change in $\bar{u}^{-1}_{max}$, for a 5% load change for three values of $\tilde{W}/\bar{W}$. That these points lie approximately along a line with negative slope is consistent with the conjecture that $\tilde{W}/\bar{W}$ may be a useful measure of the stiffening that occurs in advancing contact problems.

REFERENCES

1. Haug, E.J. and Kwak, B.M., "Contact Stress\Minimization by Contour Design," Numerical Methods in Engineering, Vol. 12, 1978, 917-930.
2. Conry, T.F., The use of mathematical programming in design for uniform load in nonlinear elastic systems, University of Wisconsin, Ph.D. Thesis, 1970.
3. Panagiotopoulos, P.D., "A nonlinear programming approach to the unilateral contact and friction boundary value problem in the theory of elasticity," Ingenieur-Archiv 44, 1975, 421-432.

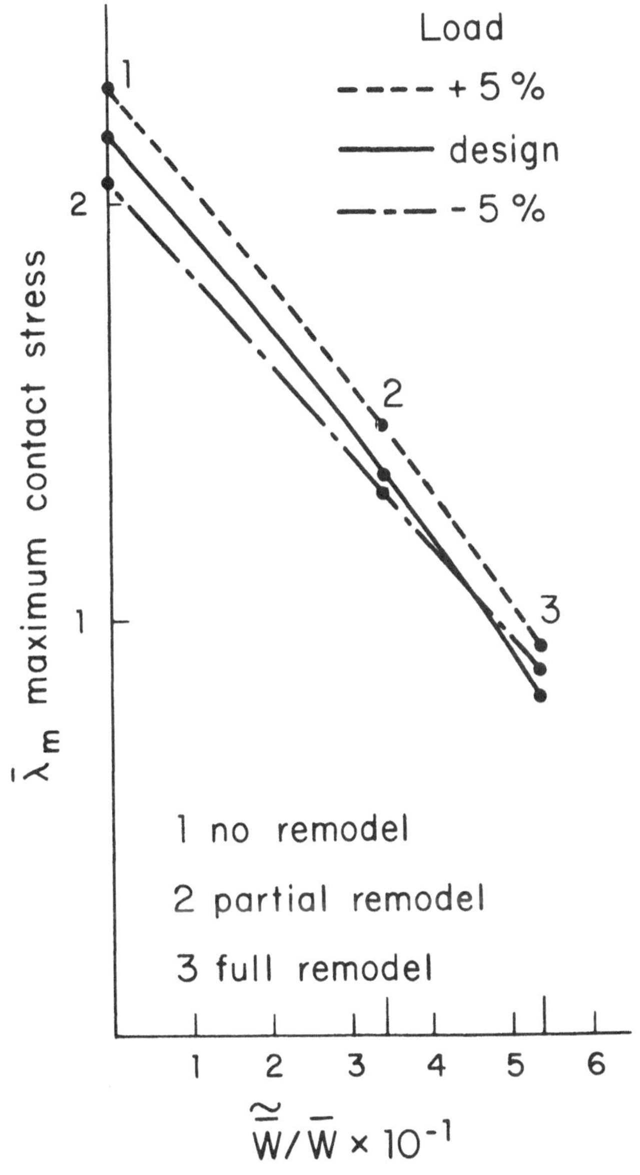

Figure 4.19 Beam Center Load $\bar{\lambda}_m$ vs $\tilde{W}/\bar{W}$. Maximum contact stress ($\bar{\lambda}_m$) for various load cases and various amounts of remodel plotted against $\tilde{W}/\bar{W}$ for clarity for a center-loaded beam on a Winkler type foundation.

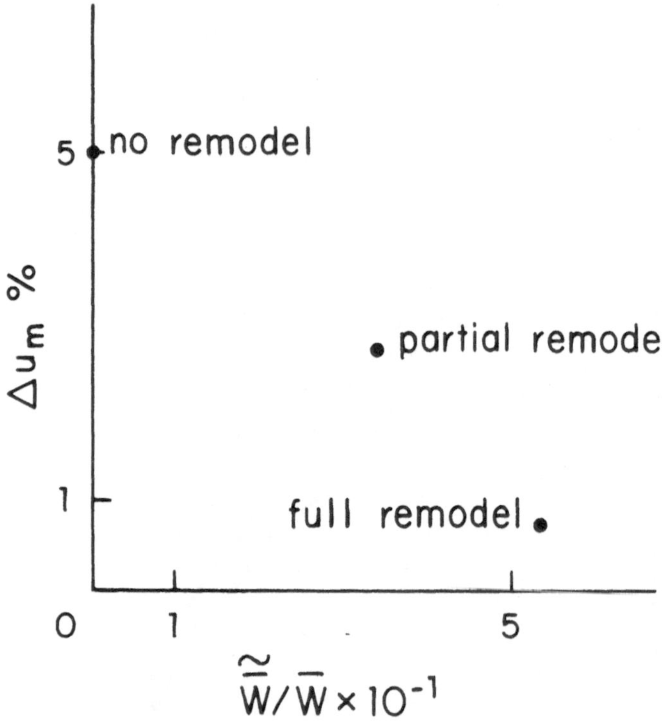

Figure 4.20 Beam Center Load Displacement Response for 5% Load Change vs $\tilde{W}/\bar{W}$. Percentage change in maximum displacement (u_m) to a 5% load change for various amounts of remodel for center-loaded beam on a Winkler type foundation.

4. Panagiotopoulos, P. D., "On the unilateral contact problem of structures with nonquadratic strain energy density," Int. J. Solids and Structures, 13, 1977, 253-261.
5. Boucher, M., "Signorini's problem in viscoelasticity," The mechanics of contact between deformable bodies, Delft University Press, 1974.
6. Fichera, G., "Unilateral Constraints in Elasticity," Handbuch der Physik, VI a/2, Springer-Verlag, 1972.
7. Kalker, J.J., "Variational principles in contact elastostatics", J. Inst. Maths. Applics., Vol. 20, 1977, 199-219.
8. Panagiotopoulos, P.D., "A Variational Inequality Approach to the Friction Problem of Structures," J. Struct. Mech., Vol. 6, No. 3, 1978, 303-318.
9. Luenberger, D.G., Optimization by vector space methods, John Wiley & Sons, New York, 1968.

10. Lanczos, C., Variational Principles in Mechanics, University of Toronto Press, 1970.
11. Dundurs, J., "Properties of Elastic Bodies in Contact," The mechanics of contact between deformable bodies, Delft University Press, 1974.
12. Prager, W. and Taylor, J.E., "Problems of Optimal Structural Design," J. of Appl. Mech., Vol. 35, 1968, 102-106.
13. Ohloff, N. and Taylor, J.E., "On Optimal Structural Remodeling," J. of Optimization Theory and Applications, Vol. 27, No. 4, April, 1979, 571-582.
14. Taylor, J.E., "Maximum Strength Elastic Structural Design," J. of Eng. Mech. ASCE, June, 1969, 653-663.
15. O'Leary, D.P., "A Generalized Conjugate Gradient Algorithm for Solving a Class of Quadratic Programming Problems," STAN-CS-77-638, Stanford Univ. Comp. Sci. Dept., 1977.
16. O'Leary, D.P., and Yang, W.H., "Elastoplastic Torsion by Quadratic Programming," Computer Methods in Applied Mech. and Eng., Vol. 16, 1978, 361-368.
17. Masur, E.F. and Mroz, Z., "Non-Stationary Optimality Conditions in Structural Design," Int. J. Solids Structures, Vol. 15, 1979, 503-512.
18. Lukasiewicz, S.A. "Optimum Design in Juncture and Contact," Colloqium #110, Contact Problems and Load Transfer in Mecahnical Assemblages, Sweden, September, 1978.

SENSITIVITY ANALYSIS FOR A CLASS OF VARIATIONAL INEQUALITIES

Jan Sokolowski

Systems Research Institute, Polish Academy of Sciences
01-447 Warszawa, ul. Newelska 6, Poland

ABSTRACT

In this paper some results concerning distributed parameter optimization for a class of problems governed by variational inequalities are given. It is shown that under some assumptions, the solution of a variational inequality is conically differentiable with respect to the design. Application of the result to control problems described by a class of variational inequalities, to obtain the necessary conditions of optimality, is discussed. A simple example of such a problem of control is presented. The variational inequality involved characterizes a contact problem.

1. INTRODUCTION

Let V be a Hilbert space, let U_{ad} be a convex, bounded, and closed subset of a Banach space U, and let $\{a_u(.,.)\}$ denote a family of bilinear forms on V that depend on the parameter $u \in U_{ad}$. Suppose that the following conditions hold:

(A1) $a_u(.,.)$ is continuous and uniformly coercive with respect to to $u \in U_{ad}$, i.e.,

$$|a_u(\phi,\psi)| \leq M \,||\phi||_V \,||\psi||_V$$

$$a_u(\phi,\phi) \geq \alpha||\phi||_V^2 \,.$$

for all ϕ and $\psi \in V$, where $\alpha > 0$ and M are constants.

(A2) $a_u(.,.)$ is differentiable with respect to the parameter $u \in U_{ad}$, i.e. for a given $u \in U_{ad}$ there exists a linear, continuous form $b_u(.;.,.)$: $U \times V \times V \to R$, i.e. a constant M_1 exists such that

$$|b_u(h;\phi,\psi)| \leq M_1 \ ||h||_U \ ||\phi||_V \ ||\psi||_V$$

for all $h \in U$ and ϕ and $\psi \in V$ and

$$\sup_{\substack{||\phi||_V \leq 1 \\ ||\psi||_V \leq 1}} |a_{u+h}(\phi,\psi) - a_u(\phi,\psi) - b_u(h;\phi,\psi)| = o(||h||_u)$$

Denote by $A_u \in L(V;\ V')$, $B_u(\cdot) \in L(U;L(V;\ V'))$, $u \in U_{ad}$, linear operators such that

$$a_u(\phi,\psi) = \langle A_u\phi,\psi\rangle$$

$$b_u(h;\phi,\psi) = \langle B_u(h)\phi,\psi\rangle$$

for all ϕ and $\psi \in V$ and $h \in U$ where V' denotes dual of V and $\langle\cdot,\cdot\rangle$ denotes duality pairing between V' and V.

Let $C \subset V$ be a closed, convex set and let $f \in V'$ be given. Consider the following variational inequality, depending on the parameter u: Find $y \in C$ such that

$$a_u(y,\phi - y) \geq \langle f,\phi - y\rangle \tag{1.1}$$

for all $\phi \in C$. It is well known that, under assumption (A1), there exists a unique solution $y = y_u(f) \in V$ to Eq. 1.1 [1]. Furthermore, it can be verified that, under assumption (A2), the solution is Lipschitz continuous with respect to u, i.e.,there exists a constant L such that

$$||y_{u_1}(f) - y_{u_2}(f)||_V \leq L||u_1 - u_2||_U \tag{1.2}$$

for all u_1 and $u_2 \in U_{ad}$.

Denote by $H = H(u,f)$, $u \in U_{ad}$, $f \in V'$, the following linear subspaced of V:

$$H(u,f) = \{\phi \in V : a_u(y_u(f),\phi) = \langle f,\phi\rangle \} \tag{1.3}$$

Let $K \subset V$ be a convex and closed subset, let $v \in K$ be given, and denote by $C_v(K)$ the cone of the form:

$$C_v(K) = \{\phi \in V: \exists\, t > 0,\ v + t\phi \in K\} \tag{1.4}$$

Let $S_v(K)$ be the closure of the cone $C_v(K)$ in V and denote $S^v = S_v(K) \cap H(u,f)$.

Definition 1.1: Let V_1 and V_2 be Banach spaces and let $F : V_1 \mapsto V_2$ be a continuous mapping. The mapping $F(\cdot)$ is said to be conically differentiable at a point $z \in V_1$ if there exists a continuous, positive homogeneous mapping $Q : V_1 \mapsto V_2$, such that in some neighborhood of $z \in V_1$,

$$F(z + tw) = F(z) + t\, Q(w) + o(t,w)$$

for all $w \in V_1$ and $t > 0$, where $\frac{1}{t}\, o(t,w) \to 0$ as $t \downarrow 0$, uniformly with respect to w on compact subsets of V_1.

Theorem 1.1: Let assumptions (A1) and (A2) be satisfied and $y = y_u(f)$ be a solution of Eq. 1.1. If the set $C_y(C) \cap H(u,f)$ is dense in $S^y = S_y(C) \cap H(u,f)$, then

$$y_{u+th}(f) = y_u(f) + t\, D(h) + o(t,h) \tag{1.6}$$

for all $h \in U$ and $t > 0$, where $D(h)$ is a unique solution of the following variational inequality: find $D(h) \in S^y$ such that

$$a_u(D(h), \phi - D(h)) \geq b_u(h; y_u(f), \phi - D(h)) \tag{1.7}$$

for all $\phi \in S^y$.

Proof: Let $u \in U_{ad}$ be a fixed parameter. Under the foregoing assumptions, the mapping

$$V' \ni f \longmapsto y_u(f) \in V \tag{1.8}$$

is conically differentiable [2], i.e.,

$$y_u(f + tw) = y_u(f) + t\, \theta(w) + o(t,w) \tag{1.9}$$

for all $w \in V'$ and $t > 0$, where for all $w \in V'$, $\theta(w) \in V$ is a unique solution of the variational inequality, find $\theta(w) \in S^y \subset V$ such that

$$a_u(\theta(w), \phi - \theta(w)) \geq \langle w, \phi - \theta(w) \rangle \tag{1.10}$$

for all $\phi \in S^y$.

Let $h \in U$ be given and let $t > 0$. Denote

$$w(t) = \frac{1}{t}\left[A_u - A_{u+th}\right] y_{u+th}(f) \tag{1.11}$$

It can be verified that

$$y_{u+th}(f) = y_u(f + tw(t)) \tag{1.12}$$

Hence, it follows from Eqs. 1.9 and 1.12 that

$$\begin{aligned} y_{u+th}(f) &= y_u(f) + t\theta(w(t)) + o(t,w(t)) \\ &= y_u(f) + t\theta(-B_u(h)\ y_u(f)) \\ &\quad + t\{\theta(w(t)) - \theta(-B_u(h)\ y_u(f))\} + o(t, w(t)) \end{aligned} \tag{1.13}$$

Denote $D(h) = \theta(-B_u(h)\ y_u(f))$ and observe that, by assumption (A2) and by Eq. 1.2 it follows that

$$\frac{1}{t}\left[A_u - A_{u+th}\right] \longrightarrow -B_u(h), \text{ in } L(V,V') \text{ as } t \downarrow 0 \tag{1.14}$$

and

$$y_{u+th}(f) \rightarrow y_u(f), \text{ in } V \text{ strongly as } t \downarrow 0 \tag{1.15}$$

Hence,

$$w(t) \longrightarrow -B_u(h)\ y_u(f), \text{ in } V' \text{ strongly as } t \downarrow 0 \tag{1.16}$$

The solution $\theta(w) \in V$ of the variational inequality Lipschitz continuous with respect to $w \in V'$. Hence Eq. 1.16 implies that

$$\theta(w(t)) \longrightarrow D(h) \text{ strongly in } V \text{ as } t \downarrow 0 \tag{1.17}$$

Thus

$$t\{\theta(w(t)) - D(h)\} + o(t,w(t)) = o(t,h) \tag{1.18}$$

which completes the proof.

Let $\bar{u} \in U_{ad}$ be given element and let $y = y_u(f)$ denote the corresponding solution to Eq. 1.1. Note that if the mapping

$$U_{ad} \ni u \rightarrow y_u(f) \in V$$

is Gateau differentiable at the point $\bar{u} \in U_{ad}$, then $S^{\bar{y}}$ is a linear subspace of V and $D(h) \in S^{\bar{y}}$ is a unique solution to the following equation:

$$a_u(D(h),\phi) = b_{\bar{u}}(h;\bar{y},\phi)$$

for all $\phi \in S^{\bar{y}}$.

2. EXAMPLE

A simple example of a parametric optimization problem is now presented, for the variational inequality characterizing [3] the problem of the deflection of two parallel membranes loaded laterally by pressures $f_i = f_i(x_1,x_2)$, i=1,2, as shown in Fig. 2.1.

To assist in the following, some notation is now introduced. Let $\Omega \subset R^2$ be an open, bounded domain with smooth boundary Γ. Denote by $H^1(\Omega)$ the Sobolev space

$$H^1(\Omega) = \{\psi \in L_2(\Omega) : \frac{\partial\psi}{\partial x_i} \in L_2(\Omega)\}$$

Let $H_0^1(\Omega)$ be the closed linear subspace of $H^1(\Omega)$,

$$H_0^1(\Omega) = \{\psi \in H^1(\Omega) : \psi|_\Gamma = 0\}$$

with the scalar product

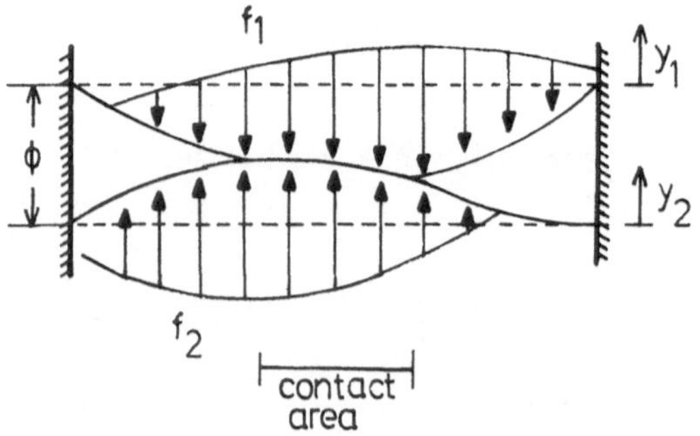

Figure 2.1. Contact Problem for Two Parallel Membranes Subjected to Transverse Pressures

$$(\xi,\psi)_{H_0^1(\Omega)} = a(\xi,\psi) = \sum_{i=1}^{2} \int_\Omega \frac{\partial \xi}{\partial x_i} \frac{\partial \psi}{\partial xi}\, dx \tag{2.1}$$

Let $E \subset \Omega$ be given and define capacity of the set E as

$$\mathrm{cap}(E) = \inf \left\{ \sum_{i=1}^{2} \int_{R^2} \left(\frac{\partial \psi}{\partial x_i}\right)^2 dx : \psi \in C_0^1(R^2), \psi \geq 1 \text{ on } E \right\} \tag{2.2}$$

The term, quasi everywhere on Ω, is defined to mean everywhere on Ω, except possibly on a set of capacity zero. Recall that any function on the Sobolev spaced $H^1(\Omega)$ can be defined [1] at all points of Ω, except possibly of a set of capacity zero, i.e., quasi everywhere on Ω. Let $\psi \in H^1(\Omega)$ be given and define a set

$$Z(\psi) = \{x \in \Omega : \psi(x) = 0\}$$

Let $u = (T_1, T_2) \in R^2$ be a parameter, and let T_{min} and T_{max} be given numbers such that $T_{max} \geq T_{min} > 0$ defines

$$U_{ad} = \{(T_1, T_2) : T_{min} \leq T_1, T_2 \leq T_{max}\} \tag{2.3}$$

Denote $V = H_0^1(\Omega) \times H_0^1(\Omega)$ and define a bilinear form on V as

$$a_u(y,\phi) = T_1\, a(y_1,\phi_1) + T_2\, a(y_2,\phi_2) \tag{2.4}$$

for all

$$y \text{ and } \phi \in V, \quad y = (y_1, y_2), \ \phi = (\phi_1, \phi_2), \text{ and } u = (T_1, T_2).$$

It can be verified that in the case of bilinear form of Eq. 2.4, assumptions (A1) and (A2) are satisfied. In particular, if $h = (\tau_1, \tau_2)$, then

$$b_u(h; y, \phi) = \tau_1\, a(y_1,\phi_1) + \tau_2\, a(y_2,\phi_2) \tag{2.5}$$

for all y and $\phi \in V$.

Let $\tilde{\phi} = \tilde{\phi}(x_1, x_2) \in L_\infty(\Omega)$ be given and define a subject $C \subset V$,

$$C = \{(\phi_1,\phi_2) \in H_0^1(\Omega) \times H_0^1(\Omega) : \phi_1(x) + \tilde{\phi}(x) \geq \phi_2(x) \text{ q.e. in } \Omega\} \tag{2.6}$$

It can be shown (see the Appendix) that in the case of the set C of Eq. 2.6, the cone $S_y(C)$ has the followiny form:

$$S_y(C) = \{(w_1,w_2) \in H_0^1(\Omega) \times H_0^1(\Omega) : w_1(x) \geq w_2(x) \text{ q.e. on } Z(y_1 + \tilde{\phi} - y_2)\} \tag{2.7}$$

Furthermore it is proved in the Appendix that is the set C has the form of Eq. 2.6, then the set $C_y(C) \cap H(u,f)$ is dense in the set $S^y = S_y(C) \cap H(u,f)$.

Let (z_1,z_2) be given functions and consider the following problem (P) of parametric optimization: minimize the cost functional

$$J(u) = \frac{1}{2} \sum_{i=1}^{2} \int_\Omega (y_i(x) - z_i(x))^2 \, dx \tag{2.8}$$

over the set U_{ad} defined in Eq. 2.3, where $y = y_u = (y_1,y_2) \in C \subset V$ is unique solution to the variational inequality

$$\sum_{i,j=1}^{2} T_j \int_\Omega \frac{\partial y_j}{\partial x_i} \left(\frac{\partial \phi_j}{\partial x_i} - \frac{\partial y_j}{\partial x_i}\right) dx \geq \sum_{j=1}^{2} \int_\Omega f_j(\phi_j - y_j) dx \tag{2.9}$$

for all $(\phi_1,\phi_2) \in C$.

With these preliminaries and notations, one can now obtain the principal result of the paper.

Theorem 2.1: Let $\hat{u} = (\hat{T}_1,\hat{T}_2)$ be an optimal solution to the problem (P). Denote by $\hat{y} = (\hat{y}_1,\hat{y}_2)$ the corresponding solution to the variational inequality of Eq. 2.9. Then, for all $u \in U_{ad}$,

$$\sum_{i=1}^{2} \int_\Omega (\hat{y}_i(x) - z_i(x)) \, D_i(u - \hat{u})(x) \, dx \geq 0 \tag{2.10}$$

where $D = (D_1,D_2) \in S^{\hat{y}} \subset V$ is the unique solution of the following variational inequality:

$$\sum_{i,j=1}^{2} \hat{T}_j \int_\Omega \frac{\partial D_j}{\partial x_i}\left(\frac{\partial \phi_j}{\partial x_i} - \frac{\partial D_j}{\partial x_i}\right) dx \geq \sum_{i,j=1}^{2} (T_j - \hat{T}_j) \int_\Omega \frac{\partial \hat{y}_j}{\partial x_i}$$

$$\cdot \left(\frac{\partial \phi_j}{\partial x_i} - \frac{\partial D_j}{\partial x_i} \right) dx \tag{2.11}$$

for all $(\phi_1, \phi_2) \in S^{\hat{y}}$.

Proof: Theorem 2.1 is a simple consequence of Thm. 1.1. Indeed it can be easily checked that the variation of the cost functional of Eq. 2.8 has the following form:

$$\delta J(u;\delta u) = \lim_{t \downarrow 0} \frac{1}{t} [J(u + t\delta u) - J(u)]$$

$$= \sum_{i=1}^{2} \int_{\Omega} (y_i(x) - z_i(x))\, D_i(\delta u)(x)\, dx, \tag{2.12}$$

for all $\delta u \in U$, where $D = (D_1, D_2)$ is a unique solution to the variational inequality of the form of Eq. 1.7, with $h = \delta u$. Since $\hat{u} \in U_{ad}$ is an optimum solution, it follows that

$$\delta J(\hat{u}\ ;\ u - \hat{u}) \geq 0 \tag{2.13}$$

for all $u \in U_{ad}$. But the above condition has exactly the form of Eq. 2.10, taking into account Eq. 2.12. This completes the proof.

Note that since the cost functional of Eq. 2.8 is continuous and the set of Eq. 2.3 is compact, it follows that there exists an optimum solution to the problem (P). For additional results relating to this class of problems, see Refs. 4, 5 and 6.

APPENDIX

Theor 1: In the example of Section 2, the set $C_y(C) \cap H(u,f)$ is dense in $S^y = S_y(C) \cap H(u,f)$.

Proof: Let $K \subset H_0^1(\Omega)$ be the following convex subset:

$$K = \{\psi \in H_0^1(\Omega) : \psi(x) \geq -\tilde{\phi}(x) \text{ on } \Omega\}$$

It can be shown [3] that

$$S_v(K) = \{\psi \in H_0^1(\Omega) : \psi(x) \geq 0, \quad \text{q.e. on } Z(v - \tilde{\phi})\}$$

Recall that $Z(v - \tilde{\phi}) = \{x \in \Omega : v(x) - \tilde{\phi}(x) = 0\}$. Since

$$C_y(C) = \{(\phi_1,\phi_2) \in H_0^1(\Omega) \times H_0^1(\Omega) : \text{there exists } t > 0$$

such that

$$y_1(x) + t\phi_1(x) + \tilde{\phi}(x) \geq y_2(x) + t\phi_2(x) \text{ on } \Omega\}$$

$$= \{(\phi_1,\phi_2) : \phi_1 - \phi_2 \in C_{y_1 - y_2}(K)\}$$

one obtains $S_y(C)$ in the form of Eq. 2.7.

On the other hand, subspace $H(u,f)$ has the form

$$H(u,f) = \{(\phi_1,\phi_2) : \sum_{i,j=1}^{2} T_j \int_\Omega \frac{\partial y_j}{\partial x_i} \frac{\partial \phi_j}{\partial x_i} dx = \sum_{j=1}^{2} \int_\Omega f_j\phi_j \, dx\}$$

Let $(\bar{\phi}_1,\bar{\phi}_2) \in S^y$ be given. Then there exists an element $\psi \in S_{y_1 - y_2}(K)$ such that $\bar{\phi}_1 = \psi + \bar{\phi}_2$. Let $\{\psi_n\} \subset C_{y_1-y_2}(K)$ be a sequence such that

$$\psi_n \to \psi, \text{ in } H_0^1(\Omega) \text{ as } n \to \infty$$

Define the sequence $\{(\phi_1^n,\phi_2^n)\} \in C_y(C) \cap H(u,f)$,

$$\phi_1^n = \psi_n + \alpha_n \bar{\phi}_2$$

$$\phi_2^n = \alpha_n \bar{\phi}_2$$

and α_n is determined from the condition $(\phi_1^n,\phi_2^n) \in H(u,f)$. It can be easily checked that

$$\left.\begin{aligned} &\alpha_n \to 1 \\ &\phi_1^n \to \phi_1, \text{ in } H_0^1(\Omega) \\ &\phi_2^n \to \phi_2 \text{ in } H_0^1(\Omega) \end{aligned}\right\} \text{ as } n \to \infty$$

which completes the proof.

REFERENCES

1. Stampacchia, G., "Variational Inequalities," Theory and Applications of Monotone Operators, Edizioni Oderisi, 1969, pp. 101-192.
2. Mignot, F., "Controle dans les Inéquations Variationelles Elliptiques," J. of Functional Analysis, Vol. 22, 1976, pp. 130-185.
3. Oden, J.T., and Kikuchi, N., "Finite Element Methods for Certain Free Boundary - Value Problems in Mechanics," Moving Boundary Problems, Academic Press, 1978, pp. 147-164.
4. Barbu, V., "Necessary Conditions for Nonconvex Distributed Control Problems Governed by Elliptic Variational Inequalities," Preprint Series in Mathematics, No. 9, Bucharest, 1979.
5. Lions, J.L., Some Aspects of the Optimal Control of Distributed Parameter Systems, SIAM, Philadelphia, 1972.
6. Saguez, C., Conditions Nécessaries d`Optimalité Pour des Problémes de Contrôle Optimal Associes á des Inéquations Variationnelles, IRIA, Raport de Recherche No. 345, 1979.